CONF-771114

ENERGY AND ENVIRONMENTAL STRESS IN AQUATIC SYSTEMS

Selected papers from a symposium held at Augusta, Georgia
November 2–4, 1977

Edited by

James H. Thorp

and

J. Whitfield Gibbons
Savannah River Ecology Laboratory and University of Georgia

Sponsored by

Savannah River Ecology Laboratory

Institute of Ecology, University of Georgia

Assistant Secretary for Environment
U. S. Department of Energy

Savannah River National Environmental Research Park

1978

Published by
Technical Information Center
U. S. Department of Energy

Library of Congress Cataloging in Publication Data

Main entry under title:

Energy and environmental stress in aquatic systems.

(DOE Symposium series; 48)
"CONF-771114"
Includes indexes.
1. Aquatic ecology—Congresses. 2. Power resources—Environmental aspects—Congresses. 3. Environmental engineering—Congresses. I. Thorp, James H. II. Gibbons, J. Whitfield, 1939- III. Savannah River Ecology Laboratory. IV. Series: United States. Dept. of Energy. DOE symposium series; 48.
QH541.5.W3E53 574.5'263 78-27913
ISBN 0-87079-115-X

Available as CONF-771114 for $15.00 from
National Technical Information Service
U. S. Department of Commerce
Springfield, Virginia 22161

DOE Distribution Category UC-11

Printed in the United States of America

December 1978

PREFACE

The United States and most industrially developed nations of the world are confronted by crucial decisions on the use of alternative sources of energy. Both short- and long-term effects on human and environmental health, and frequently on the international balance of payments, hinge on the outcome of these decisions. Too often policy is made as a reflex response to the clamor created in the public realm by industrialists, economists, or environmentalists, without giving due consideration to the overall impact and compromises associated with the choice of one fuel over another energy option. For example, many environmentalists oppose large central power plants because high temperatures increase mortality for certain aquatic organisms but do not consider the potential for increased growth rates or higher species diversity from moderate thermal inputs. In contrast, some industrialists view nuclear central power stations as the most viable energy option because the probability of an accident is minimal and the long-term efficiency is high. However, the exceedingly long-term impact of a radioactive leak and the difficulty of cleanup magnify the significance of a single "low-probability" accident. Finally, the environmental impact of nuclear power must be evaluated with respect to that of other fuels, e.g., the environmental effects of acid mine drainage and acid rainfall that can result from the procurement and conversion of coal or the effects of thermal effluents from fossil-fueled power plants.

This symposium, *Energy and Environmental Stress in Aquatic Systems*, served as a forum for discussions of the environmental effects of alternative sources of energy. This exchange of information promotes a proper perspective in which to make critical judgments affecting a country's energy and environmental policies. The symposium was ideal for comparing effects of various stressors and for enabling researchers in one discipline to become acquainted with those in another area. Because of the broad nature of the meeting, we were able to define six central topics around which the 160 oral presentations were organized. Each of the six core areas was represented by an invited speaker, who reviewed the effects of specific stressors or discussed methods of modeling stress. These speakers were encouraged to speculate on future avenues of research and to make generalizations where appropriate. Thus, each topic was addressed first by a general synthesis and then by a series of detailed, specific studies. The final 49 papers in this volume were selected from an initial group of 80 papers submitted for publication from that symposium and were included only after vigorous review by outside referees.

The theme of the symposium was environmental stress, but, as evidenced by this volume, little uniformity exists in application of the term "stress." Three of the core papers distinguished between the action and response involved in stress, two equating the action with the term "stressor" and the response with the term "stress," but another core paper described action as "stress" and response as "strain." Differentiation in terms may seem superfluous, but the dual meaning is common in the ecological literature and has implications for understanding system specificity in response to environmental fluctuations. For example, Esch and Hazen, in their core paper, characterized the response to a stressor as "either an enhancement or a diminishment in the probability of mortality, natality, or permanent change." This concept of stress as either a positive or negative response is controversial to say the least, but, if the action or stressor is defined as a neutral, time-varying input, then the response to it can be either positive, neutral, or negative depending on the characteristics of the receiving system. We believe that stress always represents a negative effect on some specific component of the system but that the ultimate response of the other components or of the overall system may be positive. For instance, species diversity or equitability may increase during periods of stressful environmental fluctuations if the average compensatory efficiency of the community members is superior to that of a species which is normally dominant in less-extreme or more-constant conditions.

Operational definitions of stress or strain frequently reflected the subdisciplines and areas of interest of the investigators. For example, Lugo, who classified environments by their "energy signatures," defined stress in terms of a drain of potential energy from a system; whereas Esch and Hazen described the homeostatic stress response of largemouth bass to thermal effluents and bacterial infection. Ulanowicz, however, was concerned with linear and nonlinear modeling of inputs and the resultant strain, in a mechanistic sense, imposed on the ecosystem. We will not provide here an additional definition of stress, but we caution the readers to consider the operational definitions carefully before comparing individual studies.

The first section of this volume is devoted to modeling stress, because ultimately the ability to model and, thus, predict the stress response is a primary raison d'etre for funding stress research. In his review paper on modeling stress, Ulanowicz classifies existing models according to whether they track local or far-field stress. He also discusses the success of various attempts at linear and nonlinear modeling of the effects on ecosystem properties of endogenous and exogenous stressors. Dwyer et al., however, contend that an "adequate description of ecosystem stability properties cannot always be derived from a linearized model of the ecosystem." They used techniques of spectral analysis to analyze phytoplankton abundance in a system exposed to time-varying inputs.

Accurate prediction of a system's response to anthropogenic stressors frequently requires an understanding of the nature of a system's response to natural perturbations. In his review of studies of stressed ecosystems, Lugo proposes that the energetic quality of both the stressor and the receiving system determines the severity of response to a perturbation. He also maintains that the stability of an ecosystem depends on the stability of the main energy source. Leffler, however, concludes from a series of microcosm experiments that "no relationships between diversity, nutrient availability, or system mass and ecosystem stability or between ecosystem resistance and resilience stabilities were demonstrated." Norse, examining the effects of biological stressors and natural physicochemical stress gradients on the distribution of tropical portunid crabs, shows that biological stress increased as physicochemical stress decreased.

The rapid proliferation in the past few years of studies of the environmental impact resulting from the procurement, conversion, burning, and disposal of coal is indicative of the reemerging importance of this energy source in many industrial nations. Gehrs concludes from a literature review of studies on organic contaminants produced by conversion of coal to hydrocarbon compounds with higher hydrogen-to-carbon ratios, that toxicological investigations of individual compounds can be made manageable by first grouping the compounds on the basis of chemical structure and then studying representative compounds. Schultz, Dumont, and Kyte demonstrate that a 2% concentration of untreated coal-conversion gasifier condensate decreased motility and increased cell lysis in a ciliate protozoan. The procurement of coal can result in acid mine pollution, which has been demonstrated to reduce severely species richness of invertebrate communities in Ohio streams (Hummon et al.) but to have a much lesser impact on Colorado streams (Ward, Canton, and Gray).

Toxicological aspects of trace metals introduced into streams by natural and anthropogenic processes have received considerable attention in recent years. Eisler reviews the effects of mercury on marine biota and suggests more severe restrictions on the discharge of mercury compounds in general and methyl mercury compounds in particular. In a separate paper, Zubarik and O'Connor discuss the uptake of mercury by Hudson River biota. Other papers examine the effects of various trace elements on survival of fish and amphibian eggs (Birge), mortality of adult Pacific oysters (Harrison and Rice), nitrification by aquatic, autotrophic bacteria (Holm and Cox), and species diversity of zooplankton communities (Marshall and Mellinger). Generalizations about the uptake and toxicity of various trace metals are difficult to make because of the specificity in action of individual elements. However, current efforts in this area center on similarities in uptake and toxicity in relation to atomic structure of the elements and to the chemical requirements of the biota being studied.

Two of the previous volumes in the Savannah River Ecology Laboratory Symposium Series on Ecological Research, and approximately half this volume, are devoted to studies of the response of organisms and communities to thermal effluents. The papers in this section of this volume are roughly grouped by their emphasis on functional (i.e., primary productivity, growth, genetics, reproduction, and behavior), structural, and toxicological aspects of thermal ecology. The diversity of these studies demonstrates that thermal effluents have a comprehensive effect on most functional and structural aspects of aquatic systems.

Difficulties in design and analysis have impeded the spread of studies on multiple and synergistic stresses, and yet, in most of our "less-than-pristine" environments, the simultaneous presence of various stressors is undoubtedly the norm rather than the exception. Vernberg's review of recent research on the effects of multiple stressors shows the reduction that can occur in an organism's zones of lethality and compatibility as a result of previous or concurrent exposure to multiple stressors. Three papers demonstrate differential effects on fish or Cladocera from multiple stressors associated with entrainment (Poje, Ginn, and O'Connor and Buikema et al.) or presence near the discharge (Cherry et al.) of fossil-fueled or nuclear power plants compared with the effects of single factors. This multiple-factor effect is also shown for the benthic macroinvertebrates in streams in Ohio (Beckett) and Pennsylvania (Bradt) which had been subjected to a variety of physical and chemical stressors as a result of channelization and urbanization.

We see a need for additional symposia that will serve as forums for comparing and contrasting the effects of a variety of stressors rather than focusing on individual perturbations. The task of selecting the energy source that best balances economic and environmental concerns in a given situation will be facilitated by such discussions. The emphasis of future meetings should differ from this one since the economic and environmental problems that industrial nations face continue to evolve with our changing world.

J. H. Thorp
J. W. Gibbons
Editors

ACKNOWLEDGMENTS

To be successful, a symposium requires the enthusiasm and unselfish assistance of numerous individuals. The major portion of effort necessary to make this particular meeting an outstanding one came from the staff and students of the Savannah River Ecology Laboratory. Contributions of time and work were given by all individuals at SREL prior to, during, and in the follow-up aspects of the symposium. We also appreciate the cooperative spirit of the Department of Energy's Savannah River Operations Office, without whose support this meeting could not have been held. In addition, we thank the individuals at DOE Headquarters for their encouragement. We especially thank the six invited speakers and numerous session chairmen for their participation in the program.

We feel a particular indebtedness to Eleanor Cato for her constant vigil to ensure attention to all details and her persistence in carrying out the plethora of jobs related to the symposium and to the publication of this volume. We thank Robert I. Nestor for his handling of the innumerable routine, but critical, exigencies that arise before and during a symposium. Once again we thank the highly competent redactory editorial staff of the DOE Technical Information Center in Oak Ridge, Tennessee. Dee Jared, Anne Goulden, and their associates successfully carried the total volume through to completion.

Finally, we are particularly grateful to those who assisted in the intensive review process through which all papers passed. We feel that many authors in the volume also owe a debt of thanks to these individuals. The following reviewed manuscripts at our request:

L. G. Abele, Florida State University, Tallahassee, FL
W. D. Adair, Duke Power Company, Huntersville, NC
S. M. Adams, Oak Ridge National Laboratory, Oak Ridge, TN
D. F. Alderdice, Pacific Biological Station, Nanaimo, British Columbia
D. E. Alston, Auburn University, Auburn, AL
A. W. Andren, University of Wisconsin, Madison, WI
N. E. Armstrong, University of Texas, Austin, TX
W. M. Barnard, State University College, Fredonia, NY
L. W. Barnthouse, Oak Ridge National Laboratory, Oak Ridge, TN
K. E. Biesinger, U. S. Environmental Protection Agency, Duluth, MN
W. J. Birge, University of Kentucky, Lexington, KY
P. C. Bradbury, North Carolina State University, Raleigh, NC
P. T. Bradt, Lehigh University, Bethlehem, PA
R. A. Brechbill, U. S. Department of Energy, Oakland, CA
J. C. Britton, Texas Christian University, Fort Worth, TX
I. M. Brook, University of Miami, Miami, FL
W. A. Brungs, U. S. Environmental Protection Agency, Duluth, MN
A. L. Buikema, Jr., Virginia Polytechnic Institute and State University, Blacksburg, VA
D. L. Bunting, University of Tennessee, Knoxville, TN
D. T. Burton, Benedict Estuarine Research Laboratory, Benedict, MD
J. Cairns, Jr., Virginia Polytechnic Institute and State University, Blacksburg, VA
W. D. Claus, Delray Beach, FL
J. L. Cooley, University of Georgia, Athens, GA
W. E. Cooper, Michigan State University, East Lansing, MI
B. C. Coull, University of South Carolina, Columbia, SC
R. A. Coulson, Louisiana State University School of Medicine, New Orleans, LA
C. C. Coutant, Oak Ridge National Laboratory, Oak Ridge, TN

F. A. Cross, National Marine Fisheries Service, Beaufort, NC

C. E. Cushing, Battelle Pacific Northwest Laboratories, Richland, WA

M. D. Dahlberg, U. S. Fish and Wildlife Service, La Crosse, WI

W. M. Darley, University of Georgia, Athens, GA

H. C. Dessauer, Louisiana State University Medical Center, New Orleans, LA

K. L. Dickson, Virginia Polytechnic Institute and State University, Blacksburg, VA

D. R. Dreesen, Los Alamos Scientific Laboratory, Los Alamos, NM

R. S. Driscoll, U. S. Forest Service, Fort Collins, CO

R. A. Drummond, U. S. Environmental Protection Agency, Duluth, MN

K. F. Ehrlich, Lockheed Center for Marine Research, Avila Beach, CA

R. Eisler, U. S. Environmental Protection Agency, Narragansett, RI

J. W. Elwood, Oak Ridge National Laboratory, Oak Ridge, TN

D. W. Engel, National Marine Fisheries Service, Beaufort, NC

D. W. Evans, Savannah River Ecology Laboratory, Aiken, SC

C. L. Fergus, Pennsylvania State University, University Park, PA

J. M. Ferris, Purdue University, West Lafayette, IN

C. B. Fliermans, E. I. du Pont de Nemours and Co., Inc., Aiken, SC

J. L. Gallagher, U. S. Environmental Protection Agency, Corvallis, OR

C. I. Gibson, Battelle Pacific Northwest Laboratory, Sequim, WA

J. M. Giddings, Oak Ridge National Laboratory, Oak Ridge, TN

J. P. Giesy, Savannah River Ecology Laboratory, Aiken, SC

J. J. Gift, Ecological Analysts, Inc., Melville, NY

R. J. Gilbert, University of Georgia, Athens, GA

J. L. Gilio, Florida Institute of Technology, Jensen Beach, FL

J. M. Glime, Michigan Technological University, Houghton, MI
D. Goodman, Scripps Institution of Oceanography, La Jolla, CA
R. A. Greig, National Marine Fisheries Service, Milford, CT
C. A. S. Hall, Cornell University, Ithaca, NY
J. Harte, Lawrence Berkeley Laboratory, Berkeley, CA
D. S. Haven, Virginia Institute of Marine Science, Gloucester Point, VA
T. C. Hazen, Wake Forest University, Winston-Salem, NC
G. R. Hendry, Brookhaven National Laboratory, Upton, NY
W. F. Hettler, National Marine Fisheries Service, Beaufort, NC
K. E. F. Hokanson, U. S. Environmental Protection Agency, Monticello, MN
R. W. Holmes, University of California, Santa Barbara, CA
T. J. Horst, Stone & Webster Engineering Corporation, Boston, MA
D. E. Hoss, National Marine Fisheries Service, Beaufort, NC
F. G. Howell, University of Southern Mississippi, Hattiesburg, MS
P. L. Hudson, U. S. Fish and Wildlife Service, Clemson, SC
W. D. Hummon, Ohio University, Athens, OH
G. R. Huntsman, National Marine Fisheries Service, Beaufort, NC
G. S. Innis, Utah State University, Logan, UT
L. D. Jensen, Ecological Analysts, Inc., Towson, MD
R. G. Kazman, Louisiana State University, Baton Rouge, LA
J. R. Kennedy, University of Tennessee, Knoxville, TN
D. L. King, Michigan State University, East Lansing, MI
R. T. Lackey, Virginia Polytechnic Institute and State University, Blacksburg, VA
H. V. Leland, U. S. Geological Survey, Menlo Park, CA
D. R. Lenat, University of North Carolina, Chapel Hill, NC
M. W. Lorenzen, Tetra Tech, Inc., Lafayette, CA

H. A. Loyacano, Clemson University, Clemson, SC
D. L. Lush, Beak Consultants Limited, Mississauga, Ontario
D. B. McDonald, University of Iowa, Iowa City, IA
R. W. McFarlane, Brown & Root, Inc., Houston, TX
I. A. McLaren, Dalhousie University, Halifax, Nova Scotia
C. P. McRoy, University of Alaska, Fairbanks, AK
J. S. Marshall, Argonne National Laboratory, Argonne, IL
J. S. Mattice, Oak Ridge National Laboratory, Oak Ridge, TN
O. E. Maughan, Oklahoma State University, Stillwater, OK
N. A. Mercando, Pennsylvania State University, Abington, PA
J. A. Mihursky, Chesapeake Biological Laboratory, Solomons, MD
P. C. Miller, San Diego State University, San Diego, CA
J. D. Milligan, Tennessee Valley Authority, Chattanooga, TN
F. J. Molz, Auburn University, Auburn, AL
R. P. Morgan, William F. Clapp Laboratories, Inc., Duxbury, MA
I. Morris, Bigelow Laboratory for Ocean Sciences, West Boothbay Harbor, ME
D. H. Nelson, University of South Alabama, Mobile, AL
S. J. Nepszy, Ontario Ministry of Natural Resources, Wheatley, Ontario
W. A. Niering, Connecticut College, New London, CT
J. M. O'Connor, New York University Medical Center, Tuxedo, NY
R. G. Otto, Johns Hopkins University, Baltimore, MD
D. S. Peters, National Marine Fisheries Service, Beaufort, NC
J. E. Pinder, III, Savannah River Ecology Laboratory, Aiken, SC
G. V. Poje, New York University Medical Center, Tuxedo, NY
T. T. Polgar, Martin Marietta Corporation, Baltimore, MD
L. R. Pomeroy, University of Georgia, Athens, GA

C. W. Reimer, Academy of Natural Sciences, Philadelphia, PA
N. L. Richards, U. S. Environmental Protection Agency, Gulf Breeze, FL
E. B. Rodgers, Tennessee Valley Authority, Decatur, AL
P. A. Sandifer, South Carolina Wildlife and Marine Resources Department, Charleston, SC
C. E. Sansbury, South Carolina Department of Health and Environmental Control, Columbia, SC
W. E. Schaaf, National Marine Fisheries Service, Beaufort, NC
C. L. Schelske, University of Michigan, Ann Arbor, MI
D. W. Schindler, Freshwater Institute, Winnipeg, Manitoba
M. J. Schneider, Battelle Pacific Northwest Laboratories, Richland, WA
P. B. Schroeder, University of Miami, Miami, FL
J. R. Schubel, State University of New York, Stony Brook, NY
F. J. Schwartz, University of North Carolina, Morehead City, NC
K. P. Sebens, Harvard University, Cambridge, MA
E. D. Seneca, North Carolina State University, Raleigh, NC
A. H. Seymour, University of Washington, Seattle, WA
J. Shapiro, University of Minnesota, Minneapolis, MN
R. R. Sharitz, Savannah River Ecology Laboratory, Aiken, SC
W. L. Shelton, Auburn University, Auburn, AL
J. M. Shick, University of Maine, Orono, ME
J. B. Shrode, Occidental College, Los Angeles, CA
L. B. Slobodkin, State University of New York, Stony Brook, NY
D. H. Smith, Stony Creek, VA
M. H. Smith, Savannah River Ecology Laboratory, Aiken, SC
J. R. Spotila, State University College, Buffalo, NY
J. B. Sprague, Guelph, Ontario
E. A. Standora, Savannah River Ecology Laboratory, Aiken, SC
R. H. Stavn, University of North Carolina, Greensboro, NC

C. E. Stephan, U. S. Environmental Protection Agency, Duluth, MN

J. S. Suffern, Oak Ridge National Laboratory, Oak Ridge, TN

R. M. Sykes, Ohio State University, Columbus, OH

M. R. Tansey, Indiana University, Bloomington, IN

K. R. Tenore, Skidaway Institute of Oceanography, Savannah, GA

J. H. Tietjen, City College of New York, New York, NY

L. J. Tilly, E. I. du Pont de Nemours and Co., Inc., Aiken, SC

F. R. Trainor, University of Connecticut, Storrs, CT

R. E. Ulanowicz, Chesapeake Biological Laboratory, Solomons, MD

T. J. Vigerstad, E. I. du Pont de Nemours and Co., Inc., Aiken, SC

J. V. Ward, Colorado State University, Fort Collins, CO

R. S. Warren, Connecticut College, New London, CT

J. R. Webster, Virginia Polytechnic Institute and State University, Blacksburg, VA

I. M. Weis, University of Windsor, Windsor, Ontario

M. O. Welch, Rice University, Houston, TX

J. L. Wilhm, Oklahoma State University, Stillwater, OK

D. R. Wiseman, College of Charleston, Charleston, SC

T. E. Wissing, Miami University, Oxford, OH

T. G. Wolcott, North Carolina State University, Raleigh, NC

D. A. Wolfe, National Oceanic and Atmospheric Administration, Boulder, CO

E. A. Woolson, U. S. Department of Agriculture, Beltsville, MD

J. C. Zieman, University of Virginia, Charlottesville, VA

R. G. Zingmark, University of South Carolina, Columbia, SC

V. Zitko, Environment Canada Biological Station, St. Andrews, New Brunswick

L. S. Zubarik, New York University Medical Center, Tuxedo, NY

CONTENTS

Modeling Stresses

Natural Stresses and Environmental Fluctuations

*Invited speaker.

Fossil-Fuel Stresses

Heavy-Metal Stresses

*Invited speaker.

Thermal Stresses

---

*Invited speaker.

*Invited speaker.

MODELING ENVIRONMENTAL STRESS

ROBERT E. ULANOWICZ
University of Maryland, Center for Environmental and Estuarine Studies, Solomons, Maryland

ABSTRACT

The word stress when applied to ecosystems is ambiguous. Stress may be low-level, with accompanying near-linear strain, or it may be of finite magnitude, with nonlinear response and possible disintegration of the system. Since there are practically no widely accepted definitions of ecosystem strain, classification of models of stressed systems is tenuous. Despite appearances, most ecosystem models seem to fall into the low-level linear response category. Although they sometimes simulate systems behavior well, they do not provide necessary and sufficient information about sudden structural changes nor structure after transition. Dynamic models of finite-amplitude response to stress are rare because of analytical difficulties. Some idea as to future transition states can be obtained by regarding the behavior of unperturbed functions under limiting strain conditions. Preliminary work shows that, since community variables do respond in a coherent manner to stress, macroscopic analyses of stressed ecosystems offer possible alternatives to compartmental models.

Unfortunately, the term stress is not used uniformly in ecological discussions. It comes to our discipline from mechanics, physiology, and psychology and brings different shades of meaning from each source. In clarifying what is meant by stress and its consequences, it is useful to refer to the meaning given to the word by nineteenth century engineers.

Stress represented "the forces or pressures exerted upon a material" (Meier, 1972). In mechanics, stress had no utility without its conjugate, strain, "a measure of the deformation brought about by the action of the stresses." The relationship between applied stress and observed strain (e.g., the elongation of a metal rod under tensile stress) was presumed linear. Twice the stress resulted in twice the strain.

The modulus of elasticity (the ratio of strain to stress) as a property of a solid is useful if we are designing a structure such as a bridge, but it is often necessary to know also the behavior of a solid system under extreme stresses. In fact, as the stress on a metal rod increases, a point is reached where the strain becomes disproportionately larger than the applied force. Not long thereafter the rod reaches the critical point; i.e., the strain is such that the rod will no longer return to its original state. Still further stress leads to increasingly disproportionate strain, culminating in a catastrophe when the rod loses its identity (yield point).

The behavior of a simple mechanical system under heavy stress differs markedly from its corresponding response to low stress. It is significantly nonlinear, and it culminates in loss of system structure. It is this response to heavy stress that is important to psychologists and physiologists, for whom stress has come to mean a "response to external or internal processes which reach those threshold levels that strain its physiological and psychological integrative capacities close to or beyond their limits" (Basowitz et al., 1955).

Stress, therefore, takes on different connotations for the engineer and the psychologist. Although it may not be obvious, this dual meaning of stress is found in ecological research. Ecologists have been slow to define and accept a useful measure of the response of the ecosystem to stress, i.e., ecological strain. Just as it is impossible to discuss mechanical stress without considering its conjugate, strain, the discussion of stress in ecological systems is fragmentary without some hypothetical measure of system strain. Before attempting a working definition of strain, however, we should consider how stress arises in ecosystem models.

Although ecosystem models may be stochastic, discrete, spatially heterogeneous, etc., much of systems analysis, following the lead of the early modelers, has concentrated on deterministic, first-order, ordinary differential equations, such as (see Lotka, 1957),

$$\dot{\mathbf{X}} = f(\mathbf{X}, \mathbf{P}, t) \tag{1}$$

where $\mathbf{X}$ is a vector of state variables, t is time, and $\mathbf{P}$ is a vector of parameters. Parameters of a model include initial conditions, fluxes into and out of the system, and characteristics of the functional form of f (such as exponents or multiplicative constants).

The external world may impinge on the ecosystem (exogenous stress) through arbitrary variations in $\mathbf{P}$ and $\mathbf{X}$; $\mathbf{X}$ may change through cropping or mass infusion of a species; and **P** may change in a number of ways. For example, the multipliers and exponents are often strongly dependent on abiotic variables, such as temperature, salinity, light, etc. These forcing functions, in turn, may possess both regular and stochastic components. The input fluxes, necessary to every living system, vary similarly.

Occasionally complex systems will exhibit the characteristics of strain without any apparent imposed stress. The term endogenous stress has been coined to describe such phenomena, but the previous discussion reveals this to be a misnomer. Nonlinear systems sometimes produce an output without any corresponding input. "Endogenous strain" would, therefore, be a more accurate descriptor for such behavior.

Attempting to provide a workable definition for strain in an ecosystem, Innis (1975) found it useful to invoke an arbitrary function of the state of the system,

$$\mathbf{H} = \mathbf{h}(\mathbf{X}, \dot{\mathbf{X}}) \tag{2}$$

to measure the deviation from some prescribed state, $\mathbf{H}^*$, characterized as unstressed. For example, $\mathbf{H}^*$ might be taken to be a stationary state, i.e.,

$$\mathbf{H}^* = \mathbf{h}(\mathbf{X}^*, \mathbf{0}) \tag{3}$$

where $\mathbf{X}^*$ is the solution of $\mathbf{f}(\mathbf{X}^*, \mathbf{P}, t) = \mathbf{0}$. Any suitably defined metric could be used to describe the distance between $\mathbf{H}$ and $\mathbf{H}^*$, i.e., the ecological strain:

$$S = \|\mathbf{H} - \mathbf{H}^*\| \tag{4}$$

As Innis remarked, whether any particular deviation is indicative of a stressed system is somewhat arbitrary and depends largely on the context of the discussion. Woodwell (1975), for example, argued against the threshold concept in ecology. In his view, any chronic stress takes its toll on the ecosystem in the form of a chronic, albeit sometimes small, deviation. The linear view of stress would be quite useful for his purposes.

In contrast, Holling (1973) cited the possibility of multiple stationary states for a given ecosystem—several $\mathbf{H}^*$, each with its own "domain of attraction" characterized by a finite deviation, S_{crit}. Deviations in excess of the critical strain can lead to transition into another domain. Furthermore, such transition may incur a change in dimensionality of the problem and/or the necessity for a new functional descriptor, $\hat{f}$, of the system dynamics. Deviations in the neighborhood of S_{crit} are signs of a stressed ecosystem in the physiological sense of the word.

Thus two classes of stress analyses are readily identifiable. In the first, there is no explicit mention of a critical deviation, a change in dimensionality, or a switch in function (topological form). The second class is identified by the prominence of at least one of these characteristics. The first class of models will be referred to as local and the second as far-field.

Unfortunately, not all ecosystem stress analyses fall neatly into these two classes. There are critics (e.g., Mann, 1975) who find compartmental modeling reductionistic. They claim that time would be better spent searching for emergent properties of the ecosystem as a whole and that these properties would serve as more reliable indicators of the response of the community to stress. To press the earlier analogy (perhaps a little far!), this is akin to observing the strain response of an assemblage of metal rods (such as a bridge truss) to various imposed stresses without being concerned with the properties of the individual members. Certainly there are characteristic dimensions or dimensionless ratios of the total structure at which strain responses are indicative of impending collapse. Such approaches to ecosystem stress will be termed macroscopic in nature.

Finally, I indulge in speculation on a principle that I believe would crystallize research on ecosystem response and, more importantly, might provide a theoretical basis to bridge the gap between ecological systems research and evolutionary theory.

LOCAL ANALYSES OF STRESS

Most of the ecosystem models in existence today were constructed to elucidate the response of the community to a stress. As I see it, most of these efforts have been local in nature. This is not to imply that local analyses are necessarily uninteresting or uninformative. In fact, some appear to be nonlinear and have been quite successful in portraying the response to exogenous stresses. For example, the highly realistic aquatic ecosystems models of Nixon and Kremer (1977), Di Toro et al. (1975), and Park, Scavia, and Clesceri (1975) are capable of predicting significant changes in response to exogenous stresses (e.g., temperature, nutrient input, and light availability).

As an analysis of stress, however, these simulations are basically local. To see this, we should appreciate that most of the exquisite architecture of the models is in their parameter specifications. Thus, with changing driving forces, the "normal" state, H*, may vary considerably according to how the instantaneous stationary state,

$$f(\mathbf{X}^*, \mathbf{P}, t) = \mathbf{0} \tag{5}$$

varies as **P** changes with exogenous stress. The behavior function, **H**, will tend to track **H*** closely. The hypothetical deviation may never be large, and the system is not stressed in the far-field sense of the word.

Nevertheless, the dramatic responses of some systems commonly referred to as stressed can be shown by a judicious choice of the functional dependence of the parameters on exogenous stress. Bierman et al. (1973), for example, chose *Chlorella* and *Microcystis* as two compartments of a nutrient-uptake model. Using separate laboratory information to describe the nutrient-uptake kinetics and the response to temperature, they investigated zones where one of the species dominated starkly. Lassiter and Kearns (1973) simulated an annual progression of six species as they dominated the phytoplankton of a hypothetical limnetic system. Falco and Mulkey (1976), using the law of mass action, anticipated the significant differential effect that pesticides can have on populations of bass and bluegills.

In these models all the information about the behavior of the system is contained in the functional form f and the parameter dependencies. Even when the population structure is predicted to be quite unbalanced (e.g., practically all blue-green algae), the system may be very near its stationary point, and, hence, by our definition it is only slightly stressed. Furthermore, the response is usually almost reversible (perhaps retrievable is a better word), meaning that, when the external influence is removed, the system returns to near its original condition.

The introductory examples of local low-stress models were purposely chosen for their nonlinear construction, but we more naturally associate low stress with linearity and, consequently, with linear models. Mathematically speaking, linear models have the form

$$\dot{\mathbf{X}} = \underline{\underline{A}}\,\mathbf{X} \tag{6}$$

where the matrix of coefficients is allowed to vary parametrically, i.e., $\underline{\underline{A}} = \underline{\underline{A}}(E, t)$. Any well-behaved nonlinear f can be approximated by a linear system in the neighborhood of a given point in phase space. The tremendous advantage of linear systems is the well-developed mathematical tools that can be brought to bear on them—especially linear stability analysis.

In terms of the effects of low-level stress on a system, there is one key question, "Will the response to the stress remain small, or will it grow to the point of disrupting the integrity of the community?" For linearized systems the procedure for answering this question is well defined and has been reviewed by May (1971). All the eigenvalues of the matrix $\underline{\underline{A}}$ must have negative real parts. If the linear ecosystem model has constant coefficients, this test will show whether the model is properly behaved. More frequently, however, the coefficients of $\underline{\underline{A}}$ vary because of exogenous driving forces. In this case the eigenvalues vary also, and, under changing conditions, it is possible that some eigenvalue will acquire a positive real part (Halfon, 1976). Thus we can map out domains of driving forces for which the systems response is possibly unstable.

Much of the literature involving linear stability theory in ecosystems has been given over to debating the question of whether diversity will better enable an ecosystem to cope with an applied stress. MacArthur (1955) suggested such a causal link, and May (1973) reviewed the use of linear stability analysis to question this hypothesis. Central to the counter argument is the observation that increasing the dimensionality and connectivity of a randomly assembled system decreases the probability that all eigenvalues will be negative. Gardner and Ashby (1970) sampled randomly constructed matrixes to illustrate this point. Others have argued that ecosystems are not randomly constructed and that constraints on the form of $\underline{\underline{A}}$ can lead to different conclusions (Roberts, 1974; McMurtrie, 1975; Saunders and Bazin, 1975; Jeffries, 1974).

The diversity—stability controversy is actually a macroscopic issue, and further discussion is best deferred to that section of this paper. What is important here is that linear stability results are neither necessary nor sufficient to determine the persistence of an ecosystem under stress.

Despite their simplicity, linear models remain a popular medium for modeling total ecosystems (Patten, 1975). In fact, there are instances where linear models seem preferable for simulating total system behavior (Patten, 1976; Ulanowicz et al., 1978). Patten's success with linear descriptions led him to propose linearity as an evolutionary design criterion (Patten, 1975)—a much criticized stance (e.g., Wiegert, 1975). Leaving philosophical considerations aside, we see that the preceding discussion of local models of stress may help illuminate why linear models are such popular tools. As stated earlier, thus far most attempts to model ecosystem response to stress have been local in nature; i.e., the system closely tracks the normal state, H*. Locally there is little difference between linear and nonlinear representations. Nonlinear representations tend to be more sensitive to parameter changes, however, and sensitivity and stability are closely related (Estberg and Patten, 1975). Thus a higher percentage of linear attempts at modeling are likely to survive into the final stages of an investigation.

Although they have served ecosystem science well, local models of stress response still leave much to be desired. The instability of a system to small stress serves as nothing more than a warning signal to the ecosystem manager. Local instability is neither necessary nor sufficient to cause a system to switch to a different configuration when subjected to a finite stress. Furthermore, the analysis reveals nothing about the future structure of the system if it should change character. Finally, in most local considerations little emphasis can be placed on endogenous strain, which may arise from finite excursions from the normal state.

FAR-FIELD STRESS ANALYSIS

The deficiencies of low-stress models are the cause of many ecological managers' suspicions of the modeling process. This point was underscored at a recent symposium on ecological modeling in a resource-management framework when three investigators—Schaaf (1975), O'Neill (1975), and Orlob (1975)—independently cited Holling's multiple steady-state hypothesis and the burning desire of ecologist and manager alike to understand more about the "collapsing" ecosystem syndrome. Managers are necessarily concerned with species changes within the systems in their charge. The conditions leading to structural changes in the system and the configuration of a collapsed or new ecosystem are matters of utmost importance in their eyes.

At the time of the symposium (1975), practically no research on the problem of switching between domains of attraction caused by finite amplitude stress was widely known. All investigators agreed on the need for theoretical research. In addition, Orlob called for controlled experimentation on collapsing ecosystems and systems subjected to low-level chronic stress.

The problem of mapping domains of attraction for an ecosystem is a formidable task. It is enough to cause a theoretical ecologist like May (1975) to remark: "I find it difficult to envision any simple number, or handful of numbers, which will quantify the resilience of a complicated natural ecosystem." There have been no major breakthroughs since May's comment to characterize the domain of attraction for a given stationary point. The task is as intriguing as it is important and formidable, however, and should continue to demand the attention of ecologists for several years to come. Nor will the problem remain of interest only to theoreticians. Lawton, Beddington, and Bower (1974) and Sutherland (1974), for example, have shown from empirical data that switching behavior occurs even among invertebrates.

Preliminary is perhaps the best word to describe the investigations on finite perturbations to date. An example of the nonreversibility of a nonlinear system is provided by McQueen (1975). His model for competition between two species of cellular slime mold exhibits two ranges of persistence in the sense of Holling. When he made the birthrate of one of the species highly dependent on climate and then shocked the model with a short burst of favorable climate, the model underwent transition from one domain to another where the favorably perturbed population was higher. Most interesting, however, was the model behavior that

> . . . suggests that a population might fluctuate for long periods of time at a low level as it tracks the lower-stable [stationary] point, but given a short burst of favorable climatic conditions it could escape and rapidly grow to an upper-stable [stationary] level. From that point on, the population will remain at a high level tracking the high-stable [stationary point] as it moves in response to changing climate. Return to a low level is only possible when negative forces increase or when climate is very unfavorable to birthrate.

We can easily envision the reverse situation occurring in a collapsing ecosystem.

Although we cannot yet say with confidence how an ecosystem will be structured after it has undergone transition, there are two notable attempts to answer this question.

In the first, Smith (1975) occupied himself with species extinctions and the realm of possible stable subsystems that a known system may possess. He defined stability in a very fundamental way (see also Ulanowicz, 1972); a system is considered stable with regard to some defined stress if none of the component species become extinct as a result of that stress. In general, when one or more species of an ecosystem is removed, either a subsystem that is itself stable (in the sense just mentioned) results or one or more of the remaining species will drive the subsystem to collapse. Smith began by enumerating all possible subsystems by dropping various combinations of state variables in turn. To test whether any of these is a stable subsystem, we must view the behavior of the system when

stressed arbitrarily near the extinction of one or more species (stress in the second sense of the word). That is, if

$$\dot{X}i = fi(X_1, X_2, \ldots, X_m, X_{m+1}, \ldots, X_n) \qquad i = 1, 2, \ldots, n \tag{7}$$

then we observe

$$\dot{X}i = fi(X_1, X_2, \ldots, X_m, 0, \ldots, 0) \qquad i = 1, 2, \ldots, n \tag{8}$$

where variables X_{m+1} through X_n were chosen (without loss of generality) as those arbitrarily driven to near extinction. Smith then listed four criteria the reduced system must satisfy to be stable. For example, a stable subsystem cannot be obtained if any one of the fi describing an extinct species has become positive as a consequence of extinction, i.e., if any

$$\dot{X}i = fi(X_1, X_2, \ldots, X_m, 0, \ldots, 0) > 0 \qquad i = m + 1, \ldots, n \tag{9}$$

He concluded his remarks by performing an analysis on a hypothetical four-species subsystem and identified three possible stable subsystems to which the original system might collapse.

Concern with environmental degradation in recent years has caused ecologists to become somewhat jaded and to focus on exogenous stress and its consequent simplification of the impacted ecosystem. In far-from-equilibrium nonlinear systems, however, endogenous strains occur which allow a chance perturbation (a mutant or migration) to flourish suddenly and to become an added dimension of the community. In the second example, Prigogine (1976) and Eigen (1971) examined this phenomenon as the crucial element in the prebiotic evolution of polymers, and Allen (1976) extended the analysis to the evolution of new populations in ecosystems. The methods used are similar to those used by Smith for collapsing systems.

Despite these interesting insights, the basic nature of structural transitions remains an enigma. This is caused in large measure by an inclination to think in terms of linear systems. It is not foreign to think of a system, an input (exogenous stress), and an output which results from that input (stability or instability), but it is discomfiting to be confronted with an output whose origin lies predominately within the system itself.

MACROSCOPIC TREATMENT OF ECOSYSTEM RESPONSE

In his presentation of evidence for multiple stable points and domains of attraction, Holling (1973) described the system from the species or population level. The enormous analytical difficulties in properly describing most multistable systems, coupled with the high dimensionality of most real ecosystems, has led a number of investigators to explore the possibility that response to stress is best described in terms of macroscopic or emergent variables. Macroscopic variables are characteristic of the ecosystem as a whole and not just parts of it. They may and often do involve some combination of lesser order variables, however. There is still no consensus as to what consitutes a proper macroscopic variable.

Several investigators have suggested semiquantitative candidates for macroscopic variables as a consequence of their empirical studies. Kerr (1974) referred to structural transitions as "emergent surprises" and believed that they can be encompassed only by macroscopic theory. He cited the particle-size spectrum of an ecosystem as a convenient indicator of stress in a community. Exogenous stress seems to always affect the larger size organisms disproportionately. Jordan, Kline, and Sasscer (1972) emphasized the ratio of recycling to input as a system variable that characterizes the recovery time of an ecosystem from a temporary stress. Golley (1974) went further; he suggested a temporal hierarchy of three system properties to describe recovery from traumatic stress. First, the system responds to restore its extensive variables (mass); second, the functional options (topological diversity) increase; and, in the final stages of return to undisturbed climax, its response time to disturbance lengthens.

Presently the reconciliation of microscopic and macroscopic properties of an ecosystem is hampered by the inability of ecological theory to provide appropriate methods for observing community properties (Kerr, 1974). Actually this hierarchical problem has always been extremely important in ecological modeling. There are many opinions on how to aggregate organisms, species, etc., into trophic compartments or functional units (see Halfon, 1978), and it is especially difficult in highly connected or "webbed" ecosystems.

To address this problem, Kemp and Homer (1977) devised a method for assigning fractions of the energy storage in a given species compartment to various trophic levels. The key to their algorithm is the matrix of partial feeding coefficients, which describes the percentage of the total input to a given species, i, that flows from another species, j. To identify the contents of the fourth trophic level, for example, we identify all pathways three steps removed from a primary producer. The fraction of the end-point species to be assigned to the fourth trophic level is the product of the partial feeding coefficients of the three links along a pathway summed over all existing three-step pathways. Operationally the transformation is calculated in a manner similar to Goh's (1975) vulnerability matrix (Ulanowicz and Kemp, 1978).

When Kemp and Homer's transformation is performed on energetics data from two comparable marsh ecosystems, one of which is impacted by thermal effluent, the results are striking. Energy flows through the lower trophic levels remain almost unchanged, but those through the higher levels fall off drastically under the thermal stress.

The observations of Kerr (1974), Golley (1974), and Kemp and Homer (1977) lead to the common conclusion that stress tends to result in more-simplified ecosystems. A great deal of debate has been devoted to the converse of this proposition, i.e., that more-diverse ecosystems are more resilient to stress. This proposition is properly macroscopic; i.e., diversity and stability are legitimate community properties. Few of the papers addressing this issue treat stability as a calculatable characteristic of the system, however. Two exceptions to this trend are Mulholland (1975) and Jorgensen and Mejer (1977).

Mulholland related the conditional entropy of an ecosystem to its buffering capacity. The conditional entropy, which comes from information theory, is the difference between the now-familiar Shannon—Wiener diversity index and the average mutual information, i.e., the amount of uncertainty about the distribution of energy from the various sources resolved by observing the behavior of the systems over a given time interval. Rutledge (1974) applied this measure to two short-grass prairie ecosystems, one under low-moisture stress. Surprisingly enough, the stressed system had higher conditional entropy (effective choice of pathways). Mulholland resolved this apparent contradiction by hypothesizing that the "ecological resilience of a system [which has not undergone transition] is maximum when conditions are harshest."

Jorgensen and Mejer defined the buffering capacity of a freshwater lake as the ratio of total phosphorous in the lake plus sediments to the steady-state value of soluble phosphorous in the lake. They found a tight correlation of this quantity with the exergy of the system (exergy is a measure of the mechanical energy equivalent of the distance from thermodynamic equilibrium).

In a third study of change in macroscopic variables in response to stress, Lane, Lauff, and Levins (1975) described the changes in several of Levins's (1968) community niche values in response to the eutrophication of a freshwater lake. Mean niche overlap, average competitive success, and mean number of organisms per unit of ecological space if no competitors were present, all increased significantly with increasing nutrient loading.

Finally, Harte and Levy (1975), borrowing from an analysis popular in physics around the turn of the century, constructed for three hypothetical ecosystems a community function whose existence ensured that the system would be stable to finite perturbations within a given domain. Briefly, if the differential equation

$$\dot{X} = f(X, P) \tag{10}$$

possesses a steady state or limit cycle, X_s, and is perturbed to $X_s + \Delta X$, it may or may not return to the neighborhood of X_s. If we can construct a function $L(\Delta X)$ (called a Liapunov function) which vanishes at the origin, is positive and monotonically increasing with ΔX in some domain about the origin, and has a negative time derivative, then the system is stable with respect to perturbations within that domain. The function L is not necessarily unique but is a conservative estimator of the stability properties of the system (and likewise a conservative estimator of the domain of stability). For certain classes of functions, f, there are standard methods for determining whether or not Liapunov functions exist. For these particular systems the question of stability is unequivocally resolved. In general, however, failure to find a Liapunov function does not imply instability of the system. The function may exist but may defy analytical description.

Despite these analytical difficulties, the Liapunov method has two things to recommend it. First, it bridges the gap between the microscopic (species level) and the macroscopic (L being a community function). Second, it offers the hope of ordinating the various domains of stability. Harte and Levy (1975) speculated that if succession is in the direction of ever more resiliency to stress, then

$$\Lambda = -\min\left(\frac{1}{2}\frac{d}{dt}\ln L\right)/\Delta X \tag{11}$$

provides a measure of the maturity of the system.

SUMMARY AND SPECULATIONS

There is a historical duality in the scientific meaning of the term stress, and this is reflected in the various models of ecosystem response to stress. Since strain, the conjugate to stress, is not well defined in ecology, discussion of the topic is difficult. If we assume a measure of system response to stress, two distinct groups of stress analysis arise, local and far-field.

I have classified most existing models of total ecosystems as local because either they are linear or they track the instantaneous stationary state closely. Analysis of local systems is quite well developed, but application of the analytical results to real ecosystems is not free from ambiguities. Some of the models exhibit considerable realism and can be used as management tools to identify exogenous stresses that might jeopardize ecosystem integrity.

The observation that ecosystems undergo sudden, radical structural changes makes a nonlinear analysis of finite amplitude stress imperative. This endeavor is bound to be wrought with analytical difficulties. Even the simplest nonlinear model can exhibit bizzare behavior (May, 1974). To analyze finite amplitude stress, the mathematical ecologist may have to enlarge his skills to include such subjects as statistical mechanics, topology, nonlinear optimization theory, and variational calculus. The progress in this field treats the feasibility of alternate stable states.

The need for a holistic approach to ecosystem response has long been recognized (Odum, 1977), but the development of a macroscopic theory remains in its preliminary stages. Some correlation of total system variables to imposed stress has been noted, but much more empiricism seems necessary before the inductive leap to fundamental principles can be made.

Despite the remoteness of holistic principles, this review brings up several questions that may indicate a fruitful approach to macroscopic laws.

First, the language used in discussing nonlinear ecosystem response [e.g., "domain of attraction," "adsorbing set" (Botkin and Sobel, 1975), "strange attractor" (May and Oster, 1976)] is intriguing. If much of the experimentation with dynamic systems leads to the recognition of attractor surfaces, why not make an effort to describe the attractor surface, both mathematically and biologically?

Second, Kerr's (1974) emphasis on reconciling microscopic and macroscopic approaches to ecosystem research deserves consideration. Is it necessary to wait until theories at both levels are well developed before the two can be related, or can a single principle bridge the gap between them?

Third, Harte and Levy's (1975) speculation on the ordination of various stable states is appealing. What ecosystems manager would not rejoice at a quantitative comparison between two ecosystem states which distinguishes the more mature?

Fourth, and most relevant to this discussion, what is a definition of total system strain, H, which can be closely related to the dynamics of the system?

Finally, how might the gulf between the systems ecologists and the evolutionary biologists be bridged? Like the fluid dynamicist who has the field representation of Newton's laws of motion from which to deduce the behavior of particular flows, most classically trained ecologists explain various species behavior in terms of Darwinian selection. Ecosystem models do not derive from any fundamental principle, however; they are a patchwork of empirical analogies and educated guesses.

No finished solutions are readily forthcoming, but I would like to speculate that all these questions can be addressed by a variational or optimizational statement. The clues that this might be so are found in the language of both the modeler and the evolutionist. Attractors can be described as points or surfaces of maximal properties. The evolutionist, in turn, is forever speaking in the superlative.

Others have hinted at an ecosystem variational principle (Glansdorf and Prigogine, 1971; Kerner, 1964; Ulanowicz, 1972). As early as 1925, Lotka suggested that living systems act to maximize the rate of energy capture. Odum and Pinkerton (1955) elaborated on this theme, and H. T. Odum treated the Lotka principle as axiomatic in many of his analyses. Energy is used in a very loose sense in these discussions. It is likely that one of the later definitions [e.g., exergy (Rant, 1956), the energetic measure of the departure of a system from the thermodynamic equilibrium state] is more suited to the descriptional task.

Thus we can envision a surface $H(X, \dot{X})$ in phase hyperspace such that any spontaneous movement along the surface maximizes the rate of energy storage (or some other suitable property). Domains of attraction are delimited by relative minimums. Attractors are points or surfaces of relative maximums. Strain in the ecosystem is defined as the distance between H and H*. Presumably H would have its zero level at thermodynamic equilibrium so that the values of the relative maximums would indicate the maturity of each domain of attraction, allowing us to compare different ecosystem structures. Like Liapunov functions, H would be a system property defined from the components.

The requisite variational principle would be an extension of evolutionary theory. That is, current evolutionary dogma would be necessary but not sufficient to explain all the phenomena the new principle would presumably encompass. Surely such extension would be objected to by many as unnecessary. Some might even complain that an ecosystem variational principle smacks of teleology, but a glance at the application of variational principles in inanimate physics shows this fear to be groundless. Ultimately, however, the burden of proof is upon the proponents of the new principle to demonstrate the unifying powers alluded to.

Viewed in the context of such a unifying theory, labors with disturbed ecosystems take on an added importance. Beyond yielding answers to temporal questions of how best to manage impacted ecosystems, they provide pieces for the unending puzzle of where we come from and where we are headed.

ACKNOWLEDGMENT

This is contribution No. 808 of the University of Maryland, Center for Environmental and Estuarine Studies.

REFERENCES

Allen, P., 1976, Population Dynamics and Evolution, in *Evolution and Consciousness: Human Systems in Transition*, E. Jantsch and C. H. Waddington (Eds.), pp. 127-130, Addison—Wesley Publishing Co., Reading, Mass.

Basowitz, H., H. Persky, S. J. Korchin, and R. R. Grinker, 1955, *Anxiety and Stress*, McGraw-Hill Book Company, New York.

Bierman, V. J., F. H. Verhoff, T. L. Poulson, and M. W. Tenney, 1973, Multi-Nutrient Dynamic Models of Algal Growth and Species Competition in Eutrophic Lakes, in *Modeling the Eutrophication Process*, pp. 89-109, Utah Water Research Laboratory, Logan.

Botkin, D. B., and M. J. Sobel, 1975, The Complexity of Ecosystem Stability, in *Ecosystems: Analysis and Prediction*, SIAM—SIMS Research Application Conference on Ecosystems, Alta, Utah, S. A. Levin (Ed.), pp. 144-150, Society for Industrial and Applied Mathematics, Philadelphia.

Di Toro, D. M., D. J. O'Connor, R. V. Thomann, and J. L. Mancini, 1975, Phytoplankton—Zooplankton—Nutrient Interaction Model for Western Lake Erie, in *Systems Analysis and Simulation Ecology*, B. C. Patten (Ed.), Vol. 3, pp. 423-473, Academic Press, Inc., New York.

Eigen, M., 1971, Self-Organization of Matter and the Evolution of Biological Macromolecules, *Naturwissenschaften*, 58: 465-522.

Estberg, G. N., and B. C. Patten, 1975, The Relation Between Sensitivity and Persistence Under Small Perturbations, in *Ecosystems: Analysis and Prediction*, SIAM—SIMS Research Application Conference on Ecosystems, Alta, Utah, S. A. Levin (Ed.), pp. 151-154, Society for Industrial and Applied Mathematics, Philadelphia.

Falco, J. W., and L. A. Mulkey, 1976, Modeling the Effect of Pesticide Loading in Riverine Ecosystems, in *Environmental Modeling and Simulation*, Proceedings of a Conference, Apr. 19—22, 1976, Cincinnati, W. R. Ott (Ed.), EPA-600/9-76-016, Environmental Protection Agency, NTIS.

Gardner, M. R., and W. R. Ashby, 1970, Connectance of Large Dynamical (Cybernetic) Systems: Critical Values for Stability, *Nature*, 228: 784.

Glansdorf, P., and I. Prigogine, 1971, *Thermodynamic Theory of Structure, Stability and Fluctuations*, John Wiley & Sons, Inc., New York.

Goh, R. S., 1975, Stability, Vulnerability and Persistence of Complex Ecosystems, *Ecol. Model.*, 1: 105-116.

Golley, F. B., 1974, Structural and Functional Properties as They Influence Ecosystem Stability, in *Proceedings of the First International Congress of*

Ecology, The Hague, Netherlands, pp. 97-102, Centre for Agricultural Publishing and Documentation, Wageningen.
Halfon, E., 1976, Relative Stability of Ecosystem Linear Models, *Ecol. Model.*, 2: 279-286.
——, 1978, *Theoretical Systems Ecology*, Academic Press, Inc., New York, in press.
Harte, J., and D. Levy, 1975, On the Vulnerability of Ecosystems Disturbed by Man, in *Unifying Concepts in Ecology*, W. H. Van Dobben and R. H. Lowe-McConnel (Eds.), pp. 208-223, Dr. W. Junk Publishers, The Hague, The Netherlands.
Holling, C. S., 1973, Resilience and Stability of Ecological Systems, *Annu. Rev. Ecol. Systemat.*, 4: 1-23.
Innis, G., 1975, Stability, Sensitivity, Resilience, Persistence. What Is of Interest? in *Ecosystems: Analysis and Prediction*, SIAM—SIMS Research Application Conference on Ecosystems, Alta, Utah, S. A. Levin (Ed.), pp. 131-139, Society for Industrial and Applied Mathematics, Philadelphia.
Jeffries, C., 1974, Qualitative Stability and Digraphs in Model Ecosystems, *Ecology*, 55(6): 1415-1419.
Jordan, C. F., J. R. Kline, and D. S. Sasscer, 1972, Relative Stability of Mineral Cycles in Forest Ecosystems, *Am. Nat.*, 106: 237-253.
Jorgensen, S. E., and H. Mejer, 1977, Ecological Buffer Capacity, *Ecol. Model.*, 3: 39-61.
Kemp, W. M., and M. Homer, 1977, Functional Analysis of Complex Estuarine Food Webs, unpublished manuscript.
Kerner, E. H., 1964, Dynamical Aspects of Kinetics, *Bull. Math. Biophys.*, 26: 151-167.
Kerr, S. R., 1974, Structural Analysis of Aquatic Communities, in *Proceedings of the First International Congress of Ecology, The Hague, Netherlands*, pp. 69-74, Centre for Agricultural Publishing and Documentation, Wageningen.
Lane, P. A., G. H. Lauff, and R. Levins, 1975, The Feasibility of Using a Holistic Approach in Ecosystem Analysis, in *Ecosystems: Analysis and Prediction*, SIAM—SIMS Research Application Conference on Ecosystems, Alta, Utah, S. A. Levin (Ed.), pp. 111-128, Society for Industrial and Applied Mathematics, Philadelphia.
Lassiter, R. R., and D. K. Kearns, 1973, Phytoplankton Population Changes and Nutrient Fluctuations in a Simple Aquatic Ecosystem Model, in *Modeling the Eutrophication Process*, pp. 131-138, Utah Water Research Laboratory, Logan.
Lawton, J. H., J. R. Beddington, and R. Bower, 1974, Switching in Invertebrate Predators, in *Ecological Stability*, M. B. Usher and M. H. Williamson (Eds.), pp. 141-158, Halsted Press, New York.
Levins, R., 1968, *Evolution in Changing Environments: Some Theoretical Explorations*, Princeton University Press, Princeton, N.J.
Lotka, A. J., 1925, *Elements of Physical Biology*, The Williams & Wilkins Company, Baltimore.
——, 1957, *Elements of Mathematical Biology*, Dover Publications, Inc., New York.
MacArthur, R. H., 1955, Fluctuations of Animal Populations, and a Measure of Community Stability, *Ecology*, 36: 533-536.
McMurtrie, R. E., 1975, Determinants of Stability of Large Randomly Connected Systems, *J. Theor. Biol.*, 50: 1-11.

McQueen, D. J., 1975, Multiple Stability and Maximum Stability in a Model Population, *Can. J. Zool.*, 53: 1844-1854.

Mann, K. H., 1975, Relationship Between Morphometry and Biological Functioning in Three Coastal Inlets of Nova Scotia, in *Estuarine Research*, L. E. Cronin (Ed.), pp. 634-644, Academic Press, Inc., New York.

May, R. M., 1971, Stability in Multi-Species Community Models, *Math. Biosci.*, 12: 59-79.

——, 1973, *Stability and Complexity in Model Ecosystems*, Princeton University Press, Princeton, N.J.

——, 1974, Biological Populations with Nonoverlapping Generations: Stable Points, Stable Cycles, and Chaos, *Science*, 186: 645-647.

——, 1975, Stability in Ecosystems: Some Comments, in *Unifying Concepts in Ecology*, W. H. Van Dobben and R. H. Lowe-McConnel (Eds.), pp. 161-168, Dr. W. Junk Publishers, The Hague, The Netherlands.

——, and G. F. Oster, 1976, Bifurcations and Dynamic Complexity of Simple Ecological Models, *Am. Nat.*, 110(974): 573-599.

Meier, R. L., 1972, Communications Stress, *Annu. Rev. Ecol. Systemat.*, 3: 289-314.

Mulholland, R. J., 1975, Stability Analysis of the Response of Ecosystems to Perturbations, in *Ecosystems: Analysis and Prediction*, SIAM—SIMS Research Application Conference on Ecosystems, Alta, Utah, S. A. Levin (Ed.), pp. 166-181, Society for Industrial and Applied Mathematics, Philadelphia.

Nixon, S. W., and J. N. Kremer, 1977, Narragansett Bay—The Development of a Composite Simulation Model for a New England Estuary, in *Ecosystem Modeling in Theory and Practice: An Introduction with Case Histories*, C. A. S. Holland and J. W. Day, Jr. (Eds.), pp. 621-673, John Wiley & Sons, Inc., New York.

Odum, E. P., 1977, The Emergence of Ecology as a New Integrative Discipline, *Science*, 195(4284): 1289-1293.

Odum, H. T., and R. C. Pinkerton, 1955, Time's Speed Regulator, the Optimum Efficiency for Maximum Output in Physical and Biological Systems, *Am. Sci.*, 43: 331-343.

O'Neill, R. V., 1975, Management of Large-Scale Environmental Modeling Projects, in *Ecological Modeling in a Resource Management Framework*, C. S. Russell (Ed.), pp. 251-282, Resources for the Future, Inc., Johns Hopkins University Press, Baltimore.

Orlob, G. T., 1975, Present Problems and Future Prospects of Ecological Modeling, in *Ecological Modeling in a Resource Management Framework*, C. S. Russell (Ed.), pp. 283-312, Resources for the Future, Inc., Johns Hopkins University Press, Baltimore.

Park, R. A., D. Scavia, and N. L. Clesceri, 1975, CLEANER: The Lake George Model, in *Ecological Modeling in a Resource Management Framework*, C. S. Russell (Ed.), pp. 49-82, Resources for the Future, Inc., Johns Hopkins University Press, Baltimore.

Patten, B. C., 1975, Ecosystems Linearization: An Evolutionary Design Problem, *Am. Nat.*, 109(969): 529-539.

——, 1976, University of Georgia, personal communication.

——, D. A. Egloff, T. H. Richardson, 1975, Total Ecosystem Model for a Cove in Lake Texoma, in *Systems Analysis and Simulation Ecology*, B. C. Patten (Ed.), Vol. 3, pp. 206-243, Academic Press, Inc., New York.

Prigogine, I., 1976, Order Through Fluctuation: Self-Organization and Social System, in *Evolution and Consciousness: Human Systems in Transition*, E. Jantsch and C. H. Waddington (Eds.), pp. 93-126, Addison—Wesley Publishing Company, Reading, Mass.

Rant, Z., 1956, Exergie, ein neues Wort fuer technische Arbeitsfaehigkeit, *Forsch. Ingenieurwes.*, 22(1): 36-37.

Roberts, A., 1974, The Stability of a Feasible Random Ecosystem, *Nature*, 251(5476): 607-608.

Rutledge, R. W., 1974, *Ecological Stability: A Systems Theory Viewpoint*, Ph.D. Thesis, Oklahoma State University, Stillwater.

Saunders, P. T., and M. J. Bazin, 1975, Stability of Complex Ecosystems, *Nature*, 256(5513): 120-121.

Schaaf, W. E., 1975, Fish Population Models: Potential and Actual Links to Ecological Models, in *Ecological Modeling in a Resource Management Framework*, C. S. Russell (Ed.), pp. 211-239, Resources for the Future, Inc., Johns Hopkins University Press, Baltimore.

Smith, D. F., 1975, Quantitative Analysis of the Functional Relationships Existing Between Ecosystem Components, *Oecologia (Berlin)*, 21: 17-29.

Sutherland, J. P., 1974, Multiple Stable Points in Natural Communities, *Am. Nat.*, 108(964): 859-873.

Ulanowicz, R. E., 1972, Mass and Energy Flow in Closed Ecosystems, *J. Theor. Biol.*, 34: 239-253.

——, D. A. Flemer, D. R. Heinle, and R. T. Huff, 1978, The Empirical Modeling of an Ecosystem, *Ecol. Model.*, 4: 29-40.

——, and W. M. Kemp, 1978, Towards Canonical Trophic Aggregations, submitted to *Am. Nat.*

Wiegert, R. G., 1975, Simulation Models of Ecosystems, *Annu. Rev. Ecol. Systemat.*, 6: 311-338.

Woodwell, G. M., 1975, The Threshold Problem in Ecosystems, in *Ecosystems: Analysis and Prediction*, SIAM—SIMS Research Application Conference on Ecosystems, Alta, Utah, S. A. Levin (Ed.), pp. 9-23, Society for Industrial and Applied Mathematics, Philadelphia.

FREQUENCY RESPONSE OF A MARINE ECOSYSTEM SUBJECTED TO TIME-VARYING INPUTS

ROBERT L. DWYER,*† SCOTT W. NIXON,* CANDACE A. OVIATT,*
KENNETH T. PEREZ,† and THEODORE J. SMAYDA*
*Graduate School of Oceanography, University of Rhode Island, Kingston, Rhode Island; and †Environmental Protection Agency, Environmental Research Laboratory, Narragansett, Rhode Island

ABSTRACT

Recent studies have drawn conclusions about the stability and linearity of ecosystem response to environmental fluctuations, on the basis of analyses of linear mathematical models of several different ecosystems. The validity of model stability estimates can be verified only by an independent method that is not based on assumptions implicit in the formulation of the model. An independent method for deriving frequency response which uses time-series data for an environmental input to and biological response from an ecosystem is presented. Time series of solar radiation, water temperature, ammonia, and phytoplankton abundance, from weekly samples taken in Narragansett Bay, Rhode Island, over a 17-year period, and time series of ammonia and chlorophyll a from a 6-month sewage-perturbation experiment in 150-liter microcosms simulating Narragansett Bay were analyzed by spectral analysis. Spectra for the environmental inputs to the bay ecosystem showed periodicities only at 1 cycle/year, whereas the spectrum for phytoplankton abundance showed additional significant periodicities at 2 and 4 cycles/year. These were judged to be harmonics resulting from a nonlinearity in the ecosystem. Cross-spectral analysis of inputs vs. phytoplankton abundance showed no evidence of linear frequency response. Inadequacies in experimental design for the microcosm experiment hindered quantitative estimation of microcosm frequency response, but the advantage of microcosms in allowing full control during input—response experiments is shown.

The last two decades have been marked by a quantum jump in the application of dynamic systems theory to the analysis of ecosystems. Mathematical modeling of ecosystems has been undertaken for two purposes, to describe mechanisms regulating the behavior of the ecosystem and to predict the effects of disturbances to the system (Wiegert, 1975). Debate concerning optimal methods for achieving these goals continues. In particular, the level of detail included in the model (e.g., the number of state variables or compartments, the form of the interaction equations, the possible inclusion of uncertainty in model structure and function, the importance of spatial and temporal heterogeneity, and methods for model validation and sensitivity analysis) has been a topic for lively discussion (Wiegert, 1975).

We have focused this study on one controversial aspect of modeling philosophy—the usefulness of limiting ecosystem models exclusively to systems of linear equations. We show that an adequate description of ecosystem stability properties cannot always be derived from a linearized model of the ecosystem.

The advantages and drawbacks of linear modeling methods were discussed in detail by Wiegert (1975), O'Neill (1975), and Bledsoe (1976). They agree that a variety of techniques exist for dealing with systems of linear equations and that linear analysis may be valuable for the range of behavior over which an ecosystem is known to respond linearly. All conclude, however, that there is little a priori justification for applying linear methods to all ecosystem studies and that the convenience of linear methods is one major reason for their use in many ecosystem studies.

One technique frequently employed is linear frequency response analysis. In this paper we evaluate the usefulness of this methodology in an instance when the linearity of ecosystem behavior is not apparent.

Child and Shugart (1972) estimated the frequency response of their linear model of magnesium cycling in a tropical forest by perturbing it with a spectrum of sinusoidal inputs. They used plots of response amplitude, normalized with respect to input amplitude, vs. frequency and of phase shift vs. frequency to provide information about linear system properties. These are the Bode plots familiar to engineers [see Fig. 7(a)] and can be used to write a transfer function for the system, which is a Laplace-transformed, linear differential-equation model. Child and Shugart (1972), Shinners (1972), Brewer (1974), or any elementary systems text will provide a more thorough discussion of Bode plots and frequency-response techniques.

Waide et al. (1974) used similar frequency domain methods to evaluate the stability and sensitivity to perturbations of a linear model of calcium cycling in the Hubbard Brook, New Hampshire, watersheds, a temperate forest. Webster, Waide, and Patten (1975) extended the linear frequency response methods used by Child and

Shugart (1972) to estimate ecosystem stability and merged with them ideas about two aspects of ecosystem stability discussed by Holling (1973). These ideas concern the ability of an ecosystem to resist displacement from a stable point and its speed of return to the stable point after perturbation.

Holling viewed ecosystems as nonlinear entities capable of reorganizing structurally and of assuming different stable states in response to perturbation. A system displaced far beyond changes caused by normal environmental fluctuations can adapt by modifying its structure or function. It may move, perhaps irreversibly, into the attractive domain of a new stable point.

In contrast, Webster, Waide, and Patten (1975) considered ecosystem stability in a more restricted sense. They viewed normal ecosystem behavior as a linear response to normally occurring environmental fluctuations (Patten, 1975). A perturbation large enough to cause a structural change as an adaptation replaces the original ecosystem with a new one, whose dynamics should be linear about a new stable point.

In this sense, Webster, Waide, and Patten (1975) redefined Holling's two aspects of stability and used two parameters from linear frequency-response analysis as estimators of normal ecosystem stability: (1) Resistance is the ability of an ecosystem to withstand displacement by an input and is estimated, inversely, by the undamped natural frequency, ω_n, of the system. (2) Resilience is the ability of the ecosystem to return to equilibrium once displaced and can be estimated by the system damping ratio, ζ. Both these parameters can be estimated numerically from Bode amplitude and phase plots calculated from linear models (see Child and Shugart, 1972; Waide et al., 1974). Webster, Waide, and Patten derived these frequency response parameters for eight different ecosystem models by analytical solution of the linear systems of equations.

Harwell, Cropper, and Ragsdale (1977), using digital and analog simulations of the same eight linear models, generated stability rankings for the ecosystems which differed from those of Webster, Waide, and Patten (1975). The discrepancies may be partially explained by the presence of nonlinearities in ecosystem response to normal fluctuations. An alternate method for quantifying ecosystem stability, one that can verify stability estimates independently of assumptions in a linear model, would be useful.

In this paper we apply a method well known in the physical sciences to evaluate the linearity of response to seasonal environmental fluctuations of one compartment in a natural ecosystem, Narragansett Bay, Rhode Island. This method, which is based on spectral analysis of time series of input—response data, does not require a quantitative compartment model for the ecosystem. Thus i is a "black-box" approach and minimizes the effects of th investigator's biases about the structure and function of the ecosys tem under study.

Since spectral frequency response analysis of ecosystem behavio in response to year-to-year fluctuations provides little informatio about its linearity or stability in response to fluctuations at highe amplitudes or frequencies of input, ecosystem experiments usin artificial inputs of the desired characteristics are required. Eisne (1971) proposed this approach to examine the dynamics of a ecosystem perturbed far from a stable point and outlined criteria fo input design which optimize the detection of response abov background noise.

Applying a time-varying input to a whole ecosystem usuall presents significant methodological problems. Using controlled replicated microcosms to simulate an ecosystem alleviates some o the difficulties. Potentially, microcosm "models" can be used t delineate the boundaries of the ecosystem's stable domain. I addition to spectral analysis of the Narragansett Bay time series, w describe the results of a first experiment using a sewage input t perturb 150-liter microcosms simulating Narragansett Bay beyon their nominal dynamics.

MATERIALS AND METHODS

Description of Data

The dynamics of phytoplankton abundance in Narragansett Bay have been described in detail (Smayda, 1976, and the literature cited therein). Smayda and his students have sampled a station in the West Passage of the bay at weekly intervals since 1959. Using this long time series of physical and biological data, we discuss solar radiation, surface water temperature, surface ammonia concentration, and total phytoplankton abundance (Fig. 1). Solar-radiation and water-temperature values are for the day of sampling and represent an essentially complete record for the period from 1959 to 1975. Averages of solar radiation and temperature weighted over the days preceding sampling would probably model phytoplankton growth processes more realistically. Complete temperature data and information to generate the appropriate weighting functions were not readily available, however, so unweighted grab measurements were used in these analyses. Surface nutrient sampling was begun in 1972. Ammonia, nitrate, nitrite, and phosphate all show the same qualitative pattern. Ammonia alone is used because its fluctuations

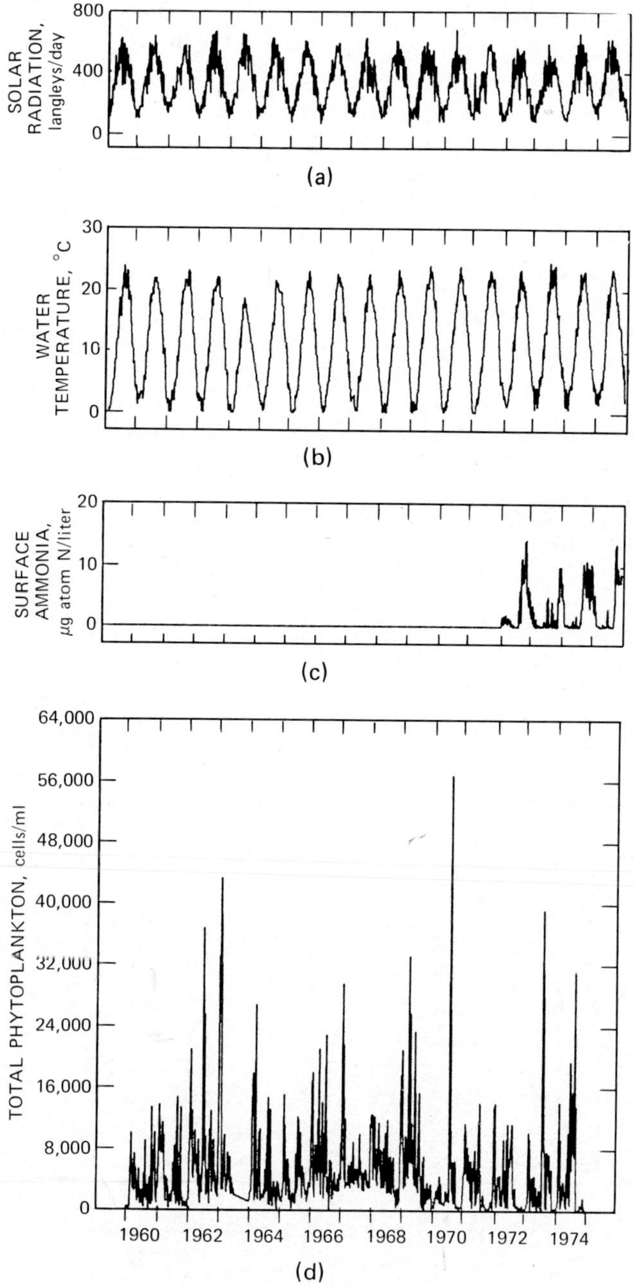

Fig. 1 Time-series data taken at weekly intervals in Narragansett Bay, Rhode Island, from 1959 to 1975. Ammonia (c) sampling was begun in 1972. Phytoplankton cell abundances (d) were counted beginning in 1960. Samples for 1975 have yet to be counted.

are most pronounced and because there is evidence that it cycles more rapidly than other nutrients in the bay. Phytoplankton abundances, by species, were counted beginning in 1960. Since samples for 1975 have yet to be processed and digitized, they are not included in the cell-count time series.

Since spectral analysis requires complete time-series data, it was necessary to interpolate missing data points linearly. For each time series the total number of interpolated points was less than 10% of the total number of data points.

Solar-radiation and temperature time series show strong deterministic oscillations at 1 cycle/year, but the periodicity for ammonia is less clear, and that for total phytoplankton abundance is undistinguishable. The frequency structure of these time series is manifest only after transformation to the frequency domain by spectral analysis.

As part of a multiphase approach to studying the Narragansett Bay ecosystem, a set of 150-liter, scaled, benthic—pelagic microcosms were developed and are being used in perturbation experiments (Nixon et al., 1978; Oviatt, Perez, and Nixon, 1977; Perez et al., 1977). Data used here are from a sewage stress—relaxation experiment described by Oviatt, Perez, and Nixon (Fig. 2). We chose ammonia and chlorophyll time series from a microcosm receiving a high sewage step input (microcosm 2, Fig. 2). Since the microcosms were not sampled at regular intervals (there was sometimes a day's delay in sampling), it was necessary to approximate the complete time series by use of linearly interpolated points (11% of the total time series). These microcosm time series are much shorter than those from Narragansett Bay, and the ammonia series was sampled twice per week.

Systems and Holistic Responses

The trend toward developing holistic measures of natural ecosystem dynamics has led to perturbation experiments on ecosystem models (Child and Shugart, 1972; Waide et al., 1974; Harwell, Cropper, and Ragsdale, 1977) in which one or more model compartments or inputs are artificially displaced. Harwell, Cropper, and Ragsdale showed that estimated model stability (e.g., in terms of natural frequency and damping ratio) may depend crucially on what is perturbed, how large the perturbation amplitude is, and what is measured as the response. Perturbations applied to real ecosystems may generate responses that are functions of the timing of application as well as of system state. This may be especially

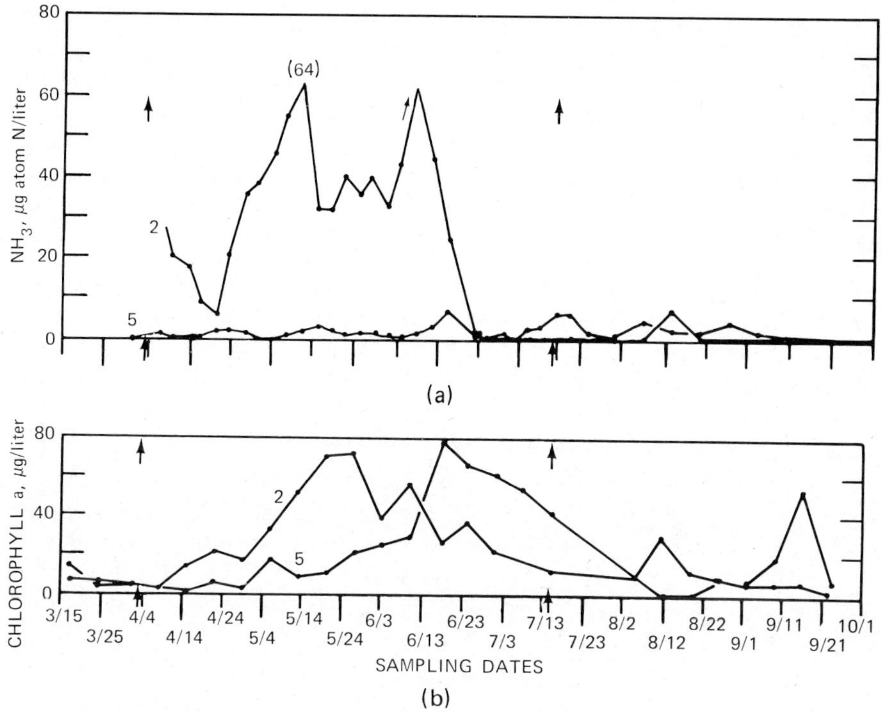

Fig. 2 Ammonia (a) and chlorophyll (b) concentrations in one of several replicate microcosms (No. 2) receiving a 3-month sewage input (between arrows). One representative control, microcosm No. 5, is also shown (adapted from Oviatt, Perez, and Nixon, 1977).

important for perturbations that temporarily or permanently disrupt feedback pathways (e.g., clear-cutting of forests).

In practice, defining system boundaries, inputs, and outputs can be troublesome. An exogenous environmental input, by definition, is one that affects, but is not affected by, system state. Solar radiation, rainfall, and temperature fit this definition. An output can be any compartment monitored by the investigator, or it can be transport of material beyond the system boundary, where it no longer modifies system functions.

Thus the operational definition of ecosystem response has depended to a great extent on what outputs researchers were equipped to measure. As yet, not a great deal of thought has been given to defining which outputs or compartments should be monitored to elucidate important aspects of ecosystem stability. The response linearization hypothesis of Patten (1975) may be correct for some types of outputs but not for others.

Defining boundaries, inputs, and outputs is obviously easier for some ecosystems. For example, the Hubbard Brook watersheds (Waide et al., 1974) have easily measurable inputs, and all material output flows through a weir where it can be monitored. Defining system boundaries in Narragansett Bay is much more difficult, and defining fluxes across an operationally established estuarine—oceanic boundary is a major research task in itself (Kremer and Nixon, 1977).

Our spectral analysis of Narragansett Bay is constrained by the availability of suitable data. The analysis is a detailed examination of the in-system and closed feedback-loop dynamics (Child and Shugart, 1972; Shinners, 1972) of the primary producer compartment of the Narragansett Bay ecosystem. We make no claims that the frequency responses we calculated represent those of the whole system. If Patten's (1975) linearization proposal is correct, however, the many nonlinear submechanisms governing this compartment (e.g., pronounced high-frequency phytoplankton species succession and saturation kinetics for light and nutrients) will be minimized in a tendency toward linearization of the frequency response of primary production.

One input to primary production, surface ammonia, is an integrating compartment in the ecosystem. It is taken up preferentially by phytoplankton in Narragansett Bay and is limiting most of the year. Through feedback mechanisms, ammonia concentration is affected by primary producers. Ammonia concentration is also modified by an exogenous input (i.e., river runoff), which carries ammonia from numerous sewage treatment plants, and by wind-driven water turbulence, which modifies nutrient regeneration by benthic sediments. Exact analysis of this problem requires a detailed signal-theory approach, which was not attempted for this analysis. Thus we must make the simplifying assumption that the effect on ammonia concentration of primary production fed back through the food web is small compared with the direct effect of ammonia on primary producers. This assumption is supported by measurements of zooplankton ammonia excretion, which show that throughout most of the year only a small portion of the primary production in the bay is supported by nitrogen that has been rapidly regenerated through the herbivores (Vargo, 1976). The assumption is even more nearly correct for the microcosm experiments, in which high levels of ammonia were added artificially. Ammonia levels in control microcosms (modified only by internal feedback mechanisms) remained much lower than those in microcosms receiving an artificial input (Fig. 2).

Spectral-Analysis Methods

Spectral analysis may be considered a statistical method for partitioning the variance of a time series of data among a range of frequencies of oscillation. Spectral analysis performs a function analogous to that of a prism breaking a beam of light into its component fractions at various wavelengths (colors) (see Fig. 3). For the interested reader without a strong mathematical background, we suggest several relatively nontechnical papers on spectral analysis (e.g., Gunnerson, 1966; Wastler, 1969; Platt and Denman, 1975). In our presentation here we minimize the mathematics underlying spectral analysis. Unless otherwise noted, all techniques, including equations incorporated into FORTRAN IV computer programs (Weisberg, 1974), are taken directly from Bendat and Piersol (1971).

Basically, spectral analysis takes time-series data sampled at regular intervals from a repeating periodic process, filters it to remove unwanted contaminating signals, removes any linear trend, tapers the two ends of the data smoothly to zero (to avoid placing a spurious artifact in the frequency spectrum), and computes the coefficients of a discrete Fourier series with one of several available numerical algorithms (see Platt and Denman, 1975). We used a Fast Fourier Transform (FFT), which yields a series of Fourier coefficients equal in length to one-half the number of data points. These coefficients yield the variance spectrum for the time series directly.

A confidence interval about the height of each spectral estimate can be calculated by a moving average of several adjacent spectral estimates. However, decreasing the width of the confidence interval necessitates broadening the frequency bandwidth resolution of the analysis. Thus, typically, an investigator must decide between estimating the heights of a few spectral peaks with great confidence or estimating more peak heights with a corresponding decrease in confidence.

Once calculated in this manner, spectral estimates for two time series connected by a causal relationship can be analyzed using cross-spectral techniques. A detailed derivation is beyond the scope of this paper, but several resulting parameters are of great interest. Coherence squared (γ^2) can be considered as a correlation coefficient between the two time series, varying from 0.0 to 1.0; i.e., as an estimate at each frequency of the proportion of the variance in one time series which can be explained in terms of variance in the other. When a significance test of the squared coherence between two time series can be shown to be statistically different from zero over a range of frequencies (Amos and Koopmans, 1963), a cause—effect or input—response relationship exists between the two variables at the

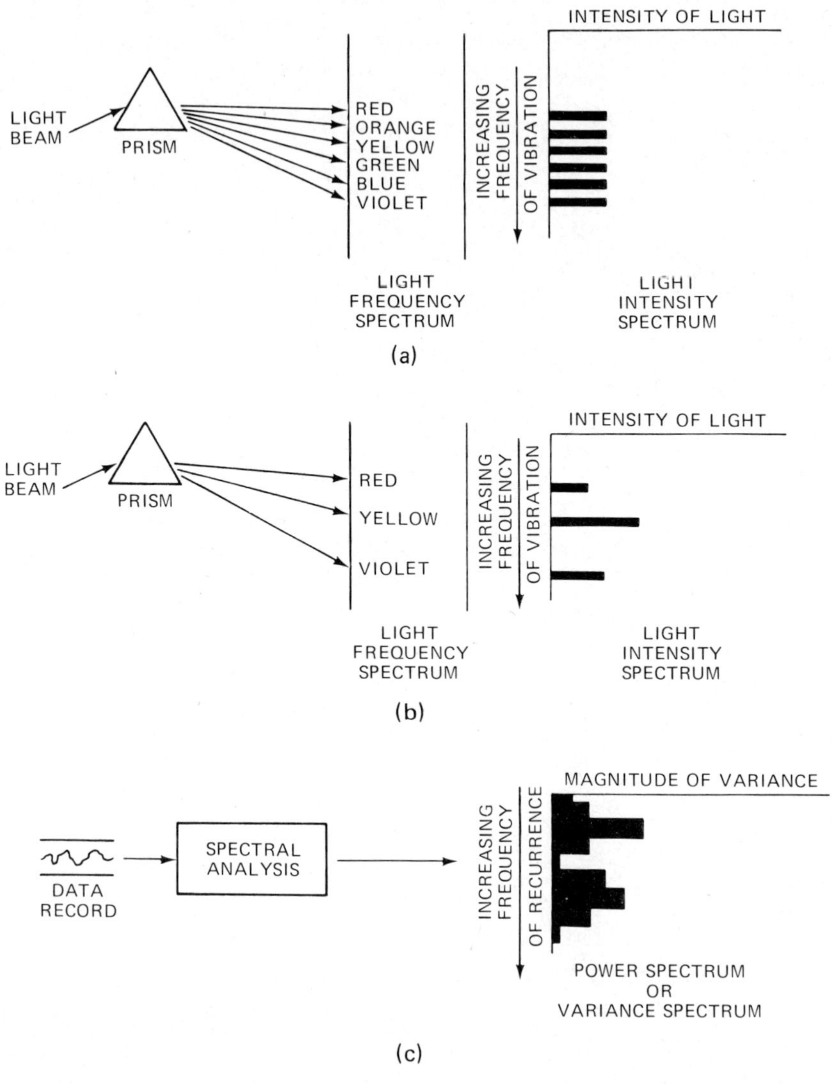

Fig. 3 Spectral analysis operating on time-series data is analogous to a prism breaking a beam of light into its colors or wavelengths (from Wastler, 1969). (a) Prismatic resolution of a beam of sunlight. (b) Prismatic resolution of a beam of light. (c) Resolution of a data record by spectral analysis.

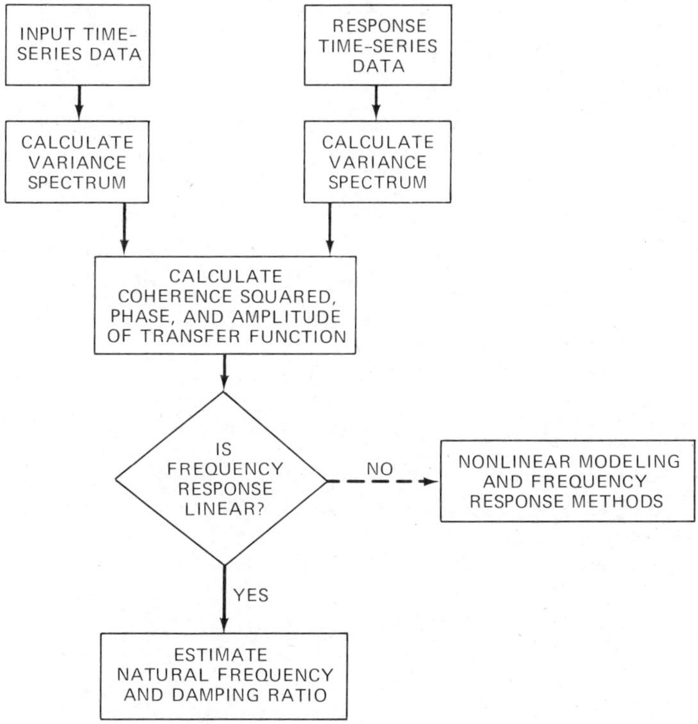

Fig. 4 Input—response spectral methodology for verifying linearity of ecosystem response and estimating frequency response of ecosystem.

significant frequencies. If squared coherence is significant over a broad band of frequencies, two more statistics can be calculated. These are the amplitude ratio and phase as functions of frequency. When plotted, these are the Bode frequency response plots described previously (see Fig. 7). Rather than being generated from a mathematical model, however, they are calculated directly from real data measured in the field or the laboratory. If the input—response transfer behaves linearly (i.e., if it has a significant squared coherence and amplitude and phase plots resembling those of linear systems), then a Laplacian transfer function model, the natural frequency (ω_n), and the damping ratio (ζ), all can be estimated graphically (Brewer, 1974). Again, we should note that this whole process (Fig. 4) represents a black-box approach in which only a qualitative cause—effect hypothesis is necessary. We use it to evaluate the linearity of the Narragansett Bay ecosystem.

RESULTS

All time-series data were tested for stationarity (constant mean and variance) and for long-term linear trends. No time series showed a significant slope over its entire length. Phytoplankton cell abundance, the most variable time series, was broken arbitrarily into six segments, each 2.5 years long, and slope, mean, and variance were calculated for each segment. Since there were no significant differences among any of the three statistics, the phytoplankton was judged to be stationary. The solar-radiation, temperature, and ammonia data were judged to be stationary by inspection (Fig. 1), as were the time series for ammonia and chlorophyll in microcosm 2 (Fig. 2).

Variance spectra were computed for the four bay time series (Fig. 5), as previously described, by modifying the Weisberg (1974) computer programs and were plotted with the graphics program of Kramer and Weisberg (1975). A 95% confidence interval, used in testing peak-height significance, is shown around phytoplankton spectral estimates [Fig. 5(d)]. The resolvable frequency bandwidth B and the number of adjacent spectral estimates over which moving averaging was done are given in the figure legend.

Variance spectra for solar radiation, temperature, and ammonia [Fig. 5(a), 5(b), and 5(c)] are similar in that they all show only one peak centered at 1 cycle/year. The flat, low-variance portion of the spectra at higher frequencies (up to 26 cycles/year) is an example of "white noise" or of a random process having variance equally distributed over all frequencies (or "colors"). White noise represents measurement errors, spatial heterogeneities, or other random errors. The errors are quite small for easily measured, relatively deterministic variables like solar radiation, temperature, and ammonia.

The variance spectrum for phytoplankton abundance [Fig. 5(d)] shows many more peaks and a high-frequency background white-noise spectrum (e.g., >10 cycles/year) of higher amplitude than the input data. In particular, spectral peaks centered at 1, 2, and 4 cycles/year all appear to have confidence intervals that do not overlap those of the background white noise or of the "valleys" on either side of the peaks. We have found no evidence to indicate that these higher frequency peaks are an artifact of our method of analysis. A tendency for spectral peaks to "leak" variance to adjacent frequency bands was almost completely suppressed by use of a 10% cosine taper window (Bendat and Piersol, 1971). We know of no environmental inputs that might pulse the Narragansett Bay ecosystem at 2 or 4 cycles/year. Other possible environmental inputs, river

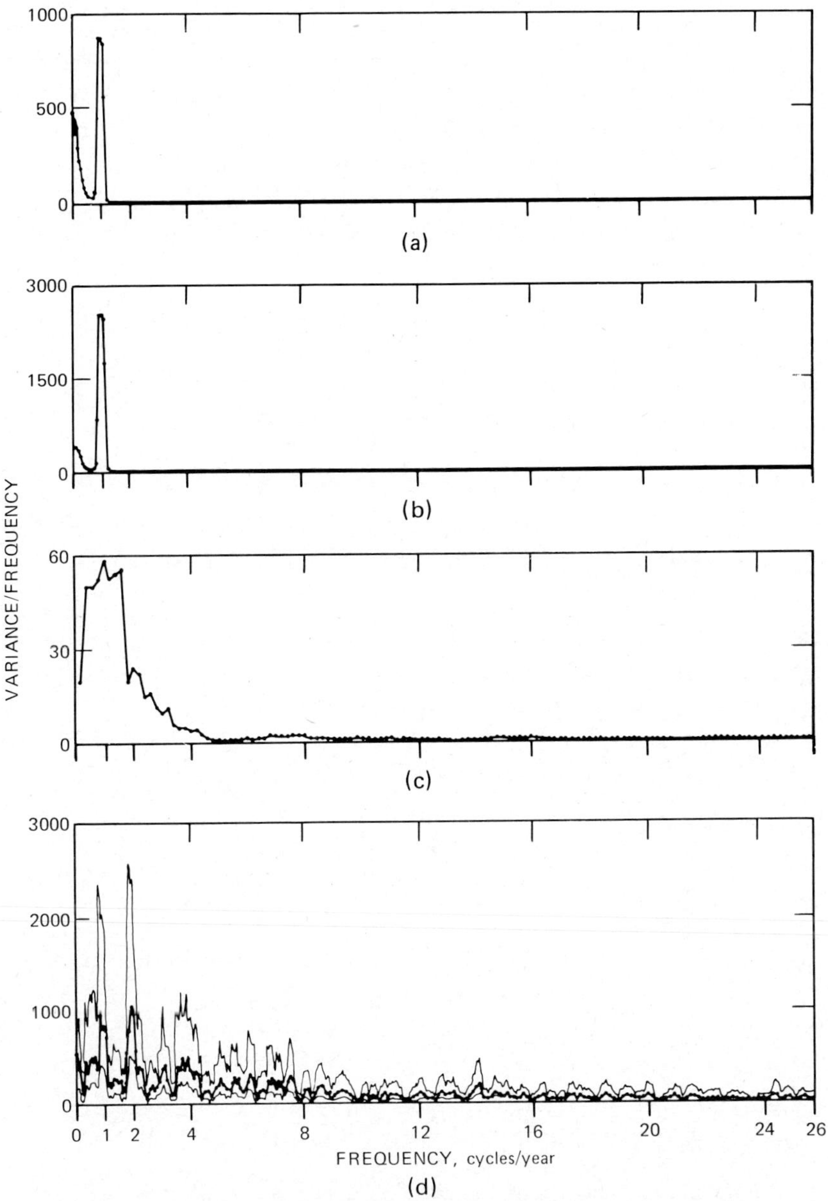

Fig. 5 Variance spectra for four Narragansett Bay time series. Each spectral estimate smoothed over seven adjacent frequency bands. (a) Solar radiation (1959–1975), B = 0.325 cycles/year. (b) Surface water temperature (1959–1975), B = 0.325 cycles/year. (c) Surface ammonia concentration (1972–1975), B = 1.31 cycles/year. (d) Phytoplankton cell abundance (1960–1974), with 95% confidence interval; B = 0.325 cycles/year.

runoff and wind-driven water turbulence being probably the most important, have not been analyzed quantitatively. River flow varies seasonally, however, with an apparently random storm-runoff component superimposed on the yearly sinusoid.

These higher frequency processes appear to be harmonics of the fundamental driving frequency of 1 cycle/year. The presence of significant oscillations in the phytoplankton data at frequencies not found in any of the major environmental inputs indicates the existence of a nonlinearity in the primary production of the ecosystem. These even harmonics are commonly observed in systems where the response is proportional to the square of an input or to a cross product of two inputs. We have not as yet found a specific mechanism in Narragansett Bay which might generate these harmonics.

Sampling of sewage input and phytoplankton response in the 6-month microcosm experiment was not designed with spectral analysis in mind. A square-wave sewage input of 3-month's duration (between arrows, Fig. 2) was attempted. One microcosm receiving sewage and one control (representing many replicates) are shown. The Fourier transformation used in spectral analysis treats this single square wave as a repeating process occurring twice per year; this is the primary spectral peak in both ammonia input and chlorophyll response (Fig. 6). Although theoretically an impulse or step input to a system can yield all the information necessary to calculate system frequency response, the actual generation of the infinite slope characteristic of impulses and step inputs is virtually impossible in the real world. Spectra for both ammonia input and chlorophyll response show secondary peaks at even multiples of 2 cycles/year and a large variance (caused by lack of stationarity) at zero frequency. These tend to mask the possible presence of true harmonics caused by ecosystem nonlinearities.

Despite our inability to ascertain through this experiment the degree of linearity in our microcosm models, we feel that input of a true sinusoid that repeats many times in the course of an experiment, combined with regular high-frequency sampling, will yield time-series data adequate for estimating ecosystem stability and linearity.

In another attempt to find linear behavior, cross-spectral analysis for the bay and microcosm time series were performed for all possible single inputs to the phytoplankton compartment. Three examples, representative of the other results, are presented here.

First, portions of the ammonia and phytoplankton time series which overlap (1972—1974) were used to compute spectra. Cross-spectral analysis of these two spectra generated estimates of

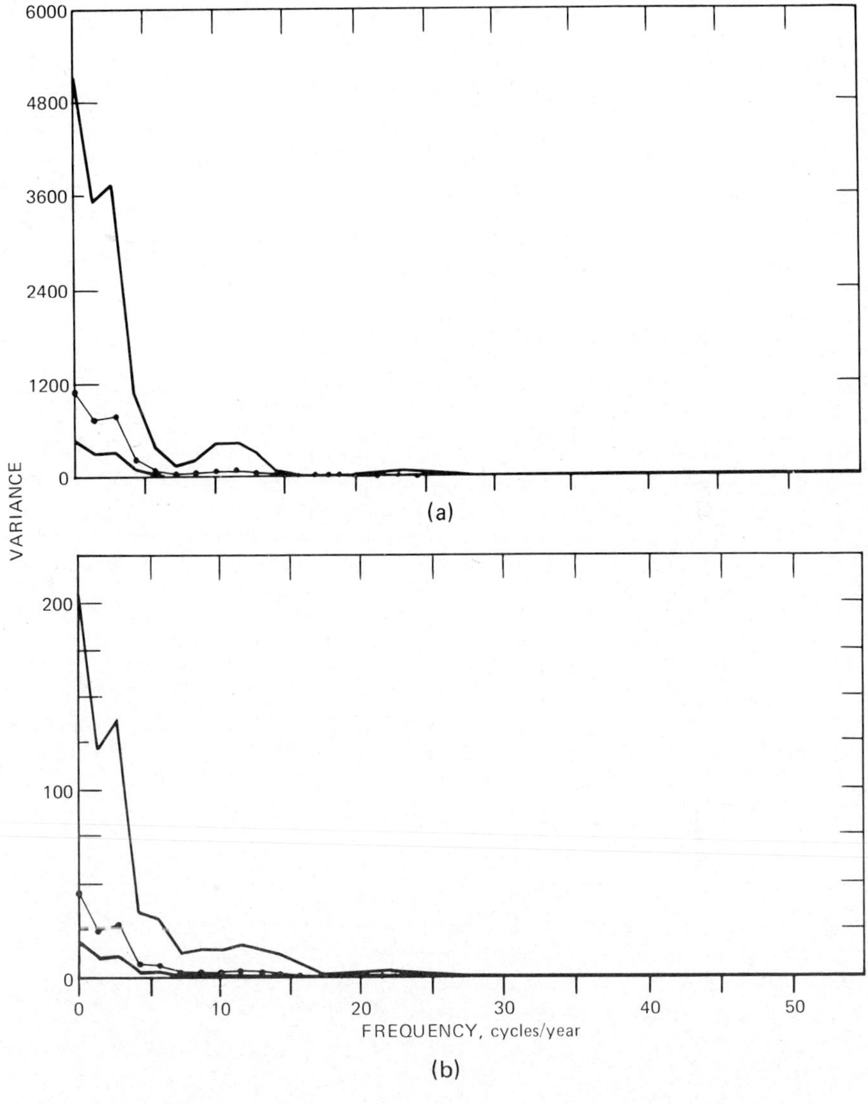

Fig. 6 Variance spectra, with 95% confidence interval, for ammonia (a) and chlorophyll (b) in microcosm 2. Each spectral estimate is smoothed over three adjacent frequency bands. B = 4.21 cycles/year. These are similar to spectra for other sewage treatment replicates. Spectra for control replicates were much less peaked.

coherence squared and the Bode amplitude and phase plots [Fig. 7(b)]. Coherence squared was not significant over most of the frequency range. Since cross-spectral frequency response analysis evaluates the linear dependence, measured by coherence squared, of response on the input, there appears to be little linear relation between the two spectra. The Bode plots show chaotic behavior at frequencies higher than 1 cycle/year; this again indicates a poor linear relation. Cross-spectral analyses of the complete solar-radiation and phytoplankton-abundance time series [Fig. 7(c)] and of the microcosm ammonia and chlorophyll time series [Fig. 7(d)] show similar random frequency responses.

Bode amplitude and phase plots show very definite forms when a linear relation exists [see Child and Shugart, 1972; Fig. 7(a)]. When calculated from time-series data from a quasilinear system, they can provide information on the minimum complexity needed in an ecosystem model, as well as estimates of system stability properties. Our inability to use them here stems from the nonlinearity of the system and from the fact that significant periodicities are represented in only a very narrow frequency range in the data (1 to 4 cycles/year). Variance at higher frequencies represents only the random portion of the data, and, by definition, the coherence between two random time series is not significant. Thus the high-frequency portions of the Bode plots represent essentially random variation.

DISCUSSION

A major difficulty in applying cross-spectral methods to data from natural ecosystems, as pointed out by the results of this study, is the fact that natural environmental inputs fluctuate only at a few frequencies. Marine ecosystems receive few stationary inputs at frequencies other than once per year, once per day, or once per tidal cycle. [Although wind-driven turbulence inputs can influence phytoplankton at very high frequencies (Platt and Denman, 1975, and the literature cited therein), these fluctuations cannot be resolved with the 1-week sampling used here.] Coherence between natural-input time series and ecological responses will be significant only at these input frequencies, regardless of what other periodicities are present in response spectra. The high-frequency portions of the spectra in this study (Figs. 5 and 6) show no significant variance other than that associated with sampling errors.

In practice, applying artificial periodic inputs to natural ecosystems presents many logistical difficulties. The use of laboratory

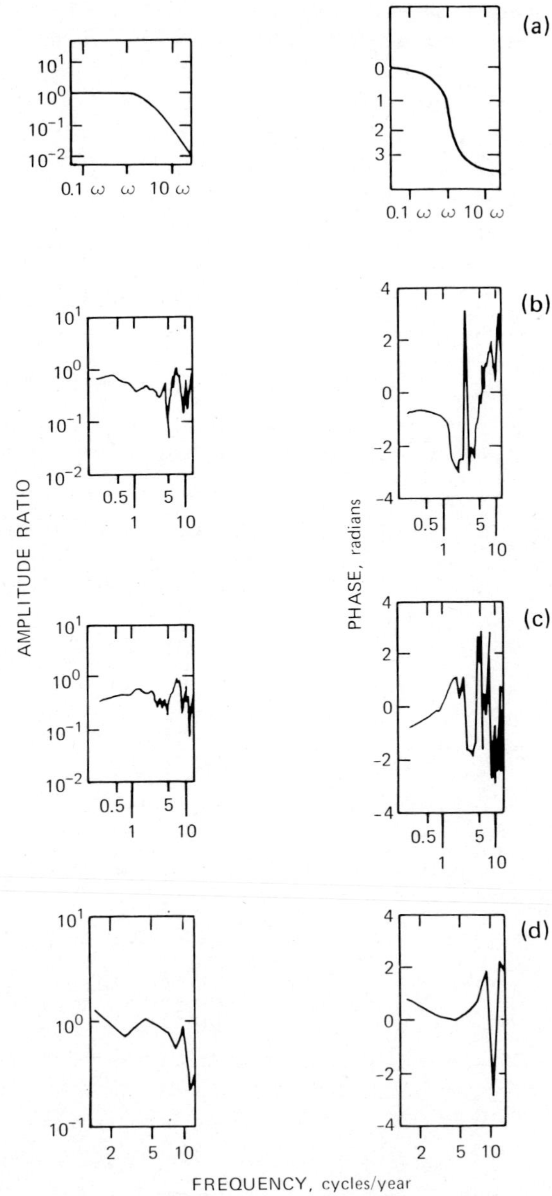

Fig. 7 Bode amplitude and phase plots generated by cross-spectral analysis of an input and a response. (a) Plots from a hypothetical second-order linear system. ω_n and ζ can be estimated graphically. (b) Plots of Narragansett Bay ammonia vs. cell abundance. (c) Plots of Narragansett Bay solar radiation vs. cell abundance. (d) Plots of microcosm 2 ammonia vs. chlorophyll.

microcosms alleviates most problems by permitting complete environmental control and making frequent sampling easy. Replicated laboratory microcosms subjected to broad-band white-noise inputs produce flatter response spectra than those produced by natural narrow-band inputs [Fig. 5(d)]. The flatter input and response spectra tend to show more coherence and, thus, are more amenable to cross-spectral frequency response analysis.

Since there is evidence of a nonlinearity in the Narragansett Bay ecosystem, we investigated the possible use of some nonlinear frequency response methods. The most promising appeared to be the describing function (Shinners, 1972; Brewer, 1974), but its application produced no useful results.

Patten (1975), in describing natural ecosystems as linear because frequencies present in the inputs are always present in the output, ignored the possible presence of additional variance at other frequencies in the response spectrum. Another example of the presence of additional variance was recently described by Ollason (1977), who performed a frequency-domain analysis of three freshwater microcosms, each subjected to a different amplitude of an incident-light sinusoid with a period of 4 days. Variance spectra were calculated for time series of the phytoplankton component and three protozoan taxa for each microcosm. The spectrum for phytoplankton in the microcosm receiving the lowest amplitude sinusoid showed variance only at the input frequency (1 cycle/4 days). Spectra for the other three components of that microcosm and all spectra for the other two microcosms showed substantial variance at frequencies other than the input frequency. Thus the characteristics of the response spectra appear to be functions of the frequency of the input sinusoid, as well as of the system compartment being monitored. We concluded that linearity of response is not a universal ecosystem property and that we must test for it with data from the natural ecosystem before we attempt to model its near-equilibrium dynamics linearly or to estimate its stability properties.

Spectral analysis is a valuable testing method. Its major drawback is the number and frequency of samples needed to resolve significant peaks in variance spectra. Many studies have already been done, however, in which useful spectra have been computed from ecological time-series data (see Platt and Denman, 1975). Rapid technological development of automated sampling equipment and phenomenal price decreases for minicomputer systems capable of automated data acquisition will make good environmental time series available to most researchers. As physical scientists have already learned, frequency-domain analyses provide versatile tools for reducing and interpreting these data.

ACKNOWLEDGMENTS

The research reported here was supported by a National Science Foundation (NSF) graduate fellowship and an Environmental Protection Agency (EPA) research traineeship to R. L. Dwyer. Support for the Narragansett Bay sampling was provided by NSF grant GA-731319X to T. J. Smayda. Support for development and experimentation on the bay microcosms was provided through EPA Environmental Research Laboratory intramural funding of K. T. Perez and a grant to S. W. Nixon and C. A. Oviatt (grant No. R803143030).

REFERENCES

Amos, D. E., and L. H. Koopmans, 1963, *Tables of the Distribution of the Coefficient of Coherence for Stationary Bivariate Gaussian Processes,* Monograph SCR-483, Sandia Corp., NTIS.

Bendat, J. S., and A. J. Piersol, 1971, *Random Data: Analysis and Measurement Procedures,* John Wiley & Sons, Inc., New York.

Bledsoe, L. J., 1976, Linear and Nonlinear Approaches for Ecosystem Dynamic Modeling, in *Systems Analysis and Simulation in Ecology,* B. C. Patten (Ed.), Vol. 4, pp. 283-298, Academic Press, Inc., New York.

Brewer, J. W., 1974, *Control Systems: Analysis, Design, and Simulation,* Prentice-Hall, Inc., Englewood Cliffs, N. J.

Child, G. I., and H. H. Shugart, Jr., 1972, Frequency Response Analysis of Magnesium Cycling in a Tropical Forest Ecosystem, in *Systems Analysis and Simulation in Ecology,* B. C. Patten (Ed.), Vol. 2, pp. 103-135, Academic Press, Inc., New York.

Eisner, E., 1971, Experiments in Ecology: A Problem in Signal Extraction, in *Statistical Ecology.* Vol. 2: *Sampling and Modeling Biological Populations and Population Dynamics,* Proceedings of the International Symposium on Statistical Ecology, New Haven, Conn., August 1969, G. P. Patil, E. C. Pielou, and W. E. Waters (Eds.), pp. 237-251, Pennsylvania State University Press, University Park.

Gunnerson, C. G., 1966, Optimizing Sampling Intervals in Tidal Estuaries, *J. Sanit. Eng. Div., Am. Soc. Civ. Eng.,* 92(SA2): 103-125.

Harwell, M. A., W. P. Cropper, Jr., and H. L. Ragsdale, 1977, Nutrient Recycling and Stability: A Reevaluation, *Ecology,* 58: 660-666.

Holling, C. S., 1973, Resilience and Stability of Ecological Systems, *Annu. Rev. Ecol. Systemat.,* 4: 1-23.

Kramer, W. P., and R. H. Weisberg, 1975, *Fortran Graphics Programs for Physical Oceanographic and Time Series Data,* Marine Technical Report 46, Graduate School of Oceanography, University of Rhode Island, Kingston.

Kremer, J. N., and S. W. Nixon, 1977, *A Coastal Marine Ecosystem: Simulation and Analysis,* Springer-Verlag, New York.

Nixon, S. W., C. A. Oviatt, J. N. Kremer, and K. T. Perez, 1978, The Use of Numerical Models and Laboratory Microcosms in Estuarine Ecosystem Analysis—Simulations of a Winter Phytoplankton Bloom, in *Marsh–Estuarine Systems Simulations,* Eighth Belle W. Baruch Institute of Marine

Biology and Coastal Research Symposium, Georgetown, S. C., January 1977, R. Dame (Ed.), University of South Carolina Press, Columbia, in press.

Ollason, J. G., 1977, Freshwater Microcosms in Fluctuating Environments, *Oikos*, 28: 262-269.

O'Neill, R. V., 1975, Dynamic Ecosystem Models: Progress and Challenges, in *Ecosystem Analysis and Prediction*, Proceedings of a SIAM—SIMS Conference, Alta, Utah, July 1—5, 1974, S. A. Levin (Ed.), pp. 280-296, Society for Industrial and Applied Mathematics, Philadelphia.

Oviatt, C. A., K. T. Perez, and S. W. Nixon, 1977, Multivariate Analysis of Experimental Ecosystems, *Helgol. Wiss. Meeresunters.*, 30: 30-46.

Patten, B. C., 1975, Ecosystem Linearization: An Evolutionary Design Problem, *Am. Nat.*, 109: 529-539.

Perez, K. T., G. Morrison, C. A. Oviatt, S. W. Nixon, and B. Buckley, 1977, The Importance of Physical and Biotic Scaling to the Experimental Simulation of a Coastal Marine Ecosystem, *Helgol. Wiss. Meeresunters.*, 30: 144-162.

Platt, T., and K. L. Denman, 1975, Spectral Analysis in Ecology, *Annu. Rev. Ecol. Systemat.*, 6: 189-210.

Shinners, S. M., 1972, *Modern Control System Theory and Application*, Addison-Wesley, Inc., Reading, Mass.

Smayda, T. J., 1976, Plankton Processes in Mid-Atlantic and Shelf Waters and Energy-Related Activities, in *Effects of Energy-Related Activities on the Atlantic Continental Shelf*, B. Manowitz (Ed.), ERDA Report BNL-50484, pp. 70-95, Brookhaven National Laboratory, NTIS.

Vargo, G. A., 1976, *The Influence of Grazing and Nutrient Excretion by Zooplankton on the Growth and Production of the Marine Diatom* Skeletonema costatum *(Greville) Cleve in Narragansett Bay*, Ph.D. Thesis, University of Rhode Island, Kingston.

Waide, J. B., J. E. Krebs, S. P. Clarkson, and E. M. Setzler, 1974, A Linear Systems Analysis of the Calcium Cycle in a Forested Watershed Ecosystem, in *Progress in Theoretical Biology*, R. Rosen and F. M. Snell (Eds.), Vol. 3, pp. 261-345, Academic Press, Inc., New York.

Wastler, T. A., 1969, *Spectral Analysis, Applications in Water Pollution Control*, Report CWT-3, Federal Water Pollution Control Administration, Cincinnati, Ohio.

Webster, J. R., J. B. Waide, and B. C. Patten, 1975, Nutrient Recycling and the Stability of Ecosystems, in *Mineral Cycling in Southeastern Ecosystems*, ERDA Symposium Series, Augusta, Ga., May 1—3, 1974, F. G. Howell, J. B. Gentry, and M. H. Smith (Eds.), pp. 1-27, CONF-740513, NTIS.

Weisberg, R. H., 1974, *The Non-Tidal Flow in the Providence River of Narragansett Bay: A Stochastic Approach to Estuarine Circulation*, Ph.D. Thesis, University of Rhode Island, Kingston.

Wiegert, R. G., 1975, Simulation Models of Ecosystems, *Annu. Rev. Ecol. Systemat.*, 6: 311-338.

APPLYING SURVIVAL CURVES TO ASSESSMENT OF FISH LARVAL ENTRAINMENT IMPACT

MICHAEL D. DAHLBERG
Ecological Sciences Division, NUS Corporation,
Pittsburgh, Pennsylvania

ABSTRACT

The equivalent-adults model for approximating numbers of fish larvae entrained at power plants and survival rates from larva to adult is assessed in regard to predicting the numbers of adults represented by entrained larvae. The accuracy of this model, in relation to other proposed approaches, depends on the proper selection of methods for calculating larval losses, lifetime fecundity, and larva-to-adult survival. Additional approaches are recommended when sufficient population data are available. Until questionable assumptions can be resolved, a reasonable approach is to calculate a possible range for the number of equivalent adults lost. These numbers can be compared to a reference of acceptability, such as year-to-year variability of stock size.

Entrainment of fish eggs and larvae is one of the major impacts of power-plant operation. Potential effects are generally assessed by translating entrained eggs and larvae to the potential number of adults they represent. The calculations are accomplished with complex models and a simplistic equivalent-adults method proposed by Horst (1975). Horst's method has been recommended by the Environmental Protection Agency (1976) and is widely applied in impact calculations. This paper examines problems encountered in applying the Horst model and discusses alternative procedures. Data for the walleye population of Oneida Lake, New York (Forney, 1976a; 1976b) are used as an example.

SURVIVAL CURVES AND ENTRAINMENT LOSSES

Constructing egg-to-adult survival curves (e.g., Fig. 1 and Table 1) is recommended as the first step in assessing larval entrainment impact. The curves in the figure illustrate a snapshot approach in which ratios of adults and earlier stages represent survival rates. Families of such curves permit graphic comparison of such variables as year-to-year differences, augmentation, and biased sampling, e.g., avoidance of sampling gear. The latter is a common source of error in lacustrine studies (Hackney, 1977) and may be responsible for the high mortality rate determined by Noble (1972) for walleye postlarvae in one bay of Oneida Lake (see Fig. 1).

The predictive value of survival curves is greatest in stable populations (Horst, 1977a). When we are restricted to specific population data, we can construct representative survival curves by averaging data acquired over several years, as in Fig. 1, or by selecting data for a typical year. Survival curves for naturally spawned (1969, 1971, and 1973) and augmented (1968, 1970, and 1972) populations of walleyes (Forney, 1976a) are presented in Fig. 1.

Egg production (13×10^9) of walleyes in Oneida Lake was estimated by Forney (1976a) from fecundity-at-age data. Walleyes hatched near May 10 (day 0 in Fig. 1) and were pelagic for approximately 40 days. The larvae occupied 971×10^6 m^3 of water, which is the upper 5.5 m (Clady, 1975) and 64% of the total lake volume. Postlarvae were transported by surface currents and yolk-sac larvae by subsurface currents (Houde and Forney, 1970).

Since larvae were not fully recruited into the plankton until the swim-up stage (day 10), numbers of larvae at hatching and survival from hatching to swim-up (13.6%) were estimated from the ratio of these stages in the augmented population. Assuming the difference in numbers of 10-day-old larvae in augmented and unstocked populations (3.192×10^7) resulted from the stocking of 2.347×10^8 1- to 3-day-old larvae (Forney, 1976a), the survival rate was 13.6%. There is a possibility that stocked larvae experience higher mortality than wild larvae (Forney, 1976a).

Populations of demersal young-of-the-year and age I juveniles were estimated from trawl studies (Forney, 1976a). Survival to adults is shown in Fig. 1 by plotting parent stock size, including all mature age classes (age IV and older), as determined by mark and recapture studies (Forney, 1976b).

Assuming a feasible open-cycle cooling-water-intake flow of 1 million gpm (5.45×10^6 m^3/day), cumulative numbers of larvae entrained were calculated and illustrated by a survival curve (Fig. 1).

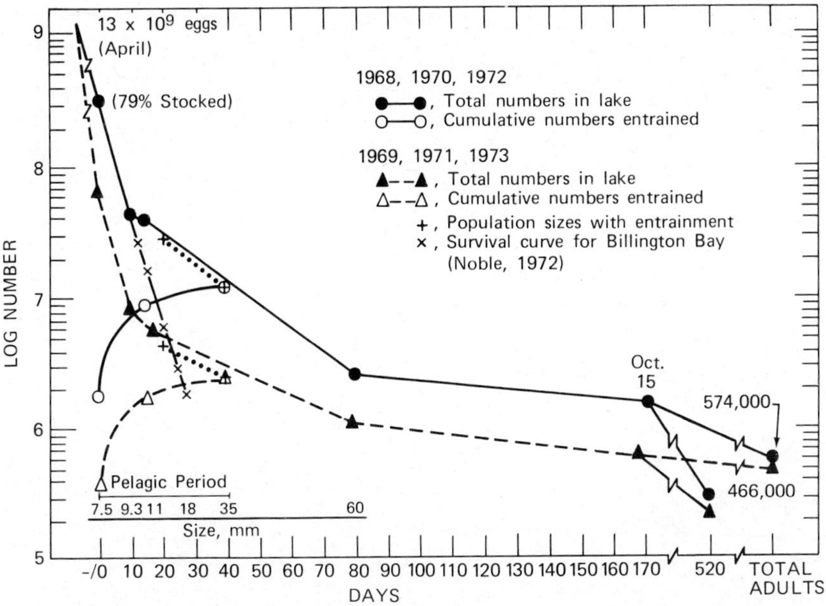

Fig. 1 **Survival curves for augmented and naturally spawned populations of walleye in Oneida Lake, New York. Sizes of entrained populations are indicated by dotted lines.**

TABLE 1

NUMBERS OF EGGS, LARVAE, AND ADULTS IN UNSTOCKED AND AUGMENTED WALLEYE POPULATIONS IN ONEIDA LAKE, NEW YORK*

	Unstocked population	**Augmented population**
Egg production†	13×10^9	13×10^9
Larval production‡	5.4775×10^7	2.8944×10^8
Swim-up larvae (10 days old)¶	7.4500×10^6	3.9368×10^7
35-mm larvae (40 days old)§	3.0×10^6	13.5×10^6
Adults	466,000	574,000

*Based on data from Forney (1976a; 1976b).

†From fecundity-at-age data.

‡From densities of swim-up larvae in net samples, at 13.6% survival from hatching to swim-up.

¶From net samples.

§Estimated from survival curve in Fig. 1.

The calculations required the assumption that entrainment is randomly exploiting the larval population, i.e., all larvae have an equal chance of being entrained. Number of larvae entrained (N_L) is

$$N_L = \sum_{i=1}^{n} d_i V_i \qquad (1)$$

where d_i is average number of larvae during the *ith* sample period and V_i is the corresponding volume of cooling water (Hackney, 1977). Calculated on a daily basis, N_L is 2.26×10^6 in the unstocked population and 12.08×10^6 in the augmented population.

Sizes of the exploited larval populations (Fig. 1) were calculated from reductions in population size:

$$L_p = \frac{V_i}{V} \times 100 \qquad (2)$$

where L_p is percent reduction of larval population, V_i is volume of water entrained (5.45×10^6 m^3/day), and V is volume of occupied water (971×10^6 m^3). Thus L_p is 0.5613%/day and 22.45% in 40 days.

EQUIVALENT-ADULTS MODEL

The equivalent-adults model was proposed by Horst (1975) to provide a simple first approximation of fish egg and larval entrainment impact in terms of potential adult loss. The basic approach is multiplying numbers of eggs and larvae entrained by egg and larva-to-adult survival rates to estimate the number of equivalent adults. The following assumptions, summarized from Horst (1977a), are made when this model is applied to larval entrainment:

1. The population is in equilibrium and has a stable age distribution.
2. Lifetime of a fish is the average age or mean generation time.
3. Numbers of males and females are equal.
4. Entrainment of larvae occurs at the time of hatching.
5. The calculated equivalent adults are distributed in proportion to the stable age distribution of adults.

Horst (1975) proposed the formula

$$N_A = N_L S_{LA} \qquad (3)$$

where N_A is potential adult loss, N_L is number of larvae entrained, and S_{LA} is survival from hatching to adult. It was proposed that S_{LA} be estimated indirectly from

$$S_{LA} = \frac{S_{EA}}{S_{EL}} \tag{4}$$

where S_{EA} (survival from egg to adult) is $2/F$ and S_{EL} is an estimated or observed survival rate from egg to hatching of larvae.

Lifetime fecundity, F, is

$$F = G \sum_{i=1}^{n} P_i E_i \tag{5}$$

where P_i is fraction of population in each age class, E_i is egg production or fecundity for each age class, and G is mean generation time in years. Since appropriate data for Oneida Lake walleyes are lacking, F is calculated for Lake Erie walleyes (Table 2). Mean generation time is estimated as 3 years in this case to correspond to the age of maximum eggs per female. An intermediate adult age, such as 5 years, can also be selected to provide possible minimum and maximum generation times (Horst, 1977b). This procedure yields S_{EA} values of 4.3728×10^{-5} and 2.6240×10^{-5}. The utility of this approach is supported by an independent calculation of egg-to-adult survival (3.5846×10^{-5}), which represents the ratio of adults and egg production in the unstocked population, as shown in Table 1.

If we limit further calculations to the unstocked population,

$$S_{LA} = \frac{S_{EA}}{S_{EL}} = \frac{4.372 \times 10^{-5}}{4.213 \times 10^{-3}} = \frac{0.00004372}{0.004213} = 0.01037$$

where S_{EL} is ratio of larva at hatching and egg production (Table 1). This value of S_{LA} is approximated by the ratio of adults to larvae at hatching (0.0085) in Table 1. The product of S_{LA} and the number of larvae entrained ($N_L = 2.26 \times 10^6$), as in Eqs. 1 and 3, yields a potential adult loss of 23,436.

The use of hatching-to-adult survival for S_{LA} is questionable since survival to adulthood increases from 1 to 15.5% through the pelagic larva period (survival rate is the ratio of adults to larvae). To make the calculation of larval entrainment losses correspond with the

S_{LA} parameter used in the Horst model, we can calculate larval losses (N_p) from

$$N_p = L_h \times L_p \tag{6}$$

where L_h is total larval production or number of larvae hatched (5.4775×10^7) and L_p is percent reduction (Eq. 2). Thus N_p is $5.4775 \times 10^7 \times 22.45\%$ or 1.2297×10^7, and the number of adults (N_A) is

$$N_A = S_{LA} \times N_p = 127{,}520 \tag{7}$$

Composition data in Table 2 indicate that N_A would be distributed among age classes as follows: age III, 0.757; age IV, 0.138; age V, 0.069; age VI, 0.014; age VII, 0.021; age VIII, 0.003.

TABLE 2

CALCULATING MEAN LIFETIME FECUNDITY AND SURVIVAL FROM EGG TO ADULT FOR WALLEYES IN WESTERN LAKE ERIE, 1955—1970*

Age	Fraction of population	Eggs per female	Weighted eggs per female
I	0.52	0	0
II	0.34	0	0
III	0.11	91,000 (X 0.86)†	8,608.6
IV	0.02	140,000	2,800.0
V	0.01	213,000	2,130.0
VI	0.002	261,000	522.0
VII	0.003	374,000	1,122.0
VIII	0.0004	364,000	145.6
Total	1.0054		15,328.2

*Data are from Busch, Scholl, and Hartman (1975) and Wolfert (1969):

Mean fecundity $(\bar{F}) = \dfrac{15{,}328}{1.0054} = 15{,}246$

Mean lifetime fecundity = Mean generation time (G) X Mean fecundity

$= 3 \times 15{,}328.2 = 45{,}984.6$

$= 5 \times 15{,}328.2 = 76{,}641.0$

Survivorship egg to adult $(S_{EA}) = \dfrac{2}{F}$

$S_{EA} = 4.3728 \times 10^{-5}$ (1 per 22,869 eggs) for G of 3

$S_{EA} = 2.6240 \times 10^{-5}$ (1 per 38,110 eggs) for G of 5

†Correction for 86% maturity.

CALCULATING DAILY POTENTIAL ADULT LOSSES

Potential adult loss can be calculated on a daily basis (N_{Ai}) through the entrainment period. Numbers of larvae entrained each day are multiplied by increasing survival rates calculated from the ratio of adults to larvae occurring each day:

$$N_{Ai} = N_{Li}S_{LAi} = N_{Li}\,\frac{A}{Li} \tag{8}$$

where N_{Li} = number of larvae entrained each day
S_{LAi} = survival rate from larva to adult
A = number of adults
Li = larval population size

However, since N_{Li}/Li is a constant (0.561%) representing the percentage of occupied waters, and presumably larvae, entrained each day, $N_{Ai} = 0.00561 \times A = 2614$ and $N_A = N_{Ai} \times 40 = 104{,}560$.

LARVAL REDUCTION METHOD

Actual reduction of a larval population caused by entrainment is less than total numbers entrained because the total numbers include larvae that would have died from natural causes by the end of the pelagic period; e.g., the number of walleye larvae entrained in 40 days is 2.26×10^6, and larval reduction is 673,500. Larval reduction is calculated by multiplying the estimated number of 40-day-old larvae (3.0×10^6) by the percent reduction, 22.45% (as described in Eq. 2).

Multiplying larval reduction by survival rate from 40-day-old larva to adult (0.1553) gives an N_A value of 104,595. The survival rate used is the ratio of adults and 40-day-old larvae (Table 1).

The simplest approach for calculating N_A is to assume that the percent reduction of adults is equivalent to the percent reduction of larvae (22.45%). This calculation yields a potential adult loss of 104,617.

DISCUSSION

Interpretations of the equivalent-adults model of Horst (1975) have generally resulted in minimal projections of potential adult loss. The primary causes are using high fecundity values, representing the maximum egg production of a species, and applying low larva-to-adult survival rates. These problems are corrected by using mean lifetime fecundities and applying appropriate combinations of survival rates and larval reductions.

Applying the survival rate for the hatching-to-adult period is reasonable when larval entrainment losses are based on the reduction of total larval production rather than the number of larvae actually entrained. This reduction in larval production can be estimated by multiplying larval production by the fraction of larvae entrained throughout the pelagic period. The results are in reasonable agreement with projections based on the assumption that the percentage reduction of adults is equivalent to the percentage of larval exploitation. Interpretations of these results should consider that the projected adult loss is distributed among all mature age classes and is in proportion to the adult age-class composition in a stable population.

With the exception of the general mode of application of the Horst model (Eqs. 1 and 3), the calculations of N_A are probably biased toward a possible maximum or worst-case estimate by the assumptions that (1) an intake randomly draws on the total larva population, (2) there is no avoidance reaction to induced intake currents, and (3) there are no compensatory responses.

Although larvae are transported by wind-generated lake currents (Houde and Forney, 1970) and by power-plant-induced currents, it is unlikely that they will be randomly recruited into an intake throughout the pelagic period. Spatial distribution and movements of larvae should be considered when possible. In Oneida Lake, postlarvae concentrate in shallow bays along the southern shore (Noble, 1972).

Avoidance of an intake probably increases with size of larvae, as does avoidance of sampling gear (Noble, 1972). Although young walleyes are pelagic up to 40 days and to 35 mm in length (Fig. 1), Noble did not capture postlarvae over 18 mm long.

Natural compensation for entrainment exploitation may occur in the exploited year class as increased survival or growth or in subsequent generations if fecundity also increases with growth. McFadden (1976) indicated that only partial compensation can be expected. High compensation is apparent in Lake Erie walleyes, which maintained their stock size despite total mortality rates of 50% in yearlings and 80% in older fish (Regier, Applegate, and Ryder, 1969). Additional growth compensation would not be expected, however, if this population had reached its physiological maximum for growth (Regier, Applegate, and Ryder, 1969) or if the supply of forage fish was low (Moyle, 1949).

Until sufficient information is available to correct for compensation and avoidance, potential adult loss should be expressed as a possible range, such as those obtained with Eqs. 3 and 7 and with the methods proposed here, e.g., 23,000 to 127,000. The calculated adult losses can then be compared with some reference for acceptability, such as year-to-year variation in population size (Environmental Protection Agency, 1976). Since the average year-to-year variability of the Oneida Lake adult walleye population was approximately 156,000 during 1957–1974 (Forney, 1976b), it appears that our calculation of potential adult loss would be acceptable.

ACKNOWLEDGMENTS

The background for this study was achieved through the preparation and review of impact statements for the Ecological Sciences Division of NUS Corporation, of which P. V. Morgan is vice president and general manager. The manuscript was reviewed by P. V. Morgan, B. C. Marcy, P. A. Dahlberg, L. K. Davis, V. R. Kranz, and H. A. Haerer of NUS Corporation; P. A. Hackney of the Tennessee Valley Authority; and T. J. Horst of Stone and Webster Engineering Corp. W. J. B. Johnson of NUS Corporation drafted the figure.

REFERENCES

Busch, W. N., R. L. Scholl, and W. L. Hartman, 1975, Environmental Factors Affecting the Strength of Walleye (*Stizostedion vitreum vitreum*) Year-Class in Western Lake Erie, 1960–70, *J. Fish. Res. Board Can.*, 32(10): 1733-1743.

Clady, M. D., 1975, *Population Dynamics of Walleye and Yellow Perch in Oneida Lake, April 1, 1970, to March 31, 1975*, Federal Aid Project F-17-R, Job I-e, New York State Conservation Department, Albany.

Environmental Protection Agency, 1976, *Development Document for Best Technology Available for the Location, Design, Construction and Capacity of Cooling Water Intake Structures for Minimizing Adverse Environmental Impact*, Sect. 316(b), P. L. 92-500, Washington, D. C.

Forney, J. L., 1976a, Year-Class Formation in the Walleye *(Stizostedion vitreum vitreum)* Population of Oneida Lake, New York, 1966–1973, *J. Fish. Res. Board Can.*, 33: 783-792.

——, 1976b, *Population Dynamics of Walleye and Yellow Perch in Oneida Lake, April 1, 1975, to March 31, 1976*, Federal Aid Project F-17-R, Job I-a, New York State Conservation Department, Albany.

Hackney, P. A., 1977, Methods for Calculating Survival Rate, Biomass Production, and Proportion Entrained of Lacustrine Ichthyoplankton, in *Proceedings of the Conference on Assessing the Effects of Power-Plant-Induced*

Mortality on Fish Populations, W. Van Winkle (Ed.), Pergamon Press, Inc., New York.

Horst, T. J., 1975, The Assessment of Impact Due to Entrainment of Ichthyoplankton, in *Fisheries and Energy Production*, S. B. Saila (Ed.), pp. 107-118, D. C. Heath & Co., Lexington, Mass.

——, 1977a, Mathematical Modeling of Power Station Impacts on Fisheries Resources of the United States, International Federation for Information Processing Working Conference on Modeling and Simulation of Land, Air and Water Resource Systems, Ghent, Belgium, preprint.

——, 1977b, Stone and Webster Engineering Corp., personal communication.

Houde, E. D., and J. L. Forney, 1970, Effects of Water Currents on Distribution of Walleye Larvae in Oneida Lake, *J. Fish. Res. Board Can.*, 27: 445-456.

McFadden, J. T., 1976, Environmental Impact Assessment for Fish Populations, in *Proceedings of the Workshop on the Biological Significance of Environmental Impacts*, R. K. Sharma, J. D. Buffington, and J. T. McFadden (Eds.), pp. 89-138, Report NR-CONF-002, U. S. Nuclear Regulatory Commission, Washington, D. C.

Moyle, J. B., 1949, Fish Population Concepts and Management of Minnesota Lakes for Sport Fishing, *Trans. North Am. Wildl. Conf.*, 14: 283-294.

Noble, R. L., 1972, Mortality Rates of Walleye Fry in a Bay of Oneida Lake, New York, *Trans. Am. Fish. Soc.*, 101: 720-723.

Regier, H. A., V. C. Applegate, and R. A. Ryder, 1969, *The Ecology and Management of the Walleye in Western Lake Erie*, Technical Report 15, Great Lakes Fisheries Commission.

Wolfert, D. R., 1969, Maturity and Fecundity of Walleyes from the Eastern and Western Basins of Lake Erie, *J. Fish. Res. Board Can.*, 26: 1877-1888.

A SIMPLE MODEL FOR ASSESSING THE POTENTIAL LOSS OF ADULT FISH RESULTING FROM ICHTHYOPLANKTON ENTRAINMENT

W. PETER SAUNDERS, JR.*
Environmental Research and Technology, Inc., Concord, Massachusetts

ABSTRACT

A mathematical model for estimating potential survival of fish eggs and larvae to reproducing adults is examined in the context of predicting the potential loss from power-station entrainment. It is demonstrated that violating one assumption of the model can result in gross underestimates of potential adult loss. High rates of natural mortality occurring during the egg and larval life stages are considered in relation to the assumption that all exploitation by the power plant occurs instantaneously at spawning or hatching. The sensitivity of potential loss estimates to various time—mortality-rate regimes is examined on the basis of the natural mortality rates observed in young life stages of several species. An alternative model is proposed, and its sensitivity to underlying assumptions is examined. This model is shown to consistently overestimate loss of equivalent adults. The magnitude of the overestimation depends on species-specific conformity of the model assumptions to actual planktonic mortality and first-year survival.

Many aquatic species that produce planktonic eggs and/or larvae have very high mortality of the young organisms. Of the many thousands of offspring initially produced, only a few individuals survive to reproductive maturity. Since we are usually familiar with the portion of a population potentially subject to exploitation by a fishery, the entrainment loss of large numbers of planktonic young may be more meaningful in the impact assessment process if considered in relation to the numbers of reproducing adults that could potentially result

*Current address: Massachusetts Cooperative Fisheries Research Unit, University of Massachusetts, Amherst, Mass.

from the entrained plankters. Horst (1975) suggested using an equivalent-adults model to translate the number of eggs and larvae entrained to an equivalent number of adults lost.

The reduced form of the model is

$$N_A = (N_E \times S_{EA}) + (N_L \times S_{LA}) \tag{1}$$

where N_A = number of adults potentially lost as a result of entrainment

N_E = number of newly spawned eggs entrained

S_{EA} = survival from egg to adult

N_L = number of newly hatched larvae entrained

S_{LA} = survival from larva to adult

and

$$S_{EA} = (Pf \times F_{\overline{x}} \times L)^{-1} \tag{1a}$$

where Pf is the proportion of the population which is female, $F_{\overline{x}}$ is the average fecundity per female per year, and L is the reproductive life expectancy of an average organism.

The term S_{LA} can be defined as

$$S_{LA} = S_{EA} \times S_{EL} \tag{1b}$$

where S_{EL} is the survival from egg to larva.

The equivalent-adults model provides an approach to the assessment of entrainment impact on a species population in instances where little information is available on the life history of the species. The only life-history information required to use the model is the fecundity of a female, the reproductive life span, and the sex ratio. A number of other life-history parameters can be incorporated into the model's calculations, however. Thus the equivalent-adults model is a valuable expedient that can be used in the impact-assessment process when data are insufficient to allow a more sophisticated approach, such as the Leslie (1945) model (Horst, 1978).

A number of assumptions implicit in its formulation must be recognized and addressed when we use the equivalent-adults model:

1. The adult population is at equilibrium, with no change in its size occurring over time.
2. The population maintains a stable age distribution.
3. The ratio of males to females is constant through time.

4. The life span of an individual is equivalent to the mean generation time (MGT) for females in the population.

5. Exploitation of planktonic eggs occurs instantaneously at spawning, and exploitation of larvae occurs instantaneously at hatching (becoming planktonic).

Horst (1978) discussed methods of dealing with departures from several of these assumptions. In many instances the method involves a simple modification of the calculated survival rate from egg to adult. Horst also discussed modifying the modeling procedure to approximate MGT.

In this paper I discuss aspects of the equivalent-adults model associated with the mortality of eggs and larvae. A conflict exists between assumption 5 and the "real-world" case; i.e., continuous exploitation of planktonic life stages by the power plant. I examine the effect of commonly observed egg and larval mortality rates on the results of the model and suggest an alternative method of computing the loss of equivalent adults which requires no additional data.

EFFECT OF EGG AND LARVAL MORTALITY ON THE RESULTS OF THE EQUIVALENT-ADULTS MODEL

I have found that very seldom are data available to allow a detailed estimate of rates of egg and larval mortality. Two cases with which I am familiar are the analysis of egg and larval mortality of the Atlantic mackerel, *Scomber scombrus* (Sette, 1943), and the analysis of larval mortality of winter flounder, *Pseudopleuronectes americanus* (Pearcy, 1962). In both these cases mortality conforms to the exponential model. My discussion here assumes that all mortality of eggs and larvae follows this model. This allows the survivorship to the end point of a life stage (e.g., hatching success) to be converted to an instantaneous mortality rate:

$$M = \frac{-\ln S}{t} \qquad (2)$$

where M is the instantaneous natural mortality rate, S is the survivorship to the end of the life stage, and t is the duration of the life stage in days.

Table 1 gives the results of a survey of daily natural mortality rates for the entrainable (planktonic) life stages of several species. The information in the table indicates that natural mortality rates for

TABLE 1

OBSERVED RATES OF NATURAL MORTALITY FOR PLANKTONIC LIFE STAGES OF FISH

Species and life stage	Duration, days	Mortality rate, %/day	Survival* (to end of stage)	Reference
Atlantic mackerel				
Egg	7	14		Sette, 1943
Larva				
Pre fin fold	28	14		Sette, 1943
During fin fold	5.5	44		Sette, 1943
Post fin fold	41	10		Sette, 1943
Egg—juvenile	85	14	0.000004	Sette, 1943
Winter flounder				
Larva				
0—25 days	25	21		Pearcy, 1962
26—53 days	28	4		Pearcy, 1962
Haddock				
Larva		10		Cushing and Walsh, 1976
Plaice				
Larva	30	5	20%/month	Cushing, 1975
Juvenile	60	2	62%/month	Cushing, 1975
Fishes (marine)				
Egg and early larva		5—10		Cushing, 1975
Cunner				
Egg	2—7	35—78	0.05	Williams, Williams, and Miller, 1973
Northern pike†				
Alevin	16	17	0.052	Franklin and Smith, 1963
	20	24	0.004	Franklin and Smith, 1963
	22	8	0.156	Monten, 1948
	35	5	0.167 (mean)	Monten, 1950
	35	3—10	0.22—0.386 (range)	Monten, 1950
Striped bass				
Egg and yolk-sac larva	10	25	0.058	Saila and Lorda, 1977
Post yolk-sac larva	24	13	0.037	Saila and Lorda, 1977
Juvenile 1	30	5	0.200	Saila and Lorda, 1977
Smallmouth bass				
Fry	90	7	0.0014	Fajen, 1975
	120	5	0.0016	Fajen, 1975

*Survival to end of life stage is provided only for those instances where it was used to back calculate mortality rate.

†Franklin and Smith (1963) and Monten (1948; 1950) dealt with hatchery-reared populations, which would, presumably, have lower mortality rates than naturally spawned fish.

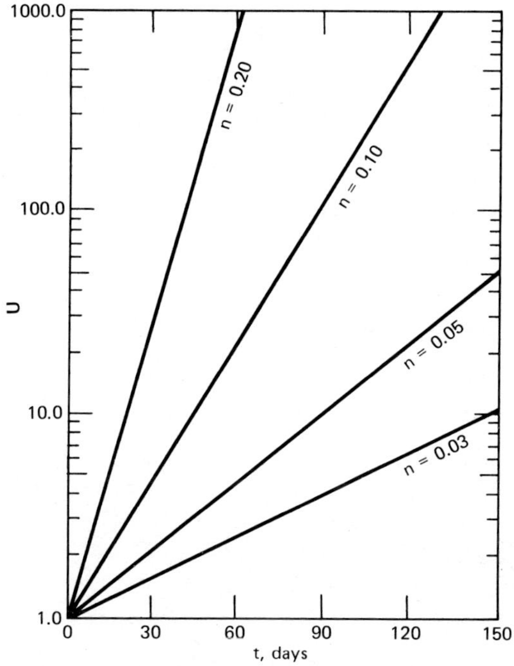

Fig. 1 Values of U, the factor by which adult loss is underestimated, for specific daily natural mortality rates (n), occurring over varying periods of entrainment (t).

planktonic eggs and larvae often exceed 5%/day. Particularly high mortality rates are evident early in the planktonic life stage or when the planktonic life stage is of short duration.

Figure 1 shows a series of curves computed by varying the duration of exploitation of planktonic life stages by the power station. These curves indicate the factor U, by which the equivalent-adults model will underestimate potential adult loss at specific natural mortality rates. In calculating this factor we assume constant vulnerability to entrainment and immediate removal (through decay, predation, or loss of buoyancy) of eggs or larvae that die naturally:

$$U = e^{Mt/2} \tag{3}$$

where M is the instantaneous natural mortality rate and t is the duration of entrainment in days.

In Eq. 3, U represents a factor that, when multiplied by the number of eggs or larvae entrained, is equivalent to the number of eggs or larvae originally produced and which would have been entrained had all entrainment taken place immediately after spawning or hatching.

At natural mortality rates of 5%/day or greater and for an entrainable life-stage duration of 30 days, the equivalent-adults model underestimates potential adult loss by a factor greater than 2 (Fig. 1). In assessing larval entrainment of winter flounder, which has a larval mortality rate of 21%/day during the first 25 days after hatching (Pearcy, 1962), the model would underestimate potential adult loss by a factor of 18.

These facts suggest that the equivalent-adults model is very sensitive to the assumption that exploitation occurs instantaneously when eggs are spawned or larvae hatch. Because the model could grossly underestimate adult loss, it may be of questionable value from a regulatory viewpoint since, in a proceeding intended to protect the environment from adverse effects, acceptable impact predictions usually must be able to assert that impact in the real-world case will not be greater than that predicted. Therefore, to assure this "conservatism," we must provide for relaxation of the assumption of immediate exploitation.

REVISION OF THE EQUIVALENT-ADULTS MODEL

When sufficient data are available to calculate mortality rates for planktonic life stages, the equivalent-adults model can be modified by incorporating U (Eq. 3) into the loss calculation. This can be accomplished by using data available in the literature or by estimating planktonic mortality rates directly from field-survey data, as discussed by Polgar (1977) and Hackney (1977). When restrictions of time or resources prevent direct estimation of U, however, I propose that an alternative form of the equivalent-adults model be used.

My model requires an estimate of survival through the first year (S_0). To relax the assumption concerning time of entrainment without introducing a number of new factors into the model, I have taken a pragmatic approach to estimating S_0. Since first-year mortality is high in comparison with mortality in older fish, I use the survival from egg to adult (S_{EA}), based on fecundity and life span (Eq. 1), as an estimator of S_0. If we then assume that all first-year mortality occurs during the planktonic life stage, a correction factor (U') can be calculated for the model:

$$U' = e^{-\ln S_{EA}/2} \qquad (4)$$

where S_{EA} is the egg-to-adult survival as defined by Eq. 1a.

This equivalent-adults model can be expressed as

$$N_A = U' [(N_E \times S_{EA}) + (N_L \times S_{LA})] \quad (5)$$

For purposes of illustration, Table 2 compares the results generated by the equivalent-adults model of Horst (1975) without providing for relaxation of the exploitation assumption; by the Horst model with relaxation based on published planktonic mortality rates, and by my version of the model for five of the species from Table 1.

SENSITIVITY ANALYSIS OF THE ALTERNATIVE MODEL

The equivalent-adults model I propose assumes (1) that all mortality occurring during the first year of life takes place during the planktonic life stage and (2) that S_{EA} approximates S_0. The sensitivity of the model to each of these assumptions was examined. Sensitivity is expressed in terms of the ratio of the predicted impact to the "actual" impact resulting when the model assumptions are not satisfied by the natural situation.

Figure 2 shows the relationship between the prediction of equivalent-adult loss based on assumption 1 and the actual loss that would result when planktonic mortality comprises only a specified portion of estimated first-year mortality. This figure indicates that the proposed model will always overestimate loss. The magnitude of the overestimation is based on the relationship

$$r = p^{-\frac{1}{2}} \quad (6)$$

where p is the proportion of first-year mortality actually occurring during the planktonic life stage(s) and r is the ratio of predicted to actual loss.

For example, if 25% of the total first-year mortality occurred during the planktonic life stages, the proposed model would overestimate loss by a factor of 2, or, if 1% of the total first-year mortality occurred during the planktonic life stages, the proposed model would overestimate loss by a factor of 10. This relationship is independent of S_0.

The effect on the model prediction resulting from the use of S_{EA} as an estimator of S_0 is similar to that described. This relationship, expressed in terms of the ratio of S_{EA} to S_0, is

$$r = \left(\frac{S_{EA}}{S_0}\right)^{-\frac{1}{2}} \quad (7)$$

where r is the ratio of predicted to actual loss.

TABLE 2

COMPARISON OF PREDICTIONS OF THREE EQUIVALENT-ADULTS MODELS

Species and life stage	Fecundity, eggs/adult	Reproductive life span, years	S_{EL}	Predictive model used*: Horst, unrelaxed	Predictive model used*: Horst, relaxed†	Predictive model used*: Saunders
Atlantic mackerel Egg and larva	200,000	10	0.35	$4 \leq N_A \leq 19$	$847 \leq N_A \leq 4{,}234$	$2{,}690 \leq N_A \leq 6{,}016$
Winter flounder Larva	250,000	10	0.35	$2 \leq N_A \leq 12$	$47 \leq N_A \leq 233$	$1{,}527 \leq N_A \leq 3{,}415$
Striped bass Egg and larva	125,000	17	0.06	$17 \leq N_A \leq 141$	$187 \leq N_A \leq 1{,}589$	$4{,}931 \leq N_A \leq 14{,}375$
Northern pike Alevin	40,000	8	0.60	$11 \leq N_A \leq 42$	$17 \leq N_A \leq 68$‡ $166 \leq N_A \leq 664$§	$3{,}240 \leq N_A \leq 6{,}481$
Smallmouth bass Fry	3,200	5	0.30	$423 \leq N_A \leq 1{,}058$	$11{,}313 \leq N_A \leq 28{,}282$	$20{,}574 \leq N_A \leq 32{,}530$

*Models estimate loss based on an entrainment of 10^6 individuals of each planktonic life stage.
†Relaxation of immediate-exploitation assumption, based on mortality rates presented in Table 1.
‡Based on highest estimated survival, from survey in Table 1.
§Based on lowest estimated survival, from survey in Table 1.

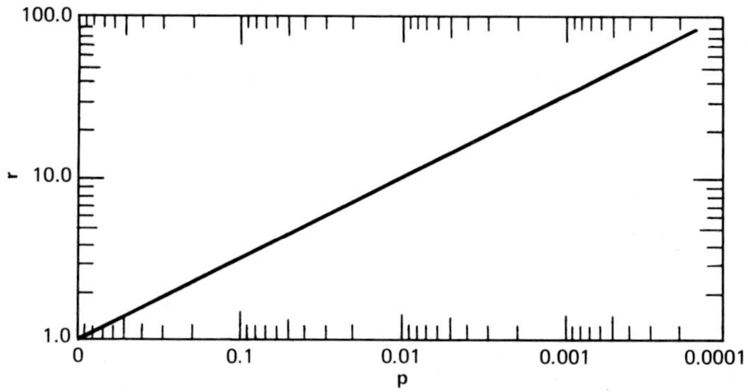

Fig. 2 The ratio (r) of predicted to actual loss when planktonic mortality comprises a specific proportion (p) of the mortality assumed by the alternative model.

The potential relationships of S_{EA} to S_0 can be examined by comparing S_{EA} to S_0 values calculated with the Leslie Matrix (Vaughan and Saila, 1976). Table 3 shows the results of comparisons for seven species [I used the method described by Horst (1978) to estimate the range of S_{EA}], along with the quantity r, the ratio of predicted to actual loss. The information in Table 3 indicates that S_{EA} is highly variable in its ability to estimate S_0, but it appears that in most cases use of S_{EA} to approximate S_0 will not result in large overestimates of impact.

DISCUSSION

In proposing a revised form of the equivalent-adults model, I have introduced several new assumptions that must be considered when using the model:

1. Survival corresponds to an exponential model during all planktonic life stages.

2. All mortality during the first year of life occurs during the planktonic life stages.

3. Vulnerability of plankton to entrainment remains constant throughout the planktonic life stages.

4. Plankton that are the victims of natural mortality are not entrained.

5. Survival from egg to adult is an approximate estimator of first-year survival, S_0.

Assumption 1 is relatively straightforward and requires no discussion, and assumptions 2 and 5 were examined in the sensitivity analysis.

TABLE 3

COMPARISON OF LESLIE MATRIX S_0 VALUES WITH ESTIMATES OF S_{EA} AND RESULTING EFFECT ON IMPACT PREDICTION

Species	S_0	S_{EA}*	S_{EA}†	Range of S_{EA}/S_0	Range of r, predicted/actual loss	References
Smallmouth bass	8.8×10^{-4}	3.2×10^{-4}	1.3×10^{-4}	0.36 to 0.15	1.66 to 2.60	‡
Cunner	2.1×10^{-4}	5.4×10^{-4}	1.5×10^{-4}	2.57 to 0.71	0.62 to 1.18	Horst, 1977
Menhaden	7.7×10^{-5}	9.6×10^{-5}	2.7×10^{-5}	1.25 to 0.35	0.90 to 1.69	Horst, 1977
Winter flounder	5.7×10^{-5}	4.0×10^{-6}	8.0×10^{-7}	0.07 to 0.01	3.77 to 8.44	Horst, 1977
Striped bass	3.5×10^{-5}	8.0×10^{-6}	9.4×10^{-7}	0.23 to 0.03	2.09 to 6.10	Saila and Lorda, 1977
Northern pike	3.3×10^{-5}	2.5×10^{-5}	6.3×10^{-6}	0.76 to 0.19	1.15 to 2.29	‡
Bluefin tuna	5.3×10^{-7}	3.9×10^{-7}	7.1×10^{-8}	0.74 to 0.13	1.17 to 2.73	Vaughan and Saila, 1976

*Based on one reproductive year.
†Based on half of the reproductive life span.
‡Calculated by Saunders.

Assumption 3 requires that the vulnerability of planktonic organisms to entrainment does not change. This is not valid since, as planktonic fish grow, swimming (avoidance) ability increases and vulnerability decreases. This would result in an adult loss lower than that predicted by the model.

Assumption 4 requires that plankton which die naturally not be entrained. Inclusion of organisms already dead in the estimated number of organisms entrained would bias the model toward an overestimation of loss. The magnitude of this increase of predicted-vs.-actual impact depends on the proportion of dead organisms entrained.

The sensitivity analysis and these observations indicate that my alternative version of the equivalent-adults model is biased toward overestimating potential adult losses. Such bias should be generally acceptable in terms of determining (in a "first-cut" approach) whether or not entrainment at a particular facility is potentially damaging to local fish populations. Use of this model will possibly generate order-of-magnitude overestimates of loss in some cases (as opposed to order-of-magnitude underestimates with the unrevised model). For this reason, I suggest using this form of the equivalent-adults model only in cases where information on planktonic mortality rates or population dynamics does not exist.

I do not consider this discussion to constitute a complete treatment of the equivalent-adults model and its application in impact-assessment situations. For a thorough discussion of other aspects of the model, see Horst (1978).

SUMMARY

An examination of one of the assumptions inherent in application of the equivalent-adults model of Horst (1975) indicated that failure to satisfy the assumption could result in underestimates of potential adult loss by a factor greater than ten. I suggest using an alternative form of the model which requires no additional data. Analysis of the assumptions of the alternative model shows that it will overestimate impact, but such overestimates are considered acceptable in using this model as a preliminary impact-assessment tool.

ACKNOWLEDGMENT

I would like to thank Tom Horst and Carol Blakely for their encouragement and comments.

REFERENCES

Cushing, D. H., 1975, *Marine Ecology and Fisheries*, Cambridge University Press, New York.

——, and J. J. Walsh, 1976, *The Ecology of the Seas*, W. B. Saunders Company, Philadelphia.

Fajen, O., 1975, Population Dynamics of Bass in Rivers and Streams, in *Black Bass Biology and Management*, R. H. Stroud and H. Clepper (Eds.), pp. 195-203, Sport Fishing Institute, Washington, D.C.

Franklin, D. R., and L. L. Smith, Jr., 1963, Early Life History of the Northern Pike, *Esox lucius* L., with Special Reference to the Factors Influencing the Numerical Strength of Year Classes, *Trans. Am. Fish. Soc.*, 92(2): 91-110.

Hackney, P. A., 1977, Methods for Calculating Natural Mortality Rate, Biomass Production, and Proportion Entrained of Lacusterine Ichthyoplankton, in *Proceedings of the Conference on Assessing the Effects of Power-Plant-Induced Mortality on Fish Populations*, W. Van Winkle (Ed.), Abstract, p. 127, Pergamon Press, Inc., New York.

Horst, T. J., 1975, The Assessment of Impact Due to Entrainment of Ichthyoplankton, in *Fisheries and Energy Production*, S. B. Saila (Ed.), pp. 107-118, D. C. Heath and Company, Lexington, Mass.

——, 1977, Effects of Power Station Mortality on Fish Population Stability in Relationship to Life History Strategy, in *Proceedings of the Conference on Assessing the Effects of Power-Plant-Induced Mortality on Fish Populations*, W. Van Winkle (Ed.), pp. 297-310, Pergamon Press, Inc., New York.

——, 1978, Mathematical Modeling of Power Station Impacts on Fisheries Resources in the United States, in *Proceedings of an International Federation for Information Processing Working Conference on Modeling and Simulation of Land, Air and Water Resources Systems*, Ghent, Belgium, 1977, North Holland Publishing Company, Amsterdam, in press.

Leslie, P. H., 1945, On the Use of Matrices in Certain Population Mathematics, *Biometrika*, 33: 183-212.

Monten, E., 1948, Undersökningar över gäddynglets biologi samt några därmed sammahängande problem, *Skr. ut. av Södra Sver. Fiskeriför.*, 1948: 3-38.

——, 1950, Studier över yngeförlusternas orsaker i fria vattan och i dammer. (II), *Skr. ut. av Södra Sver. Fiskeriför.*, 1949: 20-101.

Pearcy, W. G., 1962, Ecology of an Estuarine Population of Winter Flounder, *Pseudopleuronectes americanus* (Walbaum), *Bull. Bingham Oceanogr. Collect.*, 18(1): 5-78.

Polgar, T. T., 1977, Striped Bass Ichthyoplankton Abundance, Mortality and Production Estimation for the Potomac River Population, in *Proceedings of the Conference on Assessing the Effects of Power-Plant-Induced Mortality on Fish Populations*, W. Van Winkle (Ed.), pp. 110-126, Pergamon Press, Inc., New York.

Saila, S. B., and E. Lorda, 1977, Sensitivity Analysis Applied to a Matrix Model of the Hudson River Striped Bass Population, in *Proceedings of the Conference on Assessing the Effects of Power-Plant-Induced Mortality on Fish Populations*, W. Van Winkle (Ed.), pp. 311-332, Pergamon Press, Inc., New York.

Sette, O. E., 1943, Biology of the Atlantic Mackerel *(Scomber scombrus)* of North America. Part I, *Fish. Bull.*, 51(49): 251-358.

Vaughan, D. S., and S. B. Saila, 1976, A Method for Determining Mortality Rates Using the Leslie Matrix, *Trans. Am. Fish. Soc.*, 105(3): 380-383.

Williams, G. C., D. C. Williams, and R. J. Miller, 1973, Mortality Rates of Planktonic Eggs of the Cunner, *Tautogolabrus adspersus* (Walbaum), in Long Island Sound, in *Proceedings of a Workshop on Egg, Larval and Juvenile Stages of Fish in Atlantic Coast Estuaries,* A. L. Pacheco (Ed.), Technical Publication No. 1, Middle Atlantic Coastal Fisheries Center, National Marine Fisheries Service.

STRESS AND ECOSYSTEMS

ARIEL E. LUGO*
Department of Botany and Center for Wetlands, University of Florida, Gainesville, Florida

ABSTRACT

The literature dealing with issues of stress as it affects ecosystems is reviewed. Definitions of stress are discussed. Models and literature examples are presented to illustrate the push—pull (positive—negative) effects of most stressors and to suggest that the point of attack and the type of stressor determine the rate of response of the ecosystem. Stressors with high-quality energies (highly concentrated energy sources) that divert low-quality energy flows in a system appear to have a greater impact than stressors with low-quality energy (diluted energy sources) that impact high-quality energy flows. It is suggested that ecosystem complexity (including species diversity, physiognomy, three dimensional organization, etc.) is a function of the balance between energies that contribute to growth and organization and those that contribute to disorder. The classification of environments by their "energy signatures" (the sum of all incoming energy flows into a system and the pattern of their delivery expressed on equal energy-quality basis) is presented as the best way to arrange and analyze ecosystems hierarchically according to their capacity to develop complexity and to tolerate stress. The patterns of ecosystem response to stressors, including positive, steady-state, and declining responses and possible extinction, are discussed. It is argued that, to solve the problems of ecosystem management and the issues of environmental impact, studies and analyses must be done at the level of the ecosystem and care should be taken to quantify both the stressor and the stress with units of comparable energy quality.

Today most of the biosphere has been altered in some way by the activities of man. With human impact on ecosystems becoming more obvious and more damaging to the economy of nations, the ecologist's interests are changing from descriptions of steady-state ecosystems to studies of ecosystem perturbations and their implica-

*Current address: Council on Environmental Quality, Washington, D. C.

tions to ecosystem structure, function, and stability. Sectors of the biosphere that are relatively more sensitive to human alteration (e.g., atmospheric and aquatic systems) initially received the greatest public attention. Terrestrial systems have been studied in relation to the impacts of ionizing radiation and radionuclide contamination (Woodwell, 1970). Marine and coastal systems are now under intense study as human population centers continue to grow in coastal areas and the demands for fossil fuel increase the intensity and frequency of oil and thermal pollution incidents (Cowell, 1971; Gibbons and Sharitz, 1974; Ferguson-Wood and Johannes, 1975). Another line of research uses laboratory microcosms to study the impact of heat, radiation, oil, and other pollutants on microecosystems (Byers, 1962; Moore, 1964; Copeland, 1965; Ferens and Byers, 1972; H. T. Odum, 1974).

Platt (1965), H. T. Odum (1967), Woodwell (1970), and others suggested that ecosystem response to external impact follows certain common patterns regardless of ecosystem type or type of impact. If they are correct, it would be unwarranted to continue to duplicate certain types of studies every time a new perturbation to the biosphere is introduced by man. If there are patterns of ecosystem response to stressors, we should use this knowledge to anticipate human impact on natural ecosystems. This paper has two objectives, first, to review the information available on the response of ecosystems to stressors to determine whether there are recurrent patterns of response, and, second, to suggest some approaches for the study of stress and natural ecosystems. The synthesis highlights controversial ideas with the hope of stimulating further research on the issue.

DEFINING STRESS AND STRESSORS

The concept of stress is commonly applied to describe the behavior of systems of all kinds. Meier (1972) summarized a number of definitions of stress: Stress was initially defined as "the forces or pressures exerted upon a material." Later, biological stress was defined as "the rate of all the wear and tear caused by life." In social studies, stress characterizes "physical, social, and cultural conditions likely to be discomforting for most people living within a specific group." Stress is also viewed as "a response to external or internal processes which reach those threshold levels that strain psychological and physiological integrative capacities close to or beyond their limits." Finally, stress can be defined as "any force that pushes the functioning of a critical subsystem beyond its ability to restore homeostasis through ordinary, nonemergency adjustment processes."

Selye (1956) and Fitch and Johnson (1977) distinguished between stress and stressors. A stressor is any condition or situation that causes a system to mobilize its resources and increase its energy expenditure. Stress is the response of the system to the stressor. Responses to stressors may include adaptation or functional disorder. Adaptations allow the system to overcome the stress or to avoid it. Functional disorders can be ameliorated or can eventually lead to exhaustion and death. H. T. Odum (1967) pointed out that, although different stressors had different impacts on an ecosystem (e.g., removal of organisms, higher respiration rates, or diversion of incoming resources), they all diverted potential energy flows that otherwise could do useful work in the system. He then defined stress as a drain of calories of potential energy flow. According to his definition, stress can be measured by changes in the flows of energy in a system, disappearance of previously existing flows, or acceleration of repair work.

TYPES OF STRESSES AND STRESSORS

Regardless of how stress is defined or of the stressor involved, the concept of stress as normally used invokes an interference with the normal function of a system; its effects are most dramatically observed after certain thresholds of tolerance are exceeded; and it appears that beyond these thresholds recovery is usually difficult. When systems are stressed for short periods of time and have an opportunity to recover during periods of low stress, the stressor is said to be acute. Other systems are exposed to the effects of continuous or chronic stressors. When organisms function despite the presence of a chronic or acute stressor, the intensity and the type of stressor are said to be adaptable because they allow the systems to survive. Surviving systems are equipped to overcome the drain of potential energy and still remain competitive. Adaptability does not imply absence of an energy drain, however. All stressors are analogous to energy barriers that organisms and ecosystems must continuously overcome if they are to survive. For example, a halophyte must constantly deal with the problem of salinity even if it is adapted to salinity, just like wetlands and certain forests must overcome flooding or fire even though they have adaptations to cope with these factors. Morowitz (1968) said that the rate of protein turnover in organisms follows an exponential function of temperature which rises rapidly above 40°C for most animals. Thermophilic bacteria, however, can survive higher temperatures because their rapid rise in protein turnover occurs above 50°C. The cost of high-temperature survival in terms of protein synthesis is staggering, however, and only a few species with very low biomass and a fast turnover of this mass can survive such conditions. If the pH of the water changes, then an aquatic organism's tolerance to high temperature decreases (Brock, 1970), presumably because of a synergistic effect of one stressor on another (Vernberg and Vernberg, 1974).

Thus it appears that we need to distinguish between the stressors that are part of an ecosystem's "normal" everyday environment and those added by man or by acute events with infrequent recurrence. Both types cause energy drains, but the effects of natural stressors are not as immediately obvious to an observer as are those of additional energy drains imposed allogenically on a system.

ENERGY COST OF STRESS

The energy cost of stress to a system is probably a function of the intensity of the stressor (i.e., how much energy it drains per unit area and unit time), the multiplicative or additive effect of this energy drain on the overall function and homeostasis of the system, the frequency of its occurrence, the type of ecosystem being stressed, the condition of the system at the time of impact, the intensity of other stressors at the time of impact, the residual effects of other stressors on the system, and the frequency of return of these particular stressors. Chronic stressors, which operate for longer time periods than acute stressors, usually drain more energy from a system. Acute stressors may be very damaging, however, depending on the intensity of the stress, the time interval at which it reoccurs, and the adaptability of the stressed system. Also, the multiplicative impact of energy drains on the overall function and homeostasis of the system deserves careful consideration since a small energy drain in a component with large amplification value on the function of the ecosystem may be more damaging than a large energy drain in one with less influence on other properties of the system. An example would be the relative impact of a poison vs. that of a fire on the growth of an ecosystem.

PREDICTABLE AND UNPREDICTABLE ENVIRONMENTS

Some investigators have differentiated environments on the basis of predictability and used this difference as a measure of stress (Slobodkin and Sanders, 1969; Sanders, 1969; Colwell, 1974). They give as examples of unpredictable environments boreal and tropical sea bottoms with low dissolved-oxygen concentrations (<2 to 5% saturation) and seasonal temperature fluctuation ($\sim 5.5^\circ C$), sewer outfalls and outfalls of canneries, the edges of anoxic zones in estuaries and lakes, and deserts. Hickman (1975) considered the shallow soils of the upper slopes of mountain peaks to be unpredictable environments for plants growing on them. Bleakney (1972) suggested that acyclic extreme tides prevented the occurrence of genetic selection and behavioral adaptation in littoral communities. Ehrlich et al. (1972) documented the extinction of butterflies and the lowering of population densities in subalpine ecosystems as a result of unusual weather (a wet June followed by heavy, late snows).

These environments are obviously harsh, with factors that exhibit a relatively wide amplitude of variation, but are they really unpredictable? Slobodkin and Sanders (1969) defined an unpredictable environment as one in which "the variance of environmental properties around their mean values are relatively high and unpredictable both spatially and temporally." This definition is subjective and presents a few problems when it is used to evaluate the degree of unpredictability of natural environments. First, data are usually meager and do not allow the necessary calculations. Second, seasonal variations may appear to be unpredictable when they are, in fact, very predictable over a longer period of observation. Third, how do we know what is unpredictable to a natural system? The problem with the use of the terms predictable and unpredictable was made obvious in the discussion following the presentation of the paper by Slobodkin and Sanders at the Brookhaven symposium in 1969. Slobodkin could not agree with L. C. Cole on whether cave, hot-springs, and salt-lake ecosystems were or were not predictable environments. Yet, we would intuitively consider them very stable environments. In fact, Brock (1970) showed this to be true for hot springs, and Poulson and Culver (1969) described the environmental constancy of the cave environment.

Colwell (1974) conceptualized the term predictability as the sum of two separate components, constancy and contingency. Contingency represents the degree to which time determines a state (the value is minimal when the probability of occurrence of each state is independent of season), and constancy is a measure of sameness of state from year to year. Predictability, according to Colwell, is essentially a measure of the variation among successive periods in the pattern of a periodic phenomenon. When the variation is low, predictability is high. He argues that the same degree of predictability can have different effects on different organisms and that constancy and contingency may be as important as elements of adaptive strategies as they are as environmental constraints on evolution.

Since predictability implies anticipation, predictable environments can be anticipated, but unpredictable environments cannot. The ability to anticipate rests on the system subjected to the environment; thus the relationship between the system and its environment becomes crucial in defining the degree of predictability or unpredictability an environment represents to a system. Slobodkin and Sanders (1969) said that predictability was partly dependent on the organism and was not necessarily an environmental property. We would expect that, all other conditions being equal, the smaller and simpler the system, the greater its susceptibility to short-term environmental variability. However, miniaturization may be a successful evolutionary response to a long history of wide fluctuations. Regardless of strategy, if we assume that survival implies adaptability and that adaptability enhances anticipation of the environment, we would call the same environment an unpredictable one for the system that perishes and a predictable one for the system that survives.

Environments have also been described as constant, inconstant, cyclic, and randomly fluctuating or as being in any intermediate state. Stearns (1978) showed that it is incorrect to think in terms of just two contrasting situations (predictable or unpredictable). He found that even within one type of ecosystem (reservoirs) analysis of four variables separated 19 of them into distinct classes of variation that formed a diverse mosaic of environments. Stearns concluded that different environments could rank differently for any measure of fluctuation or stability for each significant environmental variable. Whittaker (1975) pointed out several possible characteristics of fluctuations that need consideration—quality and duration, relative amplitude of a regular fluctuation, relative irregularity, and duration of the pattern in evolutionary time. Obviously the number of points along a continuum of possible intensities of stress associated with these types of environmental situations is significant to any analysis of stress and ecosystems.

The classification of an environment in any category depends on our ability to analyze and understand data and on the completeness of the data sets. Complex time-series analyses are required to detect long-term cyclic events on data sets that before analysis appear to be random and unpredictable. The long-term cycles, although perhaps not important to organisms, may be significant to populations or whole ecosystems, as shown, e.g., by Stearns (1975) in his discussion of the effects of reservoir fluctuations on the evolution of populations of mosquitofish. Analyses of this type are becoming very common in the literature (Colwell, 1974; Sneyers, 1976; Dyer and Tyson, 1977). Findings indicate that such potential stressors as drought (Dyer and Tyson, 1977), fire (Houston, 1973), hurricanes (Thomas, 1974; Gentry, 1974), flash floods (John, 1964), storms that cause catastrophic drifts (Anderson and Lehmkuhl, 1968), or rainfall (Beatley, 1974; Colwell, 1974) have recurrent, predictable patterns to which ecosystems can adapt and which they depend on for their survival and maintenance.

Living systems "track" the environment through a number of adaptations that have bearing on environmental anticipation, e.g., storages of energy, life forms, strategies of life cycles, phenological patterns, migrations, successional recovery, alterations in rates of physiological processes, and lowering of productivity-to-biomass ratios. At the population level, e.g., Bott (1975) showed that bacteria in a stream with a fluctuating thermal event (up to 23°C/year) had adapted to a temperature 5 to 20°C higher than the mean temperature of the stream. An optimum growth temperature near the lethal temperature becomes a predictive mechanism for random temperature fluctuations; temperature increases have less likelihood of exceeding the higher tolerance limit.

Environmental fluctuations are stressful to ecosystems (Dunbar, 1960), and, as the environment becomes more variable, the cost of anticipation should increase. Anticipating a changing environment remains an essential prerequisite of life, however, if the organism or the ecosystem is to survive.

I agree with Margalef's idea (1969; 1975) that, as they develop toward the steady state, ecosystems gain information that, in part, allows them to anticipate fluctuations in their environments. At the ecosystem level an event may cause mortality in certain sectors of the system, but the system as a whole may be able to adapt to it. For example, hurricanes kill mangrove trees, but they are essential for the survival of the mangrove forest (Cintrón et al., 1978). The annual migration of the thermocline in a temperate dimictic lake creates severe problems for certain populations of bacteria, zooplankton, and phytoplankton, but, for the lake as a whole, this event is predictable and essential for the distribution of heat, nutrients, and organic fuels. Thus, if we select the whole ecosystem as a point of reference, there cannot be naturally unpredictable environments that are colonized by steady-state ecosystems. Harshness of the environment should not be confused with unpredictability.

Truly unpredictable events are, perhaps, earthquakes and many human impacts. Human interferences with natural ecosystems are unpredictable because they do not necessarily follow a recognizable pattern and they do not operate long enough to allow for adaptations to develop. Many activities of man, while nonadaptable for many natural systems, however, do select for interface ecosystems that quickly adjust to human interference and develop short-term steady states depending on continuous human intervention. These systems have many of the characteristics of the physically controlled communities described by Sanders (1969, Table 1) and are dominated by exotic or successional species.

TABLE 1

PROPERTIES OF PHYSICALLY CONTROLLED, BIOLOGICALLY ACCOMMODATED, AND STRESSED ECOSYSTEMS*

Physically Controlled
Small numbers of species per numbers of individuals
Widely fluctuating physical conditions that are not rigidly predictable
Organisms exposed to severe physiological stress
Environment of recent past history
Biologically Accommodated
Large numbers of species per numbers of individuals
Physical conditions constant and uniform for long periods of time
Environment of high predictability
Ecosystems Under Stress
Succession arrested or set back periodically
Few successional stages between pioneers and climax; autosuccession and cyclic succession common
Plants may show stress by aberrations or leaf deformations
Changes in the intensity of stressors cause shifts in species composition; generally species diversity decreases with successional time
P/R ratios of 1, >1, or <1 at steady state
Speed of succession is a function of physical environment and the point in the system where disordering energies act
Species show zonations that reflect gradients of stress; zones should not be confused with successional stages

*Data on "physically controlled" and "biologically accommodated" ecosystems are from Sanders (1969).

ENERGY BASIS OF STRESS

H. T. Odum's (1967) definition of stress as a drain of potential energy has an advantage over all other definitions because it provides a common basis (energy) on which to evaluate all types of stress in all kinds of systems. More significant, however, is that the energy required for all natural processes behaves according to well-defined laws that provide a quantitative basis for evaluating ecosystem function and response to stressors. The capacity of a system to overcome stress, then, depends on the balance between the rate of recovery of its energy stores and the magnitude of the energy drain caused by the stressor.

We could argue against Odum's definition of stress by pointing out that many drains of potential energy are not normally called stress but would be by his definition, e.g., the energy drain of nighttime respiration in plants. We would not normally identify nighttime darkness as a stressor, but shading of a plant or an ecosystem for a prolonged time period certainly would be considered one. Rogers (1977), for example, after shading a coral reef for 35 days, found significant rates of mortality and slow rates of recovery.

The health of ecosystems that are normally considered highly stressed could also be used to illustrate the futility of identifying stressors and defining what they do to a system. Salt springs, thermal springs, or deserts actually become stressed if the "stressful conditions" to which they are adapted are changed. The usefulness of the concept of stress is obviously limited if we cannot deal with the apparent subjectivity and relativity of the term.

Selye (1956) suggested that most environmental situations can become stressful to given individuals and that stress is a normal condition of anyone's environment. The realization that stressors and energy drains are part of any natural environment is an important step in generalizing about stressors and their impact on ecosystems. This is the basis for my earlier suggestion that we differentiate between background or normal stress and additional stress caused by allogenic forces. In fact, as discussed in the following section, stressors may become subsidies to certain ecosystems, and their productivity and growth may decrease if the stressors are removed (Fig. 1).

Push–Pull Model of Stress

Figure 2 is a model of the push–pull (positive–negative) effect of ordering and disordering factors in the environment (in this case, temperature) proposed by H. T. Odum (1974). The model depicts

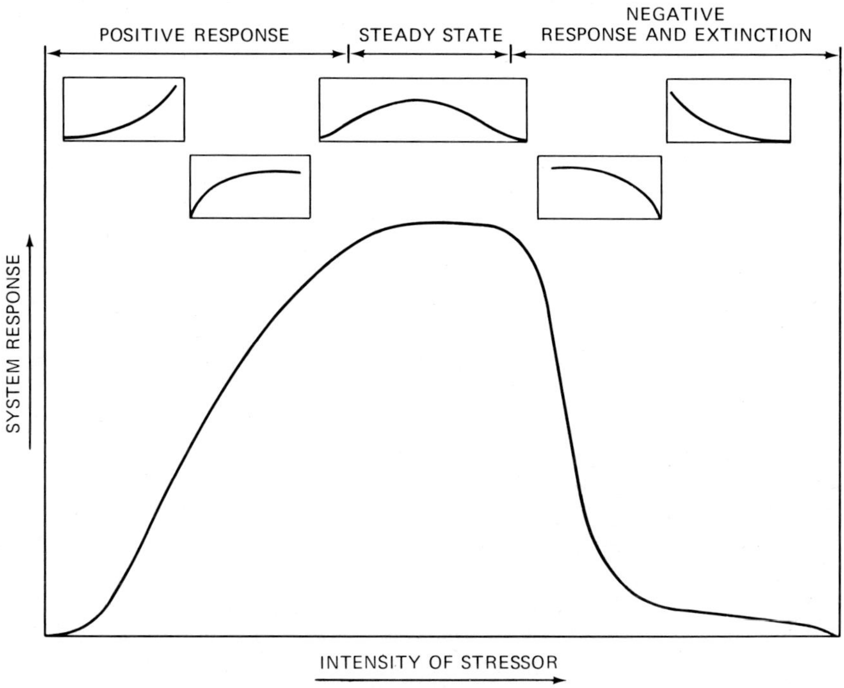

Fig. 1 Response of living systems to stressors. Similar curves of response were proposed by E. P. Odum (1971), Gibbons (1976), and Odum and Kroodsma (1976). Examples of positive response with increasing stress intensity were reported by Yentsch et al. (1974) for algal populations under temperature stress and by Carpelan (1964) for algae under salinity stress. Conditions of steady-state response are common in most stress experiments [e.g., Woodwell and Rebuck (1967) for radiation and Erdman (1966) for DDT and X rays]. Rapid declines and extinction are equally common (Woodwell and Marples, 1968; Coutant and Cox, 1976). Few studies cover the whole spectrum of response for a given environmental factor.

the well-known facts that all structures have a natural tendency to deteriorate and that life as we recognize it represents an improbable state of organization whose maintenance requires continuous energy flows. The implication is that certain drains of potential energy that we call stresses when they irreversibly alter a system are, in fact, necessary and constantly present components of all life processes. H. T. Odum (1976) suggested that disordering stimulates ordering and, thus, questioned the tendency to regard stress as bad and disorder as evil.

The push—pull model of stress proposes that stressors, such as temperature, may initially accelerate and then decelerate processes.

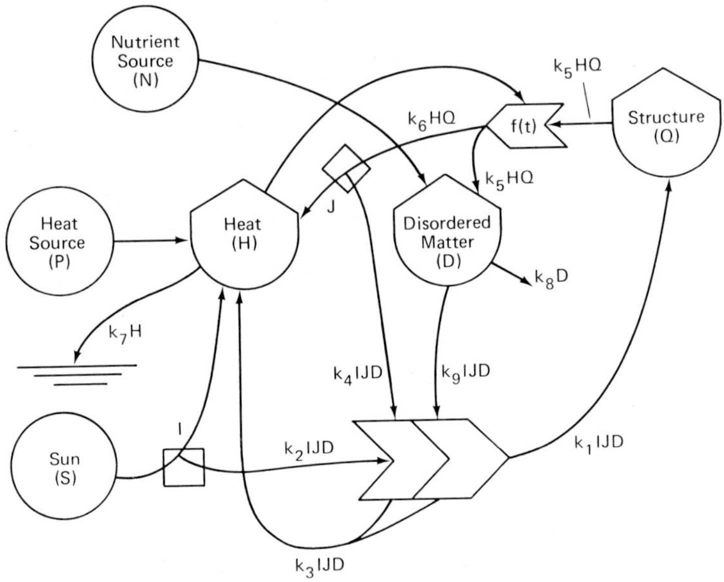

Fig. 2 Model of push—pull effects of thermal energy on ecosystems proposed by H. T. Odum (1974). Note that ordering and disordering processes depend on each other. Arrows show the direction of energy flows; tanks are state variables; the large double arrowhead represents multiplicative energy interactions; and the circle represents the energy source.

Depending on the characteristics or state of the stressed system, the energy influx of the stressor may or may not benefit the system. H. T. Odum (1974) pointed out, for example, that high-temperature stress may be useful in accelerating rates of respiration, thus causing the release of minerals for photosynthesis. This function would be a positive asset to an oligotrophic aquatic system but not to a eutrophic system. Nixon (1969) and Odum, Nixon, and DiSalvo (1970) postulated a similar role for photorespiration in oligotrophic hypersaline systems rich in organic matter. In this case, the energy drain of photorespiration was shown to benefit mineral availability in an otherwise nutrient-poor system. Disruptive stressors, such as fires, tides, floods, water currents, volcanic eruptions, and hurricanes, also have a positive role at the ecosystem level. Without a periodic disruption ecosystem growth processes stagnate as resources are immobilized by their structure. Bursts of growth and high net productivity usually follow disturbances, and rejuvenated systems replace senescent systems. To have a positive effect, the energy input to a system must be coupled with increased availability of other essential resources. Otherwise the input energies cannot be usefully converted. Table 2 summarizes literature examples of push—pull effects of different stressors.

These examples suggest that, as the inputs of energy to a system change (including energies that push and/or pull), the biotic

TABLE 2

EXAMPLES OF POSITIVE (PUSH) AND NEGATIVE (PULL) EFFECTS OF STRESSORS ON ECOSYSTEMS AND POPULATIONS*

Stressor	Pushing effect	Pulling effect
Ionizing radiation	Observed faster decomposition, leaf fall, growth, and recovery rates for a variety of dosages (Woodwell and Marples, 1968)	Can be lethal; disrupts structure and increases respiration rates; changes species composition (Ferens and Byers, 1972)
Tidal extremes	Redistribute nutrients, sediments, organic matter, and organisms (Odum, Copeland, and McMahan, 1974)	Exposes organisms to lethal conditions (Glynn, 1968)
Water flow	Brings nutrients and redistributes larvae (Anderson and Lehmkuhl, 1968); brings food and O_2 and removes toxic substances (H. T. Odum, 1955)	Removes structure; causes high energy maintenance costs in plants and animals (Anderson and Lehmkuhl, 1968; Moore, 1964)
Flooding	Removes competition; signals the beginning of phenological events (John, 1964)	Increases energy maintenance costs; temporarily decreases number of taxa and individuals (Hoopes, 1974)
Volcanic eruptions	Allow for better nutrient, moisture, and competitive environments (Eggler, 1967)	Suffocate and kill plants and animals (Eggler, 1967)
Denudation, clipping, herbivory, or defoliation	Stimulates plant productivity (Churchill et al., 1964); diversifies community at low rates (Penfound, 1964)	Removes structure and reduces diversity at high rates; causes mortality (Churchill et al., 1964)
Salinity	Allows higher gross productivity in mangroves up to about seawater strength (Hicks and Burns, 1975)	At values higher than 35‰, increases respiration rates and decreases transpiration and net productivity rates (Lugo and Snedaker, 1974)

(Table continues on following page.)

Table 2 (Cont.)

Stressor	Pushing effect	Pulling effect
Eutrophication	Stimulates primary productivity and growth of consumers (E. P. Odum, 1971)	Shifts species composition, causes anoxic conditions, and eventually reduces species diversity and complexity (E. P. Odum, 1971)
Fire	Makes nutrients and moisture more available; reduces competition (Wright, 1976)	Removes structure (Wright, 1976)
Herbicides, oil, and other poisons	Select for certain life forms over others, survivors grow faster (Wein and Bliss, 1973; Niering and Goodwin, 1974; Burk, 1977)	Reduce productivity by altering behavior and physiology, poisoning, and killing organisms (Krebs and Burns, 1977); lowers species diversity (Tomkins and Grant, 1977)
Heavy rains	Act as a clue for starting phenological events in deserts (Beatley, 1974); relieve salinity in coastal situations and redistribute nutrients (Chabreck and Palmisano, 1973)	Remove structure and may cause other stresses, such as flooding, which affect gas exchange of wetlands sediments and the turbidity in aquatic systems (Hoopes, 1974)
High temperature	Accelerates processes, particularly respiration and recycling (Smith et al., 1974)	Can be lethal; reduces species diversity (Kullberg, 1968)
Low temperature	By slowing down processes, allows for conservation of storages (Mooney and Billings, 1965)	Freezes can be lethal
Temperature amplitude	If short-term, fluctuations may stimulate metabolism (Sweeney and Schnack, 1977); increases resistance to thermal shock (Hubbs, 1964)	If amplitude is wide and change is rapid, it can be lethal (Hubbs, 1964; Sastry, 1976)
Hurricanes	Bring water, nutrients, sediments, and propagules (Chabreck and Palmisano, 1973)	Structure is removed (Craighead, 1964)
Insecticide	DDT (5—10 ppm) enhances resistance to X-ray radiation (Erdman, 1966)	Eventually lethal (Erdman, 1966)

*Literature citations are not intended to imply that these are the only or the most significant documentations of the example.

components selected for are those that somehow maximize energy flow despite the presence of stressors. This is done either by diverting all incoming energies to overcome stress or by extracting useful work from the disordering force. Energy drains are normal processes affecting living systems and should be identified as such even if they normally do not reach critical thresholds beyond which recovery is difficult.

To evaluate the relative impact of push—pull effects on ecosystems and to gain insight into the cost of recovery and ecosystem resilience, we can calculate ratios of disordering-to-ordering energies (Richey, 1970). This involves calculating the magnitude of energy

TABLE 3

RATIO OF DISORDER TO ORDER IN ECOSYSTEMS SUBMITTED TO STRESS

Ecosystem	Stressor	Disorder/order	Reference
Subtropical dry forest	Cutting	1 : 1,270	Ewel, 1971a
Subtropical dry forest	Herbicides	1 : 17	Ewel, 1971a
Subtropical moist forest	Gamma radiation	1 : 10,000	H. T. Odum, 1970
Microcosms (terrestrial)	Burning	1 : 13	Richey, 1970
Microcosms (aquatic)	Sonication and temperature	1 : 10,000	Richey, 1970
Vietnam (whole country)*	War (including bombs and herbicides)	1 : 8.7	Brown, 1977

*Reduced to fossil fuel equivalents.

flows that disrupt ecosystem structure and function and the magnitude of energy costs associated with reordering and maintaining order in the system. Table 3 summarizes literature reports on this ratio. Most values are derived from experimental situations, where the disordering input can be computed easily. Results show that disordering energies have considerable amplification and that building and maintaining order is extremely expensive relative to the cost of destruction. Obviously, the capacity of a system to regenerate depends on the availability of enough energy sources to reorganize the disordered structure. Since the availability of energy is a function of the environment, the type of environment dictates rates of recovery and degree of complexity at the steady state. In the term complexity, I include such measures of ecosystem organization as species diversity, structural organization, physiognomy, and phenological organization.

The concept of stress should, then, be applied in a hierarchical fashion relative to the various levels of biological organization. As ecologists, we are interested in stressors as they relate to life zones, individual ecosystems, and populations. Other scientists do their analysis at the organismal, cellular, or molecular and atomic levels of organization. To avoid the pitfalls of subjectivity, we need to develop appropriate criteria for evaluating the relative effects of stressors on the complexity and rate of recovery of systems at each of these levels of organization.

The Energy Signature of Ecosystems

The push—pull model (Fig. 2) suggests that the amount of structure and complexity a system is capable of building and maintaining depends on the net effect of positive and negative forces in the environment.

The sum of all incoming energy flows to a system and the pattern of their delivery expressed on an equal energy-quality basis is called the "energy signature" of the system (H. T. Odum et al., 1977). The quality of an energy source depends on its concentration (Odum et al., 1977). High-quality sources have a high energy content per unit volume, and low quality sources, like solar energy, are diluted and, thus, perform less work per unit volume. Energy signatures must be expressed on an equal energy-quality basis to show the relative ability of the components to generate work and control the system. The energy signature provides a way to categorize ecosystems according to environmental carrying capacity or capacity to deal with stress, e.g., as Odum, Copeland, and McMahan (1974) did for coastal systems. As the energy signature changes, the loss or gain of energy may become a source of stress or a subsidy to the system.

Since the actual intensity of background or normal disordering energies in a given environment may be impossible to measure, the measure of stress at the level of the life zone must be made relative to the most complex systems on earth, because they represent the best that can be done with the energy flows and resources available to all natural systems. Other life zones could then be arranged in a hierarchical and relative scale.

Perhaps the best system for classifying terrestrial and wetland ecosystems is Holdridge's life-zone system (1967) (Fig. 3). This system has worldwide application, is an objective determination of life zones based on climatic data (which approximate the energy

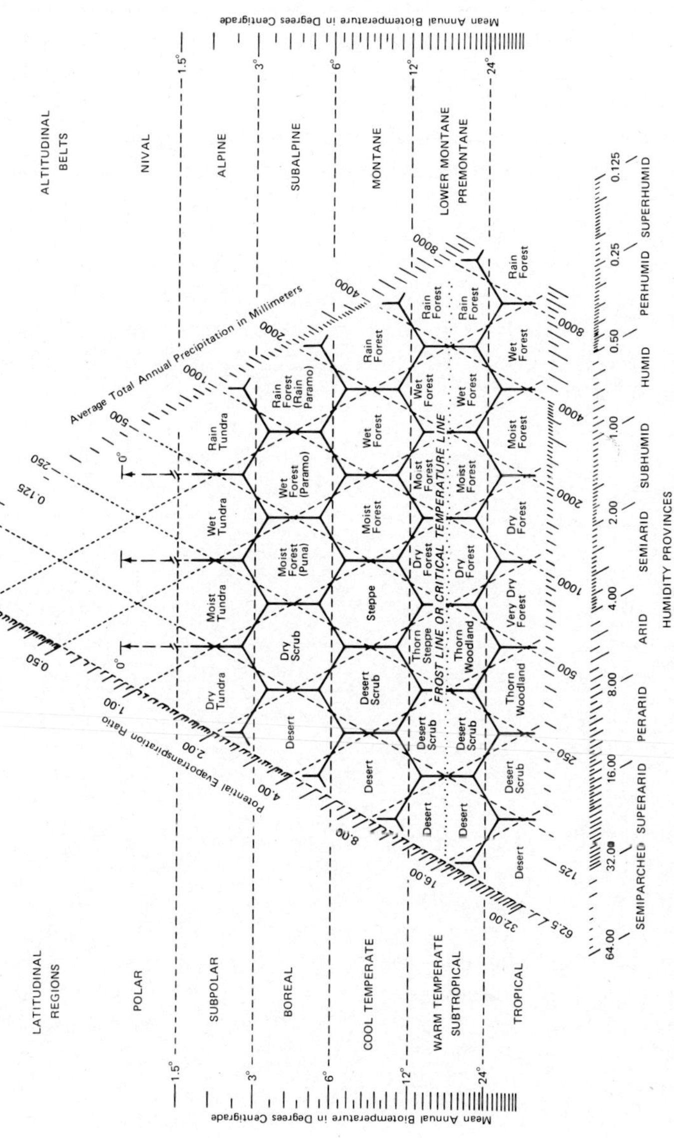

Fig. 3 Life zone classification system of Holdridge (1967). Each life zone is characterized by a specific energy signature that determines its potential to support ecosystem complexity.

signature), attempts to account for the factors that control ecosystem productivity, and has been tested widely (Holdridge et al., 1971). The life zone is a bioclimatic zone of the earth with a unique set of conditions and capacities to support biological complexity. For each climatic life zone of the world (Holdridge identifies about 120), a maximum level of ecosystem performance is possible. Performance can be measured in terms of gross energy flow or total ecosystem complexity. Holdridge (1967) has already shown that the complexity of forested terrestrial systems varies according to life zone. This complexity is equal to the product of maximum tree height (m), number of species per 0.1 ha, basal area (m^2/0.1 ha), tree density per 0.1 ha, and a constant, 10^{-3}. Life zones can be compared with one another in terms of their relative potential to support ecosystems of different complexity. The impact of additional stressors on a given type of ecosystem should vary with life zone. Ecosystems in humid life zones obviously respond differently from those in arid life zones. For example, Ewel (1977) reported slower responses to stressors (cutting and herbicides) in dry tropical climates than in humid tropical climates and described two different successional strategies for ecosystems in these two contrasting environments.

His results and those of others suggest that care should be taken in comparing ecosystems across latitudinal or climatic gradients since they should not be expected to exhibit the same speed of response to stressors. The optimism of Slobodkin and Sanders (1969), who suggest that latitude has no fundamental importance on species diversity considerations, is premature. They believe that environmental constancy is more important; however, environmental constancy per se, without adequate energy inputs, cannot sustain complexity.

The life-zone concept has not been applied to lakes and streams, but these systems are so dependent on external inputs that there is no reason not to expect a life-zone pattern within them. Since climate controls water availability and runoff, nutrient runoff from the land, and evapotranspiration, it should also influence the patterns of response of lakes and rivers.

Model of the Action of Stressors on Ecosystems

The analysis of individual ecosystems requires detailed examination of the forces acting on them. To analyze the possible ways in which stressors affect individual ecosystems, I first develop a conceptual model of an idealized unstressed ecosystem (Fig. 4). In this simple model, we can see the relationship between four state

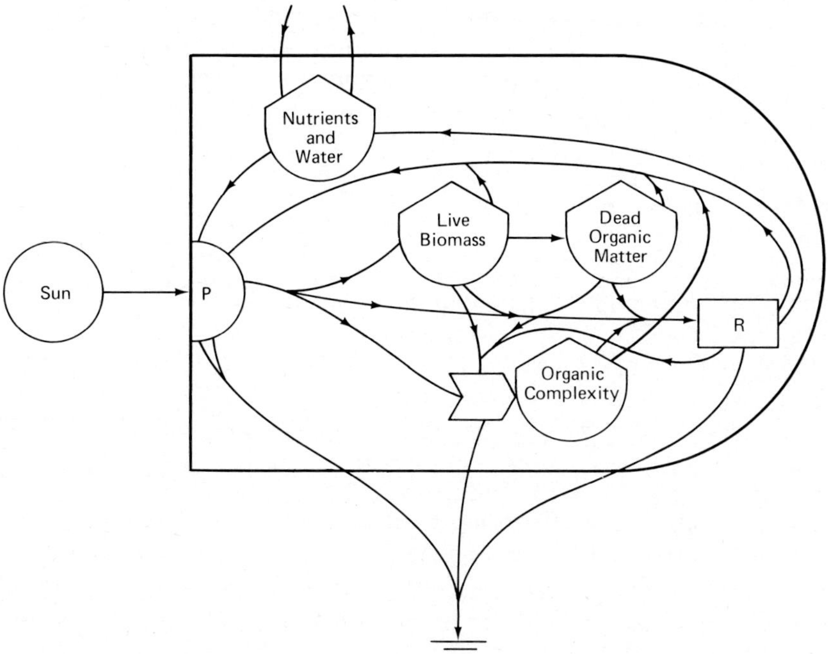

Fig. 4 Simplified model of an ecosystem without stress. The large bullet-shaped outline is the boundary of the system; P is photosynthesis; and R is respiration (for other symbols, see Fig. 2, and, for a discussion of the model, see the text). All energy inputs are eventually dissipated into degraded heat, according to the second law of thermodynamics.

variables and three ecosystem processes. The state variables are nutrient concentrations and water availability (a measure of fertility), live biomass, dead organic matter, and a hypothetical variable called organic complexity, which includes measures of species diversity and three-dimensional diversity and complexity discussed by Whittaker (1969). It is included here simply to point out that the organization of the ecosystem is an energy-demanding process and, thus, is subject to the impact of stressors. The cost of this variable is borne by each of the living components of the system, but this does not invalidate the fact that complexity of function and organization is an ecosystem property that responds to its energy signature. All state variables have a positive role in the system and provide positive feedback to its productive sector. The processes illustrated in the model are the photosynthetic conversion of solar energy into chemical energy (P), the respiratory process (R), and the many cycles and feedbacks of the system which contribute to its homeostasis.

In a system like this, there is a high cost of depreciation to support live biomass and organization. In steady-state systems the quality of energy feedbacks among ecosystem components increases relative to those of successional systems. Higher quality energy flows contribute to a higher rate of energy flow through the system as a whole. The respiration of the system increases on a unit-area basis with age because of the increasing cost of upgrading energy qualities in the system (i.e., the cost of converting low-quality forms to higher quality states).

In Fig. 5 stressors are added to the model. The energy signature of the system is subdivided into five potential sources of stress. Source 1 delivers the primary energy of the system, which in this example is solar energy for photosynthesis; source 2 diverts solar energy before it is transformed to chemical energy by plants (e.g., by shading); source 3 diverts energy after it is transformed into chemical energy but before it is incorporated into structure (e.g., by the harvest of labile sugars by a consumer, such as man); source 4 removes storages (e.g., by harvests or by drainage of water); and source 5 accelerates the respiration of the system (e.g., by changes in temperature). Each of these external sources impinges on the ecosystem at a certain rate and with a certain periodicity of delivery which depends on the type of life zone. When the energy signature changes, the change may become stressful.

The model of stress in Fig. 5 suggests that different stressors affect different functional sectors of the ecosystem and that perhaps the response of the system to these different types of drains also changes according to the type of stressor. Causal relationships must then be analyzed relative to the type of stressor, its pattern of delivery, its intensity, and the kinds of changes that it causes in the system. After a certain threshold is exceeded, the stressor causes energy flows to be diverted away from the system and diminishes the ability of the system to continue to upgrade internal energy stores. Thus the overall organization and homeostasis of the system is altered. There is flexibility in each system, however, with regard to its response to and the ultimate impact of the stressor. According to Fig. 5, more energy may be diverted into living mass, less into complexity, or more into respiration, or there may be an increase in the amount of dead organic material. The ultimate strategy probably depends on the type of stressor and where in the system it operates.

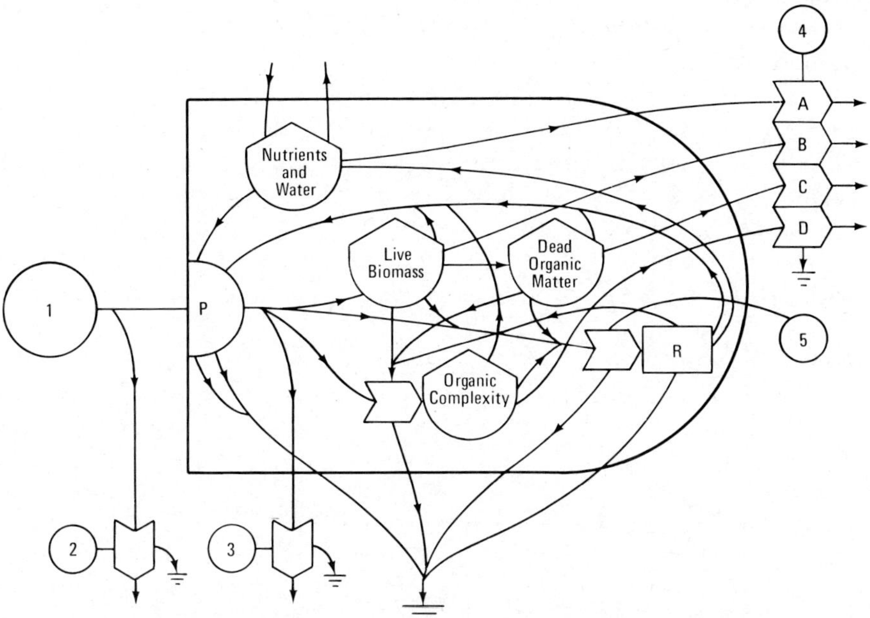

Fig. 5 Simplified model of an ecosystem influenced by a number of stressors. (For identification of symbols, see Fig. 2, and, for a discussion of the model, see the text.) Each stressor, by draining energy from a different sector of the ecosystem, affects the response in a unique way. Stressorr 4A, for example, has a stronger effect on the system than stressors 4B or 4D because it directly affects the photosynthetic process and, thus, decreases the ability of the system to restore its energy stores. Stressors 1 to 3 are equally damaging to ecosystems relative to the effects of stressors 4B to 4D and 5.

The point we need to emphasize is that, although all stressors divert potential energy away from the system, not all stressors handicap the system to the same degree, even though they drain equal amounts of energy. For example, a stressor that removes structure but does not alter the primary productivity of the system may not have as much impact as one that reduces the ability of the system to photosynthesize and, thus, to replenish its energy stores. In the first case the system will recover rapidly, but in the second recovery may be slow. This aspect of the problem has important implications for rates of recovery from stress and for management of natural ecosystems. Table 4 lists a number of naturally stressed ecosystems and classifies their respective natural stressors by type. I have tried to illustrate the rate at which the systems respond to stress

TABLE 4

EXAMPLES OF HOW THE POINT OF ACTION OF STRESSORS (ACCORDING TO FIG. 5) AFFECTS THE RATE OF ECOSYSTEM RECOVERY*

Ecosystem	Natural stressor	Natural or experimental stressors that slow down recovery rates
	Ecosystems with Fast Rates of Recovery	
Mangroves and salt marshes (Lugo and Snedaker, 1974; Cintrón et al., 1978)	Salinity (type 5)	Drainage or blocking of nutrient input (type 3), very high soil salinity (acts like type 3)
Florida sand pine scrub forest (Laessle, 1958)	Low fertility (type 3) but recovers after fire (type 4) which makes nutrients available in mesic environment	Elimination of fire or of nutrient recycling mechanisms (type 3)
Eutrophic waters (Wetzel, 1975)	Low oxygen (type 3 or 5), high water-flow rates (type 4), or no water flow (type 3 or 5)	Intensification of natural stressors of type 3
Riverine swamps (Gardner et al., 1972)	Intermittent floods (type 5)	Either drainage (type 4A) or continuous flooding (type 5), which eventually cuts plant translocation (type 4A) and lowers leaf-area index (type 2)
Springs, rivers (H. T. Odum, 1955)	Water flow (type 4)	Shading (type 2)
Coral reefs (Rogers, 1977)	High wave and water-current energy (type 4)	Shading (type 2)

Table 4 (Cont.)

Ecosystem	Natural Stressor	Natural or experimental stressors that slow down recovery rates
Ecosystems with Slow Rates of Recovery		
Granite outcrops and other arid ecosystems (Lugo and McCormick, unpublished)	Lack of water (type 3)	
Alpine and arctic ecosystems (Billings and Mooney, 1968)	Low temperature and lack of water (both act like type 3)	
Oligotrophic water (Wetzel, 1975)	Lack of nutrients (type 3)	
Bogs (Moore and Bellamy, 1973)	Continuous flooding by highly oligotrophic, low pH waters (types 3 and 5)	

*Stressors that reduce the ability of a system to photosynthesize and replenish energy stores slow down recovery rates more than stressors that act on other locations of the system. References provide relevant information on the particular ecosystems.

through successional recovery. Note that systems stressed by removal of structure or by acceleration of respiration can still function and will recover rapidly because there has been no direct impact on factors regulating primary productivity. Systems submitted to stressors that do alter their productive capacity, however, have little ability to recover or to negotiate further stress. If this is true, stressors affecting high-quality energy stores appear to have a lesser impact on a system than those affecting low energy-quality sectors of the system.

Change of Energy Signature

Examples of changes in the energy signature of an ecosystem are slow changes in climate (see Singer, 1970); changes caused in downstream sites by construction of a dam, e.g., the change in estuarine circulation patterns caused by the Aswan Dam (Sharaf El Din, 1977); and regional changes caused by alteration of water tables, e.g., in south Florida estuaries (Carter et al., 1973).

Cooper and Copeland (1973), experimenting with microcosms, simulated regional drainage changes for Trinity Bay in Texas. They found more changes in the respiration and productivity of the system than in species composition. Later, as stressors became more intense (simulated by lack of freshwater runoff or large organic-pollutant inputs), diversity index changed significantly and so did species composition. Alterations in energy flow rates before any significant changes in species composition were documented in the field by Carter et al. (1973). Apparently the substitution of species results from altered energy flows through the system. New ecosystem boundaries develop slowly in response to a different energy signature impinging on the ecosystem.

Under some circumstances a change in an energy signature could result in a new ecosystem, with higher complexity and productivity than the original system it replaced, e.g., the drainage of certain (but not all) wetlands, which creates mesic ecosystems. In this case the overall carrying capacity of the environment increases; the original ecosystem is selected against by the stressor; and the new system is more vigorous because it is better adapted to the new energy signature. J. Ewel (1978) documented such a succession in the Everglades of Florida.

Figure 6 shows as an example the energy signature of a marsh ecosystem subjected to thermal stress. It is obvious that, when energy quality is taken into consideration, tidal energy becomes the main energy source to a marsh, and fossil-fuel stress was not as significant as it appeared to be initially from its heat equivalence.

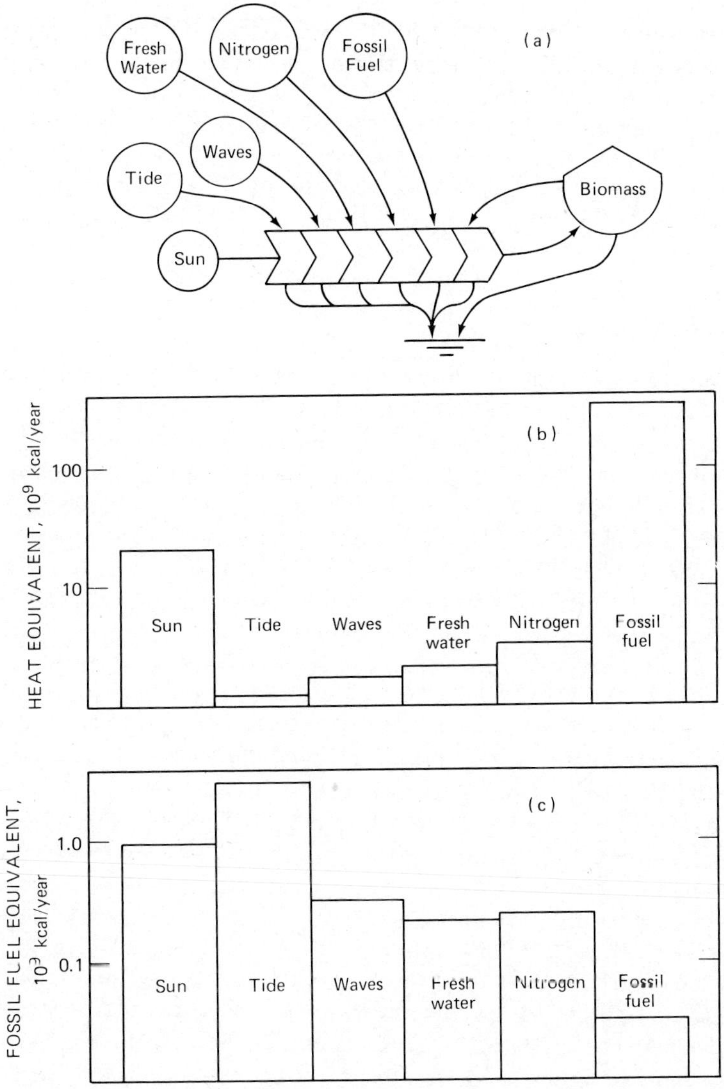

Fig. 6 Calculation of the energy signature of a marsh ecosystem under the influence of thermal pollution (a) [data from H. T. Odum et al. (1977)]. The symbols in (a) are described in Fig. 2. Note that, when all the energies are expressed on the basis of equal quality (fossil fuel equivalents), their relative importance (c) changes in comparison to calculations based on heat equivalents (b).

The change in the magnitude of an energy source when energy quality is taken into consideration raises an important question. Do energy-quality differences in stressors affect their impact on the ecosystem? For example, do poisons owe their effectiveness as stressors to their higher energy quality? The data in Table 3 do not allow for any valid generalizations, but the hypothesis could remain that stressors with high energy quality may be more effective than those with lower energy quality. Since I have already postulated the opposite effect when the quality of the stressed component is considered, it is possible that maximum stress results when a stressor with high-quality energy delivery affects a sector of the system with energy flow of low quality.

PATTERNS OF ECOSYSTEM RESPONSE TO STRESSORS

The response of cells, organisms, populations, and ecosystems to a wide variety of stressors may follow a common pattern (Fig. 1). This has been pointed out by a number of investigators (Selye, 1956; Platt, 1965; H. T. Odum, 1967; Woodwell, 1970), who also identified patterns of response. Usually, when a system is stressed, it overcomes the initial low intensity of the stressor by activating internal homeostatic mechanisms. These were described for cells, organisms, and populations by Selye (1956) and for ecosystems by Platt (1965) and Woodwell (1970). Many times low intensities of stress actually stimulate a system to performance above that of controls, probably because of higher internal physiological activity induced by the higher energy input (Fig. 1). As the intensity of the stressor increases, there is a period of steady state, when homeostatic mechanisms compensate for the energy drain, then a period of decline, and finally, if the stressor continues to intensify, a point of irreversible change. The point of collapse may be when energy reserves are exhausted or when the system reaches the limits of its adaptability. Figure 1 suggests that different stress response patterns reported in the literature may represent different sections of one single response curve that extends for the whole range of intensity of a given stressor.

Ewel (1971a), describing the pattern of ecosystem response to an intermittent stressor (Fig. 7), showed that, if the stress condition is repeated before full recovery, the system gradually loses its capacity to recover and slowly degenerates into a low-biomass steady state. This same pattern of response has been reported in salt marshes receiving chronic exposure to oil pollution (Baker, 1971) and in floodplains exposed to chronic flooding (Gardner et al., 1972).

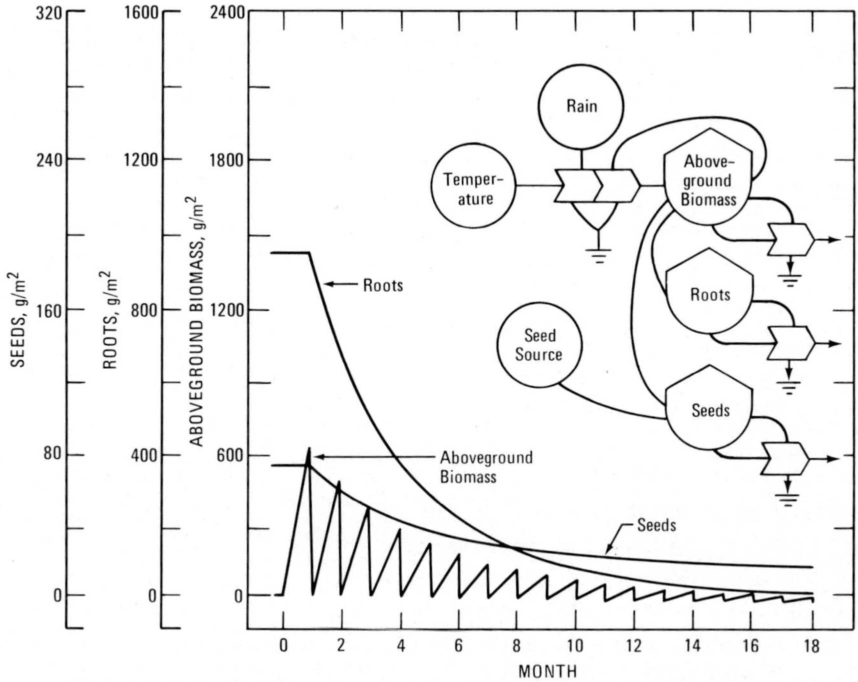

Fig. 7 Model and simulation of the effects of a chronic stressor (continuous cutting at 1-year intervals) on tropical terrestrial ecosystems [data from Ewel (1971a)]. (For identification of symbols, see Fig. 2.) Cutting is a type 4 stress (see Fig. 5). As structure is continuously removed, the system develops a steady state at a lower amount of biomass.

Stressors can also weaken systems and make them more susceptible to further stress. Woodwell and Brower (1967), for example, reported aphid population explosions induced by gamma radiation stress. The same phenomenon occurs in mangroves stressed by high salinity, frost, or alterations in drainage patterns (Lugo and Patterson Zucca, 1977). Involved in these responses are additive and/or synergistic effects among stressors which accelerate energy losses and rapidly reduce the system's capacity to negotiate more stress.

STRESS AND ECOSYSTEM COMPLEXITY

Ecologists use a variety of indexes to measure ecosystem response to stressors. Much attention has been placed on finding relationships between environmental change and species diversity or the stability of the system. [Consult the work of Baker (1970) and the volumes edited by Woodwell and Smith (1969) and Van Dobben and Lowe-McConnell (1975) for summaries of the various theories and arguments.] Sanders (1969), for example, listed the characteristics of physically controlled and biologically accommodated communities (Table 1). The idea that stressors have the effect of reducing species diversity by increasing adaptation costs is probably correct, because a stressor, by imposing energy barriers on a system, decreases its ability to support complexity. Too much emphasis has been given to the role of a stable environment, however, without also acknowledging the balance between energy subsidies and energy drains in a given environment. All the energy flows entering the system and the variety of possible ecosystem responses to stress have been ignored in such efforts as that depicted at the top of Table 1.

Slobodkin and Sanders (1969) suggested that short-term stability (a duration of about five generations) in an otherwise unpredictable environment decreased species diversity because opportunist species would temporarily gain dominance over the normal animal component of the system. If we assume that by "periods of predictable conditions" they mean a relaxation of stress, then their conclusions do not hold in theory or in practice. Theoretically, lower stress (higher predictability?) should increase the carrying capacity of an environment. Observations in rocky tidal shores (Connell, 1972; Dayton, 1971) showed that removal of dominants by a stressor results in a temporary increase in diversity. This also occurs in high-salinity environments after a storm decreases salinity stress. At these times species diversity increases for a short period but then returns to its lower value as stressful conditions return to high intensity. Thus a property of chronically stressed ecosystems is that, during their successional patterns, species diversity may decrease rather than increase (bottom of Table 1).

Conflicting arguments about the causes of ecosystem complexity can be clarified if analyses are based on the total energy flow through the system. The productivity hypothesis of Connell and Orias (1964) was criticized by Sanders (1969), who argued that highly productive systems, such as marshes, are not necessarily diverse, as inferred from the productivity theory of ecosystem diversity. Others have argued that diverse systems are not necessarily stable (e.g., May, 1975). The argument can be restated by substituting whole ecosystem analysis for generalizations based on population studies and total energy flow per unit area (the energy signature concept) for criteria of organic productivity.

Ecosystem complexity is a function of the net energy available to a system. Net energy is the balance between the energy expenditures associated with normal respiration and all incoming energies that contribute to ecosystem growth and complexity minus all incoming energies that tax the energy budget of the system. Slobodkin and Sanders (1969) asked the question, Why don't more species adapt to stress? The answer must be related to the high energy cost of certain adaptations vs. the energy limitations to which ecosystems and populations are exposed. In highly stressed environments we would expect strong selective pressure for the most energy-efficient solution to the stress problem; this adaptation would then win over less-efficient competitors. High efficiency, however, is not conducive to the fast rates of energy flow which are associated with lower efficiency systems.

Figure 8 presents a model that summarizes the proposed interactions and assumptions involved in maintaining ecosystem complexity. The first assumption is that the development and maintenance of ecosystem complexity involves energy expenditure. H. T. Odum (1970) proposed a formula to illustrate the rapid rise in the cost of ecosystem organization with increasing species diversity:

$$\frac{N^2 - N}{2} = \frac{fEaA}{K}$$

where N = number of species
f = fraction of the energy budget allocated to organization
Ea = energy input per unit area
A = total area of the system
K = daily cost of maintaining species interactions

The terms on the left side of the equation represent the number of possible interactions between species if organization is complete, and the terms on the right represent the amount of energy available for organization. Since the input of energy to a system is limited on a unit-area basis, the strategy of energy-resource allocation and the amount of work derived from energy transformations become significant factors in determining the competitive survival of a system.

H. T. Odum (1970) stated that, when a system has a low number of species, only a small amount of energy is allocated to organization and a larger fraction of its energy resource goes to other work functions; thus more energy is needed to eliminate a species and more energy is available to compensate for stress. Hickman (1975), for example, showed that energy allocation for reproduction increased with stress in *Polygonum* plants; he reported that as much as 71% of the energy stores goes into this function.

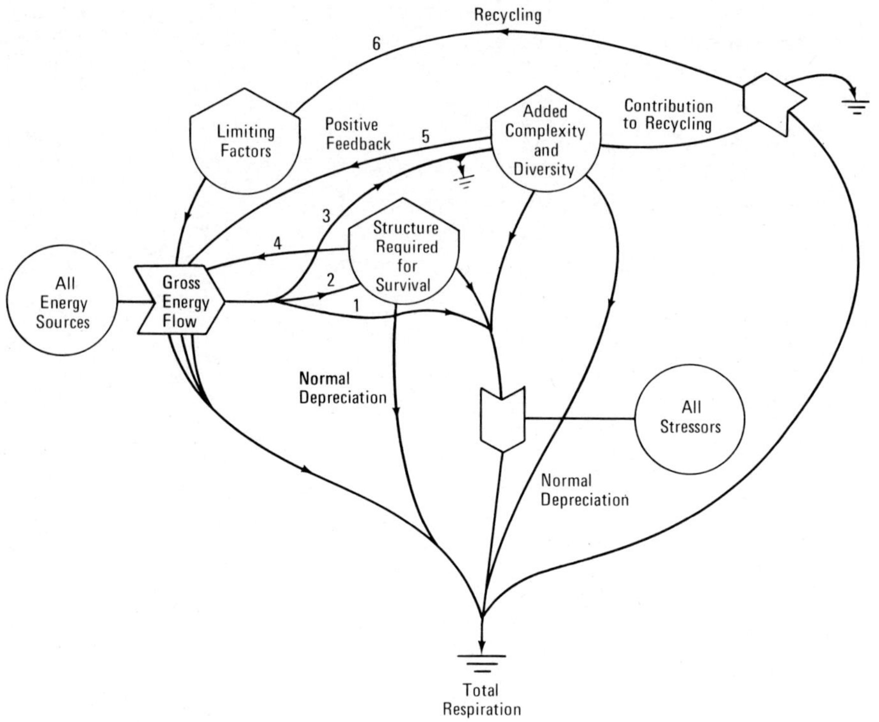

Fig. 8 Model illustrating energy flows associated with the development and maintenance of structural complexity in ecosystems. (For identification of symbols, see Fig. 2, and, for a discussion of the figure, see the text.) The gross energy flow of the system (i.e., all incoming energies that contribute to order) can be allocated to overcome disordering energies (flow 1), to develop the initial amount of structure required for survival (flow 2) and/or to develop additional complexity and species diversity (flow 3). Each of the state variables feeds back positive work that increases the ability of the system to process more energy and to use resources more efficiently. Energy allocation for structural development and complexity increases if there is net energy and if the energy investment in the feedbacks (flows 4 to 6) are offset by the increase in the energy capture and conversion efficiency of the system. As the net energy of the gross energy conversion increases (flows 1 to 3 minus flows 4 to 6), the capacity to develop more complexity and diversity also increases. Ultimately the balance between all energy sources and all stressors will determine ecosystem complexity.

A second assumption is that ecosystem complexity pays off by allowing higher efficiency in the use of energy resources and in the recycling of matter. This widely held working principle of ecology has been discussed many times (e.g., E. P. Odum, 1969; Margalef, 1969; 1975) and still makes sense to me.

The third assumption is the existence of some kind of priority development of structure in an ecosystem. Initially a minimal amount of structure necessary for survival develops. Later in successional time, if there is more net energy, complexity and diversity develop if the cost of development and maintenance do not exceed the gains that result from their feedback work. Ewel (1971b) provided evidence for the sequential development of biomass compartments in the succession of tropical ecosystems. He found that leaf biomass was the first compartment to reach steady-state values, then stems, and finally roots. The result of sequential development of compartments is a steady increase in the system's ability to use all the resources available in its surroundings. Another implication of this model is that ecosystems appear to reach functional and structural steady states early in succession, and later, with considerable lag, floristic complexity reaches its steady-state value.

In discussing ecosystem stability, we must realize that complex systems can survive only in locations with high environmental energy subsidies and low intensities of stress. An environmental change that reduces total energy flow results in a rapid decrease in structural complexity, particularly if the system loses its main power source. A decrease in complexity is interpreted as ecosystem instability. Complex cities that lose monetary subsidies, coral reefs that lose solar input energies, or complex forests growing on leached soils that lose their nutrients rapidly, lose their complexity, as expected from the behavior of the model in Fig. 8. Systems with high organic productivity but low species diversity and complexity usually export their production to other systems and, in so doing, lose the capacity to diversify. This is certainly true for mangrove forests, marshes, freshwater wetlands, some lakes, rivers, and systems stressed with high organic-matter loads. All these systems either store organics without using their potential energy or constantly export organics and receive inorganic nutrients to subsidize their primary productivity. The apparent stability of these systems is caused by the uninterrupted input of certain energy sources that represent a significant fraction of their total energy signature. Figure 6 shows evidence for this idea in relation to the role of tidal energy in maintaining stability in marshes. As long as tidal circulation continues, the marsh is capable of resisting other types of stress and recovers quickly after being subjected to a harvest or to other acute stressors. Stability is a function of stable energy input, and in most systems stability disappears when the main energy source is diverted (e.g., stress of type 1 in Fig. 5).

We have discussed how each life zone has a different background of stress which determines the potential for structural development in that locality. As more stressors are added to the background, the capacity to recover and to achieve maximum developmental potential decreases. In lowland tropical life zones, high temperatures prevail and accelerate physiological processes. For aquatic ecosystems this has serious implications for their capacity to absorb additional stress. Johannes and Betzer (1975) elaborated this point with respect to tropical marine ecosystems. Small temperature increases are very damaging to these systems; e.g., 37°C is lethal to *Thallasia* beds (Schroeder, 1975) and mangrove seedlings (Banus and Kolehmainen, 1976). These researchers also pointed out that in the tropics a higher ambient temperature and higher ambient salinity mean that a given concentration of pollutant causes a faster and greater decrease in dissolved oxygen concentration than in higher latitudes. At comparable dilution and dispersion rates, however, the pollutant concentration decreases more quickly with time and distance from the source. Thus the margin of error is smaller in tropical waters. The margin of error is equally small in ecosystems that are naturally stressed to a point near their limit of tolerance. Mangroves tolerate a more severe frost stress if they grow in low-salinity environments; they may not survive low temperatures at high soil salinities. In this case the lower salinity subsidizes high energy drains caused by frost (Lugo and Patterson Zucca, 1977).

MEASUREMENT OF STRESS AND STRESSORS AND RESEARCH NEEDS

The broad regional impact of man on the biosphere is probably why there is so much interest in studies of ecosystems and stress. Much of the research appears to ignore this broad issue, however, and instead focuses much effort on specific details that provide elegant results but do not really help society to solve the problem of coupling man to natural systems. We must document ecosystem responses to known intensities of stressors and test the ideas presented here. The researcher must take into consideration not only the intensity of the stress and the stressor but also their energy-quality characteristics relative to other energy sources impinging on the ecosystem. The points of attack in the system also require attention, and particular care should be given to the presence of feedback mechanisms and possible push—pull effects at the ecosystem level.

Table 5 summarizes some attempts at quantifying a number of stresses in natural systems and some important threshold values for specific stressors. Data of this kind, although important, are very meager. Biologists are good at counting dead or sick organisms, but little effort is made to calculate impacts per unit area or to correlate mortality with given intensities of stressors under natural conditions. Some calculations in Table 5 are based on the amount of energy actually drained from the system, but others also consider the energy the system loses when one of its components is not functioning. These unrealized energy flows are usually higher than the actual energy lost after the stressor's action. In few studies were measurements of stressor intensity and ecosystem response made (all in the same comparable units). Until researchers make these measurements, I cannot see how they will be in a position to develop a predictive capacity. For example, What is the maximum amount of energy that can be drained from an ecosystem without damaging its ability to recover? If we can learn to categorize stressors by the quality and intensity of their energy and by their point of attack in a given ecosystem and to categorize ecosystems by their energy signatures and their capacity to bounce back, then I believe we can begin to answer some of the questions about management and pollution that are being asked by those who manage natural ecosystems.

ACKNOWLEDGMENTS

I am grateful to the following colleagues for their contributions, through informal discussions, to the development of this manuscript: H. T. Odum, J. J. Ewel, S. L. Brown, C. S. Rogers, F. F. Benedict, C. A. S. Hall, S. C. Snedaker, and R. R. Twilley. None of them, however, agrees with most of this manuscript. Alma L. Lugo did most of the art work.

REFERENCES

Anderson, N. H., and D. M. Lehmkuhl, 1968, Catastrophic Drift of Insects in a Woodland Stream, *Ecology*, 49(2): 198-206.

Baker, H. G., 1970, Evolution in the Tropics, *Biotropica*, 2(2): 101-111.

Baker, J. M., 1971, Successive Spillages, in *The Ecological Effects of Oil Pollution on Littoral Communities*, E. B. Cowell (Ed.), pp. 21-32, Institute of Petroleum, London.

Banus, M. D., and S. E. Kolehmainen, 1976, Rooting and Growth of Red Mangrove Seedlings from Thermally Stressed Trees, in *Thermal Ecology II*, ERDA Symposium Series, Augusta, Ga., Apr. 2—5, 1975, G. W. Esch and R. W. McFarlane (Eds.), pp. 46-53, CONF-750425, NTIS.

TABLE 5

EXAMPLES OF CALCULATIONS OF INTENSITIES OF STRESS AND STRESSORS AND/OR OF THRESHOLDS OF ECOSYSTEM RESPONSE TO SPECIFIC STRESSORS*

Ecosystem or populations	Stressor	Intensity	Effects	Threshold	Reference
Vietnam as a whole	War	10.7% of total energy flow			Brown, 1977
Mangroves	War	40.4% of total energy flow	Extensive kills and defoliation		Brown, 1977
Estuaries	War	5.7% of total energy flow			Brown, 1977
Forests	War	7.8% of total energy flow			Brown, 1977
Agriculture	War	50% of total energy flow			Brown, 1977
Human settlements	War	12.1% of total energy flow			Brown, 1977
Eutrophic lakes	Waves		80% of end season biomass lost in submerged plants		Jupp and Spence, 1977
Mohave Desert	Rainfall			>25 mm to trigger phenology	Beatley, 1974
Volcanic slopes	Volcanic ash	1.8 m deep		<1.8 m for survival of trees; 50 cm increases growth	Eggler, 1967
Crabs *(Uca)* in salt marshes	Oil	>1000 ppm	Toxic to adults		Krebs and Burns, 1977
		100—200 ppm	Toxic to juveniles		
Coral reefs	Waves and tides	50,000 hp	Normal wear and tear on windward zones		Munk and Sargent, 1948
	Sediments			Acute dose of 200—1000 mg/cm^2 or chronic dose of 10—20 mg/liter for mortality	Rogers, 1977

Coral reefs (cont.)	Shading	5 weeks	Significantly altered structure and function of reef	Transparency < 4 m for mortality of species	Rogers, 1977
Tropical wet forest	Leaf-cutter ants' harvest	0.34 Kcal m^{-2} day^{-1}	Loss of 0.01 m^2 of leaf area/m^2, equivalent to a loss of 1.41 Kcal m^{-2} day^{-1} of productivity		Lugo et al., 1973
Granite outcrop ecosystems	Drought	Normal rainfall pattern coupled to low water storage in soil reduced growing season to about 100 days	Normal loss of unrealized productivity of about 20,000 Kcal m^{-2} growing $season^{-1}$		Lugo and J. F. McCormick, unpublished
		Additional drought caused mortality, calculated for period of 1968–1971	Plants: 506–2530 Kcal m^{-2} growing $season^{-1}$. Seeds: 66 Kcal m^{-2} growing $season^{-1}$		
Mangrove forests	Frost	–2.5 to –5.5°C	1000 Kcal m^{-2} loss as leaf biomass; 2570 Kcal m^{-2} loss as unrealized productivity		Lugo and Patterson Zucca, 1977
	Salinity			100‰; limits for growth	Cintrón et al., 1978
Oak–pine forest	Radiation			600–700 R/day increases decomposition rate	Woodwell and Marples, 1968
Lime trees	Aphids	6/leaf	Consume 19% of net primary productivity		Llewellyn, 1975
		19/tree	Consume all net primary productivity (39.74 kg/year)		

*The table is not exhaustive but shows the variety of units and methods used to report results.

Beatley, J. C., 1974, Phenological Events and Their Environmental Triggers in Mohave Desert Ecosystems, *Ecology*, 55(4): 856-863.

Billings, W. D., and H. A. Mooney, 1968, The Ecology of Arctic and Alpine Plants, *Biol. Rev.*, 43: 481-529.

Bleakney, J. S., 1972, Ecological Implications of Annual Variation in Tidal Extremes, *Ecology*, 53(5): 933-938.

Bott, T. L., 1975, Bacterial Growth Rates and Temperature Optima in a Stream with a Fluctuating Thermal Regime, *Limnol. Oceanogr.*, 20(2): 191-197.

Brock, T. D., 1970, High Temperature Systems, *Annu. Rev. Ecol. Systemat.*, 1: 191-220.

Brown, M., 1977, War, Peace, and the Computer: Simulation of Disordering and Ordering Energies in South Vietnam, in *Ecosystem Modeling in Theory and Practice: An Introduction With Case Histories*, C. A. Hall and J. W. Day (Eds.), pp. 393-417, John Wiley & Sons, Inc., New York.

Burk, C. J., 1977, A Four-Year Analysis of Vegetation Following an Oil Spill in a Freshwater Marsh, *J. Appl. Ecol.*, 14: 515-522.

Byers, R. J., 1962, Relationship Between Temperature and the Metabolism of Experimental Ecosystems, *Science*, 136(3520): 980-982.

Carpelan, L. H., 1964, Effects of Salinity on Algal Distribution, *Ecology*, 45(1): 70-77.

Carter, M. R., et al., 1973, *Ecosystems Analysis of the Big Cypress Swamp and Estuaries*, U. S. Environmental Protection Agency, Atlanta, Ga.

Chabreck, R. H., and A. W. Palmisano, 1973, The Effects of Hurricane Camille on the Marshes of the Mississippi River Delta, *Ecology*, 54(5): 1118-1123.

Churchill, G. B., H. H. John, D. P. Duncan, and A. C. Hodson, 1964, Long Term Effects of Defoliation of Aspen by the Forest Tent Caterpillar, *Ecology*, 45(3): 630-633.

Cintrón, G., A. E. Lugo, D. J. Pool, and G. Morris, 1978, Mangroves of Arid Environments in Puerto Rico and Adjacent Islands, *Biotropica*, 10(2): 110-121.

Colwell, R. K., 1974, Predictability, Constancy, and Contingency of Periodic Phenomena, *Ecology*, 55(5): 1148-1153.

Connell, J. H., 1972, Community Interactions on Marine Rocky Intertidal Shores, *Annu. Rev. Ecol. Systemat.*, 3: 169-192.

——, and E. Orias, 1964, The Ecological Regulation of Species Diversity, *Am. Nat.*, 98: 399-414.

Cooper, D. C., and B. J. Copeland, 1973, Response of Continuous-Series Estuarine Microecosystems to Point-Source Input Variations, *Ecol. Monogr.*, 43: 213-236.

Copeland, B. J., 1965, Evidence for Regulation of Community Metabolism in a Marine Ecosystem, *Ecology*, 46(4): 563-564.

Coutant, C. C., and D. K. Cox, 1976, Growth Rates of Subadult Largemouth Bass at 24 to 35.5°C, in *Thermal Ecology II*, ERDA Symposium Series, Augusta, Ga., Apr. 2—5, 1975, G. W. Esch and R. W. McFarlane (Eds.), pp. 118-120, CONF-750425, NTIS.

Cowell, E. B. (Ed.), 1971, *The Ecological Effects of Oil Pollution on Littoral Communities*, Institute of Petroleum, London.

Craighead, F. C., 1964, Land, Mangroves, and Hurricanes, *Fairchild Trop. Garden Bull.*, 19(4): 1-28.

Dayton, P. K., 1971, Competition, Disturbance, and Community Organization: The Provision and Subsequent Utilization of Space in a Rocky Intertidal Community, *Ecol. Monogr.*, 41(4): 351-389.

Dunbar, M. J., 1960, The Evolution and Stability in Marine Environments, Natural Selection at the Level of the Ecosystem, *Am. Nat.*, 94(875): 129-136.

Dyer, T. G. J., and P. D. Tyson, 1977, Estimating Above and Below Normal Rainfall Periods over South Africa, 1972—2000, *J. Appl. Meteorol.*, 16(2): 145-147.

Eggler, W. A., 1967, Influence of Volcanic Eruptions on Xylem Growth Patterns, *Ecology*, 48(4): 644-647.

Ehrlich, P. R., D. E. Breedlove, P. F. Brussard, and M. A. Sharp, 1972, Weather and the "Regulation" of Subalpine Populations, *Ecology*, 53(2): 243-247.

Erdman, H. E., 1966, Modification of Fitness in Species and Strains of Flour Beetles Due to X-Ray and DDT, *Ecology*, 47(6): 1066-1072.

Ewel, J. J., 1971a, *Experiments in Arresting Succession with Cutting and Herbicides in Five Tropical Environments*, Ph.D. Thesis, University of North Carolina, Chapel Hill.

——, 1971b, Biomass Change in Early Tropical Succession, *Turrialba*, 21: 110-112.

——, 1977, Differences Between Wet and Dry Successional Tropical Ecosystems, *Geo-Eco-Trop.*, 1(2): 103-177.

——, 1978, Department of Botany, University of Florida, personal communication.

Ferens, M. C., and R. J. Byers, 1972, Studies of a Simple Laboratory Microecosystem: Effects of Stress, *Ecology*, 53(4): 709-713.

Ferguson-Wood, E. J., and R. E. Johannes (Eds.), 1975, *Tropical Marine Pollution*, Elsevier North-Holland, Inc., New York.

Fitch, K., and P. Johnson, 1977, *Human Life Science*, Holt, Rinehart, & Winston, Inc., New York.

Gardner, G. M., D. Q. Thompson, A. E. Lugo, and D. J. Pool (Eds.), 1972, *An Environmental Assessment of Lake Ocklawaha Rodman Reservoir*, A Report of the President's Council on Environmental Quality and The Secretary of the Army, U. S. Department of Agriculture, U. S. Department of the Interior, and U. S. Department of the Army, Washington, D. C.

Gentry, R. C., 1974, Hurricanes in South Florida, in *Environments of South Florida: Present and Past.* Memoir 2, P. J. Gleason (Ed.), pp. 73-81, Miami Geological Society, Miami, Fla.

Gibbons, J. W., 1976, Thermal Alteration and the Enhancement of Species Populations, in *Thermal Ecology II*, ERDA Symposium Series, Augusta, Ga., Apr. 2—5, 1975, G. W. Esch and R. W. McFarlane (Eds.), pp. 27-31, CONF-750425, NTIS.

——, and R. R. Sharitz (Eds.), 1974, *Thermal Ecology*, AEC Symposium Series, Augusta, Ga., May 3—5, 1973, CONF-730505, NTIS.

Glynn, P. W., 1968, Mass Mortalities of Echinoids and Other Reef Flat Organisms Coincident with Midday Low Water Exposures in Puerto Rico, *Mar. Biol.*, 1(3): 226-243.

Hickman, J. C., 1975, Environmental Unpredictability and Plastic Energy Allocation Strategies in the Annual *Polygonum cascadense* (Polygonaceae), *J. Ecol.*, 63: 689-701.

Hicks, D. B., and L. A. Burns, 1975, Mangrove Metabolic Response to Alterations of Natural Freshwater Drainage to Southwest Florida Estuaries, in *Proceedings of the International Symposium on Biology and Management of Mangroves*, G. Walsh, S. Snedaker, and H. Teas (Eds.), pp. 238-255,

Institute of Food and Agricultural Sciences, University of Florida, Gainesville.

Holdridge, L. R., 1967, *Life Zone Ecology*, Tropical Science Center, San Jose, Costa Rica.

——, W. C. Grenke, W. H. Hatheway, T. Liang, and J. A. Tosi, Jr., 1971, *Forest Environments in Tropical Life Zones: A Pilot Study*, Pergamon Press, Inc., New York.

Hoopes, R. L., 1974, Flooding as a Result of Hurricane Agnes, and Its Effect on a Macrobenthic Community in an Infertile Headwater Stream in Central Pennsylvania, *Limnol. Oceanogr.*, 19(5): 853-857.

Houston, D. B., 1973, Wildfires in Northern Yellowstone National Park, *Ecology*, 54(5): 1111-1117.

Hubbs, C., 1964, Effects of Thermal Fluctuations on the Relative Survival of Greenthroat Darter Young from Stenothermal and Eurythermal Waters, *Ecology*, 45(2): 376-379.

Johannes, R. E., and S. B. Betzer, 1975, Introduction: Marine Communities Respond Differently to Pollution in the Tropics than at Higher Latitudes, in *Tropical Marine Pollution*, E. J. Ferguson-Wood and R. E. Johannes (Eds.), pp. 1-12, Elsevier North-Holland, Inc., New York.

John, K. R., 1964, Survival of Fish in Intermittent Streams, *Ecology*, 45(1): 112-119.

Jupp, B. P., and D. H. N. Spence, 1977, Limitations of Macrophytes in an Eutrophic Lake, Loch Leven. II. Wave Action Sediments, and Waterfowl Grazing, *J. Ecol.*, 65: 431-446.

Krebs, C. T., and K. A. Burns, 1977, Long-Term Effects of an Oil Spill on Populations of the Salt-Marsh Crab *Uca pugnax*, *Science*, 197: 484-487.

Kullberg, R. G., 1968, Algal Diversity in Several Thermal Spring Effluents, *Ecology*, 49(4): 751-755.

Laessle, A. M., 1958, The Origin and Successional Relationship of Sandhill Vegetation and Sand-Pine Scrub, *Ecol. Monogr.*, 28: 361-387.

Llewellyn, M., 1975, The Effect of the Lime Aphid (*Eucallipterus tiliae* L.) (Aphididae) on the Growth of the Lime (*Tilia* x *vulgaris Hayne*). II. The Primary Production of Saplings and Mature Trees, the Energy Drain Imposed by the Aphid Populations and Revised Standard Deviations of Aphid Population Energy Budgets, *J. Appl. Ecol.*, 12: 15-23.

Lugo, A. E., E. G. Farnworth, D. Pool, P. Jerez, and G. Kaufman, 1973, The Impact of the Leaf Cutter Ant *Atta colombica* on the Energy Flow of a Tropical Wet Forest, *Ecology*, 54(6): 1292-1301.

——, and C. Patterson Zucca, 1977, The Impact of Frost on Mangrove Ecosystems, *Trop. Ecol.*, 18(2): 149-161.

——, and S. C. Snedaker, 1974, The Ecology of Mangroves, *Annu. Rev. Ecol. Systemat.*, 5: 39-64.

Margalef, R., 1969, Diversity and Stability: A Practical Proposal and a Model of Interdependence, in *Diversity and Stability in Ecological Systems*, Brookhaven Symposia in Biology, No. 22, May 26—28, 1969, G. M. Woodwell and H. H. Smith (Eds.), USAEC Report BNL-50175, pp. 25-37, Brookhaven National Laboratory, NTIS.

——, 1975, Diversity, Stability and Maturity in Natural Ecosystems, in *Unifying Concepts in Ecology*, W. H. van Dobben and R. H. Lowe-McConnell (Eds.), pp. 151-160, Dr. W. Junk, B. V. Publishers, The Hague, The Netherlands.

May, R. M., 1975, Stability in Ecosystems: Some Comments, in *Unifying Concepts in Ecology*, W. H. van Dobben and R. H. Lowe-McConnell (Eds.), pp. 161-168, Dr. W. Junk, B. V. Publishers, The Hague, The Netherlands.

Meier, R. L., 1972, Communications Stress, *Annu. Rev. Ecol. Systemat.*, 3: 289-314.

Mooney, H. A., and W. D. Billings, 1965, Effects of Altitude on Carbohydrate Content of Mountain Plants, *Ecology*, 46(5): 750-751.

Moore, I. J., 1964, Effect of Water Currents on Freshwater Snails *Stagnicola palustris* and *Physa propinqua*, *Ecology*, 45(3): 558-564.

Moore, P. D., and D. J. Bellamy, 1973, *Peatlands*, Springer-Verlag, Inc., New York.

Morowitz, H. J., 1968, *Energy Flow in Biology*, Academic Press, Inc., New York.

Munk, W., and M. C. Sargent, 1948, Adjustment of Bikini Atoll to Ocean Waves, *Trans. Am. Geophys. Union*, 29: 855-860.

Niering, W. A., and R. H. Goodwin, 1974, Creation of Relatively Stable Shrublands with Herbicides: Arresting "Succession" on Right-of-Way and Pastureland. *Ecology,* 55(4): 784-795.

Nixon, S., 1969, *Characteristics of Some Hypersaline Ecosystems*, Ph.D. Thesis, University of North Carolina, Chapel Hill.

Odum, E. P., 1969, The Strategy of Ecosystem Development, *Science*, 164: 262-270.

——, 1971, *Fundamentals of Ecology*, 3rd ed., W. B. Saunders Company, Philadelphia.

——, and R. L. Kroodsma, 1976, The Power Park Concept: Ameliorating Man's Disorder with Nature's Order, in *Thermal Ecology II*, ERDA Symposium Series, Augusta, Ga., Apr. 2—5, 1975, G. W. Esch and R. W. McFarlane (Eds.), pp. 1-9, CONF-750425, NTIS.

Odum, H. T., 1955, Trophic Structure and Productivity of Silver Springs, Florida, *Ecol. Monogr.*, 27: 55-112.

——, 1967, Work Circuits and Systems Stress, in *Symposium on Primary Productivity and Mineral Cycling in Natural Ecosystems*, H. E. Young (Ed.), pp. 81-138, University of Maine Press, Orono.

——, 1970, Summary: An Emerging View of the Ecological Systems at El Verde, in *A Tropical Rain Forest*, H. T. Odum and R. F. Pigeon (Eds.), USAEC Report TID-24270, pp. I-191—I-289, U. S. Atomic Energy Commission, NTIS.

——, 1974, Energy Cost—Benefit Models for Evaluating Thermal Plumes, in *Thermal Ecology*, AEC Symposium Series, Augusta, Ga., May 3—5, 1973, J. W. Gibbons and R. R. Sharitz (Eds.), pp. 628-649, CONF-730505, NTIS.

——, 1976, Macroscopic Minimodels of Man and Nature, in *Systems Analysis and Simulation in Ecology*, B. C. Patten (Ed.), Vol. IV, pp. 249-280, Academic Press, Inc., New York.

——, B. J. Copeland, and E. McMahan (Eds.), 1974, *Coastal Ecological Systems of the United States*, 4 vols., The Conservation Foundation, Washington, D. C.

——, S. W. Nixon, and L. H. DiSalvo, 1970, Adaptations for Photoregenerative Cycling, in *The Structure and Function of Freshwater Microbial Communities,* J. Cairns (Ed.), pp. 1-29, Research Division Monograph 3, Virginia Polytechnic Institute and State University, Blacksburg.

——, W. Kemp, M. Sell, W. Boynton, and M. Lehman, 1977, Energy Analysis and the Coupling of Man and Estuaries, *Environ. Manag.*, 1(4): 297-315.

Penfound, W. T., 1964, Effects of Denudation on the Productivity of Grassland, *Ecology*, 45(4): 838-845.

Platt, R. B., 1965, Ionizing Radiation and Homeostasis of Ecosystems, in *Ecological Effects of Nuclear War*, G. M. Woodwell (Ed.), USAEC Report BNL-917, pp. 39-60, Brookhaven National Laboratory, NTIS.

Poulson, T. L., and D. C. Culver, 1969, Diversity in Terrestrial Cave Communities, *Ecology*, 50(1): 153-158.

Richey, J. E., 1970, *The Role of Disordering Energy in Microcosms*, M.S. Thesis, University of North Carolina, Chapel Hill.

Rogers, C. S., 1977, *The Response of a Coral Reef to Sedimentation*, Ph.D. Thesis, University of Florida, Gainesville.

Sanders, H. L., 1969, Benthic Marine Diversity and the Stability-Time Hypothesis, in *Diversity and Stability in Ecological Systems*, Brookhaven Symposia in Biology, No. 22, May 26—28, 1969, G. M. Woodwell and H. H. Smith (Eds.), USAEC Report BNL-50175, pp. 71-81, Brookhaven National Laboratory, NTIS.

Sastry, A. N., 1976, Effects of Constant and Cyclic Temperature Regimes on the Pelagic Larvae Development of a Brachyuran Crab, in *Thermal Ecology II*, ERDA Symposium Series, Augusta, Ga., Apr. 2—5, 1975, G. W. Esch and R. W. McFarlane (Eds.), pp. 81-87, CONF-750425, NTIS.

Schroeder, P. B., 1975, *Thermal Stress in* Thalassia testudinum, Ph.D. Thesis, University of Miami, Miami, Fla.

Selye, H., 1956, *The Stress of Life*, McGraw-Hill Book Company, New York.

Sharaf El Din, S. H., 1977, Effect of the Aswan High Dam on the Nile Flood and on the Estuarine and Coastal Circulation Pattern Along the Mediterranean Egyptian Coast, *Limnol. Oceanogr.*, 22(2): 194-207.

Singer, S. F. (Ed.), 1970, *Global Effects of Environmental Pollution*, American Association for the Advancement of Science, Symposium, Dallas, December 1968, D. Reidel Publishing Co., Hingham, Mass.

Slobodkin, L. B., and H. L. Sanders, 1969, On the Contribution of Environmental Predictability to Species Diversity, in *Diversity in Ecological Systems*, Brookhaven Symposia in Biology, No. 22, May 26—28, 1969, G. M. Woodwell and H. H. Smith (Eds.), USAEC Report BNL-50175, pp. 82-95, Brookhaven National Laboratory, NTIS.

Smith, W. H. B., H. McKellar, D. L. Young, and M. E. Lehman, 1974, Total Metabolism of Thermally Affected Coastal Systems on the West Coast of Florida, in *Thermal Ecology*, AEC Symposium Series, Augusta, Ga., May 3—5, 1973, J. W. Gibbons and R. R. Sharitz (Eds.), pp. 475-489, CONF-730505, NTIS.

Sneyers, R., 1976, Application of Least Squares to the Search for Periodicities, *J. Appl. Meteorol.*, 15(4): 387-393.

Stearns, S. C., 1975, *A Comparison of the Evolution and Expression of Life History Traits in Stable and Fluctuating Environments:* Gambusia affinis *in Hawaii*, Ph.D. Thesis, University of British Columbia, Vancouver, B. C.

——, 1978, On the Stable—Fluctuating Dichotomy in Evolutionary Biology, in press.

Sweeney, B. W., and J. A. Schnack, 1977, Egg Development, Growth and Metabolism of *Sigara alternata* (Say) (Hemiptera:Corixidae) in Fluctuating Thermal Environments, *Ecology*, 58(2): 265-277.

Thomas, T. M., 1974, A Detailed Analysis of Climatological and Hydrological Records of South Florida with Reference to Man's Influence upon Ecosystem Evolution, in *Environments of South Florida: Present and Past.*

Memoir 2, P. J. Gleason (Ed.), pp. 82-122, Miami Geological Society, Miami, Fla.

Tomkins, D. J., and W. F. Grant, 1977, Effects of Herbicides on Species Diversity of Two Plant Communities, *Ecology*, 58(2): 398-406.

Van Dobben, W. H., and R. H. Lowe-McConnell (Eds.), 1975, *Unifying Concepts in Ecology*, Dr. W. Junk, B. V. Publishers, The Hague, The Netherlands.

Vernberg, F. J., and W. B. Vernberg, 1974, Synergistic Effects of Temperature and Other Environmental Parameters on Organisms, in *Thermal Ecology*, AEC Symposium Series, Augusta, Ga., May 3—5, 1973, J. W. Gibbons and R. R. Sharitz (Eds.), pp. 94-99, CONF-730505, NTIS.

Wein, R. W., and L. C. Bliss, 1973, Experimental Crude Oil Spills on Arctic Plant Communities, *J. Appl. Ecol.*, 10: 671-682.

Wetzel, R. G., 1975, *Limnology*, W. B. Saunders Company, Philadelphia.

Whittaker, R. H., 1969, Evolution of Diversity in Plant Communities, in *Diversity and Stability in Ecological Systems,* Brookhaven Symposia in Biology, No. 22, May 26—28, 1969, G. M. Woodwell and H. H. Smith (Eds.), USAEC Report BNL-50175, pp. 178-196, Brookhaven National Laboratory, NTIS.

——, 1975, The Design and Stability of Some Plant Communities, in *Unifying Concepts in Ecology*, W. H. van Dobben and R. H. Lowe-McConnell (Eds.), pp. 169-181, Dr. W. Junk, B. V. Publishers, The Hague, The Netherlands.

Woodwell, G. M., 1970, Effects of Pollution on the Structure and Physiology of Ecosystems, *Science*, 168: 429-433.

——, and J. H. Brower, 1967, An Aphid Population Explosion Induced by Chronic Gamma Irradiation of a Forest, *Ecology*, 48(4): 680-683.

——, and T. G. Marples, 1968, The Influence of Chronic Gamma Irradiation on Production and Decay of Litter and Humus in an Oak—Pine Forest, *Ecology*, 49(3): 456-465.

——, and A. L. Rebuck, 1967, Effects of Chronic Gamma Radiation on the Structure and Diversity of an Oak—Pine Forest, *Ecol. Monogr.*, 37(1): 53-69.

——, and H. H. Smith (Eds.), 1969, *Diversity and Stability in Ecological Systems*, Brookhaven Symposia in Biology, No. 22, May 26—28, 1969, USAEC Report BNL-50175, Brookhaven National Laboratory, NTIS.

Wright, R. F., 1976, The Impact of Forest Fire on the Nutrient Influxes to Small Lakes in Northwestern Minnesota, *Ecology*, 57(4): 649-663.

Yentsch, C. S., C. M. Yentsch, L. R. Strube, and I. Morris, 1974, Influence of Temperature on the Photosynthetic Efficiency in Natural Populations of Marine Phytoplankton, in *Thermal Ecology*, AEC Symposium Series, Augusta, Ga., May 3—5, 1973, J. W. Gibbons and R. R. Sharitz (Eds.), pp. 508-517, CONF-730505, NTIS.

ECOSYSTEM RESPONSES TO STRESS IN AQUATIC MICROCOSMS

JOHN W. LEFFLER*
Institute of Ecology, University of Georgia, Athens, Georgia

ABSTRACT

Ecosystem stress is defined as any change in exogenous inputs which produces an output significantly different from the system's normal response. Any discussion of stress requires a description of the state space in which the ecosystem normally functions. This normal operating range can be either static or dynamic, depending on the time scale of interest. Ecosystem stress responses can be discussed in terms of the system's relative stability. Hypotheses regarding such responses can be empirically evaluated by use of microcosm models. Experiments using aquatic microcosms to evaluate the stabilizing effects of species diversity and nutrient availability are described as examples of this technique. Contrary to previously proposed hypotheses, no relationships between diversity, nutrient availability, or system mass and ecosystem stability or between ecosystem resistance and resilience stabilities were demonstrated. Data from microcosm studies illustrate the importance of specifying the nature of the perturbation, the parameters used to characterize the system, and the prior history of the system in discussions of an ecosystem's response to stress.

Although stress is a term with an obvious meaning, it is difficult to define precisely. Dictionaries provide definitions such as "the action on a body of any system of balanced forces whereby strain or deformation results" and "the internal resistance or reaction of an elastic body to the external forces applied to it" (Barnhart and Stein, 1964). Medical textbooks consider stress to be any force that elicits increased levels of epinephrine, glucocorticoids, blood glucose, blood coagulation, arterial pressure, and cellular metabolism (Guyton, 1971; Smith, Jones, and Hunt, 1972). In most definitions of stress, it

*Present address: Department of Biology, Ferrum College, Ferrum, Virginia.

is the response of the organism or system which is used as a criterion for identifying a specific force as a stress. Certain characteristic parameters must display abnormal behavior before a system can be considered stressed. At an organismal level, epinephrine, glucocorticoids, and blood glucose are such parameters. To identify stress at an ecosystem level, we must monitor appropriate ecosystem-level parameters. Ecosystem stress studies should consider some characterization of the total community metabolism, the nutrient-cycling mechanisms, and the structural properties of the system. Deviations of such parameters from some normal condition can be used to identify exogenous inputs as stresses and to describe ecosystem responses to those stresses.

THE CONCEPT OF ECOLOGICAL STRESS

As used here, stress is defined as any deviation in input variables which produces an output different from the system's normal response. Obviously this definition depends on the description of the normal, unstressed state. The region of state space within which an ecosystem normally functions is called its normal operating range. The dashed lines in Fig. 1 delineate such a region for the very simple

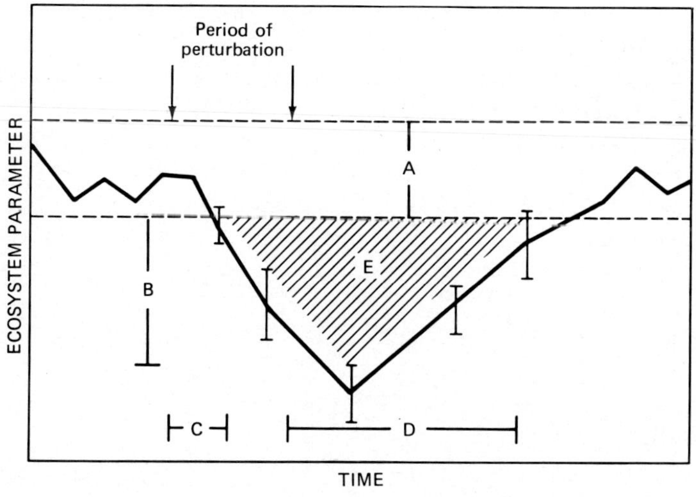

Fig. 1 Diagram of five measures of relative stability. Dashed lines are boundaries of the normal operating range. Letters refer to quantitative stability measures: A, constancy; B, resistance; C, response time; D, resilience; and E, total relative stability.

case of an ecosystem in a nonoscillating steady state. This range can be calculated as confidence bounds about the mean of all preperturbation observations, assuming the system is in steady state. The width of the region or the coefficient of variation of the observations can be regarded as a measure of the relative constancy of the ecosystem.

Description of an ecosystem's normal operating range provides a standard for judging whether a particular input should be considered a stress. The criterion is whether the input causes the system's trajectory to vary significantly from its normal operating range. Since both the normal operating range and the system trajectory can be described statistically, desired confidence levels can be placed on such deviations. On the basis of this discussion and analysis, it is obvious that the same qualitative input may or may not be a stress. Temperature, for example, is an inevitable input to all ecosystems. A normal operating range can be defined under conditions of temperature fluctuations that are typical for the system. Temperature can become a stress, however, if the quantitative level of this input assumes a value that causes system response to exceed the normal operating bounds.

The description of the normal operating range is also extremely observer dependent. The width of such a region is a function of the characteristics used to describe the ecosystem and of the ability of the observer to discriminate differences in these parameters. For example, an algal culture can be classified on the basis of "greenness" or in terms of milligrams of chlorophyll per liter, depending on the observer. Of perhaps greater importance is the choice of a time scale for describing the normal operating range. Since the rate of change of dissolved oxygen in aquatic systems typically displays diel variation, an hourly description gives an oscillating normal operating range and permits a dynamic analysis. In this case, an input can become a stress on the system only at certain times in the daily cycle. Depending on the observer's objective, the same system can be described on a daily time scale. This gives a more static and much broader normal operating range at all time points. Such a concept is flexible enough to be applied to ecosystems undergoing oscillations, seasonal changes, developmental sequences, or even cyclic succession. Similarly, the width of the normal operating range need not remain constant at all times. It is plausible, for example, to believe that early seral stages possess broader operating ranges than climax systems.

It becomes evident that the concept of ecological stress cannot be discussed in a general sense. Both the change of system inputs and the properties of the system itself determine the ecosystem's stress response. The stress of a specific ecosystem cannot be discussed without reference to the nature of the input changes, the system's past and present states, and the observer's characterization of the system, ability to detect changes in the system, and his time scale. These considerations not only require that the concept of ecological stress be discussed in precise terms but also provide a practical approach to assessing stress in a specific ecosystem.

An ecosystem's stress response can be described in several ways. A system's ability to resist the effects of exogenous input changes can be measured by the magnitude of its deflection from the normal operating range (Webster, Waide, and Patten, 1975) and by the time required for an initial displacement to occur (Hurd and Wolf, 1974). The first characteristic can be referred to as resistance and the second response time. The trajectory of the stressed system must be statistically defined so that significant deviations from the normal operating range can be assessed. After a system has been displaced from its normal operating range, it may or may not return to its original, statistically defined condition. If it fully recovers to its original normal operating range, the time required is a measure of its resilience to the stress (Webster, Waide, and Patten, 1975). (These concepts are depicted in Fig. 1.)

Since resistance and resilience are not mutually exclusive properties, another measure is necessary to describe the divergence of the perturbed trajectory from that of the unstressed system. The integral difference between the normal operating range and the perturbed trajectory can be defined as total relative stability (Webster, 1975), which is illustrated by the shaded area in Fig. 1. This reflects the total impact of a stress on an ecosystem and is a function of both the specific system and the altered input. Ecosystems can be compared by dividing resistance and total relative stability measures by the mean of the observations used to define the normal operating range. The resilience measure can be made independent of resistance by dividing the recovery time by the relative resistance. This model provides a means of identifying stresses and of describing an ecosystem's stress response.

Many hypotheses have been proposed to describe system characteristics that influence an ecosystem's stress response or stability. The best known of these has been the hypothesis that increased diversity leads to increased stability [see the review by Goodman (1975)]. Large abiotic nutrient reserves have also been proposed as a major influence on an ecosystem's response to stress. Pomeroy (1975) suggested that increased nutrient availability leads to increased resistance and resilience, but Webster, Waide, and Patten

(1975) hypothesized that resistance would increase and resilience would decrease. They believed that resistance is a function of the mass of the system and resilience is related to functional dynamics, especially turnover rates. Thus resistance and resilience might be considered inverse concepts. Other investigators have discussed the effect of environmental constancy on ecosystem stability. Systems adapted to less variable environments often exhibit high levels of constancy stability but very poor resistance and resilience in the event of a major perturbation (Copeland, 1970; Jernelöv and Rosenberg, 1976; Larsen, 1974). Estuaries, on the other hand, typically exhibit low constancy stability because of their highly variable environment but high levels of resistance and resilience (Boesch, 1974).

THE MICROCOSM APPROACH TO ECOSYSTEM THEORY

Ecologists have had difficulty in testing the validity of hypotheses that describe an ecosystem's response to stress. Most of these propositions were derived either by an intuitive synthesis of field observations or by mathematical simulation. Rigorous experimental evaluation is rare because of problems typical of all ecosystem-level studies, e.g., long time scales, lack of replication, difficulty of ecosystem parameter measurement, lack of control over extraneous environmental conditions and system histories, and economic feasibility. One means of testing the generality of a hypothesis is to use microcosms as living analogs of ecosystems. I have used several types of aquatic microecosystems to evaluate hypotheses regarding ecosystem stress responses. Microcosms can be designed to meet all requirements of standard ecosystem definitions (Odum, 1971) and to overcome the examples cited of experimental difficulties in field studies. By designing microcosm systems in which extraneous variables can be controlled, we can focus on the specific concept to be evaluated. No effort is made to duplicate any specific natural ecosystem; the microcosms are merely models of general ecosystem properties, such as energy flow through a trophic structure, nutrient cycling, and species diversity. In addition, the hypothetical factors influencing stability can be modeled by subjecting microcosms to different treatment levels of a factor, e.g., nutrient availability. Ecosystems developed under each treatment level can be subjected to identical stresses, and their respective responses can be compared.

The goal of such experiments is not to extrapolate from microcosm results to all ecosystems but to determine whether a hypothesis derived by nonempirical methods is empirically falsifiable. This approach is subject to the appropriateness of the experimental treatments for simulating mechanisms postulated by the hypothesis. Negation of a hypothesis by microcosm experiments limits its generality and, at least, indicates that caution is advisable in extrapolating the hypothesis to all types of ecosystems. Several examples of this procedure will illustrate the microcosm approach.

The Diversity–Stability Hypothesis

The diversity—stability hypothesis states that increasing diversity within an ecosystem increases the system's stability. The first step in evaluating this proposal by microcosm experiments was to define operationally and to clarify the statement to be tested. Diversity was defined simply as the total number of species within each microcosm. Each of the five stability measures in Fig. 1 were evaluated. Thus the hypothesis predicted that microcosms with greater numbers of species should have greater constancy stability and that under stressed conditions they should have greater resistance, response time, resilience, and total relative stability.

Four types of microcosms, each containing a different number of taxonomic groups, including algae, protozoans, metazoans, and bacteria, were established in 500-ml cotton-stoppered Erlenmeyer flasks. System A contained 10 species; system B, 17; system C, 21; and system D, 25. Each system was developed from stock cultures that were originally derived intact from natural sources but had been grown in flasks in the laboratory for a minimum of 6 months. Thus the organisms had coevolved histories and were adapted to the laboratory environment. Sterile techniques ensured against invasion by additional taxa.

Fifty-four replicates of each system were used in the study. All microcosms were maintained under identical environmental conditions. After inoculation the systems underwent a 12-week succession to allow each to attain a steady-state condition. The normal operating range and constancy stability were determined for each of the four systems by sacrificing unperturbed control microcosms during weeks 13 to 18. Other microcosms were stressed by raising their temperatures from 24 to 45°C during weeks 14 to 16. These systems were sacrificed at intervals from weeks 13 to 40 to monitor their response to the stress. Each microcosm was characterized by its net day production, night respiration, production-to-respiration (P/R) ratio, microscopic counts of dominant organisms, and nutrient agar plate counts of bacterial types. An element distribution index (EDI) was used to characterize the cycling of phosphorus and iron. This is an isotope technique that is sensitive to changes in the size of

TABLE 1

STABILITY RANK ORDERING OF FOUR DIVERSITY LEVELS IN AN AQUATIC MICROCOSM EXPERIMENT*

Parameter	Measure of stability			
	Constancy	Resistance	Resilience	Total relative stability
Net day production	A-B-C-D	A-D-C-B	(C,D)-B-A	D-C-B-A
Night respiration	A-D-C-B	D-A-B-C	(A,B,C,D)	B-A-D-C
P/R ratio	A-D-B-C	A-B-C-D	(B,C)-D-A	C-B-D-A
Iron EDI	A-C-D-B	B-A-C-D	(A,B,C,D)	B-A-C-D
Phosphorus EDI	C-D-A-B	D-A-(B,C)	(B,C)-A-D	B-D-C-A
Species composition				
Microscopic counts	C-D-A-B	A-D-C-B†	B-(A,C)-D†	‡
Bacterial counts	A-C-B-D			

*System A contained 10 species; system B, 17; system C, 21; and system D, 25. The predicted ordering for all stability measures from least to most stable is A-B-C-D. Parentheses indicate ties in ranking.

†Microscopic and bacterial counts were combined.

‡Data were not determined.

nutrient storage compartments and the rates of nutrient movement through them. Details of these experiments are described elsewhere (Leffler, 1977).

Microcosm behavior was plotted for each of the four types of systems and was analyzed according to the criteria described in Fig. 1. Constancy, resistance, response time, resilience, and total relative stability of each parameter were calculated for each microcosm type to rank each of the four diversity levels from most stable to least stable. Results of this analysis are shown in Table 1. Observed rankings seldom correlated with predicted rankings. No significant relationship between taxonomic diversity and any of the five stability measures could be demonstrated. These results were interpreted to refute the generality of the diversity—stability hypothesis. This does not deny that under some circumstances diversity and stability may be related, but confident estimates of ecosystem stability based on counting taxonomic units do not appear justified by these results.

The Relation Between Nutrient–Energy Subsidies and Ecosystem Stability

Several researchers have proposed that large abiotic nutrient reserves, together with such energy subsidies as tidal action, may affect ecosystem stability (e.g., Pomeroy, 1975; Webster, Waide, and Patten, 1975). They suggest that increased nutrient availability would increase the constancy and resistance of ecosystems. As mentioned previously, through intuitive syntheses, Pomeroy predicted increased resilience, whereas Webster, Waide, and Patten postulated an inverse relationship between resistance and resilience on the basis of mathematical simulations. I have evaluated these hypotheses experimentally with aquatic microcosms (Leffler, manuscript in preparation).

Four-liter, open, flow-through systems were used in the study. All microcosms initially contained identical chemical mediums, and each was inoculated from the same stocks. Cross-seeding improved replication. Three levels of nutrient subsidy and three levels of energy subsidy were used. Nutrients were added in the form of Taub's #36 microcosm medium (Taub and Dollar, 1964). Concentrations of the inflow solutions were set at one-tenth, normal, and ten times the normal strength of this medium. Turnover time for each system was 7 days. Energy subsidies were provided by Teflon stir bars and magnetic stirrers, which operated constantly, for 2 hr on alternating days, and not at all. Each combination of nutrients and energy was replicated four times, for a total of 36 microcosms. After inoculation, all systems experienced a 90-day succession and were then monitored at steady state for an additional 90 days. A thermal stress was induced by increasing the temperature from 22 to 40°C for 48 hr and then returning it to normal. Stress response was monitored for 75 days. At the end of this period, new normal operating ranges were described for each system by further observation for 70 days. Another thermal stress of the same magnitude but of 7-days duration was induced. The microcosms' responses were followed for 50 days.

A variety of parameters characterized ecosystem dynamics—net day production, night respiration, P/R ratio, chlorophyll a, particulate matter, heterotrophic bacterial numbers, and inflow-to-outflow ratios of ammonia, nitrate, phosphorus, and magnesium. The responses of the microcosms to the thermal stresses were analyzed as described for the diversity experiment. Nonparametric tests were used to determine the significance of the ordering of nutrient treatments, energy treatments, and nutrient—energy combinations for the five measures of stability for each parameter. Few consistent generalizations could be drawn from this analysis. Effects of energy subsidies were seldom significant for any parameter or stability measure. Constancy stability was enhanced for all chemical inflow-to-outflow ratios by increasing nutrient subsidies and by increasing the level of nutrient—energy combinations.

All parameters were combined, and rankings of each treatment level were assessed by Friedman nonparametric analyses of variance and by sign tests ($P \leqslant 0.200$) (Siegel, 1956). These results are shown in Table 2. Energy subsidies were not significantly related to any type of stability. Constancy stability and to a lesser extent resilience

TABLE 2

EFFECTS ON ECOSYSTEM STABILITY OF INCREASING NUTRIENT AND ENERGY SUBSIDIES IN AN AQUATIC MICROCOSM EXPERIMENT*

Measure of stability	Nutrient	Energy	Nutrient + energy
Constancy	H-M-L ($P < 0.01$)	NS	NS
Resistance	NS	NS	L-H ($P = 0.105$)
Response time	L-M-H ($P = 0.04$)	NS	L-H ($P = 0.019$)
Resilience	H-(M,L) ($P = 0.19$)	NS	NS
Total relative stability	NS	NS	NS

*Based on summation of all measured parameters. Friedman analyses of variance and the sign test were used to detect significance at the $P \leqslant 0.200$ level. Abbreviations H, M, and L are high, medium, and low, respectively; NS is not significant. The predicted rankings from most to least stable are H-M-L or H-L. Parentheses indicate ties in ranking.

stability increased with increasing nutrient subsidies, but response times decreased. Increased levels of nutrient—energy combinations led to decreases in resistance stability and response time. Table 3 summarizes the behavior predicted by the hypotheses for large abiotic nutrient reserves and the empirically determined results of the microcosm experiments. The observed increase in constancy stability with increasing nutrients verified the predictions of both Pomeroy (1975) and Webster, Waide and Patten (1975). This suggests that nutrient availability may play an important role in ecosystem dynamics. It is obvious, however, that the mechanisms by which nutrient reserves and energy subsidies affect ecosystem stability are poorly understood, as evidenced by contradictions between observations and hypotheses for the four types of ecosystem stress response.

TABLE 3

SUMMARY OF PREDICTED AND OBSERVED EFFECTS OF LARGE ABIOTIC NUTRIENT RESERVES ON ECOSYSTEM STABILITY*

	Predicted		
Measure of stability	**Pomeroy (1975)**	**Webster, Waide, and Patten (1975)**	**Observed**
Constancy	Increase	Increase	Increase
Resistance	Increase	Increase	No effect
Response time	†	Increase	Decrease
Resilience	Increase	Decrease	Increase
Total relative stability	Increase	†	No effect

*Observations are based on the results of the nutrient—energy subsidy experiment described in Table 2 and in text.

†Not discussed.

OTHER INFLUENCES ON ECOSYSTEM STRESS RESPONSE

Mass of the System

Many other factors may influence an ecosystem's response to stress. Data derived from the previously described experiments has been used to evaluate some of these influences. Golley (1974) and Webster, Waide, and Patten (1975) related resistance and resilience of ecosystems to nutrient storages represented by standing-crop biomass within the system. They proposed that a large biomass would damp fluctuations induced by the external environment and, thus, confer greater resistance to stresses impinging on the system. Webster, Waide, and Patten also suggested that resilience has an inverse relationship to resistance. The biomass-derived inertia that increases a system's resistance to stress would also slow its recovery from a stress, thereby decreasing its resilience.

On the basis of these hypotheses, I would predict that resistance and resilience rankings of treatments in the nutrient—energy experiments would be correlated with the amount of particulate matter in the microcosms of those treatments. The stability rankings for each of the nine treatments were correlated with the mean amount of particulate matter in the microcosms under steady-state conditions by use of Kendall's rank correlation coefficient. The probability of a significant correlation ($P \leqslant 0.100$) was also calculated. The binomial test was used to detect trends across all parameters. The only significant correlation was a positive relationship between particulate matter and constancy stability during the first perturbation ($P = 0.020$). I must conclude that a generalization that attempts to explain an ecosystem's stress response as a function of its mass is not justified by these results. Neither resistance nor resilience showed any tendencies toward such a correlation. This contradicts the hypotheses of Golley (1974) and Webster, Waide, and Patten (1975). Harwell, Cropper, and Ragsdale (1977) reported similar conclusions based on mathematical arguments.

Relationships Among Types of Stability

The relationships among the five different measures of stability were also investigated. Kendall's rank correlations were calculated for each measured parameter and each pairing of stability results based on the stability ordering of treatments for both the diversity and the nutrient—energy studies. Few significant correlations ($P \leqslant 0.100$) were discovered. Again, binomial tests across all parameters from both studies were used to detect significant ($P \leqslant 0.100$) trends. No such relationships were found. The most highly correlated ($P = 0.132$) pair of stabilities was resistance and total relative stability. This was not unexpected considering their definitions. No significant relationship between resistance and resilience was discernible based on all 22 parameters. This apparently contradicts the generality of the inverse relationship postulated by Webster, Waide, and Patten.

It has been suggested that an ecosystem's ability to resist and to recover from a stress may be a function of habitat predictability (O'Neill, 1976). Such ecosystems as estuaries, which are subjected to highly variable environments, typically exhibit low constancy stability but high levels of resistance and resilience (Boesch, 1974; Copeland, 1970; Larsen, 1974). It may be worthwhile to consider whether a system's response to stress can be predicted by monitoring its variability under unstressed steady-state conditions. An inverse relationship between constancy stability and the perturbation response stabilities might be postulated. Correlation results, however, indicate that such a generalization is not justified, at least in terms of the two microcosm studies considered. Constancy stability fails to correlate with any of the measures that characterize an ecosystem's stress response.

EVALUATION OF AN ECOSYSTEM'S STRESS RESPONSE

The examples discussed demonstrate how an empirical approach can be used to evaluate general theories of ecosystem stress response. Microcosms can be used to model ecological processes common to all ecosystems. They provide the advantage of controlling many extraneous variables that complicate attempts at rigorous field experiments. Studies of an ecosystem's response to stress must also be of sufficient duration for dynamics to be observed. This factor, as well as ease of replication, make microcosms a practical and economically feasible approach for studying the theory of stress ecology. A researcher must be aware, however, of the many factors that influence interpretation of a system's response to stress. Foremost is the validity of the microcosm as an accurate analog of the natural systems of primary concern. In addition, such factors as the type of perturbation, the parameters used to characterize the system, and the historical development of the system must be considered in discussing the stress response of any ecosystem.

Nature of the Stress

Although theoreticians tend to generalize about ecosystem dynamics, it is not surprising that the specific type of stress to which a system is subjected will greatly influence its response. Figure 2 shows the differing responses of replicate microcosms of the 21-species system from the diversity experiment after being subjected to several types of perturbations. Each trajectory represents two microcosms receiving the same perturbation. The ordinate scale is a subjective evaluation of microcosm health based on a grading system from 1 to 4, with plus (+) or minus (—) indicating variation within a category; the categories are 1, yellow and brown colors dominant, much debris, metazoans dead; 2, green color dominant but yellow and brown colors common, debris present; 3, normal, no different from controls; and 4, dark green color, suggesting high productivity, structure similar to controls. Stresses on the various microcosms included increasing the pH to 8.8 by adding sodium hydroxide (curve A in the figure); introducing 10 amphipods (*Gammarus* sp.) into microcosms (curve B); increasing the temperature from 22 to 35°C for 2 weeks and then returning it to normal (curve C); and increasing the temperature to 40°C for 1 week and then returning it to normal (curve D). Although the qualitative nature of the stress is obviously important, curves C and D illustrate the effect of different intensities of the same perturbation. The

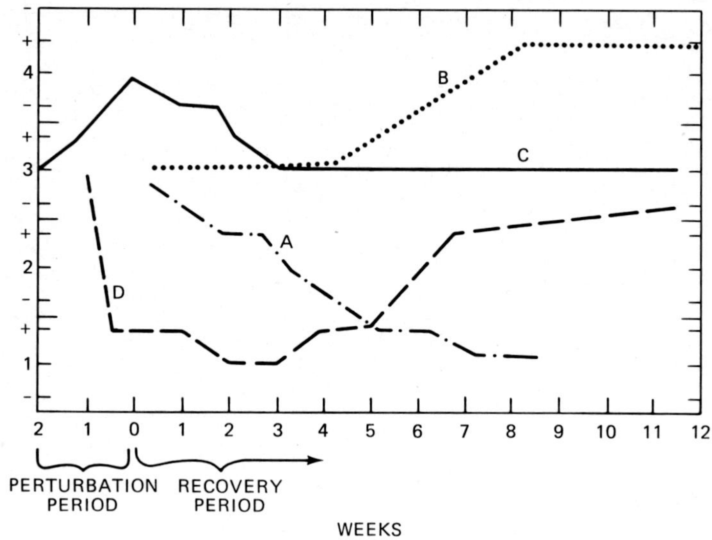

Fig. 2 Effects of different perturbations on replicate microcosms as described by a subjective evaluation of microcosm health. The stresses illustrated are A, pH of system increased by addition of sodium hydroxide; B, 10 amphipods introduced to each system; C, temperature increased from 22 to 35°C for 2 weeks; and D, temperature increased from 22 to 40°C for 1 week. (See text for discussion of the evaluation.)

duration, the frequency of occurrence, and the synergistic effects of different stresses also must be considered.

Parameters Characterizing the Ecosystem

Understanding ecosystem behavior depends on the parameters used to characterize the system, but this is seldom pointed out in discussions of ecosystem dynamics. This was obvious in the diversity study illustrated by Table 1, where stability rankings of the four treatments varied greatly depending on which parameter was considered. The relative sensitivity of each parameter is also worth noting, especially if microcosm experiments continue to be used for studies of ecosystem dynamics. To evaluate parameter sensitivity, I examined data from the nutrient—energy subsidy studies. For each parameter and each of the nine treatments, coefficients of variation of the steady-state observations were used to rank the treatments in terms of constancy stability. The median coefficient of variation from the first and second perturbations combined was calculated for every parameter. A comparison of these numbers led to the rankings in Table 4. The P/R ratio, night respiration, and magnesium input-to-output ratio were the least variable parameters during steady state. Ammonia, nitrate, and chlorophyll a fluctuated most. These rankings, of course, are partially a function of the specific methods

TABLE 4

RANKING OF PARAMETER SENSITIVITIES FOR STEADY-STATE VARIABILITY, RESISTANCE TO STRESS, AND RECOVERY FROM STRESS*

Parameter	Number of cases	Steady-state variability (least to most)	Resistance (least to most)	Resilience (least to most)
Net day production	18	4	4	5
Night respiration	18	2	5	3
P/R ratio	18	1	2	6
Chlorophyll a	18	7	8	2
Particulate matter	18	6	9	4
Ammonia input-to-output ratio	18	9	7	1
Nitrate input-to-output ratio	9	8	1	9
Phosphorus input-to-output ratio	9	5	3	8
Magnesium input-to-output ratio	9	3	6	7

*The ranking is based on all treatments from both perturbations in the nutrient—energy subsidy experiment. The median coefficient of variation of steady-state observations, the median relative resistance, and the median resilience for these treatments were compared for each parameter to arrive at a ranking.

used in this study. The median relative resistance and median relative resilience, similarly calculated for each parameter, were compared. The nitrate, P/R ratio, and phosphorus parameters were most sensitive to stress (i.e., least resistant), and particulate matter, chlorophyll a, and ammonia were least sensitive. Ammonia, chlorophyll a, and night respiration recovered most slowly, but nitrate, phosphorus, and magnesium returned rapidly to their normal operating ranges. Differences in response of various parameters should be considered during the design of a microcosm experiment.

System History

A third consideration in the assessment of an ecosystem's stress response is the effect of history. We might expect that prior exposure to stress would alter a system's future response to similar disturbances. Data from the nutrient—energy subsidy experiment were examined for such evidence. Coefficients of variation of the steady-state observations were compared by treatment for each of the two perturbations to evaluate the increase or decrease in steady-state variability before and after the first perturbation of each of the nine treatments. The binomial test was used to detect whether a significant ($P \leqslant 0.100$) change in constancy stability had occurred. As shown in Table 5, only the P/R ratio was changed significantly; the perturbation caused a decrease in steady-state constancy. Similar procedures allowed comparisons of resistance, resilience, and total relative stability after the first and second stress periods. The occurrence of a prior disturbance caused night respiration to be less resistant to a second perturbation, but the P/R ratio became more resistant. Resilience and total relative stability were also affected for

TABLE 5

EFFECTS OF PRIOR PERTURBATION ON AN ECOSYSTEM'S RESPONSE TO STRESS IN THE MICROCOSM NUTRIENT—ENERGY SUBSIDY EXPERIMENT*

Parameter	Constancy	Resistance	Resilience	Total relative stability
Net day production	NS	NS	NS	—
Night respiration	NS	—	+	—
P/R ratio	—	+	—	—
Chlorophyll a	NS	NS	NS	NS
Particulate matter	NS	NS	NS	NS
Ammonia input-to-output ratio	NS	NS	+	NS

*The coefficient of variation of steady-state observations, the relative resistance, the relative resilience, and the total relative stability of each of the nine treatments were compared for the first and second perturbations. Significant trends for all treatments were detected with the binomial test at the $P \leqslant 0.100$ level. Plus (+) indicates that the stability of the microcosms was greater for the second stress than for the first; minus (—) indicates that the stability was less for the second stress than for the first; and NS is no significant change.

certain parameters. It is perhaps worth noting that the total relative stability of all three metabolic parameters was less for the second stress than for the first.

For each of the nine treatments, median coefficients of variation among replicate microcosms were also compared before and after perturbations. The binomial test again indicated significant ($P \leqslant 0.100$) trends. The thermal stress increased replicate variation for net day production but reduced variation for the chlorophyll a, phosphorus, and magnesium parameters. Although no general trends are obvious from these studies, we must certainly consider the influence of prior events (i.e., history) when assessing an ecosystem's stress response.

SUMMARY

Most of the current theory that seeks to describe and predict the response of ecosystems to stress conditions has been developed by mathematical or intuitive reasoning. I attempted to clarify the stress concept and advocated an empirical approach complementing these methodologies. This approach requires that strict operational definitions be developed to clarify theoretical concepts. Relations between concepts must be expressed as falsifiable hypotheses. Five measures of relative stability can be used to quantify a system's response to stress. Microcosms are proposed as living models of large natural ecosystems. Since in any type of model (mathematical, verbal, or physical) some behavior is specific to the model, results must be extrapolated cautiously. Microcosms are potentially valuable tools, however, both conceptually and pragmatically, for developing and testing theories of ecosystem dynamics.

The two microcosm experiments briefly described to illustrate how such models can be used failed to support several hypotheses that were derived by mathematical or intuitive approaches. The relative stability of these experimental systems was not related simply to their species diversity, nutrient—energy subsidies, or mass. Constancy stability, however, was directly related to nutrient availability. No relationship could be demonstrated between resistance and resilience of stressed microcosms or between their steady-state variability and any measure of their perturbed behavior. These empirically derived results suggest that some hypotheses that seek to describe an ecosystem's stress response are not correct as generalities and require further clarification.

Several factors influence an ecosystem's response to stress. Among these are the type of perturbation, the parameters used to characterize the ecosystem, and the history of the system. Each factor must be rigorously specified before predictions about an ecosystem's stress response can be made. This recognition greatly limits our ability to generalize about the perturbed behavior of ecological systems. Data from microcosm studies were used to illustrate the possible effects of these three factors. Empirical evaluation of hypotheses concerning an ecosystem's response to stress is sorely needed. The rigorous design and specificity of analysis required limit generalization of results but may enhance understanding of ecosystem dynamics when used in conjunction with mathematical and intuitive methods.

ACKNOWLEDGMENTS

I thank G. T. Auble, M. C. Chase, F. B. Golley, J. V. Nabholz, and C. L. Thomas for their critical comments on earlier drafts of this paper. Microcosm research was supported by grant BMS-75-19765 from the National Science Foundation.

REFERENCES

Barnhart, C. L., and J. Stein (Eds.), 1964, *The American College Dictionary*, p. 1197, Random House, Inc., New York.

Boesch, D. F., 1974, Diversity, Stability and Response to Human Disturbance in Estuarine Ecosystems, in *Proceedings of the First International Congress on Ecology*, The Hague, Sept. 8—14, 1974, pp. 109-114, Centre for Agricultural Publishing and Documentation, Wageningen, The Netherlands.

Copeland, B. J., 1970, Estuarine Classification and Responses to Disturbances, *Trans. Am. Fish. Soc.*, 99: 826-835.

Golley, F. B., 1974, Structural and Functional Properties as They Influence Ecosystem Stability, in *Proceedings of the First International Congress on Ecology*, The Hague, Sept. 8—14, 1974, pp. 97-102, Centre for Agricultural Publishing and Documentation, Wageningen, The Netherlands.

Goodman, D., 1975, The Theory of Diversity—Stability Relationships in Ecology, *Q. Rev. Biol.*, 50: 237-266.

Guyton, A. C., 1971, *Textbook of Medical Physiology*, 4th ed., pp. 700-701, W. B. Saunders Co., Philadelphia.

Harwell, M. A., W. P. Cropper, Jr., and H. L. Ragsdale, 1977, Nutrient Recycling and Stability: A Reevaluation, *Ecology*, 58: 660-666.

Hurd, L. E., and L. L. Wolf, 1974, Stability in Relation to Nutrient Enrichment in Arthropod Consumers of Old-Field Successional Ecosystems, *Ecol. Monogr.*, 44: 465-482.

Jernelöv, A., and R. Rosenberg, 1976, Stress Tolerance of Ecosystems, *Environ. Conserv.*, 3: 43-46.

Larsen, P. F., 1974, Structural and Functional Responses of an Oyster Reef Community to a Natural and Severe Reduction in Salinity, in *Proceedings of the First International Congress on Ecology*, The Hague, Sept. 8—14, 1974, pp. 80-85, Centre for Agricultural Publishing and Documentation, Wageningen, The Netherlands.

Leffler, J. W., 1977, *A Microcosm Approach to an Evaluation of the Diversity—Stability Hypothesis*, Ph.D. Thesis, University of Georgia, Athens.

Odum, E. P., 1971, *Fundamentals of Ecology*, 3rd ed., p. 8, W. B. Saunders Co., Philadelphia.

O'Neill, R. V., 1976, Ecosystem Persistence and Heterotrophic Regulation, *Ecology*, 57: 1244-1253.

Pomeroy, L. R., 1975, Mineral Cycling in Marine Ecosystems, in *Mineral Cycling in Southeastern Ecosystems*, ERDA Symposium Series, Augusta, Ga., May 1—3, 1974, F. G. Howell, J. B. Gentry, and M. H. Smith (Eds.), pp. 209-223, CONF-740513, NTIS.

Siegel, S., 1956, *Nonparametric Statistics for the Behavioral Sciences*, McGraw-Hill Book Company, New York.

Smith, H. A., T. C. Jones, and R. D. Hunt, 1972, *Veterinary Pathology*, 4th ed., pp. 186-187, Lea & Febiger, Philadelphia.

Taub, F. B., and A. M. Dollar, 1964, A *Chlorella—Daphnia* Food Chain Study: The Design of a Compatible Chemically Defined Culture Medium, *Limnol. Oceanogr.*, 9: 61-74.

Webster, J. R., 1975, *Analysis of Potassium and Calcium Dynamics in Stream Ecosystems on Three Southern Appalachian Watersheds of Contrasting Vegetation*, Ph.D. Thesis, University of Georgia, Athens.

——, J. B. Waide, and B. C. Patten, 1975, Nutrient Recycling and the Stability of Ecosystems, in *Mineral Cycling in Southeastern Ecosystems*, ERDA Symposium Series, Augusta, Ga., May 1—3, 1974, F. G. Howell, J. B. Gentry, and M. H. Smith (Eds.), pp. 1-27, CONF-740513, NTIS.

PHYSICOCHEMICAL AND BIOLOGICAL STRESSORS AS DISTRIBUTIONAL DETERMINANTS OF CARIBBEAN AND TROPICAL EASTERN PACIFIC SWIMMING CRABS

ELLIOTT A. NORSE
Department of Zoology, University of Iowa, Iowa City, Iowa

ABSTRACT

Zones where influences of terrestrial—limnetic and marine ecosystems overlap provide natural physicochemical stress gradients for groups that evolved at either end. Portunid crabs, originally a marine group, now occur from equable offshore marine biotopes to stressful rivers and streams. Field studies on population and community structure in coastal areas of the eastern Pacific (Colombia, Panama, and Mexico) and the Caribbean (Jamaica, Florida, Curaçao, and Colombia) and experiments on physiological tolerances and biological interactions of Caribbean portunids showed that *Euphylax robustus* in the Pacific and *Portunus*, *Arenaeus*, and *Cronius* spp. in both oceans occur predominantly farther offshore, deeper, or in higher salinities than *Callinectes* spp., which are more euryhaline than members of the other genera. Among *Callinectes*, high-temperature and desiccation tolerances are much less important than hyposalinity tolerances in permitting species to penetrate estuaries and shallow areas, and likelihood of severe dilution directly or indirectly limits distributions as stress increases. Seaward distributional limits result from increased interference and exploitation competition, predation, parasitism, decreasing food quantity and quality, and the interactions of these factors. Thus biological stress increases as physicochemical stress decreases. Distributions of populations along gradients are envisioned as responses to risk of mortality from physicochemical and biological stressors, with abundance peaks corresponding to zones of minimum combined risk.

When a species belonging to a group that evolved in marine environments leaves the sea for land or freshwater, it has an opportunity to exploit new resources and leave behind old problems.

There is an admission price, however; the species must have morphological, physiological, behavioral, or life-historic adaptations to abiotic and biological conditions that are increasingly alien and stressful with increasing departure from the sea. If the group were in some way preadapted for ecological expansiveness (weedy), repeated invasions of the gradient extending from marine dominated to terrestrially dominated aquatic climates would lead either to one species' eliminating the others or to partitioning of the gradient. This paper considers one group, bottom-dwelling swimming crabs of the family Portunidae, which has repeatedly invaded this gradient. Observations are given on species distributions, factors limiting portunid incursions into estuaries, and factors resulting in coexistence rather than in monopolization by a single species. Some of the principles illustrated should be valid for other groups inhabiting natural physicochemical gradients as well.

The tropical American coasts have a small share of the world's ~300 portunid species (Norse, 1977), but they do have most of the species (members of the genus *Callinectes*) which are known or believed to invade estuaries. The taxonomy of American portunids, although less fluid than that of portunids from the Indo-West Pacific, has been complicated by variations in areal distributions, abundance, temporal fluctuations, and intensities of collecting. Many species are sexually dimorphic and dichromatic, have allometric growth and vary geographically, and, therefore, require considerable experience to identify with confidence. In fact, newly described *Portunus* (Holthuis, 1959; 1969) and *Callinectes* spp. (Williams, 1966; Taissoun, 1972) have been added to the tropical west Atlantic fauna in recent years. Unfortunately, *C. similis* Williams is still misidentified as *C. ornatus* Ordway or *C. danae* Smith by some scientists, and I failed to realize fully that *C. maracaiboensis* Taissoun, ostensibly a Lake Maracaibo area endemic, was common in my main study area (Norse, 1975; 1977). This taxonomic uncertainty has confused our already limited view of the group's ecology, but identification of eastern Pacific portunids and the world's *Callinectes* has been greatly facilitated by the monographs of Garth and Stephenson (1966) and Williams (1974).

After studying the taxonomy of *Callinectes*, I visited Florida, Jamaica, Curaçao, both the Caribbean and Pacific coasts of Colombia, and the Pacific coasts of Mexico and Panama (Fig. 1) to delimit the ecological distributions of demersal portunids and the roles of environmental factors contributing to them. This paper synthesizes some of the results I have discussed in greater detail elsewhere (Norse, 1975; 1977; in preparation; Norse and Estevez, 1977).

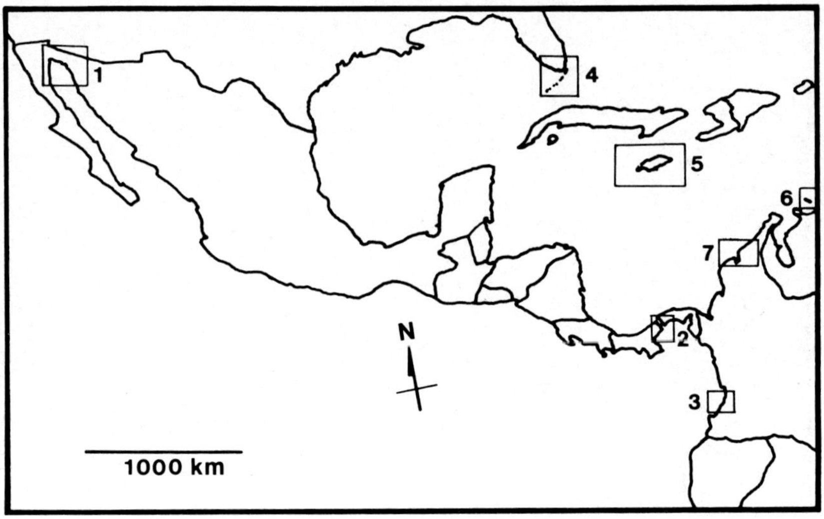

Fig. 1 The study areas in the east Pacific: (1) Northern Gulf of California, Mexico; (2) Panama; and (3) Colombia; and the west Atlantic: (4) Miami area and Florida keys; (5) Jamaica; (6) Curaçao; and (7) Colombia.

METHODS

Distributions

On Colombia's Pacific coast, I worked with Mario Estevez, a biologist then a member of the Colombian Marine Fisheries Development Project, using the Colombian government's shrimp trawler *Inderena* to sample in the lower reaches of four large rivers, Guafuí, Timbiquí, Saija, and Naya, and on the adjacent continental shelf. We planned four transects, each having seven trawls from as far upriver as nets could be operated to the 20-fathom (37-m) contour on the shelf (upriver, midriver, river mouth, and at 9, 18, 27, and 37 m), but two of the transects were only partially completed. To assure sufficient depth for shrimp-trawl operation and to minimize the effects of running tides on crab movements, we made all estuarine trawls within 1 hr of high slack tide. We did shelf trawls during daylight hours irrespective of tides. We trawled without tickler chains at equal speeds for equal periods to cover equal areas of bottom. At the midpoint of most trawls, we took surface temperatures and water samples, but the water samples were lost before salinities were determined. Supplemental eastern Pacific collections were made by a shrimp trawler in Mexico and by hand netting and trapping in Colombia, Mexico, and Panama.

Jamaica was the main Caribbean study area. I sampled lower reaches of rapidly flowing through nearly stagnant lotic waters, a coastal salt pond, periodically exposed mangrove mud flats, sea grass and algal beds, and featureless bottoms in both estuarine and nonestuarine bays and lagoons, and among coral reefs, by wading, snorkeling, and scuba diving. I used a gill net and traps at one site but trapped and/or hand netted at all others. By standardizing the trapping procedure (using the same trap designs, trapline pattern, bait, and period throughout), I was able to compare crab densities (in number per trap-hr) among sites. Temperatures and water samples were taken from bottom waters in the middle of each site. Most salinities were measured with a refractometer, but a few were measured by titration. I made supplemental collections in Jamaica, in the Miami area and the middle Florida keys, on the Colombian coast between the Ciénaga Grande de Santa Marta and Bahía Concha, and in Curaçao.

At all sites the physical conditions and presence of other species were noted, and, in quantitative Jamaican and Pacific Colombian sites, crabs were sexed, and their carapace lengths were measured immediately after capture.

Physiological Tolerances

The hyposalinity tolerances of the common portunids of Jamaica and the Florida keys were tested by acclimating them for at least 30 hr in running seawater ($\overline{X}$ = 34.5‰ in Jamaica and 36.25‰ in the keys), then placing them into 10, 25, 37.5, or 50% seawater, and noting survival after 24 hr.

High-temperature tolerances of Jamaican *Callinectes* were tested by acclimating them for at least 21 days in the Discovery Bay Marine Laboratory's running seawater system ($\overline{X}$ = 28.9° C), placing them in 39.0°C aerated seawater for 60 min, replacing them in acclimation-temperature running seawater for 24 hr, and examining for survival.

Desiccation tolerances were tested by acclimating Jamaican *Callinectes* for at least 30 hr in running seawater, weighing one of each species and replacing it in seawater to recover water lost during weighing, then placing each crab in an individual compartment in a wire-mesh cage in a breezy but shaded corridor between buildings, and noting time until death and lethal weight losses for each species.

Interspecific Agonism

Male Jamaican *Callinectes* of varied known sizes were placed in a fenced-off "arena" in a concrete seawater table, and each crab's retreats from approach, display by, or contact with each individual of the other species were tallied. When agonism in the arena diminished, increasing concentrations of conch, fish, and crab extracts were added to increase agonistic activity. Only repeatedly decisive results were considered a victory for one of the combatants.

AREA DESCRIPTIONS

The Pacific Colombian coastal state of Cauca has an extensive depositional shoreline. Rainfall is very heavy, particularly on the western range of the Andes, which peak about 80 km inland. Many mangrove-lined rivers pour enormous amounts of freshwater into the turbid sea overlying the gently sloping continental shelf. The Pacific coast of Panama (the coastal areas of the Canal Zone and Taboga Island) has considerable rain but is much less affected by upland rainfall and has less freshwater mixing with the turbid shelf waters than the Colombian coast. The northern Gulf of California, in the San Felipe, Baja California Norte, and Puerto Peñasco, Sonora, areas of Mexico, has clearer, slightly hypersaline waters over the gently sloping shelf, but there is very little freshwater input in this desert area.

In contrast, inshore areas in Jamaica have varied aquatic climates. The island is mountainous and fairly wet, but, where permanent and temporary streams are few, coastal waters are oceanic in character, particularly on the steeply sloping, coral-reef-fringed north coast. The Caribbean Colombian coast has adjacent areas of markedly different aerial and aquatic climate. The area of Ciénaga Grande de Santa Marta, a large estuarine lake, is a fairly wet, depositional coast. To the east, near the city of Santa Marta, the edges of a very tall, isolated mountain massif slope steeply into the sea. Since the aerial climate is semiarid, coastal waters are largely oceanic. Curaçao is a geologically complex semiarid island; oceanic waters abut both coasts, which are indented by protected, turbid bays that lack permanent freshwater input. In the fairly wet Miami area, the Oleta River, which has intermittent flow, empties into Biscayne Bay, a large, culturally stressed shallow bay. Shore areas may be estuarine, but the central and eastern parts usually have little seawater dilution. The keys are emergent points on the eastern edge of the Florida plateau. Although they receive moderate rainfall, their small size, lack of topographic relief, and geologic composition (permeable limestones) prevent continuing existence of hyposaline biotopes (for more detailed descriptions, see Norse, 1975, and Norse and Estevez, 1977).

RESULTS AND DISCUSSION

Since 1969, I have examined and measured the *Callinectes* collections in the Allan Hancock Foundation of the University of Southern California, the Harvard Museum of Comparative Zoology, and the American Museum of Natural History and have collected and/or observed in the field about 8000 portunids belonging to 26 tropical American species (Table 1). Two species are not primarily demersal; *Euphylax dovii* Stimpson is an epipelagic species that enters inshore waters to spawn (Norse and Fox-Norse, 1977), and *Portunus sayi* (Gibbes) lives in floating *Sargassum* patches that occasionally drift ashore (Williams, 1965). Several species were too rarely found or locally distributed to permit any but tentative conclusions about their distributions. Some species are commonly associated with particular plant communities or substrate types, but, except to note that most occur mainly on soft bottoms, I will concentrate on distributions with regard to climatic and biological stressors.

From upriver to continental-shelf Pacific Colombian waters, surface temperatures increased steadily from 26.3 to 28.4°C. This small change would not be likely to cause great changes in species compositions of demersal crabs, but rather indicates the change in the ratio of cooler freshwater to warmer seawater. The surface conditions, which directly affect benthos only in the shallows, reflect deeper bottom conditions, depending on the hydrographic characteristics of the estuaries (Gibbs, 1970). The most obvious major climatic difference between river water and shelf water is salinity.

Distributions of portunids and other groups reflect this salinity gradient. Portunids comprised nearly 99% of the total of benthic brachyuran crabs. Table 2 gives the species compositions (mean relative abundances) of demersal portunids along the transects. *Callinectes toxotes* Ordway dominates the freshest sites but is replaced by *C. arcuatus* Ordway, which is then replaced by *Euphylax robustus* A. Milne Edwards and *Portunus asper* (A. Milne Edwards) in increasingly equable biotopes. I found a few *C. arcuatus* inshore in Panama, but, in the northern Gulf of California, there is a clear ecological separation between *Callinectes* spp. Tidal channels draining salt marshes and shallow areas of both bays and the open gulf are

TABLE 1

DATA BASE FOR ASSESSING DISTRIBUTIONS OF AMERICAN PORTUNIDS

East Pacific

Mexico, Northern Gulf of California		Panama		Colombia	
Callinectes arcuatus	~2000	*Callinectes arcuatus*	~8	*Euphylax dovii**	1543
Callinectes bellicosus	~500			*Callinectes arcuatus*	~900
Portunus xantusii minimus	~15			*Callinectes toxotes*	~300
Euphylax robustus	1			*Euphylax robustus*	292
				Portunus asper	46
				Arenaeus mexicanus	7
				Cronius ruber	1

West Atlantic

Miami area		Florida keys		Jamaica		Colombia		Curaçao	
Callinectes ornatus	~60	*Callinectes ornatus*	~300	*Callinectes marginatus*	~280	*Callinectes danae*	~60	*Callinectes danae*	~150
Callinectes sapidus	~40	*Portunus spinimanus*	~80	*Callinectes exasperatus*	~250	*Callinectes bocourti*	~50	*Portunus anceps*	~40
Callinectes exasperatus	21	*Portunus depressifrons*	~70	*Callinectes ornatus*	~235	*Callinectes sapidus*	~6	*Callinectes exasperatus*	~30
Callinectes similis	17	*Callinectes marginatus*	~50	*Callinectes danae*	~200	*Portunus spinimanus*	3	*Callinectes marginatus*	~20
Callinectes danae	~8	*Callinectes exasperatus*	~15	*Callinectes sapidus*	~115	*Callinectes marginatus*	2	*Callinectes maracaiboensis*	7
Callinectes marginatus	1	*Callinectes sapidus*	~5	*Callinectes maracaiboensis*	~100	*Callinectes ornatus*	1	*Callinectes bocourti*	1
Portunus spinimanus	1	*Cronius tumidulus*	4	*Callinectes bocourti*	~70	*Portunus spinicarpus*	1	*Callinectes ornatus*	1
		Portunus sebae	3	*Portunus sayi**	~50	*Portunus anceps*	1		
		Portunus gibbesii	2	*Portunus sebae*	27				
				Arenaeus cribrarius	15				
				Portunus anceps	7				
				Callinectes similis	3				
				Portunus depressifrons	1				
				Cronius tumidulus	1				
				Total portunids	~8000				
				Total demersal portunids	>6000				

*A nondemersal species.

TABLE 2

SPECIES COMPOSITION (%) OF DEMERSAL PORTUNIDS ALONG PACIFIC COLOMBIAN TRANSECTS

Species	Upriver	Midriver	River mouth	Continental shelf 9 m	18 m	27 m	37 m
Callinectes toxotes	94.6	43.5	34.0		0.9		
Callinectes arcuatus	5.4	56.5	64.3	85.7	43.2	31.5	
Euphylax robustus				6.1	49.9	61.3	87.4
Portunus asper				7.6	5.3	7.2	12.6
Arenaeus mexicanus				0.6	0.7		
Cronius ruber			1.7				

strongly dominated (>95%) by *C. bellicosus* (Stimpson), which was completely absent from offshore trawl hauls in which I found great numbers of *C. arcuatus*, some *Portunus xantusii minimus* Rathbun, an *E. robustus*, and a handful of nonportunid brachyurans. Brusca (1973) noted that *E. robustus* can be abundant in shrimp trawl hauls in the northern gulf, presumably in deeper waters than I sampled. Thus the Colombian and northern gulf gradients, although ~5000 km apart, show striking similarities. The most terrestrially influenced biotopes are dominated by different species (*C. toxotes* vs. *C. bellicosus*), but these are replaced to seaward by *C. arcuatus*, a *Portunus* sp. (*P. asper* vs. *P. xantusii minimus*), and *E. robustus*.

Although the Caribbean has more portunid species than the east Pacific, salinity distributions (aside from the absence of extant Caribbean *Euphylax*) in tropical and warm temperate areas are similar at the generic level. *Callinectes* monopolize terrestrially dominated waters, and *Arenaeus*, *Portunus*, *Cronius*, and fully aquatic brachyurans in many other families make, at most, limited incursions into hyposaline waters (Table 3). Among Jamaican *Callinectes*, *C. maracaiboensis*, *C. bocourti* A. Milne Edwards, and *C. sapidus* Rathbun dominate the freshest sites. All seven common species occur in the brackish range, *C. exasperatus* (Gerstaecker) being the most abundant. In waters usually diluted slightly below seawater values, *C. marginatus* (A. Milne Edwards), *C. danae*, and *C.*

exasperatus dominate. *Callinectes ornatus*, *C. marginatus*, and *C. exasperatus* dominate biotopes having virtually undiluted seawater. Similar patterns exist in regions having long-lasting hyposaline environments (e.g., Miami; Ciénaga Grande de Santa Marta, Colombia). In areas lacking permanent hyposaline waters (e.g., the keys; areas near Santa Marta, Colombia; and Curaçao), however, the portunid species composition changes with the degree of exchange with the sea. Thus the back bays most tenuously connected with the sea have only the *Callinectes* that occur below 10‰ in wetter regions.

TABLE 3

SPECIES COMPOSITION OF JAMAICAN PORTUNIDS IN FOUR SALINITY CATEGORIES

Species	Freshwater (0—10.0‰)	Brackish water (10.1—30.0‰)	Near marine (30.1—33.0‰)	Marine (33.1—35.5‰)
Callinectes maracaiboensis	41.0	10.0	3.8	0.4
Callinectes bocourti	34.5	10.4	0.4	0.7
Callinectes sapidus	22.5	15.2	4.5	1.4
Callinectes exasperatus	2.0	29.6	22.8	25.2
Callinectes danae		16.0	22.4	1.8
Callinectes marginatus		16.6	30.8	26.4
Callinectes ornatus		2.4	13.9	32.6
Callinectes similis				0.2
Arenaeus cribrarius			1.5	0.3
Portunus sebae				8.9
Portunus anceps				1.5
Portunus depressifrons				0.3
Cronius tumidulus				0.3

Upstress Distributional Determinants

The simplest explanation for these patterns is that species distributions respond to the likelihood of severe dilution. Some crabs must either avoid or be killed in climates having constant (always fresh or brackish) or occasional (e.g., back bays, where heavy rains cause normally marine salinities to plummet) severe dilution. Such patterns imply marked differences among species' tolerances of dilution, differences my tests confirmed. Combined results from Jamaica and the Florida keys give the following order of euryhalinity: *C. maracaiboensis* = *C. bocourti* = *C. sapidus* ⩾ *C. exasperatus* > *C. danae* > *C. marginatus* > *C. ornatus* > *P. depressifrons* ⩾ *P. spinimanus* ⩾ *A. cribrarius* ⩾ *P. sebae.* Using data from experiments in which *Callinectes* spp. were transferred to 10% seawater, I found that the (arcsin transformed) proportions surviving are very strongly correlated with the species lowest salinity records ($r = -0.989$, with 5 df; $P < 0.001$). In contrast, their high-temperature tolerances are only weakly correlated with their highest temperature records ($r = 0.397$; not significant), and I could discern no marked relationship between species desiccation tolerances and their potential danger from desiccation. Unfortunately, I was unable to do planned experiments on anoxia tolerances, but current evidence indicates that upstress distributional limits are largely controlled directly (by mortality) or indirectly (e.g., by the depressing effects of physiological stress on competitive abilities) by the likelihood of severe dilution.

There is a strong resemblance between taxonomic and ecophysiological relationships in portunids from both oceans, undoubtedly resulting from Pliocene interoceanic connection. Among Pacific Colombian crabs, *C. toxotes*, whose closest Caribbean relatives are *C. maracaiboensis* and *C. bocourti*, dominates the freshest estuarine sites; *C. arcuatus*, whose closest relative is *C. danae*, dominates intermediate salinities; whereas *Portunus asper* and *Arenaeus mexicanus* (Gerstaecker), whose congeners are among the less euryhaline Caribbean portunids, occur only in equable shelf waters.

Downstress Distributional Determinants

Since all species can occur in marine salinities and all Caribbean species can survive in undiluted seawater in the laboratory, climatic conditions do not directly limit downstress distributions. What brings about the observed patterns of serial replacement in both oceans? Possibilities include interference and exploitation competition, predation, parasitism, and food quantity and quality. Some investigators (e.g., MacArthur, 1972) have stressed the role of competition; others (e.g., Connell, 1975) have shown that predation and parasitism further limit distributions within the organisms' physicochemically limited fundamental niche space.

Interference Competition

To determine whether interference competition could prevent seaward expansion of crabs from hyposaline waters, I needed data on relative agonistic abilities, size structures, and relative abundances of species. I characterized the Caribbean species by salinity as freshwater (*C. maracaiboensis*, *C. bocourti*, and *C. sapidus*), brackish water (*C. exasperatus* and *C. danae*), and marine (*C. marginatus* and *C. ornatus*). I tested the relative agonistic abilities of the species and calculated the proportion of wins between combatants in adjacent salinity groups in size-ratio classes between 0.5 and 1.9. I arcsin transformed these proportions and regressed them against size ratios, from the highest ratio, where a salinity group won none of the interactions, to the lowest ratio, where it won all. By solving for a 0.5 probability, I obtained a "combative equals ratio," the size ratio at which adjacent salinity groups have equal probabilities of winning agonistic interactions. Since crab stages range from a few millimeters to 50 or 75 mm, I used the method of Schoener (1969) to obtain mean sizes of adult males (the mean of the largest third) for comparisons between species groupings. Relative abundances are in numbers per trap-hour.

A comparison of mean sizes, sizes of combatively equal crabs, and densities (Table 4) shows that, although freshwater and brackish-water crabs closely approach combative equality, the much greater population densities in brackish waters could prevent the seaward expansion of freshwater crabs. Using a model of slightly greater complexity (Norse, in preparation) in which I calculated the total numbers of losses which would be suffered by various-sized freshwater crabs as they moved through increasing salinities, I estimated that brackish water poses an "agonistic barrier" of considerable magnitude (there were up to three times as many losses per unit distance traveled as in freshwater). This could account for both the abundance and size structure of freshwater crabs along the gradient.

In contrast, only marine crabs (which are less abundant than brackish water crabs) of well above average size are combatively equal to average brackish water crabs. Such large marine crabs are so few that they probably do not agonistically confine them to brackish water, nor does agonism seem to prevent the seaward expansion of

TABLE 4

MEAN SIZES OF ADULT MALE *Callinectes*, COMBATIVE EQUALS OF MEAN-SIZED MALES, AND ABUNDANCES WHERE EACH SALINITY GROUP IS DOMINANT

	Freshwater	Brackish
Mean sizes, mm	68.3	56.5
Combative equals, mm	68.3	57.3
Number/trap-hour	0.42	0.78
	Brackish	**Marine**
Mean sizes, mm	56.5	45.7
Combative equals, mm	56.5	51.7
Number/trap-hour	0.78	0.57

the Pacific Colombian *C. toxotes*. It is much larger than *C. arcuatus* ($\overline{X}$ carapace length for adult males is 87.3 vs. 49.0 mm, respectively). If they were agonistically equivalent to their closest Caribbean relatives (freshwater *Callinectes* vs. *C. danae*), *C. arcuatus* would have to average 71.6 mm long to equal average *C. toxotes* combatively. If they were agonistically equivalent to the most mismatched Caribbean species (freshwater *Callinectes* vs. *C. marginatus*), *C. arcuatus* would have to average 67.4 mm to have an equal chance of winning against *C. toxotes*. Since the maximum size of *C. arcuatus* (55.5 mm) is far less than either of these, interference competition apparently cannot explain downstress limits of either Jamaican brackish-water or Pacific Colombian freshwater crabs.

Exploitation Competition and Predation

Although I took quantitative data only on Pacific Colombian and Jamaican brachyuran crabs, I repeatedly observed that numbers of species in groups likely to be competitors and predators of portunids (e.g., other reptant decapods, stomatopods, gastropods, cephalopods, echinoderms, and fishes) increase sharply as the likelihood of severe dilution decreases. In Jamaica the number of portunid species quadruples from the freshest to the most equable marine waters. In contrast, I found no obvious rise in the number of individuals along the gradient, although population size and structure of many species have undoubtedly been affected by human activities. As MacArthur (1972) suggested, an increase in diffuse competition (i.e., number of species) on a resource gradient increases the chance that one species sandwiched between others will be eliminated.

By analogy, what can be called "diffuse predation" also increases to seaward. Activities of some crab predators (wading birds, crab-eating raccoons in Colombia, and crocodilians in Florida, Jamaica, and Colombia) are not greatly affected by the salinity of the water containing their prey. Distributions of many other predators are affected by salinity, however. Man, American eels, and *Callinectes* are the only other known *Callinectes* predators in Jamaican fresh waters. More predator species occur in brackish waters, however, including ariid catfishes in both oceans in Colombia; snook in both oceans in Colombia, in Florida, and in Curaçao; tarpon and gray and schoolmaster snappers in the Caribbean; and some sciaenids in Pacific Colombia. Still more occur in marine waters; these include many sciaenid species along mainland coasts and moray eels, balistids, snappers, grunts, goatfishes, serranids, and sea turtles in all areas where they have not been fished out. In fact, Randall (1967) found that the majority of 212 Caribbean inshore and reef fish species eat crabs. From these observations, it appears that predation (and exploitation competition since many of these predators consume foods that would otherwise be taken by portunids) steadily increases to a maximum in lagoons and coral reefs, probably because many predators are orthostenohaline. This trend may not be so simple in some areas, however. Freshwaters on Caribbean islands, which are not easily reached by oligostenohaline fishes and decapod crustaceans, are depauperate in comparison with analogous mainland communities. On the continents both downstress and upstress distributions of freshwater *Callinectes* may be limited by competition and predation from diverse marine and freshwater stenohaline communities.

In addition to studying extrinsic evidence of predation pressure from observed instances of predation and numbers of predator species and individuals, we can assess its impact on a species by examining the intrinsic responses that have evolved to thwart predation. Several of these responses indicate that predation pressure is more intense in higher salinities.

All *Callinectes* are more or less countershaded, although ventral surfaces of crabs in some environments may be covered with reddish or black films. Countershading is almost ubiquitous among demersal animals but has a unique denouement in *Callinectes*. In advanced brachyurans the abdomen is usually folded beneath the body. In male and immature female *Callinectes*, the actual dorsal abdominal surface and surrounding ventral sternites are indistinguishably light colored. When a female undergoes the molt of puberty, her abdomen darkens, becoming virtually identical in color to the dorsal surface of her carapace. When she spawns, the egg mass distends the abdomen from her ventral surface so that it is held behind her, as in a lobster, thereby maintaining countershading. This must have particular adaptive value in higher salinities where females spawn and hatch their eggs.

The upper surfaces of Caribbean portunids' carapaces match substrata to varying degrees. The freshwater *Callinectes* can be found on widely different substratum colors and textures. Coloration in *C. bocourti* (bright reddish brown and dull green tones) and in *C. maracaiboensis* (highly variable, ranging from ashen gray to dark chocolate brown) is cryptic in some environments but somewhat conspicuous in others. In contrast, portunids from higher salinities occur on fewer substrate types and, with one notable exception, are better camouflaged. For example, ground colors in *Portunus anceps* (Saussure), *P. depressifrons* (Stimpson), and *Arenaeus cribrarius* (Lamarck) are very light (white, gray, and tan), with disruptive markings that render them extremely inconspicuous against the light calcareous sediments on and in which they live. The species from the most equable climates I studied, marine lagoons and coral reefs, has almost completely eschewed cryptic coloration. *Portunus sebae* (H. Milne Edwards), apparently a strictly nocturnal species, has two unmistakable white-ringed black ocelli on the posterior half of its carapace; this is startle coloration, which it shares with some coral-reef octopuses and stomatopods and with terrestrial moths and mantids. These ocelli have probably evolved as deterrents to goatfishes and triggerfishes, which "blow" into reefal sands as they forage during the day.

As Stephenson (1962) stated, swimming crabs are also diggers, and those I observed in the field spend a large fraction of their time buried in soft substrata, usually with only their eyes, antennae, and antennules protruding. Not surprisingly, their eyes are cryptically colored to varying degrees. Eyes of freshwater *Callinectes* have dark and light barred or checked patterns, which make them somewhat more conspicuous than the light-colored, often stippled eyes of species from higher salinities, including *P. sebae*. Demersal portunids have ridged, pilose, or granulose carapaces that retain sediments when the crabs have partly or completely emerged from the substratum, thereby disrupting their form and permitting a degree of camouflage even on substrates that their carapace colors do not match closely. As in carapace and eye coloration, there is a trend in carapace texturing among species along the gradient. Freshwater

Callinectes have sparse granules of low relief (*C. sapidus*) or large areas devoid of granulation (*C. bocourti* and *C. maracaiboensis*). *Callinectes, Arenaeus, Portunus,* and *Cronius* spp. from higher salinities have carapaces adorned by more even, higher, or denser granules; strong ridges; dense pile; or depressions.

Portunids possess another antipredation mechanism, spininess, for which variation in the degree of development may reflect similar variation in predation pressure. Spines may stimulate the release of improperly seized crabs or, failing that, may prevent the predator from swallowing the crab. Indeed, Burnett and Snyder (1954) found starved eider ducks with blue crabs lodged in their throats. With a few exceptions, American portunids in equable climates are spinier, having longer, sharper, and/or more spines, reaching maximum development in deepwater species such as *P. spinicarpus* (Stimpson).

Yet despite extrinsic and intrinsic evidence that predation pressure (at least for islands) increases as physicochemical stress decreases, regions with greatly differing human fishing pressures have similar portunid replacement sequences along salinity gradients. This may imply that *Callinectes* have evolved proximate antipredation behaviors that still operate despite relaxed predation pressure (e.g., avoidance of higher salinities or waters with lower concentrations of terrigenous organics). Such behaviors may not be selected out because portunids seem to have long larval lives (Costlow and Bookhout, 1959; Bookhout and Costlow, 1974; 1977), and adults in one place may have hatched hundreds of kilometers distant, where predation pressures were not relaxed. Thus, in some areas, down-stress distributions may be intrinsically limited in response to predation pressures that no longer exist.

Parasitism

Parasites can be divided into two functional groups: endoparasites, which receive the benefits of host osmoregulation (if present) and may be unaffected by salinity fluctuations of the host's environment, and ectoparasites (in the broad sense, epibionts that diminish the host's health), which must be as eurytopic as their hosts to survive. Some *Callinectes* in Pacific Colombian river-mouth sites had carapaces fouled with *Balanus*, but heavier incrustations of *Balanus, Chelonibia* (another balanomorph barnacle), bryozoans, and sabellid polychaetes occurred on *Callinectes* in shelf waters. Marine portunid species were much freer of external fouling organisms. Parasites living on the gills are also ectoparasites. I have found stalked (lepadomorph) barnacles of the genus *Octolasmis* on the gills of all common Jamaican and eastern Pacific *Callinectes*, particularly adult females, which tend to occur in higher salinities; in some cases gill chambers were clogged with hundreds. Stenohaline marine portunids, however, usually have few or no *Octolasmis* on their gills (De Turk, 1940; Humes, 1941; Walker, 1974). In low salinities euryhaline crabs have few external or gill foulers, but, when they enter higher salinities, they become susceptible to fouling. Marine species, however, are less vulnerable, presumably because of morphological, physiological, behavioral, or life-historic adaptations for preventing larval settling. Crabs with fouled carapaces and limbs suffer higher energetic maintenance costs and diminished walking and swimming abilities. Those with gill foulers must compete with the parasites for oxygen, are prevented from cleaning foreign matter (sand grains, ciliates, nemerteans, bryozoans) from their gills, further diminishing their oxygen uptake, and probably have weakened gill discs, which provide an avenue for entry of pathogenic bacteria and protozoa into their circulatory systems. Not surprisingly, crabs having heavily fouled gills are noticeably less active than less-fouled conspecifics.

Food Availability

Both quantitative and qualitative aspects of food change along the gradient in terrestrial influence on aquatic climate. Available data show that *Callinectes* have varied diets (Darnell, 1958; Tagatz, 1968; Estevez, 1972), and V. Fox-Norse and I observed that they have a surprisingly varied repertoire of feeding behavior. They hunt prey as diverse as polychaetes and mullet, scavenge for carrion, scrape aufwuchs from sea-grass blades and rocks, eat detritus-rich sediments, and even filter feed when they encounter dense concentrations of zooplankton (Norse, 1975; Fox-Norse, in preparation). In the Caribbean study areas, water clarity permits autochthonous planktonic and benthic primary production, but much benthic macrophyte productivity and allochthonous material from terrestrial plants overhanging streams enter detritus-based food webs (Fenchel, 1970; Odum and Heald, 1972). On the Pacific Colombian coast, since estuaries and inshore shelf waters appear too turbid to allow major planktonic and benthic plant production, most energy input is in the form of wood, bark, and leaves from mangroves and other trees. Since *Callinectes* feed largely on detritus or detritus-feeding organisms, their productivity should be a function of the amount of available detritus. The trawls collected animals, whose total biomass did not vary greatly along transects, and macroscopic terrigenous detritus, which far outweighed animal biomass in fresher sites but decreased rapidly, constituting perhaps only 1% of the hauls in deeper shelf sites. If the amount of smaller, edible detritus is proportional to that of macrodetritus, the most climatically stressful environments have superabundant food for their inhabitants, but species are increasingly likely to be food-limited to seaward. Although upriver species can afford to be inefficient at finding or assimilating food, selection should favor increased efficiency among organisms in climatically more equable but trophically poorer waters. Downstress distributional limits may correspond to the points where an individual's maintenance costs equal its income.

Energy intake is affected by the availability, as well as the overall quantity of potential food. In lagoons and among coral reefs, predation pressures have stimulated the evolution of highly developed antipredation mechanisms in benthic invertebrates (Bakus, 1964). Thus potential prey are usually out of reach (e.g., they bore into hard substrates), hard to find (cryptic), or in some way unpleasant to would-be predators (e.g., they have startling or warning coloration, sharp spines, envenomating apparatus, toxicity, or noxiousness). In informal experiments (Norse, 1975), I found that normally exposed reefal animals were immune from predation by *Callinectes*, but estuarine organisms were not. Thus an estuarine crab on a coral reef, like the Ancient Mariner, could be surrounded by food resources but unable to use them.

We have seen correlative evidence supporting alternate hypotheses that downstress portunid distributions are limited by interference competition, exploitation competition, predation, parasitism, and food-resource availability. Interactions of these factors could also be limiting since, as climatic stress decreases, eurytopic portunids increasingly become host to ectoparasites. This raises their maintenance costs while diminishing their abilities to compete for increasingly scarce food resources, to win agonistic contests against relatively stenotopic portunids, and to escape increasingly intense predation. Unfortunately, if my observations are correct, we cannot yet reject any of the hypotheses. Even if only one of the major categories were correlated with changes in climatic stress and portunid distributions, we still would not know the precise limiting factors. Rather, the hypotheses generated by these observations should be tested rigorously so that we can continue to approach an understanding of the determinants of organismic distributions.

The *Callinectes* spp. whose crab stages can live in the freshest waters at the edges of both oceans are more than transients but less than permanent residents; they are catadromous (Norse, 1975; 1977; Norse and Estevez, 1977). Although this skews their distributions to seaward, the species that reach peak abundances along rather than at

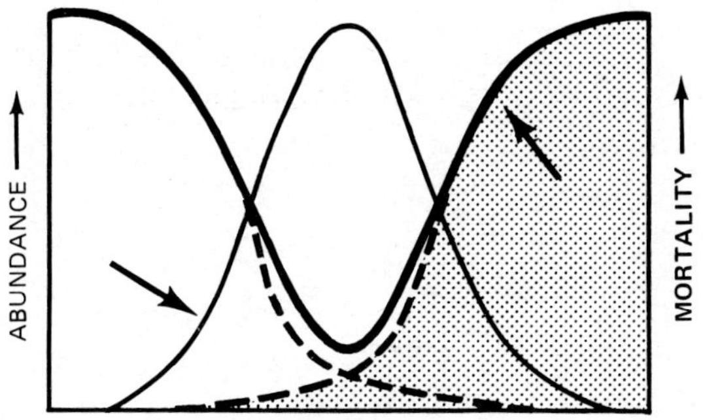

Fig. 2 Effect of risk of mortality (solid bold line) on abundance (thin line) of a species inhabiting a portion of a physicochemical gradient. Sources of risk of mortality: abiotic stressors (white area under bold line); biological stressors (shaded area under bold line).

the ends of the gradient otherwise have Gaussian abundance curves, a feature they share with very different organisms, e.g., some trees (Whittaker, 1967) and birds (Terborgh, 1971) inhabiting other gradients. The model shown in Fig. 2 could account for these characteristic distribution patterns in groups that originated at one end of a physicochemical gradient. As the risk of mortality from usual or unusual weather decreases, risk from various biological stressors increases. The abundance of the species in question mirrors the combined risk of mortality from both abiotic and biological stressors, with peak abundance corresponding to minimum combined risk of mortality. The abundance curves of species result from actual selective mortality if they lack life-history stages that can discriminate among habitats. In species having the potential for such behavior, fitness is increased by avoidance of risky portions of the gradient, and abundance curves result from both habitat selection and mortality. The positions of species along the gradient, then, reflect the relative degrees to which they have specialized toward coping with physicochemical vs. biological stressors.

ACKNOWLEDGMENTS

I have profitted greatly from the essential balance of knowledge, wisdom, and honesty provided by John S. Garth, Gerald J. Bakus, William Stephenson, Alan D. Havens, Hugh Dingle, and Virginia Fox-Norse. Many people helped in the field and laboratory, particularly Heather Copland, Mario Estevez, Virginia Fox-Norse, Alan D. Havens, Paul Sammarco, Avril Siung, Michael Weber, and Bernd Werding. I am indebted to the directors and staffs of the following institutions for providing field, laboratory, and library facilities and taxonomic reference collections: the University of Arizona—University of Sonora Cooperative Marine Station; the Smithsonian Tropical Research Institution; Colombia's Institute for the Development and Management of Natural Resources; Pigeon Key Marine Environmental Station, University of Miami; Discovery Bay Marine Laboratory, University of the West Indies; the Caraibisch Marien-Biologisch Instituut (CARMABI); the Instituto Colombo-Alemán de Investigaciones Cientificas "Punta de Betin"; the Harvard University Museum of Comparative Zoology; the American Museum of Natural History; the Allan Hancock Foundation of the University of Southern California; and the University of Iowa. Donald Doumakes, Virginia Fox-Norse, and James H. Thorp gave valuable criticism of the manuscript. Funds were provided, in part, by the Allan Hancock Foundation, University of Southern California; the Organization for Tropical Studies; the Society of Sigma Xi; the Theodore Roosevelt Memorial Fund, American Museum of Natural History; National Science Foundation grant BMS 74-22718 to Hugh Dingle; and National Institute of Mental Health grant (5 T01 10641-10) to the Neurobehavioral Sciences Program, University of Iowa. This is contribution No. 157, Discovery Bay Marine Laboratory.

REFERENCES

Bakus, G. J., 1964, The Effects of Fish-Grazing on Invertebrate Evolution in Shallow Tropical Waters, *Occ. Pap. Allan Hancock Fdn.*, 27: 1-29.

Bookhout, C. G., and J. D. Costlow, 1974, Larval Development of *Portunus spinicarpus* Reared in the Laboratory, *Bull. Mar. Sci.*, 24: 20-51.

——, 1977, Larval Development of *Callinectes similis* Reared in the Laboratory, *Bull. Mar. Sci.*, 27: 704-728.

Brusca, R. C., 1973, *Handbook to the Common Intertidal Invertebrates of the Gulf of California*, University of Arizona Press, Tucson.

Burnett, F. L., and D. E. Snyder, 1954, Blue Crabs as Starvation Food of Oiled American Eiders, *Auk*, 71: 315-316.

Connell, J. H., 1975, Some Mechanisms Producing Structure in Natural Communities: A Model and Evidence from Field Experiments, in *Ecology and Evolution of Communities*, M. L. Cody and J. M. Diamond (Eds.), pp. 460-490, Harvard University Press, Cambridge, Mass.

Costlow, J. D., and C. G. Bookhout, 1959, The Larval Development of *Callinectes sapidus* Rathbun Reared in the Laboratory, *Biol. Bull. (Woods Hole, Mass.)*, 116: 373-396.

Darnell, R. M., 1958, Food Habits of Fishes and Larger Invertebrates of Lake Ponchartrain, Louisiana, An Estuarine Community, *Publ. Inst. Mar. Sci., Univ. Tex.*, 5: 353-416.

De Turk, W. E., 1940, *The Parasites and Commensals of Some Crabs at Beaufort, North Carolina*, Ph.D. Thesis, Duke University, Durham, N. C.

Estevez, M., 1972, Estudio preliminar sobre la biología de dos especies alopátricas de cangrejos brachyrhyncha del Pacifico Colombiano, *Bol. Museo Mar* (U. Bogota Jorge Tadeo Lozano, Fac. Cienc. Mar.), 4: 1-17.

Fenchel, T., 1970, Studies on the Decomposition of Organic Detritus Derived from the Turtle Grass *Thalassia testudinum, Limnol. Oceanogr.*, 15: 14-20.

Garth, J. S., and W. Stephenson, 1966, Brachyura of the Pacific Coast of America Brachyrhyncha: Portunidae, *Allan Hancock Monogr. Mar. Biol.*, 1: 1-154.

Gibbs, R. J., 1970, Circulation in the Amazon River Estuary and Adjacent Atlantic Ocean, *J. Mar. Res.*, 28: 113-123.

Holthuis, L. B., 1959, The Crustacea Decapoda of Suriname (Dutch Guiana), *Zool. Verh. (Leiden)*, 44: 1-296.

——, 1969, *Portunus binoculus*, N. Sp., A New Deep-Water Swimming Crab from the Caribbean Region (Crustacea, Decapoda, Brachyura), *Bull. Mar. Sci.*, 19: 409-427.

Humes, A. G., 1941, Notes on *Octolasmis mülleri* (Coker), A Barnacle Commensal on Crabs, *Trans. Am. Microsc. Soc.*, 60: 101-103.

MacArthur, R. H., 1972, *Geographical Ecology: Patterns in the Distribution of Species*, Harper and Row Publishers, Inc., New York.

Norse, E. A., 1975, *The Ecology of Blue Crabs, Genus* Callinectes *(Brachyura: Portunidae) in the Caribbean*, Ph.D. thesis, University of Southern California.

——, 1977, Aspects of the Zoogeographic Distribution of *Callinectes (Brachyura: Portunidae), Bull. Mar. Sci.*, 27: 440-447.

——, and M. Estevez, 1977, Studies on Portunid Crabs from the Eastern Pacific. I. Zonation Along Environmental Stress Gradients from the Coast of Colombia, *Mar. Biol.*, 40: 365-373.

——, and V. Fox-Norse, 1977, Studies on Portunid Crabs from the Eastern Pacific. II. Significance of the Unusual Distribution of *Euphylax dovii, Mar. Biol.*, 40: 374-376.

Odum, W. E., and E. J. Heald, 1972, Trophic Analyses of An Estuarine Mangrove Community, *Bull. Mar. Sci.*, 22: 671-738.

Randall, J. E., 1967, Food Habits of Reef Fishes of the West Indies, *Stud. Trop. Oceanogr.*, 5: 665-847.

Schoener, T. W., 1969, Size Patterns in West Indian *Anolis* Lizards. I. Size and Species Diversity, *Syst. Zool.*, 18: 368-401.

Stephenson, W., 1962, Evolution and Ecology of Portunid Crabs, with Especial Reference to Australian Species, in *The Evolution of Living Organisms*, G. W. Leeper (Ed.), pp. 311-327, Melbourne University Press, Melbourne, Victoria, Australia.

Tagatz, M. E., 1968, Biology of the Blue Crab *Callinectes sapidus* Rathbun in the St. Johns River, Florida, *Fish. Bull.*, 67: 17-33.

Taissoun, E., 1972, Estudio comparativo, taxonómico y ecológico entre los cangrejos (Dec. Brachyura. Portunidae), *Callinectes maracaiboensis* (nueva especie), *C. bocourti* (A. Milne Edwards) y *C. rathbunae* (Contreras) en el Golfo de Venezuela, Lago de Maracaibo y Golfo de Mexico, *Bol. Cent. Invest. Biol., Univ. Zulia*, 6: 1-45.

Terborgh, J., 1971, Distribution on Environmental Gradients: Theory and a Preliminary Interpretation of the Distributional Patterns in the Avifauna of the Cordillera Vilcabamba, Peru, *Ecology*, 52: 23-40.

Walker, G., 1974, The Occurrence, Distribution and Attachment of the Pedunculate Barnacle *Octolasmis müller*i (Coker) on the Gills of Crabs, Particularly the Blue Crab, *Callinectes sapidus*, *Biol. Bull. (Woods Hole, Mass.)*, 147: 678-689.

Whittaker, R. H., 1967, Gradient Analysis of Vegetation, *Biol. Rev. Cambridge Philos. Soc.*, 42: 207-264.

Williams, A. B., 1965, Marine Decapod Crustaceans of the Carolinas, *Fish. Bull.*, 65: 1-298.

——, 1966, The Western Atlantic Swimming Crabs *Callinectes ornatus*, *C. danae* and a New Related Species (Decapoda, Portunidae), *Tulane Stud. Zool. Bot.*, 13: 83-93.

——, 1974, The Swimming Crabs of the Genus *Callinectes* (Decapoda: Portunidae), *Fish. Bull.*, 72: 685-798.

EFFECTS OF FLUCTUATING FLOW RATES AND WATER LEVELS ON CHIRONOMIDS: DIRECT AND INDIRECT ALTERATIONS OF HABITAT STABILITY

ALAN P. COVICH, WILLIAM D. SHEPARD, ELIZABETH A. BERGEY, and CARYN S. CARPENTER
Department of Zoology, University of Oklahoma, Norman, Oklahoma

ABSTRACT

Regulation of stream flow-through water storage by hydroelectric dams and flood-control impoundments may alter benthic macroinvertebrate communities that are adapted to distinct seasonal shifts in water levels. Isolating these changes in community structure from changes caused by alterations in water quality is not possible currently, because data on effects of fluctuating flow in uncontaminated streams are infrequently reported. Studies from February through August 1977 on the Grand River and two of its tributaries in Oklahoma were conducted at seven stations. Of the 122 taxa of macroinvertebrates collected, 24 were genera of chironomids. *Glyptotendipes* dominated most locations in terms of density and biomass. Fluctuations in water level and water flow resulting from reservoir discharges were found to influence chironomid populations in both the discharge-receiving river and its two tributaries. Decreased density, biomass, and numbers of genera characterized the most intensely fluctuating sites. Indirect effects on substrate size, composition, and movement, as well as direct washing out of organisms from the substrate, may have caused the relative declines in abundance.

Many studies of macroinvertebrate community structure emphasize analyses of changing population densities and species diversities because these biological changes are known to reflect long-term changes in many physical and chemical parameters (Cummins, 1975; Edwards, Hughes, and Read, 1975; Gaufin, 1973; Goodnight, 1973; Hynes, 1970; Isom, 1971). Despite the recognized importance of fluctuations in flow rates and water levels for "catastrophic drift" (Waters, 1972) of aquatic invertebrates, data on fluctuations of natural lakes and streams are infrequently reported. Thus base-line data for evaluating potential natural stresses in uncontaminated, unmanaged watersheds are lacking. Although ranges and mean discharge rates are usually reported in studies of managed reservoirs, the frequency of the fluctuations and their impacts on benthic macroinvertebrates in downstream waters is often not monitored. The concern expressed by Wolman (1971) is still unresolved:

> Many measurements of biological effects are done during low and summer flows where measurement is easy, organisms often flourish, and concentrations of various substances in the flow are high. The effect of winter flow . . . and the special significance to the flora and fauna of periodic floods are not well documented. Significantly, however, among the most common trends in river management is the progressive regulation of flow through the provision of storage. Conceivably regulation rather than pollution alone may have the most far-reaching effects on the character of many river systems . . . .

Several investigators have called attention to some direct and indirect effects of flow rates and water-level changes on benthic macroinvertebrates (Baxter, 1977; Cairns et al., 1971; Trotsky and Gregory, 1974; Ward, 1976). For example, Neel (1963) reported that "reduction of winter stages below normal unregulated levels is common and often reduces the stream's carrying capacity for many forms of life . . . . Daily fluctuations in reservoir releases, which are often occasioned by power peaking, discourage littoral stream life and cause many readaptations." Minshall and Winger (1968) noted that "irrigation and reservoir management practices tend to cause periodic fluctuation on local stream levels, particularly during the summer months, when stream flows are already minimal." Ward and Short (1978) compared benthic communities from streams with constant and widely fluctuating flow regimes in the Rocky Mountains. They found that low standing-crop biomass values were associated with pronounced water-level fluctuations and that streams with regulated water flows had very different species compositions and lower species diversities than streams with natural flow regimes. In this report, we discuss the effects of varied flow rates and water levels on the chironomid community downstream from a large reservoir during a 7-month period.

STUDY AREA

Our seven sample stations (Fig. 1) were in the Grand (Neosho) River drainage basin. The watershed, located in Mayes County, Oklahoma, includes two tributaries (Pryor Creek, a permanent

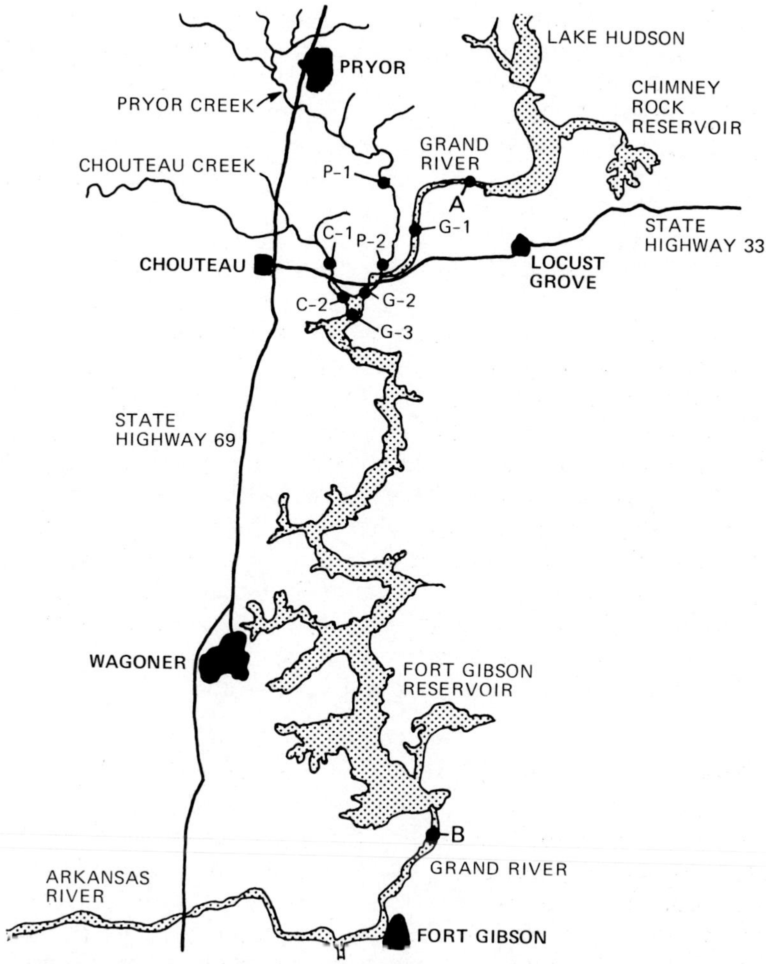

Fig. 1 Map of study area showing locations of seven sampling stations in the Grand River drainage basin, Mayes County, Oklahoma. U. S. Geological Survey gauging stations are located at two sites (A and B) along the Grand River. During period of maximal water storage in Fort Gibson Reservoir, water flows upstream in Chouteau Creek (C-1 and C-2) and Pryor Creek (P-1 and P-2), and the flow rate on the Grand River is slow. During periods of water release from both Fort Gibson Reservoir and Lake Hudson, downstream flow rates rapidly increase at all stations. Station G-1 is subject to frequent daily fluctuations in water level and flow rates as a result of releases from Lake Hudson.

stream, and Chouteau Creek, an intermittent stream) of the Grand River, which flows south into the Arkansas River. Upstream from the study area, the Grand River is impounded by Lake Hudson and, downstream, by Fort Gibson Reservoir. Because the reservoir system is used to produce hydroelectric power (Lake Hudson generates 100 MW, and Fort Gibson produces 45 MW), a large amount of storage capacity is needed. The major river flow occurs in winter and spring, whereas the peak power demand occurs during the summer months (Fredrich and Beard, 1975).

Most of the drainage area is used for pasture (~50%) and rangeland (~10%). Forest covers less than 25%; cropland is ~10%, and the remaining 5% is urbanized. Heavy rainfall is followed by rapid runoff because natural vegetative cover is sparse.

The U. S. Geological Survey collects data on mean daily discharge 4.0 km downstream from Lake Hudson (Fig. 1, point A) and 1.8 km downstream from Fort Gibson Reservoir (Fig. 1, point B). From March through August 1977, the maximum discharge occurred in late June below Lake Hudson and in early July below Fort Gibson Reservoir (Fig. 2). During periods when releases from Fort Gibson did not match those from Lake Hudson and other inflows, we observed upstream flow in Pryor and Chouteau creeks. Although we made no attempt to measure flow rate, we did note rapid increases in water level (up to 2 m at station G-1) after managed releases from upstream and downstream reservoirs.

METHODS

Beginning in February 1977, three prelabeled Hester—Dendy samplers (Hester and Dendy, 1962) were placed at each of the seven sample stations. Samplers were suspended ~1 m below floats tied to emergent trees and stumps. After a period of approximately 1 month, allowed for colonization by invertebrates, the samplers were removed, and another set was exposed. The Hester—Dendy samplers were carefully raised, placed immediately into plastic bags, and washed with 70% isopropyl alcohol to preserve the specimens.

Organisms were sorted under a dissecting microscope and identified to the generic level. The techniques described by Mason (1973), Bryce and Hobart (1972), and Beck (1975) were used for mounting head capsules for microscopic identification. Specimens were deposited in the permanent collection of the Stovall Museum of Science and History, University of Oklahoma.

Results were expressed as density (number of individuals per square meter of sampler), biomass (total weight of each taxon

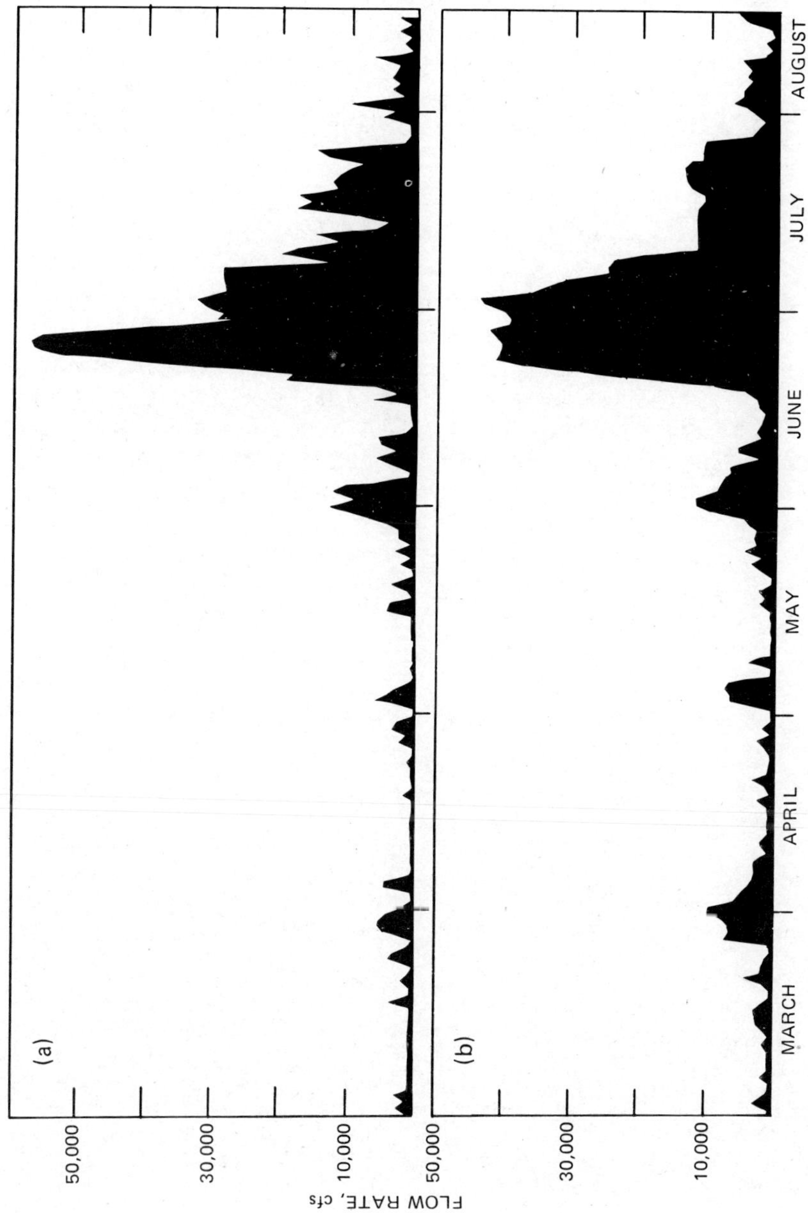

Fig. 2 Fluctuations in discharge at two gauging stations in the Grand River watershed (data from USGS) from March through early August 1977.

per sampler), and diversity. Diversity was measured by the Shannon—Weaver index ($\bar{d}$) as described by Wilhm (1972) and Weber (1973). Pupae were not used in diversity calculations because of difficulties in identification. Biomass was determined by weighing ovendried (4 hr at 100°C) specimens on an electrobalance (Cahn Instruments, model G). Measured biomass was an underestimation because head capsules were removed for identification before drying and weighing.

RESULTS

Of the 122 taxa of macroinvertebrates collected, 64 were insects. Twenty-four genera in three subfamilies of the Chironomidae were collected. The most represented subfamily was the Chironominae, with 15 genera. Of these 15, ten were very widespread, being found at all seven sample stations. In contrast, five genera were found at only one station (Table 1). Station C-2 had the highest number of genera; rarer genera were restricted to one or two stations on Pryor and Chouteau creeks.

The dominant chironomid, in terms of both density and biomass, was *Glyptotendipes*, which comprised over half the chironomids collected throughout the sampling period, except at station G-1 (Fig. 3). *Dicrotendipes* was the second most abundant genus at five of the sampling stations, but *Ablabesmyia* was more common at the Pryor Creek stations. *Glyptotendipes* and *Dicrotendipes* are both in the subfamily Chironominae, and *Ablabesmyia* is in the subfamily Tanypodinae.

Station G-1 showed the greatest increase in water level and flow rate when water was released from Lake Hudson because a low-water dam is located immediately below this station and the river channel is constricted. The density and biomass values were consistently low at station G-1 (Figs. 4 and 5) in comparison with other stations on the Grand River, especially during May and June, when discharge rates and water-level fluctuations were maximal. Station G-2, which was somewhat sheltered, being located near an island among emergent dead trees, frequently showed a higher chironomid density than did the more exposed stations, G-1 and G-3. Station G-2 also had the highest biomass values of any Grand River station in early June before maximal discharge (Fig. 4).

Although the chironomid densities of the Chouteau Creek stations (C-1 and C-2) were higher than those of Pryor Creek stations (P-1 and P-2), the patterns of chironomid density throughout the sampling period were quite similar for the two tributaries (Figs. 6

TABLE 1

TOTAL GENERIC DISTRIBUTION AND ABUNDANCE PER STATION*

	Station						
	G-1	G-2	G-3	C-1	C-2	P-1	P-2
Glyptotendipes	1	1	1	1	1	1	1
Dicrotendipes	1	1	1	1	1	2	2
Ablabesmyia	3	3	3	3	1	1	1
Polypedilum	3	3	3	4	3	3	3
Procladius	4	3	3	3	3	3	3
Psectrocladius	3	3	3	3	3	4	3
Tribelos	3	3	3	3	3	4	3
Chironomus	3	3	4	3	4	3	3
Micropsectra	3	3	3	4	4	4	3
Orthocladius	4	3	4	4	3	4	3
Cricotopus	3	3	3		3		3
Parachironomus	4	4	4	3	4	4	4
Trissocladius	4	3		4		3	3
Cryptochironomus	4	4	4	4	3	4	
Rheotanytarsus				4	4	3	4
Einfeldia		4	4		4		4
Metriocnemus		4		3	4		
Pseudochironomus					4		4
Phaenopsectra				4			4
Clinotanypus					4		
Kiefferulus				4			
Endochironomus					4		
Goeldichironomus					4		
Trichocladius					4		
Total genera	14	16	14	17	21	14	17
Total samples	12	13	12	11	11	12	12

*Dominance classes, based on abundance in total samples: 1, dominant (1000+); 2, abundant (500 to 1000); 3, common (50 to 500); 4, uncommon (1 to 50).

and 7). The upstream sites (C-1 and P-1) had lower densities than those downstream (C-2 and P-2). Both density and biomass reached their highest peaks in June at station C-2 (Figs. 5 and 6), where the downstream flow was minimal and the stream was very broad and shallow. At all times density and biomass values were relatively high at C-1 and C-2. No sampling was conducted at Chouteau or Pryor creeks in August because time for sorting the samples collected earlier became limited.

Another important difference among the stations was the month of maximum density. Maximum density peaks occurred in May at

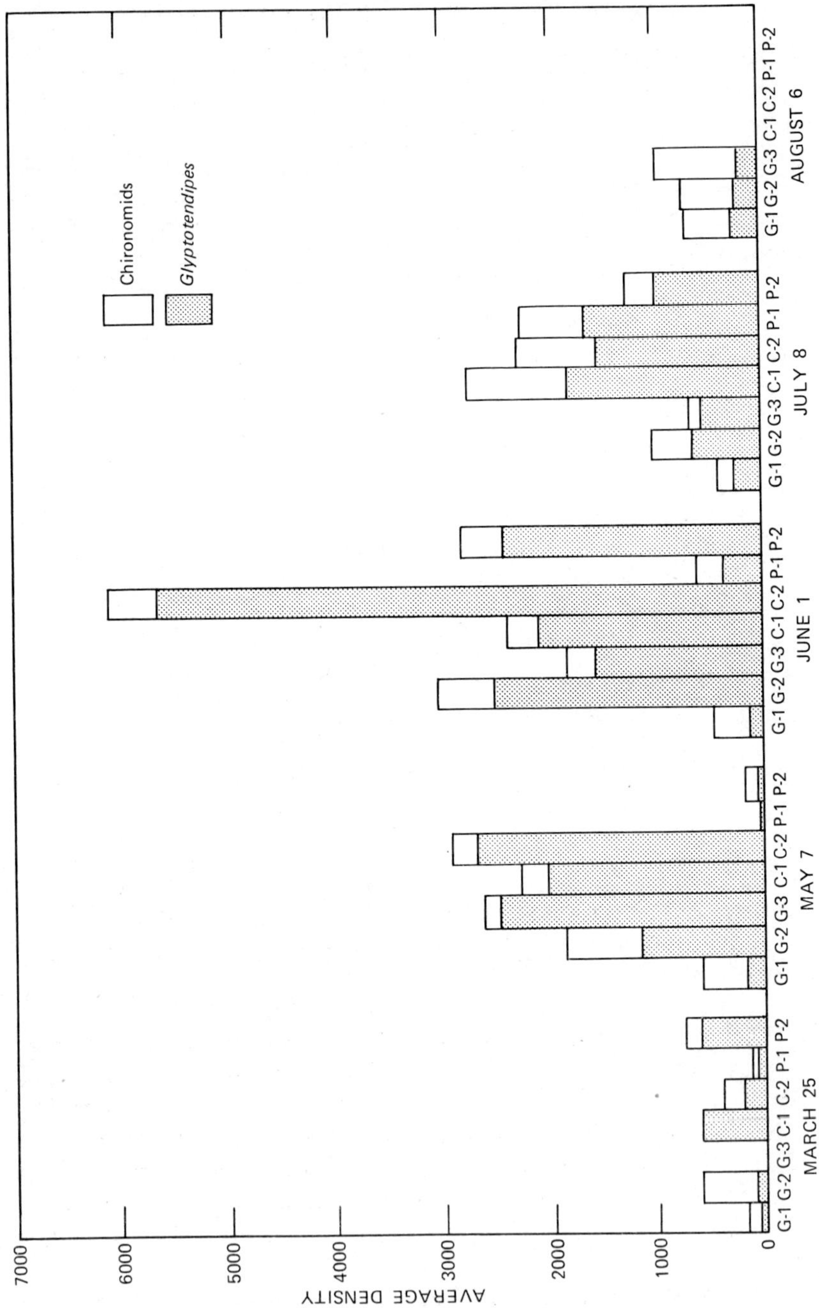

Fig. 3 Average densities of *Glyptotendipes* and all other chironomids from each station.

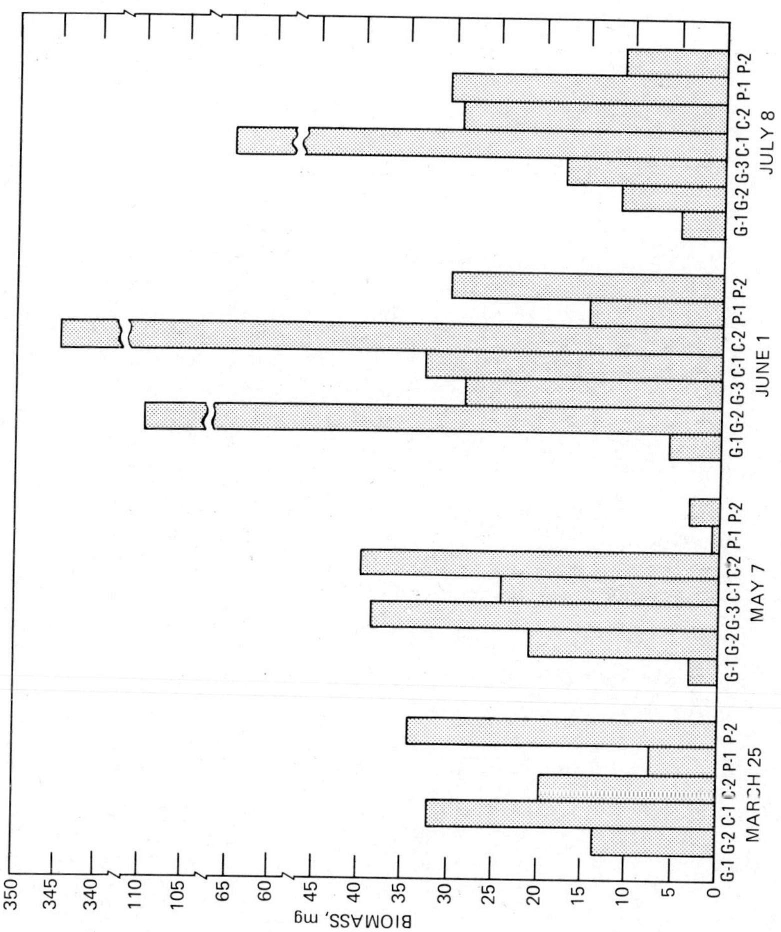

Fig. 4 Average biomass (milligrams dry weight) per triplicate Hester–Dendy, multiplate sampler during March, May, June, and July 1977. (No data were collected from station G-3 in March.)

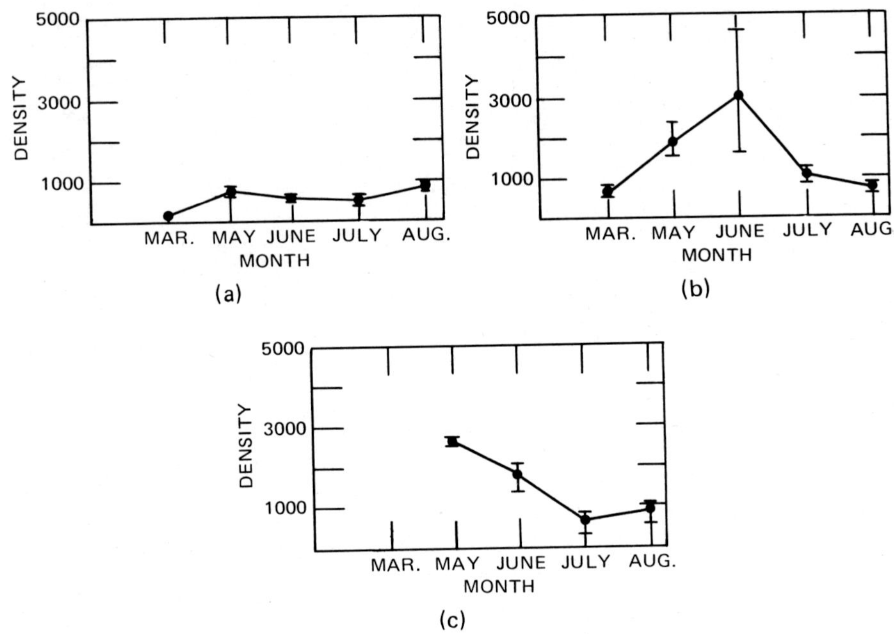

Fig. 5 Density values for stations G-1 (a), G-2 (b), and G-3 (c), based on averages of triplicate samples. Because of sampler losses only duplicate samples were available at G-1 and G-2 in August. Bars indicate range of data; black dots denote means.

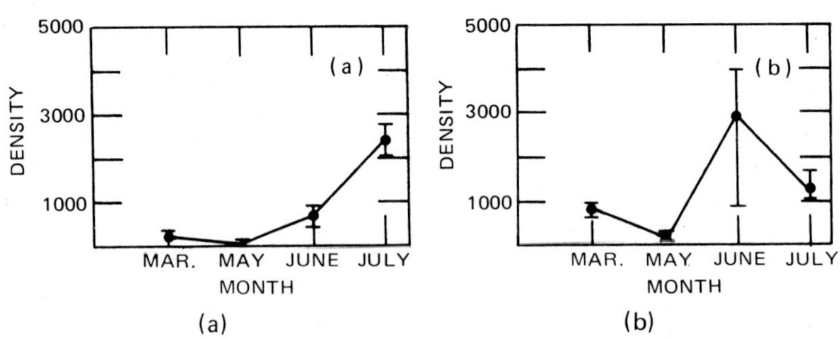

Fig. 6 Density values for stations P-1 (a) and P-2 (b), based on averages of triplicate samples at all stations.

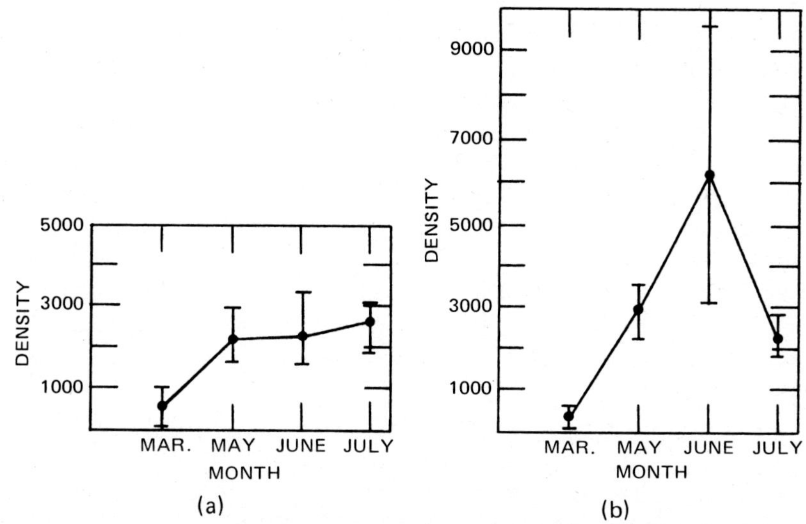

Fig. 7 Density values for stations C-1 (a) and C-2 (b), based on averages of triplicate samples, except for use of duplicates in March.

G-3; June at C-2, P-2, and G-2; July at C-1 and P-1; and August at G-1. Thus the station with relatively slower and more uniform flow had earlier peak densities.

Chironomid diversity ($\bar{d}$) was relatively low throughout the sampling period; values ranged from 0.34 to 2.67. Only at station G-1 did diversity remain relatively uniform, but these values were strongly influenced by the relatively low densities of chironomids collected at this station.

The patterns of generic distributions resemble those of chironomid density. The exposed Grand River sites generally contained fewer taxa (G-1 and G-3 both had only 14 genera) than did the more protected station (G-2 had 16 genera) and most of the tributary stations. The stations on Chouteau Creek had high generic diversity, with the maximum number (21 genera) occurring at C-2. Again, as for most values of density and biomass, the numbers of taxa were lower at the upstream stations in both Pryor and Chouteau creeks.

DISCUSSION

The high density of *Glyptotendipes* at most sampling stations is not surprising when we consider the ecological distribution of the genus (Beck, 1977; Weber, 1973). Paine and Gaufin (1956) found *Glyptotendipes* to be typical of organically enriched streams, whereas Kimerle and Anderson (1970) reported that *G. barbipes* was the dominant species on waste-stabilization lagoons. In Oklahoma this genus occurs where summer oxygen levels are lowered by high rates of organic productivity, e.g., in Arbuckle Lake, as reported by Parrish and Wilhm (1978). Thus the genus appears to be well adapted to the relatively nutrient-rich slow-flowing water found in most of our sampling area. In contrast to four of our seven stations, Harrold (1978) found approximately equal numbers of *Glyptotendipes* and *Ablabesmyia* in his Hester—Dendy samplers in the Knife River in North Dakota.

Although the species of *Glyptotendipes* from the Grand River are not yet identified, at least three species *(G. libiferous, G. meridionalis,* and *G. barbipes)* are known to be opportunistic in their rapid invasion of new reservoirs (Aggus, 1971; Patterson and Fernando, 1969; 1970). Presumably the species that dominate the Grand River share this trait and are readily able to colonize samplers placed in habitats with fluctuating flow rates and water levels.

A general pattern emerges when the chironomid populations are compared with the relative degree of water-flow and water-level fluctuations at each of the seven sampling sites. First, under conditions of extreme fluctuation (i.e., at station G-1, where the Grand River is constricted), chironomid density, biomass, and numbers of genera present were all low and were relatively constant throughout our study period. A smaller degree of fluctuation occurs in a second group of stations. Stations G-3, which is on a relatively exposed, straight section of the Grand River, and C-1 and P-1, where the relative narrowness and shallowness of the two upstream tributaries make them more susceptible to water-level fluctuations than the downstream stations, have an intermediate level of chironomid densities. These densities also showed a successive series of changes (increasing in the tributaries and decreasing at G-3) during the sampling period. In addition, stations P-1 and C-1 have fewer genera than the downstream stations on the same streams.

The remaining stations (G-2, P-2, and C-2), which comprise the third group, have a smaller degree of water-flow and water-level fluctuations. The two downstream tributary stations (P-2 and C-2) are located where the streams meet the Grand River. Both areas are wide and fairly deep. Each of these stations has a high chironomid biomass and the highest densities and largest number of genera in each of their respective river or stream systems. Also, each has a June peak in chironomid density, indicating a possible minimal effect of washout by high discharge rates. High recruitment from hatching eggs in May and June, followed by late summer emergence of other cohorts of chironomids, would be characteristic of this pattern of changing density. The two downstream tributary stations (P-2 and C-2) appear to have a combination of both genera typical of the larger, slower flowing Grand River and genera typical of small, fast-flowing streams.

The relatively high biomass and density of chironomid larvae in most of our study area apparently result from a combination of relatively nutrient-rich, stable substrate and slow-flowing waters. It is difficult to distinguish natural fluctuations of flow rates and water levels from managed fluctuations in this reservoir system. Other investigators have noted significant stresses on chironomid populations as a result of both natural and regulated flow regimes. For example, Spence and Hynes (1971) concluded that, relative to stations farther downstream or upstream, increased numbers of chironomids result from greater availability of detritus and uniform flow below Shand Dam in the Grand River in Ontario. Unregulated streams may also be affected strongly by flow-rate and water-level fluctuations that occur after rapid runoff in foothill streams (Siegfried and Knight, 1977) or during flooding after heavy rainfall (Hoopes, 1974).

Various types of adaptations give distinct advantages to particular species in habitats with fluctuating waters. The end result is a high standing-crop biomass of only a few dominant species and a general pattern of low diversity. These frequent disruptions in the physical and chemical nature of the habitat are analogous to observations in plankton communities that high species diversity results from a "contemporaneous disequilibrium" of the open waters (Richerson, Armstrong, and Goldman, 1970). Benthic habitats exposed to intensely fluctuating waters are apparently characterized by low diversity, however, because of the high degree of seasonal and daily unpredictability in flow and because of the relative newness of these types of frequently exposed and inundated substrates. Fisher and LaVoy (1972) aptly termed these substrates freshwater "intertidal" biotopes and pointed out that a high degree of "pulse stability" is potentially attainable in some new habitats. Currently the managed freshwater "intertidal beaches" appear unstable in comparison with the naturally regulated streams and are dominated by a very restricted number of taxa. The dominant chironomids generally have high fecundity and are well adapted for dispersal both as larvae and as airborne adults. Thus they can be viewed as typical "r-selected" species (Baxter, 1977).

CONCLUSION

Fluctuations in water level and water flow resulting from reservoir discharge were found to influence the chironomid populations in both the discharge-receiving river and its tributaries (at least as far upstream as water backed up from a second reservoir pool). The greater the degree of fluctuations, the greater the effects appeared to be, so that decreasing density, biomass, and numbers of genera characterized the most intensely fluctuating sites.

A single chironomid genus, *Glyptotendipes*, clearly dominated all seven stations. In our study, however, we were unable to identify specific mechanisms responsible for this dominance. With the establishment of adequate base-line data, we should be able to follow community development in selected habitats and to learn a great deal about relationships between dominance and habitat stability.

REFERENCES

Aggus, L. R., 1971, Summer Benthos in Newly Flooded Areas of Beaver Reservoir During the Second and Third Years of Filling, 1965—1966, in *Reservoir Fisheries and Limnology*, G. E. Hall (Ed.), Special Publication No. 8, pp. 139-152, American Fisheries Society, Bethesda, Md.

Baxter, R. M., 1977, Environmental Effects of Dams and Impoundments, *Annu. Rev. Ecol. Systemat.*, 8: 255-283.

Beck, W. M., 1975, Chironomidae, in *Keys to the Water Quality Indicative Organisms of the Southeastern United States*, F. K. Parrish (Ed.), Report EPA-657-695/5483, Environmental Protection Agency, GPO.

——, 1977, *Environmental Requirements and Pollution Tolerance of Common Freshwater Chironomidae*, Report EPA-600/4-77-024, Environmental Protection Agency, GPO.

Bryce, D., and A. Hobart, 1972, The Biology and Identification of Larvae of the Chironomidae (Diptera), *Entomol. Gaz.*, 23: 175-217.

Cairns, J., J. S. Crossman, K. L. Dickson, and E. E. Herricks, 1971, The Recovery of Damaged Ecosystems, *Assoc. Southeast. Biologists Bull.*, 18: 79-106.

Cummins, K. W., 1975, Macroinvertebrates, in *River Ecology*, B. A. Whitton (Ed.), pp. 170-198, University of California Press, Berkeley.

Edwards, R. W., B. D. Hughes, and M. W. Read, 1975, Biological Survey in the Detection and Assessment of Pollution, in *The Ecology of Resource Degradation and Renewal*, M. J. Chadwick and G. T. Goodman (Eds.), pp. 139-156, John Wiley & Sons, Inc., New York.

Fisher, S. G., and A. LaVoy, 1972, Differences in Littoral Fauna Due to Fluctuating Water Levels Below a Hydroelectric Dam, *J. Fish. Res. Board Can.*, 29: 1472-1476.

Fredrich, A. J., and L. R. Beard, 1975, Complementary Use of Hydro and Thermal Power, in *Water Management by the Electric Power Industry*, E. F. Gloyna, H. H. Woodson, and H. R. Drew (Eds.), pp. 65-74, University of Texas Press, Austin.

Gaufin, A. R., 1973, Use of Aquatic Invertebrates in the Assessment of Water Quality, in *Biological Methods for the Assessment of Water Quality*, J. Cairns and K. L. Dickson (Eds.), pp. 96-116, American Society for Testing and Materials, Philadelphia.

Goodnight, C. J., 1973, The Use of Aquatic Macroinvertebrates as Indicators of Stream Pollution, *Trans. Am. Microsc. Soc.*, 92: 1-13.

Harrold, J. F., Jr., 1978, Relation of Sample Variations to Plate Orientation in the Hester—Dendy Plate Sampler, *Prog. Fish-Cult.*, 40: 24-25.

Hester, F. E., and J. S. Dendy, 1962, A Multiple-Plate Sampler for Aquatic Macroinvertebrates, *Trans. Am. Fish. Soc.*, 91: 420-421.

Hoopes, R. L., 1974, Flooding, as a Result of Hurricane Agnes, and Its Effects on a Macrobenthic Community in an Infertile Headwater Stream in Central Pennsylvania, *Limnol. Oceanogr.*, 19: 853-857.

Hynes, H. B. N., 1970, *Ecology of Running Waters*, University of Toronto Press, Toronto, Ont.

Isom, B. G., 1971, Effects of Storage and Mainstream Reservoirs on Benthic Macroinvertebrates in the Tennessee Valley, in *Reservoir Fisheries and Limnology*, G. E. Hall (Ed.), Special Publication No. 8, pp. 179-191, American Fisheries Society, Bethesda, Md.

Kimerle, R. A., and N. H. Anderson, 1970, Production and Bioenergetic Role of the Midge *Glyptotendipes barbipes* (Staeger) in a Waste Stabilization Lagoon, *Limnol. Oceanogr.*, 16: 646-659.

Mason, W. T., 1973, *An Introduction to the Identification of Chironomid Larvae*, Environmental Protection Agency, Report EPA-758-495/1237, Environmental Protection Agency, GPO.

Minshall, G. W., and P. V. Winger, 1968, The Effect of Reduction in Stream Flow on Invertebrate Drift, *Ecology*, 49: 580-582.

Neel, J. K., 1963, The Impact of Reservoirs, in *Limnology in North America*, D. G. Frey (Ed.), pp. 575-593, University of Wisconsin Press, Madison.

Paine, G. H., and A. R. Gaufin, 1956, Aquatic Diptera as Indicators of Pollution in a Midwestern Stream, *Ohio J. Sci.*, 56: 291-304.

Parrish, J. H., and J. Wilhm, 1978, Relationship Between Physiochemical Conditions and the Distribution of Benthic Macroinvertebrates in Arbuckle Lake, *Southwest. Nat.*, 23: 135-144.

Patterson, C. G., and C. H. Fernando, 1969, The Macro-Invertebrate Colonization of a Small Reservoir in Eastern Canada, *Verh. Int. Verein. Limnol.*, 17: 126-136.

——, and C. H. Fernando, 1970, Benthic Fauna Colonization of a New Reservoir with Particular Reference to the Chironomidae, *J. Fish. Res. Board Can.*, 27: 213-232.

Richerson, P., R. Armstrong, and C. R. Goldman, 1970, Contemporaneous Disequilibrium, A New Hypothesis to Explain the Paradox of the Plankton, *Proc. Nat. Acad. Sci. U. S. A.*, 67: 1710-1714.

Siegfried, C. A., and A. W. Knight, 1977, The Effects of Washout in a Sierra Foothill Stream, *Am. Midl. Nat.*, 98: 200-206.

Spence, J. A., and H. B. N. Hynes, 1971, Differences in Benthos Upstream and Downstream of an Impoundment, *J. Fish. Res. Board Can.*, 28: 35-43.

Trotsky, H. M., and R. W. Gregory, 1974, The Effects of Water Flow Manipulation Below a Hydroelectric Power Dam on the Bottom Fauna of the Upper Kennebec River, Maine, *Trans. Am. Fish. Soc.*, 103: 318-324.

Ward, J. V., 1976, Effects of Flow Patterns Below Large Dams in Stream Benthos: A Review, in *Instream Flow Needs Symposium*, J. F. Orsborn and C. H. Allman (Eds.), Vol. II, pp. 235-253.

——, and R. A. Short, 1978, Macroinvertebrate Community Structure of Four Lotic Habitats in Colorado, U. S. A., *Verh. Int. Verein. Limnol.*, in press.

Waters, T. F., 1964, Recolonization of Denuded Stream Bottom Areas by Drift, *Trans. Am. Fish. Soc.*, 93: 311-315.

——, 1972, The Drift of Stream Insects, *Annu. Rev. Entomol.*, 17: 253-272.

Weber, C. I., 1973, Macroinvertebrates, in *Biological Field and Laboratory Methods for Measuring the Quality of Surface Waters and Effluents*, Report 670/4-73-001, pp. 1-38, Environmental Protection Agency, GPO.

Wilhm, J., 1972, Graphic and Mathematical Analyses of Biotic Communities in Polluted Streams, *Annu. Rev. Entomol.*, 17: 223-252.

Wolman, M. G., 1971, The Nation's Rivers, *Science*, 174: 905-918.

ENVIRONMENTAL IMPLICATIONS OF COAL-CONVERSION TECHNOLOGIES: ORGANIC CONTAMINANTS

C. W. GEHRS
Environmental Sciences Division, Oak Ridge National Laboratory,
Oak Ridge, Tennessee

ABSTRACT

The Department of Energy is currently supporting development of more than 20 major coal-conversion facilities, with many more in the exploratory and predesign stage. All coal-conversion processes utilize pyrolysis (destructive heating) to produce hydrocarbons enriched in their hydrogen-to-carbon ratio from the parent coal. This entails the use of high temperature, often high pressure, and a reducing atmosphere. The products are easier to handle than the parent coal and are usable in a greater variety of ways. The organic components may reach the aquatic environment, however, from product usage, through leaks, or in plant effluents. This paper briefly describes the conversion technologies, identifies the types of organic contaminants expected to be released, summarizes the literature on environmental effects, and outlines a research strategy for addressing the potential environmental ramifications of organic contaminants released by coal-conversion technologies.

The United States contains more than 50% of the world's known reserves of coal. Although 90% of the total reserves of fossil fuels in the United States are coal, only 20% of the energy produced results from the use of this fuel. This relatively small reliance on coal as an energy source is expected to change drastically over the next few decades. Consumption is expected to increase from 600 tonnes of coal per year in 1975 to double that amount in the early 1980s and potentially to quadruple by the year 2020. Although coal is being considered a major source of energy over the next half century, its use is not without potential health and environmental ramifications. As early as 1750, Sir Percival Pott reported a high incidence of scrotal cancer among chimney sweeps in London (Henry, 1946). He hypothesized that the disease was associated with particulate material clinging to the clothing and the skin of the individuals. Butlin (1892) strengthened the hypothesis when he found a much higher rate of scrotal cancer in chimney sweeps working in countries consuming coal than in countries where wood was the primary fossil fuel. Although coal has been used for several centuries, it was not until the 1950s that a large-scale epidemiological investigation provided valuable data (MacMahon, Pugh, and Ipsen, 1960). In 1952, smog over London resulted in several thousands of deaths related to emphysema. This smog was correlated with coal consumption.

Unfortunately, quantitative and qualitative determinations of the environmental responses to coal are more difficult. Among the few exceptions have been several investigations relating acid mine drainage to decreases in fish and benthic communities downstream from the source (Butler et al., 1973; Huckabee, Goodyear, and Jones, 1975). The best-known incidences of environmental insult resulting from coal consumption on a regional scale is acid precipitation. In this respect, the Scandinavian countries offer the greatest documentation. Decreases in productivity of forests (Tamm, 1976) and of aquatic environments (Schofield, 1976) were associated with decreases in the pH of rainfall. The changes in pH of rainfall were directly associated with coal consumption in the Ruhr Valley of central Europe and in England. Several recent studies in the eastern United States have postulated similar responses. For example, Shriner, McLaughlin, and Baes (1977) found acid precipitation with pH as low as 3.2 in certain episodes in eastern Tennessee. Since buffering capacities of soils and waters in the eastern United States are similar to those in the Scandinavian countries, it is feasible that problems similar to those identified in the Scandinavian countries have already occurred in the eastern United States.

COAL-CONVERSION TECHNOLOGIES

The United States is committed to the development of coal-conversion technologies, partly as a result of concern for acid precipitation but also from a necessity to produce usable and transportable fuels to replace dwindling supplies of crude oil and natural gas. Coal conversion is a generic term referring to any of 20 some processes whereby coal is changed into a liquid, gaseous, or solid product relatively free of sulfur and inorganic particulate materials.

To produce usable gaseous or liquid hydrocarbons from coal, we must increase the atomic ratio of hydrogen to carbon (Fig. 1), and

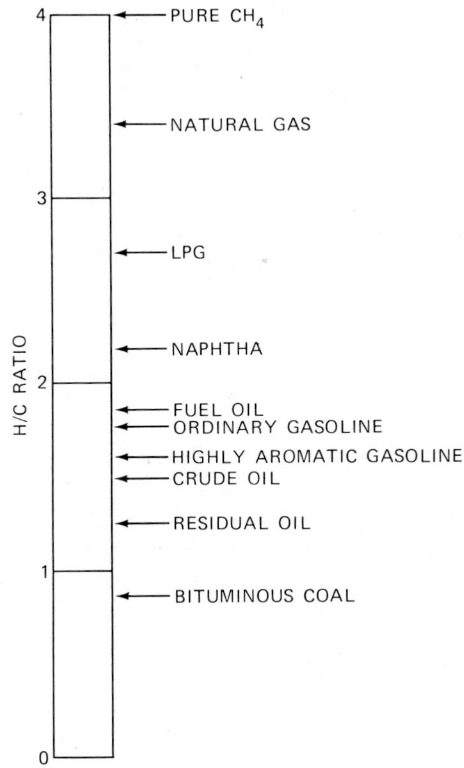

Fig. 1 Hydrogen-to-carbon ratios for various hydrocarbon fuels. (Data from Wiser, 1973.)

the impurities and potential pollutants present in the feed coal must be removed. The hydrogen-to-carbon ratio can be enriched in several ways, e.g., by removing the volatile hydrogen-rich organic content through pyrolysis, adding extra hydrogen to the carbon contained in the coal, or producing CO and H_2 from coal and catalytically reacting the molecules to form methane and/or higher hydrocarbons (Richmond, Reichle, and Gehrs, 1976). In general, the greater the enrichment, the higher are the energy costs.

Although more than 20 processes for converting coal are being developed (Energy Research and Development Administration, 1977), sufficient similarities exist to list only five general types (Fig. 2), including two for gasification and three for liquefaction. Coal gasification can produce either low- or high-Btu gas. Low-Btu gasification produces a fuel usable at a stationary source, such as a power-generating facility. Processes for converting coal to low-Btu

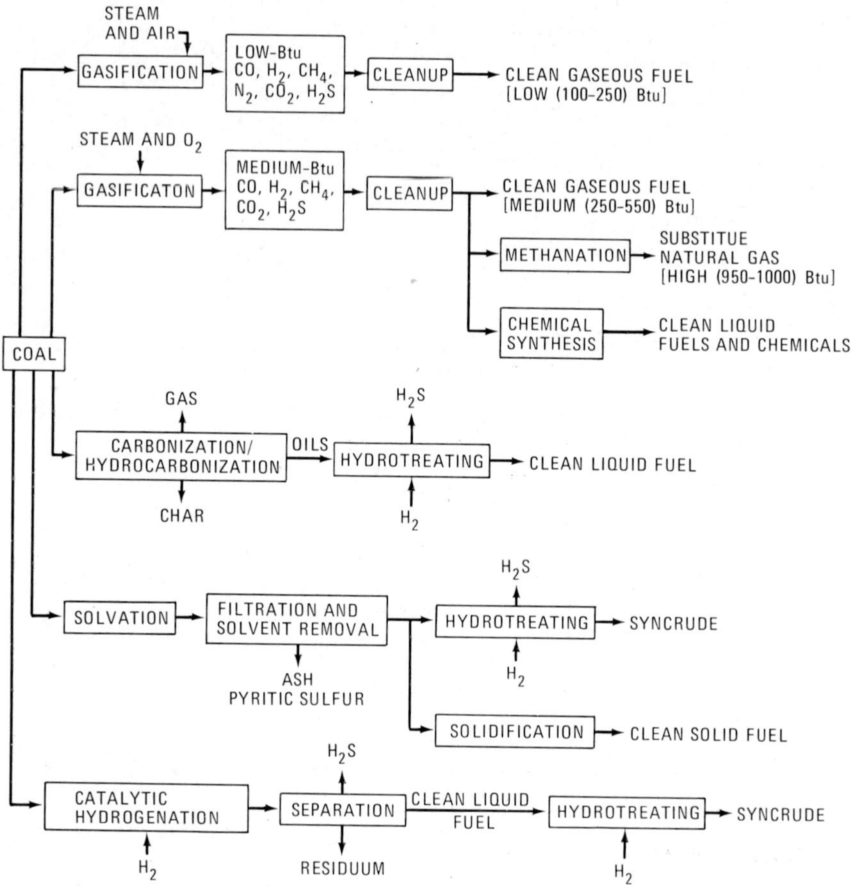

Fig. 2 Schematic of five basic approaches for converting coal to gaseous or liquid fuels. (Data from Richmond, Reichle, and Gehrs, 1976.)

fuel are currently being used in several countries. In the United States attempts are being made to demonstrate the use of caking coals (the major type in the eastern United States) and high temperature. In contrast, high-Btu gas can be produced from gasification with oxygen and upgrading of the product. This allows transport of the gas to diverse sources and ultimate substitution for natural gas as a home-heating source.

Coal liquefaction processes are in the early exploratory stages. Three general approaches are being pursued, pyrolysis or carbonization, dissolution, and hydrogenation. All rely on high temperature, often high pressure, and a reducing atmosphere to produce liquid from coal. These are also the three major factors in the production of potentially hazardous organic substances during the conversion process. The products of coal liquefaction can be used directly as boiler fuel in stationary power-generating stations or upgraded and refined for use as gasoline or petroleum feedstock.

The primary source of contaminants reaching the aquatic environment will be aqueous effluents, which will be enriched in organic compounds through the process of product hydrotreating (Fig. 2). Leachates arising from the ultimate disposal of solid wastes can also contaminate water (Gehrs, 1978), as can accidental spills and disruption of plant operation.

No large-scale coal-conversion facilities currently exist in the United States, and the smaller units that are now operating do so intermittently (and, thus, have atypical effluents) with no or little waste treatment. Although these facts pose problems for ecologists attempting to determine potential contamination from coal conversion, there is also a unique opportunity to aid in developing an environmentally acceptable technology(ies) if the ecologist produces relevant, timely data.

HISTORIC DATA ON ENVIRONMENTAL HAZARDS OF COAL CONVERSION

Although no large coal-conversion facilities are presently operating in the United States, Union Carbide operated a 300-ton/day experimental facility at Institute, W. V., in the early 1950s. The results of this experiment, which included both medical observations and animal testing, are reported in a series of papers by Sexton (1960) and Weil and Condra (1960). They suggest the need for health and environmental concerns related to coal conversion. Despite the use of an aggressive industrial hygiene program, the incidence of skin cancer among the approximately 364 workers was between 16 and 37 times that reported in the general population (Sexton, 1960).

Whole animal studies on various process and product streams also revealed carcinogenic potential (Weil and Condra, 1960). These results showed that most of the carcinogenic activity was found in the heavy oils (boiling point $\geqslant 260°C$). More recently, Bingham (1975) found the products of three liquefaction processes currently under development to be carcinogenic in animals, and Hueper (1956) identified the heavy ends as more active than the light ends. Caution must be taken in interpreting these data, however, since the products were from small facilities that might not be indicative of commercial operations. Nevertheless, sufficient data exist to cause concern.

Unfortunately, ecological data are lacking with respect to either the products or the effluents from the Carbide facility or from the surrounding environment.

The United States, through the Department of Energy, is actively supporting the development of more than 20 processes for converting coal to liquid, gaseous, or solid fuels. Sufficient data exist to necessitate research concerning the possible health and environmental consequences of these technologies. Several unique characteristics of the level of development of the processes limit the types of research that can be conducted (Gehrs and Wells, 1977). The small size of the existing facilities, coupled with changes in mode of operating, makes conducting field studies around such units relatively useless. Furthermore, evaluation of effluents from a specific plant is tenuous because of lack of effluent treatment, little steady-state operation, and basic questions concerning similarity of effluent composition after process scale up. Engineers are unsure that a 20,000-ton/day facility will produce the same effluents as a scaled-down replica using only 50 tons/day of coal. Finally, the chemical composition of effluent streams is extremely complex, with hundreds, perhaps thousands, of compounds in each effluent (Shults, 1976). Developing meaningful data on the ecological hazards of these effluents is further complicated by the myriad of potential interactions that may occur between the various chemical compounds (i.e., synergism, antagonism, etc.). These unique attributes of currently existing coal-conversion facilities are not listed here to suggest that environmental research is futile; they are presented rather to emphasize that developing a meaningful environmental research program requires awareness and understanding of these limitations and a delineation of the purpose or goals of the research. Ecological research related to coal conversion has three major goals:

1. To aid the development of amenable coal-conversion technologies
2. To develop a data base for assessing the effects of coal-conversion technologies on biological communities
3. To determine the form, source, and potential concentrations of trace contaminants that may reach man and other biota from environmental releases of coal-conversion effluents

The approach chosen is determined by the specific goal of the research activity.

The remainder of this paper addresses the scientific approaches which have been used to assess the hazards of a particular technology and which can be used to address coal conversion. It includes a description and a discussion of two major approaches that might be taken (including their strengths and weaknesses), a summary of the environmental data available on the various organic components of coal-conversion effluents, and a suggested approach that we have adopted for assessing the environmental implications of coal-conversion effluents.

Historically, the potential environmental hazards of aqueous effluents have been evaluated by one of two approaches: (1) identifying and subsequently evaluating all the various effluent components or (2) testing the whole effluent.

The first approach produces the data necessary to attain the second and third major goals of ecological research. The types of research involved in this approach include, in addition to determining sublethal effects and the mechanisms of effects, determining the ultimate fate of a compound within the environment and the critical environmental pathways that lead to exposure of various biota. This type of research can be referred to as mechanistic. It is essential in developing predictive capabilities. Unfortunately, when the number of compounds in an effluent stream is large, identifying all of them and testing each individually and in concert becomes too costly and time consuming. Such is the case in coal conversion, where perhaps thousands of compounds may be present in a particular effluent stream (Shults, 1976).

The second approach, testing the whole effluent, enables us to determine the acute toxicity of various effluent streams. Such data reveal the relative toxicity of the streams although they do not allow determination of what component(s) is(are) most toxic, and this information is extremely important in assessing treatability and developing control technology. This approach is particularly useful for coal conversion where, as mentioned earlier, a great many processes are currently under development. Thus, if he can produce and deliver relevant data within the time frame the developing technology follows, the ecologist has a unique opportunity to help determine the ultimate form of the technology(ies). For example, if decisions on whether to pursue a technology are to be made in early 1980, then ecological-effects data pertaining to the technology must be developed before that time.

The toxic components of complex chemical mixtures (primarily petroleum and synthetic petroleum) have been determined in two ways. The most common method uses boiling point to separate a mixture into smaller, similar fractions. Moore and Dwyer (1974), reviewing the literature on the toxicity of aromatic hydrocarbons to

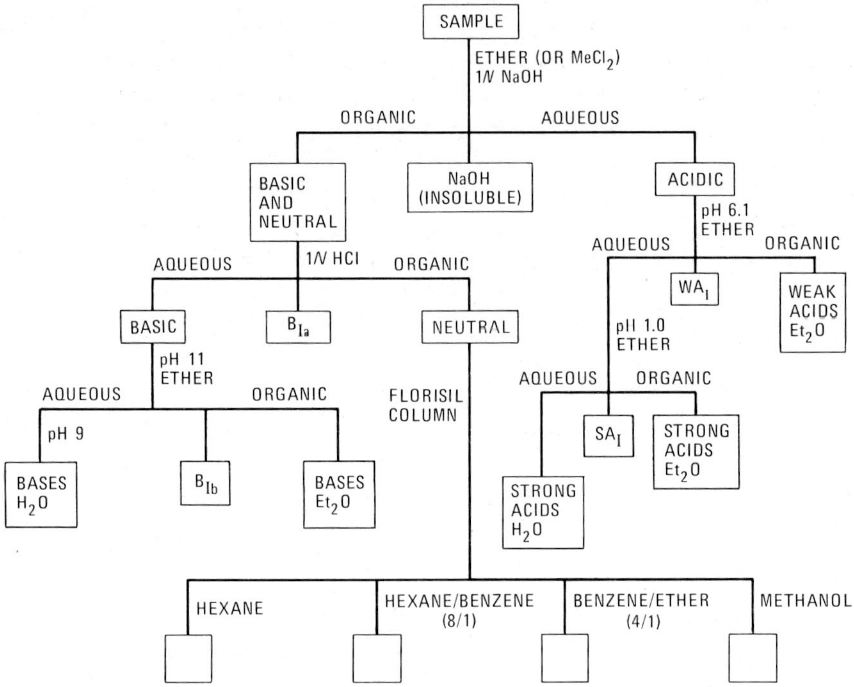

Fig. 3 Separation and fractionation process used to obtain materials for environmental testing. (Data from Rubin et al., 1976.)

aquatic organisms, found an inverse association between toxicity and boiling point. Legore (1974) concurred with this relationship and suggested that solubility was the primary factor.

More recent investigations (Gehrs, 1975; Parkhurst, Gehrs, and Rubin, 1978) have begun to separate complex mixtures on the basis of chemical similarities of compounds (Fig. 3). Rubin et al. (1976) adapted this chemical fractionation approach, which was developed for evaluating tobacco smoke, to aqueous samples. The scheme makes use of the differential solubility of various compounds to produce from three to more than 20 semidiscrete and reproducible fractions. Mutagenesis testing (Epler et al., 1978) has been used to identify the more active of these fractions, with subsequent subfractionation and testing allowing eventual determination of problematic compounds.

Parkhurst, Gehrs, and Rubin (1978) used a similar approach with *Daphnia magna* as the test species. They found (Table 1) that the neutral fraction, which comprised only 7% of the soluble materials in

TABLE 1

CONCENTRATION AND TOXICITY OF VARIOUS COMPONENTS IN AQUEOUS EFFLUENT FROM THE SRC PROCESS*

Fraction	% of total dissolved fractions of SRC effluent	48-hr LC_{50}, mg/liter	Fraction concentration (mg/liter) at 48-hr LC_{50} concentration		Contribution to whole effluent toxicity,† %	
			Reconstituted effluent	SRC effluent	Reconstituted effluent	SRC effluent
Acid	28.5	29.5	15.2	15.6	51.5	51.3
Base	3.0	45.8	1.6	1.7	3.5	3.5
Neutral	7.1	9.0	3.8	3.9	42.2	41.9
Inorganic	61.3	1000‡	32.7	33.6	3.0	3.3

*The effluents were stored for substantial time at room temperature. Consequently, results cannot be assumed directly applicable to the SRC process. The samples do, however, show the validity of the approach. Adapted from Parkhurst, Gehrs, and Rubin, 1978.

†The contribution of a fraction to the toxicity of the whole effluent is given by the expression:

$$\frac{Am/Ai}{(Am/Ai) + (Bm/Bi) + \ldots + (Zm/Zi)}$$

where A, B . . . , Z, are the fraction, i is the toxicity (LC_{50}) of the individual fraction, and m is the fraction concentration at the LC_{50} concentration of the whole effluent.

‡No mortality occurred at highest concentration in test series; 1000 mg/liter is assumed to be a very conservative estimate of the 48-hr LC_{50}.

the effluent, contributed 42% of the total toxicity. The acid fraction, which was almost 29% of the effluent, contributed about 52% of the toxicity. Preliminary characterization work showed naphthalenes, phenanthrene, and anthracene in the neutral fraction and phenols and cresols in the acid fraction. Such an approach provides useful information (identification of active components) to the design engineer and control technologist through a process of selective elimination.

Parkhurst, Gehrs, and Rubin (1978) also compared the toxicity of an unfractionated effluent with a reconstituted effluent (various fractions mixed together in same concentration as initial effluent). That they found no significant difference in results suggests the validity of fractionation and toxicity testing as an initial step in evaluating complex chemical mixtures.

Both approaches have several weaknesses, however. The identities and concentrations of the specific compounds being delivered are unknown. Interactions are masked. Hence, developing predictive information from these types of studies is not possible. Using chemical fractionation rather than boiling point to separate complex mixtures appears to have several advantages, the most important being that separation is into similar groups (phenolics, aromatics, etc.) rather than across groups by boiling point. By combining the fractionation–subfractionation scheme with subsequent molecular weight profiles, we can obtain a rough estimate of the amount of the various classes, as well as the distribution (by weight) within the various classes. Using the data developed by Herbes, Southworth, and Gehrs (1977) (discussed in the following section) enables initial prediction of both potential effects and causative agents.

Neither of these screening approaches provides the information needed to understand mechanisms of effect, however, and, hence, neither permits predictive capabilities. Predictions can be made only by using specific compounds in studies of transport, fate, and effects.

LIQUID EFFLUENT COMPOSITION

No data are available on the exact composition of effluents that might be released from coal-conversion facilities. Herbes, Southworth, and Gehrs (1977) used data from the coking industry as a surrogate to estimate potential effluent components. They arranged the compounds into five groups on the basis of chemical structure and used available data on anticipated concentrations in liquid effluents, plus efficacy of waste-water treatment methods, to calculate expected concentrations released to the environment.

TABLE 2

CONCENTRATIONS AND WASTE-WATER TREATMENT EFFICIENCIES OF MAJOR CONSTITUENTS OF COAL-CONVERSION EFFLUENTS*

Constituent	Anticipated effluent concentration, mg/liter	Waste-water treatment removal efficiency, %	Expected levels in final effluent, mg/liter
Phenols	10^4	99.9+	<1
Aromatic amines	10^2–10^3	30–50	50–700
Monoaromatic hydrocarbons	10^1–10^2	90+	1–10
Thiophenes	10^1	?	1–10
Polycyclic hydrocarbons	10^{-1}–10^0	30–80	0.02–0.7

*Adapted from Herbes, Southworth, and Gehrs, 1977.

Identities of classes and their anticipated levels in untreated and treated wastes are given in Table 2. Phenols will be the predominant class of chemicals present in the untreated effluents, with levels as high as 6000 mg/liter being measured (Wender, 1975). Chemical stripping of most of the phenolics (to retrieve them for use as chemical feedstocks) will be followed by biological treatment, which will reduce levels to $\leq$1 mg/liter. Concentrations of unsubstituted aromatic hydrocarbons present in untreated waste waters will be limited primarily by water solubility, with the higher molecular weight polycyclic aromatic hydrocarbons constituting lesser amounts. Microbial degradation rates of aromatic hydrocarbons are inversely related to molecular weight as reflected in the estimated removal efficiencies of 90+% and 30 to 80% for monoaromatic and polycyclic aromatic hydrocarbons, respectively (Table 2). Incorporating nitrogen (aromatic amines) or sulfur (thiophenes) into the basic aromatic ring structure tends to increase solubility and decrease microbial degradation rates as compared with unsubstituted aromatics. Thus these two classes of compounds are expected to comprise a significant percentage of the total organics contained in treated, aqueous, coal-conversion effluents.

POTENTIAL TOXICITY OF COAL-CONVERSION EFFLUENT CLASSES

The toxicity of phenolics to aquatic life is better defined and understood than is the toxicity of other organic classes expected to be present in aqueous effluents. Paralysis, loss of equilibrium, increased respiration rates, and increased swimming rates are responses observed in fish exposed to phenols. Trout and salmon are killed by phenol at levels of 3 to 5 mg/liter, but rough fish appear 5 to 10 times less sensitive (Albersmeyer and Erichsen, 1959). Phenolic toxicity is inversely related to dissolved oxygen content (Anonymous, 1961) and water hardness (Anonymous, 1962) and directly related to water temperature (Bucksteeg, Thiele, and Stoltzel, 1955). Available data reveal that fish are more sensitive to phenolics than are other aquatic organisms (McKee and Wolf, 1963). Although phenolics make up the largest fraction of organics anticipated in untreated aqueous wastes, the well-developed methodologies for their removal and degradation, coupled with regulations regarding quantities that can be released to surface waters, suggest that they will not be a significant hazard to aquatic environments.

Data on the toxicity of phenolics to aquatic life are plentiful, whereas little information on arylamines is available. Arylamines constitute the second largest quantity (class) of organics in untreated effluents. After treatment, undiluted aqueous effluents might contain $\geqslant$50 mg/liter of arylamines, constituting more than 90% of the organic load arising from this class. Herbes, Southworth, and Gehrs (1977), summarizing the available data on acute toxicity of arylamines to aquatic life, found a good correlation between molecular weight and 96-hr LC_{50} values. A 40-unit increase in molecular weight resulted in a tenfold increase in acute toxicity (Fig. 4). Parkhurst (1977) conducted static tests on the fathead minnow *(Pimephales promelas)* for several arylamines in preparation for chronic exposure studies. He obtained 96-hr LC_{50} values of ~1.5, 24.7, and 7.0 mg/liter for quinoline, 2-methylquinoline, and 2,6-dimethylquinoline, respectively.

Chronic effects of arylamines on fish have not been studied, but Vasilenko et al. (1972) investigated mammalian responses to sublethal exposures of aniline. Cyanosis, anemia, and neurological disorders were observed in these studies. Epler et al. (1977) found that several of the quinolines are mutagenic agents when administered to microbial systems. The relatively high concentrations of arylamines expected to be released from coal-conversion facilities (as a result of poor treatability) suggest the possibility that acute and chronic effects might occur in surrounding environments.

The unsubstituted aromatic hydrocarbons, in particular the polycyclic aromatic hydrocarbons (PAH), have received the greatest interest with respect to their occurrence in coal-conversion liquid

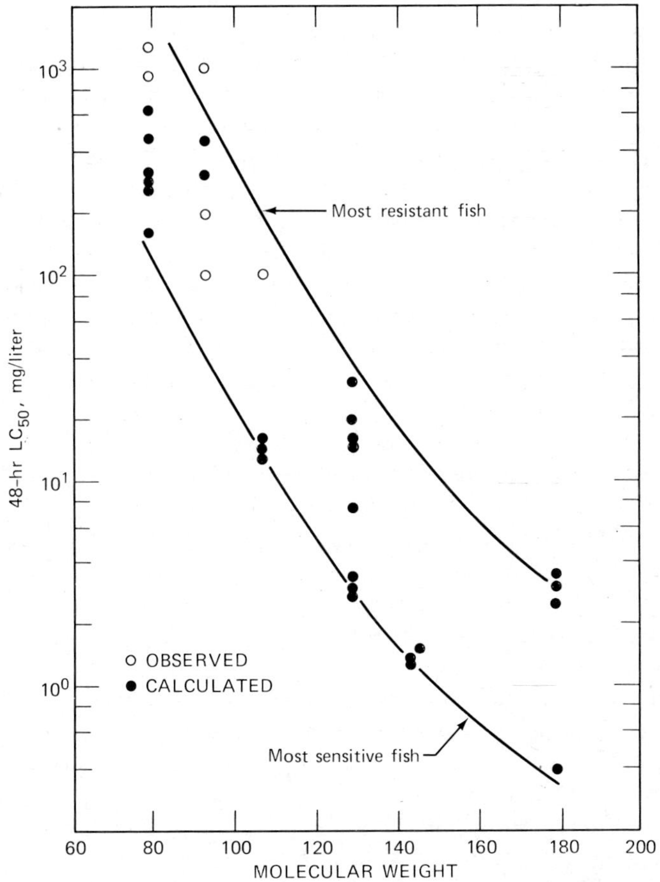

Fig. 4 48-hr LC_{50} values of arylamines to fish. (Data from Herbes, Southworth, and Gehrs, 1977.)

effluents. Members of this class, such as benzo(a)pyrene, are known carcinogens, and levels in coal-conversion products are expected to be substantially higher than in petroleum products (Shults, 1976). A preponderance of the biological activity (with mutagenesis as the test parameter) is associated with the PAH fraction of the liquid effluents (Epler et al., 1978; Epler et al., 1977; Guerin et al., 1976).

Data relative to the toxicity of aromatic hydrocarbons to aquatic organisms are also sparse. Moore and Dwyer (1974), in a review of literature on the effects of oil on marine organisms, generalized that the lower the boiling point of the compound the greater the toxicity and suggested that the relative toxicity might be associated with solubility. Compounds with low boiling points generally are more soluble in water and, hence, more available to be transported across membranes. Legore (1974) noted a good direct relationship between boiling point and relative toxicity, however. He found that naphthalene, a diaromatic hydrocarbon, was considerably more toxic to oyster larvae *(Crassostrea gigas)* than benzene, ethylbenzene, b-propylbenzene, isopropylbenzene, and ortho-, meta-, and para-xylenes, aromatic hydrocarbons with lower boiling points. Naphthalenes have been cited as the most toxic water-soluble petroleum fraction (Anderson et al., 1974). Concentrations of naphthalenes in effluents are expected to be below levels reported to cause acute toxicity to adult fish (15 to 25 mg/liter) (McKee and Wolf, 1963).

Chronic effects from aromatic hydrocarbon effluents are a distinct possibility. Anderson et al. (1974) found increased respiration rates in marine crustacea exposed to 5 to 10 mg/liter of aromatics. Soluble petroleum aromatics (0.01 to 0.1 mg/liter) disrupted social behavior in the lobster *(Homarus americanus)* and altered attraction of the snail *(Nassarius obsoletus)* to food at concentrations as low as 0.1 μg/liter (Bresch et al., 1972; Jacobson and Boylan, 1973).

Data on the toxicity of thiophenes (sulfur-substituted aromatic hydrocarbons) are almost nonexistent. Thiophene is 33% more toxic to sunfish than benzene (Jones, 1964), and thiophene and 2-methylthiophene are more toxic to mammals than their benzene analogs. Since we can expect substantial quantities of thiophenes in aqueous effluents from coal-conversion facilities, investigations into their ecological impacts are required.

Summarizing the data on the toxicity of the various effluent classes leads to several generalities. First, insufficient data are available to predict the potential for acute toxicity which might result from the various effluent chemical classes, except phenols, and the data that are available have resulted primarily from the testing of specific compounds, not "real world" effluent mixtures. Second, the information base for evaluating potential chronic effects is an abyss. If we are willing to accept the inherent weaknesses of the existing data base (it is sparse, does not evaluate interactions, etc.), we find that two apparently opposite generalities have been formulated with respect to toxicity. Moore and Dwyer (1974) suggested that the toxicity of aromatic hydrocarbons is inversely associated with boiling point and, hence, with solubility. Herbes, Southworth, and Gehrs (1977) hypothesized a direct correlation between molecular weight (which relates in a general manner to boiling point and solubility) and acute toxicity within each of the various classes of compounds

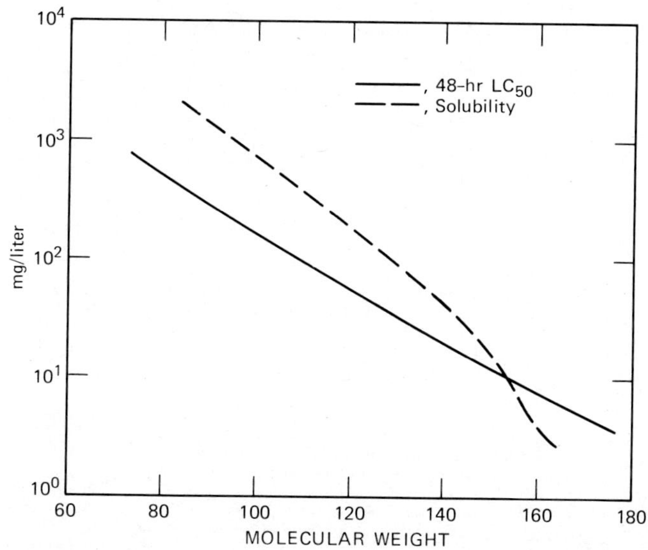

Fig. 5 Theoretical solubility and LC_{50} values plotted against molecular weights for a series of aromatic hydrocarbons.

potentially present in coal-conversion effluents. This divergence in interpretation might be explained if solubility becomes the limiting factor in reaching concentrations in water necessary for toxicity to be expressed. This is shown in Fig. 5, a theoretical plot of solubility and LC_{50} vs. molecular weight. To the right of the junction of the two curves, the materials would not be toxic because sufficient quantities would not be in solution. The finding of the preponderance of toxicity in materials with low boiling points (which transects all the chemical classes) quite possibly results from insufficient quantities in the water columns.

Certain flaws or, more appropriately, certain real-world phenomena detract from this oversimplification. These include:

1. Differential sensitivity of biological systems to the same molecule
2. Differential response of biological systems to various isomers of the same compound
3. Interactive phenomena, both toxicological and physio-chemical, already alluded to
4. Transport and transformation of materials in the aquatic environment

For example, quinoline, an aromatic amine, has an LC_{50} of 1.5 mg/liter in the fathead minnow and 17.5 mg/liter in the bluegill sunfish (Parkhurst, Gehrs, and Rubin, 1978), and benzo(a)pyrene is a potent mutagen and carcinogen, whereas benzo(e)pyrene is virtually inactive.

The complexity of developing a program for evaluating the environmental hazards of coal conversion is obvious. It is apparent that such a program should incorporate toxicity screening of effluents and their chemical fractions, but it is less apparent how to develop an approach that will produce the data necessary for predictive capabilities (research using individual compounds).

Generating a manageable approach to such a large number of contaminants requires adopting the same philosophy that produced the screening approach, i.e., look first for the greatest commonalities among the various compounds. This allows us to arrange compounds in groups that are most similar in certain parameters (e.g., structure).

Herbes, Southworth, and Gehrs (1977), adopting such a philosophy and arranging the compounds according to structure and molecular weight, were able to conduct research of a predictive nature for all coal conversion effluents by using approximately 30 compounds rather than the thousands present. This is clearly a more manageable number.

SUMMARY

In summary, the term coal conversion refers to any of a group of processes designed to produce liquid, gaseous, or clean-burning solid fuels from coal. All the processes use high temperature and high pressure (often in a reducing atmosphere) to produce the fuel and give rise to a myriad of organics that may ultimately find their way into aquatic environments. Hundreds, perhaps thousands, of individual compounds may be present in aqueous effluents. Investigation of the potential for environmental impact of these effluents can be approached in one of two ways, each of which possesses weaknesses. The two approaches are testing the whole complex effluent or testing individual compounds. The first provides rapid data on toxicity but has little value for predictive purposes since neither the identities nor concentrations of materials are known. The second, testing individual compounds, provides the necessary data for developing predictions. To study all the various compounds (including isomers of each) and all their combinations with respect to effects, transport, transformation, fate, and food chain kinetics and including all the environmental factors that might affect these parameters would require a substantial percentage of the Gross National Product and more time than is available. Obviously the necessary tactic is to use certain components from each of the two approaches.

Testing complex mixtures, coupled with chemical fractionation, allows us to determine acute toxicity and also to identify active components of the effluent. Investigations of individual compounds can be made manageable by grouping the compounds on the basis of chemical structure.

ACKNOWLEDGMENT

The research reported here was sponsored by the Division of Biomedical and Environmental Research, U. S. Department of Energy, under contract W-7405-eng-26 with Union Carbide Corporation. Environmental Sciences Division Publication No. 1194, Oak Ridge National Laboratory.

REFERENCES

Albersmeyer, W., and L. V. Erichsen, 1959, Investigations on the Effects of Tar Constituents in Waste Waters, *Z. Fisch. Deren Hilfswiss.*, 8(1/3): 29.

Anderson, J. W., J. M. Neff, B. A. Cox, H. E. Tatum, and G. M. Hightower, 1974, The Effects of Oil on Estuarine Animals: Toxicity, Uptake, and Depuration, Respiration, in *Pollution and Physiology of Marine Organisms*, F. J. Vernberg and W. B. Vernberg (Eds.), Academic Press, Inc., New York.

Anonymous, 1961, Some Effects of Pollution on Fish. *Notes on Water Pollution*, No. 13 (June), Department of Scientific and Industrial Research, England (from McKee and Wolf, 1963).

Anonymous, 1962, Drinking Water Standards. Title 42—Public Health, Chap. 1, Public Health Service, Department of Health, Education, and Welfare. Part 72—Interstate Quarantine, *Fed. Regist.* (Washington, D.C.), 2152 (Mar. 6, 1962).

Bingham, E., 1975, *Carcinogenic Investigations of Oils from Fossil Fuels*, National Institute of Environmental Health Sciences, Grant ES OU159-08, Kettering Laboratory, University of Cincinnati, Cincinnati, Ohio.

Bresch, H., R. Spielhoff, U. Mohr, and H. Barkemyer, 1972, Use of the Sea Urchin Egg for Quick Serum Testing of the Biological Activities of Substances. I. Influence on Fractions of a Tobacco Smoke Condensate on Early Development, *Proc. Soc. Exp. Biol. Med.*, 141: 747-752.

Bucksteeg, W., H. Thiele, and K. Stoltzel, 1955, The Effect on Fish of Toxic Substances in Waste Waters, *Vom Wasser*, 22: 194.

Butler, R. L., E. L. Cooper, J. K. Crawford, D. C. Hales, W. G. Kimmel, and C. C. Wagner, 1973, *Fish and Food Organisms in Acid Mine Waters of Pennsylvania*, Ecological Research Series, Publication No. EPA-R3-73-032, Environmental Protection Agency, Washington, D.C.

Butlin, H. T., 1892, Cancer of the Scrotum in Chimney Sweeps and Others, *Br. Med. J.*, 1: 1341-1346.

Energy Research and Development Administration, 1977, *Fossil Energy Research Program, FY 1978*, ERDA Report ERDA 77-33, GPO.

Epler, J. L., et al., 1978, Analytical and Biological Analyses of Test Materials from the Synthetic Fuel Technologies. I. Mutagenicity of Crude Oils Determined by the *Salmonella typhimurium*/Microsomal Activation System, *Mutat. Res.*, in press.

——, W. Winton, T. Ho, F. W. Larimer, T. K. Rao, and A. A. Hardigree, 1977, Comparative Mutagenesis of Quinolines, *Mutat. Res.*, 39: 285-296.

Gehrs, C. W., 1975, in *Coal Technology Program Quarterly Progress Report for the Period Ending Sept. 30, 1975*, pp. 61-62, USAEC Report ORNL-5093, Oak Ridge National Laboratory, NTIS.

——, 1978, in *Coal Technology Program Quarterly Progress Report for the Period Ending Dec. 31, 1977*, DOE Report ORNL-5387, Oak Ridge National Laboratory, NTIS.

——, and R. M. Wells, 1977, Environmental Transport and Effects, in *Sampling Strategy and Characterization of Potential Emissions from Synfuel Production*, Symposium and Workshop, Austin, Tex., June 6, 1976, ERDA Report CONF-760602, p. 79, NTIS.

Guerin, M. R., W. H. Griest, C.-h. Ho, and W. D. Shults, 1976, Chemical Characterization of Coal Conversion Pilot Plant Materials, in *Proceedings of the Third Environmental Protection Conference*, Chicago, Ill., Sept. 23—26, 1975, ERDA Report CONF-750967, Energy Research and Development Administration, NTIS.

Henry, S. A., 1946, *Cancer of the Scrotum in Relation to Occupation*, Oxford University Press, London.

Herbes, S. E., G. R. Southworth, and C. W. Gehrs, 1977, Organic Contaminants in Aqueous Coal Conversion Effluents: Environmental Consequences and Research Priorities, in *Trace Substances in Environmental Health—X*, D. D. Hemphill (Ed.), University of Missouri Press, Columbia.

Huckabee, J. W., C. P. Goodyear, and R. D. Jones, 1975, Acid Rock in the Smokies: Unanticipated Impact on Aquatic Biota of Road Construction in Regions of Sulfide Mineralization, *Trans. Am. Fish. Soc.*, 104(4): 672-684.

Hueper, W. C., 1956, Experimental Carcinogenesis Studies on Hydrogenated Coal Oils. II. Fisher—Tropsch Oils, *Ind. Med. Surg.*, 25: 456-459.

Jacobson, S. M., and D. B. Boylan, 1973, Effects of Seawater Soluble Fraction of Kerosene on Chemotaxis in a Marine Snail, *Nassarius obsoletus*, *Nature*, 241: 213.

Jones, J. R. E., 1964, *Fish and River Pollution*, Butterworth & Co., Ltd., London.

Legore, R. S., 1974, *The Effect of Alaskan Crude Oil and Selected Hydrocarbon Compounds on the Embryonic Development of the Pacific Oyster,* Crassostrea gigas, Ph.D. Thesis, University of Michigan, Ann Arbor.

McKee, J. E., and H. W. Wolf (Eds.), 1963, *Water Quality Criteria*, 2nd ed., Publication No. 3-A, The Resources Agency of California State Water Quality Control Board, Sacramento.

MacMahon, B., T. F. Pugh, and J. Ipsen, 1960, *Epidemiologic Methods*, Little, Brown & Company, Boston.

Moore, S. F., and R. L. Dwyer, 1974, Effects of Oil on Marine Organisms: A Critical Assessment of Published Data, *Water Res.*, 8: 819-827.

Parkhurst, B. R., 1977, Oak Ridge National Laboratory, personal communication.

——, C. W. Gehrs, and I. B. Rubin, 1978, Chemical Fractionation with Acute Toxicity Testing for Identifying the Toxic Components of Complex Aqueous Effluents, in *Proceedings of the 2nd Annual Symposium on*

Aquatic Toxicology, Cleveland, Ohio, Oct. 31—Nov. 1, 1977, American Society for Testing Materials, in press.

Richmond, C. R., D. E. Reichle, and C. W. Gehrs, 1976, *Balanced Program Plan.* Volume IV. *Coal Conversion, Analysis for Biomedical and Environmental Research,* ERDA Report ORNL-5123, Oak Ridge National Laboratory, NTIS.

Rubin, I. B., M. R. Guerin, A. A. Hardigree, and J. L. Epler, 1976, Fractionation of Synthetic Crude Oils from Coal for Biological Testing, *Environ. Res.,* 12: 358-365.

Schofield, C. L., 1976, Acid Precipitation: Effects on Fish, *Ambio,* 5: 228-230.

Sexton, R. J., 1960, The Hazards to Health in the Hydrogenation of Coal, *Arch. Environ. Health,* 1: 181-233.

Shriner, D. S., S. B. McLaughlin, and C. F. Baes, Jr., 1977, *Character and Transformation of Pollutants from Major Fossil Fuel Energy Sources,* ERDA Report ORNL-TM-5919, Oak Ridge National Laboratory, NTIS.

Shults, W. D. (Ed.), 1976, *Preliminary Results: Chemical and Biological Examination of Coal-Derived Materials,* ERDA Report ORNL/NSF/EATC-18, Oak Ridge National Laboratory, NTIS.

Tamm, C. O., 1976, Acid Precipitation: Biological Effects in Soil and on Forest Vegetation, *Ambio,* 5: 235-238.

Vasilenko, N. M., et al., 1972, Toxic Effects of Aniline, *Vrach. Delo,* 1972(8): 132-134 (In Russian).

Weil, C. S., and N. I. Condra, 1960, The Hazards to Health in the Hydrogenation of Coal. II. Carcinogenic Effect of Materials on the Skin of Mice, *Arch. Environ. Health,* 1: 187-193.

Wender, I., 1975, *Quarterly Technical Progress Report,* ERDA Report PERC/QTR-75/1, Pittsburgh Energy Research Center.

Wiser, W. H., 1973, Coal Characteristics and Coal Conversion Processes, Pennsylvania State University, Oct. 29—Nov. 3.

THE STREAM ENVIRONMENT AND MACROINVERTEBRATE COMMUNITIES: CONTRASTING EFFECTS OF MINING IN COLORADO AND THE EASTERN UNITED STATES

JAMES V. WARD,* STEVEN P. CANTON,* and LAWRENCE J. GRAY†
*Department of Zoology and Entomology, Colorado State University, Fort Collins, Colorado, and †Department of Zoology, Arizona State University, Tempe, Arizona

ABSTRACT

Studies were conducted year-round on a Colorado stream that receives drainage from a coal mine to assess the potential response of macroinvertebrate communities to mining activities in the western United States. Species composition, diversity, and standing crop were examined, and results are compared with similar studies conducted in eastern states. Generally low values of sulfate and iron, highly-buffered waters, and low levels of toxic substances characterized the Colorado stream and applied, in general, to many streams in the western energy-development region. Moderate inputs of soluble salts increased abundance of macroinvertebrates without significant changes in community structure or other discernible indications of stressed conditions. This is attributed to the relatively soft waters above the mine and the protection afforded by a buffer strip between the mine spoils and the stream. Increased salinity, sedimentation, and water depletion are major problems, but, with proper environmental considerations, mining in the West may not have the severe impacts on stream biota which characterize many eastern mining regions.

The western United States has vast expanses of coal. Kauffman and Schaefer (1977) indicated that "70 percent of all known, high-grade, low-sulphur, strippable coal deposits are located in this area, as are all of the country's major high-grade oil shale reserves." They estimated that the next 10 to 25 years will be a developmental period for the energy resources of the western United States.

Coal-bearing strata underlie 28% of the state of Colorado (Landis, 1964). Coal deposits are of the Upper Cretaceous and early Tertiary periods, as are the majority of surface mineable coal reserves of the Rocky Mountains and the Northern Great Plains (McWhorter, Skogerboe, and Skogerboe, 1975).

The ecological impacts on aquatic ecosystems of mining and processing activities differ in many ways in the West from those in mining regions of the eastern states. Yet, although there is a plethora of literature documenting the effects of mining on aquatic biota in the eastern United States, there are relatively few data on problems associated with mining and resulting effects on aquatic organisms in the West.

A year-round study was conducted on a Colorado stream that receives drainage from a strip coal mine. The objectives were (1) to compare macroinvertebrate communities and environmental conditions at stream locations differentially affected by mining activities and (2) to contrast the results with those of studies of eastern streams receiving coal-mine drainage.

METHODS AND SITE DESCRIPTION

Trout Creek, in the Upper Colorado River Basin of northwestern Colorado (Fig. 1), receives groundwater and surface runoff from a strip coal mine (Edna Mine). The area of the watershed above and including the mine totals approximately 1.1×10^4 ha; the area of mine spoils (597 ha) is slightly more than 4% of the total watershed area upstream from the mine (McWhorter, Skogerboe, and Skogerboe, 1975). The upper portion of the basin, which lies in Routt National Forest, is well vegetated with aspen and conifers. The middle portion of the watershed is a mixture of forests and farmland. Lower reaches are more xeric, with sagebrush and other shrubs and grasses predominating. Agricultural practices (primarily grazing) variously affect middle and especially lower portions of the basin. At elevations ranging from 2160 to 2100 m, the stream is bordered on the east by the Edna Mine. Trout Creek progressively flows past spoils from mining about 30 years ago, spoils 20 to 30 years old, and an area of current mining activity.

Sampling stations on Trout Creek were located on rubble riffles above, adjacent to, and below the mine spoils at sites from which water chemistry data had been collected in a previous study (McWhorter, Skogerboe, and Skogerboe, 1975). An additional sampling station (TC-1) was established upstream from a mine shaft (see Fig. 1) even though the shaft is horizontal and there was no evidence of seepage entering the stream.

Macroinvertebrates were sampled monthly from July 1975 through June 1976. High water or inclement weather precluded

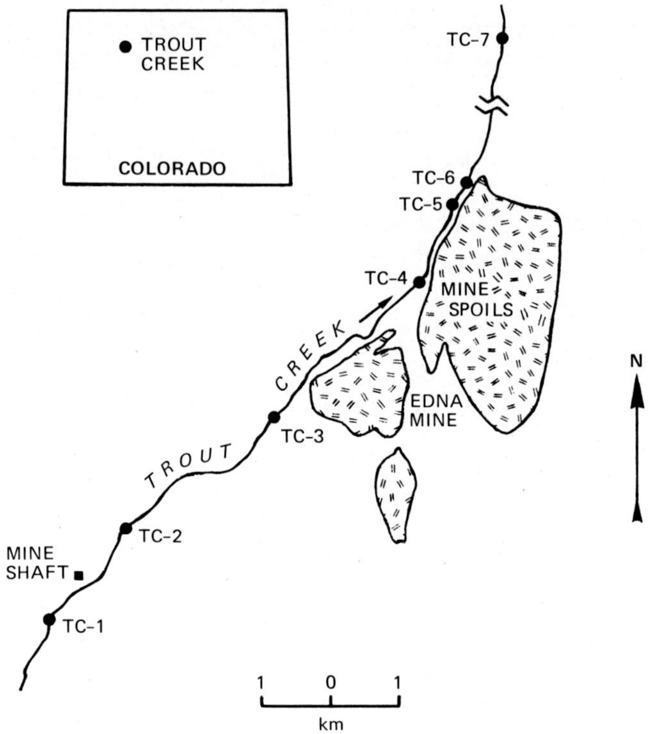

Fig. 1 Map of Trout Creek showing sampling stations and mine spoils. Station TC-7 is about 3 km downstream from TC-6.

sampling at one or two sites on three occasions. A 929-cm^2 Surber sampler with net mesh apertures of ~700 μm was used to take five or more replicate samples at each site. The organisms in each sample were enumerated separately, and numbers were pooled for diversity index calculations. The Shannon—Weaver index with logarithms to base 2 (bits per individual) was used to calculate macroinvertebrate diversity. Analysis of variance (f-distribution) was calculated on log transformations of raw density data to allow use of parametric tests (Elliott, 1973).

RESULTS

A total of 88 macroinvertebrate taxa were identified from the seven sampling stations on Trout Creek. A list of taxa, detailed methods, and site descriptions are given in Canton and Ward (1978).

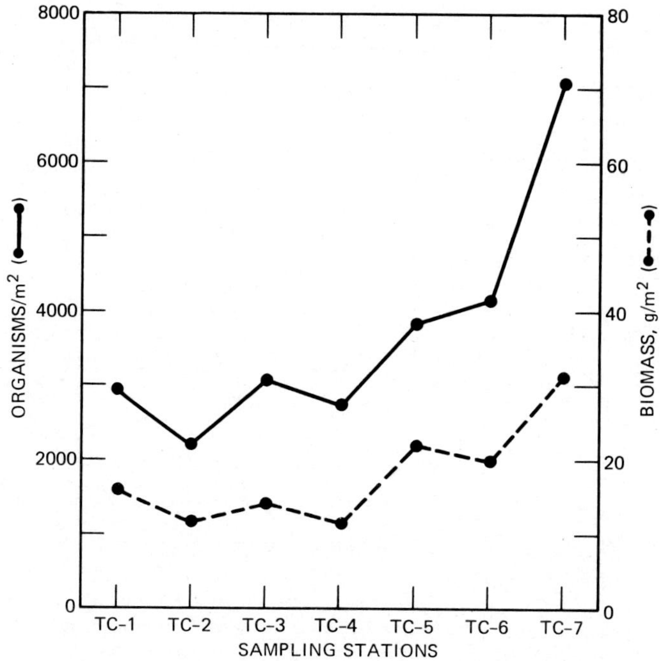

Fig. 2 Downstream changes in macroinvertebrate annual mean density and biomass values on Trout Creek.

Rather than decreasing, as expected, density and biomass exhibited a general increase downstream (Fig. 2). Differences between stations were significant ($P < 0.01$) for total numbers of organisms collected during the study but were not significant on certain collecting dates (June, July, and January). Significant differences ($P < 0.01$) between dates reflect periods of emergence and recruitment, effects of spring runoff, and onset of winter conditions.

Although density increased downstream, similar numbers of taxa occurred at all stations (Fig. 3). Shannon—Weaver index values also indicated no longitudinal pattern. Median values at all stations were between 3.0 and 4.0, the "normal" range defined by Wilhm (1970).

Five orders of aquatic insects (Trichoptera, Diptera, Ephemeroptera, Plecoptera, and Coleoptera) comprised well over 90% of the mean annual density and biomass of macroinvertebrates at all sampling stations. Eight taxa at each station represented 55 to 83% of the density, and most of these taxa were among the most abundant at many stations. The mayfly *Baetis* was among the top eight at all stations. Others abundant at most stations included the

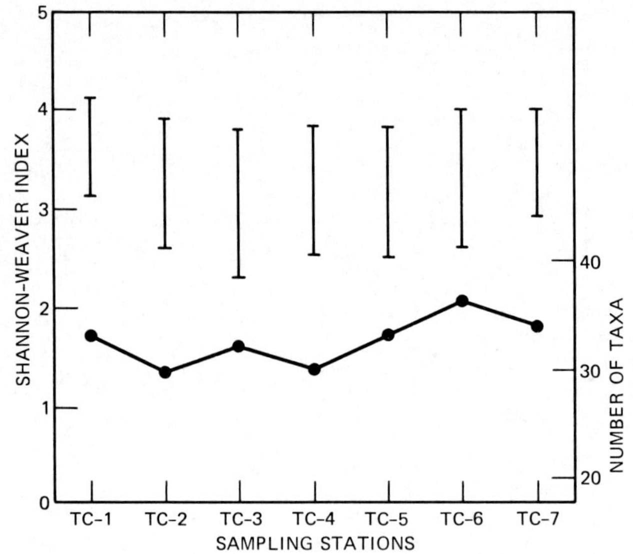

Fig. 3 Annual range of Shannon—Weaver index values (I) and the mean number of macroinvertebrate taxa per collection date (●—●) for sampling stations on Trout Creek.

caddisflies *Agapetus* and *Brachycentrus*, the mayfly *Rhithrogena*, and the elmid beetle *Optioservus*. Caddisflies were the most abundant order by density at all Trout Creek sampling stations (at TC-7 trichopterans and dipterans each comprised 33% of the mean density) and also comprised the largest biomass (at TC-2 trichopterans and ephemeropterans each comprised 28% of the mean biomass). Plecopterans comprised less than 10% of the macroinvertebrate numbers but were much more important gravimetrically. The reverse was true for coleopterans, which were primarily small riffle beetles. The relative importance of the five insect orders was similar at all stations and exhibited no longitudinal trend. Other orders of insects and noninsects were generally rare, and differences in their distribution patterns cannot be reliably relegated to anything except the sampling inadequacies inherent in studies of rocky streams (Hynes, 1970). Even at the generic and specific level, many of the common organisms were widely distributed.

DISCUSSION

Streams of the eastern United States which receive acid mine drainage are characterized by lowered diversity and density and greatly modified macroinvertebrate species composition (Appalachian Regional Commission, 1969; Roback and Richardson, 1969; Dills and Rogers, 1974; Herricks and Cairns, 1974).

Herricks and Cairns (1974) studied macroinvertebrate communities at locations above and below acid mine drainage in Pennsylvania streams (Table 1). Control stations were dominated by ephemeropterans, plecopterans, and odonates; dipterans (mainly chironomids) and *Hydropsyche* were the common organisms at a station receiving acid mine drainage.

In an Alabama stream an isopod *(Lirceus)* comprised 43% of the benthos at unpolluted stations, whereas acid stations were dominated by chironomids, ceratopogonids and megalopterans (Dills and Rogers, 1974). Reduced density and diversity were reported for polluted stations (Table 1).

These eastern studies contrast sharply with our study of Trout Creek. Despite the fact that coal strip mining began about 30 years ago and is presently being conducted along the creek, macroinvertebrate communities adjacent to and below the mine spoils indicated no discernible detrimental effect from mining activities. The number of taxa was similar above, adjacent to, and below the mine. Standing crop increased rather than decreased downstream, and taxonomic composition was remarkably similar throughout the stream section studied. Shannon—Weaver index values gave no indication of a stressed macroinvertebrate community at any of the sites on Trout Creek.

Chemical Conditions and Macroinvertebrates

Wentz (1974), referring to the effects of mine drainage on the water quality of streams in Colorado, indicated that "approximately 450 miles (724 kilometers) of streams in 25 different areas are adversely affected by metal-mine drainage. Coal-mine drainage is not a problem, apparently because of the low sulfur content of Colorado coal."

Table 2 compares the water quality of Trout Creek with criteria for acid mine drainage. In streams in western energy-development areas, acidity is typically undetectable, and pH values generally range from 6.5 to 8.5, with over 85% of all pH values falling between 7.0 and 8.0 (Skogerboe, 1976). Stream waters are well buffered, and major inputs of acid or base would be required to shift the pH one or more units. The chemoautotrophic bacteria, which greatly speed the production of sulfuric acid and ferric hydroxide, are effective only at low pH, whereas bacteria that oxidize ferrous iron at higher pH are relatively unimportant in acid formation (Wentz, 1974).

TABLE 1

MACROINVERTEBRATE DIVERSITY, DENSITY, AND COMPOSITION AT STATIONS ABOVE AND BELOW MINE DRAINAGE ON TROUT CREEK AND IN STREAMS IN THE EASTERN UNITED STATES

Number of taxa*	Diversity index (median $\bar{d}$)	Organisms/m^2	Predominant groups†	Stream system	Reference‡
(33)	3.7	2930	T, E, D, P, C	Trout Creek (TC-1)	This study
(36)	3.5	4124	T, E, D, P, C	Trout Creek (TC-6)	
(37)	3.6	2518	E, P, O	Indian Creek, Penn. (Station 1, control)	Herricks and Cairns (1974)
(6)	2.2	low	T, D	Indian Creek, Penn. (Station 3)	
55	2.9	high	I (E, P, O)§	Cane Creek, Ala. (unpolluted stations)	Dills and Rogers (1974)
24	1.8	low	D, M	Cane Creek, Ala. (polluted stations)	

*Numbers in parentheses are mean numbers of taxa per collection date. Other values are total taxa.

†Abbreviations are T, Trichoptera; E, Ephemeroptera; D, Diptera; C, Coleoptera; P, Plecoptera; O, Odonata; I, Isopoda; and M, Megaloptera.

‡In some cases values were calculated from data of studies cited.

§Odonata, Plecoptera, and Ephemeroptera implied.

TABLE 2

CRITERIA FOR ACID MINE DRAINAGE
COMPARED WITH WATER QUALITY OF TROUT CREEK

Water quality characteristic*	Acid mine drainage†	Trout Creek‡	
		TC-3	TC-6
Acidity	>3.0	0.0	0.0
Alkalinity	0	89.2 (41–166)	102.4 (24–205)
pH	<6.0	7.6 (6.5–8.2)	7.6 (6.8–8.3)
Iron	>0.5	0.17 (<0.02–1.10)	0.12 (0.04–0.54)
Sulfate	>250	10.6 (5–20)	80.6 (12–250)
Total hardness	>250	115.9 (43–270)	169.3 (67–418)
Suspended solids	>250	5.3 (0–24)	9.8 (0–62)
Dissolved solids	>500	110.4 (40–176)	223.0 (52–509)

*All values except pH are in milligrams per liter.

†Herricks and Cairns (1974).

‡Single values are annual means; ranges at sites above and below mine spoils are in parentheses.

Much of the stress on macroinvertebrates associated with acid mine drainage in eastern streams can be attributed to low pH, per se, although ferric hydroxide precipitation and heavy metals also have adverse effects. Napier and Hummon (1976) found that mayflies did not occur in streams which had otherwise recovered from acid mine drainage but which maintained low pH values. They concluded that low pH was primarily responsible for the elimination of mayflies.

From the results of a survey of Colorado sites receiving drainage from mines, Wentz (1974) tabulated the percentage of sample sites where concentrations of 15 heavy metals and pH exceeded stream criteria for fish and other aquatic organisms (criteria were drawn from various cited sources). Only one station associated with coal mining, the drain from an adit of an abandoned mine, exceeded these criteria and only for iron and manganese. Wentz found no macroinvertebrates in the drain, the bottom of which had a bright-orange coating.

Concentrations of most heavy metals are low in Trout Creek and in western energy-development areas generally (Skogerboe, 1976). The solubilities of many heavy metals associated with acid production processes in eastern mining areas are limited by hydroxide or carbonate precipitation processes. An examination of the solubility equilibria of metals showed that most are least soluble from pH 7.0 to 8.0, the range within which over 85% of all observed pH values in western energy areas lie (Skogerboe, 1976). The well-buffered nature

TABLE 3

IONIC COMPOSITION (mg/liter), SPECIFIC CONDUCTANCE, AND pH OF EDNA MINE SPOILS

Ion, mg/liter	Edna Mine spoils*
SO_4^{2-}	820
Na^+	41
Mg^{2+}	270
Ca^{2+}	52
Cl^-	15
HCO_3^-	280
K^+	24
Specific conductance, μmhos/cm	3967
pH	8.2

*Saturated paste analyses (McWhorter, Skogerboe, and Skogerboe, 1975).

of most western streams would require major inputs of acid or base to change the solubilities of most metals appreciably. In addition, Cairns (1976) stressed that hard, well-buffered waters reduce the adverse effects to biota of many toxic materials.

There is some evidence that many elements contained in the mine spoils do not reach Trout Creek in large quantities. Sediment from runoff channels in the mine contained higher concentrations of all heavy metals analyzed, except mercury, than did sediment from Trout Creek (Skogerboe, 1976); this "implies that those elements contained in the runoff are largely removed from solution, perhaps by precipitation, before the runoff reaches the creek." The presence of a strip of unmined land between the mine spoils and the creek may considerably reduce the input of potentially toxic substances to the stream.

Soluble salts from the overburden are considered by some to be the most significant potential pollutant of streams from strip mines in the West (McWhorter, Skogerboe, and Skogerboe, 1975). Edna Mine spoils contain large amounts of certain ions (Table 3). Precipitation in more mesic regions is sufficient to leach out soluble salts as they form, but, in the arid and semiarid regions of the western United States, only the top few centimeters of soil are adequately leached. Surface mining exposes fresh surfaces and increases leaching of soluble salts; this results in an approximate doubling of total dissolved solids in Trout Creek (Table 2). Increases in hardness and soluble salts may, within limits, result in increased productivity, especially in western streams, which often originate in mountainous regions of insoluble crystalline rock. Downstream increases in specific conductance in Trout Creek were correlated with increases in the density ($r = 0.9$) and biomass ($r = 0.8$) of macroinvertebrates.

Sedimentation and Macroinvertebrates

Coal-mine drainage often results in the production of ferric hydroxide, which is insoluble and forms a yellow—orange precipitate on the stream substrate (Wentz, 1974). This did not occur in Trout Creek, which had relatively low iron concentrations. Inorganic sediment from erosion of surfaces exposed by mining may also cover the substrate. Apart from any toxic effects, sedimentation decreases substrate heterogeneity, fills interstices with silt, may severely reduce algal populations, and directly affects the benthos (Ward, 1976). The presence of a strip of unmined land between the mine spoils and Trout Creek, in combination with the semiarid climate of the region, is apparently largely responsible for the maintenance of a heterogeneous and relatively silt-free substrate.

Climate, Groundwater, and Stream Flow

Climatic differences between western and eastern mining regions undoubtedly modify effects on stream ecosystems. Differences in leaching efficiency between xeric and mesic regions have already been referred to. Herricks and Cairns (1974) stressed the importance of relationships between stream flow and mine drainage. The apparent lack of adverse effects of Edna Mine drainage on Trout Creek macroinvertebrates may result in part from a hydrologic situation in which most substances enter the stream during a relatively short period associated with high stream flow. For example, over 80% of the total salt input from the mine occurs during April, May, and early June.

In summary, effects of mining on stream ecosystems in the western United States are different in many ways from those in eastern states. Acid mine drainage and ferric hydroxide precipitates are rarely associated with coal mines in western energy-development areas, partly because of the low sulfate content of western coal. Alkaline, highly buffered waters prevent the formation of acid conditions, even where sulfate values are high, and reduce the solubility of heavy metals. Water depletion, sedimentation, and increased salinity are the major potential problems associated with western mining.

The Trout Creek study shows that mining in the West need not be as seriously damaging to the stream ecosystem as mining in the eastern states. Although the Trout Creek basin had been mined for about 30 years, macroinvertebrates exhibited no definitive indications of stressed conditions in the stream section studied. Although Edna Mine spoils are extensive, we should emphasize that the headwaters of Trout Creek provide a source of organisms for recolonization of affected areas downstream. The mine spoils adjacent to Trout Creek are 30 to over 100 m from the stream channel. This buffer zone between the mine spoils and the stream apparently reduces sediment input and the quantities of other potentially harmful substances. It is postulated that physical, chemical, and biological processes occurring in the soil of the buffer zone reduce the toxicity of certain substances and retard the movement of potentially harmful substances into the stream channel. This is an area of research which merits further investigation.

ACKNOWLEDGMENTS

We wish to thank R. K. Skogerboe, Department of Chemistry, Colorado State University, for suggestions regarding the manuscript and for supplying unpublished chemical data. This research was funded in part by the U. S. Environmental Protection Agency, Environmental Research Laboratory, Duluth, Minn. (Research Grant No. R803950), and by a National Science Foundation Energy Traineeship awarded to S. P. Canton. This paper is based in part on a thesis submitted by S. P. Canton in partial fulfillment of requirements for a M.S. degree in zoology at Colorado State University. The Natural Resource Ecology Laboratory, Colorado State University, provided support facilities and coordinated research activities.

REFERENCES

Appalachian Regional Commission, 1969, *Acid Mine Drainage in Appalachia*, Washington, D.C.

Cairns, John, Jr., 1976, Heated Waste-Water Effects on Aquatic Ecosystems, in *Thermal Ecology II*, ERDA Symposium Series, Augusta, Ga., Apr. 2–5, 1975, G. W. Esch and R. W. McFarlane (Eds.), pp. 32-38, CONF-750425, NTIS.

Canton, S. P., and J. V. Ward, 1978, *Environmental Effects of Western Energy Development*. Part II. The Aquatic Macroinvertebrates of Trout Creek, Colorado, EPA Ecological Research Series, in press.

Dills, G., and D. T. Rogers, Jr., 1974, Macroinvertebrate Community Structure as an Indicator of Acid Mine Pollution, *Environ. Pollut. (London)*, 6: 239-262.

Elliott, J. M., 1973, *Some Methods for the Statistical Analysis of Samples of Benthic Invertebrates,* Scientific Publication No. 25, Freshwater Biological Association, Ambleside, England.

Herricks, E. E., and J. Cairns, Jr., 1974, *Rehabilitation of Streams Receiving Acid Mine Drainage,* Virginia Water Resources Research Center Bulletin 66, Virginia Polytechnic Institute and State University, Blacksburg.

Hynes, H. B. N., 1970, *The Ecology of Running Waters,* University of Toronto Press, Toronto, Ont.

Kauffman, K. O., and R. R. Schaefer, 1977, Water for Energy—Western United States, *Fisheries,* 2: 6-13.

Landis, E. R., 1964, Coal, in *Mineral and Water Resources of Colorado,* pp. 35-45, U. S. Geological Survey, U. S. Government Printing Office, Washington, D.C.

McWhorter, D. B., R. K. Skogerboe, and G. V. Skogerboe, 1975, *Water Quality Control in Mine Spoils, Upper Colorado River Basin,* EPA Report 670/2-75-048, Environmental Protection Agency, National Environmental Research Center, Cincinnati.

Napier, S., Jr., and W. D. Hummon, 1976, Survival of Mayfly Larvae Under Mine Drainage Conditions, *Int. Revue Gesam. Hydrobiol.,* 61: 677-682.

Roback, S. S., and J. W. Richardson, 1969, The Effects of Acid Mine Drainage on Aquatic Insects, *Proc. Acad. Nat. Sci. Philadelphia,* 121: 81-98.

Skogerboe, R. K., 1976, Monitoring of Water Quality Characteristics of Energy Development Field Sites, in *Toxic Effects on the Aquatic Biota from Coal and Oil Shale Development: Progress Report—Year 1,* pp. 349-374, Natural Resource Ecology Laboratory, Colorado State University, Fort Collins, and Fisheries Bioassay Laboratory, Montana State University, Bozeman.

Ward, J. V., 1976, Effects of Flow Patterns Below Large Dams on Stream Benthos: A Review, in *Proceedings, Instream Flow Needs Symposium,* Boise, Id., J. F. Orsborn and C. H. Allman (Eds.), pp. 235-253, American Fisheries Society, Bethesda, Md.

Wentz, D. A., 1974, *Effect of Mine Drainage on the Quality of Streams in Colorado, 1971—72,* Colorado Water Resources Circular No. 21, Colorado Water Conservation Board, Denver.

Wilhm, J., 1970, Range of Diversity Index in Benthic Macroinvertebrate Populations, *J. Water Pollut. Control Fed.,* 42: R221-R224.

MEIOFAUNAL ABUNDANCE IN SANDBARS OF ACID MINE POLLUTED, RECLAIMED, AND UNPOLLUTED STREAMS IN SOUTHEASTERN OHIO

WILLIAM D. HUMMON, WAYNE A. EVANS, MARGARET R. HUMMON, FRANCIS G. DOHERTY, ROBERT H. WAINBERG, AND WILLIAM S. STANLEY
Department of Zoology and Microbiology, Ohio University, Athens, Ohio

ABSTRACT

In October and November 1976, a collection was made at each of two sites along seven streams in Athens, Vinton, and Hocking counties, Ohio, during a period of stable weather. Streams were chosen to include watersheds with no mining and with varying histories of mining and reclamation. Meiofaunal sampling followed a regular pattern. Four stations were sampled along a cross-stream transect; four depths were sampled at each station, and triplicate subsamples were taken at each depth, yielding 48 2.7-cm^3 sediment subsamples per site collected. Physical—chemical parameters measured for stream and sediment at each site included temperature, pH, conductivity, carbonate alkalinity, sulfates, calcium and total hardness, manganese, iron, dissolved oxygen, biochemical oxygen demand, and stream flow. Meiofauna were narcotized with 1% $MgCl_2$, extracted by multiple decantation, and tallied and identified to major taxon with multiple whole Sedgwick—Rafter cell counts.

The number of major taxa recovered at the 14 sites ranged from 3 to 11; geometric mean abundance of total meiofauna per coring station expressed as organisms per 10 square centimeters of surface ranged from 21 to 443. Diptera, one of 16 taxa observed, total numbers of taxa present, and geometric means of total meiofauna were significantly correlated (negatively) with values for compensated noncarbonate conductivity from the 14 sites.

In a dendrogram of $S_{H'}$ similarity analysis, the unpolluted streams along with several sites with a past history of mining formed an eight-site complex, showing H' taxon diversity values of 1.6 to 2.8 with 40 to 80% in common. The remaining sites, all with a past history of mining, formed two groups. One group of four sites, dominated by rotifers, showed H' values of 1.2 to 1.6 with 36 to 52% in common. The second group of two sites, dominated by nematodes, showed H' values of 1.3 to 1.5 with 46% in common.

Streams and other waters of areas with a history of mining and with sulfur-rich coal are often polluted with acid mine drainage. This pollutant continues to form and be washed into streams long after mining has ceased and may affect the fauna of such waters after the area has been vegetatively reclaimed (Napier and Hummon, 1976). Iron pyrite (FeS), present in the coal, is oxidized in the presence of water and oxygen and, catalyzed by bacteria, forms sulfuric acid. In addition to exhibiting lowered pH, waters affected by acid mine pollution contain various dissolved minerals, including iron, aluminium, zinc, copper, and manganese (Hill, 1968; Massey and Barnhisel, 1972) and are characterized by high conductivity and an absence of buffering capacity (Parsons, 1968; Dills and Rogers, 1974; Faucon and Hummon, 1976).

The effect of acid mine drainage on the occurrence of stream macroinvertebrates is fairly well documented (Parsons, 1968; Roback and Richardson, 1969; Warner, 1971; Koryak, Shapiro, and Sykora, 1972; Nichols and Bulow, 1973; Dills and Rogers, 1974; Napier and Hummon, 1976). However, relatively little attention has been given to the effect of mine acid on stream microinvertebrates or meiofauna (Lackey, 1939; Hummon, 1977).

The single best measure of the degree of acid mine pollution in a natural body of water and of the effect of this pollution on flora or fauna has generally been considered to be pH (Parsons, 1956; Bennett, 1969; Warner, 1971). But recently it has been suggested that total conductivity (Dills and Rogers, 1974) or certain combinations of conductivity and carbonate concentration (Faucon and Hummon, 1976) may be critical.

This paper contains results from a continuing study of the relationship between natural and acid mine-polluted water and invertebrate fauna, particularly the meiofauna. Our report (1) introduces results on the abundance and taxon diversity of meiofaunal elements in natural and polluted stream ecosystems and (2) develops preliminary criteria by which the state of acid mine stream pollution and recovery can be assessed with respect to its meiobenthic fauna.

MATERIAL AND METHODS

During an extended period of stable, fair weather, 14 sites, two each along seven streams in Athens, Vinton, and Hocking counties, Ohio (Table 1), were sampled once over a month-long period (Oct. 23 to Nov. 20, 1976). The streams were chosen to represent a variety of conditions. Raccoon Creek and Sandy Run are highly acid mine-polluted streams. Minkers Run, subject of stripping and

TABLE 1
LOCATIONS OF SITES SAMPLED, BASED ON THE OHIO RIVER SURVEY

Stream	Site	County	Quarter of quarter sect.	Section	Township	Range	Comments
Long Run	Upper	Athens	NE of NE	31	T5N	R13W	Off county road 25, 2.6 km SE of junction with Co. 90
	Lower	Athens	SE of SE	30	T4N	R13W	150 m above stream crossing beneath Co. 25
Minkers Run	Upper	Athens	SE of SE	22	T12N	R15W	Off USFS road, 1.1 km NW of junction with Co. 1
	Lower	Athens	SW of SE	16	T12N	R15W	10 m below stream crossing beneath Co. 1
Raccoon Creek	Upper	Hocking	NW of NE	9	T12N	R16W	150 m above stream crossing beneath Co. 325
	Lower	Hocking	SE of SW	9	T12N	R16W	20 m above stream crossing beneath Co. 324
Tick Ridge	Upper	Hocking	NW of SE	33	T12N	R16W	400 m below ridge forming headwaters of unnamed creek
	Lower	Hocking	SE of SE	33	T12N	R16W	400 m further downstream
Strouds Run	North	Athens	SE of SE	30	T5N	R13W	10 m above trail exiting campground off Twp. 212
	West	Athens	NE of NE	35	T5N	R13W	Off parking lot along Co. 20, just west of Dow Lake
Sandy Run	Upper	Vinton	NE of NE	11	T11N	R16W	150 m below stream from Big Four Hollow crossing Ohio 278
	Lower	Vinton	NW of SE	9	T11N	R16W	15 m above stream crossing beneath Twp. 4
Margaret Creek	South	Athens	SE of SE	32	T8N	R14W	Beneath bridge where stream crosses Ohio 681
	West	Athens	SW of SW	35	T8N	R14W	20 m above stream crossing beneath Co. 80

vegetative reclamation in the 1950s, is a chronically acid mine-polluted stream (Napier and Hummon, 1976), particularly the lower site, which is located just below an old, but extensive, tipple zone. Long Run is situated in an area in which there was some stripping in the 1950s; judging by its macrofauna (Feldner and Stanley, 1976), however, the stream may have recovered from some of the adverse effects of acid mine effluents. The unnamed stream adjacent to Tick Ridge receives water from a 13- to 14-ha area that was stripped during the 1970s and reclaimed according to current practices. Margaret Creek and Strouds Run are streams in watersheds with no history of coal mining. Five of the seven streams were sampled at upstream (upper) and downstream (lower) sites. For the two other streams, Margaret Creek and Strouds Run, sites were chosen on different branches.

Both sites for a given stream were collected on the same day. Sites were located at least 400 m, usually 1 km or more, apart. Each site was sampled for meiofauna following a regular pattern. A transect was oriented perpendicular to stream flow, running from an emergent sand bar into deeper water covering submerged but continuous sandy substratum. Samples were taken from four coring stations, whose surface elevations were located at +2 cm (bar station), −3 cm (stream-edge station), and −10 cm and −20 cm (midstream stations) with respect to the stream surface.

Four sediment cores were taken at each coring station. A 20-cm^3 disposable plastic syringe (2.68-cm^2 surface) with its tip cut off operated as a suction corer. Three sediment cores were fractioned, and the 0 to 1, 1 to 2, 3 to 4, and 6 to 7-cm core depths were used for faunistic analysis. Thus there were 4 stations × 4 sample depths per station, or 16 samples, × 3 subsample replicates per sample, or a total of 48 subsamples. Replicates from eight samples were treated individually as splits (8 × 3 – 24), and those from the other eight samples were combined and treated as grouped subsamples (8 × 1 = 8). This reduced by one-third the number of entities to be examined. Selection of those samples in which subsample replicates would be grouped was made by a stratified random process, such that two of four samples per station and two of four samples per depth would be included, based on a 4 × 4 matrix of stations and depths. The fourth sediment core was reserved for sedimentary analysis.

Physical—chemical data were also collected from the stream and from interstitial water allowed to seep into a hole dug in the sandbar adjacent to the stream. Air, stream water, sediment surface, and interstitial water temperatures were measured with a laboratory thermometer. Light transmission was measured in stream water with a Beckman EV-4 Envirotrans meter. Conductivity and pH were measured in stream and interstitial water with YSI 33 S-T-C and Orion 404 Specific Ion meters. Water samples from both stream and hole were placed in plastic bottles and BOD (biochemical oxygen demand) bottles for further chemical analysis in the laboratory. The azide modification of the Winkler technique was used on site for preliminary fixation of dissolved oxygen. A twig, meterstick, and stopwatch were used to measure flow rates in shallow zones having rather even depths. A tape and level were used to make a bank-to-bank stream and channel profile along the transect line.

Samples were returned to the laboratory within 3 hr of collection, where fauna were narcotized with 1% $MgCl_2$ and extracted from one jar into an identical jar by multiple decantation, using 1% $MgCl_2$ as the decantation fluid throughout. Samples were fixed with enough 100% Formalin containing rose bengal to make a 10% final solution, capped with the numbered cap from the original jar, and mixed thoroughly by swirling. Sieving was omitted, since even a 38-μm sieve results in a serious loss of small animals, including rotifers, nematodes, and gastrotrichs.

Concurrently with the extraction and fixation of faunistic samples, bottles for BOD determination were placed in a constant-temperature chamber without light at 20°C. Duplicate titrations were made for methyl orange (total) alkalinity with bromocresol green—methyl red indicator and for dissolved oxygen. As soon as time allowed, tests for manganese, iron, and sulfates were conducted with Hach chemical tests modified for use with a Bausch and Lomb Spectronic 20. Calcium and total hardness were determined with unmodified Hach chemical tests. After 5 days, BOD samples were fixed and titrated for residual dissolved oxygen.

Extracted and fixed faunistic samples were analyzed under a Wild M-8 stereomicroscope at 50 and 100× magnification using multiple Sedgwick—Rafter (S—R) cell counts. Eight whole-cell counts were used for split samples and twelve were used for grouped samples. Material was prepared for the S—R cells by first decanting about two-thirds of the supernatant fluid. The residual material was swirled to determine its suspended load of organic material, silt, and clay. If the suspended load was great enough to impede counting efficiency, fluid was added back to dilute the suspended load and increase the counting efficiency. The fluid was swirled to randomize its contents; then the S—R cells were filled by a large-bore pipette, and the remaining fluid volume was noted. Fauna were enumerated to major taxon, and the resulting number of organisms in each taxon was multiplied by a correction factor to convert to organisms per cubic centimeter of sediment:

$$\begin{aligned}(\text{Organisms per cm}^3\text{ sed.}) = {} & (\text{organisms counted from S—R cells}) \\ & \times (\text{total fluid, ml, from which S—R cells were filled}) \\ & \times [(\text{amount of sediment, cm}^3\text{, represented by sample}) \\ & \times (\text{number ml counted in S—R cells})]^{-1}\end{aligned}$$

Parenthetically, the procedure provides generous flexibility for use with any predominately sandy sediments. Under light suspended loads it allows for the actual counting of 70 to 90% of the fauna contained in the sample. The procedure also gives reliable results with meiofauna of marine sandy sediments, requiring only two alterations: (1) the concentration of $MgCl_2$ should be increased to 6 to 7% and (2) the observation magnification can be reduced by one-half owing to the generally larger sizes of the marine meiofauna.

Faunistic abundance per 10 square centimeters of surface ($\times 10^3$ = abundance per square meter of surface) was estimated separately for each taxon and coring station. Density per cubic centimeter of sediment for individually analyzed samples or subsamples was multiplied for each depth fraction analyzed by an appropriate factor (0 to 1 cm depth by 1.0, 1 to 2 cm depth by 1.5, 3 to 4 cm depth by 2.5, and 6 to 7 cm depth by 3.5) to convert estimate from partial to complete cores, and then by 10 to convert from 1 to 10 square centimeters of surface. The sum of taxon abundances at a given coring station was taken as the abundance of total meiofauna at that station.

Geometric mean abundance per station and 95% confidence limits were then calculated for each taxon and site, using log (X) or (X + 1) transformations as appropriate. The resulting values were correlated with a provisional three-parameter index of the physical—chemical environment. Of the three parameters in the index—noncarbonate conductivity tempered by high oxygen saturation and stream flow—the first two represent faunally weighted means of stream-water and interstitial-water values. For the faunal weighting factor, the fauna in the upper centimeter of sediment was thought to be most responsive to stream-water values and the fauna of the remaining depths to be most responsive to interstitial-water values. Hence, for a given site, stream-water values were multiplied by the decimal fraction of abundance occurring in the upper centimeter of sediment, and the interstitial-water values were multiplied by the decimal fraction of remaining abundance. Noncarbonate conductivity is given in micromhos per centimeter, and oxygen saturation is given as a decimal fraction, rather than as percent. The third parameter is the cube root of cubic decimeters of stream flow per second, or decimeters per second.

In an alternate way of viewing meiofaunal relationships between sites, the H′ (Lloyd, Zar, and Karr, 1968) taxon diversity was calculated for the arithmetic aggregate of station abundances at each site. It is proper to use the H′ diversity measure with respect to taxa so long as (1) the results are compared with those of other studies using the same taxa and sampling—analytic scheme and (2) no inferences are made with respect to H′ generic or species diversity. Faunal abundances from all 14 sites were then subjected to an S_H' similarity analysis (Hummon, 1974), based on shared taxon diversity, and the results were assembled in a dendrogram using the unweighted mean, pair—group method.

RESULTS

Values for pH were lowest at Raccoon Creek and Sandy Run, the two streams most heavily polluted by acid mine effluents (Table 2). In both streams the lower site showed slightly moderated effects of mine acid. Minkers Run was the only other stream having pH values consistently below 7.0, with the lower site in this case showing slightly increased effects of mine acid, probably associated with the nearby tipple zone. Interstitial waters generally showed slightly lower pH values than the adjacent stream waters, the major exception being Sandy Run (upper site) where interstitial water was higher than stream water.

Total conductivity was greatest at Raccoon Creek and least at Tick Ridge, which lay just below the recently stripped and reclaimed area (Table 2). Intermediate values, in descending order, were Minkers Run, Sandy Run (upper site), Long Run, Strouds Run, Sandy Run (lower site), and Margaret Creek. Total conductivity of interstitial waters was similar to or lower than that of stream waters, the trend being most accentuated at Sandy Run (lower site), but with major reversals at Minkers Run and Margaret Creek (south site). Carbonate conductivity was absent or minimal at Raccoon Creek, Sandy Run, and Minkers Run (lower site), low at Tick Ridge and in the stream water at Minkers Run (upper site), and high in the interstitial water at Margaret Creek (south site). Other stream sites had intermediate

TABLE 2

STREAM-STUDY SITES, ARRANGED BY DATE OF COLLECTION, AND PHYSICAL—CHEMICAL—BIOLOGICAL DATA IMPORTANT TO EACH SITE OR TO THE DEVELOPMENT OF THE COMPENSATED NONCARBONATE CONDUCTIVITY INDEX

Stream and date (1976) sampled	Site, type of water sampled*	Temp., °C	pH	BOD	Faunal weight'g factor	Conductivity, μmho/cm at 25°C				Dissolved O_2		Stream flow	
						Total	− carb.	= noncarb.	(A) Weighted noncarb.	Dec. satn.	(B) Weighted dec. satn.	dm^3/sec	(C) $(dm^3/sec)^{1/3}$
Long Run, Oct. 23	U, S	6.0	7.7	1.8	0.677	566	277	289	256	0.96	0.73	20.0	2.71
	U, I	6.0	7.6	2.6	0.323	428	240	188		0.25			
	L, S	6.0	7.8	1.8	0.864	610	196	414	397	0.96	0.88	21.0	2.76
	L, I	5.0	7.4	1.1	0.136	515	229	286		0.37			
Minkers Run, Oct. 24	U, S	8.0	6.7	2.3	0.247	657	17	640	702	0.94	0.56	22.8	2.84
	U, I	8.5	6.6	1.5	0.753	918	195	723		0.43			
	L, S	7.5	6.3	5.5	0.409	716	6	710	936	1.13	0.74	54.7	3.80
	L, I	9.0	6.2	2.2	0.591	1096	3	1093		0.47			
Raccoon Creek, Nov. 6	U, S	5.0	3.0	3.3	0.406	2142	0	2142	1847	1.05	0.83	5.3	1.74
	U, I	3.5	2.9	2.6	0.594	1645	0	1645		0.68			
	L, S	4.5	3.2	2.0	0.153	1466	0	1466	1319	0.93	0.60	38.1	3.36
	L, I	4.0	3.2	3.8	0.847	1292	0	1292		0.54			
Tick Ridge, Nov. 7	U, S	6.0	7.4	4.3	0.254	192	30	162	158	0.85	0.46	0.3	0.67
	U, I	5.5	6.8	2.8	0.746	195	38	157		0.32			
	L, S	5.0	7.3	2.7	0.349	169	19	150	145	0.99	0.54	0.6	0.84
	L, I	4.0	7.3	1.6	0.651	160	18	142		0.30			
Strouds Run, Nov. 13	N, S	4.0	7.6	2.2	0.264	345	211	134	138	0.94	0.78	6.0	1.82
	N, I	2.0	7.5	4.9	0.736	318	178	140		0.72			
	W, S	1.0	7.4	3.4	0.313	360	187	173	170	0.95	0.74	10.6	2.20
	W, I	1.0	7.6	5.9	0.687	355	187	168		0.64			
Sandy Run, Nov. 14	U, S	2.0	3.0	3.0	0.546	949	0	949	760	0.91	0.83	6.0	1.82
	U, I	1.0	3.9	7.1	0.454	533	0	533		0.73			
	L, S	2.5	4.3	3.1	0.794	334	0	334	280	0.97	0.83	6.3	1.85
	L, I	1.0	4.2	3.4	0.206	78	5	73		0.29			
Margaret Creek, Nov. 20	S, S	2.5	7.2	2.4	0.370	278	143	135	209	1.11	0.60	9.7	2.13
	S, I	2.0	6.9	1.6	0.630	683	430	253		0.30			
	W, S	5.5	7.3	1.1	0.458	282	121	161	149	0.93	1.01	5.0	1.71
	W, I	5.0	7.4	3.6	0.542	273	135	138		1.07			

*Key to abbreviations. Site: U, upper; L, lower; N, north; W, west; S, south. Type of water: S, stream water; I, interstitial water.

values. Subtracting carbonate from total conductivity and applying the faunal weighting factors (which show, incidentally, that only at Sandy Run and Long Run were the majority of animals in the upper layer of sediment) gives the numerator values (Table 2, column A) of the preliminary physical—chemical index.

Dissolved oxygen treated as decimal saturation showed the expected constancy in stream water ($\bar{x} = 0.97$, SD = 0.08, n = 14) but a wide variability in interstitial water ($\bar{x} = 0.51$, SD = 0.24, n = 14) (Table 2). Application of faunal weighting factors resulted in one of two parts (Table 2, column B) of the index's denominator. The BOD showed wide variability both in stream water ($\bar{x} = 2.8$, SD = 1.1, n = 14) and in interstitial water ($\bar{x} = 3.2$, SD = 1.8, n = 14). Little pattern was visible from the standpoint of faunal weighting factors or faunal abundance.

Stream flow showed more than a 100-fold difference between sites, though no site carried a large volume of water (Table 2). Minkers Run had the largest flow, more than doubling its volume between the upper and lower sites; Tick Ridge had the smallest flow. Raccoon Creek showed the largest difference in flow between sites, and, relative to the amount of flow, Long Run and Sandy Run showed the least. Taking the cube root of flow rate resulted in the second part (Table 2, column C) of the index's denominator. Light transmission in stream waters was high at the time of most collections; this indicated stable conditions and a lack of erosional runoff due to rainfall. Only Minkers Run (upper site, 60%; lower site, 77%) showed less than 100% transmission.

Iron ranged from trace amounts at Tick Ridge, Strouds Run, and Margaret Creek to 2.2 ± 0.5 ppm at Raccoon Creek; manganese ranged from trace amounts at Strouds Run to 19.6 ± 2.1 ppm at Raccoon Creek; and sulfate ranged from 72 ± 7 ppm at Tick Ridge, Strouds Run, and Margaret Creek to 1055 ± 375 ppm at Raccoon Creek (Table 2). Calcium and total hardness were lowest at Tick Ridge (51 ± 7 and 85 ± 7 ppm) and highest at Raccoon Creek (327 ± 46 and 650 ± 57 ppm). Little pattern was noted among the streams having intermediate values. Only at Raccoon Creek was there a consistent relationship between stream and interstitial waters with respect to these five chemical parameters, the values of interstitial waters being 0.67 ± 0.08 times those observed in stream waters.

The compensated noncarbonate conductivity, A/(B × C), values for each site are shown in Table 3 in decreasing order of magnitude together with the faunistic data with which they were correlated. The number of taxa encountered, ranging from 3 at Raccoon Creek (upper site) to 10 to 11 at Strouds Run and Margaret Creek, showed a significant negative correlation with compensated non-carbonate conductivity ($P < 0.01$). Geometric mean abundance per station of total meiofauna and Diptera likewise showed significant negative correlations ($P < 0.01$ and $P < 0.05$), but for no other taxon was this the case.

The fauna was dominated by Nematoda, Rotifera, and Diptera, with occasionally high counts of Oligochaeta (Margaret Creek, south site; Strouds Run, west site), Gastrotricha (Tick Ridge, lower site; Strouds Run, west site), and Cyclopoida copepods (Strouds Run, north site). There was a total of 16 taxa (listed in Table 3 in decreasing order of $G\bar{x}$ abundance per station, summed over all sites), of which six (Turbellaria, Harpacticoida copepods, Pelecypoda, Ostracoda, Collembola, and Ephemeroptera) were observed at but three or four sites each and three (Cladocera, Gastropoda, and Plecoptera) were found at but one site each.

The two Long Run sites showed greatest taxon similarity, with 86% shared diversity (Fig. 1). The following sites joined at about the same level (53 to 54%): Minkers upper with the two Long Run sites, Margaret Creek west with south, Strouds west with Tick lower, and Raccoon upper with Sandy lower. At a slightly lower level (49%), the latter two pairs were joined, respectively, by Strouds north and Sandy upper. Already, at 49% similarity, four of the seven stream pairs have joined one another. At 46% similarity, Margaret Creek joined the Minkers upper—Long Run group, and it in turn was joined at 40% similarity by the Strouds Run—Tick lower group to form one major complex of predominantly unpolluted streams. Also, at 46% similarity, Raccoon lower and Minkers lower joined. This group joined the central complex at 35%, just 1% less than the point at which Tick upper joined the Raccoon upper—Sandy Run group. Finally, the latter group joined the enlarged complex at the 30% similarity level to complete the series.

Sites were arranged along the top of the figure, within the confines of the axes provided by their sequential joining, in such a manner as to maximize the $S_{H'}$ values between adjacent members. A pattern emerges with two interesting features. One is that the H' values for sites as arranged along the figure form a sine wave, with but two reversals, increasing from a minimum on the left to a maximum just right of center to a minimum again on the right. The second feature is that this arrangement portrays a bipolarity of faunal dominance, from rotifer-dominated faunae on the left to nematode-dominated faunae on the right, with the high diversity—low dominance faunae in the center.

TABLE 3

DATA FROM STREAM STUDY SITES FOR COMPENSATED NONCARBONATE CONDUCTIVITY AND GEOMETRIC MEAN ($G\bar{x}$) ABUNDANCE OF MEIOFAUNAL TAXA PRESENT

(Correlations between index and faunistic parameters are included)

Stream, site	Compensated noncarbonate conductivity [A/(B × C)]	No. of taxa	$G\bar{x}$ abundance (and 95% confidence limits) per station (n = four stations per site) expressed as organisms per 10 sq. cm of surface					
			Total meiofauna	Nematoda	Rotifera	Diptera	Oligochaeta	Gastrotricha
Raccoon Cr., U	1279	3	22.8(9–60)	2.3(0–30)	11.9(2–57)	3.2(0–19)		
Raccoon Cr., L	654	6	21.2(8–57)	6.7(0–102)	4.7(1–18)	0.6(0–3)		0.4(0–3)
Tick Ridge, U	518	7	159.6(34–751)	43.8(29–66)	68.5(3–1371)		0.3(0–2)	6.8(0–85)
Sandy Run, U	504	6	40.2(15–108)	1.6(0–15)	10.5(0–167)	18.3(10–34)		0.5(0–4)
Minkers Run, U	445	9	95.1(21–422)	42.3(11–162)	11.1(3–41)	12.6(1–159)	2.4(0–31)	
Minkers Run, L	333	8	213.8(122–374)	152.1(73–318)	27.4(11–67)	1.6(0–14)	1.6(0–17)	0.3(0–2)
Tick Ridge, L	319	9	122.6(82–184)	36.1(12–110)	32.4(19–54)	2.4(0–12)	1.0(0–9)	24.8(9–68)
Sandy Run, L	182	6	186.0(102–340)	15.4(10–25)	144.8(82–257)	12.4(4–37)	1.6(0–14)	
Margaret Cr., S	164	11	134.9(97–188)	36.2(18–74)	7.3(1–48)	21.7(16–30)	21.8(7–69)	0.3(0–2)
Long Run, L	163	7	173.3(76–397)	104.5(37–293)	5.9(0–59)	27.8(6–131)	4.6(0–35)	0.6(0–7)
Long Run, U	129	7	138.1(49–388)	84.4(31–230)	6.4(0–67)	23.3(5–116)	2.5(0–13)	0.4(0–4)
Strouds Run, W	105	10	443.2(222–887)	52.4(22–127)	174.3(95–318)	16.5(6–48)	93.4(25–343)	44.4(8–254)
Strouds Run, N	97	10	284.7(220–368)	76.0(22–261)	91.8(39–215)	16.1(5–49)	14.0(2–91)	0.8(0–10)
Margaret Cr., W	87	10	254.1(110–588)	111.0(43–284)	24.5(7–90)	24.7(11–58)	12.5(1–106)	4.3(0–47)
Correlation coefficient with A/(B × C)		–0.759*	–0.666*	–0.529	–0.340	–0.608†	–0.350	–0.229

TABLE 3 (Continued)

Stream, site	Compensated noncarbonate conductivity [A/(B × C)]	No. of taxa	$G\bar{x}$ abundance (and 95% confidence limits) per station (n = four stations per site) expressed as organisms per 10 sq. cm surface								
			Cyclopoida	Tardigrada	Turbellaria	Harpacticoida	Pelecypoda	Ostracoda	Collembola	Ephemeroptera	Other
Raccoon Cr., U	1279	3									
Raccoon Cr., L	654	6	0.3(0–2)					0.4(0–4)			
Tick Ridge, U	518	7	0.3(0–2)		0.3(0–2)			0.9(0–6)			
Sandy Run, U	504	6	0.4(0–3)		0.4(0–3)						
Minkers Run, U	445	9	4.9(0–58)			2.8(0–44)	1.6(0–18)	0.4(0–3)	0.9(0–13)		
Minkers Run, L	333	8	8.9(2–35)			1.6(0–53)			0.3(0–2)		
Tick Ridge, L	319	9	0.7(0–9)	10.1(3–31)	0.3(0–2)						Plecoptera, 0.3(0–2)
Sandy Run, L	182	6	0.9(0–13)		4.3(0–39)						
Margaret Cr., S	164	11	7.6(0–68)	12.0(3–46)	2.6(0–38)	0.6(0–8)				0.3(0–2)	Cladocera, 1.3(0–10)
Long Run, L	163	7	2.5(0–39)		1.6(0–15)						
Long Run, U	129	7	3.4(0–67)		1.3(0–10)						
Strouds Run, W	105	10	10.7(6–19)	5.7(0–55)	7.0(0–55)		0.3(0–2)			0.3(0–2)	
Strouds Run, N	97	10	31.0(8–127)		5.4(2–13)		0.9(0–5)	0.5(0–4)		0.2(0–2)	
Margaret Cr., W	87	10	4.8(0–23)	10.4(6–20)	25.5(2–214)				0.3(0–2)		Gastropoda, 0.6(0–7)
Correlation coefficient with A/(B × C)			−0.422	−0.352	−0.416	+0.029	−0.083	+0.145	−0.006	−0.379	

*$P_{0.01}$ (12 df) = 0.661 (0.01 > P); from N. T. J. Bailey, *Statistical Methods in Biology*, Appendix 4, The English Universities Press, 1959.
†$P_{0.05}$ (12 df) = 0.532 (0.05 > P); from N. T. J. Bailey, *Statistical Methods in Biology*, Appendix 4, The English Universities Press, 1959.

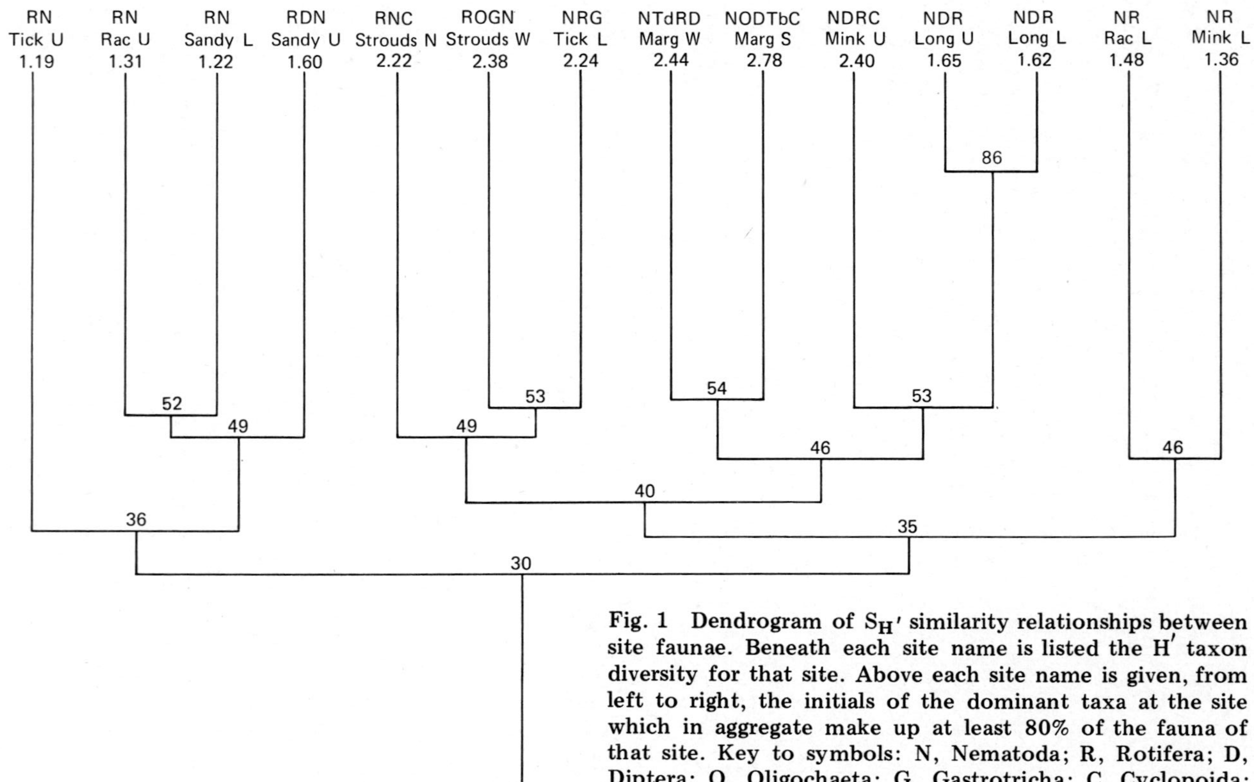

Fig. 1 Dendrogram of $S_{H'}$ similarity relationships between site faunae. Beneath each site name is listed the H' taxon diversity for that site. Above each site name is given, from left to right, the initials of the dominant taxa at the site which in aggregate make up at least 80% of the fauna of that site. Key to symbols: N, Nematoda; R, Rotifera; D, Diptera; O, Oligochaeta; G, Gastrotricha; C, Cyclopoida; Tb, Turbellaria; and Td, Tardigrada.

DISCUSSION

It was important to collect all sites during stable, fair weather undisturbed by temporary rains, winter runoff, or spring floods. The three parameters in the physical—chemical index are easily changed by an influx or spate of water, which would also change the observed abundance of meiofaunal taxa since the sampling scheme is closely tied to the water level.

Total conductivity is a blackbox-like measure that in no way specifies which ions are present, in what proportions, or involving what sorts and amounts of chemical complexing. What we have done in proposing the compensated noncarbonate conductivity index is to exclude from consideration those ions which are measured in titrating alkalinity. It can be said at present that, in the autumn season under relatively stable weather conditions, the index provides a relationship between certain physical—chemical parameters of the streams surveyed and the number of meiofaunal taxa and geometric mean abundance per station of total meiofauna and Diptera within those same streams.

Based on levels of compensated noncarbonate conductivity, Long Run is more closely associated with Strouds Run and Margaret Creek, the two unpolluted streams, than with Raccoon Creek and Sandy Run upper, the two polluted streams. On the other hand, both sites on Long Run have fewer than the expected number of taxa, and the lower site has greater than expected abundance of total meiofauna. Tick Ridge upper is most closely associated with the polluted streams, although it has a greater than expected number of taxa and total meiofaunal abundance and no observed Diptera. Minkers Run, Tick Ridge lower, and Sandy Run lower occupy intermediate positions, with the Sandy Run site having fewer than the expected number of taxa and both the lower sites at Minkers and Sandy Runs having greater than expected abundances of total meiofauna.

Results of the similarity analysis form a pattern, not greatly at variance with results of the correlation analysis. Long Run again is more closely associated with the unpolluted than with the polluted streams, with Minkers upper and Tick lower occupying comparable positions. Raccoon Creek and Sandy upper, along with Tick upper, are located at the ends of the array, but joining them in this case are Sandy lower and Minkers lower.

On balance, considering Strouds Run and Margaret Creek to be unpolluted and Raccoon Creek and Sandy upper to be polluted, the following comments on the other seven sites can be made. The Long

Run sites appear to have undergone the greatest amount of recovery, although they lack the number of taxa one would expect of an unpolluted stream. Tick lower and Minkers upper would be next in order from unpolluted to polluted, both having a moderate load of compensated noncarbonate conductivity and lacking the abundance of meiofauna expected of an unpolluted stream but both showing greatest similarity of the meiofaunal taxa present with the meiofauna of unpolluted streams or those sites well along in the recovery process. Next in order would be Minkers lower and Sandy lower, both of which have abundant meiofauna, an intermediate to depressed number of taxa, and a moderate to less than moderate load of compensated noncarbonate conductivity. The meiofauna of each of these two sites is most similar, each in its own way, to one of the Raccoon Creek sites. Finally, and most anomalous, is the Tick upper site which, aside from the low measured total conductivity, gives some indications that it may have been adversely affected by the recently stripped and reclaimed area at its headwaters.

ACKNOWLEDGMENTS

This research was supported by a grant from the Office of Water Research and Technology, U. S. Department of Interior, through the Ohio Water Resources Center. We wish to thank George Billy, Ohio Department of Natural Resources, Athens; Roger Jewell, Wayne National Forest, Athens; and James Tong, Department of Chemistry, Ohio University, Athens, for consultations regarding the nature and extent of acid mine pollution in southeastern Ohio. We are particularly indebted to Roger Jewell for suggesting the Raccoon Creek and Tick Ridge collecting sites. Our thanks go to Deborah Zmarzly, Lynn Sarles, and Stuart Arkett for aiding in various of the field collections.

REFERENCES

Bailey, N. T. J., 1959, *Statistical Methods in Biology*, The English Universities Press, Ltd., London.

Bennett, H. D., 1969, Algae in Relation to Mine Water, *Castanea*, 34: 306-328.

Dills, G., and D. T. Rogers, 1974, Macroinvertebrate Community Structure as an Indicator of Acid Mine Pollution, *Environ. Pollut. (London)*, 6: 239-262.

Faucon, A., and W. D. Hummon, 1976, Effects of Mine Acid on Longevity and Reproductive Rate of the Gastrotricha *Lepidodermella squammata* (Dujardin), *Hydrobiologia*, 50: 265-269.

Feldner, M. E., and W. S. Stanley, 1976, Macrofaunal Survey of Long Run, unpublished.

Hill, R. D., 1968, *Mine Drainage Treatment: State of the Art and Research Needs,* U. S. Department of the Interior, Federal Water Pollution Control Administration, Cincinnati, Ohio.

Hummon, W. D., 1974, S_H': A Similarity Index Based on Shared Species Diversity, Used to Assess Temporal and Spatial Relations Among Intertidal Marine Gastrotricha, *Oecologia (Berlin),* 17: 203-220.

——, 1977, Meiobenthos of the Mississippi Headwaters, *Am. Zool.,* 17: 869 (abstract).

Koryak, M., M. A. Shapiro, and J. L. Sykora, 1972, Riffle Zoobenthos in Streams Receiving Acid Mine Drainage, *Water Res.,* 6: 1239-1247.

Lackey, J. B., 1939, Aquatic Life in Waters Polluted by Acid Mine Waste, *U. S. Public Health Rep.,* 54: 740-746.

Lloyd, M., J. H. Zar, and J. R. Karr, 1968, On the Calculation of Information—Theoretical Measures of Diversity, *Am. Midl. Natur.,* 79: 257-272.

Massey, H. F., and R. I. Barnhisel, 1972, Copper, Nickel and Zinc Release from Acid Coal Spoil Material of Eastern Kentucky, *Soil Sci.,* 13: 207-212.

Napier, S., Jr., and W. D. Hummon, 1976, Survival of Mayfly Larvae Under Mine Acid Conditions, *Int. Rev. Gesam. Hydrobiol.,* 61: 677-682.

Nichols, L. E., and F. J. Bulow, 1973, Effects of Acid Mine Drainage on the Stream Ecosystem of the East Fork of the Obey River, Tennessee, *J. Tenn. Acad. Sci.,* 48: 30-39.

Parsons, J. D., 1956, The Effects of Acid Strip Mine Pollution on the Ecology of a Central Missouri Stream, *Diss. Abstr.,* 16: 1301-1302.

——, 1968, The Effects of Acid Strip-Mine Effluents on the Ecology of a Stream, *Arch. Hydrobiol.,* 65: 25-50.

Roback, S. S., and J. W. Richardson, 1969, The Effects of Acid Mine Drainage on Aquatic Insects, *Proc. Acad. Nat. Sci. Philadelphia,* 12: 81-107.

Warner, R. W., 1971, Distribution of Biota in a Stream Polluted by Acid-Mine Drainage, *Ohio J. Sci.,* 71: 202-215.

CYTOTOXICITY OF UNTREATED COAL-CONVERSION GASIFIER CONDENSATE

T. WAYNE SCHULTZ,*† JAMES N. DUMONT,‡ and LOLA M. KYTE‡
*University of Tennessee, Knoxville, Tennessee, and Oak Ridge Graduate School of Biomedical Sciences, Oak Ridge, Tennessee;
‡Biology Division, Oak Ridge National Laboratory, Oak Ridge, Tennessee

ABSTRACT

The possible environmental effects of an untreated gasifier condensate (filtered product water from the Synthane gasification process) are examined. Populations of the ciliate *Tetrahymena pyriformis* were exposed to varying concentrations of condensate, and their behavior, respiration, cytology, and growth rates were investigated. Concentrations of 1% and less cause little if any behavioral (shape and motility) changes, but concentrations of 2% and greater decrease motility and increase cell lysis. The condensate causes a nonlinear, dose-dependent reduction in oxygen consumption. At concentrations of less than 2%, no alteration in respiration is noted over a 300-min exposure. At all concentrations tested, the most striking cytological alterations are in the mitochondrial matrixes, which become more electron dense. Cell membranes are also disrupted, and mucocysts are discharged. Population growth is reduced by concentrations of product water as low as 0.2% and is completely inhibited by 1% concentration. The density of test populations plateaus at values inversely related to concentration. Pure phenolic compounds elicit similar responses.

For many years gas and oil have been the major energy sources in the United States, but, with the depletion of domestic reserves and increasing dependence on foreign sources for these fuels, coal, which accounts for 75% of the U. S. fossil-fuel resources, will be relied on more and more as a source of energy. To be of greater industrial and commercial value, however, coal must be converted into gas and oil.

†Current address: Department of Biology, Pan American University, Edinburg, Texas.

One of the more promising coal-to-gas schemes is the Synthane gasification process, which encompasses basically four steps: (1) pretreatment; (2) gasification; (3) purification; and (4) methanation to produce a medium-Btu gas (for a review of this process, see Forney et al., 1974). A potential effluent is the contaminated condensate from the gasifier (Forney et al., 1974). Chemical analyses of this aqueous effluent (Forney et al., 1974; Schmidt, Sharkey, and Friedel, 1974; Ho, Clark, and Guerin, 1976) show it to be a potential environmental and health hazard since it contains large quantities of phenolic compounds.

The ciliate *Tetrahymena pyriformis* has been used as a test organism and model to study the effects of trace elements (Carter and Cameron, 1973; Tingle, Paulet, and Cameron, 1973) and organic compounds (Gray and Kennedy, 1974; Schultz and Dumont, 1977). It provides a rapid and inexpensive means of analyzing several parameters of cellular effects of potential pollutants. This paper reports some lethal and sublethal effects of untreated laboratory-scale Synthane gasifier condensate on *Tetrahymena*.

MATERIALS AND METHODS

General

Tetrahymena pyriformis, strain GL-C, were grown in axenic cultures in a semidefined proteose peptone medium (Schultz and Dumont, 1977). Cultures were grown without shaking in 500-ml Erlenmeyer flasks containing 100 ml of medium in a 28°C water bath. Stationary-growth-phase cultures (3 to 4 days old, with 90,000 cells/ml) were used throughout the investigation. The Synthane process condensate was provided by the Pittsburgh Energy Research Center, Pittsburgh, Pa.

Behavior

The behavior of test cells was observed with a phase-contrast microscope immediately after the test solutions were added, and the general reactions and appearance of the cells were recorded. Particular attention was paid to changes in shape and motility, and the percent of cells altered was estimated at various times.

Oxygen Consumption

The effect of the gasifier condensate on the respiratory rate of *Tetrahymena* was examined in a respirometer (Gilson Medical Electronics), with the temperature stabilized at 28°C. Reaction flasks with a capacity of 15 ml, each containing 4.5 ml of cell suspension, were used. A paper wick saturated with 15% KOH solution was placed in the center well to absorb free CO_2, and 0.5 ml of condensate at a concentration 10 times that desired for final testing was placed in the sidearm of the reaction vessel and later was mixed with the sample. Experiments were performed with final concentrations of 1 through 5%. Experimental and control samples were monitored at 15-min intervals for a total of 300 min, and total oxygen consumed (μ liter) was recorded for each time point. The immediate effect of gasifier condensate on respiration was determined with a biological oxygen monitor (Yellow Springs Instrument Company, Inc.) equipped with a platinum—oxygen electrode. The temperature was stabilized at 28°C, and the rate of oxygen consumption (%/min) was recorded. For each test a 3-ml aliquot of cells was aerated for 3 min in the sample chamber before recording the data. The 0.3-ml aliquots of 10x condensate or sterile medium were added directly to the sample chambers, which contained either the cell suspension or sterile medium, and the samples were continuously stirred. Oxygen consumption was recorded before, during, and immediately after the test solutions were added to the chambers.

Electron Microscopy

The test animals were removed from the condensate solution at intervals of 15, 30, 60, 120, and 360 min for electron microscopy and were fixed according to the technique of Kennedy and Richardson (1969). Fixed cells were dehydrated in ethanol and propylene oxide and then infiltrated with and embedded in Epon (Shell Chemical Company). Thin sections stained with uranyl acetate and lead citrate were viewed with an electron microscope (Hitachi) operated at 75 kV.

Population Growth and Density

The effects of gasifier condensate on growth rates and population densities of *Tetrahymena* were measured spectrophotometrically. Optically matched 18- by 150-mm glass culture tubes were used, and experiments were performed with final concentrations of 0.1, 0.2, 0.4, 0.6, 0.8, and 1% condensate. Each tube, containing 10 ml of medium, was inoculated with 0.2 ml of log-growth-phase culture and maintained in a water bath at 28°C. Absorbance at 540 nm was used to estimate population density. The period of exponential growth was determined for each test concentration, and the best line for the data (Y is absorbance, and X is time) was fitted by the least-squares method of linear regression. The slopes were tested for homogeneity using analysis of covariance at the 5% level ($P = 0.05$).

Pure Compounds

Pure phenolic components known to be present in gasifier condensate (3 methyl; 2,6-dimethyl; 3,5-dimethyl; and 4-ethyl) were also examined for their effects on the behavior, respiration, and cytology of *Tetrahymena*.

RESULTS

Behavior

Little, if any, behavioral alteration was caused by 1% toxicant. Reduction of motility and increased incidence of alteration in cell shape and lysing was directly related to concentrations of from 2 to 5%.

The general scheme of morphological alteration, although a continuum, can be subdivided into three phases: (1) the normal pear-shaped cells become rounded posteriorly; (2) the rounded cells become completely spherical; (3) the spherical cells become swollen. Concomitant with shape changes are alterations in the contractile vacuole. For example, the rate of discharge decreased and the volume of the vacuole increased with increased concentrations and/or increased exposure to condensate. Table 1 summarizes morphological and motility changes in test populations.

Oxygen Uptake

Gasifier condensate caused a nonlinear, dose-dependent reduction in oxygen consumption (Fig. 1). Oxygen uptake by *Tetrahymena* exposed to 2% or less product water was not different from controls. Respiration of cells exposed to 3% toxicant continued at a normal rate for the initial 90 min of exposure but then decreased with time and plateaued after 300 min. Respiration of cells treated with 5% condensate fell quickly and then leveled off after 120 min. Uptake of oxygen by ciliates exposed to 4% toxicant was intermediate to that for those exposed to 3 and 5%. The reduction of culture respiratory rates was correlated with the number of viable cells in the culture at any given time. For example, we can see from Table 1 that, at 180 min in 5% condensate, ~75% of the individuals in the population had lysed. Figure 1 indicates that, at the same concentration and time, the respiratory rate was reduced to about

TABLE 1

SUMMARY OF MORPHOLOGICAL AND MOTILITY CHANGES IN *Tetrahymena* EXPOSED TO SYNTHANE PRODUCT WATER

Synthane concentration, %	Exposure, min	General appearance*	Motility†
2	30	Slight alterations in shape	100% normal
	60	<5% lysed; 60% phase I and II	50% reduced
	120	10% lysed; 15% phase II and III	15% reduced (many feeding on detritus)
	180	10% lysed; <10% phase III	100% normal
3	30	Occasional lysing; <10% phase I and II	100% normal
	60	10% lysed; 25% phase I to III	90% reduced
	120	15–20% lysed; 50% phase I to III	5% immobile; 90% reduced
	180	<25% lysed; 70% phase I to III	10% immobile; 85% reduced
5	30	25% phase I and II	99% reduced
	60	10% lysed; 85% phase II and III	50% reduced; 50% immobile
	120	50% lysed; 50% phase III	25% reduced; 75% immobile
	180	75% lysed; 25% phase III	100% immobile

*Percent of initial population expressing trait; remainder have normal shape.
†Percent of existing population expressing trait; remainder have normal motility.

35% of that of the control. Thus, at least with this toxicant, respiratory rates undoubtedly reflected changes in population densities rather than alterations in the respiration of individuals. We should note that the slow, continued uptake of oxygen, even up to 240 min in 5% condensate, was probably caused by respiratory activity of detritus (mitochondria) from lysed cells.

The following results were obtained in tests designed to monitor changes in oxygen utilization and/or saturation levels immediately after addition of toxicant to the oxygen monitor (during the first 3 min). Introducing 0.3 ml of full-strength gasifier condensate into the sample chamber containing the cell suspension caused an

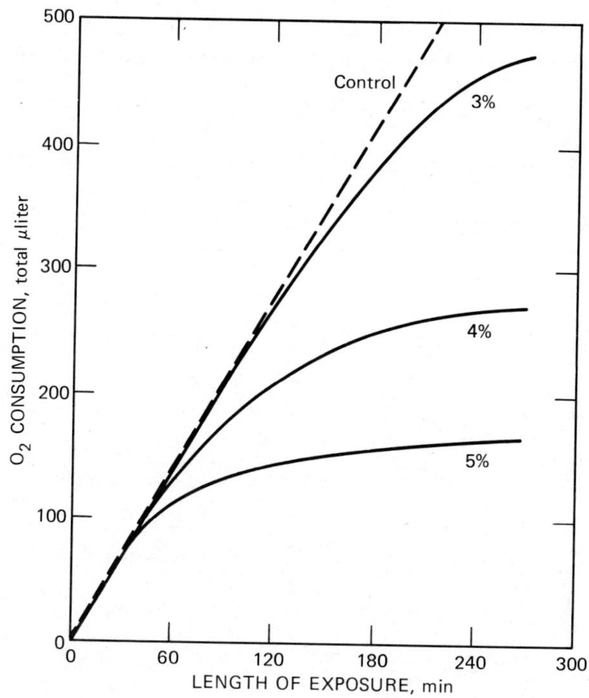

Fig. 1 Effect of gasifier condensate on respiration of *Tetrahymena* at 28°C.

immediate 5- to 7-unit decrease in the percent-saturation reading (Fig. 2, curve A). The total time elapsed from the addition of toxicant to reestablishment of a linear decrease in percent saturation (which we refer to as reaction time) was ~30 sec. In this case introducing half-strength condensate caused an immediate 1-unit decrease in the percent saturation (Fig. 2, curve B); the reaction time was less than 15 sec. The concentration of test solution added affected the slope (rate of respiration) of the reestablished oxygen-consumption line, as well as the reaction time and percent saturation. Adding 0.3 ml of sterile medium to the cell suspension caused a very slight decrease in the rate of oxygen uptake, less than 0.5%/min (Fig. 2, curve C). Sterile medium injected into sterile medium had no effect (Fig. 2, curve F). Full-strength condensate introduced into sterile medium caused a decrease in saturation of about 6 to 8 units, with a reaction time of less than 15 sec (Fig. 2, curve D). Half-strength condensate added to sterile medium caused only a 3% decrease within 9 sec (Fig. 2, curve E). These tests showed that, although the addition of gasifier condensate initially reduced the oxygen content of the medium, respiration was also reduced.

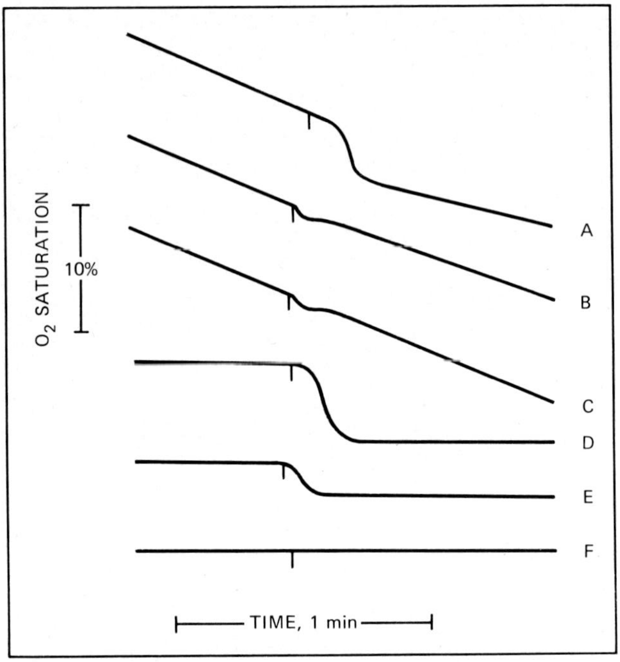

Fig. 2 Immediate effects on respiration of adding gasifier product water at 28°C, measured with an oxygen electrode with constant stirring. A, Undiluted Synthane product water added to cell suspension (final concentration, 10%). B, 50% Synthane product water added to cell suspension (final concentration, 5%). C, Sterile medium added to cell suspension. D, Undiluted Synthane product water added to sterile medium (final concentration, 10%). E, 50% Synthane product water added to sterile medium (final concentration, 5%). F, Sterile medium added to sterile medium.

Cytology

Elliott and Kennedy (1973) reviewed the fine structure of *Tetrahymena.* A series of three unit membranes enclose each cell; the innermost is the plasma membrane, and the outer two form the pellicle (Fig. 3a). Located immediately underneath the plasma membrane are numerous small mucocysts. The matrix of mature mucocysts displays a crystalline lattice pattern (Fig. 3b). The circular mitochondrial matrix varies with different fixation techniques; Elliott and Kennedy (1973) stated ". . . with good preservation the matrix is densely granular."

The most prominent cytological alterations of *Tetrahymena* exposed to gasifier condensate occurred in the mitochondria and the pellicular and plasma membranes. For all concentrations tested, the first cytological indication of alteration was observed in the first 15 min, during which time there was a marked increase in the electron density of the mitochondrial matrixes (Fig. 4). The increased matrix density remained unaltered over the 6-hr test period in toxicant concentrations as low as 1% (Fig. 4a).

Disruption of the cellular membranes occurred concomitantly with mitochondrial alteration. Initially the pellicular membrane split or separated (Fig. 4b), and the disruption of the plasma membrane and cell lysis, which followed, were directly related to the concentration of the toxicant and the length of exposure, at least up to ~240 min. An indirect indication of lysis was the presence of cellular

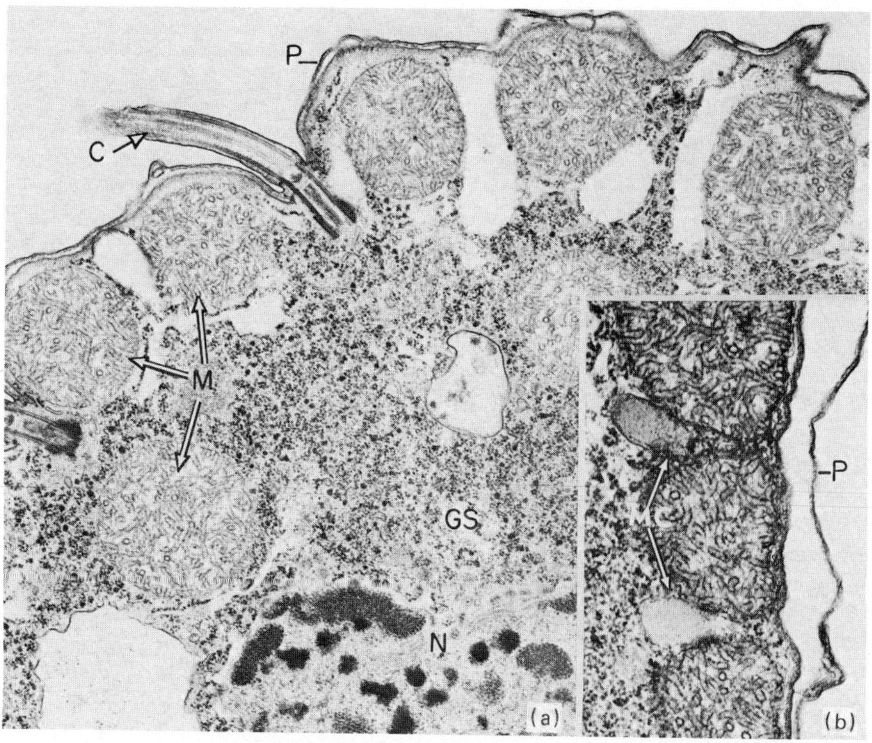

Fig. 3 Electron micrographs of *Tetrahymena pyriformis*, strain GL-C (3- to 4-day-old cultures). (a) Section of a portion of a control cell showing cilium (C); pellicle (P); mitochondria (M), with tubular cristae and granular matrices; electron-dense glycogen stores (GS); and a segment of the macronucleus (N) (16,000×). (b) Higher magnification of the periphery of a control cell revealing pellicular membrane organization, a mitochondrion, and mature mucocysts (MC) with internal crystalline lattice (16,400×).

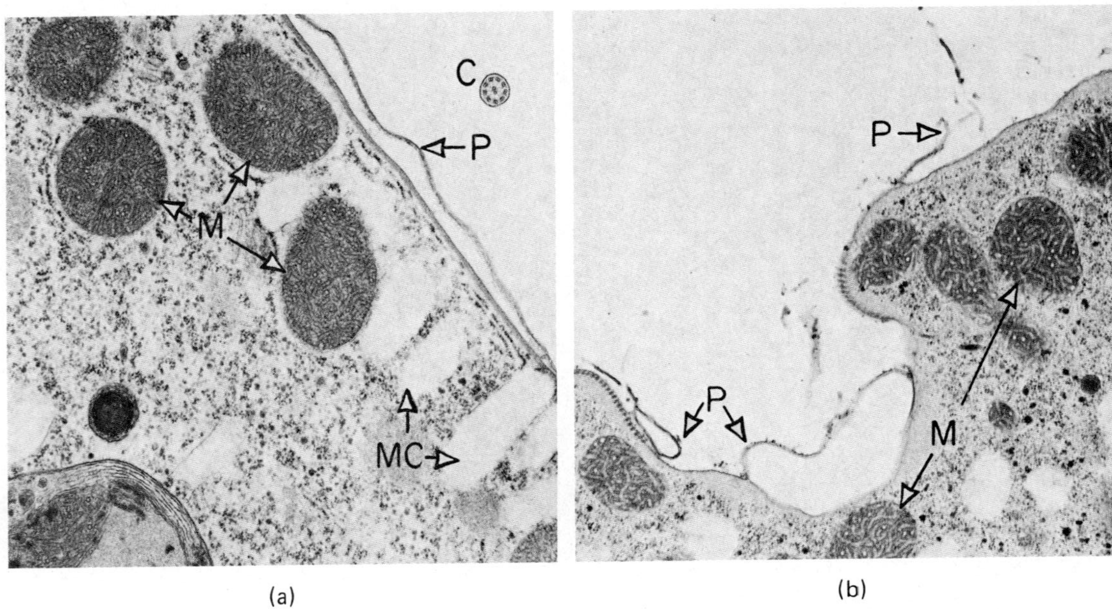

Fig. 4 Electron micrographs of *Tetrahymena Pyriformis*, strain GL-C (3- to 4-day old cultures). (a) Portion of the periphery of a cell exposed to 1% gasifier condensate for 360 min showing cilium (C), pellicle (P), several mucocysts (MC), mitochondria (M) with electron-dense matrices, glycogen stores (14,500X). (b) Portion of a cell exposed to 4% gasifier product water for 15 min. The mitochondria are electron dense, and the pellicle has split and separated from the cell (14,500X).

detritus in the food vacuoles of ciliates remaining after exposure to the toxicant.

Test cells also discharged mucocysts (Fig. 4a) within the first 15 min for all concentrations tested. All mucocysts appeared to be affected since no crystalline lattice was seen in mucocysts of any cells exposed to concentrations greater than 1%. Cells exposed to 1% toxicant or less retained a few mucocysts with a crystalline matrix, however, even after 6 hr of exposure. The apparent decrease in glycogen content with increased length of exposure suggested for *Tetrahymena* exposed to phenol (Schultz and Dumont, 1977) was not as apparent for cells exposed to gasifier condensate.

Population Growth

Population growth of axenic cultures of *Tetrahymena* grown at 28°C was measured spectrophotometrically as absorbance at 540 nm. Growth of cultures exposed to condensate concentrations of 0.2 to 1% for 72 hr showed an inverse relationship to concentration, with the mean rate during exponential growth being significantly affected ($P < 0.05$) by concentration (Fig. 5). In the presence of 0.2% toxicant, cells grew at a somewhat slower rate than controls, but in 0.4% they grew at about half the rate of controls. These cultures plateaued at 48 hr at about 66% of the final control value. In 0.6% toxicant, populations grew at an even slower rate and plateaued within 36 hr at ~33% of the final control value. Growth of cultures exposed to 0.8% condensate plateaued in ~30 hr at an absorbance less than 25% of that of 72-hr controls. Gasifier condensate at 1% concentration inhibited population growth during the 72 hr tested.

Pure Compounds

The behavioral, respiratory, and cytological results for the four pure phenolic derivatives paralleled results described by Schultz and Dumont (1977) for pure phenol and those given here for gasifier condensate. We should note, however, that, although *Tetrahymena* detoxified or otherwise altered phenol (as indicated by return to normal shape and motility within 8 hr), cells exposed to methyl- and ethyl-substituted phenols did not recover as rapidly, exhibiting shape and motility alterations for as long as 24 hr after exposure.

DISCUSSION

Although the gasifier condensate water used in this investigation is not a final effluent, there is a real need for biomedical and

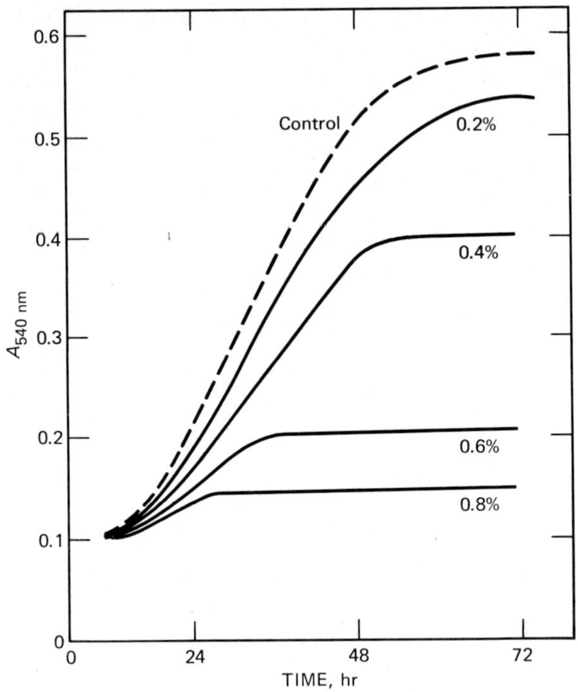

Fig. 5 Effect of gasifier product water on population growth of *Tetrahymena* at 28°C. Slopes of the mean log-phase growth lines (based on three replicate determinations) for control populations and for populations exposed to 0.2, 0.4, 0.6, and 0.8% condensate are 0.013, 0.013, 0.006, 0.004 and 0.002, respectively. Analyses of covariance were performed to test these slopes. Control and 0.2% populations are not significantly different, but control and 0.2% populations and the slopes for 0.4, 0.6, and 0.8% populations differ significantly ($P < 0.05$).

environmental screening of such untreated product streams to ensure water quality and to avoid potential health hazards since the coal-conversion process is advancing from bench-scale and pilot-plant facilities to demonstration plants using 5000 tons/day and eventually to commercial-level plants using 25,000 tons/day.

The major organic components of the Synthane process gasifier condensate are phenol, which alone accounts for 35% by weight of the total organics present (Ho, Clark, and Guerin, 1976), and its methyl-substituted derivatives.

The most obvious effects of gasifier condensate on *Tetrahymena* were changes in shape and discharge of its mucocysts. Similar shape alterations and mucocyst discharge have been observed in ciliates exposed to a variety of compounds, including mercuric chloride (Tingle, Paulet, and Cameron, 1973), nontobacco cigarette-smoke residue (Gray and Kennedy, 1974), and phenol (Schultz and Dumont, 1977). These responses undoubtedly represent generalized reactions to a wide variety of adverse environmental conditions. Concomitant with these events was the cessation of contractile vacuole function; the contractile vacuole filled but ceased to discharge and became severely enlarged. We interpreted this response as indicative of the action of the phenolic components since pure phenol (Schultz and Dumont, 1977) and the alkane-substituted phenols examined in this study all elicit a similar response.

The cytological effect of the toxicant on mitochondria is similar to mitochondrial matrix changes noted in animals exposed to as little as 10 mg/liter of phenol (Schultz and Dumont, 1977). Unlike the phenol-treated animals, however, in which the matrix density returned to normal after 240 min, the matrix density of mitochondria of *Tetrahymena* treated with 1% gasifier condensate did not return to normal, possibly because methyl-substituted phenols, which have been shown to elicit the same mitochondrial response, were present in the sample. Changes in the configuration of mitochondria have also been observed in media-grown exponential-growth-phase *Tetrahymena* transferred to inorganic salts (Nilsson, 1970) or exposed to higher temperatures (Nilsson, 1976) or to dimethyl sulfoxide (Nilsson, 1974). Hamburger and Zeuthen (1957; 1960) and Skriver and Nilsson (1974) reported a reduction in the rate of respiration after these treatments. Schwab-Stey, Schwab, and Krebs (1971) described three configurations of mitochondria isolated from *Tetrahymena* (orthodox, intermediate, and condensed) but were unable to correlate the configurational states with specific energy stages or physiological conditions of the animal. The orthodox configuration corresponds to the normal or control type (see Fig. 3), and the condensed form is similar to mitochondria in cells exposed to gasifier condensate (see Fig. 4). In our study the configurational change corresponds to reduced respiratory activity, but other physiological parameters that may also contribute to such mitochondrial transformations are unknown at this time.

A discussion of the significance of the effects on respiratory rates of *Tetrahymena* cultures of longer exposure (up to 240 min) to gasifier condensate must be tempered with the knowledge that during these experiments the density of the population was reduced because of cell lysis. Thus the respiratory rates correlated with the number of viable cells in the culture at any given time. The immediate decrease in oxygen saturation after addition of toxicant to the cultures probably represents an interaction of the toxicant and the medium rather than a respiratory effect (see Fig. 2). After the sharp initial decline of oxygen saturation, however, the oxygen consumption rate declined again; this indicated a reduction in respiration of the population.

Population growth rates and final stationary-phase population densities were sensitive to lower concentrations of product water than were either cytological changes or respiration rates. The exponential growth rates (slopes) for cultures exposed to as little as 0.2% toxicant were significantly reduced from those of controls. That this was not observed for cells grown in the presence of 5 to 25 mg/liter of pure phenol (Schultz and Dumont, 1977) suggests that some component of the gasifier condensate other than phenol affected the rate of population growth (cell division). On the other hand, the lengthened lag phase and the premature onset of stationary growth at higher concentrations is consistent with data obtained from the growth rates of phenol-treated cells (Schultz and Dumont, 1977). Since the cessation of population growth at suboptimum levels is characteristic of *Tetrahymena* cultures exposed to a variety of compounds (Blum, Kirshner, and Utley, 1966; Satir, 1967; Meyer et al., 1972; Schultz and Dumont, 1977), it is difficult to determine precise modes of action that bring about this effect.

Finally, attempts to relate the observed effects of gasifier condensate to one or a group of its specific identified components focus attention immediately on the fact that very high concentrations of phenol (2100 mg/liter) and methylated phenols (3120 mg/liter) were present in the sample (Ho, Clark, and Guerin, 1976). Although effects were almost certainly elicited by many of the other components present, it is interesting to compare the data presented here with data previously obtained for pure phenol or some methyl phenols. For example, the point at which respiration plateaued in populations exposed to 5% condensate (263 mg/liter total phenols) was comparable to that in populations exposed to 100 mg/liter phenol (Schultz and Dumont, 1977), 400 mg/liter 3-methyl phenol, 300 mg/liter 2,6- and 3,5-dimethyl phenol, and 200 mg/liter 4-ethyl phenol. There is no assurance that phenol per se and/or the other phenolic compounds present were primarily responsible for this effect. Nonetheless, they were the major organic components, and we can be reasonably confident that they had the greatest influence on the parameters measured. Although correlation of toxicity with π and σ of selected phenolic compounds has been successful (Kopperman, Carlson, and Caple, 1974), the compounds that the investigators chose for testing had no *ortho* substitution; the Hammett equation is valid only for *meta* and *para* substitution (Jaffe, 1953). Partition coefficients per se may be of value for indicating relative toxicities, however. We have found, for example, that, within certain groups of related compounds, there is a direct correlation between the partition coefficient and the toxicity of a compound, but elaboration of this concept must await further investigations.

ACKNOWLEDGMENTS

We acknowledge the excellent technical assistance provided by Patricia Miller and Carole Richter. We are also indebted to the Pittsburgh Energy Research Center and especially P. M. Yavorsky for the Synthane condensate samples used in this investigation.

When this work was done, T. W. Schultz was a postdoctoral investigator supported by subcontract No. 3322 from the Biology Division, Oak Ridge National Laboratory, to the University of Tennessee.

REFERENCES

Blum, J. J., N. Kirshner, and J. Utley, 1966, The Effect of Reserpine on Growth and Catecholamine Content of *Tetrahymena*, *Mol. Pharmacol.*, 2: 606-613.

Carter, J. W., and I. L. Cameron, 1973, Toxicity Bioassay of Heavy Metals in Water Using *Tetrahymena pyriformis*, *Water Res.*, 7: 951-961.

Elliott, A. M., and J. R. Kennedy, 1973, Morphology of *Tetrahymena*, in *The Biology of* Tetrahymena, A. M. Elliott (Ed.), pp. 58-87, Dowden, Hutchinson and Ross, Inc., Stroudsburg, Pa.

Forney, A. J., W. P. Haynes, S. J. Gasior, G. E. Johnson, and J. P. Strokey, 1974, *Analyses of Tars, Chars, Gases and Waters Found in Effluents from the Synthane Process*, Technical Progress Report No. 76, U. S. Bureau of Mines, Department of the Interior.

Gray, J. P., and J. R. Kennedy, Jr., 1974, Ultrastructure and Physiological Effects of Non-Tobacco Cigarettes on *Tetrahymena*, *Arch. Environ. Health*, 28: 283-291.

Hamburger, K., and E. Zeuthen, 1957, Synchronous Division in *Tetrahymena pyriformis* as Studied in an Inorganic Medium. The Effect of 2,4-dinitrophenol, *Exp. Cell Res.*, 13: 443-453.

——, and E. Zeuthen, 1960, Some Characteristics of Growth in Normal and Synchronized Populations of *Tetrahymena pyriformis*, *C. R. Trav. Lab. Carlsberg*, 32: 1-18.

Ho, C. H., B. R. Clark, and M. R. Guerin, 1976, Direct Analysis of Organic Compounds in Aqueous By-Products from Fossil Fuel Conversion Processes: Oil Shale Retorting, Synthane and Gasification and COED Coal Liquefaction, *J. Environ. Sci. Health, Part A*, 11: 481-489.

Jaffe, H. H., 1953, A Reexamination of the Hammett Equation, *Chem. Rev.*, 53: 191-261.

Kennedy, J. R., and S. H. Richardson, 1969, Fine Structure of *Vibrio cholerae* During Toxin Production, *J. Bacteriol.*, 100: 1393-1401.

Kopperman, H. L., R. M. Carlson, and R. Caple, 1974, Aqueous Chlorination and Ozonation Studies. I. Structure—Toxicity Correlations of Phenolic Compounds to *Daphnia magna*, *Chem.-Biol. Interact.*, 9: 245-251.

Meyer, R. R., C. R. Boyd, D. C. Rein, and S. J. Keller, 1972, Effects of Ethidium Bromide on Growth and Morphology of *Tetrahymena pyriformis*, *Exp. Cell Res.*, 70: 233-237.

Nilsson, J. R., 1970, Cytolysomes in *Tetrahymena pyriformis* GL. II. Reversible Degeneration, *C. R. Trav. Lab. Carlsberg*, 38: 107-121.

——, 1974, Effects of DMSO on Vacuole Formation, Contractile Vacuole Function, and Nuclear Division in *Tetrahymena pyriformis* GL, *J. Cell Sci.*, 16: 36-47.

——, 1976, Physiological and Structural Studies on *Tetrahymena pyriformis* GL, *C. R. Trav. Lab. Carlsberg*, 40: 215-355.

Satir, B., 1967, Effect of Actinomycin D on Cultural Growth Phases and on the Pattern of Total RNA Synthesis in *Tetrahymena*, *Exp. Cell Res.*, 48: 253-262.

Schmidt, C. E., A. G. Sharkey, Jr., and R. A. Friedel, 1974, *Mass Spectrometric Analysis of Product Water from Coal Gasification*, Technical Progress Report No. 86, U. S. Bureau of Mines, Department of the Interior.

Schultz, T. W., and J. N. Dumont, 1977, Cytotoxicity of Synthetic Fuel Products on *Tetrahymena pyriformis*. I. Phenol, *J. Protozool.*, 24: 164-172.

Schwab-Stey, H., D. Schwab, and W. Krebs, 1971, Electron Microscopic Examination of Isolated Mitochondria of *Tetrahymena pyriformis*, *J. Ultrastruct. Res.*, 37: 82-93.

Skriver, L., and J. R. Nilsson, 1974, Oxygen Uptake and Food Vacuole Formation in *Tetrahymena*, *J. Protozool.*, 21: 462.

Tingle, L. E., W. A. Paulet, and I. L. Cameron, 1973, Sublethal Cytotoxic Effects of Mercuric Chloride on the Ciliate *Tetrahymena pyriformis*, *J. Protozool.*, 20: 301-304.

AQUATIC TOXICOLOGY OF TRACE ELEMENTS OF COAL AND FLY ASH

WESLEY J. BIRGE
T. H. Morgan School of Biological Sciences, University of Kentucky, Lexington, Kentucky

ABSTRACT

Aquatic bioassays were performed on 22 coal elements, with developmental and juvenile stages of fish and amphibians used as test organisms. For more sensitive test species, LC_{50} values of 0.1 ppm or less were observed for 15 trace elements, and LC_1 values ranged down to 0.1 to 0.2 ppb for mercury and silver. Studies also were performed on the aqueous leaching characteristics and toxicological properties of coal-produced fly ash. A 52-kg sample of precipitator-collected fly ash was subjected to continuous flow-through washing for 2000 consecutive hours in a bench-scale ash-settling pond. Chemical characteristics of the simulated effluent compared closely with those recorded for actual ash-settling ponds. During the first 500 hr of operation, conductivity averaged 690 μmhos/cm, and a mean of 0.56 g/liter was observed for total dissolved solids. Gradual decreases were observed thereafter. Effluent pH averaged 4.3 during the first 500 hr but approached the level of influent water (7.1 to 7.9) by 2000 hr. Maximum effluent concentrations detected for seven selected metals were 126 ppm Al, 766 ppb Zn, 518 ppb Cu, 500 ppb Cd, 370 ppb Ni, 87 ppb Hg, and 8 ppb Ag. Mean concentrations for the first 500 hr of elution were 32.6 ppm Al, 350 ppb Zn, 156 ppb Cd, 155 ppb Ni, 110 ppb Cu, 2.1 ppb Ag, and 1.8 ppb Hg. All metals except mercury reached maximum levels within 500 hr, and concentrations declined thereafter. Mean mercury levels increased to 27 and 15 ppb for the second and third 500-hr elution intervals. After 522 and 1033 hr of elution time, continuous-flow treatment with undiluted ash effluent produced 100% mortality of frog and sunfish eggs. A 0.1 dilution at 1033 hr resulted in 42% hatchability of sunfish eggs. After 1775 hr of continuous leaching, undiluted ash effluent and 0.1 and 0.01 dilutions gave survival frequencies of 57, 76, and 88% for goldfish eggs, compared with 92% for controls. Metals analyzed for undiluted effluent administered to goldfish were well below LC_1 determinations, except aluminum, which was present at the LC_{50} level.

With the increasing dependence on coal as a national energy source, there is a serious need to investigate further the effects of coal production and use on environmental health. Numerous recent studies emphasize the extent to which coal utilization has grown beyond our ability to identify fully the hazardous trace elements in coal, quantify their release rates into the environment, and define their biological and health-related effects (Ayer, 1974; Yavorsky and Akhtar, 1974; Babu, 1975; Vaughan et al., 1975). Approximately two-thirds of the over 60 elements that occur in coal have been detected as environmental pollutants (Vitez, 1976). Although toxicological data are largely incomplete, most of the elements found in coal and other fossil fuels are known to have at least some toxic effects on animal species (National Academy of Sciences—National Academy of Engineering, Committee, 1973; Smith, Ferguson, and Carlson, 1975; Vitez, 1976).

The annual combustion of 600 million tons of coal constitutes the major source of environmental contamination with coal trace elements (Abel and Rancitelli, 1975; Bolton et al., 1975; Sheibley, 1975). The main sources of water pollution are power plants, which dispose of more than 50 million tons/year of bottom ash and precipitated fly ash (Rubin and McMichael, 1974; Chu, Nicholas, and Ruane, 1975). Large quantities of water are used in sluicing ash residues to settling ponds, and pond effluents contain toxic metals that affect the quality of receiving waters (Theis, 1975; Hildebrand, Cushman, and Carter, 1976). For each 1000-MW capacity, sluicing-water requirements for Tennessee Valley Authority (TVA) power stations average 11.5 million gal/day or 4.2 billion gal/year (Environmental Protection Agency, 1974; Chu, Nicholas, and Ruane, 1975). Current projections indicate that the rate of coal combustion will double by the mid 1980s (Vaughan et al., 1975). Therefore, a better understanding of the aquatic toxicology of coal-derived contaminants is essential if we are to maximize coal use and also institute safeguards necessary to maintain reasonable and proper environmental health.

In this study, aquatic bioassays were performed to establish a comparative toxicological ranking for 22 coal elements, identify those which may be particularly hazardous to aquatic ecosystems, and provide quantitative data for use in further evaluations of environmental standards and pollution-abatement technology. In addition, a bench-scale settling pond was developed to simulate fly-ash effluents and to investigate the aqueous leaching of toxic elements. Continuous flow embryo—larval bioassays were used for in situ monitoring of ash effluents to provide direct toxicological evaluations on complex suites of trace elements.

MATERIALS AND METHODS

Aquatic Bioassays

Semistatic embryo–larval bioassays were performed on the goldfish *(Carassius auratus)*, the rainbow trout *(Salmo gairdneri)*, and the narrow-mouthed toad *(Gastrophryne carolinensis)* with rapid-scan procedures previously described (Birge and Just, 1975). Eggs were exposed to coal elements from fertilization through 4 days posthatching, giving treatment periods of 7 days for toads and goldfish and 28 days for trout. Test water and toxicant were renewed at regular 12-hr intervals. Mean water hardness, with standard error, was 195 ± 5.4 ppm $CaCO_3$ for goldfish and toad stages and 104 ± 2.0 for trout. Test water pH averaged 7.4 ± 0.1. Dissolved oxygen was maintained near saturation by continuous, moderate aeration. Other chemical and physical characteristics of the reconstituted test water were described by Birge and Black (1977). Water temperature was 13.0 ± 0.5°C for trout eggs and 22.0 ± 1.0°C for other species.

Test organisms were examined daily to tabulate frequencies of mortality and teratogenesis. Control adjusted LC_1 and LC_{50} values were calculated for combined test responses by log probit analysis (Daum, 1969). Anomalous survivors were counted as lethals. Control eggs were cultured simultaneously with experimentals and under identical conditions, except for omission of toxic coal elements. Minimum sample size was set at 150 eggs per culture.

The 22 coal elements and test compounds selected for bioassay analysis are given in Table 1. Depending on the degree of anticipated toxicity, exposure concentrations were initiated at 10 to 100 ppm and continued at two- to tenfold dilutions until survival of experimental animals equaled or approached that observed for controls. Each coal element was administered at 10 to 14 exposure levels. Elemental concentrations of test water were monitored by atomic absorption spectrophotometry with a model 503 Perkin–Elmer unit equipped with an HGA-2100 graphite furnace and a mercury analyzer (Perkin–Elmer Corp., 1973).

Aqueous Leaching of Fly Ash

A bench-scale Plexiglas settling pond was designed to investigate the aqueous leaching characteristics of precipitator-collected fly ash obtained from a local 1000-MW coal-fired power plant. A 52-kg sample of dry ash was deposited in an 88.2-liter settling chamber. A Gilson Minipuls II peristaltic pump (Gilson Medical Electronics, Inc.) provided a continuous flow of water over the ash bed at a rate of 1

TABLE 1

COAL TRACE ELEMENTS SELECTED FOR BIOASSAY EVALUATIONS

Trace element	Bioassay test compound	Concentration* in coal, ppm
Aluminum	$AlCl_3$	10,440—12,900
Antimony	$SbCl_3$	0.50—1.26
Arsenic	$NaAsO_2$	4.45—14.02
Cadmium	$CdCl_2$	0.47—2.52
Cobalt	$Co(NO_3)_2$	2.90—9.57
Chromium	CrO_3	13.75—18.00
Copper	$CuSO_4$	8.30—15.16
Germanium	GeO_2	1.00—6.59
Lanthanum	$LaCl_3$	3.80—10.00
Lead	$PbCl_2$	4.90—34.78
Manganese	$MnCl_2$	33.80—49.40
Mercury	$HgCl_2$	0.12—0.20
Molybdenum	Na_2MoO_4	5.00—7.54
Nickel	$NiCl_2$	16.00—21.07
Silver	$AgNO_3$	0.03—0.12
Selenium	Na_2SeO_4	2.08—2.20
Strontium	$SrCl_2$	10.00—23.00
Thallium	$TlCl_3$	0.29—2.00
Tin	$SnCl_2$	0.03—4.79
Tungsten	Na_2WO_4	0.10—3.00
Vanadium	V_2O_5	28.50—32.71
Zinc	$ZnCl_2$	46—272

*The majority of values are from Ruch, Gluskoter, and Shimp (1974), Fulkerson et al. (1975), and Carter (1975) and represent means for multiple coal samples taken largely from western Kentucky and southern Illinois. Lower means for Ag, Tl, Sn, and W are from Lloyd (1976), and the upper mean for Ag is from Vaughan et al. (1975).

liter/hr, giving a detention time of 42 hr. Water was discharged from the settling chamber into an overflow-equipped effluent reservoir. The ash-to-water ratio and the detention time were calculated to approach conditions observed for a local ash-settling pond (Freeman and Birge, 1978). Also, the detention time was in good agreement with times reported for a number of TVA ash ponds (Chu, Ruane, and Steiner, 1976). Influent and effluent water samples were taken at 1- to 2-day intervals for 2000 hr of continuous operation to observe changes in water-quality parameters. Determinations were made on pH, conductivity, alkalinity, and total dissolved solids, and analyses were performed for seven selected metals (Ag, Al, Cd, Cu, Hg, Ni, and Zn). Alkalinity and total dissolved solids were determined according to standard methods (American Public Health Association, 1975), and metals were analyzed by atomic absorption spectrophotometry. Fly ash displayed good settling characteristics, and effluent water was essentially free of ash particulates. The fly-ash bed compacted sufficiently to impede interstitial percolation, limiting water movement primarily to surface flow. Through the first 770 hr of operation, influent water was distilled and deionized and had a conductivity less than 0.25 μmhos/cm and a pH of 6.8. Total dissolved solids and trace metals were not detectable. During the remainder of the leaching period, the settling chamber was supplied with carbon-filtered tap water, which had a pH of 7.1 to 7.9, conductivity of 141 to 252 μmhos/cm, alkalinity of 54 to 70 ppm $CaCO_3$, and total dissolved solids of 0.19 to 0.24 g/liter. Water temperatures ranged from 24.4 to 26.0°C.

Bioassay Monitoring of Fly-Ash Effluent

Continuous-flow bioassays were performed on the simulated ash effluent to evaluate toxicological properties of the aqueous leachates. Eggs of the goldfish *(Carassius auratus)*, redear sunfish *(Lepomis microlophus)*, leopard frog *(Rana pipiens)*, and Fowler's toad *(Bufo fowleri)* were used as test organisms. Full-strength effluent and serial dilutions thereof were perfused continuously through 300-ml egg chambers at flow rates of 200 to 300 ml/hr. Effluent dilutions of 0.1, 0.01, 0.001, and 0.0001 were achieved with a proportional diluter (Freeman and Birge, 1978). Exposure was maintained from fertilization through hatching, and results were expressed as percent survival (hatchability). Hatching times averaged 1.5 days for Fowler's toad, 2.5 days for the leopard frog, and 3 days for sunfish and goldfish. Minimum sample size was set at 100 eggs. Control egg chambers received the same influent water as that supplied to the simulated ash-settling pond. Bioassays were initiated after 522, 1033, and 1775 hr of continuous aqueous leaching of the original 52-kg fly-ash sample.

RESULTS

Embryo–Larval Bioassays

Fish and amphibian eggs were exposed to each of 22 selected coal elements (Table 1) from fertilization through 4 days posthatching, giving treatment periods of 28 days for trout and 7 days for the narrow-mouthed toad and goldfish. Probit-derived LC_{50} and

LC_1 values expressed in parts per million and parts per billion, respectively, are summarized in Table 2. In order of decreasing toxicity, on the basis of LC_{50} determinations, the 12 elements most lethal to trout were Hg, Ag, La, Ge, Ni, Cu, Cd, V, Tl, Pb, Cr, and Sr. The LC_{50} values were 0.005, 0.01, and 0.02 for Hg, Ag, and La, respectively; 0.05 for Ge and Ni; 0.09 for Cu; 0.13 for Cd; and 0.16 to 0.20 for V, Tl, Pb, Cr, and Sr. The calculated LC_1's for the more toxic elements were 0.2 for Hg and Ag, 0.4 for Ge, 0.6 for Ni, 0.8 for La, 1.8 for Cu, 2.5 for Pb, and 6.0 to 6.1 for Sr and Cd (Table 2).

The goldfish was the least sensitive of the three test species. The 12 elements most toxic to goldfish eggs were Ag, Hg, Al, Cd, As, Cr, Co, Pb, Ni, Sn, Zn, and V. The LC_{50} values were 0.03, 0.12, 0.15, 0.17, 0.49, 0.66, 0.81, and 1.66 for the first eight, respectively; 2.14 for Ni and Sn; 2.54 for Zn; and 4.60 for V. The LC_1 values obtained for these metals ranged from as low as 0.4 and 0.6 ppb for Al and Ag to 400 ppb for Zn. Certain coal elements (e.g., Ag and Hg) were more toxic to fish embryos; others (e.g., Al, Cd, Ge, and Pb) exhibited considerable toxicity to posthatched juveniles.

In bioassays with the narrow-mouthed toad, the 12 most lethal elements were Hg, Ag, Zn, Cr, Pb, Cd, Cu, As, Ge, Co, Ni, and Al. The LC_{50} values were 0.001, 0.01, 0.01, and 0.03 ppm for the first four elements, respectively; 0.04 ppm for Pb, Cd, Cu, and As; and 0.05 ppm for Ge, Co, Ni, and Al. The calculated LC_1's ranged only from 0.1 and 0.6 ppb for Hg and Ag to 1.6 and 3.2 ppb for Cd and Pb.

The LC_{50} values were averaged for all animal species (Table 3) to provide a simplified toxicological index for the 22 elements. This mean toxicity index provided a convenient ranking, consisting of three general toxicity groups. Group 1 included ten highly toxic elements with mean LC_{50} values below 1 ppm; group 2 included nine elements with LC_{50} values of 1 to 5 ppm; and group 3 included three elements with an LC_{50} range of 20 to 47 ppm.

The selected coal elements were also ranked according to a most sensitive species index (Table 3) based on median lethal concentrations determined for the animal species exhibiting highest sensitivity to each of the 22 elements. The LC_{50} values ranged from 0.001 ppm Hg to 2.90 ppm W. Elements with approximately the same LC_{50} concentration were further differentiated on the basis of LC_1 values.

Aqueous Leaching of Fly Ash

A 52-kg sample of precipitator-collected fly ash was subjected to continuous washing for 2000 hr at a flow rate of 1 liter/hr.

TABLE 2

COAL-ELEMENT LC_1 AND LC_{50} VALUES WITH 95% CONFIDENCE LIMITS

Element and animal species	LC_{50}, ppm	Confidence limit Lower	Confidence limit Upper	LC_1, ppb	Confidence limit Lower	Confidence limit Upper
Aluminum						
Trout	0.56	0.40	0.70	256	52.7	371
Goldfish	0.15	0.02	0.82	0.4	0.0	5.6
Toad	0.05	0.04	0.08	2.3	0.7	4.8
Antimony						
Trout	0.58	0.34	0.92	28.6	4.6	72.2
Goldfish	11.3	3.99	55.0	111	0.1	663
Toad	0.30	0.18	0.51	3.8	0.7	10.7
Arsenic						
Trout	0.54	0.42	0.67	39.7	15.5	71.6
Goldfish	0.49	0.39	0.61	15.5	7.5	26.6
Toad	0.04	0.02	0.07	1.6	0.2	4.4
Cadmium						
Trout	0.13	0.10	0.18	6.1	1.8	12.9
Goldfish	0.17	0.13	0.21	15.0	4.4	19.2
Toad	0.04	0.03	0.05	1.6	0.9	2.5
Chromium						
Trout	0.18	0.07	0.31	19.1	0.4	56.5
Goldfish	0.66	0.40	1.10	8.1	1.5	22.1
Toad	0.03	0.03	0.04	1.0	0.6	1.5
Cobalt						
Trout	0.47	0.38	0.58	34.2	13.8	60.8
Goldfish	0.81	0.27	2.27	6.8	0.0	42.6
Toad	0.05	0.02	0.08	0.9	0.3	2.0
Copper						
Trout	0.09	0.05	0.15	1.8	1.0	4.5
Goldfish	5.20	4.13	6.41	299	101	571
Toad	0.04	0.03	0.05	1.0	0.3	1.3
Germanium						
Trout	0.05	0.03	0.07	0.4	0.1	0.7
Goldfish	5.60	1.76	7.84	143	2.7	567
Toad	0.05	0.03	0.08	1.2	0.1	3.8
Lanthanum						
Trout	0.02	0.01	0.04	0.8	0.0	2.7
Goldfish	60.4	30.3	105	1987	136	6503
Toad	0.29	0.19	0.43	7.5	2.3	16.2
Lead						
Trout	0.18	0.10	0.32	2.5	0.2	8.1
Goldfish	1.66	0.85	3.05	14.6	1.4	53.0
Toad	0.04	0.02	0.07	3.2	0.1	9.0
Manganese						
Trout	2.91	1.85	4.37	388	66.2	800
Goldfish	8.22	2.39	24.6	21.5	0.1	182
Toad	1.42	0.84	2.40	3.0	0.5	9.8

(Table continues on following page.)

TABLE 2 (Continued)

Element and animal species	LC_{50}, ppm	Confidence limit Lower	Confidence limit Upper	LC_1, ppb	Confidence limit Lower	Confidence limit Upper
Mercury						
Trout	0.005	0.004	0.005	0.2	0.1	0.3
Goldfish	0.12	0.10	0.14	14.3	8.2	21.2
Toad	0.001	0.001	0.002	0.1	0.0	0.3
Molybdenum						
Trout	0.73	0.30	1.40	22.3	1.2	83.4
Goldfish	60.0	7.94	92.2	39.3	0.9	261
Toad	0.96	0.58	1.60	3.1	0.6	9.5
Nickel						
Trout	0.05	0.04	0.06	0.6	0.2	1.2
Goldfish	2.14	1.19	3.63	55.8	7.9	160
Toad	0.05	0.03	0.09	0.4	0.0	1.5
Selenium						
Trout	4.18	2.82	5.82	79.5	17.5	202
Goldfish	8.78	7.23	10.6	506	267	805
Toad	0.09	0.08	0.15	5.0	0.7	7.3
Silver						
Trout	0.01	0.01	0.02	0.2	0.1	0.4
Goldfish	0.03	0.02	0.03	0.6	0.4	0.9
Toad	0.01	0.01	0.03	0.6	0.0	2.0
Strontium						
Trout	0.20	0.10	0.38	6.0	0.3	19.9
Goldfish	8.58	2.11	21.2	45.3	0.0	396
Toad	0.16	0.12	0.21	2.4	1.0	4.6
Thallium						
Trout	0.17	0.09	0.30	8.4	0.7	24.8
Goldfish	7.00	1.94	9.96	52.5	1.1	266
Toad	0.11	0.09	0.14	2.4	1.1	4.5
Tin						
Trout	0.40	0.23	0.67	15.5	2.1	42.5
Goldfish	2.14	0.36	3.45	68.8	0.0	390
Toad	0.09	0.08	0.13	1.7	0.3	4.5
Tungsten						
Trout	15.61	6.71	31.98	828	14.2	2810
Goldfish	120	92.3	156	345	4.7	2139
Toad	2.90	2.44	3.50	10.7	4.5	21.2
Vanadium						
Trout	0.16	0.07	0.30	6.9	0.3	22.8
Goldfish	4.60	0.51	9.10	55.2	0.0	116
Toad	0.25	0.13	0.44	7.4	0.6	23.0
Zinc						
Trout	1.06	0.75	1.39	20.0	5.7	33.2
Goldfish	2.54	1.59	4.18	400	26.5	500
Toad	0.01	0.01	0.04	0.6	0.0	2.2

TABLE 3

COMPARATIVE TOXICITY OF COAL ELEMENTS TO FISH AND AMPHIBIAN EMBRYO—LARVAL STAGES

Mean toxicity index*		Most sensitive species index			
Element	LC_{50}, ppm	Element	Species	LC_{50}, ppm	LC_1, ppb
Toxicity group 1		Mercury	Toad	0.001	0.1
Silver	0.02	Silver	Trout	0.01	0.2
Mercury	0.04	Zinc	Toad	0.01	0.6
Cadmium	0.11	Lanthanum	Trout	0.02	0.8
Aluminum	0.25	Chromium	Toad	0.03	1.0
Cobalt	0.29	Copper	Toad	0.04	1.0
Arsenic	0.36	Cadmium	Toad	0.04	1.6
Chromium	0.45	Arsenic	Toad	0.04	1.6
Lead	0.62	Lead	Toad	0.04	3.2
Nickel	0.75	Nickel	Trout	0.05	0.6
Tin	0.88	Cobalt	Toad	0.05	0.9
Toxicity group 2		Germanium	Toad	0.05	1.2
Zinc	1.20	Aluminum	Toad	0.05	2.3
Vanadium	1.67	Tin	Toad	0.09	1.7
Copper	1.78	Selenium	Toad	0.09	5.0
Germanium	1.90	Thallium	Toad	0.11	2.4
Thallium	2.43	Strontium	Toad	0.16	2.4
Strontium	2.98	Vanadium	Trout	0.16	6.9
Antimony	4.07	Antimony	Toad	0.30	3.8
Manganese	4.18	Molybdenum	Trout	0.73	22.3
Selenium	4.35	Manganese	Toad	1.42	3.0
Toxicity group 3		Tungsten	Toad	2.90	10.7
Lanthanum	20.25				
Molybdenum	20.56				
Tungsten	47.17				

*LC_{50} values at 4 days posthatching averaged for three species, narrow-mouthed toad, goldfish, and rainbow trout.

Detention time was 42 hr, and water-quality parameters were plotted and averaged for each of four 500-hr elution intervals. Total dissolved solids, conductivity, and pH showed marked decreases during the first 100 hr of elution time (Fig. 1). Total dissolved solids decreased from 2.2 g/liter at 18 hr to 0.5 g/liter at 94 hr, averaging 0.56 g/liter for the first 500 hr. Conductivity (μmhos) decreased from 2400 at 20 hr to 900 at 94 hr and continued to drop slowly to 200 at 500 hr. The sharpest decline was observed for pH, which decreased from 7.7 at 1 hr to about 4.0 at 18 hr, averaging 4.3

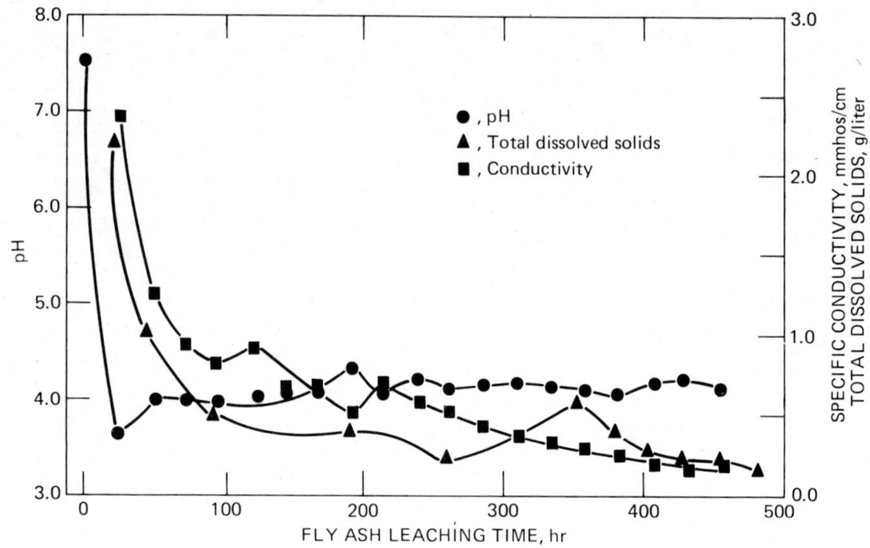

Fig. 1 Changes in fly-ash effluent with leaching time.

for the first elution interval. As seen in Fig. 1, changes in these effluent parameters were most pronounced during the initial 4 days of ash-leaching time, presumably correlating with the period during which leachable components were most rapidly removed from the fly ash. Mean values for the first 500 hr are given in Table 4.

Midway in the second elution interval, at 770 hr, the influent source was changed to carbon-filtered tap water. This action was taken to determine whether influent water of higher pH and greater buffering capacity would alter the leaching process. Initially, conductivity and total dissolved solids of the effluent rose proportionately with increases observed for the new influent water, but values for these parameters declined steadily over the third and fourth elution intervals, closely approaching those obtained for influent water by 2000 hr (see Materials and Methods section). After the change to influent tap water, effluent pH for the second elution interval increased steadily from 4.5 to 7.1. A gradual increase continued thereafter, and, during the last two elution intervals, pH ranges of 7.1 to 7.9 and 7.0 to 7.7 were recorded for influent and effluent water, respectively. Although total alkalinity was not determined during the first 1000 hr, ranges for the third and fourth elution intervals were 32 to 55 and 46 to 62 ppm $CaCO_3$, compared with 54 to 69 ppm for influent tap water. After 2000 hr of continuous washing of the original fly-ash sample, influent and

TABLE 4

CHARACTERISTICS OF SIMULATED ASH-POND EFFLUENT

Characteristic	Simulated effluent*	TVA range†
Total dissolved solids, g/liter	0.56 ± 0.17	0.14–0.52
pH	4.3 ± 0.1	4.4–11.3
Conductivity, μmhos/cm	690 ± 80	242–855
Alkalinity,‡ mg/liter $CaCO_3$	43 ± 3	40–154
Ag, μg/liter	2.1 ± 0.9	
Al, mg/liter	32.6 ± 6.1	1.4–7.2
Cd, μg/liter	156 ± 35	1–37
Cu, μg/liter	110 ± 26	10–310
Hg, μg/liter	1.8 ± 0.5	0.2–38
Ni, μg/liter	155 ± 16	31–1100
Zn, μg/liter	350 ± 33	30–1510
Input, ml/hr	969 ± 10	
Output, ml/hr	897 ± 18	
% evaporation	8 ± 2	

*Mean ± standard error for initial 500 hr of continuous operation.

†Range of mean values for 14 TVA ash ponds (Chu, Ruane, and Steiner, 1976).

‡Alkalinity determined for the third 500-hr elution interval.

effluent water did not differ substantially in pH, alkalinity, conductivity, or total dissolved solids.

Effluent concentrations for the seven selected metals monitored through 1500 hr are summarized in Table 5. Maximum concentrations, which in most instances were observed during the first 100 hr of elution time, were 126 ppm Al, 766 ppb Zn, 518 ppb Cu, 500 ppb Cd, 370 ppb Ni, 87 ppb Hg, and 8 ppb Ag. Mean concentrations for the first 500 hours were 32.6 ppm Al, 350 ppb Zn, 156 ppb Cd, 155 ppb Ni, 110 ppb Cu, 2.1 ppb Ag, and 1.8 ppb Hg. The 500-hr elution patterns for Al, Cu, Ni, and Zn are illustrated in Fig. 2. Elevated concentrations observed at about 300 hr correlated with mechanical disturbances that temporarily facilitated water filtration through the fly-ash bed. Concentrations for all metals except mercury continued to decline progressively with further leaching time, resulting in mean values for the third elution interval of 540 ppb Al, 61.4 ppb Zn, 33.6 ppb Ni, 25.7 ppb Cd, 4.1 ppb Cu, and 0.2 ppb Ag (Table 5). Mercury fluctuated from 0.3 to 7.4 ppb during the first 500 hr but increased substantially thereafter, with mean values of 27.4 and 14.9 ppb for the second and third elution intervals. However, the mercury level dropped markedly toward the end of the third elution period, averaging 2.6 ± 0.7 ppb after 1360 hr. Metals were not detected in

TABLE 5

METAL CONCENTRATIONS FOR FLY-ASH EFFLUENT

Element	Maximum concentration, ppb	Mean concentration for three elution intervals, ppb*			Influent concentration, ppb
		0–500 hr	500–1000 hr	1000–1500 hr	
Ag	8	2.1 ± 0.9	0.2 ± 0.2	0.2 ± 0.2	0.0
Al	126,000	32,600 ± 6,100	1,570 ± 260	540 ± 80	230 ± 29
Cd	500	156 ± 35	93.8 ± 24.8	25.7 ± 11.3	1.7 ± 0.8
Cu	518	110 ± 26	14 ± 1.8	4.1 ± 0.8	3.5 ± 0.9
Hg	87	1.8 ± 0.5	27.4 ± 11.7	14.9 ± 5.3	0.0
Ni	370	155 ± 16	58 ± 10	33.6 ± 2.3	3.3 ± 1.4
Zn	766	350 ± 33	106 ± 20	61.4 ± 4.1	4.9 ± 3.1

*Mean values with standard errors were based on analyses taken at 1- to 3-day intervals.

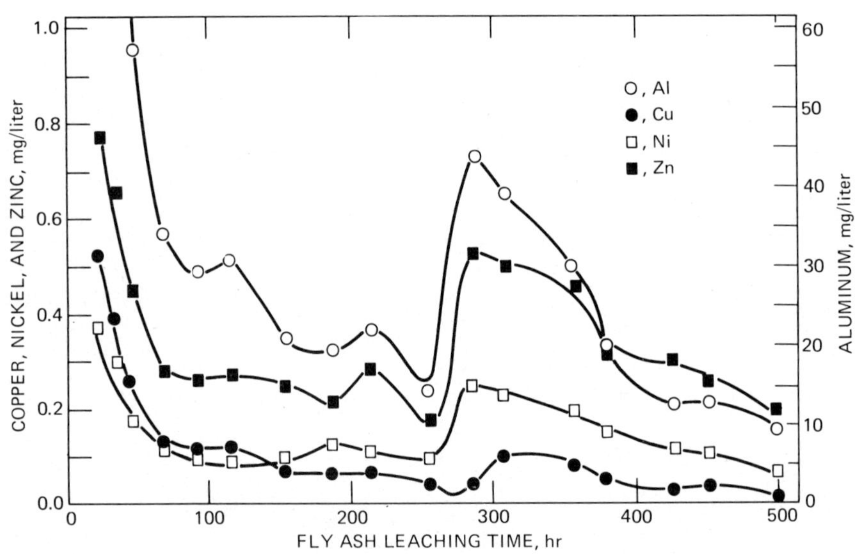

Fig. 2 Leaching patterns for fly-ash metals.

the distilled–deionized influent water used for the first 770 hr, and background values for the carbon-filtered tap water are given in Table 5.

Bioassay Analysis of Fly-Ash Effluent

Four sets of embryo–larval bioassays were performed on fly-ash effluent by use of a continuous-flow system. Tests were initiated after 522, 1033, and 1775 hr of continuous aqueous leaching of the original 52-kg sample of precipitator-collected fly ash. At 522 hr tests were conducted on eggs of the leopard frog and Fowler's toad with undiluted ash effluent. Frog eggs suffered rapid and complete mortality, and a hatching frequency of 46% was observed for Fowler's toad. Survival was 97 to 99% for control populations treated with the same influent water source used to supply the fly-ash leaching chamber (Table 6). Bioassays were initiated at 1033 hr on eggs of the redear sunfish. Undiluted effluent produced complete mortality, and 0.1 and 0.01 dilutions gave survival frequencies of 42 and 90%, which closely approached control survival. In tests with goldfish eggs conducted at 1775 hr, survival averaged 57, 76, and 88% for undiluted effluent and 0.1 and 0.01 dilutions, respectively. Control survival was 92%.

Effluent metal concentrations observed for the amphibian bioassays approximated mean values given for the second elution interval (Table 5). Although ash toxicants produced a near-LC_{50} response for toad eggs, the exposure period was limited to only 1.5 days. In addition, developmental stages of Fowler's toad are highly resistant to trace metals, compared with other amphibian and piscine species (Birge, 1976). Animal species used for the initial toxicological

TABLE 6

EMBRYO—LARVAL BIOASSAYS ON FLY-ASH EFFLUENT

Species	Elution interval, hr	Exposure time, days	Bioassay solution	Percent survival at hatching
Leopard frog	522—582	2.5	Ash effluent	0
(Rana pipiens)			Control	97
Fowler's toad	522—558	1.5	Ash effluent	46
(Bufo fowleri)			Control	99
Redear sunfish	1033—1105	3.0	Ash effluent	0
(Lepomis			Diluted effluent	
microlophus)			0.1	42
			0.01	90
			0.001	93
			0.0001	95
			Control	89
Goldfish	1775—1847	3.0	Ash effluent	57
(Carassius			Diluted effluent	
auratus)			0.1	76
			0.01	88
			Control	92

characterization of coal elements were not available for the first two sets of effluent bioassays. Sunfish eggs, however, have the same hatching time (exposure period) as goldfish and generally exhibit similar sensitivity when used in aquatic bioassays (Birge, Black, and Westerman, 1978). During the exposure period for sunfish eggs, mean effluent metal concentrations, with standard errors, were 0.4 ± 0.2 ppb Ag, 1070 ± 230 ppb Al, 72.0 ± 43.0 ppb Cd, 5.5 ± 1.8 ppb Cu, 20.6 ± 4.8 ppb Hg, 31.5 ± 1.4 ppb Ni, and 70.0 ± 6.7 ppb Zn. At the 0.1 dilution, which gave 42% survival for sunfish eggs (Table 6), all analyzed metals except aluminum were well below goldfish LC_1 values. Aluminum was present at approximately two-thirds of the LC_{50} value. Although the effluent was not analyzed for all possible toxicants, this correlation tends to support application of the toxicological index given for coal elements (Tables 2 and 3). Before the goldfish bioassays were initiated, effluent metal concentrations had dropped to 0 ppb Ag, 160 ± 10 ppb Al, 1.5 ± 1.5 ppb Cd, 4.5 ± 1.8 ppb Cu, 3.8 ± 0.7 ppb Hg, 23.0 ± 3.0 ppb Ni, and 44.5 ± 8.0 ppb Zn. These values were all below LC_1's calculated for goldfish, except aluminum, which was present at about the LC_{50} level (Table 2). The undiluted effluent gave 57% survival.

DISCUSSION

The embryo–larval bioassays reported in Table 2 demonstrate the high toxicity of numerous inorganic coal elements to aquatic biota. Depending on the animal species, LC_{50} values of 0.1 ppm or less were observed for 15 coal elements, and calculated LC_1's ranged down to 0.1 to 0.2 ppb for mercury and silver. Tungsten was the least toxic element in all cases, with LC_{50} values ranging from 2.90 ppm for the toad to 120 ppm for the goldfish. When the test data were averaged, the increasing order of sensitivity of animal species was goldfish, trout, and toad.

The order of toxicity of the 22 elements, as determined by LC_{50} values, varied somewhat for embryo–larval stages of the three species. Only Ag, Cd, Cr, Hg, Ni, and Pb occurred among the 12 most toxic elements for all three, but Al, As, Co, Cu, Ge, V, and Zn were included in this group for two species (Table 2). Of particular interest were the consistent extreme toxicity of mercury and silver to developmental stages of all species and certain selective responses, such as the high relative toxicity of aluminum to goldfish, germanium and lanthanum to trout, and selenium and zinc to the toad.

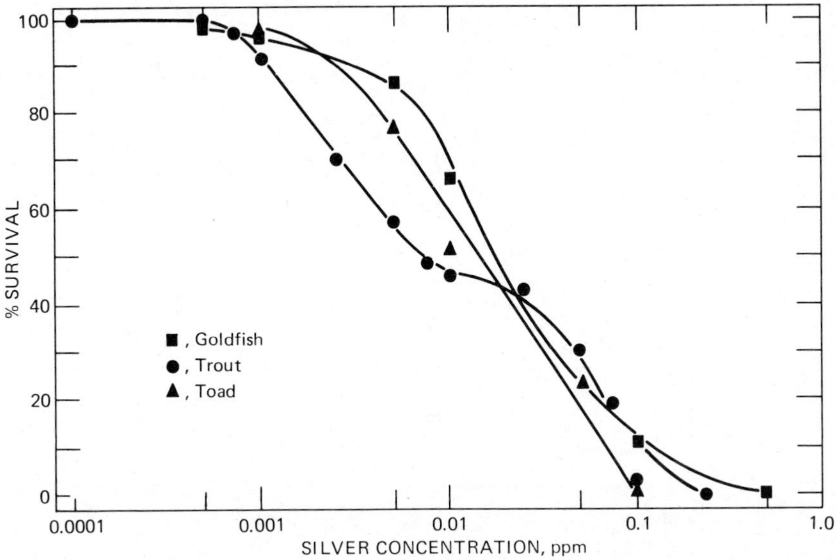

Fig. 3 Effects of silver on embryo—larval stages.

Several patterns of response were discernible concerning the differential sensitivity of the three test organisms. Highest uniformity was obtained for silver, which gave an exceptionally narrow range of LC_{50} values (0.01 to 0.03 ppm). As seen in Fig. 3, this relationship held for the full range of exposure concentrations. Germanium produced similar effects on the two most sensitive test animals, trout and toads, but was substantially less toxic to the goldfish (Fig. 4). This same pattern was given by Cu, Mo, Ni, Sb, Sr, Tl, and V. A still more heterogeneous response occurred for Pb (Fig. 5), Hg, and certain other elements (e.g., La, Sn, and W). Considering the response patterns summarized in Figs. 3 to 5, it is probable that the diversity of aquatic species affected by pollution would increase in the order of Pb, Ge, and Ag. Although elements such as Pb and Ge likely would affect fewer species, these toxicants probably would contribute to an ecological imbalance of aquatic biota. As noted, Se, Zn, and certain other elements (e.g., As and Co) were more selective for the toad, and, on the basis of LC_{50} values, the toad was the most sensitive species for 17 of the 22 elements. This suggests that amphibians may constitute particularly sensitive target sites for coal contaminants. For example, goldfish LC_{50} determinations for selenium and zinc exceeded those for the toad about 100 and 250 times, respectively.

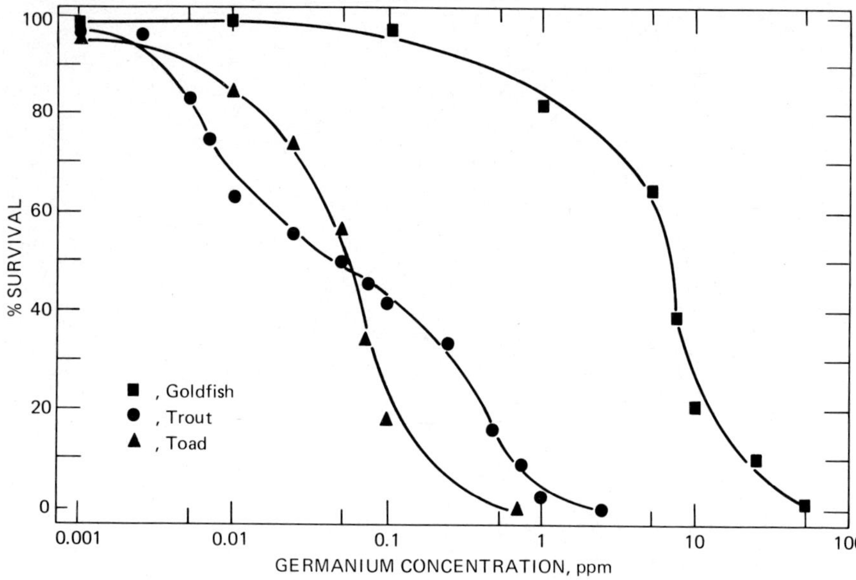

Fig. 4 Effects of germanium on embryo—larval stages.

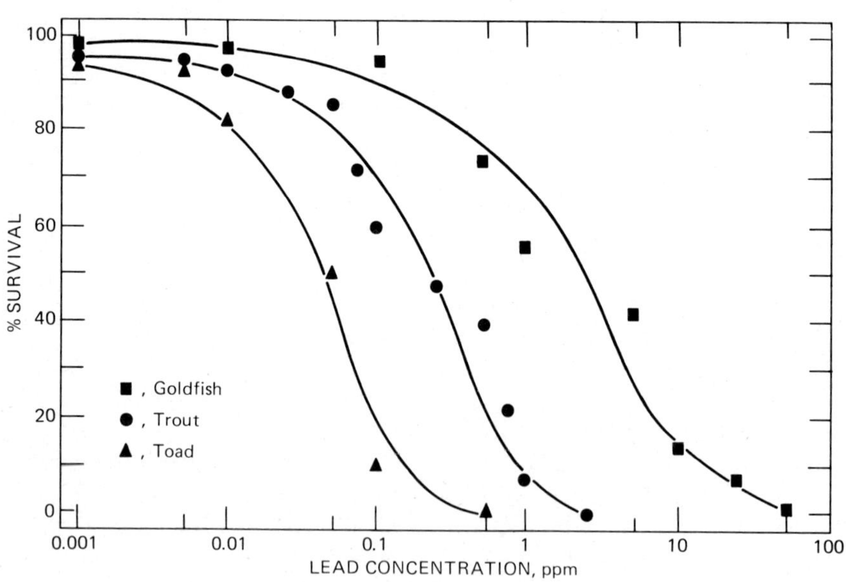

Fig. 5 Effects of lead on embryo—larval stages.

The heterogeneity of response observed for the three animal species somewhat complicates application of the bioassay data on coal elements to impact assessment and pollution-abatement technology. Therefore, the mean LC_{50} and sensitive species indexes (Table 3) were developed to provide a simplified data base for energy and environmental engineers. The sensitive-species ranking for coal elements was used to delineate upper limits of toxicity observed for the 66 independent bioassays, and the mean index summarized average test responses. Despite some notable exceptions (e.g., Al and La), the toxicological orders given in the two indexes were generally similar. Principal differences in relative order were attributed to elements exhibiting disproportionate selective toxicity for a particular animal species. The only extreme disparity involved lanthanum, for which mean and sensitive species LC_{50} values differed by three orders of magnitude. We should note that several recent publications review additional bioassay data for some of the trace metals found in coal (National Academy of Sciences—National Academy of Engineering, Committee, 1973; Vaughan et al., 1975; Environmental Protection Agency, 1976).

In the fly-ash leaching study, characteristics of the simulated effluent were compared with those recorded for 14 TVA ash ponds. As seen in Table 4, good agreement was obtained for all test parameters except aluminum and cadmium, but concentrations for these metals were within TVA ranges early in the third elution interval. The high initial values for aluminum and cadmium may have resulted from use of distilled influent water, which contributed to low pH in the simulated ash pond. These and other results indicate that the quality of effluent water may be improved somewhat by regulating certain parameters (e.g., pH and alkalinity) of influent water used for ash sluicing.

Appreciable metal leaching continued, however, even after 770 hr, when the change was made in influent water (Table 5). The resulting suite of toxic metals produced lethality of test organisms through 1775 to 1847 hr of continuous elution time (Table 6). Effluent metal concentrations (Table 5) were compared to freshwater guidelines (National Academy of Sciences—National Academy of Engineering, Committee, 1973; Environmental Protection Agency, 1976) to further evaluate potential effects of fly-ash leaching on aquatic biota. Through 1775 hr, mercury remained well above the limit of 0.05 ppb, and aluminum exceeded the 100-ppb level considered deleterious to growth and survival of fish. Cadmium was over the trout standard of 0.4 to 1.2 ppb for 1775 hr and exceeded the maximum limit for other aquatic species (4 to 12 ppb) for 1050 to 1435 hr. On the basis of the Environmental Protection Agency's (EPA) application factor (0.01) and the trout, toad, and goldfish embryo–larval LC_{50} values, nickel and zinc concentrations exceeded recommended levels for all test species through 1775 hr, and silver was above acceptable limits for 1050 to 1266 hr. Copper, with an application factor of 0.1, exceeded calculated concentrations for trout and toad through 1000 to 1500 hr but was over the goldfish limit for only 22 hr. Using embryo–larval rather than adult LC_{50} values resulted in more stringent limits for Ag, Cu, Ni, and Zn. However, freshwater standards should permit adequate protection for sensitive life-cycle stages. Except for copper, the suggested EPA application factors appeared acceptable for embryo–larval stages. On the basis of data in Table 2, 0.01 to 0.05 of LC_{50} determinations gave values that generally fell within or near 95% confidence limits for LC_1's. In comparison with the EPA value of 0.1, a more suitable application factor for copper was found to be 0.05 for goldfish and trout and 0.01 for the toad.

Since combined toxicological effects of complex suites of trace metals are difficult to quantify by existing hazard-assessment criteria, direct bioassay monitoring was used to provide further characterization of ash effluent. As noted, after 1033 hr of continuous elution, undiluted ash effluent produced 100% mortality of sunfish eggs, and survival of goldfish eggs was reduced to 57% when exposure was initiated at 1775 hr. A 0.1 dilution produced an approximate LC_{50} for sunfish, and 0.01 gave essentially control-level survival for both species. When median survival was obtained, concentrations of all monitored metals except aluminum were at or below goldfish LC_1 values, and aluminum was present at about the LC_{50} level (Table 2). Effluent dilutions that gave control-level survival did not contain any monitored metals at concentrations exceeding goldfish LC_1 values. Although ash effluent was not analyzed for all possible toxicants, results obtained by direct effluent monitoring were in good agreement with the independent embryo–larval bioassays for coal elements. Also, trace metals present at or below the probit LC_1's did not exert any overt synergistic effects. In addition, results indicate that continuous-flow embryo–larval test systems are highly suitable for in situ toxicological monitoring of complex coal effluents.

Although not intended to serve in lieu of actual field studies, simulated ash ponds can be used to characterize aqueous leaching processes and to evaluate ash effluents for potential environmental hazards. Test parameters can be manipulated individually to determine effects on metal elution rates, and such model systems can be particularly useful in comparing ash residues of coal from different formations. The chemical composition of bottom and fly ash is highly variable, depending on the source of the coal used, combustion conditions, and such factors as the efficiency of emission-control equipment (Moulton, 1973; Chu, Nicholas, and Ruane, 1975; Cooper, 1975). In addition to differences in ash composition, numerous physical and chemical factors may affect the leaching of trace elements and the final composition of ash-pond water. These factors include the quantity of water used for sluicing; its temperature, pH, and hardness; and various performance characteristics of the settling pond. As noted by Chu, Nicholas, and Ruane (1975), the effects of such variables on the quality of ash-pond effluents are not sufficiently understood. It is known, however, that a number of coal-derived inorganic elements reach appreciable concentrations in ash-pond waters. Since 1973, TVA has analyzed for 17 trace elements in quarterly grab samples from bottom ash, fly ash, and combined ash ponds, and the results have been summarized by Chu and co-workers (Chu, Nicholas, and Ruane, 1975; Chu, Krenkel, and Ruane, 1976). Discharges from fly-ash ponds were reported to contain up to 7.3 ppm Al, 0.3 ppm Ba, 0.04 ppm Cd, 0.1 ppm Cr, 0.3 ppm Cu, 0.08 ppm Pb, 13.4 ppm Mn, 1.1 ppm Ni, and 1.5 ppm Zn. Ranges for a number of these metals are summarized in Table 4.

Other investigators also have considered various problems associated with fly-ash disposal (Guthrie, Cherry, and Rodgers, 1974; Theis, 1975; Holland et al., 1975). Theis (1975) indicated that the production of metal leachates and alterations of pH and dissolved oxygen may affect receiving waters. He also demonstrated significant release rates for trace metals when fly ash was dispersed in distilled water. Holland et al. (1975) investigated the environmental effects of trace elements in the pond disposal of ash and flue-gas desulfurization sludge. Samples of ash and sludge from five generating stations were studied by simulated ponding. In general, concentrations of aqueous leachates were low, but Ba, B, Cr, Hg, and Se exceeded EPA guidelines for public water supplies. However, these investigators did not compare their findings with EPA standards for freshwater biota, which generally are more stringent, and they did not consider the combined toxic effects of the resulting metal mixtures. Guthrie, Cherry, and Rodgers (1974) evaluated the impact on biota in waters receiving ash-basin effluent from a coal-fired power plant. Bacterial, plant, and animal diversities were reduced at sites affected by ash effluents. Abiotic water parameters affected by ash-basin effluents included temperature, turbidity, dissolved oxygen, and pH. Concentrations of coal-ash leachates (e.g., Cd, Cr, Cu, Hg, and Zn) were lowest in effluent water, somewhat greater in aquatic biota, and highest in benthos. This indicated accumulation of these toxicants in biomass and bottom sediment.

The results given here show clearly that a substantial number of minor and trace elements of coal and fly ash are highly toxic to aquatic organisms. Many are leachable from ash residues at concentrations that prove lethal to fish and amphibian embryo—larval stages and other organisms. Since annual coal utilization in the United States may reach 1 billion tons or more in the near future (Vaughan et al., 1975), it remains essential to characterize more fully the toxic properties of coal-derived contaminants, ascertain their release rates, and determine their pathways of exchange within and ultimate effects upon aquatic ecosystems.

ACKNOWLEDGMENTS

I should like to acknowledge efforts devoted to bioassay evaluation by A. G. Westerman and J. A. Black and contributions by R. A. Freeman and J. E. Hudson to the fly-ash-leaching studies. I am most grateful to B. A. Ramey for preparation of the manuscript and figures. The research reported here was supported by the Institute for Mining and Minerals Research, Lexington, Kentucky (grant number 7576-EZ).

REFERENCES

Abel, K. H., and L. A. Rancitelli, 1975, Major, Minor and Trace Element Composition of Coal and Fly Ash as Determined by Instrumental Neutron Activation Analysis, in *Trace Elements in Fuel*, S. P. Babu (Ed.), Advances in Chemistry Series, No. 141, pp. 118-138, American Chemical Society, Washington, D. C.

American Public Health Association, 1975, *Standard Methods for the Examination of Water and Wastewater*, 14th ed., Washington, D. C.

Ayer, F. A. (Comp.), 1974, *Environmental Aspects of Fuel Conversion Technology*, Symposium Proceedings, St. Louis, Mo., May 13—15, 1974, Report EPA-650/2-74-118, Environmental Protection Agency, NTIS.

Babu, S. P. (Ed.), 1975, *Trace Elements in Fuel*, Advances in Chemistry Series, No. 141, American Chemical Society, Washington, D. C.

Birge, W. J., 1976, *Effects of Metals on Embryogenesis and Use of Vertebrate Embryos as Sensitive Indicators of Environmental Quality*, NSF(RANN) Technical Report, Grant No. AEN 74-08768 A01, National Science Foundation, Washington, D. C.

——, and J. A. Black, 1977, *A Continuous Flow System Using Fish and Amphibian Eggs for Bioassay Determinations on Embryonic Mortality and Teratogenesis*, Report EPA-560/5-77-002, Environmental Protection Agency.

——, J. A. Black, and A. G. Westerman, 1978, Embryo—Larval Bioassays with Goldfish and Redear Sunfish, unpublished data.

——, and J. J. Just, 1975, *Bioassay Procedures Using Developmental Stages as Test Organisms*, Research Report No. 84, U. S. Department of the Interior, Washington, D. C.

Bolton, N. E., et al., 1975, Trace Element Mass Balance Around a Coal-Fired Steam Plant, in *Trace Elements in Fuel*, S. P. Babu (Ed.), Advances in Chemistry Series, No. 141, pp. 175-187, American Chemical Society, Washington, D. C.

Carter, J. A., 1975, Trace Element Composition of Coal-Derived Materials (NSF—RANN), in *Coal Technology Program Quarterly Progress Report No. 1 for the Period Ending December 31, 1974*, pp. 66-69, ERDA Report ORNL-5026, Oak Ridge National Laboratory, NTIS.

Chu, T. J., P. A. Krenkel, and R. J. Ruane, 1976, Characterization and Reuse of Ash Pond Effluents in Coal-Fired Power Plants, paper presented at 49th Annual Water Pollution Control Federation Conference, Minneapolis, Minn., Oct. 3—8, 1976.

——, W. R. Nicholas, and R. J. Ruane, 1975, Complete Reuse of Ash Pond Effluents in Fossil-Fueled Power Plants, paper presented at 68th Annual Meeting of the American Institute of Chemical Engineers, Los Angeles, Calif., Nov. 16—20, 1975.

——, R. J. Ruane, and G. R. Steiner, 1976, Characteristics of Wastewater Discharges from Coal-Fired Power Plants, paper presented at the 31st Annual Purdue Industrial Waste Conference, Purdue University, West Lafayette, Ind., May 4—6, 1976.

Cooper, H. B., Jr., 1975, The Ultimate Disposal of Ash and Other Solids from Electric Power Generation, in *Water Management by the Electric Power Industry*, E. F. Gloyna, H. H. Woodson, and H. R. Drew (Eds.), pp. 183-195, Water Resources Symposium No. 8, Center for Research in Water Resources, The University of Texas, Austin.

Daum, R. J., 1969, A Revision of Two Computer Programs for Probit Analysis, *Bull. Entomol. Soc. Am.*, 16: 10-15.

Environmental Protection Agency, 1974, *Steam Electric Power Generating Point Source Category*, Report EPA-440/1-74 029-a, group 1, Environmental Protection Agency, GPO.

——, 1976, *Quality Criteria for Water*, Washington, D. C.

Freeman, R. A., and W. J. Birge, 1978, Aqueous Leaching of Toxic Metals from Coal-Produced Fly Ash, unpublished data.

Fulkerson, W., et al., 1975, Allen Steam Plant Study, in *Energy Division Annual Progress Report for Period Ending December 31, 1974*, pp. 77-82, ERDA Report ORNL-5030, Oak Ridge National Laboratory, NTIS.

Guthrie, R. K., D. S. Cherry, and J. H. Rodgers, 1974, The Impact of Ash Basin Effluent on Biota in the Drainage System, in *Mid-Atlantic Industrial Waste Conference*, Vol. 7, pp. 17-43, Drexel University, Philadelphia, Pa.

Hildebrand, S. G., R. M. Cushman, and J. A. Carter, 1976, The Potential Toxicity and Bioaccumulation in Aquatic Systems of Trace Elements Present in Aqueous Coal Conversion Effluents, in *Trace Substances in Environmental Health—X*, D. D. Hemphill (Ed.), pp. 305-313, University of Missouri Press, Columbia, Mo.

Holland, W. F., K. A. Wilde, J. L. Parr, P. S. Lowell, and R. F. Pohler, 1975, *The Environmental Effects of Trace Elements in the Pond Disposal of Ash and*

Flue Gas Desulfurization Sludge, prepared by Radian Corporation, Austin, Tex., for the Electric Power Research Institute, Palo Alto, Calif.

Lloyd, W. G., 1976, *Mercury in Texas Coal Samples*, unpublished report, Institute for Mining and Minerals Research, University of Kentucky, Lexington.

Moulton, L. K., 1973, Bottom Ash and Boiler Slag, in *Information Circular No. 8640*, U. S. Bureau of Mines, Pittsburgh, Pa.

National Academy of Sciences—National Academy of Engineering Committee on Water Quality Criteria, 1973, *Water Quality Criteria 1972*, GPO.

Perkin—Elmer Corp., 1973, *Analytical Methods for Atomic Absorption Spectrophotometry*, Norwalk, Conn.

Rubin, E. S., and A. McMichael, 1974, Some Implications of Environmental Regulatory Activities on Coal Conversion Processes, in *Environmental Aspects of Fuel Conversion Technology*, Symposium Proceedings, St. Louis, Mo., May 13—15, 1974, pp. 69-90, Report EPA-650/2-74-118, Environmental Protection Agency, NTIS.

Ruch, R. R., H. J. Gluskoter, and N. F. Shimp, 1974, Distribution of Trace Elements in Coal, in *Environmental Aspects of Fuel Conversion Technology*, Symposium Proceedings, St. Louis, Mo., May 13—15, 1974, Report 650/2-74-118, Environmental Protection Agency, NTIS.

Sheibley, D. W., 1975, Trace Elements by Instrumental Neutron Activation Analysis for Pollution Monitoring, in *Trace Elements in Fuel*, S. P. Babu (Ed.), Advances in Chemistry Series, No. 141, pp. 98-117, American Chemical Society, Washington, D. C.

Smith, I. C., T. L. Ferguson, and B. L. Carson, 1975, Metals in New and Used Petroleum Products and By-Products—Quantities and Sequences, in *The Role of Trace Metals in Petroleum*, T. F. Yen (Ed.), pp. 123-148, Ann Arbor Science Pubs., Inc., Ann Arbor, Mich.

Theis, T. L., 1975, The Potential Trace Metal Contamination of Water Resources Through the Disposal of Fly Ash, paper presented at 2nd National Conference on Complete Water Reuse, American Institute of Chemical Engineers and Environmental Protection Agency, Chicago, Ill., May 4—8, 1975.

Vaughan, B. E., et al., 1975, *Review of Potential Impact on Health and Environmental Quality from Metals Entering the Environment as a Result of Coal Utilization*, ERDA file No. NP-20585, Battelle Pacific Northwest Labs., Richland, Wash.

Vitez, B., 1976, Trace Elements in Flue Gases and Air Quality Criteria, *Power Eng.*, January: 56-60.

Yavorsky, P. M., and S. Akhtar, 1974, Environmental Aspects of Coal Liquefaction, in *Environmental Aspects of Fuel Conversion Technology*, Symposium Proceedings, St. Louis, Mo., May 13—15, 1974, pp. 325-330, Report EPA-650/2-74-118, Environmental Protection Agency, NTIS.

MERCURY CONTAMINATION STANDARDS FOR MARINE ENVIRONMENTS

RONALD EISLER
Environmental Protection Agency, Environmental Research Laboratory, Narragansett, Rhode Island

ABSTRACT

Selected technical literature on biological and ecological effects of mercury compounds on marine and estuarine biota is reviewed. Potential and actual hazards to public health through marine vectors are considered. Within this framework, approaches for establishing mercury contamination standards in saline environments are presented.

Toxicological aspects of mercury and mercury compounds in coastal and offshore environments as a result of anthropogenic or natural processes have been extensively reviewed elsewhere (D'Itri, 1972; Friberg and Vostal, 1972; Gavis and Ferguson, 1972; Harriss, 1971; Holden, 1973; Jernelöv, Landner, and Larsson, 1975; Keckes and Miettinen, 1972; Newberne, 1974). Most of these authorities agree on five points. First, forms of mercury with relatively low toxicity can be transformed into forms with very high toxicity through biological and other processes. Second, uptake of mercury directly from seawater or through biomagnification in marine food chains returns mercury to man in concentrated form. Third, mercury uptake may result in genetic changes. Fourth, elevated levels of mercury in some marine fishes, such as tuna or swordfish, emphasize the complexity of both natural mercury cycles and man's impact on those cycles. Finally, man's use of mercury should be curtailed because, in contrast to some other pollutants, the difference between tolerable natural background levels of mercury and harmful levels in the environment is exceptionally small.

This paper considers recent material on mercury effects in saline waters and recommends useful criteria and promising research approaches for incorporating mercury contamination standards that will protect marine products of commerce, their food organisms, and their predators, including man.

HISTORICAL REVIEW

Laboratory Studies

Comparative Toxicity, Survival, and Biotic and Abiotic Modifiers

In general, salts of mercury and its organic compounds have been shown in short-term bioassays to be more toxic to marine organisms than are salts of other heavy metals. To oyster embryos, for example, mercury salts were more toxic than were Ag, Cu, Zn, Ni, Pb, Cd, As, Cr^{+6}, Mn, or Al (Calabrese et al., 1973); to clam embryos, mercury was the most toxic metal tested, followed by Ag, Zn, Ni, and Pb, in that order (Calabrese and Nelson, 1974). Toxicity bioassays of 168 hr duration with salts of Hg, Cd, Cr^{+6}, Ni, and Zn on adults of representative marine fishes, crustaceans, bivalve and gastropod molluscs, annelids, and echinoderms confirmed that mercury was consistently the most toxic metal in this series (Eisler and Hennekey, 1977). Similar results are reported for marine algae (Berland et al., 1976); marine fungi (Schneider, 1972); sea urchin eggs (Kobayashi, 1971); larvae of marine molluscs and crustaceans (Connor, 1972); some species of marine polychaete worms (Reish et al., 1976); freshwater annelids, insects, and gastropods (Rehwoldt et al., 1972); crustaceans (Cabejszek and Stasiak, 1960); and fish (Weir and Hine, 1970).

Acute toxicity values of mercury compounds for various marine species are summarized in Table 1. These values are similar to those reported for freshwater groups (McKim, 1977). Differences in species resistance to mercury compounds (Hendrick and Everett, 1965; Schweiger, 1957; Wisely and Blick, 1967) may account for some of the variability in test results shown in Table 1, but many additional factors are known to affect or modify the survival time of mercury-stressed marine biota, sometimes by one order of magnitude or more. Abiotic modifiers include the chemical form of mercury administered (Boetius, 1960; Boney, 1971; Boney and Corner, 1959; Boney, Corner, and Sparrow, 1959; Corner and Sparrow, 1957; Ellis, 1947; Fang, 1973; MacLeod and Pessah, 1973; Middaugh and Rose, 1974); cations other than mercury present in the medium (Barnes

TABLE 1

ACUTE TOXICITY OF MERCURY COMPOUNDS TO VARIOUS SPECIES OF MARINE BIOTA*

Species	Concentration	Effect	Reference
Algae			
18 spp.	10–25	LC_{100} (17 days)	Berland et al., 1976
8 spp.	12–8,000	LC_{50} (0.5-hr exposure plus 168-hr postexposure	Boney, 1971; Boney and Corner, 1959
5 spp.	60	LC_{100}	Ukeles, 1962
Laminaria hyperboria	2,500	LC_{100} (24 hr)	Hopkins and Kain, 1971
Coelenterata			
Eirene viridula	300	LC_{100} (3 hr)	Karbe, 1972
Echinodermata			
Acanthocidaris crassispina eggs	18–46	LC_{100} (48 hr)	Kobayshi, 1971
Asterias forbesi	10	LC_0 (168 hr)	Eisler and Hennekey, 1977
Asterias forbesi	20	LC_{50} (168 hr)	
Asterias forbesi	125	LC_{100} (168 hr)	
Annelida			
Neanthes arenaceodentata	22–100	LC_{50} (96 hr)	Reish et al., 1976
Neanthes arenaceodentata	17–90	LC_{50} (28 days)	
Capitella capitata	14–<100	LC_{50} (96 hr)	
Capitella capitata	100	LC_{50} (28 days)	
Nereis virens	25	LC_0 (168 hr)	Eisler and Hennekey, 1977
Nereis virens	60	LC_{50} (168 hr)	
Nereis virens	125	LC_{100} (168 hr)	
Ophryotrocha labronica	1,000	LC_{50} (0.5 hr)	Brown and Ahsanullah, 1971
Mollusca			
Mya arenaria	1	LC_0 (168 hr)	Eisler and Hennekey, 1977
Mya arenaria	4	LC_{50} (168 hr)	
Mya arenaria	30	LC_{100} (168 hr)	
Mercenaria mercenaria embryos	2.5	LC_0 (48 hr)	Calabrese and Nelson, 1974
Mercenaria mercenaria embryos	4.8	LC_{50} (48 hr)	
Mercenaria mercenaria embryos	7.5	LC_{100} (48 hr)	
Crassostrea virginica embryos	1.0	LC_0 (48 hr)	Calabrese et al., 1973
Crassostrea virginica embryos	5.6	LC_{50} (48 hr)	
Crassostrea virginica embryos	8.0	LC_{100} (48 hr)	

(Table continues on following page.)

Table 1 (Continued)

Species	Concentration	Effect	Reference
Mollusca (continued)			
Crassostrea virginica	50	LC_{50} (19 days)	Kopfler, 1974
Argopecten irradians	89	LC_{50} (96 hr)	Nelson et al., 1976
Nassarius obsoletus	100	LC_0 (168 hr)	Eisler and Hennekey, 1977
Nassarius obsoletus	700	LC_{50} (168 hr)	
Nassarius obsoletus	5,000	LC_{100} (168 hr)	
Congeria leucophaeta	1,000	LC_{20} (48 hr)	Dorn, 1974
Congeria leucophaeta	10,000	LC_{100} (48 hr)	
Cardium edule	10,000	LC_{50} (48 hr)	Portmann, 1968
Rangia cuneata	6,300—40,000	LC_{50} (48 hr)	Olson and Harrel, 1973
Crustacea			
Tigriopus japonicus	6	LC_{100} (72 hr)	D'Agostino and Finney, 1974
Paneus setiferus postlarvae	17	LC_{50} (96 hr)	Green et al., 1976
Pagurus longicarpus	10	LC_0 (168 hr)	Eisler and Hennekey, 1977
Pagurus longicarpus	50	LC_{50} (168 hr)	
Pagurus longicarpus	125	LC_{100} (168 hr)	
Petrolisthes armatus	50—64	LC_{50} (96 hr)	Roesijadi et al., 1974
Sesarma sp.	60	LC_{50} (24 hr)	Okubo and Okubo, 1962
Pandalus montague	100	LC_{50} (48 hr)	Portmann, 1968
Uca pugilator larvae	180	LC_{100} (24 hr)	DeCoursey and Vernberg, 1972
Uca pugilator	1,000	LC_{60} (28 days)	Weis, 1976
Nitocra spinopes	600	LC_{50} (24 hr)	Barnes and Stanbury, 1948
Carcinus maenus	1,000	LC_{50} (48 hr)	Portmann, 1968
Balanus balanoides	1,000	LC_{90} (48 hr)	Clarke, 1947
Crangon crangon	6,000	LC_{50} (48 hr)	Portmann, 1968
Artemia salina	1,000	LC_{50} (25 hr)	Brown and Ahsanullah, 1971
Artemia salina	21,000—50,000	LC_{50} (24 hr)	Okubo and Okubo, 1962

(Table continues on following page.)

Table 1 (Continued)

Species	Concentration	Effect	Reference
Teleosts			
Mugil cephalus	500	LC_{58} (3 hr)	Middaugh and Rose, 1974
Mugil cephalus	50	LC_{0} (3 hr)	
Fundulus heteroclitus	100	LC_{0} (168 hr)	Eisler and Hennekey, 1977
Fundulus heteroclitus	800	LC_{50} (168 hr)	
Fundulus heteroclitus	1,000	LC_{100} (168 hr)	
Fundulus heteroclitus	1,000	LC_{100} (168 hr)	Weis and Weis, 1976
Fundulus heteroclitus	100	LC_{100} (264 hr)	
Fundulus heteroclitus	860	LC_{0} (96 hr)	Klaunig, Koepp, and McCormick, 1975
Fundulus heteroclitus	2,000	LC_{50} (96 hr)	
Fundulus heteroclitus	4,600	LC_{100} (96 hr)	

*Concentrations are in micrograms per liter (ppb) of total mercury added at start of the experiment which were lethal to specified percentage within the indicated time frame.

and Stanbury, 1948; Corner and Sparrow, 1956; Gray, 1974; Gray and Ventilla, 1973; Hunter, 1949; Lask et al., 1971); anion concentration in the medium (Amend, Yasutake, and Morgan, 1969; Lask et al., 1971); complexing agents in solution (Boney and Corner, 1959; Eisler et al., 1972; Nishikawa and Tabata, 1969; Tabata and Nishikawa, 1969); presence of surface active agents (Calamari and Marchetti, 1973; Eisler et al., 1972); rate at which mercury was administered (Corner and Rigler, 1957; Gillespie, 1972; Horn and Katz, 1946); temperature of the water (Amend, Yasutake, and Morgan, 1969; Ballard and Oliff, 1969; Boney and Corner, 1959; Clemens and Sneed, 1958; Jones, 1973; Lask et al., 1971; MacLeod and Pessah, 1973; Portmann, 1968; Rodgers et al., 1951; Vernberg, DeCoursey, and O'Hara, 1974; Vernberg and O'Hara, 1972; Vernberg and Vernberg, 1972); seasonal differences, as distinct from temperature (Vernberg, DeCoursey, and O'Hara, 1974); dissolved oxygen content (Amend, Yasutake, and Morgan, 1969); pH (Lask et al., 1971); and salinity of the medium (Binet and Nicolle, 1940; Boetius, 1960; Hunter, 1949; Jones, 1973; 1975; Olson and Harrel, 1973; Pyefinch and Mott, 1948; Roesijadi et al., 1974; Schneider, 1972; Vernberg, DeCoursey, and O'Hara, 1974; Vernberg and Vernberg, 1972). Biological factors modifying the mercury survival response of marine organisms include: age of test organisms (Connor, 1972; Green et al., 1976; Okubo and Okubo, 1962; Portmann, 1968; Reish et al., 1976; Vernberg, DeCoursey, and O'Hara, 1974; Wilson and

Connor, 1971); sex (Vernberg, DeCoursey, and O'Hara, 1974; Vernberg and Vernberg, 1972); general health of animal (Portmann, 1968; Shealy and Sandifer, 1975); previous exposure to organic compounds (Boney and Corner, 1959); or to mercury (Corner and Sparrow, 1956; Foster and Olson, 1951; Gillespie, 1972; Green et al., 1976); moulting stage of crustaceans (Wilson and Connor, 1971); and presence of parasites (Clemens and Sneed, 1958).

It was not possible to document a specific trend for any individual modifier. Some studies showed relatively short survival times at comparatively high temperatures, and others demonstrated unchanged survival times for some species at a Δt of 10°C or more. Most studies, however, showed that previous exposure to mercury salts was in some way associated with increased sensitivity to mercury and other stressors; shrimp were an exception. Pre-exposure of shrimp for 57 days to 1.0 μg Hg/liter did not affect LC_{50} values (Green et al., 1976). There was general agreement that organomercury compounds were more toxic than inorganic compounds; that chelating agents in combination with mercury salts produced less-than-additive toxicity but other compounds, especially salts of copper, lead, or zinc act synergistically to increase toxicity; and that salinity stress, especially abnormally low salinities, reduced significantly the survival time of mercury-exposed organisms. The evidence suggests that species adapted to a fluctuating estuarine environment could be more vulnerable to the added stresses of heavy metal pollution, including mercury, than species inhabiting more uniformly stable environments (Jones, 1973).

Sublethal and Latent Effects

Mercury adversely affects various metabolic processes essential to normal growth, development, reproduction, and general well-being of marine and estuarine biota. Typically, metabolic effects are manifested at mercury concentrations that are orders of magnitude below those producing death. Morphological variations, for example, occur in dinoflagellates at 1.0 μg Hg/liter (Kayser, 1976), and teratogenic effects, such as the abnormal growth of multiple eyes and tentacles, are found in marine gastropods exposed only as embryos for 3 to 24 days to 10.0 μg Hg/liter (Reinhart and Myers, 1975). Exposure of shrimp larvae for 48 hr to high sublethal mercury levels is associated with reduced survival at post larval stage, delayed molting, extended development time, and morphological deformities (Shealy and Sandifer, 1975). At 1/100th of the lethal concentration, mercury disrupts avoidance-learning responses in fish (Weir and Hine, 1970), and, in fact, mercury is more effective in reducing these conditioned responses than are salts of arsenic, lead, or selenium (Weir and Hine, 1970). Altered feeding behavior and appetite reduction during and after exposure to mercury salts are documented for teleosts (Klaunig, Koepp, and McCormick, 1975; Rodgers et al., 1951; Salzinger et al., 1973) and mussels (Dorn, 1976).

Mercury-induced growth inhibition is reported for several marine species. The sediment-living, ciliate protozoan *Cristigera* exhibited reduction in growth during immersion in 2.5 to 5.0 μg Hg/liter (Gray, 1974; Gray and Ventilla, 1973); synergistic effects of Hg—Cu and Hg—Pb mixtures on *Cristigera* growth were also recorded (Gray and Ventilla, 1971). Growth of the colonial, marine hydroid *Campanularia flexuosa* was depressed during immersion for 11 days in 1.6 to 1.7 μg Hg/liter; however, growth stimulation occurred in 1.0 μg Hg/liter between days 3 and 7 (Stebbing, 1976). Kelps and other species of algae exhibited growth inhibition during immersion in < 5.0 to 350.0 μg Hg/liter (Berland et al., 1976; Boney, 1971; Clendenning and North, 1959; Kayser, 1976; Sick and Windom, 1975); similar results were observed for various bacteria (Ben-Bassat et al., 1972). Photosynthesis was reduced in several species of marine alga during immersion in 0.06 to 10.0 μg Hg/liter; pronounced effects were consistently induced by organomercury compounds at concentrations of 1.0 μg/liter (Harriss, White, and MacFarlane, 1970; Hopkins and Kain, 1971; Nuzzi, 1972). Recovery of algae to control growth levels after initial mercury-induced inhibition was also reported (Davies, 1974).

Mercury disrupts various reproductive processes of aquatic organisms, including fecundity, egg deposition, and hatching frequency. Concentrations between 32 and 92 μg Hg/liter seriously impair development of sea urchin embryos (Okubo and Okubo, 1962; Waterman, 1937); higher concentrations (1,000 μg/liter) were necessary to produce a similar effect on marine protozoa (Persoone and Uyttersprot, 1975). Flagellar contractility of sea urchin sperm was inhibited during immersion in 20 to 200 μg Hg/liter (Young and Nelson, 1974); however, mercury was exceeded in this effect by Cd, Cu, Zn, and Fe (Morisawa and Mohri, 1974). Teleost zygotes fertilized and incubated in water containing at least 0.2 μg Hg/liter exhibit a reduced hatching rate (Kihlstrom, Lundberg, and Hulth, 1971). The number of teleost eggs deposited decreases when the concentration of organomercury compounds is 1.0 μg Hg/liter or greater (Kihlstrom, Lundberg, and Hulth, 1971). Eggs fertilized in mercury-free water and then allowed to develop under various mercury regimes exhibited increased hatch at 10.0 μg Hg/liter but reduced or negligible hatch at 50.0 μg Hg/liter (Kihlstrom and Hulth,

1972). Mercury also inhibits zoospore activity and subsequent development of sporangia in marine fungi (Schneider, 1972).

Mercury interferes with enzyme production or activity in marine teleosts, including brain cholinesterase (Abou-Donia and Menzel, 1967), blood transaminases and dehydrogenases (Christensen, 1971), hemopoietic Na—K—ATPase activities (Renfro et al., 1974), and liver aminolevulinate dehydrases and other enzymes (Jackim, 1973; Jackim, Hamlin, and Sonis, 1970). Low levels, 5.0 μg Hg/liter, increased allantoise formation in marine annelids, but higher levels prevent formation (May and Brown, 1973). Enzyme activities may have potential as an early warning system of mercury-induced stress. For example, a concentration of 0.17 μg Hg/liter, one-tenth the level causing growth inhibition of marine hydroids, inhibited the lysosomal hydrolase staining reaction in that group (Moore and Stebbing, 1976).

Histopathological damage was observed among marine teleosts subjected to 500 μg Hg/liter; damage was most severe in olfactory organs and in the lateral line system (Gardner, 1975). An observed dose-related increase in mantle tentacle epithelial cells of mercury-treated clams (Fowler, Wolfe, and Hettler, 1975) suggests that histological studies might also make it possible to predict impending mercury stress.

Mercury affects oxygen consumption and ventilation rates of aquatic animals. Increased respiration of the mollusc *Congeria leucophaeta* was observed during immersion in 10.0 μg Hg/liter for 48 hr (Dorn, 1974). Concentrations of 3.0 μg Hg/liter caused a significant increase in "cough frequency" of trout, and this may alter oxygen consumption in that species (Drummond, Olson, and Batterman, 1974). Larvae of the fiddler crab *Uca pugilator* exposed to 1.8 μg Hg/liter for 24 hr showed a marked reduction in swimming ability and reduction in general metabolic processes, including oxygen consumption (DeCoursey and Vernberg, 1972). Marine teleosts exposed for 60 days to 10.0 μg Hg/liter exhibited elevated respiration rates, and, at 5.0 μg/liter, there were increases in plasma proteins and decreases in plasma osmolality (Calabrese et al., 1975).

Studies with antifouling compounds suggest that mercury is relatively ineffective in preventing metamorphosis of barnacles attached to glass plates (Clarke, 1947). Some mariculture studies have indicated that low concentrations of mercury salts for short periods have either no measurable effect or beneficial, therapeutic, or prophylactic applications (Green et al., 1976; Rodgers et al., 1951; Rucker and Amend, 1969). The preponderance of studies, however, seems to demonstrate that mercury adversely affects several vital life processes of marine organisms at concentrations as low as 0.2 and 1.0 μg/liter. In addition to the deleterious effects previously listed on behavior, growth, reproduction, histology, respiration, and enzyme production, there are a variety of mercury-related effects that are imperfectly understood. These include increased mucous production in fish (Gardner, 1975; Lask et al., 1971); initiation of diuresis in amphipods (Lockwood and Inman, 1975); a rise in urine nitrogen levels of mercury-poisoned crabs (Corner, 1959); altered osmoregulatory ability of estuarine isopods (Jones, 1975); and negative phototaxy and increased operculate movements of teleosts (Klaunig, Koepp, and McCormick, 1975). Increased efficiency of mercury transport mechanisms has also been noted in crabs exposed at comparatively high temperatures (Vernberg and O'Hara, 1972).

Uptake, Retention, and Translocation

Rapid accumulation of mercury, especially organomercury compounds, by various species of marine biota, primarily teleosts and molluscs, is well documented (Bligh, 1972; Boulton and Hetling, 1972; Cunningham and Tripp, 1975a; 1975b; Davies, 1974; Fang, 1973; Fowler, Wolfe, and Hettler, 1975; Hannerz, 1968; Hasselrot, 1968; Hibiya and Oguri, 1961; Johnels et al., 1967; Kramer and Neidhart, 1975; Lask et al., 1971; Laumond et al., 1973; MacLeod and Pessah, 1973; McKone et al., 1971; Middaugh and Rose, 1974; Nelson et al., 1976; Olson, Bergman, and Fromm, 1973; Pentreath, 1976a; 1976b; Rucker and Amend, 1969). Bioaccumulation of mercury and its compounds from seawater can be modified by many factors. These include chemical form of mercury administered (Cunningham and Tripp, 1975a; Hannerz, 1968; Kramer and Neidhart, 1975; Pentreath, 1976a; 1976b); mode of administration (Jarvenpaa, Tillander, and Miettinen, 1970); presence of complexing agents in medium (Kramer and Neidhart, 1975); initial concentration (Kramer and Neidhart, 1975; Sick and Windom, 1975); exposure time (Sick and Windom, 1975); presence of selenium (Fowler and Benayoun, 1976); salinity of medium (Vernberg, DeCoursey, and O'Hara, 1974); water temperature (Cunningham and Tripp, 1975b; Vernberg, DeCoursey, and O'Hara, 1974); age of organism (Beckett and Freeman, 1974; Cunningham and Tripp, 1975b; Glooschenko, 1969; Hannerz, 1968); biological surface area (Sick and Windom, 1975); variability in detoxication mechanisms (Davies, 1976); sexual condition (Cunningham and Tripp, 1975b); tissue specificity (Cunningham and Tripp, 1975a; Fowler, Wolfe, and Hettler, 1975; Hannerz, 1968; Pentreath, 1976a; 1976b; Vernberg, DeCoursey, and O'Hara, 1974); presence of mercury-resistant strains of bacteria

(Colwell and Nelson, 1975; Colwell et al., 1976); and accumulation after death (Glooschenko, 1969). Two studies, both on adult American oysters, *Crassostrea virginica*, are worth emphasizing. Cunningham and Tripp (1973) held oysters in seawater containing 10 μg Hg/liter as mercuric acetate. After 45 days the whole-body mercury concentration of exposed animals was 28,000 μg/kg wet weight. The mercury concentration in control oysters was always less than 20 μg/kg wet weight. The mercury concentration in exposed oysters dropped to 18,000 μg/kg by day 60, probably because of spawning. At day 60, oysters were transferred to mercury-free seawater for 160 days. During the first 18 days, levels declined to 15,000 μg Hg/kg, but thereafter no further decline occurred. It was concluded that oysters can concentrate 10 μg Hg/liter by a factor of 2800 and that total self-purification was not achieved over a 6-month cleansing period. Kopfler (1974) found that continuous exposure to 1 μg Hg/liter in any of the three mercury compounds tested caused oysters to concentrate mercury rapidly in their tissues far in excess of 0.5 mg/kg wet weight, i.e., at levels potentially hazardous to humans ingesting these oysters, according to the action guideline established by the Food and Drug Administration. Details of Kopfler's study follow:

> Accumulation of mercury compounds by oysters was determined in two experiments, each utilizing three groups of 100 adult oysters. In the first experiment, conducted between 0 and 10°C, mercury levels were maintained at 50 μg/l with flow rate adjusted to one liter per oyster per hour. In the second experiment, mercury levels were reduced to one μg/l, the water temperature varied between 25 and 35°C, and flow rates maintained at 2 l/oyster/hr because of the increased temperature. Controls were maintained for each study. In the first experiment, the administration of organic mercury compounds was terminated after 19 days because many of the oysters in the groups receiving salts of either methylmercury or phenylmercury were dead or moribund. Oysters classified as moribund exhibited slow, incomplete valve closure when disturbed. When oysters which survived 19 days of exposure to methylmercury and phenylmercury were placed in flowing seawater, about half in each group died within a week with all oysters in both groups dead within 14 days. Oysters exposed to inorganic mercuric chloride exhibited no apparent ill effects over a 42-day period of exposure to 50 μg/l of Hg. Mean mercury levels in experimentals after one week was 1000X over controls. Mercury values in experimental oysters at 7 days, in mg Hg/kg wet wt, ranged between 15 and 25; for controls this was 0.02. Copper and zinc levels were depressed in flesh from all three groups of oysters after one week exposure to mercury; in the two groups exposed to organomercury, Cu and Zn declined over the 19-day exposure. Copper and zinc levels in the oysters exposed to mercuric chloride began to increase during the third week and continued to do so until they were essentially the same as control values. In the second experiment methyl- and phenylmercury were concentrated

to essentially the same degree, while inorganic mercury was concentrated about 4 times less over a 74-day exposure. For organomercury compounds, values in mg Hg/kg wet weight were about 10 at 20 days and 30 at 60 days (a final concentration factor of about 30,000). For inorganic mercuric chloride, values were about 2 at 20 days and about 10 at 60 days.

Mercury is also accumulated from seawater by other marine organisms. In radiotracer studies, clams (*Tapes decussatus*) contained 10 times more ^{203}Hg per unit weight than the medium within 24 hr (Unlu, Heyraud, and Keckes, 1972). Two species of marine alga, *Chaetoceros galvestonensis* and *Phaeodactylum tricornutum*, contained 7.4 and 2.4 g/kg of mercury, respectively, when cultured in media containing 100 μg Hg/liter (*Chaetoceros*) or 50 μg Hg/liter (*Phaeodactylum*) (Hannan et al., 1973).

Mercury can also be accumulated through food webs. Bacteria have a demonstrable effect on mercury accumulation in food chains that include filter feeders. Thus whole oyster accumulation of mercury more than doubled in the presence of mercury-resistant strains of *Pseudomonas* spp. (Colwell and Nelson, 1975; Colwell et al. 1976; Sayler, Nelson, and Colwell, 1975). Sarcophagous flies feeding on mercury-contaminated fish muscle accumulated 3 times more mercury than their food; this was especially pronounced when the fish flesh contained >0.5 mg Hg/kg wet weight (Nuorteva and Hasanen, 1972). A similar pattern was noted among wild fish feeding on mercury-contaminated fingerlings (Rucker and Amend, 1969). In the food chain algae-detritus—worm—fish (prey)—fish (predator), mercury had a long biological half-life, >1000 days in predatory fish compared with about 55 days in prey fish (Huckabee and Blaylock, 1972). The transfer efficiency of inorganic mercury from prey to predator fish was about 40% but from worm to prey fish, only 12%; worms assimilated 60% of the inorganic mercury contained in the algae-detritus. Huckabee and Blaylock concluded that food-chain uptake can account for a significant percentage of the mercury body burden in fish. In an algae-to-copepod food link, however, the crustacea showed no impairment of egg laying or egg development and no retention of mercury in tissues, eggs, or feces (Parrish and Carr, 1976). No progressive mercury concentration was observed in a clam-to-eel transfer (Tsuruga, 1963). When dogfish meal containing up to 2.3 mg total Hg/kg, of which 1.9 mg/kg was methyl mercury, replaced low-mercury fish rations used in salmon culture, the flesh of the salmon contained >0.5 mg/kg within 240 days (Spinelli and Mahnken, 1976). When dogfish meal comprised less than half the diet, however, mercury levels in salmon flesh were <0.5 mg/kg. Similar results were observed among sablefish fed dogfish meal (Kennedy and Smith, 1972).

Skeletal muscle of teleosts appears to function as a reservoir for methyl mercury (Giblin and Massaro, 1973; Johnels et al., 1967; MacLeod and Pessah, 1973; Middaugh and Rose, 1974; Pentreath, 1976b; Weisbart, 1973). Maximum concentration factors of radiomercury were reached in skeletal muscle, brain, and lens after 34, 56, and >90 days, respectively; maximum values were reached in most other tissues and organs in about 7 days (Giblin and Massaro, 1973). One pathway by which anadromous fishes accumulate mercury from a medium is through the gills; up to 90% of the mercury taken up on the gills is subsequently bound to erythrocytes within 40 min (Olson, Bergman, and Fromm, 1973; Olson and Fromm, 1973).

The retention time of mercury in marine organisms depends on several factors. Weisbart (1973) found that the gill, heart, and swim-bladder tissues of fish lost mercury at rates faster than the whole organism, but some tissues, including brain, liver, muscle, and kidney, showed no significant decrease with time. In clams radiomercury seems to have a biological half-life of 10 days if accumulated via the food chain but only 5 days if taken up from the water (Unlu, Heyraud, and Keckes, 1972). Retention times of ^{203}Hg in most marine species were comparatively lengthy, ranging from 267 days for the fish *Serranus scriba* to 1000 days for mussels, *Mytilus galloprovincialis* (Miettinen, Heyraud, and Keckes, 1972; Miettinen et al., 1969; 1972). Miettinen and coworkers found that phylogenetically related species follow a similar pattern of methyl mercury excretion, with biological half-life depending on water temperature and mode of entry into organisms; it is longer after intramuscular injection than after peroral administration. There is some evidence that mercury excretion rates follow a biphasic or polyphasic pattern (Burrows and Krenkel, 1973; Giblin and Massaro, 1973; Tillander, Miettinen, and Koivisto, 1972; Weisbart, 1973). Tillander, Miettinen, and Koivisto found that, in seals, *Pusa hispada*, the fastest excreted component took 20 days for 50% elimination and the slowest excreted mercury fraction, which comprised 45% of all mercury, 500 days for 50% elimination.

Field Investigations

Residues in Water and Sediments

Mercury levels in coastal seawater from various locations ranged from 0.021 to 0.078 μg/liter (Dehlinger et al., 1973); concentrations were lower in continental slope water, ranging from 0.010 to 0.041 μg/liter. Mercury levels in water were several orders of magnitude higher in the vicinity of chlor—alkali plants and similar industrial operations (Kazantzis, 1971). Mercury levels in sediments varied widely and tended to be elevated in the neighborhood of sewer outfalls (Klein and Goldberg, 1970), sludge disposal areas, and especially areas impacted by mercury wastes from industrial operations, such as Minamata Bay, Japan (Irukayama, 1967). Vucetic, Vernberg, and Anderson (1974) reported values ranging from 130 to 1500 μg/kg of mercury in sediments in the Adriatic Sea, and Williams and Weiss (1973) found 390 μg/kg in pelagic clays collected 430 km southeast of San Diego, Calif. It is reported that mercury in seawater exists almost entirely bound to suspended particles (Jernelöv et al., 1972), that the surface area of sediment granules is instrumental in determining final mercury content (Renzoni, Bacci, and Falciai, 1974), that conversion and transformations occur in the surface layer of the sediment or on suspended organic particles in the water (Dean, 1972; Fagerstrom and Jernelöv, 1972; Jernelöv et al., 1972), and, finally, that mercury-containing sediments would require many decades to purge themselves naturally to background levels (Langley, 1973).

Mercury—sediment—water interactions affect uptake by marine life. Bivalves can accumulate mercury directly from seawater, and molluscs had higher concentrations of the metal in turbulent waters than in clear waters (Raymont, 1972). Further, marine organisms feeding in direct contact with sediment have higher overall mercury levels than those feeding above the sediment—water interface (Klemmer, Unninayer, and Ukubo, 1976).

Residues in Biota

Table 2 summarizes the ranges of concentrations of mercury in field collections of marine flora and fauna. These data support the findings of other investigators (Cocoros, Cahn, and Siler, 1973; Huckabee and Blaylock, 1972; Jernelöv and Lann, 1971; Knauer and Martin, 1972; Skei, Saunders, and Price, 1976; Stickney et al., 1975), who demonstrated that the efficiency of mercury transfer through natural marine food chains among lower trophic levels is comparatively low. Higher trophic levels, such as fishes, fish-eating birds, and mammals, however, exhibit marked mercury amplification. The variability of concentrations is partly explainable in terms of collection locale; some field collections were taken from areas where human activities have raised the mercury content in the aquatic environment above natural levels, thus producing a significant increase in the mercury content of endemic fauna (Hearnden, 1970; Johnels et al., 1967; Kazantzis, 1971; Kleinert and Degurse, 1972;

TABLE 2

RANGE IN CONCENTRATIONS OF MERCURY (μg/kg wet weight) REPORTED FROM FIELD COLLECTIONS OF MARINE BIOTA

Sample	Concentration	Reference
Algae	40—2,000	Greig, Wenzloff, and Shelpuk, 1975; Haug, Melsom, and Omang, 1974; Leatherland and Burton, 1974; Lulic and Strohal, 1974
Higher aquatic plants	10—8,700	Gardner et al., 1975; Johnson and Braman, 1975; Lindberg and Harriss, 1974; Windom, 1973
Porifera	330—1,580	Leatherland and Burton, 1974; Williams and Weiss, 1973
Coelenterata	70—860	Leatherland and Burton, 1974; Leatherland et al., 1973
Tunicata	60—570	Leatherland and Burton, 1974; Leatherland et al., 1973
Echinodermata	280—400	Williams and Weiss, 1973
Annelida	350	Leatherland and Burton, 1974
Mollusca	1—4,440	Anderlini, 1974; Burton and Leatherland, 1971; Craig, 1967; Greig, Wenzloff, and Shelpuk, 1975; Hussain and Bleiler, 1973; Irukayama et al., 1962a; 1962b; Laumond et al., 1973; Leatherland and Burton, 1974; Lulic and Strohal, 1974; Stenner and Nickless, 1975; Thibaud, 1973; Vucetic, Vernberg, and Anderson, 1974; Williams and Weiss, 1973; Young, 1974
Crustacea	5—2,520	DeClerk, Vanderstoppen, and Vyncke, 1974; Eftekhari, 1975; Evans, Bails, and D'Itri, 1972; Greig, Wenzloff, and Shelpuk, 1975; Holden and Topping, 1972; Johnson and Braman, 1975; Leatherland et al., 1973; Raeder and Snekvik, 1949b; Skei, Saunders, and Price, 1976; Somayajulu and Rama, 1972; Stenner and Nickless, 1975; Vucetic, Vernberg, and Anderson, 1974; Williams and Weiss, 1973
Insecta	221—630	Kim, Chu, and Barron, 1974

(Table continues on following page.)

Renzoni, Bacci, and Falciai, 1974; Windom, Taylor, and Stickney, 1973; Wobeser et al., 1970; Zitko et al., 1971).

Inshore marine biota often contain higher mercury concentrations than the same or similar species collected offshore (Dehlinger et al., 1973; Jones, Jones, and Stewart, 1972; Westoo, 1969). In

Table 2 (Continued)

Sample	Concentration	Reference
Elasmobranchs	24—2,080	Childs and Gaffke, 1973; Childs, Gaffke, and Crawford, 1973; Forrester, Ketchen, and Wong, 1972; Gardner et al., 1975; Leatherland et al., 1973; Peterson, Kalwe, and Sharp, 1973; Windom et al., 1973
Teleosts	10—6,300	Alexander et al., 1973; Arima and Umemoto, 1976; Barber, Vijayakumar, and Cross, 1972; Beasley, 1971; Beckett and Freeman, 1974; Benson et al., 1976; Childs and Gaffke, 1973; Cocoros, Cahn, and Siler, 1973; Cugurra and Maura, 1976; DeGoeij and Zegers, 1971; Establier, 1975a; Freeman and Horne, 1973; Gardner et al., 1975; Greig, Wenzloff, and Pearce, 1976; Greig, Wenzloff, and Shelpuk, 1975; Greve and Wit, 1971; Johnson and Braman, 1975; Kamps, Carr, and Miller, 1972; Leatherland and Burton, 1974; Lulic and Strohal, 1974; Peterson, Kalwe, and Sharp, 1973; Portmann, 1972; Raeder and Snekvik, 1949a; 1949b; Renzoni, Bacci, and Falciai, 1974; Shultz et al., 1976; Stenner and Nickless, 1975; Stickney et al., 1975; Suzuki, Miyama, and Toyama, 1973; Ui and Kitamura, 1971; Vucetic, Vernberg, and Anderson, 1974; Williams and Weiss, 1973; Windom et al., 1973
Birds	50—65,000	Benson et al., 1976; Berg et al., 1966; Koivusaari et al., 1976
Mammals	40—387,000	Anas, 1971; 1974; Buhler, Claeys, and Mate, 1975; Gaskin et al., 1972; 1973; 1974; Holden, 1975; Jones et al., 1976; Kim, Chu, and Barron, 1974; Koeman et al., 1973; Roberts, Heppleston, and Roberts, 1976; Robertson et al., 1972; Sergeant and Armstrong, 1973; Tillander, Miettinen, and Koivisto, 1972

Sweden, marine fish caught near shore often showed comparatively high methyl mercury levels, with many values in the range from 5,000 to 10,000 μg/kg wet weight (Ackefors, Lofroth, and Rosen, 1970; Westoo, 1969); however, levels above 1,000 μg/kg in Swedish fish are usually caused by industrial discharges of mercury compounds (Westoo, 1969).

Several observations can be made regarding mercury residues in teleosts collected in the field. First, mercury tends to concentrate in the edible flesh of finfish, with older fish containing more mercury per unit weight than young fish (Alexander et al., 1973; Barber, Vijayakumar, and Cross, 1972; Cross et al., 1973; Cumont et al., 1972; DeClerk, Vanderstoppen, and Vyncke, 1974; Evans, Bails, and D'Itri, 1972; Forrester, Ketchen, and Wong, 1972; Giblin and Massaro, 1973; Greichus, Greichus, and Emerick, 1973; Hannerz, 1968; Johnels and Westermark, 1969; Johnels et al., 1967; Kleinert and Degurse, 1972; Knight and Herring, 1972; Nuorteva and Hasanen, 1971; 1975; Peterson, Kalwe, and Sharp, 1973; Scott and Armstrong, 1972; Svansson, 1975; Taylor and Bright, 1973; Vermeer, 1972). Second, most of the mercury in fish flesh was in the organic form, primarily methyl mercury (Fukui et al., 1973; Kamps, Carr, and Miller, 1972; Koeman et al., 1973; Peterson, Kalwe, and Sharp, 1973; Rissanen, Erkama, and Miettinen, 1972; Westoo, 1966; 1969; 1973; Zitko et al., 1971). Third, levels of mercury in muscle from adult tunas, billfishes, sharks, and other carnivores were higher than those in young fishes with a shorter food chain; this indicates associations among predatory behavior, longevity, and mercury accumulation (Forrester, Ketchen, and Wong, 1972; Jernelöv, 1972; Peakall and Lovett, 1972; Peterson, Kalwe, and Sharp, 1973; Ratkowsky, Dix, and Wilson, 1975; Saila, 1971; Ui, 1972; Walter, June, and Brown, 1973). Finally, elevated levels of mercury in wide-ranging oceanic fish are not solely the consequence of anthropogenic activities but also result from natural concentrations as well (Greig, Wenzloff, and Pearce, 1976; Miller et al., 1972; Schultz et al., 1976).

Birds that feed on aquatic animals exhibit markedly increased concentrations of mercury in comparison with terrestrial raptorial species (Johnels and Westermark, 1969; Karppanen and Henriksson, 1970). Top trophic-level predators, such as cormorants and pelicans, demonstrated mercury concentration factors in their flesh of 14- and 6-fold, respectively, over prey fish (Greichus, Greichus, and Emerick, 1973). Mercury levels in aquatic and fish-eating birds were elevated in the vicinity of chlor—alkali plants where mercury is used as a catalyst (Fimreite, 1974; Fimreite et al., 1971); these differences were detectable up to 300 km from the chlor—alkali plant (Fimreite and Reynolds, 1973). Eggs of fish-eating birds, including herons and grebes, collected near mercury point-source discharges contained abnormally high levels of mercury; 29% of eggs contained more than 0.5 mg/kg wet weight, and 9% contained more than 1.0 mg/kg (Faber and Hickey, 1973). Seasonal variations in mercury levels occur in livers of aquatic birds. Higher levels were observed in winter, when birds were exclusively estuarine, and drastically lower levels were observed in summer when birds migrate to inland Arctic and sub-Arctic breeding grounds (Parslow, 1973).

Diet is an important concentrating mechanism in marine mammals. Gray seals and harbor seals, which feed on large fish and cephalopods, contained 10 times more mercury than harp seals, which feed on small pelagic fish and crustacea (Sergeant and Armstrong, 1973). Age and tissue specificity also influence mercury residue levels. Liver from older sea lions contained more mercury per unit weight than that from younger specimens (Anas, 1974; Holden, 1975; Jones et al., 1976; Sergeant and Armstrong, 1973). Some adult seals found dead had high levels of mercury in the brain in comparison with levels in various species poisoned by exposure to mercury compounds (Koeman et al., 1973). Mercury and selenium concentration in livers of marine mammals seem to be positively correlated (Koeman et al., 1973; 1975; Martin et al., 1976). Koeman and co-workers (1973) suggested that selenium protects these species against mercury poisoning by completely binding to subcellular S sites, the presumed location of mercury's toxic action.

Three points are worth emphasizing at this juncture. First, mercury discharged into rivers, bays, or estuaries as metallic mercury, inorganic divalent mercury, phenyl mercury, or alkoxyalkylmercury can all be converted to methyl mercury compounds by natural processes (Jernelöv, 1969). Second, organomercury complexes are rapidly accumulated in tissues with high lipid content (Wood, 1973). Finally, mercury-resistant strains of bacteria which have been developed or discovered may have application in mercury mobilization or fixation from mercury-contaminated waters to the extent that polluted areas become innocuous (Colwell and Nelson, 1975; Nelson et al., 1973; Vosjan and Van der Hoek, 1972).

Minamata

Any review of mercury hazards in the marine environment should include the Minamata Bay incident in southwestern Kyushu, Japan. This extensively documented case (Fujiki, 1963; Irukayama, 1967; Irukayama et al., 1961; 1962a; 1962b; Kiyoura, 1963; Kurland, Faro, and Siedler, 1960; Matida and Kumada, 1969; Matida et al., 1972; Takevchi, 1972; Tsubaki et al., 1967) reveals effects upon man of chronic discharges of low-level methyl mercury wastes into coastal waters. The source of mercury was waste discharged from an acetaldehyde plant into Minamata Bay beginning in 1952.

Several years later, the mercury levels in sediments near the plant outfall were about 2010 mg/kg wet weight; this decreased sharply with distance from the plant. Sediments in the bay contained between 0.4 and 3.4 mg Hg/kg wet weight. Concentrations of mercury in fish, shellfish, and other food decreased with increasing distance from the point of effluence and appeared to reflect sediment mercury levels. Late in 1953, a severe neurological disorder was recognized among inhabitants of the Minamata Bay region. By 1956, the outbreak had reached epidemic proportions; 111 cases of poisoning were found by the end of 1960; and 41 deaths were reported as of August 1965. The poisoning was caused by eating fish and shellfish from Minamata Bay and the neighboring sea, except for 19 congenital cases in children born of mothers who had eaten the same diet. In addition to humans, cats and water fowl living near the bay succumbed to the disease. Experimental cats and rats fed fish and shellfish collected from the bay developed the same symptoms as animals spontaneously affected. Symptoms included cerebellar ataxia, constriction of visual fields, dysarthria, and, in congenital cases, disturbance of physical and mental development. Pathological findings included regressive changes in the cerebellum and the cerebral cortexes. The clinico-pathological features resembled alkyl mercury poisoning. Abnormal mercury content (i.e., >30.0 mg Hg/kg wet weight) was measured in fish, shellfish, and muds from the bay and in organs of necropsied humans and cats that had succumbed to the disease. Years after the waste-discharge situation was corrected, fish and shellfish still contained levels of mercury hazardous to human health. Mercury levels in Minamata fish and shellfish were significantly higher than levels shown in Table 2 for teleosts and molluscs; note also that the high levels shown for mammals in Table 2 are from liver.

Potential Health Hazards

The dangerous mercury level in human blood is about 2000 ng/g; the normal background level in about 80% of people without occupational exposure to mercury is <5 ng/g (Saha, 1972). Since methyl mercury has a biological half-life of 70 to 200 days in man (Lofroth, 1970), it is not unexpected that mutagenic and teratogenic effects of mercury have been reported at levels well below those associated with acute poisoning (Kazantzis, 1971; Keckes and Miettinen, 1972).

The findings of elevated blood mercury levels, increased human chromosome breakage, and easy passage of mercury through placental membranes among people who regularly eat fish containing

1 to 17 mg Hg/kg wet weight suggest to some health authorities that the currently accepted level in the United States [0.5 mg Hg/kg wet weight (500 μg/kg)] does not have an appreciable level of safety for pregnant women (Establier, 1975b; Peakall and Lovett, 1972; Skerfving, Hansson, and Lindstem, 1970). In Sweden, the official limit for mercury in fish is 1.0 mg/kg wet flesh. Human risk in consumption of mercury-contaminated fish within this guideline is considered negligible in that country (Berglund and Berlin, 1969); however, some Scandinavian scientists recommend that fish consumption should be limited to one meal a week (Lofroth, 1970).

Selected Studies

Mercury retention by teleost species is variable. In one study, when northern pike from a lake that was heavily contaminated with mercury were transplanted to waters relatively free of mercury, only 30% of the mercury in their muscle was eliminated in 1 year (Lockhart et al., 1972). A study by Amend (1970) of juvenile sockeye salmon indicated that fish treated repeatedly with mercurials during their freshwater phase accumulated and retained high levels of mercury for several months, but, after 4 years at sea, the returning salmon contained normal levels of mercury in all tissues examined.

A number of investigators report that mercury from point-source discharges, including sewer outfalls and chlor—alkali plants, was taken up by sediments; sediment levels were eventually reflected by an increased mercury content of epibenthic fauna (Dehlinger et al., 1973; Hoggins and Brooks, 1973; Klein and Goldberg, 1970; Klemmer, Luoma, and Lau, 1973; Parsons, Bawden, and Heath, 1973; Takevchi, 1972). Analysis of the effluent from the Hyperion sewer outfall in Los Angeles showed a mercury content slightly below 1 μg/liter (Klein and Goldberg, 1970). Concentrations of mercury in sediments near this outfall were as high as 820 μg/kg but decreased with increasing distance from the outfall; mercury levels in epibenthic fauna, including crabs, whelks, and scallops, were also highest near the discharge and lowest tens of kilometers away.

A mercury budget for estuaries along the Georgia coast indicates that the dominant salt-marsh plant, *Spartina alterniflora*, exerts a strong control on mercury migration (Gardner et al., 1975; Windom et al., 1976; Windom, 1973). Mercury entered the estuary primarily in solution, delivering ~1.5 mg annually to each square meter of salt marsh. Annual uptake of mercury by *S. alterniflora* alone was about 0.7 mg/m^2 salt marsh. Mangrove vegetation plays a similarly important role in mercury cycling in the Florida everglades (Lindberg and Harriss, 1974; Tripp and Harriss, 1976).

RECOMMENDATIONS

On the basis of these studies, it appears that establishing marine mercury-contamination standards requires abatement of existing discharges, development of technologies to decontaminate polluted waterways, monitoring criteria for protecting human health, and research to modify existing standards, as well as to predict impending mercury stress in marine ecosystems. Specific recommendations are listed.

1. Prohibit the sale, manufacture, and use of methyl mercury compounds. All evidence presented to date implicates methyl mercury as the most toxic chemical form of mercury. Its demonstrated deleterious effects to marine biota and human health mandate passage of legislation banning all environmental uses of methyl mercury compounds. In the decade since Sweden prohibited agricultural use of methyl mercury for seed coatings and pesticides, bird populations have begun to recover and the amounts of mercury in many types of food have decreased (Johnels and Westermark, 1969).

2. Restrict use of other mercurials. Investigations are presently in progress in several countries on the geochemistry and biogeochemistry of mercury, but much remains to be learned about its distribution, its conversion from one form to another, and its transport on a global basis. It is well known, however, that inorganic mercury compounds can be converted in nature to organomercury compounds, which are more readily transported and accumulated in tissues. Accordingly, legislation is needed which will curtail environmental use of all mercury compounds, especially organomercurials.

3. Develop alternative technologies. About 90% of the mercury consumed annually is used for industrial purposes. Mercury is used as a catalyst in chlor—alkali plants, as a slimicide in pulp and paper mills, as a chemical reagent in the plastics industry, in pharmaceuticals and paints, in power plants in special heat engines, in metal-refinement operations by amalgamation, and in the manufacture of electric switches, batteries, and lamps. Chlor—alkali plants probably represent the greatest point-source hazard at this time; typical plants contain more than 500,000 kg of liquid mercury and discharge several kilograms daily to the environment. In the absence of better controls, this loss could be averted by developing a new technology that precludes mercury compounds. This technology is urgently required and should be awarded the highest priority.

4. Remove mercury from contaminated waterways. Thousands of kilometers of our inland waterways and some coastal areas are contaminated with mercury to the extent that fish and other organisms taken from these waters are unsafe to eat (Anonymous, 1971). This process shows little sign of abatement. Accordingly, high priority should be given to research and development technology to remove mercury from existing contaminated areas. These could include bacterial strains and other methods to solubilize mercury from sediments, with subsequent removal and reclamation from waterways via biological or abiotic vectors. One interim alternative would be to feed mercury-contaminated scrapfish and seafoods as partial diet supplements to migratory anadromous fishes. Removal by dredging of mercury-contaminated sediments and storage at designated locations is another alternative, but numerous problems are associated with this approach as currently practiced.

5. Develop monitoring criteria. Preventing mercury feedback into human diets is of paramount importance. To achieve this goal, we must refine existing mercury-residue standards for waste waters, sediments, and biota, with due consideration for geographical location and economic aspects of the resource.

It is emphasized at this time that mercury in marine waters is usually bound to bottom materials, proteinaceous matter, plankton, and higher organisms. Even in water heavily polluted by mercury discharges, only a fraction is found free in solution, except near the point of discharge. The best available evidence indicates that total mercury in waste waters, dissolved plus particulate, emanating from the pipe terminus should not exceed 1.0 μg/liter; however, I believe two additional restrictions on pipe terminus concentrations are warranted to achieve a reasonable margin of safety. First, outfall volume should not exceed one-fiftieth of the mean daily waterflow of the receiving water. Second, the receiving waters upstream of the terminus (or in direction of positive tidal flow, if applicable) should contain less than 0.02 μg/liter total mercury, dissolved plus particulate. The final recommended value of 0.0392 μg total Hg/liter after mixing is about 2.5 times lower than the currently recommended value of less than or equal to 0.10 μg/liter proposed by the National Academy of Sciences (1973); however, the 0.10 μg/liter does not consider mixing processes.

Because benthic animals sometimes reflect the mercury levels of their sediment substrate, it is recommended that sediments in outfall environs should not exceed 500 μg total Hg/kg. This value is less than the 750 μg/kg set by the Environmental Protection Agency (1977) for solid wastes disposed offshore.

For biological monitoring, two groups of indicator organisms are recommended, filter-feeding bivalve molluscs and piscivorous fishes.

Total mercury levels in soft tissues from bivalves or fish axial musculature should not exceed 500 μg/kg wet weight (0.5 ppm). This is the current limit in comestibles set by the Food and Drug Administration. Until additional evidence becomes available it is also suggested that expectant mothers limit to 200 g/week their dietary intake of seafood items containing total mercury concentrations in the 0.3- to 0.5-ppm range.

6. Conduct research. One of the highest priorities should be assigned to refining the current analytical chemistry methods for detecting mercury. Specifically, procedures must be developed for accurate, precise, and inexpensive determination of mercury at concentrations as low as 0.02 μg/liter.

An equally high priority could be assigned to laboratory studies on lethal, sublethal, and latent effects of mercurials on selected marine indicator species during lifetime exposure. These data do not exist at present. Tests must be conducted in flowing seawater and should, at a minimum, accurately assess mercury effects on survival, growth, behavior, reproduction, accumulation, and histopathology. Results of these studies would provide a basis for a more reasonable assessment of safe levels of mercury in water and would probably result in some modification of the proposed 0.0392 μg/liter value recommended here.

Field and laboratory investigations of interaction effects of mercury with other constituents in the environment should be conducted. The significance of the apparent protective effect of selenium in marine mammals is a case in point. At present, data on field studies of mercury cycling and effects of mixed wastes containing mercury on community structure are scarce or absent. These studies should be conducted, with a major effort being assigned to roles played by aquatic mammals, fish-eating birds, predatory fishes, higher plants, and mercury-resistant strains of bacteria. Establishment of a nation-wide monitoring program on mercury levels in coastal sediments appears to be a necessary prerequisite for identifying heavily contaminated areas and selecting appropriate sites for field investigations.

One of the most important research needs is to develop an early warning system in selected indicator organisms that will forecast generalized stress among marine communities, with emphasis on response parameters peculiar to mercury-induced stress. Within the past decade, multiparametric approaches have been used extensively by industrial toxicologists and others who work with poisonous substances; however, application of these approaches to marine communities is rare and difficult. Typically, mercury-induced stress would be evident long before physiological or morphological damage occurs, would be manifested at the cellular or subcellular level, and, if allowed to remain unabated, would ultimately cause death or large-scale changes in community structure. Biochemical responses are probably most sensitive at present, but behavioral and histological responses should not be discounted as possible early indicators.

7. *Apply recommendations to other contaminants.* With minor modifications, the preceding recommendations could become protocols for other heavy metals in coastal ecosystems and, with additional modifications, could be useful in abatement, decontamination, and research programs now in progress on many groups of synthetic chemicals routinely discharged into the marine environment.

ACKNOWLEDGMENT

I am obligated to fellow staff members of the Environmental Research Laboratory, Narragansett, for their many constructive comments on this manuscript, especially Gerald Pesch, Frank Lowman, Daniel O'Neill, Glen Thompson, Earl Davey, Gerald Zaroogian, John Gentile, Sandra Marburg, and Brian Melzian. All opinions and data interpretations expressed herein are mine and do not necessarily reflect those of the Environmental Protection Agency or any other regulatory agency.

REFERENCES

Abou-Donia, M. B., and D. B. Menzel, 1967, *Comp. Biochem. Physiol.*, 21: 99.

Ackefors, H., G. Lofroth, and C. G. Rosen, 1970, *Ocean. Mar. Biol. Annu. Rev.*, 8: 203.

Alexander, J. E., J. Foehrenbach, S. Fisher, and D. Sullivan, 1973, *NY Fish Game J.*, 20: 147.

Amend, D. F., 1970, *Prog. Fish-Cult.*, 32: 192.

——, W. T. Yasutake, and R. Morgan, 1969, *Trans. Am. Fish. Soc.*, 98: 419.

Anas, R. E., 1971, in *Mercury in the Western Environment*, Workshop, Portland, Ore., Feb. 25–26, 1971, D. R. Buhler (Ed.), p. 91, Environmental Health and Science Center, Oregon State University, Corvallis.

——, 1974, *Fish. Bull.*, 72: 133.

Anderlini, V., 1974, *Arch. Environ. Contam. Toxicol.*, 2: 253.

Anonymous, 1971, *Sport Fishing Inst. Bull.*, 221: 4.

Arima, S., and S. Umemoto, 1976, *Bull. Jpn. Soc. Sci. Fish.*, 42: 931.

Ballard, J. A., and W. D. Oliff, 1969, *Water Res.*, 3: 313.

Barber, R. T., A. Vijayakumar, and F. A. Cross, 1972, *Science*, 178: 636.

Barnes, H., and F. A. Stanbury, 1948, *J. Exp. Biol.*, 25: 270.

Beasley, T. M., 1971, *Environ. Sci. Technol.*, 5: 634.

Beckett, J. S., and H. C. Freeman, 1974, Special Scientific Report 675, p. 154, U. S. Department of Commerce.
Ben-Bassat, D., G. Shelef, N. Gruner, and H. I. Shuval, 1972, *Nature*, 240(5375): 43.
Benson, W. W., D. W. Brock, J. Gabica, and M. Loomis, 1976, *Bull. Environ. Contam. Toxicol.*, 15: 543.
Berg, W., A. Johnels, B. Sjostrand, and T. Westermark, 1966, *Oikos*, 17: 71.
Berglund, F., and M. Berlin, 1969, in *Chemical Fallout, Current Research on Persistent Pesticides*, M. W. Miller and G. G. Berg (Eds.), p. 423, Charles C Thomas, Publisher, Springfield, Ill.
Berland, B. R., D. J. Bonin, V. I. Kapkov, S. Y. Maestrini, and D. P. Arlhac, 1976, *C.R. Acad. Sci., Paris*, t. 282, Ser. D: 633.
Binet, L., and P. Nicolle, 1940, *C.R. Soc. Biol., Paris*, 134: 562.
Bligh, E. G., 1972, *J. Inst. Can. Sci. Technol. Aliment.*, 5: A6.
Boetius, J., 1960, *Medd. Komm. Havunders., Copenhagen*, 3(N.S.)(4): 93.
Boney, A. D., 1971, *Mar. Pollut. Bull.*, 2: 69.
——, and E. D. S. Corner, 1959, *J. Mar. Biol. Assoc. U.K.*, 38: 267.
——, E. D. S. Corner, and B. W. P. Sparrow, 1959, *Biochem. Pharmacol.*, 2: 37.
Boulton, P., and L. J. Hetling, 1972, Technical Paper 19, New York State Department of Environmental Conservation, Albany.
Brown, B., and M. Ahsanullah, 1971, *Mar. Pollut. Bull.*, 2: 182.
Buhler, D. R., R. R. Claeys, and B. R. Mate, 1975, *J. Fish. Res. Board Can.*, 32: 2391.
Burrows, W. D., and P. A. Krenkel, 1973, *Environ. Sci. Technol.*, 7: 1127.
Burton, J. D., and T. M. Leatherland, 1971, *Nature*, 231: 440.
Cabejszek, I., and M. Stasiak, 1960, *Roczn. Panst. Zakl. Hig.*, 11: 533.
Calabrese, A., R. S. Collier, D. A. Nelson, and J. R. MacInnes, 1973, *Mar. Biol.*, 18: 162.
——, and D. A. Nelson, 1974, *Bull. Environ. Contam. Toxicol.*, 11: 92.
——, F. P. Thurberg, M. A. Dawson, and D. R. Wenzloff, 1975, in *Sublethal Effects of Toxic Chemicals on Aquatic Animals*, J. H. Koeman and J. J. T. W. A. Strik (Eds.), p. 15, Elsevier-North Holland Publishing Co., New York.
Calamari, D., and R. Marchetti, 1973, *Water Res.*, 7: 1453.
Childs, E. A., and J. N. Gaffke, 1973, *Fish. Bull.*, 71: 713.
——, J. N. Gaffke, and D. L. Crawford, 1973, *Bull. Environ. Contam. Toxicol.*, 9: 276.
Christensen, G. M., 1971, *Chem.-Biol. Interact.*, 4(1971/72): 351.
Clarke, G. L., 1947, *Biol. Bull.*, 92: 73.
Clemens, H. P., and K. E. Sneed, 1958, *Prog. Fish-Cult.*, 20: 147.
Clendenning, K. A., and W. J. North, 1959, in *Proceedings of the First International Conference on Waste Disposal in Marine Environments*, Berkeley, Calif., E. A. Pearson (Ed.), p. 82, Pergamon Press, Inc., New York.
Cocoros, G., P. H. Cahn, and W. Siler, 1973, *J. Fish. Biol.*, 5: 641.
Colwell, R. R., and J. D. Nelson, Jr., 1975, EPA Rept. 600/3-75-007, Environmental Research Laboratory, Narragansett, R.I., NTIS.
——, G. S. Sayler, J. D. Nelson, Jr., and A. Justice, 1976, in *Environmental Biogeochemistry*. Vol. 2. *Metals Transfer and Ecological Mass Balances*, J. D. Nriagu (Ed.), p. 437, Ann Arbor Science Publishers, Ann Arbor, Mich.
Connor, P. M., 1972, *Mar. Pollut. Bull.*, 3: 190.
Corner, E. D. S., 1959, *Biochem. Pharmacol.*, 2: 121.

——, and F. H. Rigler, 1957, *J. Mar. Biol. Assoc. U.K.*, 37: 85.
——, and B. W. Sparrow, 1956, *J. Mar. Biol. Assoc. U.K.*, 35: 531.
——, and B. W. Sparrow, 1957, *J. Mar. Biol. Assoc. U.K.*, 36: 459.
Craig, S., 1967, *J. Am. Osteopath. Assoc.*, 66: 1000.
Cross, F. A., L. H. Hardy, N. Y. Jones, and R. T. Barber, 1973, *J. Fish. Res. Board Can.*, 30: 1287.
Cugurra, F., and G. Maura, 1976, *Bull. Environ. Contam. Toxicol.*, 15: 568.
Cumont, G., C. Viallex, H. Lelievre, and P. Bobenrieth, 1972, *Rev. Int. Oceanogr. Med.*, 28: 95.
Cunningham, P. A., and M. R. Tripp, 1973, *Mar. Biol.*, 20: 14.
——, and M. R. Tripp, 1975a, *Mar. Biol.*, 31: 321.
——, and M. R. Tripp, 1975b, *Mar. Biol.*, 31: 311.
D'Agostino, A., and C. Finney, 1974, in *Pollution and Physiology of Marine Organisms*, F. J. Vernberg and W. B. Vernberg (Eds.), p. 445, Academic Press, Inc., New York.
Davies, A. G., 1974, *J. Mar. Biol. Assoc. U.K.*, 54: 157.
——, 1976, *J. Mar. Biol. Assoc. U.K.*, 56: 39.
Dean, R. B., 1972, Environmental Protection Agency, PB-213-692:1, NTIS.
DeClerk, R., R. Vanderstoppen, and W. Vyncke, 1974, *Ocean Manage.*, 2: 117.
DeCoursey, P. J., and W. B. Vernberg, 1972, *Oikos*, 23: 241.
DeGoeij, J. J. M., and C. Zegers, 1971, *TNO Nieuws*, 26: 400.
Dehlinger, P., W. F. Fitzgerald, S. Y. Feng, D. F. Paskausky, R. W. Garvine, and W. F. Bohlen, 1973, University of Connecticut, Marine Science Institute, Groton, Conn.
D'Itri, F. M., 1972, Technical Report 23, Institute of Water Research, Michigan State University, Lansing.
Dorn, P., 1974, *Bull. Environ. Contam. Toxicol.*, 12: 86.
——, 1976, *Bull. Environ. Contam. Toxicol.*, 15: 714.
Drummond, R. A., G. F. Olson, and A. R. Batterman, 1974, *Trans. Am. Fish. Soc.*, 103: 244.
Eftekhari, M., 1975, *Sci. Peche*, 250: 9.
Eisler, R., G. R. Gardner, R. J. Hennekey, G. LaRoche, D. F. Walsh, and P. P. Yevich, 1972, *Water Res.*, 6: 1009.
——, and R. J. Hennekey, 1977, *Arch. Environ. Contam. Toxicol.*, 6: 315.
Ellis, M. M., 1947, Special Scientific Report 42, U.S. Fish and Wildlife Service.
Environmental Protection Agency, 1977, *Federal Register*, 42(7): 2477.
Establier, R., 1975a, *Invest. Pesq.*, 39: 509.
——, 1975b, *Invest. Pesq.*, 39: 249.
Evans, R. J., J. D. Bails, and F. M. D'Itri, 1972, *Environ. Sci. Technol.*, 6: 901.
Faber, R. A., and J. J. Hickey, 1973, *Pestic. Monit. J.*, 7: 27.
Fagerstrom, T., and A. Jernelöv, 1972, *Water Res.*, 6: 1193.
Fang, S. C., 1973, *Arch. Environ. Contam. Toxicol.*, 1: 18.
Fimreite, N., 1974, *J. Wildl. Manage.*, 38: 120.
——, W. N. Holswroth, J. A. Keith, P. A. Pearce, and I. M. Gruchy, 1971, *Can. Field-Nat.*, 85: 211.
——, and L. M. Reynolds, 1973, *J. Wildl. Manage.*, 37: 62.
Forrester, C. R., K. S. Ketchen, and C. C. Wong, 1972, *J. Fish. Res. Board Can.*, 29: 1487.
Foster, R. F., and P. A. Olson, 1951, *Prog. Fish-Cult.*, 13: 129.
Fowler, B. A., D. A. Wolfe, and W. F. Hettler, 1975, *J. Fish. Res. Board Can.*, 32: 1767.

Fowler, S. W., and G. Benayoun, 1976, *Mar. Biol.*, 37: 59.
Freeman, H. C., and D. A. Horne, 1973, *J. Fish. Res. Board Can.*, 30: 454.
Friberg, L., and J. Vostal, 1972, CRC Press, Cleveland, Ohio.
Fujiki, M., 1963, *Kumamoto Igakkai Zasshi*, 37: 494.
Fukui, S., K. Tanaka, S. Kanno, and T. Ukita, 1973, *Adv. Water Pollut. Res., Proc. Int. Conf., 6th*, 1972: 819.
Gardner, G. R., 1975, in *The Pathology of Fishes*, W. E. Ribelin and G. Migaki (Eds.), p. 657, University of Wisconsin Press, Madison.
Gardner, W. S., H. L. Windom, J. A. Stephens, F. E. Taylor, and R. R. Stickney, 1975, in *Mineral Cycling in Southeastern Ecosystems*, ERDA Symposium Series, Augusta, Ga., May 1—3, 1974, F. G. Howell, J. B. Gentry, and M. H. Smith (Eds.), p. 268, CONF-740513, NTIS.
Gaskin, D. E., R. Frank, M. Holdrinet, K. Ishida, C. J. Walton, and M. Smith, 1973, *J. Fish. Res. Board Can.*, 30: 471.
——, K. Ishida, and R. Frank, 1972, *J. Fish. Res. Board Can.*, 29: 1644.
——, G. J. D. Smith, P. W. Arnold, M. V. Louisy, R. Frank, M. Holdrinet, and J. W. McWade, 1974, *J. Fish. Res. Board Can.*, 31: 1235.
Gavis, J., and J. F. Ferguson, 1972, *Water Res.*, 6: 989.
Giblin, F. J., and E. J. Massaro, 1973, *Toxicol. Appl. Pharmacol.*, 24: 81.
Gillespie, D. C., 1972, *J. Fish. Res. Board Can.*, 29: 1035.
Glooschenko, W. A., 1969, *J. Phycol.*, 5: 224.
Gray, J. S., 1974, in *Pollution and Physiology of Marine Organisms*, F. J. Vernberg and W. B. Vernberg (Eds.), p. 465, Academic Press, Inc., New York.
——, and R. J. Ventilla, 1971, *Mar. Pollut. Bull.*, 2: 39.
——, and R. J. Ventilla, 1973, *Ambio*, 2: 118.
Green, F. A., Jr., J. W. Anderson, S. R. Petrocelli, B. J. Presley, and R. Sims, 1976, *Mar. Biol.*, 37: 75.
Greichus, Y. A., A. Greichus, and R. J. Emerick, 1973, *Bull. Environ. Contam. Toxicol.*, 9: 321.
Greig, R. A., D. R. Wenzloff, and J. B. Pearce, 1976, *Mar. Pollut. Bull.*, 7: 185.
——, D. Wenzloff, and C. Shelpuk, 1975, *Pestic. Monit. J.*, 9: 15.
Greve, P. A., and S. L. Wit, 1971, *TNO Nieuws*, 26: 395.
Hannan, P. J., P. E. Wilkniss, C. Patouillet, and R. A. Carr, 1973, Report 7628, U. S. Naval Research Laboratory, Washington, D.C.
Hannerz, L., 1968, *Rep. Inst. Freshwat. Res. Drottningholm*, 48: 120.
Harriss, R. C., 1971, *Biol. Conserv.*, 3: 279.
——, D. B. White, and R. B. MacFarlane, 1970, *Science*, 170(3959): 736.
Hasselrot, T. B., 1968, *Rep. Inst. Freshwat. Res. Drottningholm*, 48: 102.
Haug, A., S. Melsom, and S. Omang, 1974, *Environ. Pollut. (London)*, 7: 179.
Hearnden, E. H., 1970, *Fisher. Can.*, 22(10): 3.
Hendrick, R. D., and T. R. Everett, 1965, *J. Econ. Entomol.*, 58: 958.
Hibiya, T., and M. Oguri, 1961, *Bull. Jpn. Soc. Sci. Fish.*, 27: 996.
Hoggins, F. E., and R. R. Brooks, 1973, *NZ J. Mar. Freshwater Res.*, 7: 125.
Holden, A. V., 1973, *J. Food Technol.*, 8: 1.
——, 1975, *Rapp. P.-v. Reun. Cons. Int. Explor. Mer*, 169: 353.
——, and G. Topping, 1972, *Proc. R. Soc. Edinburgh, Sect. B*, 71(14): 189.
Hopkins, R., and J. M. Kain, 1971, *Mar. Pollut. Bull.*, 2: 75.
Horn, W. M. V., and M. Katz, 1946, *Science*, 104: 557.
Huckabee, J. W., and B. G. Blaylock, 1972, in *Metal Ions in Biological Systems*, S. K. Dhar (Ed.), p. 125, Plenum Publishing Corporation, New York.

Hunter, W. R., 1949, *J. Exp. Biol.*, 26: 113.
Hussain, M., and E. L. Bleiler, 1973, *Mar. Pollut. Bull.*, 4: 44.
Irukayama, K., 1967, *Adv. Water Pollut. Res., Proc. Int. Conf., 3rd*, 1966: 153.
——, M. Fujiki, F. Kai, and T. Kondo, 1962a, *Kumamoto Igakkai Zasshi*, 15: 1.
——, F. Kai, M. Fujiki, and T. Kondo, 1962b, *Kumamoto Igakkai Zasshi*, 15: 57.
——, T. Kondo, F. Kai, and M. Fujiki, 1961, *Kumamoto Igakkai Zasshi*, 14: 158.
Jackim, E., 1973, *J. Fish. Res. Board Can.*, 30: 560.
——, J. M. Hamlin, and S. Sonis, 1970, *J. Fish. Res. Board Can.*, 27: 383.
Jarvenpaa, T., M. Tillander, and J. K. Miettinen, 1970, *Suomen Kem., B.*, 43: 439.
Jernelöv, A., 1969, in *Chemical Fallout, Current Research on Persistent Pesticides*, M. W. Miller and G. G. Berg (Eds.), p. 68, Charles C Thomas, Publisher, Springfield, Ill.
——, 1972, *The Changing Chemistry of the Oceans*, Proceedings of the 20th Nobel Symposium, Lerum, Sweden, Aug. 16—20, 1971, D. Dyrssen and D. Jagner (Eds.), p. 161, John Wiley & Sons, Inc., New York.
——, R. Hartung, P. B. Trost, and R. R. Bisque, 1972, in *Environmental Mercury Contamination*, R. Hartung and B. D. Dinman (Eds.), p. 167, Ann Arbor Science Publishers, Ann Arbor, Mich.
——, L. Landner, and T. Larsson, 1975, *J. Water Pollut. Control Fed.*, 47: 810.
——, and H. Lann, 1971, *Oikos*, 22: 403.
Johnels, A. G., and T. Westermark, 1969, in *Chemical Fallout, Current Research on Persistent Pesticides*, M. W. Miller and G. G. Berg (Eds.), p. 221, Charles C Thomas, Publisher, Springfield, Ill.
——, T. Westermark, W. Berg, P. I. Persson, and B. Sjostrand, 1967, *Oikos*, 18: 323.
Johnson, D. L., and R. S. Braman, 1975, *Deep-Sea Res.*, 22: 503.
Jones, A. M., Y. Jones, and W. D. P. Stewart, 1972, *Nature*, 238(5360): 164.
Jones, D., K. Ronald, D. M. Lavigne, R. Frank, M. Holdrinet, and J. F. Uthe, 1976, *Sci. Total Environ.*, 5: 181.
Jones, M. B., 1973, *Estuarine Coastal Mar. Sci.*, 1: 425.
——, 1975, *Mar. Biol.*, 30: 13.
Kamps, L. R., R. Carr, and H. Miller, 1972, *Bull. Environ. Contam. Toxicol.*, 8: 273.
Karbe, L., 1972, *Mar. Biol.*, 12: 316.
Karppanen, E., and K. Henriksson, 1970, *Nord. Med.*, 84(35): 1097.
Kayser, H., 1976, *Mar. Biol.*, 36: 61.
Kazantzis, G., 1971, *Int. J. Environ. Stud.*, 1: 301.
Keckes, S., and J. J. Miettinen, 1972, in *Marine Pollution and Sea Life*, M. Ruivo (Ed.), p. 276, Fishing Trading News (Books), Ltd., London.
Kennedy, W. A., and M. S. Smith, 1972, Technical Report 309, Fisheries Research Board of Canada, Ottawa.
Kihlstrom, J. E., and L. Hulth, 1972, *Bull. Environ. Contam. Toxicol.*, 7: 111.
——, C. Lundberg, and L. Hulth, 1971, *Environ. Res.*, 4: 355.
Kim, K. C., R. C. Chu, and G. P. Barron, 1974, *Bull. Environ. Contam. Toxicol.*, 11: 281.
Kiyoura, R., 1963, *Int. J. Air Water Pollut.*, 7: 459.
Klaunig, J., S. Koepp, and M. McCormick, 1975, *Bull. Environ. Contam. Toxicol.*, 14: 534.
Klein, D. H., and E. D. Goldberg, 1970, *Environ. Sci. Technol.*, 4: 765.

Kleinert, S. J., and P. E. Degurse, 1972, Technical Bulletin 52, Department of Natural Resources, Madison, Wisc.
Klemmer, H., S. N. Luoma, and L. S. Lau, 1973, *Gov. Rep. Announce.*, 73(10): 76.
Klemmer, H. W., C. S. Unninayer, and W. I. Ukubo, 1976, *Bull. Environ. Contam. Toxicol.*, 15: 454.
Knauer, G. A., and J. H. Martin, 1972, *Limnol. Oceanogr.*, 17: 868.
Knight, L. A., Jr., and J. Herring, 1972, *Pestic. Monit. J.*, 6: 103.
Kobayashi, N., 1971, *Publ. Seto Mar. Biol. Lab.*, 18: 379.
Koeman, J. H., W. H. M. Peeters, C. H. M. Koudstaal-Hol, P. S. Tijoe, and J. J. M. de Goeij, 1973, *Nature*, 245: 385.
——, W. S. M. van de Ven, J. J. M. de Goeij, P. S. Tijoe, and J. L. van Haaften, 1975, *Sci. Total Environ.*, 3: 279.
Koivusaari, J., I. Nuuja, R. Palokangas, and M. L. Hattula, 1976, *Bull. Environ. Contam. Toxicol.*, 15: 235.
Kopfler, F. C., 1974, *Bull. Environ. Contam. Toxicol.*, 11: 275.
Kramer, H. J., and B. Neidhart, 1975, *Bull. Environ. Contam. Toxicol.*, 14: 699.
Kurland, T., S. N. Faro, and H. Siedler, 1960, *World Neurol.*, 1: 370.
Langley, D. G., 1973, *J. Water Pollut. Control Fed.*, 45: 44.
Lask, D. J., C. E. McKone, R. G. Young, and C. A. Bache, 1971, *Environ. Sci. Technol.*, 5: 1138.
Laumond, F., M. Neuburger, B. Donnier, A. Fourcy, R. Bittel, and M. Aubert, 1973, *Rev. Int. Oceanogr. Med.*, 31-32: 47.
Leatherland, T. M., and J. D. Burton, 1974, *J. Mar. Biol. Assoc. U.K.*, 54: 457.
——, J. D. Burton, F. Culkin, M. J. McCartney, and R. J. Morris, 1973, *Deep-Sea Res.*, 20: 679.
Lindberg, E., and C. Harriss, 1974, *Mar. Pollut. Bull.*, 5: 93.
Lockhart, W. L., J. F. Uthe, A. R. Kenney, and P. M. Mehrle, 1972, *J. Fish. Res. Board Can.*, 29: 1519.
Lockwood, A. P. M., and C. B. E. Inman, 1975, *Comp. Biochem. Physiol.*, 52(2C): 75.
Lofroth, G., 1970, Bulletin 4, Natural Science Research Council, Ecological Research Committee, Stockholm, Sweden.
Lulic, S., and P. Strohal, 1974, *Rev. Int. Oceanogr. Med.*, 33: 119.
Lunde, G., 1973, *J. Sci. Food Agric.*, 24: 1029.
McKim, J., 1977, Environmental Protection Agency, Duluth, Minn., personal communication.
McKone, C. E., R. G. Young, C. A. Bache, and D. J. Lisk, 1971, *Environ. Sci. Technol.*, 5: 1138.
MacLeod, J. C., and E. Pessah, 1973, *J. Fish. Res. Board Can.*, 30: 485.
Martin, J. H., et al., 1976, *Mar. Biol.*, 35: 91.
Matida, Y., and H. Kumada, 1969, *Tansuiku Suisan Kenkyusho Kenkyu Hokoku*, 19: 73.
——, H. Kumada, S. Kimura, Y. Saiga, T. Nose, M. Yokote, and H. Kawatsu, 1972, *Tansuiku Suisan Kenkyusho Kenkyu Hokoku*, 21: 197.
May, D. R., and G. W. Brown, 1973, *Research in Fisheries*, Annual Report, Fisheries, p. 66, University of Washington, Seattle, Wash.
Middaugh, D. P., and C. L. Rose, 1974, *Water Res.*, 8: 173.
Miettinen, J. K., M. Heyraud, and S. Keckes, 1972, in *Marine Pollution and Sea Life*, M. Ruivo (Ed.), p. 295, Fishing Trading News (Books), Ltd., London.
——, M. Tillander, K. Rissanen, V. Miettinen, and Y. Ohmomo, 1969, *Proceedings of the 9th Japanese Conference on Radioisotopes*, Tokyo, p. 474.

Miettinen, V., E. Blankenstein, K. Rissanen, M. Tillander, J. K. Miettinen, and M. Valtonen, 1972, in *Marine Pollution and Sea Life,* M. Ruivo (Ed.), p. 298, Fishing Trading News (Books), Ltd., London.
Miller, G. E., P. M. Grant, R. Kishore, F. J. Steinkruger, F. S. Rowland, and V. P. Guinn, 1972, *Science,* 175: 1121.
Moore, M. N., and A. R. D. Stebbing, 1976, *J. Mar. Biol. Assoc. U.K.,* 56: 995.
Morisawa, M., and H. Mohri, 1974, *Exp. Cell Res.,* 83: 87.
National Academy of Sciences, 1973, in *Water Quality Criteria 1972,* Report R3-73-033, p. 252, Environmental Protection Agency, Washington, D.C.
Nelson, D. A., A. Calabrese, B. A. Nelson, J. R. MacInnes, and D. R. Wenzloff, 1976, *Bull. Environ. Contam. Toxicol.,* 16: 275.
Nelson, J. D., W. Blair, F. E. Brinckman, R. R. Colwell, and W. P. Iverson, 1973, *Appl. Microbiol.,* 26: 321.
Newberne, P. M., 1974, *Crit. Rev. Food Technol.,* 4: 311.
Nishikawa, K., and K. Tabata, 1969, *Tokaiku Suisan Kenkyusho Kenkyu Hokoku,* 58: 233.
Nuorteva, P., and E. Hasanen, 1971, *Ann. Zool. Fenn.,* 8: 331.
——, and E. Hasanen, 1972, *Ann. Zool. Fenn.,* 9: 23.
——, and E. Hasanen, 1975, *Ann. Zool. Fenn.,* 12: 247.
Nuzzi, R., 1972, *Nature,* 237(5349): 38.
Okubo, K., and T. Okubo, 1962, *Tokaiku Suisan Kenkyusho Kenkyu Hokoku,* 32: 131.
Olson, K. R., H. L. Bergman, and P. O. Fromm, 1973, *J. Fish. Res. Board Can.,* 30: 1293.
——, and P. O. Fromm, 1973, *J. Fish. Res. Board Can.,* 30: 1575.
——, and R. C. Harrel, 1973, *Contrib. Mar. Sci.,* 17: 9.
Parrish, K. M., and R. A. Carr, 1976, *Mar. Pollut. Bull.,* 7: 90.
Parslow, J. L. F., 1973, *Environ. Pollut. (London),* 5: 295.
Parsons, T. R., C. A. Bawden, and W. A. Heath, 1973, *J. Fish. Res. Board Can.,* 30: 1014.
Peakall, D. B., and R. J. Lovett, 1972, *Bioscience,* 22: 20.
Pentreath, R. J., 1976a, *J. Exp. Mar. Biol. Ecol.,* 25: 51.
——, 1976b, *J. Exp. Mar. Biol. Ecol.,* 24: 121.
Persoone, G., and G. Uyttersprot, 1975, *Rev. Int. Oceanogr. Med.,* 37-38: 125.
Peterson, C. L., W. L. Kalwe, and G. D. Sharp, 1973, *Fish. Bull.,* 71: 603.
Portmann, J. E., 1968, *Meeresunters.,* 17: 247.
——, 1972, *Aquaculture,* 1: 91.
Preston, A., 1973, *Nature,* 242(5393): 95.
Pyefinch, K. A., and J. C. Mott, 1948, *J. Exp. Biol.,* 25: 296.
Raeder, M. G., and E. Snekvik, 1949a, *K. Nor. Vidensk. Selsk.,* 21(25): 102.
——, and E. Snekvik, 1949b, *K. Nor. Vidensk. Selsk.,* 21(25): 105.
Ratkowsky, D. A., T. G. Dix, and K. C. Wilson, 1975, *Aust. J. Mar. Freshwater Res.,* 26: 223.
Raymont, J. E. G., 1972, *Proc. R. Soc. (London), Ser. B,* 180(1061): 451.
Rehwoldt, R., L. W. Menapace, B. Nerrie, and D. Alessandrello, 1972, *Bull. Environ. Contam. Toxicol.,* 8: 91.
Reinhart, K., and T. D. Myers, 1975, *Chesapeake Sci.,* 16: 286.
Reish, D. J., and J. M. Martin, F. M. Piltz, and J. Q. Word, 1976, *Water Res.,* 10: 299.
Renfro, J. L., B. Schmidt-Neilson, D. Miller, D. Benos, and J. Allen, 1974, in *Pollution and Physiology of Marine Organisms,* F. J. Vernberg and W. B. Vernberg (Eds.), p. 101, Academic Press, Inc., New York.

Renzoni, A., E. Bacci, and L. Falciai, 1974, *Rev. Int. Oceanogr. Med.*, 35-36: 159.
Rissanen, K., J. Erkama, and J. K. Miettinen, 1972, in *Marine Pollution and Sea Life*, M. Ruivo (Ed.), p. 289, Fishing Trading News (Books), Ltd., London.
Roberts, T. M., P. B. Heppleston, and R. D. Roberts, 1976, *Mar. Pollut. Bull.*, 7: 194.
Robertson, D. E., L. A. Rancitelli, J. C. Langford, and R. W. Perkins, 1972, USAEC Report BNWL-SA-1791, Battelle Pacific Northwest Laboratories, Richland, Wash.
Rodgers, E. O., B. H. Hazen, S. B. Friddle, and S. F. Sneiszko, 1951, *Prog. Fish-Cult.*, 13: 71.
Roesijadi, G., S. R. Petrocelli, J. W. Anderson, B. J. Presley, and R. Sims, 1974, *Mar. Biol.*, 27: 213.
Rucker, R. R., and D. F. Amend, 1969, *Prog. Fish-Cult.*, 31: 197.
Saha, J. G., 1972, in *Radiotracer Studies of Chemical Residues in Food and Agriculture,* Symposium Proceedings, Vienna, 1971, p. 81, STI/PUB/332, International Atomic Energy Agency, Vienna.
Saila, S. B., 1971, *Community Health,* 3: 129.
Salzinger, K., S. P. Fairhurst, S. J. Freimark, and F. D. Wolkoff, 1973, *J. Environ. Sys.*, 3: 27.
Sayler, G. S., J. D. Nelson, Jr., and R. R. Colwell, 1975, *Appl. Microbiol.*, 30: 91.
Schneider, J., 1972, *Mar. Biol.*, 16: 214.
Schweiger, G., 1957, *Arch. Fischereiwiss.*, 8: 54.
Scott, D. P., and F. A. J. Armstrong, 1972, *J. Fish. Res. Board Can.*, 29: 1685.
Sergeant, D. E., and F. A. J. Armstrong, 1973, *J. Fish. Res. Board Can.*, 30: 843.
Shealy, M. H., and P. A. Sandifer, 1975, *Mar. Biol.*, 33: 7.
Schultz, C. D., D. Crear, J. E. Pearson, J. B. Rivers, and J. W. Hylin, 1976, *Bull. Environ. Contam. Toxicol.*, 15: 230.
Sick, L. V., and H. L. Windom, 1975, in *Mineral Cycling in Southeastern Ecosystems,* ERDA, Symposium Series, Augusta, Ga., May 1—3, 1974, F. G. Howell, J. B. Gentry, and M. H. Smith (Eds.), p. 239, CONF-740513, NTIS.
Skei, J. M., M. Saunders, and N. B. Price, 1976, *Mar. Pollut. Bull.*, 7: 34.
Skerfving, S., K. Hansson, and J. Lindstem, 1970, *Arch. Environ. Health,* 21: 133.
Somayajulu, B. L. K., and Rama, 1972, *Curr. Sci. (India),* 41: 207.
Spinelli, J., and C. Mahnken, 1976, *J. Fish. Res. Board Can.*, 33: 1771.
Stebbing, A. R. D., 1976, *J. Mar. Biol. Assoc. U.K.*, 56: 977.
Stenner, R. D., and G. Nickless, 1975, *Mar. Pollut. Bull.*, 6: 89.
Stickney, R. R., H. L. Windom, D. B. White, and F. E. Taylor, 1975, in *Mineral Cycling in Southeastern Ecosystems,* ERDA Symposium Series, Augusta, Ga., May 1—3, 1974, F. G. Howell, J. B. Gentry, and M. H. Smith (Eds.), p. 257, CONF-740513, NTIS.
Suzuki, T., T. Miyama, and C. Toyama, 1973, *Bull. Environ. Contam. Toxicol.*, 10: 347.
Svansson, A., 1975, Report 1, Institute of Marine Research, Fishery Board, Goteborg, Sweden.
Tabata, K., and K. Nishikawa, 1969, *Tokaiku Suisan Kenkyusho Kenkyu Hokoku*, 58: 255.
Takevchi, T., 1972, in *Environmental Mercury Contamination*, R. Hartung and B. D. Dinman (Eds.), p. 79, Ann Arbor Science Publishers, Ann Arbor, Mich.

Taylor, D. D., and T. J. Bright, 1973, Report TAMU-SG-73-208, Department of Oceanography, Texas A&M University, College Station, Tex.
Thibaud, Y., 1971, *Sci. Peche,* 209: 1.
——, 1973, *Sci. Peche,* 221: 1.
Tillander, M., J. K. Miettinen, and I. Koivisto, 1972, in *Marine Pollution and Sea Life,* M. Ruivo (Ed.), p. 303, Fishing Trading News (Books), Ltd., London.
Tripp, M., and R. C. Harris, 1976, in *Environmental Biogeochemistry.* Vol. 2. *Metals Transfer and Ecological Mass Balances,* J. D. Nriagu (Ed.), p. 489, Ann Arbor Science Publishers, Ann Arbor, Mich.
Tsubaki, T., et al., 1967, *Proc. Jpn. J. Med.,* 6: 132.
Tsuruga, H., 1963, *Bull. Jpn. Soc. Sci. Fish.,* 29: 403.
Uchida, M., K. Hirakawa, and T. Inoue, 1961, *Kumamoto Igakkai Zasshi,* 14: 171.
Ui, J., 1972, *The Changing Chemistry of the Oceans,* Proceedings of the 20th Nobel Symposium, Lerum, Sweden, Aug. 16—20, 1971, D. Dyrssen and D. Jagner (Eds.), p. 171, John Wiley & Sons, Inc., New York.
——, and S. Kitamura, 1971, *Mar. Pollut. Bull.,* 2: 56.
Ukeles, R., 1962, *Appl. Microbiol.,* 10: 532.
Unlu, M. Y., M. Heyraud, and S. Keckes, 1972, in *Marine Pollution and Sea Life,* M. Ruivo (Ed.), p. 292, Fishing Trading News (Books), Ltd., London.
Vermeer, K., 1972, *Can. Field-Nat.,* 86: 123.
Vernberg, W. B., P. J. DeCoursey, and J. O'Hara, 1974, in *Pollution and Physiology of Marine Organisms,* F. J. Vernberg and W. B. Vernberg (Eds.), p. 381, Academic Press, Inc., New York.
——, and J. O'Hara, 1972, *J. Fish. Res. Board Can.,* 29: 1491.
——, and J. Vernberg, 1972, *Fish. Bull.,* 70: 415.
Vosjan, J. H., and G. J. Van der Hoek, 1972, *Neth. J. Sea Res.,* 5: 440.
Vucetic, T., W. B. Vernberg, and G. Anderson, 1974, *Rev. Int. Oceanogr. Med.,* 33: 75.
Walter, C. M., F. C. June, and H. G. Brown, 1973, *J. Water Pollut. Control Fed.,* 45: 2203.
Waterman, A. J., 1937, *Biol. Bull.,* 73: 401.
Weir, P. A., and C. H. Hine, 1970, *Arch. Environ. Health,* 20: 45.
Weis, J. S., 1976, *Fish. Bull.,* 74: 464.
Weis, P., and J. S. Weis, 1976, *Bull. Environ. Contam. Toxicol.,* 16: 197.
Weisbart, M., 1973, *Can. J. Zool.,* 51: 143.
Westoo, G., 1966, *Acta Chem. Scand.,* 20: 2131.
——, 1969, in *Chemical Fallout, Current Research on Persistent Pesticides,* M. W. Miller and G. G. Berg (Eds.), p. 75, Charles C Thomas, Publisher, Springfield, Ill.
——, 1973, *Science,* 181: 567.
Williams, P. M., and H. V. Weiss, 1973, *J. Fish. Res. Board Can.,* 30: 293.
Wilson, K. W., and P. M. Connor, 1971, *Coop. Res. Rep., Int. Counc. Explor. Sea*, 1971/E8: 343.
Windom, H., W. Gardner, J. Stephens, and F. Taylor, 1976, *Estuarine Coastal Mar. Sci.,* 4: 579.
——, R. Stickney, R. Smith, D. White, and F. Taylor, 1973, *J. Fish. Res. Board Can.,* 30: 275.
——, F. Taylor, and R. Stickney, 1973, *J. Cons. Int. Explor. Mer*, 35: 18.
Windom, H. L., 1973, American Society of Civil Engineers, Waterway Harbors and Coastal Engineering Division, 99(WW2): 257.

Wisely, B., and R. A. P. Blick, 1967, *Aust. J. Mar. Freshwater Res.*, 18: 63.
Wobeser, G., N. O. Nielsen, R. H. Dunlop, and F. M. Atton, 1970, *J. Fish. Res. Board Can.*, 27: 830.
Wood, J. M., 1973, *Rev. Int. Oceanogr. Med.*, 31-32: 7.
Young, D. R., 1974, Southern California Coastal Water Research Project, El Segundo, Calif.
Young, L. G., and L. Nelson, 1974, *Biol. Bull.*, 147: 236.
Zitko, V., B. J. Finlayson, D. J. Wildish, J. M. Anderson, and A. C. Kohler, 1971, *J. Fish. Res. Board Can.*, 28: 1285.

A RADIOISOTOPIC STUDY OF MERCURY UPTAKE BY HUDSON RIVER BIOTA

LOIS S. ZUBARIK and JOSEPH M. O'CONNOR
New York University Medical Center, Institute of Environmental Medicine, A. J. Lanza Research Laboratories, Tuxedo, New York

ABSTRACT

Planktonic organisms from the Hudson River were exposed to various forms of mercury (^{203}Hg) to evaluate the role of estuarine forage organisms in the kinetics of toxic metal transport in aquatic systems. Mercury accumulation was expressed as concentration per unit time. Concentration factors ranged from 10^2 to 10^6 times that in the filtered river water. Mercury-203 uptake was greater in microzooplankton and algae than in macrozooplankton and fish larvae. The amphipod *Gammarus* sp. was tested to determine if differences in mercury uptake could be related to the form of mercury or to changes in environmental conditions. Exposure to two inorganic forms of mercury (mercuric nitrate and mercuric chloride) and two organic forms of mercury (methyl mercury chloride and phenyl mercuric acetate) showed no differences in concentrations of the four forms of mercury after a 1-day exposure. Concentration of the organic forms by *Gammarus* sp. was three times greater than that of inorganic mercury compounds after exposure for 1 week. Studies of mercury uptake throughout the year showed that uptake of all the mercury compounds tested increased during the summer months, but uptake of organic compounds increased to a greater extent than that of inorganic compounds. Temperature change was shown to be an important variable for determining the degree of uptake; however, no single environmental parameter adequately explained the seasonal fluctuations in mercury uptake.

Levels of mercury in excess of 0.5 mg/liter in freshwater fish have been reported in Japan (Fujiki and Fujima, 1973), Sweden (Johnels and Westermark, 1969; Berglund et al., 1971), Norway (Underdal and Hastein, 1971), Finland (Miettinen, 1969), the United States (Wallace et al., 1971), and Canada (Bligh, 1970). Marine plankton may show concentrations of mercury from 0.2 to 25 ppm dry weight

(Skei, Saunders, and Price, 1976; Williams and Weiss, 1973), and estuarine organisms in the southeastern United States, including blue crabs and many species of fish, have average mercury levels above 0.5 ppm. Larger predators, such as lemon and bull sharks, exhibited mercury levels of 4 to 10 ppm (Gardner et al., 1975).

Despite abundant data on mercury levels in aquatic systems, it is the consensus that insufficient data exist on the mechanisms of mercury accumulation and cycling (Gavis and Ferguson, 1972; D'Itri, 1973; Gardner et al., 1975; Lindberg, Andrew, and Harriss, 1975; Matsunaga, 1975; Tsai, 1975; Stickney et al., 1975; Wallace et al., 1971; Friberg and Vostal, 1972). Measuring mercury levels in organisms is important in assessing the degree of an environmental problem, but survey and monitoring data frequently cannot be used for making predictions about mercury cycling because the conditions under which uptake and accumulation occurred are not known.

Most uptake and accumulation studies performed to date have been on fish, marine mammals, and larger invertebrates (D'Itri, 1973; Miettinen, 1969; Olson, Bergman, and Fromm, 1973; MacLeod and Pessah, 1972; Guarino and Anderson, 1976; Hannerz, 1968). Often there seems to be no clear reason for the high mercury levels frequently found in fish. Confusion still exists over the relative importance of mercury availability for direct uptake from water vs. biomagnification through food chains. To understand aquatic cycling of mercury, we must determine uptake and accumulation of mercury by forage organisms as well as commercially important fish and shellfish.

There are few investigations of bioaccumulation by smaller invertebrates, few studies on mercury uptake by any estuarine biota, and no studies on mercury uptake by Hudson River biota. Our study was carried out to evaluate the role of estuarine organisms in mercury dynamics and to determine the accumulation of mercury compounds by Hudson River biota. A broad spectrum of planktonic organisms, from algae to fish larvae, were selected for study of bioconcentration under natural conditions. The organisms were chosen on the basis of their abundance in the river and their importance as forage. The amphipod *Gammarus* sp. was studied in detail, to acertain if there was a difference in uptake of four mercury compounds (mercury chloride, mercury nitrate, methyl mercury chloride, and phenyl mercuric acetate) and to determine if fluctuations in selected environmental parameters might relate to changes in mercury uptake.

MATERIALS AND METHODS

Collection and Maintenance of Organisms

Zooplankton and algae were collected from midchannel and shore locations in the Hudson River 66 km north of New York City. Zooplankton (see Table 2) were collected on the day preceding or on the morning of an experiment with a 0.5-m-diameter nylon net having a mesh size of either 76 or 500 μ. River water and algae were collected by filling 4-liter plastic buckets with surface water. Algae were subsequently concentrated with a nannoplankton net (18-μ mesh). *Gammarus* sp. collected from *Myriophyllum spicatum* gathered from the Hudson River were placed in aquariums containing Hudson River water together with *M. spicatum* and maintained in culture at ambient river temperature.

Before use in an experiment, the larger zooplankton (*Crangon, Gammarus, Amnicola, Neomysis,* and *Chiridotea*) were maintained in 380-liter aquariums filled with Hudson River water, and small zooplankton (*Eurytemora, Acartia, Daphnia, Leptodora, Cypris, Mesocyclops, Monoculoides, Chaoborus* nauplii and cyclopoid copepodids) were kept in 800-ml battery jars filled with Hudson River water. All holding vessels were placed in water troughs filled with flowing river water and were illuminated on a 12 : 12 light—dark cycle. Organisms were fed ad libitum before the experiment.

Striped bass (*Morone saxatilis*) larvae from Hudson River stock were obtained from the Consolidated Edison Hudson River fish hatchery at Verplanck, N. Y. Larvae were maintained in holding facilities at the hatchery before the experiment and were brought to the New York University laboratory at Indian Point for acclimation over a 2-day period to Hudson River water. This was accomplished by serial dilution of hatchery water with Hudson River water. Striped bass larvae were maintained at ambient Hudson River temperatures and fed brine shrimp nauplii.

Accumulation Studies

Organisms were taken from holding facilities and placed in individual culture bowls. The numbers of organisms added to each bowl varied according to species. Generally 150 to 300 microzooplankton, 30 to 50 macrozooplankton, or 6 to 10 fish larvae were placed in the experimental containers. Each bowl contained 100 ml, or 300 ml in the case of fish larvae, of Hudson River water. One milliliter of radioisotope carrier (~0.1 μCi) was injected with an

Eppendorf pipettor directly into Hudson River water containing the test organisms. The four mercury compounds used [mercuric nitrate, $^{203}Hg(NO_3)_2$; mercuric chloride, $^{203}HgCl_2$; methyl mercury chloride, $^{203}HgCH_3Cl$; and phenyl mercuric acetate, $\phi\,^{203}HgCOOCH_3$] were purchased from New England Nuclear. Dilutions to appropriate radioactive concentrations (0.1 to 1 μCi) were made in deionized water. The final concentrations of stable mercury added with the radioisotope was always less than mercury concentrations reported for Hudson River water (0.1 ppb).

At predetermined intervals organisms were removed from the water by collection in netted cages, killed in hot water, and rinsed with clean Hudson River water. There was no evidence that the technique caused breakage of cells of the organisms used. They were then air dried overnight and placed in scintillation vials filled with 15 ml of phase combining system solubilizer (PCS) (Amersham/Searle Corp.) and 5 ml of water.

The test water was filtered immediately through 0.45-μ Millipore filters when an experiment was terminated. Five milliliters of the filtered river water and the filter paper containing the suspended particulate matter from the test water were placed in separate liquid scintillation vials containing 15 ml of PCS. The filter paper was air dried overnight and then weighed. To avoid yellowing of the sample, we added the cocktail to the filter paper just before counting. The samples were counted on Nuclear Chicago Mark I or Isocap/300 liquid scintillation counters. Both instruments were calibrated with a primary standard from Amersham/Searle Corp. (126 mCi/g ± 2.9%). Quenching was corrected for by the channels ratio technique on the Mark I and by an external standard on the Isocap (Wang and Willis, 1965; Chase and Rabinowitz, 1967). Samples were counted for 10 min or longer (if necessary) to give a counting error of less than 3%. Experiments generally consisted of three replicates.

All counts were corrected for decay and background, and the counts were converted to disintegrations per minute. Concentration factors were obtained by dividing the disintegrations per minute per gram of organism by the disintegrations per minute per gram of filtered test water.

Food organisms were labeled as if they were being tested for uptake from water except that, instead of being killed at the end of the experiment, they were washed in filtered Hudson River water, recollected, and added to the test culture. Food organisms were always supplied to test cultures in excess amounts.

During each experiment, salinity, dissolved oxygen, conductivity, alkalinity, and temperature were measured according to procedures in *Standard Methods* (American Public Health Association, 1971). Algal cell counts and carbon values were obtained from ongoing ecological studies of the Hudson River at Indian Point (New York University Medical Center, 1976; P. Storm, New York University, unpublished data). Turbidity, phosphate, and nitrogen levels were obtained from Lawler, Matusky, and Skelly Engineers (1976).

Statistical Analysis

All experimental data were converted to natural logarithms before analysis. Bartlett's test for homogeneity of variance and the Kolmogorov—Smirnov test for normality were run to ensure that assumptions of the analysis of variance (ANOVA) were met. Once differences were shown, the Scheffe test compared each experiment for significant differences. Multiple regression analysis (Nie et al., 1975) was used as an inferential tool for evaluating the relationship between mercury uptake and fluctuations in the chemical and physical parameters of Hudson River water.

RESULTS

Both fish larvae and zooplankton took up mercury directly from water and from food. In most cases mercury uptake from filtered river water was similar in degree to uptake from whole river water and from labeled algae, detritus, and bacteria. Presence of sediment reduced the level of mercury in organisms (Table 1).

Comparing the concentration factors for several Hudson River species shows that each species is able to concentrate all four forms of mercury to a considerable extent over the water phase. After a 1-day exposure, mercury concentration in organisms was 10^2 to 10^6 times that in the filtered river water (Tables 2 and 3). In general, Hudson River algae and microzooplankton showed much higher concentration factors (50,000 to 1,000,000) than macrozooplankton (1,000 to 30,000) and fish (1,000 to 10,000). Of the crustacea, the filter feeders (e.g., *Eurytemora affinis*, *Acartia tonsa*, and *Daphnia pulex*) show greater concentrations of mercury taken up from river water than carnivorous copepods (e.g., cyclopoid copepods) or detrital feeders (e.g., *Gammarus* sp.).

Gammarus sp. was extensively studied to determine if there was greater uptake of organic mercury than of inorganic mercury. One-way ANOVA comparing uptake of four mercury compounds [$Hg(NO_3)_2$, $HgCl_2$, $HgCH_3Cl$, $\phi HgCOOCH_3$] showed no differences after 1 day (68 df, $P > 0.05$), but there were differences at 1 week

TABLE 1

INORGANIC MERCURY UPTAKE IN HUDSON RIVER WATER FOR DIFFERENT LABELING CONDITIONS*

Experiment	Organism	Mean dis min^{-1} g^{-1} X 10^3 ± SE
1. Labeled river water	*Gammarus* sp.	1,130 ± 560 (3)
Labeled filtered river water	(*G. daiberi* and *G. tigrinus*)	880 ± 110 (3)
2. Labeled river water	*Neomysis*	510 ± 220 (4)
Labeled filtered river water	*americana*	680 ± 130 (2)
3. Labeled river water	*Morone saxatilis*	8,040 ± 230 (4)
Labeled filtered river water		8,200 ± 4,400 (2)
4. Labeled river water	*Gammarus* sp.	4,500 ± 2,000 (3)
Labeled leaves		1,090 ± 240 (3)
5. Labeled filtered river water	*Gammarus* sp.	620 ± 160 (3)
Labeled algae, detritus, and bacteria		710 ± 110 (3)
Labeled algae		270 ± 47 (3)
6. Labeled river water	*Gammarus* sp.	2,190 ± 390 (3)
Labeled algae, detritus, and bacteria		2,200 ± 990 (3)
7. Labeled river water	*Gammarus* sp.	510 ± 270 (3)
Labeled algae, detritus, and bacteria		700 ± 600 (3)
8. Labeled river water	*Gammarus* sp.	150,000 ± 30,000 (10)
Labeled *Gammarus*		12,000 ± 2,100 (10)
9. Labeled river water	*Gammarus* sp.	13,220 ± 670 (3)
1-in. mud layer		99 ± 14 (3)
10. Labeled river water	*Gammarus* sp.	87,000 ± 13,000 (3)
Labeled river water (1 day), then nonlabeled river water (1 day)		126,000 ± 27,000 (3)
11. Labeled river water	*Morone saxatilis*	8,040 ± 200 (4)
Labeled copepods		10,300 ± 5,000 (2)
Labeled copepods, then 1 day in nonlabeled river water		4,400 ± 2,000 (2)

*Numbers in parentheses are numbers of replicates.

TABLE 2

MERCURY CONCENTRATION FACTORS FOR SELECTED AQUATIC BIOTA AFTER 1-DAY EXPOSURE TO INORGANIC MERCURY COMPOUNDS*

Biota	$Hg(NO_3)_2$	$HgCl_2$
Phytoplankton		
Bacillariophyta		
Melosira sp.		1,100,000 ± 170,000 (2)
Synedra sp.		220,000 ± 33,000 (2)
Chrysophyta		
Scenedesmus sp.		93,000 ± 36,000 (2)
Hudson River algae, detritus, and bacteria	49,000 (1)	92,000 (1)
Crustacea		
Cladocera		
Daphnia pulex	62,000 ± 11,000 (3)	200,000 ± 130,000 (5)
Leptodora kindti		200,000 (1)
Ostracoda		
Cypris sp.†		58,000 (1)
Copepoda		
Nauplii	9,800 ± 1,200 (2)	
Copepods (cyclopoid copepodids)	3,700 ± 920 (2)	2,830 ± 310 (2)
Mesocyclops edax	194,000 ± 14,000 (2)	
Eurytemora affinis	104,000 ± 32,000 (7)	230,000 ± 160,000 (7)
Acartia tonsa	61,000 ± 23,000 (2)	14,000 ± 13,000 (2)
Isopoda		
Chiridotea almyra†		2,336 ± 30 (2)
Amphipoda		
Gammarus sp. (*G. daiberi* and *G. tigrinus*)	1,440 ± 400 (14)	1,490 ± 520 (23)
Monoculodes edwardsi		1,330 ± 780 (2)
Mysidacea		
Neomysis americana	133 (1)	95 (1)
Decapoda		
Crangon septemspinosa†	18,000 (1)	
Mollusca		
Gastropoda		
Amnicola limosa		412 ± 40 (3)
Ichthyoplankton		
Percichthyidae		
Morone saxatilis		
Eggs		58 ± 25 (2)
Larvae	1,300 ± 770 (4)	7,600 ± 5,900 (5)

*Numbers in parentheses are numbers of experiments.

†Concentration factors are based on uptake after exposure of 5 to 9 days.

TABLE 3

MERCURY CONCENTRATION FACTORS FOR SELECTED AQUATIC BIOTA AFTER 1-DAY EXPOSURE TO ORGANIC MERCURY COMPOUNDS*

Biota	$^{203}HgCH_3Cl$	$\phi^{203}HgCOOCH_3$
Phytoplankton		
Bacillariophyta		
Melosira sp.	201,600 ± 1,200 (2)	
Synedra sp.	390,000 ± 72,000 (2)	
Chrysophyta		
Scenedesmus sp.		
Hudson River algae, detritus, and bacteria	12,000 ± (1)	12,000 (1)
Crustacea		
Cladocera		
Daphnia pulex	240,000 ± 220,000 (3)	63,000 (1)
Leptodora kindti	140,000 (1)	
Ostracoda		
Cypris sp.†		
Copepoda		
Nauplii	15,000 ± 4,300 (2)	
Copepods (cyclopoid copepodids)	9,800 ± 7,000 (2)	15,000 ± 4,500 (2)
Mesocyclops edax	360,000 ± 200,000 (2)	
Eurytemora affinis	250,000 ± 190,000 (6)	105,000 ± 46,000 (6)
Acartia tonsa	540,000 ± 220,000 (2)	180,000 ± 170,000 (2)
Isopoda		
Chiridotea almyra†		
Amphipoda		
Gammarus sp. (*G. daiberi* and *G. tigrinus*)	4,900 ± 2,600 (22)	2,660 ± 950 (13)
Monoculodes edwardsi		
Mysidacea		
Neomysis americana	2,800 ± 700 (2)	1,700 ± (1)
Decapoda		
Crangon septemspinosa†	53,000 ± 15,000 (2)	4,039 (1)
Insecta		
Diptera		
Chaoborus sp.	14,200 ± 1,000 (2)	
Mollusca		
Gastropoda		
Amnicola limosa		
Ichthyoplankton		
Percichthyidae		
Morone saxatilis		
Eggs	620 ± 220 (10)	
Larvae	5,200 ± 410 (3)	10,000 ± 29,000 (2)

*Numbers in parentheses are numbers of experiments.
†Concentration factors are based on uptake after exposure of 5 to 9 days.

(28 df, $P < 0.05$). Uptake of methyl mercury was greater than uptake of mercuric nitrate. There were no uptake differences between mercuric nitrate and mercuric chloride or between methyl mercury and phenyl mercury. Combining uptake of the two inorganic forms and the two organic forms after a 1-day exposure showed that mean uptake of inorganic compounds was less than that of organic compounds, but this was not statistically significant. Organic mercury uptake was three times greater than inorganic uptake after 1 week ($P \leqslant 0.007$; Table 4).

TABLE 4

MEAN MERCURY CONCENTRATION FACTORS AND 95% CONFIDENCE LIMITS (L_1 L_2) FOR *Gammarus* sp.

Exposure time	Mercury type					
	Inorganic			Organic		
	Mean	L_1	L_2	Mean	L_1	L_2
1 day	688	475	996	1,153	736	1,798
1 week	2,562	1,540	4,260	8,196	4,065	16,526

Rates of uptake over time were rapid for both methyl mercury and mercuric chloride (Figs. 1 and 2). Of the total mercury taken up by *Gammarus* after 1 week, almost one-half the radioactivity of organic mercury and one-third the radioactivity of inorganic mercury were taken up during the first day. Studies of mercury uptake over periods longer than a week show a decline in mercury per gram of organism between days 10 and 19 (Fig. 3).

Further studies were undertaken with *Gammarus* sp. to determine what changing conditions might explain the variations in concentration factors shown by one species. Studies of mercury uptake in conjunction with seasonally changing estuarine variables show that the difference between inorganic and organic uptake is increased during the summer months (Fig. 4). There was a strong positive correlation between uptake and temperature for both types of mercury compounds and an inverse correlation of mercury uptake with total organic carbon. Positive correlations existed with phosphates and algal cell concentration, but there were peaks in these values when there were no peaks in mercury uptake. No single environmental parameter adequately explains the seasonal fluctuations in mercury uptake (Fig. 4).

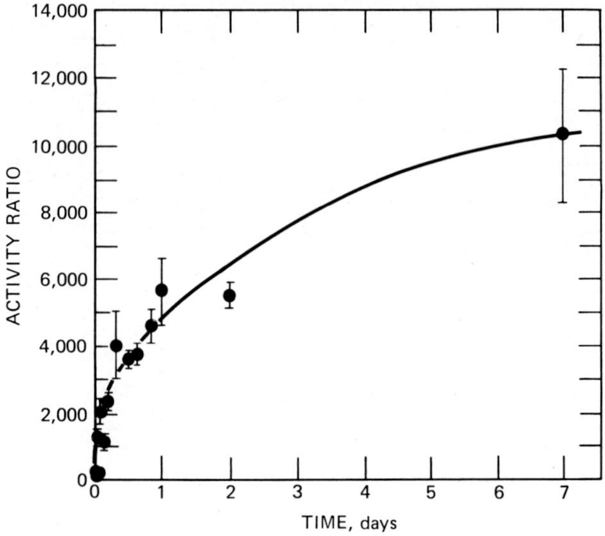

Fig. 1 Methyl mercury uptake by *Gammarus* sp. over a period of 1 week. Each point represents the mean of three replicates. Activity ratio = dis min^{-1} g^{-1} organism per dis min^{-1} g^{-1} in the water at the end of the experiment.

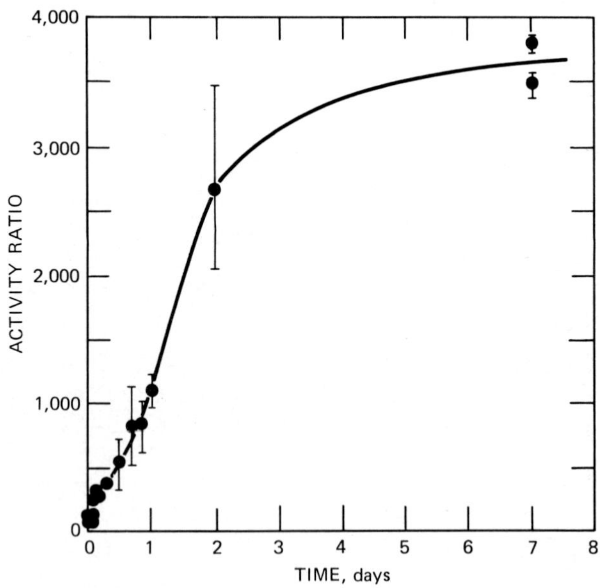

Fig. 2 Mercuric chloride uptake by *Gammarus* sp. over a period of 1 week. Each point represents the mean of three replicates.

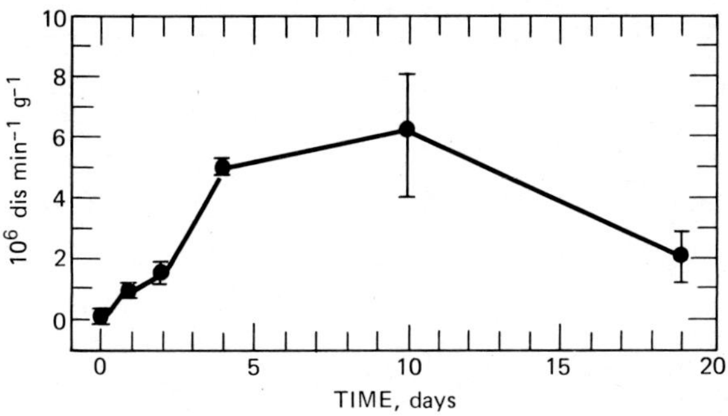

Fig. 3 Long-term accumulation and loss of mercuric chloride by *Gammarus* sp. Each point represents the mean uptake of 15 to 20 organisms.

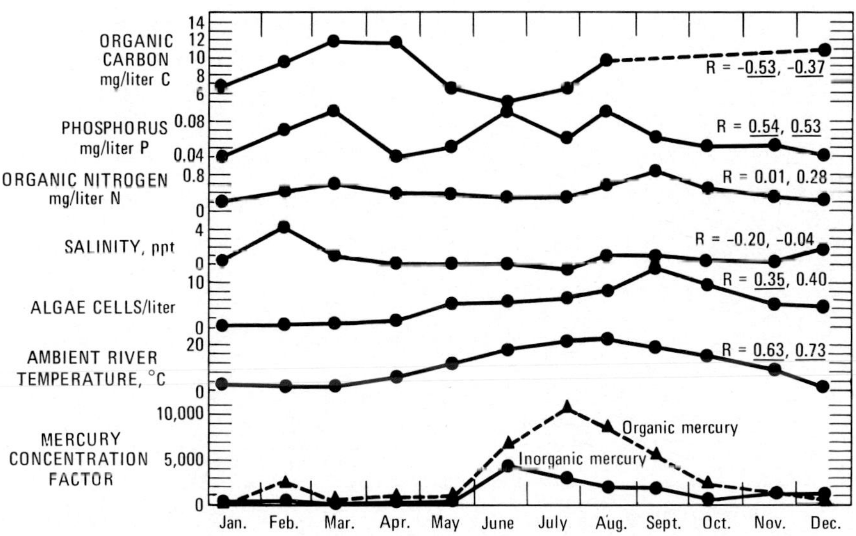

Fig. 4 Correlation between mercury concentration factors (inorganic and organic) for *Gammarus* sp. and seasonal fluctuations in physical and chemical parameters of Hudson River water during 1975. Values of R are correlation coefficients for inorganic (first value given) and organic (second value) mercury concentration factors with the variable graphed. Coefficients underlined are significant ($P < 0.005$).

A stepwise multiple regression was done to determine which parameters might best explain variability in mercury uptake. Ambient river temperature accounted for the largest percent of variability for both inorganic and organic mercury uptake. For organic uptake the variance contributed by changes in algal cell concentration and salinity was also significant. Inorganic uptake was apparently related to several factors other than temperature (Table 5). The significant concentration factors (CF) for inorganic and organic uptake produced the following predictive equations:

$$CF_{inorganic} = 2.3 + 0.07(T) - 2.2(ON) - 0.055(OC) + 48(TB) + 0.0002(COND) \qquad R^2 = 0.65 \tag{1}$$

$$CF_{organic} = 2.4 + 0.07(T) + 0.024(SAL) - 0.06(AC/liter) - 0.02(OC) \qquad R^2 = 0.75 \tag{2}$$

where T = temperature
ON = organic nitrogen
OC = organic carbon
TB = turbidity
COND = conductivity
SAL = salinity
AC/liter = algal cells per liter

TABLE 5

SIGNIFICANT VARIABLES USED IN PREDICTION EQUATIONS FOR INORGANIC AND ORGANIC MERCURY UPTAKE

Concentration factor	Variables	Probability F value
Inorganic 66 cases	Temperature (T)	<0.000
	Organic nitrogen (ON)	0.027
	Organic carbon (OC)	0.043
	Turbidity (TB)	0.001
	Conductivity (COND)	0.003
Organic 66 cases	Temperature (T)	<0.000
	Algal cells per liter (AC/liter)	<0.000
	Salinity (SAL)	<0.000
	Organic carbon (OC)	0.097

DISCUSSION

A variety of Hudson River forage organisms, ranging from algae to fish larvae, accumulated mercury compounds from the test environment. The concentrations in organisms ranged from 10^3 to 10^6 times that in the test medium. In all cases tested, uptake of mercury occurred via food, as well as directly from water. The greatest relative concentrations (10^6) over the external medium occurred among algae; the least (10^3) occurred among the larger planktonic crustacea (e.g., *Gammarus*) and striped bass larvae. Mercury was transferred to striped bass larvae from copepods labeled with mercury. Thus mercury concentrations in forage organisms and the conditions that affect their mercury uptake are relevant to predictions about mercury cycling. However, the concept of biomagnification along a food chain is misleading in aquatic systems where bioaccumulation occurs from water as well as food.

Mercury uptake in Hudson River organisms was found to vary on a seasonal basis within the same species or species group, showing a strong positive correlation with temperature. MacLeod and Pessah (1973), in their studies of *Salmo gairdneri* (rainbow trout), and Heit and Fingerman (1977), in their studies of *Procambarus clarki* and *Faxonella clypeata* (crayfishes), also found that mercury accumulation was strongly dependent on temperature. This suggests that several possible mechanisms are operating to cause a gradual increase in mercury content. First, since poikilothermic organisms generally show an increase in metabolic rate as temperature increases, the increased bioaccumulation during warm months may be due simply to a greater rate of uptake of mercury-contaminated food. Increased metabolism should also result in more rapid rates of mercury clearance, however, but this is not borne out by our results. Nonetheless, the potential for bioaccumulation of mercury compounds during warm months is greater than during cool months. Field studies should be carried out in mercury-contaminated environments to verify this hypothesis and its several ramifications: (1) that the concentration of mercury compounds per gram of organism increases during warm periods, (2) that uptake and retention of mercury is greater during the summer, and (3) that an increase in the absolute quantity of mercury compounds available for direct uptake and food-chain transfer occurs in warm environments.

The form of mercury available to organisms affects uptake rates and the environmental impact of mercury on aquatic systems. Experiments with *Gammarus* showed a greater uptake of organic over inorganic forms of mercury, but only after prolonged (1-week) exposures. Initial uptake of both forms was extremely rapid, however; this indicates that perhaps the hypothesized "necessary" conversion of inorganic mercury to a methylated form before uptake by organisms is not required. Mercury methylation by natural bacterial assemblages could not convert mercuric nitrate or mercuric chloride to an organic form rapidly enough to generate the rates of uptake observed in these experiments (Langley, 1973). These results are confirmed by those of Olson, Bergman, and Fromm (1973), who studied the rainbow trout; Prabhu and Hamdy (1977), who studied *Lebistes reticulatus* (guppy); and Mellinger (1973), who studied *Margaritifera margaretifera* (freshwater mussel). Kramer and Neidhart (1975) found that mercury uptake by fish *(Poecilia reticulata)* in a methyl mercury chloride solution was about four times as fast as in a mercuric nitrate solution.

Variations in concentration factors for the same species can be related to the conditions under which bioconcentration of mercury took place. The facts that the presence of sediments decreases uptake by *Gammarus* sp. and that there is an inverse correlation between mercury uptake and total organic carbon levels indicate that mercury concentration or type of mercury compound are not the only factors controlling mercury uptake by aquatic organisms. Environmental parameters may affect the availability of mercury for uptake and accumulation. Tsai, Boush, and Matsumura (1975) reported that the pH of the ambient water has a great effect on the translocation of inorganic mercury to fish. In the area of the Hudson River studied, pH varies annually by only 0.6 of a unit (7.2 to 7.8). Therefore, it was impossible to assess the effect of pH on mercury uptake in this study. Lindberg, Andrew, and Harriss (1975) reported that transport and deposition of mercury in the estuarine environment is largely controlled by interaction with natural organic matter. Cranston (1976) stated that the process controlling accumulation and distribution of mercury in estuarine sediments is related to the amount of fine-grain sediments.

The experiments reported here address only the effects of relationships among species, environmental parameters, and chemical form on the uptake of mercury from water and food. The impact related to the community and to seasonal differences in mercury uptake was not studied directly. The relationship between temperature and mercury uptake identified in our study, however, justifies some speculation on the stress mercury compounds may cause on aquatic systems. First, since the toxicity of organic mercury compounds is greater than that of inorganic and since organic compounds can be accumulated to greater levels during periods of high temperature, we might propose that mercury contaminated environments are most stressed during the summer months. Even at sublethal concentrations, stress responses to mercury intoxication, if at all related to total body burden, must be greater if accumulated quantities are greater during periods of high temperature.

The absence of intensive study on the physiological and biochemical effects of metals on planktonic organisms, which may act as the major vectors for mercury transport to organisms used as food by human populations, has led to the unfortunate circumstance wherein the environmental impact of mercury is based on surveys of concentrations in fish flesh and the results of relatively few laboratory bioassay experiments. Our experiments attest to the need for further study of the behavior of mercury compounds under conditions that simulate natural conditions to ascertain in a valid ecological context the true meaning of mercury-related stress on aquatic ecosystems.

ACKNOWLEDGMENTS

The research reported here was funded in part by grant No. ES-00260 from the National Institute of Environmental Health Sciences.

REFERENCES

American Public Health Association, 1971, *Standard Methods for the Examination of Water and Wastewater*, 13th ed., Washington, D. C.

Berglund, F., et al., 1971, Methyl Mercury in Fish, a Toxicologic-Epidemiologic Evaluation of Risks, Report from an Expert Group, *Nord. Hyg. Tidskr., Suppl.* 4.

Bligh, E. G., 1970, *Mercury and the Contamination of Freshwater Fish*, Freshwater Institute MS Report 1088, Fisheries Research Board of Canada, Ottawa.

Chase, G. D., and J. L. Rabinowitz, 1967, *Principles of Radioisotope Methodology*, 3rd ed., Burgess Publishing Company, Minneapolis, Minn.

Cranston, R. E., 1976, Accumulation and Distribution of Total Mercury in Estuarine Sediments, *Estuarine Coastal Mar. Sci.*, 4: 695-700.

D'Itri, F. M., 1973, Mercury in the Aquatic Ecosystem, in *Bioassay Techniques and Environmental Chemistry*, G. Glass (Ed.), pp. 3-70, Ann Arbor Science Publishers, Ann Arbor, Mich.

Friberg, L., and J. Vostal, 1972, *Mercury in the Environment*, 2nd ed., CRC Press, Cleveland, Ohio.

Fujiki, M., and S. Fujima, 1973, The Pollution of Minamata Bay and the Neighboring Sea by Factory Waste Water Containing Mercury, in *Research Conference on New Methodology on Ecological Chemistry*, M. Goto,

F. Coulston, and F. Korte (Eds.), pp. 217-229, International Academic Printing Co., Tokyo.

Gardner, W. S., H. I. Windom, J. A. Stephens, F. E. Taylor, and R. R. Stickney, 1975, Concentrations of Total Mercury and Methyl Mercury in Fish and Other Coastal Organisms: Implications to Mercury Cycling, in *Mineral Cycling in Southeastern Ecosystems,* ERDA Symposium Series, Augusta, Ga., May 1—3, 1974, F. G. Howell, J. B. Gentry, and M. H. Smith (Eds.), pp. 268-278, CONF-740513, NTIS.

Gavis, J., and J. F. Ferguson, 1972, The Cycling of Mercury Through the Environment, *Water Res.,* 6: 989-1008.

Guarino, A. M., and Anderson, J. B., 1976, Tissue Distribution of [^{14}C] Methyl Mercury in the Lobster, *Homarus Americanus, J. Toxicol. Environ. Health,* 2: 13-24.

Hannerz, Lennart, 1968, Experimental Investigations on the Accumulation of Mercury in Water Organisms, *Rep. Inst. Freshwat. Res., Drottningholm,* 48: 120-176.

Heit, M., and M. Fingerman, 1977, The Influences of Size, Sex and Temperature on the Toxicity of Mercury to Two Species of Crayfishes, *Bull. Environ. Contam. Toxicol.,* 18: 572-580.

Johnels, A. G., and T. Westermark, 1969, Mercury Contamination of the Environment in Sweden, in *Chemical Fallout,* M. W. Miller and G. Berg (Eds.), pp. 221-239, Charles C Thomas, Publisher, Springfield, Ill.

Kramer, H. J., and B. Neidhart, 1975, The Behavior of Mercury in the System Water—Fish, *Bull Environ. Contam. Toxicol.,* 14(6): 699-704.

Langley, D. G., 1973, Mercury Methylation in an Aquatic Environment, *J. Water Pollut. Control Fed.,* 45: 44-51.

Lawler, Matusky, and Skelly Engineers, 1976, *1975 Hudson River Aquatic Ecology Studies—Bowline Point and Lovett Generating Stations,* a report to Orange and Rockland Utilities, Pearl River, N. Y.

Lindberg, S. E., A. W. Andrew, and R. C. Harriss, 1975, Geochemistry of Mercury in the Estuarine Environment, in *Estuarine Research,* L. E. Cronin (Ed.), pp. 64-107, Academic Press, Inc., New York.

MacLeod, J. C., and E. Pessah, 1973, Temperature Effects on Mercury Accumulation, Toxicity, and Metabolic Rate in Rainbow Trout (*Salmo gairdneri*), *J. Fish. Res. Board Can.,* 30: 485-492.

Matsumura, F., Y. G. Doherty, F. Furukawa, and G. M. Boush, 1975, Incorporation of Hg into Methylmercury in Fish Liver: Studies on Biochemical Mechanisms in Vitro, *Environ. Res.,* 10: 224-235.

Matsunaga, K., 1975, Concentration of Mercury by Three Species of Fish from Japanese Rivers, *Nature,* 257: 49-50.

Mellinger, Peter J., 1973, The Comparative Metabolism of Two Mercury Compounds as Environmental Contaminants in the Freshwater Mussel, *Margaritifera margaritifera,* in *Proceedings of University of Missouri's 6th Annual Conference on Trace Substances in Environmental Health,* June 13—15, 1972, D. D. Hemphill (Ed.), pp. 173-180, University of Missouri Press, Columbia, Mo.

Miettinen, K., 1969, Methylmercury as a Food Chain Problem in Scandinavia, in *Chemical and Toxicological Aspects of Environmental Quality,* International Symposium, Munich, pp. 155-156, G. Thieme, Stuttgart.

New York University Medical Center, 1976, *Effects of Entrainment by the Indian Point Power Plant on Biota in the Hudson River Estuary: A Progress Report for 1975,* to the Consolidated Edison Company of New York, Inc.

Nie, N. H., C. H. Hull, J. G. Jenkins, K. Steinbrenner, and D. H. Bent, 1975, *Statistical Package for the Social Sciences*, 2nd ed., McGraw-Hill Book Company, New York.

Olson, K. R., H. L. Bergman, and P. O. Fromm, 1973, Uptake of Methyl Mercuric Chloride and Mercuric Chloride by Trout: A Study of Uptake Pathways into the Whole Animal and Uptake by Erythrocytes in Vitro, *J. Fish. Res. Board Can.*, 30: 1293-1299.

Prabhu, N. V., and M. K. Hamdy, 1977, Behavior of Mercury in Biosystems. 1. Uptake and Concentration in Food-Chain, *Bull. Environ. Contam. Toxicol.*, 18: 409-417.

Skei, J. M., M. Saunders, and N. B. Price, 1976, Mercury in Plankton from a Polluted Norwegian Fjord, *Mar. Pollut. Bull.*, 7: 34-35.

Stickney, R. R., H. L. Windom, D. B. White, and F. E. Taylor, 1975, Heavy-Metal Concentrations in Selected Georgia Estuarine Organisms with Comparative Food-Habit Data, in *Mineral Cycling in Southeastern Ecosystems*, ERDA Symposium Series, Augusta, Ga., May 1—3, 1974, F. G. Howell, J. B. Gentry, and M. H. Smith (Eds.), pp. 257-267, CONF-740513, NTIS.

Tsai, Shan-Ching, G. M. Boush, and F. Matsumura, 1975, Importance of Water pH in Accumulation of Inorganic Mercury in Fish, *Bull. Environ. Contam. Toxicol.*, 13(2): 188-193.

Underdal, B., and T. Hastein, 1971, Mercury in Fish and Water from a River and a Fjord in the Krager Region, South Norway, *Oikos*, 22: 101-105.

Wallace, R. A., W. Fulkerson, W. D. Schults, and W. S. Lyon, 1971, *Mercury in the Environment: The Human Element*, USAEC Report ORNL-NSF-EP-1, Oak Ridge National Laboratory, NTIS.

Wang, C. H., and D. L. Willis, 1965, *Radiotracer Methodology in Biological Science*, Prentice-Hall, Inc., Englewood Cliffs, N. J.

Williams, P. M., and H. V. Weiss, 1973, Mercury in the Marine Environment: Concentration in Sea Water and in a Pelagic Food Chain, *J. Fish. Res. Board Can.*, 30(2): 293-295.

IMPACT OF ARSENICALS ON NITRIFICATION IN AQUEOUS SYSTEMS

HARVEY W. HOLM and MARILYN F. COX
Environmental Research Laboratory, Environmental Protection Agency,
College Station Road, Athens, Georgia

ABSTRACT

The impact of both arsenate and cacodylic acid (at arsenic concentrations of 0, 0.1, 1, 10, 100, and 1000 mg/liter) on mixed populations of nitrifiers in model aqueous systems containing ammonia was determined by measuring levels of ammonia and nitrite over a 24-day incubation period. Arsenate decreased the rate of oxidation of ammonia by *Nitrosomonas* only at high concentrations (100 and 1000 mg/liter) of arsenic; low levels of arsenate (0.1, 1, and 10 mg As/liter) had no effect on the oxidation rate, in comparison with arsenic-free controls. The oxidation of nitrite to nitrate by *Nitrobacter* was affected by all concentrations of arsenic added as arsenate; low concentrations (0.1, 1, and 10 mg/liter) delayed the oxidation of nitrite and high concentrations (100 and 1000 mg/liter) inhibited the process. The only impact of cacodylic acid on nitrification occurred at 1000 mg As/liter. The oxidation of ammonia by *Nitrosomonas* was delayed by the arsenical, but the *Nitrobacter* population was not affected. Although cacodylic acid is not toxic to the nitrification process, its degradation product, arsenate, can inhibit nitrification if it is in an available form. This inhibition of the *Nitrobacter* population may promote the accumulation of nitrite in the environment.

An understanding of the impact of pollutants on nutrient cycling in aquatic ecosystems is important to decision makers in industry and to surveillance and enforcement personnel in local, state, and federal government. These individuals need to know (1) how a pollutant affects an environment, (2) what the consequence of an altered environment might be, and (3) how the environment influences the fate of a pollutant.

Economic factors prevent the use of individual field studies to answer these questions for every pollutant. Instead, smaller laboratory-associated studies must be completed and the data ultimately extrapolated to a field situation. To meet this end, we have focused much of our laboratory's research on the fate and impact of pollutants in different environmental systems. Our environmental system studies largely involve mixed microbial populations that are important in the nutrient cycling process.

For this study, we investigated the impact of two arsenicals, cacodylic acid (hydroxydimethylarsine oxide) and sodium arsenate, on the nitrification process in aqueous systems. Cacodylic acid, which is representative of the organic arsenicals, is used in the cotton-growing states as a defoliant (Versar, Inc., 1976). Sodium arsenate was chosen because it is a common metabolic product of the organic arsenicals (Woolson and Kearney, 1973).

Oxidation of nitrogen from a reduced state, such as ammonia, to a more oxidized state, such as nitrite and nitrate, is called nitrification (Hardy and Holsten, 1972). These nitrogen transformations are usually mediated by two types of chemosynthetic, autotrophic bacteria. Ammonia oxidizers, typified by *Nitrosomonas*, get energy for growth by oxidizing ammonia to nitrite; nitrite oxidizers of the *Nitrobacter* type complete the process by oxidizing nitrite to nitrate.

The objectives of this research were to develop simple systems to study the impact of pollutants on nitrification and to determine whether arsenicals, represented by sodium arsenate and cacodylic acid, significantly alter nitrification rates. These objectives were met by the following steps: (1) Mixed populations of nitrifiers were cultured in an aqueous medium; (2) the reproducibility and reliability of the system was determined; and (3) the impact of arsenicals on nitrification was determined by measuring the oxidation of ammonia and nitrite.

MATERIALS AND METHODS

Growth Medium

Nitrification rates were determined in a chemically defined medium containing micronutrients and macronutrients. In a final volume of 1000 ml, the filter-sterilized micronutrient stock contained EDTA—Na_2, 500 mg; $FeSO_4 \cdot 7H_2O$, 200 mg; $ZnSO_4 \cdot 7H_2O$, 10 mg; $MnCl_4 \cdot 4H_2O$, 3 mg; H_3BO_3, 30 mg; $CoCl_2 \cdot 6H_2O$, 20 mg; $CuCl_2 \cdot 2H_2O$, 1 mg; $NiCl_2 \cdot 6H_2O$, 2 mg; and $Na_2MoO_4 \cdot 2H_2O$, 3 mg. Each liter of the chemically defined medium contained NH_4Cl,

38.2 mg; K_2HPO_4, 1.1 mg; $MgCl_2$, 5.7 mg; $MgSO_4 \cdot 7H_2O$, 1.9 mg; $CaCl_2 \cdot 2H_2O$, 4.4 mg; $NaHCO_3$, 150 mg; and micronutrient stock, 10 ml. The medium was sterilized by filtration, and 100-ml aliquots were added to sterilized 250-ml Erlenmeyer flasks.

Appropriate amounts of sodium arsenate or cacodylic acid were added to the medium before filter sterilization. Reagent grade chemicals were used to prepare the growth medium. Mixed nitrifying populations (containing both *Nitrosomonas* and *Nitrobacter* types) from local garden soil were used as inocula for the systems. Although the functional populations were not identified, oxidation of ammonia to nitrate indicated that both *Nitrosomonas* and *Nitrobacter* were present.

Ammonia and nitrite concentrations in each flask were measured at selected intervals during the incubation period using a Technicon Autoanalyzer unit (AAII). Ammonia was measured using the automated colorimetric phenate method (Technicon Corp., 1971). Nitrite concentrations were measured using the automated cadmium reduction method (Environmental Protection Agency, 1974).

Experimental Design

Enrichment of nitrifying populations was completed by adding 1 g of garden soil to 250-ml Erlenmeyer flasks containing 100 ml of the growth medium. The flasks were incubated in the dark at 25°C at a shaking rate of 125 rpm. At 2- to 3-day intervals, samples were taken, diluted, and analyzed for ammonia and nitrite. After populations had oxidized all the ammonia to nitrate, 1-ml inocula were transferred into fresh media, and the sampling cycle was repeated. Inocula from the second and third transfers were used for the arsenic studies.

The reproducibility of the nitrification system was measured in seven replicate flasks, each receiving a 5-ml inoculum from a culture transferred twice previously. Biological activity was determined within each flask by measuring the oxidation of ammonia to nitrate over a 20-day incubation period.

The impact of both sodium arsenate and cacodylic acid (at arsenic concentrations of 0, 0.1, 1, 10, 100, and 1000 mg/liter) was measured on mixed cultures of nitrifiers in aqueous systems containing 10 mg/liter nitrogen as ammonia. Duplicate flasks were set up for each treatment, and nitrification rates were determined in each flask by measuring levels of ammonia and nitrite over a 24-day period.

The hours required for the oxidation of ammonia by *Nitrosomonas* and nitrite by *Nitrobacter* to a nitrogen concentration of 1 mg/liter were the quantities chosen for statistical analysis. Data were analyzed using analysis of variance and Tukey's ω procedure (Steel and Torrie, 1960).

RESULTS

Nitrifying populations were easily cultured from the garden-soil inoculum. For the initial enrichment, ammonia was oxidized to nitrite by *Nitrosomonas* organisms within 2 weeks, and the nitrite formed in the first step was oxidized to nitrate by *Nitrobacter* within 3 weeks. In successive transfers of 1-ml inocula, however, the rate of oxidation of ammonia to nitrate was significantly increased. By the second transfer, all the ammonia was oxidized to nitrite within 1 week. The entire nitrification process, involving the oxidation of 10 mg/liter ammonia nitrogen to 10 mg/liter nitrate nitrogen, was shortened from 24 days to 10 days by the third transfer.

No counts of nitrifiers were made as the ammonia was being oxidized to nitrate. *Nitrosomonas* probably outnumbered *Nitrobacter* in each flask, however, because the oxidation of ammonia to nitrite provides more energy for growth of cells than does the oxidation of nitrite to nitrate (Tuffey, 1973; Curtis, Durrant, and Harmon, 1975).

These nitrification systems provide reproducible results. Data in Table 1 were collected from seven replicate flasks, each receiving a 5-ml inoculum from a culture transferred twice previously. These results demonstrate that aqueous systems in flasks can be used to study the impact of soluble pollutants on nitrification. An estimate of homogeneity of data, the coefficient of variation (CV), which is 100(standard deviation/mean), calculated from the data in Table 1 shows that these systems have CV values of 20% or less when the ammonia or nitrite nitrogen concentration is above 1 mg/liter.

Using these test systems, we observed that *Nitrosomonas* organisms are not sensitive to arsenicals. At low levels of arsenate (10 mg As/liter), the rate of ammonia oxidation was identical to that of the arsenic-free controls (Fig. 1). The rate of ammonia oxidation was significantly different from the control only at 1000 mg As/liter added as arsenate, determined with Tukey's ω procedure ($\alpha = 0.05$). These results show that arsenate should not have an inhibiting action on *Nitrosomonas* in the natural environment because of the high concentrations necessary before any effects are seen.

Likewise, cacodylic acid should not have a significant direct effect on *Nitrosomonas. Nitrosomonas* oxidized ammonia at normal rates in media containing as much as 100 mg As/liter as cacodylic

TABLE 1

CONCENTRATIONS OF NITROGEN IN AQUEOUS NITRIFICATION SYSTEMS*

Incubation, days	NH_4^+—N, mg/liter	NO_2^-—N, mg/liter
0		
1	9.35† (S = 0.38)	
3	9.23 (S = 0.38)	0.04† (S = 0.01)
6	8.42 (S = 0.36)	0.71 (S = 0.13)
8	5.14 (S = 0.97)	4.35 (S = 0.78)
10	0.12 (S = 0.22)	8.93 (S = 0.51)
13		7.81 (S = 0.46)
15		6.06 (S = 0.82)
17		4.79 (S = 0.99)
20		0.43 (S = 0.49)

*The growth medium containing 10 mg/liter NH_4^+—N was inoculated with 5 ml of a mixed nitrifier population. Abbreviation S is standard deviation of seven flasks.

†Mean of seven observations.

acid. There was a small decrease in the oxidation rate at concentrations of 1000 mg As/liter (Fig. 2).

Nitrobacter-type organisms used in this study are sensitive to arsenate (Fig. 3). Apparently there is a progressively longer lag phase in the arsenic concentration range from 0.1 to 10 mg/liter, but the only treatment statistically different from the control (based on analysis of variance, $\alpha = 0.05$) is 10 mg/liter. Because only one point from each oxidation curve was used for the statistical analysis and only two replicate curves were obtained for each treatment, the analyses for differences are probably conservative. At very high levels of arsenate (100 and 1000 mg As/liter), *Nitrobacter* was inhibited, and nitrite accumulated.

Cacodylic acid does not affect *Nitrobacter*. Arsenic concentrations as high as 1000 mg/liter as cacodylic acid did not alter the oxidation rate of nitrite (Fig. 4).

DISCUSSION

The technique we adapted for this study is useful for studying the impact of pollutants on specific functional groups of bacteria. In this study we enriched the medium for chemosynthetic autotrophs important in the nitrification process. There are several advantages to

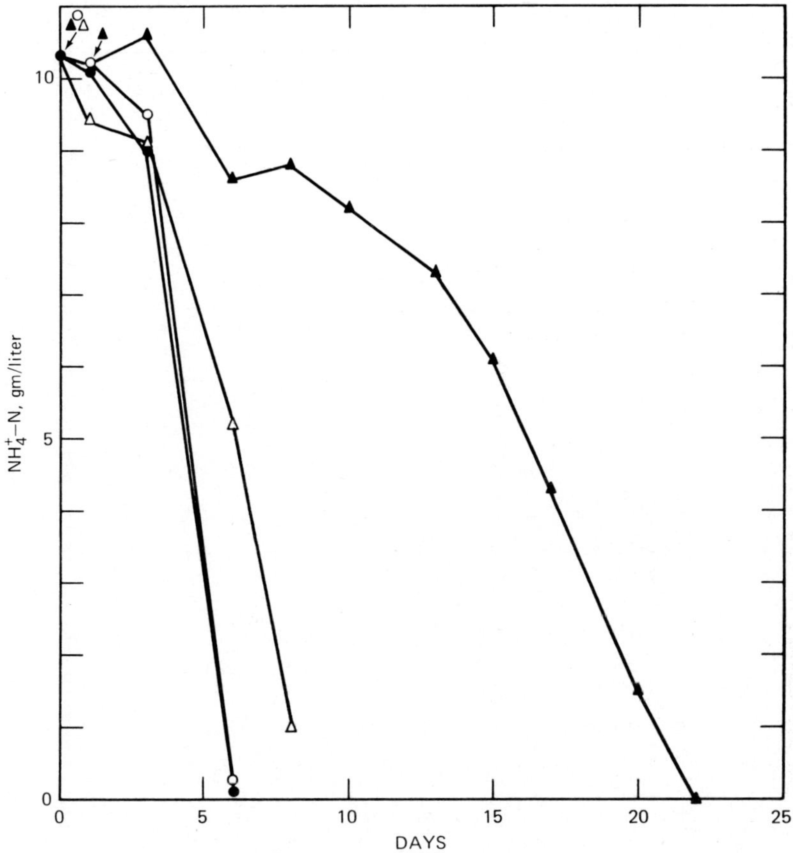

Fig. 1 Impact of arsenate on *Nitrosomonas*. ●, no arsenic; ○, 10 mg As/liter; △, 100 mg As/liter; and ▲, 1000 mg As/liter.

this approach. First, mixed cultures growing in a selective medium provide a larger genetic pool than do pure cultures. This large genetic pool may allow the system to adapt to the toxicant much as populations do in nature. Second, enrichment of the mixed population assures the researcher uniform cultures to be used for a given set of toxicity tests (Table 1). This assurance usually is not present in studies of functional groups using natural samples or colonized slides. Third, monitoring ecosystem function is usually faster and easier than measuring specific bacterial populations. For example, the study by Fliermans, Bohlool, and Schmidt (1974), using fluorescent antibodies for counting nitrifiers, is very time and labor consuming.

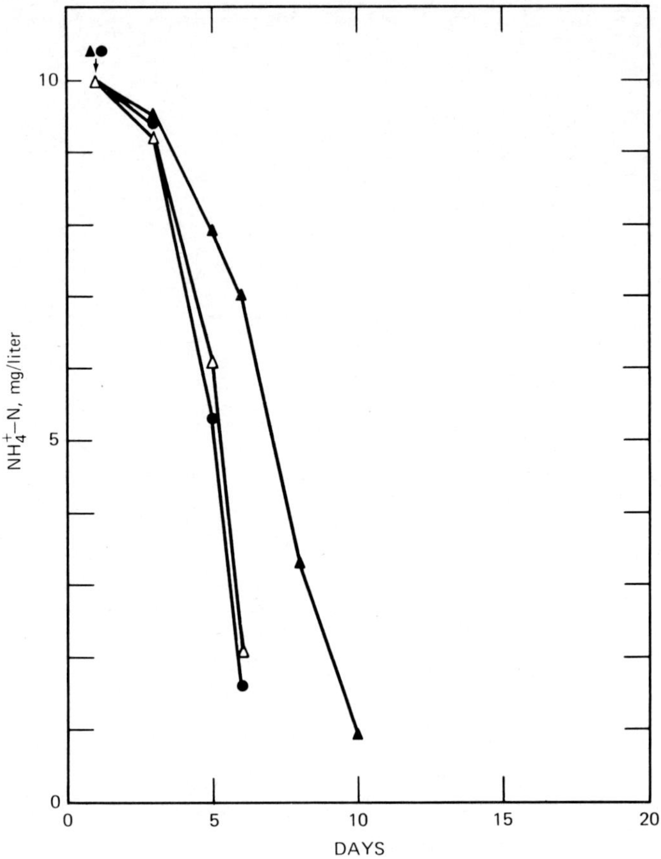

Fig. 2 Impact of the organic arsenical, cacodylic acid, on *Nitrosomonas.* **●, no arsenic; △, 100 mg As/liter; and ▲, 1000 mg As/liter.**

Our approach also has some disadvantages. First, selection of new variants or species may occur, but this is not monitored. In fact, the selection process may have occurred during the enrichment phase of the study since the nitrification rate increased on successive transfer. This phenomenon may also have occurred when populations of *Nitrobacter* were exposed to arsenicals (Fig. 3). Second, interactions of populations from different segments of the nutrient cycle are not possible with enrichment techniques. This contrasts with work by Laveglia and Dahm (1974), who used soil systems to monitor the activity of several functional groups of bacteria after the test systems were dosed with pollutants.

Inhibition of *Nitrosomonas* and, consequently, inhibition of the total nitrification process, can be economically and ecologically

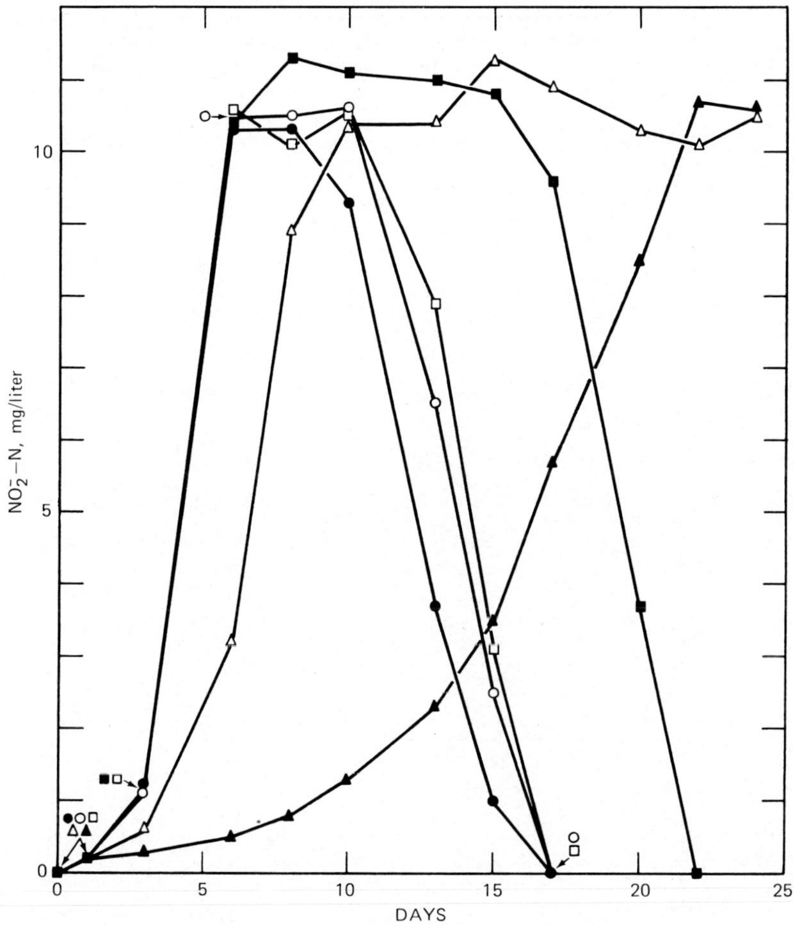

Fig. 3 Impact of arsenate on *Nitrobacter*. ●, no arsenic; ○, 0.1 mg As/liter; □, 1 mg As/liter; ■, 10 mg As/liter; △, 100 mg As/liter; and ▲, 1000 mg As/liter.

desirable. The agricultural industry prolongs the availability of nitrogen fertilizer by applying N-SERVE [2-chloro, 6-(trichloromethyl) pyridine], an inhibitor specific for *Nitrosomonas* (Campbell and Aleem, 1965). When the oxidation of ammonia to nitrite is inhibited, the nitrogen remains in a relatively immobile form, ammonia, but it is still available for plants. Pesticides such as lindane, Malathion, and Baygon also inhibit nitrification at this first stage (Garretson and San Clemente, 1968). It may also be desirable to control nitrification in streams and lakes because it contributes a significant oxygen demand in some waters. Theoretically, 4.57 mg of

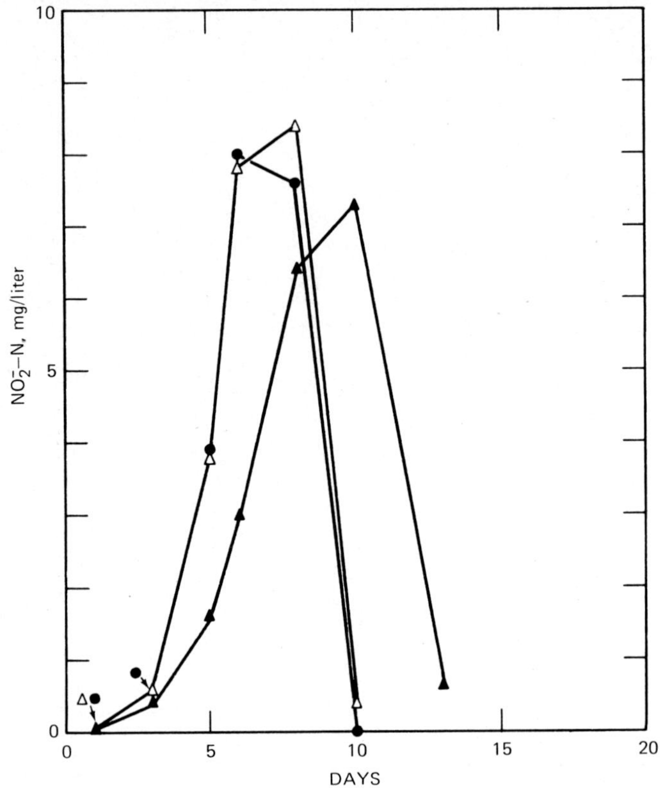

Fig. 4 Impact of the organic arsenical, cacodylic acid, on *Nitrobacter*. ●, no arsenic; △, 100 mg As/liter; and ▲, 1000 mg As/liter.

molecular oxygen are required to oxidize 1 mg of ammonia nitrogen to nitrate nitrogen (Young, 1969).

Our results show that arsenate inhibits the second step of nitrification to a greater degree than it does the first stage. This was reflected in a delay in the oxidation of nitrite at low concentrations of arsenate; at high levels no nitrite was oxidized within 25 days. The mechanism of this inhibition was not studied. Button and Dunker (1971), however, showed that arsenate interrupts phosphate metabolism in some microorganisms. Torstensson (1974) reported that MCPA, 2,4,5-T, and Linuron inhibit *Nitrobacter* and cause an accumulation of nitrite.

Nitrite in the environment is undesirable because of its toxicity to biota (National Academy of Sciences, 1973) and its possible implication in the formation of carcinogenic nitrosamines (Lijinsky and Epstein, 1970; Elespuru and Lijinsky, 1973; Eisenbrand, Ungerer, and Preussmann, 1975).

Generally organic arsenicals are reported to be between 10 and 100 times less toxic to biota than inorganic arsenicals (Versar, Inc., 1976). Our results showed this to be the case. Although arsenic in the form of arsenate increased the lag phase of *Nitrobacter* at concentrations as low as 0.1 and 1 mg/liter, cacodylic acid affected the nitrifiers only at arsenic concentrations of 100 mg/liter and above. Bollen, Norris, and Stowers (1977), studying forest floors and forest soils treated with either cacodylic acid or MSMA, concluded that these organic arsenicals should not have a significant adverse effect on nitrogen metabolism of forest-floor and soil microorganisms.

In aerobic systems, cacodylic acid is readily degraded to arsenate (Woolson and Kearney, 1973). This fact, coupled with our observation of arsenate toxicity towards nitrifiers, points toward a potential temporary buildup of nitrite in environments receiving high levels of the nontoxic arsenical.

In summary, our study showed (1) that these nitrification systems provide reproducible results, (2) that *Nitrobacter*-type organisms used in these systems are sensitive to arsenate, and (3) that cacodylic acid is not directly toxic to nitrifiers. Field investigations should be completed to test whether arsenicals inhibit *Nitrobacter* in field situations.

DISCLAIMER

Mention of tradenames of commercial products does not constitute endorsement or recommendation for use.

REFERENCES

Bollen, W. B., L. A. Norris, and K. L. Stowers, 1977, Effect of Cacodylic Acid and MSMA on Nitrogen Transformations in Forest Floor and Soil, *J. Environ. Qual.*, 6: 1-3.

Button, D. K., and S. S. Dunker, 1971, *Biological Effects of Copper and Arsenic Pollution*, Report No. R71-8, Institute of Marine Science, University of Alaska, College.

Campbell, N. E. R., and M. I. H. Aleem, 1965, The Effect of 2-Chloro, 6-(Trichloromethyl) Pyridine on the Chemoautotrophic Metabolism of Nitrifying Bacteria, *Antonie van Leeuwenhoek J. Microbiol. Serol.*, 31: 137-144.

Curtis, E. J. C., K. Durrant, and M. M. Harmon, 1975, Nitrification in Rivers in the Trent Basin, *Water Res.*, 9: 255-268.

Eisenbrand, G., O. Ungerer, and R. Preussmann, 1975, The Reaction of Nitrite with Pesticides. II. Formation, Chemical Properties and Carcinogenic

Activity of the N-Nitroso Derivative of N-Methyl-1-Naphthyl Carbomate (Carbaryl), *Food Cosmet. Toxicol.*, 13: 365-367.

Elespuru, R. K., and W. Lijinsky, 1973, The Formation of Carcinogenic Nitroso Compounds from Nitrite and Some Types of Agricultural Chemicals, *Food Cosmet. Toxicol.*, 11: 807-817.

Environmental Protection Agency, 1974, *Methods for Chemical Analysis of Water and Wastes,* Publication No. EPA-625/6-74-003, Washington, D.C.

Fliermans, C. B., B. B. Bohlool, and E. L. Schmidt, 1974, Autecological Study of the Chemoautotroph Nitrobacter by Immunofluorescence, *Appl. Microbiol.*, 27: 124-129.

Garretson, S. L., and C. L. San Clemente, 1968, Inhibition of Nitrifying Chemolithotrophic Bacteria by Several Insecticides, *J. Econ. Entomol.*, 61: 285-288.

Hardy, R. W. F., and R. D. Holsten, 1972, Global Nitrogen Cycling: Pools, Evolution, Transformations, Transfers, Quantitation and Research Needs, in *The Aquatic Environment: Microbial Transformations and Water Management Implications,* symposium sponsored by EPA Office of Water Programs Operations, October 1972, pp. 87-132, U.S. Government Printing Office, Washington, D.C., Stock No. 5501-00615.

Laveglia, J., and P. A. Dahm, 1974, Influence of AC92,100 (Counter) on Microbial Activities in Three Iowa Surface Soils, *Environ. Entomol.*, 3: 528-533.

Lijinsky, W., and S. S. Epstein, 1970, Nitrosamines as Environmental Carcinogens, *Nature*, 225: 21-23.

National Academy of Sciences, 1973, *Water Quality Criteria 1972, Washington, D.C.*, Publication No. EPA-R3-73-003, prepared for Environmental Protection Agency, Washington, D.C.

Steel, R. G. D., and J. H. Torrie, 1960, *Principles and Procedures of Statistics,* 1st ed., pp. 109-110, McGraw-Hill Book Company, Inc., New York.

Technicon Corp., 1971, *Industrial Method, AAII,* Report No. 98-70W, Tarrytown, N.Y.

Torstensson, L., 1974, Effects of MCPA, 2,4,5-T, Linuron and Simazine on Some Functional Groups of Soil Microorganisms, *Swedish J. Agric. Res.*, 4: 151-160.

Tuffey, T., 1973, *The Detection and Study of Nitrification in Streams and Estuaries,* Ph.D. Thesis, Rutgers University, New Brunswick, N.J.

Versar, Inc., 1976, *Technical and Microeconomic Analysis, Task III—Arsenic and Its Compounds,* Publication No. 560/6-76-016, prepared for Environmental Protection Agency, Washington, D.C.

Woolson, E. A., and P. C. Kearney, 1973, Persistence and Reactions of ^{14}C-Cacodylic Acid in Soils, *Environ. Sci. Technol.*, 7: 47-50.

Young, J. C., 1969, Chemical Methods for Nitrification Control, in *Proceedings of the 24th Industrial Waste Conference,* Purdue University, Lafayette, Ind., May 6—8, 1969, pp. 1090-1102.

COPPER SENSITIVITY OF ADULT PACIFIC OYSTERS

F. L. HARRISON and D. W. RICE, JR.
Environmental Sciences Division, Lawrence Livermore Laboratory,
University of California, Livermore, California

ABSTRACT

Sensitivity of the Pacific oyster, *Crassostrea gigas*, to copper was evaluated in experiments of about 2-weeks duration. Groups of 8 to 10 animals were exposed to copper concentrations ranging from 100 to 1300 μg/liter in a high-volume flow-through bioassay system. Copper concentrations in the oysters and the major chemical form of copper in the water were determined. In addition, experiments were performed with ^{64}Cu to determine accumulation rates and distribution. The family of mortality curves of time vs. concentration for oysters was unlike that for other organisms in that the greatest mortality at 96 hr occurred in animals at intermediate rather than high copper concentrations. However, typical mortality curves were obtained by correcting the duration of copper exposure of each oyster by a factor related to the amount of time the shells were open. Significant mortalities occurred at copper concentrations of 200 μg/liter and higher. Analysis of the data yields an LC_{50} at 48 hr of 650 μg/liter, an LC_{50} at 96 hr of 430 μg/liter, and an incipient LC_{50} of 230 μg/liter. Distributions of ^{64}Cu in oysters after a 24-hr exposure differed in animals maintained in water containing high and low concentrations of copper. The flux of copper into tissues was related to copper concentrations in the water; this indicated that there was little metabolic regulation.

Shellfish resources of estuarine and marine ecosystems can be harmed by increased levels of heavy metals in water. Of the heavy metals commonly released as a result of man's activities, copper is of special interest because of the known sensitivity of aquatic organisms to it (Becker and Thatcher, 1973). Oysters accumulate copper (Galtsoff, 1964; Pringle et al., 1968; Shuster and Pringle, 1969;

Drifmeyer, 1974; Okazaki, 1976) and are harvested commercially in large numbers as an important shellfish consumed by man.

Although a considerable amount of information exists on quantities of copper in oysters from pristine and polluted waters (Brooks and Rumsby, 1965; Establier, 1972; Windom and Smith, 1972; Thrower and Eustace, 1973; Establier and Pascual, 1974; Ayling, 1974; Boyden and Romeril, 1974; Cronin et al., 1974; Ratkowsky et al., 1974; Huggett, Cross, and Bender, 1975; Mackay et al., 1975; Frazier, 1975; 1976), very little data are available on the sensitivity or accumulation rates of copper in the adult oyster. Previously reported LC_{50} values at 96 hr range from 500 to 1900 μg Cu/liter (Galtsoff, 1932; Fujiya, 1960; Okazaki, 1976). Neither Galtsoff nor Fujiya determined the concentrations and chemical forms of copper to which the oysters were exposed; Okazaki analyzed the bioassay water for total copper in his experiments.

Our studies were initiated to determine the effects on oysters of increased levels of copper in water. Under controlled laboratory conditions we determined the effects on mortality rates, on copper levels in the soft tissues, and on accumulation rates and tissue distributions of ^{64}Cu. Amounts and major chemical forms of copper in the bioassay waters and amounts of copper in the oysters were measured.

MATERIAL AND METHODS

Experimental Population

Oysters, *Crassostrea gigas*, were obtained from the Coast Oyster Company beds in Arcata Bay, the shallow north arm of Humboldt Bay, Calif. They were shipped by air to San Francisco, transported immediately by truck to Lawrence Livermore Laboratory, and then maintained for 7 to 10 days in a 10,000-gal recirculating seawater system before testing. Water in this system was continuously refiltered and was kept between 10 and 12°C by refrigeration. The water for the bioassay system was collected before the experiment and stored in a 10,000-gal underground tank. Water in the circulating system and in the storage tank was obtained from the University of California Marine Station at Bodega Bay, Calif. This water is pumped from the sea off the open coast in an area that receives little anthropogenic input. The water contains low levels of trace metals, dissolved organics, and particulate material. Although the oysters were reported to be 3 years old, they varied considerably in size. Soft body masses were typically between 30 and 60 g. Oysters were scrubbed before random selection for placement in the bioassay system.

Toxicity Experiments

Bioassay System

Animals were exposed to copper in a flow-through bioassay system. The exposure solutions were introduced by gravity from a mixing chamber into 10-liter 5- by 40- by 48-cm plastic trays containing 10 oysters placed hinge side up at known locations. Filtered seawater and $CuCl_2$ solutions were pumped into the mixing chambers at predetermined rates to provide the desired copper concentrations. The volume of water in the exposure tray was maintained constant with an outflow siphon. The flow of seawater from the trays was 6.6 liters/hr. Replacement of 90% of the seawater in the tray required approximately 3.4 hr.

Gentle streams of bubbles from outlets arranged in a 5- by 15-cm grid at the bottom of the trays aerated and mixed the seawater in the exposure trays. All exposure trays were immersed in a water bath whose temperature was monitored and maintained at 12.8 ± 1°C. Continuous illumination was provided because we observed that some oysters responded to abrupt changes in light intensity by closing the valves of their shells. The salinity and pH of the bioassay water were 31.1‰ and 8.08 ± 0.05, respectively.

During the 10- to 12-day experiments, observations were made at 4-hr intervals from 8:00 a.m. to 12:00 p.m. (no observations were made at 4:00 a.m.). At each observation the amount of valve separation of the shell was recorded. Gaping oysters (having a separation of ~1 cm between the valves) were gently tickled with a blunt, soft, plastic probe to check for muscular responses. Animals that did not respond were considered dead and were removed from the bioassay system. At the termination of the experiment, oysters alive in the 230 μg Cu/liter and control trays were sacrificed.

Copper Analyses of Oysters

Oysters were weighed intact; the bodies were removed from the shells; and the soft tissues were weighed before freezing for later dissection. To prepare for copper analyses, we dissected the animals while they were still partially frozen into gills, digestive glands and stomachs, and remaining tissues. Tissues were dried at 100°C, ashed at 450°C, and wet ashed in a mixture of HCl and HNO_3 for copper analysis on an atomic absorption spectrometer.

Copper Analyses of Bioassay Water

Total and ionic copper concentrations in the bioassay water were measured at 1- to 3-day intervals during the test period. Water was collected either by siphoning directly from the trays over a 5-min interval (grab sample) or by continuously pumping from the tray at a low rate over a 2- to 3-day interval (integrated sample). Both total and ionic copper concentrations generally were measured in grab samples, but only total copper was measured in integrated samples since they were acidified to pH 1 during collection.

Total copper in samples containing copper concentrations $\geqslant$400 μg/liter was determined by direct aspiration of seawater into the flame of the atomic absorption spectrometer and in samples with copper concentrations $\geqslant$10 μg/liter by direct injection into a graphite furnace.

Ionic copper, operationally defined as the fraction retained by Chelex 100 resin (Bio-Rad Laboratories), was determined by the method of Riley and Taylor (1968). Eluants from the columns were analyzed directly in the flame or in the graphite furnace.

Accumulation of ^{64}Cu

Groups of oysters were exposed in refrigerated aquariums to ^{64}Cu-labeled seawater containing concentrations of approximately 1.0, 800, or 2400 μg Cu/liter. Sixteen oysters were maintained at each concentration, and four animals were sampled from each system after 1, 5, 10, and 24 hr of exposure.

The initial amount of ^{64}Cu in the water at all three copper concentrations was 300 pCi/liter. Standards of ^{64}Cu were made when the isotope was added to the aquarium water. At regular intervals concentrations of ^{64}Cu in aquarium waters were compared with the standards. Additions of ^{64}Cu were made to aquarium waters when the ^{64}Cu level was lower than expected from decay alone so that changes in the water reflected only radioactive decay.

To prepare for ^{64}Cu analysis, we weighed oysters whole, removed the soft tissues from the shells, and dissected them into mantles, gills and labial palps, digestive glands and stomachs, gonads and intestines, adductor muscles, body fluids, and remains. Each tissue was weighed and counted in a gamma well counter.

All counts of water and tissues were corrected for background and counting efficiency and then for decay to the time the experiment began.

RESULTS

Toxicity Experiments

The family of mortality curves from our first two experiments with oysters was difficult to interpret because increased copper concentrations in the water did not consistently result in increased mortality rates; 100% mortality of the oysters occurred first in the water containing 500 rather than 900 or 1500 μg Cu/liter. Since oysters can prevent the entrance of water into their mantle cavities by tight closure of the valves of the shell, they might have reduced their exposure to copper in this way. In experiments 3 and 4, oysters were observed at 4-hr intervals from 8:00 a.m. to 12:00 p.m. so that the exposure of each oyster could be corrected for the difference between the time the animal was first observed with its shell open and the time it was introduced into the water and also by the ratio of the number of times the shell was seen opened to the number it was seen closed (Table 1). Typical mortality curves were generated with adjusted exposure times (Fig. 1), and, in all further measures of toxicity, we used adjusted exposure time. A logit instead of a probit analysis was performed on the time-vs.-mortality data (American Public Health Association, 1975). Differences between the slopes of the lines fitted to the curves and the intercepts of the lines were tested using a t statistic (Snedecor and Cochran, 1967). The slopes were significantly different from the controls ($P < 0.01$), but the intercepts were not.

Sensitivity to copper sometimes differed between the two groups of oysters in experiments 3 and 4. Such differences are expected when oysters are collected at different times, the number of animals tested at each concentration is small, and the exposure conditions and concentrations are not exactly identical (Table 1).

The concentration vs.-time-to-50%-mortality data from experiments 3 and 4 were used to generate a toxicity curve (American Public Health Association, 1975). Acute lethality stopped at concentrations below about 230 μg Cu/liter (Fig. 2). In this group of oysters, the LC_{50} value (lethal concentration for 50% of the test population) ±95% confidence interval at 48 hr was 650 ± 100 μg Cu/liter; at 96 hr, 430 ± 60 μg Cu/liter; and at 168 hr, 30 ± 100 μg Cu/liter. The incipient lethal or threshold concentration is defined as the concentration that kills 50% of the population during an exposure sufficiently long for acute lethal action to cease (Sprague, 1969). In this group of oysters, the estimated incipient lethal concentration (incipient LC_{50}) was 230 ± 100 μg Cu/liter.

TABLE 1

CONDITIONS OF TOXICITY TESTS OF COPPER TO ADULT OYSTERS

	Bioassay test system					
Condition	1	2	3	4	5	6
Experiment 3						
Copper						
Total,* μg/liter	3 (control)	180	310	580	680	1050
Range, μg/liter		160–190	270–360	480–620	640–700	1000–1100
Ionic, %		84 ± 5	94 ± 3	86 ± 6	89 ± 6	76 ± 8
Oysters						
Number	10	10	9	8	8	8
Ratio of opened to closed	0.59	0.66	0.72	0.64	0.43	0.32
LT_{50}						
$\bar{x}$		164	163	59	49	12
95% CI†		±47	±59	±12	±15	±6
Experiment 4						
Copper						
Total,* μg/liter	1 (control)	230	440	510	760	1300
Range, μg/liter		210–270	400–600	410–570	750–760	1100–1400
Ionic, %		76 ± 5	74 ± 10	86 ± 6	83 ± 2	78 ± 2
Oysters						
Number	10	10	10	10	10	9
Ratio of opened to closed	0.60	0.74	0.79	0.82	0.56	0.48
LT_{50}						
$\bar{x}$			95	64	24	18
95% CI†			±7	±6	±10	±5

*Mean values of six to eight determinations.
†In calculating the 95% confidence intervals, we assumed the error mean square to be 1.

At each test concentration an LT_{50} value (time required for 50% mortality of the test population) was determined. The LT_{50} values ranged from 164 hr at the lowest concentration to 18 hr at the highest concentration (Table 1).

Concentrations of total and ionic copper in the water were relatively constant during the test period. Deviations from the average concentrations (Table 1) can be attributed to (1) small changes in inflow rates resulting from conformational changes in the tubing of the pumping system, (2) changes in rates of accumulation of copper from the water by the oysters, and (3) analytical errors. The ionic fraction was generally greater than 75% of the total amount of copper present (Table 1).

Copper Concentrations in Oysters

Oysters maintained at increased levels of copper in the water contained more copper than those in control groups (Table 2). The

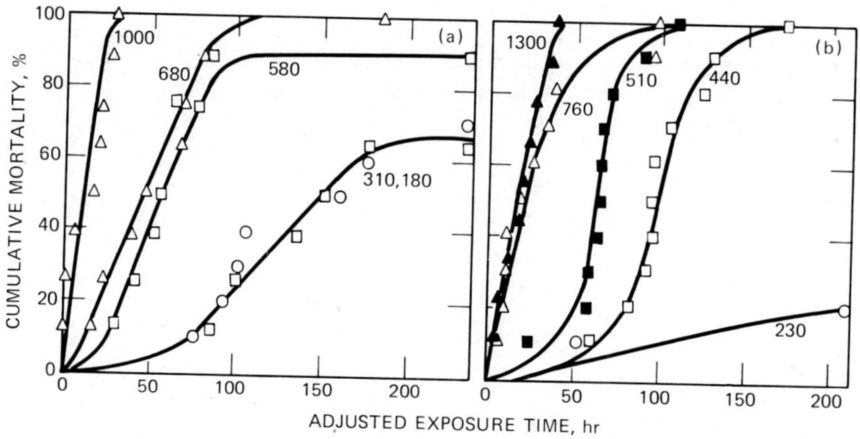

Fig. 1 Cumulative mortality of adult oysters continuously exposed to copper. (a) Experiment 3. (b) Experiment 4. Exposure time was adjusted by the ratio of amount of time the shells were opened to the time closed. Numbers on the curves are copper concentrations in micrograms per liter.

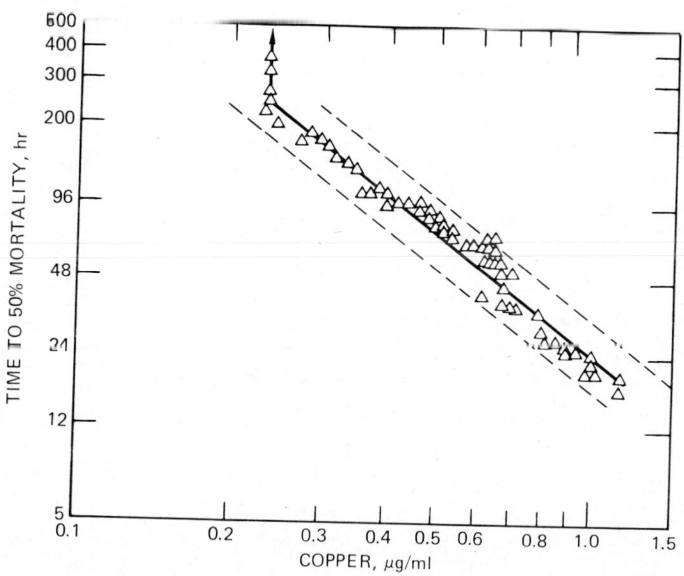

Fig. 2 Toxicity curve for adult oysters continuously exposed to copper based on data from experiments 3 and 4 with adjusted exposure times. The least square equations for this line are $Y = 6.3595 - 1.6619X$ and $r^2 = 0.9645$. The dashed lines (= = = =) indicate ±95% confidence limits around the x value.

TABLE 2

AVERAGE COPPER CONCENTRATIONS* IN WATER AND OYSTERS FROM EXPERIMENT 4

Copper in bioassay water, μg/liter	Whole body	Gills	Digestive gland and stomach	Remainder of body
1 (control)				
μg/g	19.4	65.0	42.0	15.3
total μg	1,790	426	131	1,230
230				
μg/g	221	368	382	195
total μg	11,400	1,530	1,250	8,650
440				
μg/g	73.8	95.2	90.9	70.8
total μg	3,380	404	237	2,740
510				
μg/g	68.4	99.9	85.7	63.3
total μg	3,510	515	251	2,750
760				
μg/g	57.6	111	50.8	53.3
total μg	2,310	404	126	1,870
1300				
μg/g	49.6	112.0	66.2	42.4
total μg	2,770	592	164	1,990

*Fractional standard deviation (SD/mean) generally ranged from 0.2 to 0.4 for the concentrations expressed in micrograms per gram and from 0.4 to 0.6 for the total amounts.

increases were not linear with concentrations in the water, however; the highest levels were found in the oysters in water containing 230 μg Cu/liter. The quantities of copper contained in oysters held in seawater with copper concentrations greater than 230 μg/liter did not differ significantly from each other.

The quantities of ^{64}Cu (decay corrected) in the total soft tissues at the end of the 24-hr exposure were different from oysters in each of the three copper concentrations. The highest concentrations of ^{64}Cu were detected in animals held in water with the lowest copper concentrations (Table 3). Some regulation of copper accumulation is suggested; the ^{64}Cu concentrations would have been similar in the animals from the three test systems if the amount of copper accumulated per unit time was directly proportional to the copper concentration in the water (no homeostasis).

The ^{64}Cu in the tissues of oysters placed in each of the three copper concentrations increased with time, except in the digestive

TABLE 3

CONCENTRATIONS OF ^{64}Cu IN OYSTER TISSUE (AVERAGE pCi/mg wet weight)

Tissue	Copper concentration, μg/liter											
	1 (control), hr				800, hr				2400, hr			
	1	5	10	24	1	5	10	24	1	5	10	24
Total soft tissue	0.61	2.1	4.0	4.6	0.38	1.4	2.6	3.3	0.03	0.54	0.66	1.5
Mantles	0.27	1.5	8.2	9.7	0.46	1.1	2.7	7.3	0.02	0.58	0.98	2.9
Gills	1.7	3.9	14.0	20.0	1.20	2.8	5.8	14.0	0.02	1.6	2.8	7.3
Digestive glands and stomachs	11.0	27.0	27.0	26.0	0.34	1.9	3.4	7.3	0.02	0.69	0.99	3.8
Adductor muscles	0.12	0.35	1.9	2.0	0.22	0.49	2.8	5.0	0.01	0.48	0.72	2.3
Gonads and intestines	0.40	1.4	4.2	4.9	0.27	0.76	3.1	4.1	0.06	0.57	0.66	1.9
Body fluids	0.03	0.08	0.02	0.02	0.15	0.32	0.23	0.15	0.02	0.34	0.05	0.07
Remains	0.03	0.26	0.05	0.05	0.09	0.15	0.09	0.07	0.01	0.08	0.02	0.04

glands and stomachs of control animals (Table 3). The highest ^{64}Cu concentrations in tissue were in the digestive glands and stomachs of control oysters and in the gills of oysters at 800 and 2400 μg Cu/liter. Although ^{64}Cu concentrations in the gills of control animals and of those at 800 μg Cu/liter were similar after the first hour, concentrations in the digestive glands were more than 20 times higher in the controls than in oysters at 800 μg Cu/liter. These results indicate that a much greater fraction of the ^{64}Cu-labeled copper reaching the gills is transported to the digestive glands and stomachs of animals at low rather than high copper concentrations. However, even though the influx of ^{64}Cu into digestive glands and stomachs is greater in controls, the influx of stable copper is less. The specific activity was 800 times greater in controls than in oysters at 800 μg Cu/liter, whereas the ^{64}Cu in digestive glands and stomachs was only 20 times greater in control animals than in those at 800 μg Cu/liter.

The accumulation rates of ^{64}Cu in the soft tissues of oysters decreased with increasing concentrations of copper in the water (Table 3). The average and the maximum amounts of ^{64}Cu increased linearly for the first 10 hr at all three exposure concentrations. The average rates of accumulation in the oysters in 1.0 (control), 800, and 2400 μg Cu/liter were 390, 260, and 60 pCi g^{-1} wet weight hr^{-1}, respectively. The maximum rates at all three concentrations were about 1.7 times the average rates. Since the initial specific activity of the water was known, calculation of the accumulation rates of stable copper was possible. The rates of accumulation (μg Cu/hr) of a hypothetical oyster whose wet weight was 50 g were 0.25 for the control, 32 at 800 μg Cu/liter, and 23 at 2400 μg Cu/liter.

DISCUSSION

The LC_{50} that we obtained for *C. gigas* at 96 hr (430 μg Cu/liter) is higher than that obtained for many other aquatic species (Becker and Thatcher, 1973) and considerably lower than the 1900 μg Cu/liter for oysters reported by Fujiya (1960), but it is in the same range as the 500 μg/liter obtained by Okazaki (1976). Fujiya used a static system and did not determine the final copper concentration in the bioassay water. We found in our preliminary experiments that oysters could remove a significant fraction of the copper from test waters. When the flow rate was 1.5 liters/hr and the copper concentration was ⩽500 μg/liter, the outflow concentration was about half that of the inflow. When the copper concentration was between 500 and 1200 μg/liter, the outflow concentration was about three-fourths that of the inflow. The actual copper concentrations to which the animals were exposed in Fujiya's experiments were probably lower than he reported.

The greatest mortality at 96 hr in our experiments and in those of Okazaki (1976) occurred at intermediate rather than at high copper concentrations. Our observations indicate that this response resulted from the oysters' remaining open for shorter periods in the water containing high concentrations of copper. In an early experiment one oyster maintained at 1200 μg Cu/liter was closed during each observation period for 14 days. When sacrificed at the end of the experiment, the oyster appeared normal, and its heart was beating regularly. We observed during preliminary experiments that oysters will stop circulating seawater when copper is added to it at levels ⩾1000 μg/liter. Establier and Pascual (1974) noted increased incidence of shell closure in *C. angulata* at high copper concentrations. This response of oysters may serve as protection from episodic releases of copper. Moreover, it suggests that oysters may not be the best biological indicators of copper pollution even though they have been used as such in the past (Huggett, Bender, and Stone, 1973). They may accumulate copper only when the concentration in the water is at low and intermediate levels. Since accumulation may not be continuous, the copper content of oysters may not reflect the total amount of copper in their environment.

Mussels, like oysters, can detect copper in their environment. *Mytilus edulis* avoids the detrimental consequences of exposure to copper by closing its shell valves (Davenport, 1977). Davenport suggested that mussels and other animals possessing similar closure mechanisms are of doubtful use as a biological pollutant-monitoring system.

The toxic response in some animals is attributed to the concentration of ionic copper in the water (Sunda and Guillard, 1976). In our experiments the ionic fraction (as operationally defined) was greater than 75% of the total in all test systems. If oysters are sensitive to only the ionic form, however, the LC_{50} values for the toxic copper form should be lower than that reported.

The impact of past releases on oyster populations can be approximated by comparing the concentrations of copper in past releases to concentrations known to elicit a toxic response in oysters. Two important sources of copper in marine systems are municipal wastes and power-plant effluents. A survey of 108 municipal waste discharges along the Atlantic coast reported concentrations of 20 to 5900 μg Cu/liter; 50% of the effluents contained more than 100 μg Cu/liter (Mytelka et al., 1973). These represent the total copper concentrations in the waste at the point of discharge (before dilution) and do not indicate the fraction that might have been in a toxic form. A year-long survey at a single Pacific coastal power plant showed copper values in the intake and discharge waters to vary from 0.1 to 5.8 μg Cu/liter (Dorband et al., 1976). These values are in contrast to the 1800 μg Cu/liter (before dilution) discharged in the initial surge of water from the cooling system during startup at another power plant (Warrick, Sharp, and Friedrich, 1975). Until recently the California State Water Resources Control Board permitted concentrations in discharges of 200 μg Cu/liter for half a month and 300 μg/liter for 10% of a month (California State Water Resources Control Board, 1972). Based on an analysis of acute toxicity, chronic toxicity, and reported seawater concentrations, the board has proposed amendments to the water quality control plan for ocean waters of California which have a water-quality objective of 5 μg Cu/liter in the receiving water (Klapow, 1978). This objective should adequately protect adult oysters because the estimated incipient lethal concentration for adults is 230 μg Cu/liter.

Organism mortalities resulting from heavy metals may be caused by (1) excessive amounts of a metal in tissues, (2) excessive rates of influx into tissues, or (3) inhibition of a vital metabolic process. In the first case, if the influx of copper into an organism is greater than its efflux, increased levels result; lethal amounts may accumulate in all tissues or in a special storage depot. We tested for the first possibility by copper analyses on the experimental animals and for the second possibility with ^{64}Cu-uptake experiments.

The results of copper analyses on test oysters show that the highest concentrations of copper were in oysters at 230 μg Cu/liter rather than at the higher concentrations in seawater. Also, levels of copper in oysters held in water containing more than 230 μg Cu/liter did not increase with increasing concentrations in the water but were relatively constant. The duration of exposure was not always the same for oysters held at the same and at different concentrations, however. For example, exposure for oysters in water containing 440 μg Cu/liter ranged from 3.9 to 8.3 days. The concentrations in this group of animals did not increase consistently with increased exposure. The durations of exposure and the tissue concentrations (Cu/g wet weight) were: 3.9 days, 37.4 μg; 4.2 days, 48.8 μg; 5.2 days, 92.1 and 86.5 μg; 5.9 days, 82.5 μg; 6.0 days, 77.6, 90.4, and 107 μg; 6.5 days, 48 μg; and 8.3 days, 68.2 μg. At 230 μg Cu/liter, the two animals that died at 8.8 and 9.4 days contained 60.0 and 96.0 μg Cu/g wet weight, respectively. Copper may be lost from animals near death as a result of permeability changes. The hearts of many oysters held in water containing high concentrations of copper were observed beating when they were removed from the shell; no significant difference in copper content was found in animals with and without a firm heartbeat. Permeability changes may occur before the heart is affected, however. Consequently, the results of the copper analysis do not demonstrate conclusively that mortality was caused by an excessive amount of copper in the tissues.

Results of the ^{64}Cu experiment indicate that the flux of copper is much greater in oysters maintained at 800 and 2400 μg Cu/liter than in the controls in seawater. If the concentration of copper in oyster tissues were regulated, the flux should remain constant with changes in copper concentration in seawater. Since this did not occur, strict homeostatic control probably does not take place. The decreased rate of ^{64}Cu accumulation with increased copper concentration does suggest a small amount of regulation of copper intake, however. Whether this is the result of changes in transport across membranes or changes in the circulation of seawater by the oyster is not known.

Coombs (1974) reported that copper in oyster tissues occurs in at least two different forms, one a readily solubilized component and the second firmly bound to tissues. No information is available on which fraction is related to mortality.

ACKNOWLEDGMENTS

The research reported here was performed under the auspices of the U. S. Nuclear Regulatory Commission, research order No. 60-76-144, and the U. S. Department of Energy, contract No. W-7405-ENG-48.

We thank Jack Dawson, Rose Carrillo, and James Alexander for their assistance in this investigation and the Coast Oyster Company, Eureka, Calif., for supplying the oysters.

Reference to a company or product name does not imply approval or recommendation of the product to the exclusion of others that may be suitable.

REFERENCES

American Public Health Association, 1975, *Standard Methods for the Examination of Water and Wastewater*, 14th ed., Washington, D. C.

Ayling, G. M., 1974, Uptake of Cadmium, Zinc, Copper, Lead, and Chromium in the Pacific Oyster, *Crassostrea gigas*, Grown in the Tamar River, Tasmania, *Water Res.*, 8: 729-738.

Becker, C. D., and T. O. Thatcher, 1973, *Toxicity of Power Plant Chemicals to Aquatic Life*, USAEC Report WASH-1249, U. S. Atomic Energy Commission, GPO.

Boyden, C. R., and M. G. Romeril, 1974, A Trace Metal Problem in Pond Oyster Culture, *Mar. Pollut. Bull.*, 5: 75-78.

Brooks, R. R., and M. G. Rumsby, 1965, The Biogeochemistry of Trace Element Uptake by Some New Zealand Bivalves, *Limnol. Oceanogr.*, 10: 521-527.

California State Water Resources Control Board, 1972, *Water Quality Control Plan, Ocean Waters of California*, Table B, Resolution No. 72-45, Sacramento.

Coombs, T. L., 1974, The Nature of Zinc and Copper Complexes in the Oyster *Ostrea edulis*, *Mar. Biol.*, 28: 1-10.

Cronin, L. G., D. W. Pritchard, J. R. Schubel, and J. A. Sherk, 1974, *Metals in Baltimore Harbor and Upper Chesapeake Bay and Their Accumulation by Oysters*, Chesapeake Bay Institute, The Johns Hopkins University, and Chesapeake Biological Laboratory, University of Maryland, Baltimore.

Davenport, J., 1977, A Study of the Effects of Copper Applied Continuously and Discontinuously to Specimen of *Mytilus edulis* (L.) Exposed to Steady and Fluctuating Salinity Levels, *J. Mar. Biol. Assoc. U. K.*, 57: 63-74.

Dorband, W. R., J. C. Van Olst, J. M. Carlberg, and R. F. Ford, 1976, *Effects of Chemicals in Thermal Effluent on* Homarus americanus *Maintained in Aquaculture Systems*, Contribution No. 15, Center for Marine Studies, San Diego State University, San Diego, Calif.

Drifmeyer, J. E., 1974, Zn and Cu Levels in the Eastern Oyster, *Crassostrea virginica*, from the Lower James River, *J. Wash. Acad. Sci.*, 64: 292-294.

Establier, R., 1972, Copper Content of Oysters (*Crassostrea angulata*) Along the Huelva Coast, *Invest. Pesq.*, 36: 293-296.

——, and E. Pascual, 1974, Copper, Iron, Manganese, and Zinc in Oysters (*Crassostrea angulata*) of the Gulf of Cadiz, *Invest. Pesq.*, 38: 371-385.

Frazier, J. M., 1975, The Dynamics of Metals in the American Oyster, *Crassostrea virginica*. I. Seasonal Effects, *Chesapeake Sci.*, 16: 162-175.

——, 1976, The Dynamics of Metals in the American Oyster, *Crassostrea virginica*. II. Environmental Effects, *Chesapeake Sci.*, 17: 188-197.

Fujiya, M., 1960, Studies on the Effects of Copper Dissolved in Seawater on Oysters, *Bull. Jpn. Soc. Sci. Fish.*, 26: 462-468.

Galtsoff, P. S., 1932, The Life in the Ocean from a Biochemical Point of View, *J. Wash. Acad. Sci.*, 22: 246-257.

——, 1964, *The American Oyster* Crassostrea virginica *(Gmelin)*, Fishery Bulletin 64, U. S. Fish and Wildlife Service, GPO.

Huggett, R. J., M. E. Bender, and H. D. Stone, 1973, Utilizing Metal Concentration Relationships in the Eastern Oyster (*Crassostrea virginica*) to Detect Heavy Metal Pollution, *Water Res.*, 7: 451-460.

——, F. A. Cross, and M. E. Bender, 1975, Distribution of Copper and Zinc in Oysters and Sediments from Three Coastal-Plain Estuaries, in *Mineral Cycling in Southeastern Ecosystems*, ERDA Symposium Series, Augusta, Ga., May 1—3, 1974, F. G. Howell, J. B. Gentry, and M. H. Smith (Eds.), pp. 224-238, CONF-740513, NTIS.

Klapow, L. A., 1978, California State Water Resources Control Board, Sacramento, personal communication.

Mackay, N. J., R. J. Williams, J. L. Kacprzac, M. N. Kazacos, A. J. Collins, and E. H. Auty, 1975, Heavy Metals in Cultivated Oysters (*Crassostrea com-*

mercialis, Saccostrea cucullata) from the Estuaries of New South Wales, *Aust. J. Mar. Freshwater Res.*, 26: 31-46.

Mytelka, A. I., J. S. Czachor, W. B. Guggino, and H. Golub, 1973, Heavy Metals in Wastewater and Treatment Plant Effluents, *J. Water Pollut. Control Fed.*, 45: 1859-1864.

Okazaki, R. K., 1976, Copper Toxicity in the Pacific Oyster *Crassostrea gigas*, *Bull. Environ. Contam. Toxicol.*, 16: 658-664.

Pringle, B. H., D. E. Hissong, E. L. Katz, and S. T. Mulawka, 1968, Trace Metal Accumulation by Estuarine Mollusks, *J. Sanit. Eng. Div., Am. Soc. Civ. Eng.*, 94: 455-475.

Ratkowsky, D. A., S. J. Thrower, I. J. Eustace, and J. Olley, 1974, A Numerical Study of the Concentration of Some Heavy Metals in Tasmanian Oysters, *J. Fish. Res. Board Can.*, 31: 1165-1171.

Riley, J. P., and D. Taylor, 1968, Chelating Resins for the Concentration of Trace Metals from Seawater and Their Analytical Use in Conjunction with Atomic Absorption Spectrophotometry, *Anal. Chim. Acta*, 40: 479-485.

Shuster, C. N., Jr., and B. H. Pringle, 1969, Trace Metal Accumulation by the American Eastern Oyster, *Crassostrea virginica*, *Proc. Nat. Shellfish Assoc.*, 59: 91-103.

Snedecor, G. W., and W. C. Cochran, 1967, *Statistical Methods*, 6th ed., Iowa State University Press, Ames.

Sprague, J. B., 1969, Measurement of Pollution Toxicity to Fish. I. Bioassay Methods for Acute Toxicity, *Water Res.*, 3: 793-831.

Sunda, W., and R. R. L. Guillard, 1976, The Relationship Between Cupric Ion Activity and the Toxicity of Copper to Phytoplankton, *J. Mar. Res.*, 34: 311-329.

Thrower, S. J., and I. J. Eustace, 1973, Heavy Metal Accumulation in Oysters Grown in Tasmanian Waters, *Food Technol. Aust.*, 25: 546-553.

Warrick, J. W., S. C. Sharp, and S. J. Friedrich, 1975, *Chemical, Biological, and Corrosion Investigations Related to the Testing of the Diablo Canyon Unit 1 Cooling Water System*, Report No. 7333-129-75, Pacific Gas and Electric Company, San Ramon, Calif.

Windom, H. L., and R. G. Smith, 1972, Distribution of Iron, Magnesium, Copper, Zinc, and Silver in Oysters Along the Georgia Coast, *J. Fish. Res. Board Can.*, 29: 450-452.

AN IN SITU STUDY OF CADMIUM STRESS IN A NATURAL ZOOPLANKTON COMMUNITY

J. S. MARSHALL and D. L. MELLINGER
Ecological Sciences Section, Radiological and Environmental Research Division, Argonne National Laboratory, Argonne, Illinois

ABSTRACT

The effects of elevated cadmium concentrations on a natural zooplankton community were studied in situ in Green Bay, Lake Michigan, during the summer of 1976, with polyethylene carboys as enclosures. The results of five experiments showed that cadmium concentrations as low as 5 μg/liter, the lowest concentration tested, caused large and highly significant effects ($P < 0.001$) on both functional and structural attributes of the zooplankton community within 9 days. The effects of enclosure up to 15 days were mostly insignificant. Functional responses, as indicated by changes in the different species' rates of increase, showed highly significant differences ($P < 0.001$) among total Cladocera, Calanoida, and Cyclopoida, but no differences among species in each of these major groups. Structural responses included reduced total abundance, species diversity (3 indexes), and community similarity (2 indexes). The results indicate that added cadmium concentrations much lower than 5 μg/liter probably would cause detectable effects within 9 days and that incubations up to 15 days are feasible.

The need for experimental studies of pollutant stress in whole communities and ecosystems, as opposed to lower levels of organization, has been emphasized recently by several ecologists (Barrett, Van Dyne, and Odum, 1976; Jernelov and Rosenberg, 1976; Fisher and Wurster, 1974). The usefulness of in situ experiments in studies of toxic stress in a marine ecosystem was shown by Reeve et al. (1976), who concluded that the zooplankton community was probably the component most affected. Although it has been possible to maintain nearly normal zooplankton communities in enormous enclosures (18,000 m^3) up to 22 months in a small lake (Smyly, 1976), the use of large enclosures in the Great Lakes has been limited by their high unit cost and susceptibility to storm damage (Schelske and Stoermer, 1972).

Cadmium has long been recognized as a highly toxic element, but only recently has concern been expressed over the impact of anthropogenic emissions of cadmium into the environment. Man's activities now appear to be contributing more cadmium to streams than is contributed by natural processes (Fleischer et al., 1974). Unfortunately, the concentrations and distribution of cadmium in the Great Lakes remain poorly known. The average concentration of cadmium in Lake Ontario is 0.09 μg/liter (Chau et al., 1970). In the heavily polluted, extreme southern portion of Green Bay, Lake Michigan, cadmium concentrations ranged from 1.0 to 1.4 μg/liter during 1973 (Wiersma et al., 1974). Recent analyses at Argonne National Laboratory indicate, however, that the concentration of cadmium in the main body of Lake Michigan is less than 0.1 μg/liter (Tisue, 1976).

The purpose of our study was to determine the short-term effects of increased cadmium concentrations on the zooplankton component of the Lake Michigan ecosystem. We used an in situ method designed to overcome some of the limitations of large enclosures for experiments in large lakes. Following guidelines recommended by Barrett, Van Dyne, and Odum (1976) for ecosystem stress studies, we studied both functional and structural attributes of the zooplankton component under seminatural conditions and evaluated the relative sensitivities of different indexes of stress.

MATERIALS AND METHODS

Field Methods

During the summer of 1976, five in situ cadmium-enrichment experiments were conducted in northern Green Bay, Lake Michigan. Two sampling and incubation stations were located where the bottom depth is approximately 30 m, 1 km northwest of the larger of the Sister Islands off Sister Bay, Wisconsin. Water was collected from different depths in the lower half of the epilimnion (6 to 12 m) with a nonmetallic, 16-liter Kemmerer bottle and composited in a large mixing tub. In each experiment, two to four opaque polyethylene carboys (8- or 20-liter capacity) for each of five levels of added $CdCl_2$ were filled gradually with water from the mixing tub by adding small portions in turn to each carboy. Two to four additional carboys, which were not to be incubated, were filled at the same time. When all the carboys were nearly filled, those to be incubated were assigned prescribed cadmium additions at random. The concentrations of added cadmium in the five experiments were 0, 50, 100, 150, and 200 μg/liter in experiments 1 to 3; 0, 25, 50, 75, and 100 μg/liter in experiment 4; and 0, 5, 10, 20, and 40 μg/liter in experiment 5. After the cadmium was added, each carboy was capped, thoroughly shaken, and finally filled completely through a small plughole in the cap to exclude all air bubbles. This last step is very important because of the propensity of small cladocerans, such as *Bosmina*, to become trapped at air—water interfaces.

For in situ incubations the carboys were suspended in the epilimnion at depths of 2 to 12 m in the sampling area. Inclusive dates of the incubation periods for the five experiments were 1, July 22—26; 2, July 27—August 1; 3, August 20—27; 4, August 27—September 11; and 5, September 11—20. At the end of each incubation, the zooplankton in each carboy were removed by filtering through a nylon net with apertures of either 153 μm (experiments 1 to 3) or 85 μm (experiments 4 and 5), narcotized with "club soda" (Gannon and Gannon, 1975), and preserved in 4% Formalin.

Laboratory Methods

The zooplankton were identified with the aid of keys, illustrations, and descriptions in Edmondson (1959) and Torke (1974) by use of binocular compound and dissecting microscopes. The identification categories in each entire sample were enumerated in an open-top, chambered counting cell (Gannon, 1971) with a binocular dissecting microscope. Special care was taken to count only specimens that showed no signs of decomposition.

Calculations and Definitions of Symbols

As in the work of Edmondson (1968), rates of increase, r, were calculated from the equation

$$r = \frac{\ln N_t - \ln \overline{N}_i}{\Delta t} \qquad (1)$$

where N_t is the terminal number of individuals of a given identification category in an incubated sample, $\overline{N}_i$ is the average initial number in the unincubated samples, and Δt is the duration of incubation in days.

Three indexes of species diversity were calculated according to Hill (1973):

$$N_0 = \text{total number of identification categories} \tag{2}$$

$$N_1 = \exp(-\Sigma\, P_i \ln P_i) \tag{3}$$

$$N_2 = \frac{1}{\Sigma P_i^2} \tag{4}$$

where P_i is the proportional abundance of the different identification categories in a given sample.

Two indexes of community similarity, S and S , were calculated according to Whittaker (1972):

$$S = \frac{2C}{A + B} \tag{5}$$

where A is the number of identification categories in the control samples, B is the number of categories in any single sample, and C is the number common to both; and

$$S' = 1 - 0.5\, \Sigma\, |\Delta P_i| \tag{6}$$

where ΔP_i is the difference between the proportional abundance of a given identification category in any single sample and its average proportional abundance in the controls.

RESULTS

The effects of cadmium on functional attributes of the community, as indicated by the relationships among average rates of increase ($\bar{r}$) and added cadmium (micrograms per liter) for the predominant zooplankton groups and component species, were more consistent and much more pronounced than the effects of enclosure, which are discussed later. The linear regression coefficients (slopes) for relationships among average rates of increase and added cadmium (0 to 50 μg/liter) for the three major groups of crustacean zooplankton in each of the five experiments are shown in Table 1. Friedman's analysis of variance by ranks (Siegel, 1956) indicates that the differences between slopes for cladocerans, calanoid copepods, and cyclopoid copepods are highly significant ($P < 0.001$). The mean regression coefficients for the Cladocera, Calanoida, and Cyclopoida

TABLE 1

INCUBATION (EXPOSURE TIME) AND LINEAR REGRESSION COEFFICIENTS FOR RELATIONSHIPS AMONG RATES OF INCREASE AND ADDED CADMIUM FOR MAJOR GROUPS OF CRUSTACEANS FOR IN SITU EXPERIMENTS IN LAKE MICHIGAN*

Experiment	Exposure time, days	Cladocera	Calanoida	Cyclopoida
1	4	−0.0212	−0.0049	−0.0004
2	5	−0.0202	−0.0057	−0.0005
3	7	−0.0190	−0.0084	−0.0006
4	15	−0.0059	−0.0031	−0.0018
5	9	−0.0075	−0.0071	−0.0015

*Added cadmium ranged from 0 to 50 μg/liter.

are −0.0131 ± 0.0032, −0.0058 ± 0.0009, and −0.0010 ± 0.0003, respectively (mean ± standard errors).

The effects of cadmium on different species or other categories within each major crustacean group in experiment 5 are represented by the linear regression equations and corresponding lines fitted to the mean values, $\bar{r}$ (Fig. 1). The data from which these values were calculated are summarized in Table 2; summaries of the data from experiments 1 to 4 are available (Marshall and Van Reken, 1977). Differences in apparent sensitivity among species or other categories within each major crustacean group, as indicated by the differences in the slopes (steepness) of the lines in Fig. 1, are quite small in comparison with the differences among the major groups. The immature copepodites of both calanoid and cyclopoid copepods appear to be no more sensitive to cadmium than the adults are. Furthermore, the mean number of copepod nauplii in samples at each added cadmium level (Table 2) was not significantly different ($P > 0.1$) from that of the controls.

At higher ranges of added cadmium (> 50 μg/liter) in experiments 3 and 4, the relationships between $\bar{r}$ and cadmium are better represented by curves (not shown) whose slopes decrease with increasing cadmium. This is most pronounced in Cladocera and is noticeable even for the 0 to 20 μg Cd/liter range in the data for Bosminidae plus Daphnidae in Fig. 1. Even in this range the data are fitted better by an exponential function. In other words, the sensitivity of Cladocera in the range of 0 to 5 μg Cd/liter is probably greater than that indicated by the slope of the line in Fig. 1.

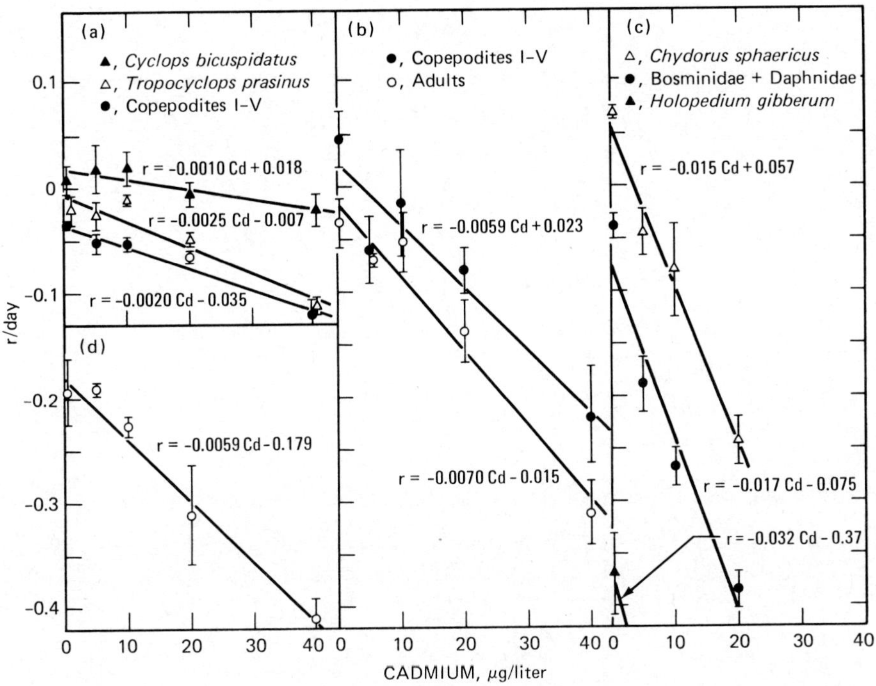

Fig. 1 Effects of cadmium on rate of increase, r, of (a) cyclopoid copepods, (b) calanoid copepods, (c) cladocerans, and (d) large rotifers.

The mean numbers of nauplii, several rotifer species, and one large protozoan in the samples for experiment 5 are shown in Table 2. The effect of added cadmium on the large rotifers is shown in Fig. 1(d) by the linear regression equation and the corresponding line fitted through the average rates of increase, $\bar{r}$, for the large rotifers in experiment 5. The apparent cadmium sensitivity of these rotifers, as indicated by the slope of the line, is comparable to that of calanoid copepods. Differences among mean numbers of *Ceratium hirundinella* in incubated controls and unincubated samples indicated a pronounced effect of enclosure on this protozoan. The mean numbers remaining in incubated samples containing different levels of added cadmium were not significantly different from the controls (Table 2).

The effects of cadmium on structural attributes of the crustacean zooplankton community are exemplified by decreases in total abundance, relative abundance of the major groups, species diversity,

TABLE 2

NUMBER OF ZOOPLANKTON PER LITER (x 10) IN A CADMIUM-ENRICHMENT EXPERIMENT WITH IN SITU INCUBATION IN LAKE MICHIGAN*

Zooplankton	Unincubated samples	Incubated samples with added cadmium, μg/liter				
		0	5	10	20	40
Cladocera						
Leptodora kindtii	0.4 ± 0.4	0.4 ± 0.4	0	0	0	1 ± 1
Holopedium gibberum	49 ± 6	1 ± 1	0.4 ± 0.4	0	0	0
Bosmina longirostris	37 ± 3	39 ± 3	22 ± 8	18 ± 1	5 ± 1	5 ± 1
Eubosmina coregoni	144 ± 13	125 ± 9	3 ± 1	4 ± 1	2 ± 1	1 ± 1
Daphnia retrocurva	87 ± 10	42 ± 4	22 ± 6	1 ± 1	1 ± 1	4 ± 2
D. galeata mendotae	13 ± 1	10 ± 5	7 ± 2	5 ± 2	3 ± 1	0.3 ± 0.3
D. longiremis	2 ± 2	3 ± 3	0.4 ± 0.4	0	0	0
Ceriodaphnia lacustris	15 ± 1	11 ± 4	3 ± 1	1 ± 1	0	0.3 ± 0.3
Chydorus sphaericus	50 ± 1	94 ± 5	35 ± 7	26 ± 10	4 ± 2	8 ± 2
Copepoda: Cyclopoida						
Cyclops vernalis	2 ± 1	3 ± 1	5 ± 2	5 ± 3	3 ± 1	7 ± 1
C. bicuspidatus thomasi	37 ± 9	41 ± 5	47 ± 11	45 ± 6	36 ± 4	32 ± 4
Mesocyclops edax	6 ± 1	15 ± 2	14 ± 2	9 ± 1	9 ± 2	11 ± 2
Tropocyclops prasinus	47 ± 1	40 ± 5	38 ± 5	43 ± 2	31 ± 2	17 ± 1
Copepodites I–V	231 ± 14	169 ± 2	146 ± 14	144 ± 8	128 ± 6	95 ± 13
Copepoda: Calanoida						
Eurytemora affinis	2 ± 1	2 ± 1	3 ± 2	1 ± 0	1 ± 1	1 ± 1
Diaptomus spp.†	9 ± 1	12 ± 1	6 ± 1	9 ± 3	5 ± 2	0.4 ± 0.4
Copepodites I–V	14 ± 1	15 ± 3	8 ± 3	9 ± 4	5 ± 1	1 ± 1
Copepoda						
Nauplii	244 ± 10	115 ± 21	138 ± 34	150 ± 26	120 ± 4	80 ± 10
Rotifera						
Keratella cochlearis	172 ± 11	31 ± 8	35 ± 3	22 ± 1	9 ± 3	4 ± 1
Asplanchna priodonta	54 ± 4	5 ± 3	5 ± 3	4 ± 2	0.4 ± 0.4	0
Ploesoma spp.	15 ± 1	18 ± 3	11 ± 3	11 ± 3	12 ± 3	2 ± 1
Polyarthra spp.	25 ± 1	5 ± 3	2 ± 1	1 ± 1	1 ± 1	1 ± 1
Trichocerca spp.	16 ± 4	0	0	0	0	0
Protozoa						
Ceratium hirundinella	1183 ± 104	7 ± 2	1 ± 1	2 ± 2	10 ± 3	11 ± 3

*The experiment was conducted Sept. 11–12, 1976; values are mean ± standard error.
†*D. oregonensis*, *D. minutus*, and *D. ashlandi*.

and community similarity. The results for experiment 5 are illustrated in Fig. 2.

Changes in total abundance of crustacean zooplankton (minus copepod nauplii) and relative abundances of the major crustacean groups are shown in Fig. 2(a). The relationship between total abundance and added cadmium is represented by the uppermost line, the slope of which decreases with increasing cadmium. Thus total abundance is most sensitive to cadmium in the lowest segment of the range (0 to 5 μg/liter). Total abundance at 5 μg Cd/liter was reduced by 42% from that in the controls, and the difference is highly significant ($P < 0.002$); whereas, from 5 to 10 μg/liter, it was reduced only 11% further.

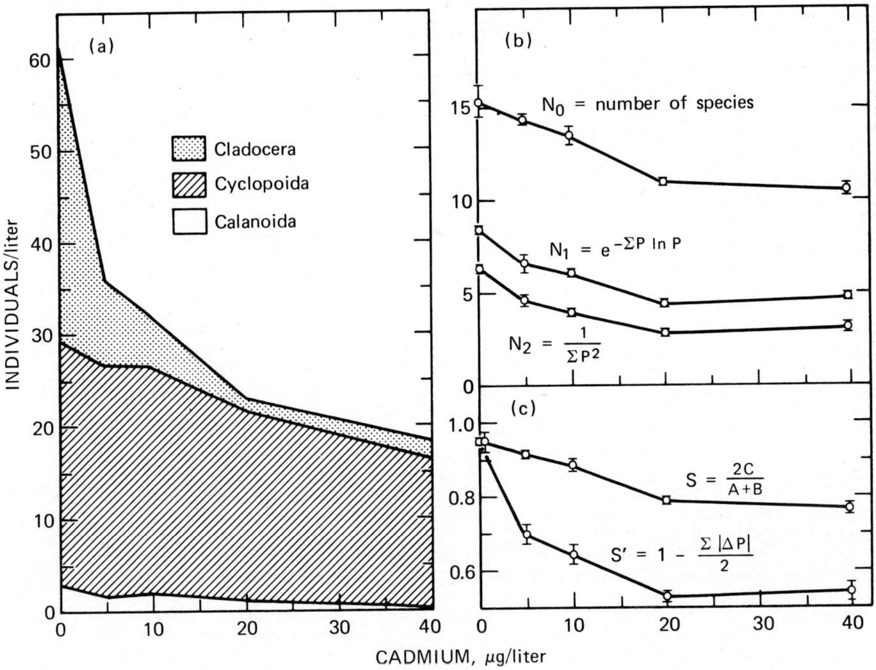

Fig. 2 Effects of cadmium on (a) abundance of major zooplankton groups, (b) species diversity, and (c) community similarity. (See text for definitions of symbols.)

Changes in relative abundance of the major crustacean groups were primarily an increase in the percentage of cyclopoid copepods and a decrease in that of the cladocerans. Again, these changes were greatest in the lowest portion of the overall range. From 0 to 5 μg Cd/liter, the relative abundance of cyclopoid copepods increased from 43 to 70%, cladocerans decreased from 52 to 25%, and calanoid copepods remained at 4 to 5%.

The effects of cadmium on species diversity, as indicated by the mean values of Hill's (1973) diversity numbers in the range of 0 to 40 μg Cd/liter, are shown in Fig. 2(b). The mean number of identification categories, $\overline{N}_0$, decreased with increasing cadmium although the means at 5 and 10 μg/liter are not significantly different from the controls. The mean diversity numbers $\overline{N}_1$ and $\overline{N}_2$ also decreased with increasing cadmium but were more sensitive than $\overline{N}_0$ in the range of 0 to 5 μg Cd/liter. The values of $\overline{N}_1$ and $\overline{N}_2$ at 5 μg Cd/liter are both significantly different ($P < 0.05$) from the controls.

Changes in community similarity, as measured by indexes S and S′, are shown in Fig. 2(c). Both S and S′ decreased with increasing cadmium, but S′ was by far the more sensitive index, especially in the range of 0 to 5 μg/liter. The mean value of S′ at 5 μg/liter is reduced to 73% of that for the controls, and the difference is highly significant ($P < 0.001$); whereas the mean value of S at 5 μg/liter is not significantly different.

The effects of enclosure, in contrast to the effects of cadmium, were mostly insignificant. This is shown by the results of t-tests of the significance of differences between mean numbers in the incubated controls and those in unincubated samples taken either at the end of experiments 1, 3, and 5 (48 tests) or at the beginning of each experiment (80 tests). These results are summarized in Table 3. Both sets of tests indicated that most of the differences (58 to 62%) were not significant ($P > 0.05$). The two sets of tests also indicated that 35 to 37% of the significant differences were caused by larger populations in the incubated controls as compared with the unincubated samples. Most of the significant differences, furthermore, are accounted for by relatively few species. *Holopedium gibberum*, *Eubosmina coregoni*, and immature copepods account for 54 to 68% of the significant decreases, and *Chydorus sphaericus* and *Cyclops bicuspidatus thomasi* account for 57% of the significant increases (Table 3).

DISCUSSION

The results of this study showed that added cadmium concentrations as low as 5 μg/liter caused pronounced effects on both functional and structural attributes of a Lake Michigan zooplankton community in 9 days, but the effects of enclosure were relatively minor up to 15 days. The results indicate further that added cadmium concentrations less than 5 μg/liter probably would cause detectable effects within 9 days and that incubations up to 15 days are feasible.

The responses of species' r to environmental changes are, collectively, a fundamental functional attribute of a plankton community because these responses affect mineral-cycling and energy-flow pathways by determining the absolute and relative abundances of the different species. In this study the observed effects of cadmium on values of r for the different species were probably caused mostly by direct toxicity because the effect in all cases was a reduction of r. Furthermore, the duration of the experiments was probably too short for secondary effects to manifest themselves. Hurlbert (1975) defined secondary effects as those which take place in an ecosystem following and as a result of

TABLE 3

RESULTS OF T-TESTS OF DIFFERENCES BETWEEN MEAN NUMBERS OF VARIOUS CRUSTACEANS IN INCUBATED CONTROL SAMPLES AND UNINCUBATED SAMPLES TAKEN AT INITIATION AND TERMINATION OF IN SITU INCUBATION*

	Incubated controls vs. unincubated initial samples (80 tests)			Incubated controls vs. unincubated terminal samples (48 tests)		
Crustacean	0	—	+	0	—	+
Cladocera						
Holopedium gibberum	1	4	0	1	2	0
Bosmina longirostris	3	2	0	2	1	0
Eubosmina coregoni	2	3	0	1	2	0
Daphnia longiremis	4	1	0	3	0	0
D. galeata mendotae	5	0	0	3	0	0
D. retrocurva	2	3	0	2	1	0
Ceriodaphnia lacustris	4	0	1	3	0	0
Chydorus sphaericus	2	0	3	1	0	2
Copepoda: Cyclopoida						
Cyclops vernalis	3	0	2	2	0	1
C. bicuspidatus thomasi	4	0	1	1	0	2
Mesocyclops edax	3	0	2	2	1	0
Tropocyclops prasinus	5	0	0	0	2	1
Copepodites I—V	0	4	1	1	2	0
Copepoda: Calanoida						
Eurytemora affinis	4	0	1	2	1	0
Diaptomus spp.†	5	0	0	2	0	1
Copepodites I—V	3	2	0	2	1	0

*The number of tests with no significant differences are shown as 0; those with a significant difference ($P < 0.05$) are shown as — for decrease or + for increase of the incubated control populations.

†*D. oregonensis*, *D. minutus*, *D. ashlandi*, and *D. siciloides*.

direct effects on growth, survival, or reproduction of the sensitive species. He also stated that toxicity is the parsimonious explanation for any population decline that is demonstrated to result from exposure.

The observed differences in apparent cadmium toxicity to the different major groups of zooplankton and the similar sensitivities of species within each major group are in agreement with the results of previous studies of heavy-metal toxicity to zooplankton. McIntosh and Kevern (1974) found that in copper-treated ponds both cladoceran and rotifer populations were more sensitive to copper than the cyclopoid copepods. In a laboratory study of the acute toxicity of cadmium and other heavy metals to three species of crustacean zooplankton, Baudouin and Scoppa (1974) found that the cladoceran *Daphnia hyalina* was considerably more sensitive than the two copepod species and that the calanoid *Eudiaptomus padanus* was more sensitive than the cyclopoid *Cyclops abyssorum*. They found only minor differences in the rank orders of the metals' toxicities to these species, and their results for *D. hyalina* were in good agreement with those reported for *D. magna* by Biesinger and Christensen (1972). The observed similarity of cadmium effects on different species of Cladocera agrees with the findings of Winner and Farrell (1976) that four species of *Daphnia* did not differ in their sensitivity to copper.

The observed effects of cadmium on structural attributes of a Lake Michigan zooplankton community are qualitatively similar to those found in studies of long-term effects of other stresses in natural communities (Woodwell, 1970). The long-term effects of cadmium on the Lake Michigan zooplankton community would probably be more pronounced than the effects observed in this study, however, because heavy-metal toxicity tends to increase with chronic exposure. For example, chronic cadmium toxicity to laboratory populations of *Daphnia galeata mendotae* increases with time, and 23 weeks are required to establish equilibrium relationships (Marshall, 1978).

The various indexes of community structure differed considerably in their sensitivity to cadmium. Total zooplankton abundance, the most sensitive index, was reduced by 42% in 9 days at 5 μg Cd/liter, whereas the most sensitive indexes of species diversity and community similarity, N_2 and S', were reduced by only 29 and 27%, respectively. The diversity number N_2 was the most sensitive index of species diversity because it is the one most heavily weighted by the more abundant species (Hill, 1973). This also explains the different sensitivities of the two indexes of similarity, S and S'. The indexes of similarity account for changes in relative proportions of each species, whereas the diversity number accounts only for presence or absence of species. Although total abundance was the most sensitive index of community structure in our study, species diversity or community similarity would probably be more sensitive to long-term effects of chronic, low-level cadmium stress because total zooplankton abundance would probably be partially restored, given time, by more tolerant species.

There are few previous studies of heavy-metal effects on aquatic ecosystems with which to compare our results, and none dealing with the effects of cadmium. The most relevant study of heavy metal stress is that of Reeve et al. (1976) on the short-term effects of copper on a marine zooplankton community in enclosed water columns. Reduced abundance of total zooplankton was apparently the most sensitive index of copper stress, just as it was for cadmium in our study. Reeve and co-workers were not able to detect a statistically significant reduction of either Shannon's index (H) or percentage similarity (S′) caused by copper concentrations up to 50 μg/liter because of pronounced enclosure effects. Total phytoplankton, as measured by chlorophyll a, increased in the copper-treated enclosures because of reduced zooplankton grazing (Reeve et al., 1976). The effects of various heavy metals and other pollutants on functional and structural changes in freshwater algae and protozoan communities include reduced total abundance and increased relative abundance of tolerant species (Cairns, Lanza, and Parker, 1972). The effects of copper have probably been studied more than those of any other heavy metal because it is frequently used as an aquatic herbicide, but there are few studies even of copper effects on zooplankton (McIntosh and Kevern, 1974). Heavy metals associated with acid precipitation have been implicated in the reduction of biomass and species diversity of phytoplankton, zooplankton, and fish communities in lakes chronically exposed to these inputs (Gorham, 1976). The well-documented reductions of abundance and species diversity of zooplankton and fish communities in the LaCloche Mountain lakes near the mining and smelting region of Sudbury, Ontario, appear to be caused primarily by acid stress rather than by the associated heavy metals (Beamish, 1976; Sprules, 1975).

Reduced biomass and diversity appear to be common symptoms of stress caused by many kinds of disturbances, at least in terrestrial ecosystems (Woodwell, 1970). Not all stresses have the same effects, however. Eutrophication is frequently referred to as a stress, yet its effects are just the opposite of those of oligotrophication (Hutchinson, 1973), a process that is claimed to occur in lakes subjected to excessive acid loading (Grahn, Hultberg, and Lander, 1974). Thus different stress syndromes are more likely to be found for different classes of stressors, e.g., toxic substances. Universally toxic substances, such as cadmium or strong acids, are even more likely to have similar and predictable effects than are more-selective toxicants, such as pesticides (Hurlbert, 1975).

The effects of heavy metals and other toxic substances on aquatic communities in the Great Lakes have been rated low in importance relative to other stresses, including uncontrolled commercial fishing, erosion and nutrient loading, invading species, and stream destruction and shoreline restructuring (Regier and Hartman, 1973). Certainly any reduction of plankton abundance which may have been caused by toxic substances in recent decades has been masked by dramatic effects of eutrophication and fish predation. In Lake Michigan, alewife predation has had a dominant influence on changes in the zooplankton community (Wells, 1970) although eutrophication may have been responsible for counteracting the effects of alewife predation in Green Bay (Gannon, 1972). Eutrophication has not only increased the total abundance of Lake Michigan phytoplankton (Beeton, 1965) but has also changed the species composition (Schelske and Stoermer, 1972). As long as such dramatic changes are taking place in the Great Lakes, there seems to be little chance of detecting the effects of toxic substances by direct observation (monitoring), except in the vicinity of strong local sources.

The results of this study indicate that the use of relatively small enclosures overcomes some of the limitations of large enclosures for short-term experiments in large lakes. For studies of pollutant effects on zooplankton, at least for short-term effects, the in situ method used in this study has several advantages over other methods. Recent work indicates that effects of enclosure in opaque polyethylene carboys in longer experiments become excessive, but effects of enclosure in translucent carboys incubated for 3 weeks at optimal depths are not much greater than effects of large enclosures of various kinds (Marshall and Mellinger, 1978a). Longer in situ experiments still do not appear to be feasible in the Great Lakes, however. Therefore, future determinations of relationships between short- and long-term stress effects in laboratory populations of species that live in these lakes (Marshall, 1978) and the same relationships in zooplankton communities of small, experimental lakes (Marshall and Mellinger, 1978b) will be useful in assessing the long-term significance of the results of in situ experiments in large lakes.

ACKNOWLEDGMENTS

The research reported here was performed under the auspices of the U. S. Department of Energy.

We thank J. A. Zischke and R. Van Reken for assistance in various phases of the laboratory and field work. Helpful reviews of an earlier version of this paper were provided by A. L. Buikema and W. T. Edmondson.

REFERENCES

Barrett, G. W., G. M. Van Dyne, and E. P. Odum, 1976, Stress Ecology, *Bioscience*, 26: 192-194.

Baudouin, M. F., and P. Scoppa, 1974, Acute Toxicity of Various Metals to Freshwater Zooplankton, *Bull. Environ. Contam. Toxicol.*, 12: 745-751.

Beamish, R. J., 1976, Acidification of Lakes in Canada by Acid Precipitation and the Resulting Effects on Fishes, *Water, Air, Soil Pollut.*, 6: 501-514.

Beeton, A. M., 1965, Eutrophication of the St. Lawrence Great Lakes, *Limnol. Oceanogr.*, 10: 240-254.

Biesinger, K. E., and G. M. Christensen, 1972, Effects of Various Metals on Survival, Growth, Reproduction, and Metabolism of *Daphnia magna, J. Fish. Res. Board Can.*, 29: 1691-1700.

Cairns, J., Jr., G. R. Lanza, and B. C. Parker, 1972, Pollution Related Structural and Functional Changes in Aquatic Communities with Emphasis on Freshwater Algae and Protozoa, *Proc. Acad. Nat. Sci. Philadelphia*, 124: 79-124.

Chau, T. K., V. K. Chawla, H. F. Nicholson, and R. A. Vollenweider, 1970, Distribution of Trace Elements and Chlorophyll a in Lake Ontario, *Proc., Conf. Great Lakes Res.*, 13: 659-672.

Edmondson, W. T. (Ed.), 1959, *Fresh-Water Biology*, 2nd ed., John Wiley & Sons, Inc., New York.

——, 1968, A Graphical Model for Evaluating the Use of the Egg Ratio for Measuring Birth and Death Rates, *Oecologia (Berlin)*, 1: 1-37.

Fisher, N. S., and C. F. Wurster, 1974, Impact of Pollutants on Plankton Communities, *Environ. Conserv.*, 1: 189-190.

Fleischer, M., et al., 1974, Environmental Impact of Cadmium: A Review by the Panel on Hazardous Trace Substances, *Environ. Health Perspect.*, 1974: 253-323.

Gannon, J. E., 1971, Two Counting Cells for the Enumeration of Zooplankton Micro-Crustacea, *Trans. Am. Microsc. Soc.*, 90: 486-490.

——, 1972, Effects of Eutrophication and Fish Predation on Recent Changes in Zooplankton Crustacea Species Composition in Lake Michigan, *Trans. Am. Microsc. Soc.*, 91: 82-84.

——, and S. A. Gannon, 1975, Observations on the Narcotization of Crustacean Zooplankton, *Crustaceana (Leiden)*, 28: 220-224.

Gorham, E., 1976, Acid Precipitation and Its Influence upon Aquatic Ecosystems—An Overview, *Water, Air, Soil Pollut.*, 6: 457-481.

Grahn, O., H. Hultberg, and L. Lander, 1974, Oligotrophication—A Self-Accelerating Process in Lakes Subjected to Excessive Supply of Acid Substances, *Ambio*, 3: 93-94.

Hill, M. O., 1973, Diversity and Evenness: A Unifying Notation and Its Consequences, *Ecology*, 54: 427-432.

Hurlbert, S. H., 1975, Secondary Effects of Pesticides on Aquatic Ecosystems, *Residue Rev.*, 58: 81-148.

Hutchinson, G. E., 1973, Eutrophication, *Am. Sci.*, 61: 269-279.

Jernelov, A., and R. Rosenberg, 1976, Stress Tolerance of Ecosystems, *Environ. Conserv.*, 3: 43-46.

McIntosh, A. W., and N. R. Kevern, 1974, Toxicity of Copper to Zooplankton, *J. Environ. Qual.*, 3: 166-170.

Marshall, J. S., 1978, Population Dynamics of *Daphnia galeata mendotae* as Modified by Chronic Cadmium Stress, *J. Fish. Res. Board Can.*, 35: 461-469.

——, and D. L. Mellinger, 1978a, Effects of Cadmium Enrichment on a Lake Michigan Plankton Community, in *Radiological and Environmental Research Division Annual Report*, Part 3, January—December 1977, ANL-77-65(Part 3), Argonne National Laboratory, NTIS, in press.

——, and D. L. Mellinger, 1978b, Zooplankton Responses to Cadmium Enrichment in Large Enclosures in Canada's Experimental Lakes Area Lake 223, in *Radiological and Environmental Research Division Annual Report*, Part 3, January—December 1977, ANL-77-65(Part 3), Argonne National Laboratory, NTIS, in press.

——, and R. Van Reken, 1977, A Method for the In Situ Study of Pollutant Effects on Natural Zooplankton Communities, in *Radiological and Environmental Research Division Annual Report*, Part 3, January—December 1976, ANL-76-88(Part 3), pp. 26-30, Argonne National Laboratory, NTIS.

Reeve, M. R., G. D. Grice, V. R. Gibson, M. A. Walter, K. Darcy, and T. Ikeda, 1976, A Controlled Environmental Pollution Experiment (CEPEX) and Its Usefulness in the Study of Large Marine Zooplankton Under Toxic Stress, in A. P. M. Lockwood (Ed.), *Effects of Pollutants on Aquatic Organisms*, pp. 145-162, Cambridge University Press, New York.

Regier, H. A., and W. L. Hartman, 1973, Lake Erie's Fish Community: 150 Years of Cultural Stresses, *Science*, 180: 1248-1255.

Schelske, C. L., and E. F. Stoermer, 1972, Phosphorus, Silica and Eutrophication in Lake Michigan, in *Nutrients and Eutrophication Special Symposia*, G. E. Likens (Ed.), Vol. 1, pp. 157-171, American Society of Limnology and Oceanography, Gaithersburg, Md.

Siegel, S., 1956, *Nonparametric Statistics*, McGraw-Hill Book Company, Inc., New York.

Smyly, W. J. P., 1976, Some Effects of Enclosure on the Zooplankton in a Small Lake, *Freshwater Biol.*, 6: 241-251.

Sprules, G. W., 1975, Midsummer Crustacean Zooplankton Communities in Acid-Stressed Lakes, *J. Fish. Res. Board Can.*, 32: 389-395.

Tisue, G. T., 1976, Argonne National Laboratory, personal communication.

Torke, B. G., 1974, An Illustrated Guide to the Identification of Planktonic Crustacea of Lake Michigan with Notes on Their Ecology, Special Report No. 17, Center for Great Lakes Studies, The University of Wisconsin, Milwaukee.

Wells, L., 1970, Effects of Alewife Predation on Zooplankton Populations in Lake Michigan, *Limnol. Oceanogr.*, 15: 556-565.

Whittaker, R. A., 1972, Evolution and Measurement of Species Diversity, *Taxon*, 21: 223-251.

Wiersma, J. H., I. K. Iskandar, L. J. Schwartz, R. W. Lanz, and C. C. Weber, 1974, The Effect of a Fossil Fuel Power Plant's Cooling System on the Water Quality of Green Bay, Wisconsin, *Proc., Conf. Great Lakes Res.*, 17: 535-543.

Winner, R. W., and M. P. Farrell, 1976, Acute and Chronic Toxicity of Copper to Four Species of *Daphnia*, *J. Fish. Res. Board Can.*, 33: 1685-1691.

Woodwell, G. M., 1970, Effects of Pollution on the Structure and Physiology of Ecosystems, *Science*, 168: 429-433.

THERMAL ECOLOGY AND STRESS: A CASE HISTORY FOR RED-SORE DISEASE IN LARGEMOUTH BASS

GERALD W. ESCH and TERRY C. HAZEN
Department of Biology, Wake Forest University, Winston—Salem, North Carolina; and Savannah River Ecology Laboratory, Aiken, South Carolina

ABSTRACT

The stress concept is analyzed and related to individual, population, and ecosystem levels of biological organization. Red-sore disease, produced by the gram-negative bacterium *Aeromonas hydrophila*, is discussed in terms of its relationship to stress. An effort is made to relate seasonal changes in red-sore disease to environmentally induced alterations in the physiology and behavior of largemouth bass. A hypothesis is proposed to explain these interactions in terms of stress.

ANALYSIS OF THE STRESS CONCEPT

The operative word in the title of this symposium is stress, yet we question whether many investigators have anything more than an intuitive notion of what it means. To establish a common ground and to set the stage for a discussion of our own work in this area, we feel we should first briefly discuss the concept of stress and consider its application at various levels of biological organization. We hope that the discussion of stress at individual and ecosystem levels of organization is not misconstrued as an effort to rediscover the wheel. The following conceptualization of stress represents our amalgamation of the ideas and notions presented by Selye (1950; 1956), Brett (1958), Slobodkin (1967), Odum (1969), Cairns (1976), and Gibbons (1976).

Traditional Perspective

Historically the word stress has been used by biomedical scientists to describe a somewhat vague array of physiological, morphological, and biochemical responses by an individual organism to an even more vague and less defined group of etiologies. Thus there is more certainty about how an individual organism manifests stress than there is about what causes the response to occur.

Selye (1950) described the response of an individual organism to stressor input as a succession of physiological and biochemical reactions to which he collectively referred as the general adaptation syndrome (or GAS). Selye separated the GAS into three parts, the alarm stage, the resistance stage, and the exhaustion stage.

The alarm stage begins with stressor input, which promotes the release of epinephrine into the blood vascular system from the adrenal glands and increases the activity of the sympathetic portion of the autonomic nervous system (Fig. 1). The combined action of epinephrine and the autonomic nervous system then produces a wide range of physiological and biochemical changes in the respondent

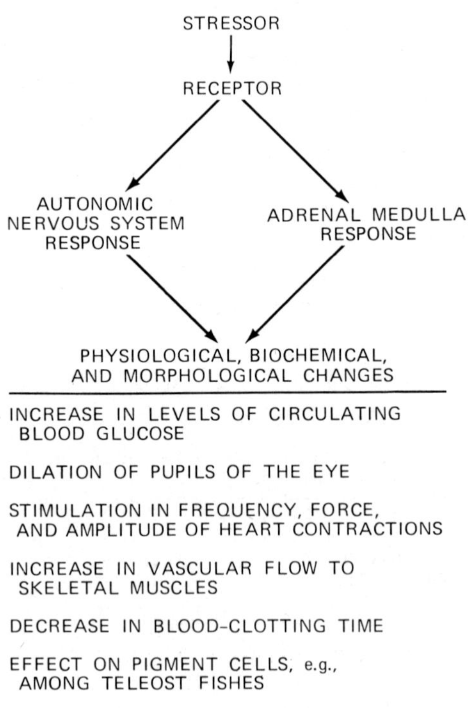

Fig. 1 Alarm stage of the general adaptation syndrome.

organism, including increasing the levels of circulating blood glucose, dilating the pupils of the eye, stimulating the frequency and force of heart contractions, etc. These kinds of responses are clearly effective in preparing an individual to respond defensively or offensively to the threat of external attack or provocation. It is appropriate that the alarm stage is sometimes called the fright—flight—fight response.

The second phase of the GAS is the resistance stage. Very soon after the alarm stage has run its course, the activity of the autonomic nervous system is diminished, and the release of epinephrine from the adrenal medulla also slows down. If the stressor persists in time, however, there is a substantial rise in the level of circulating corticosteroids produced by cells in the adrenal cortex (Fig. 2). The cortical cells are stimulated by adrenocorticotrophic hormone (ACTH), which, in turn, is produced by the pituitary gland. The stimuli for discharge of ACTH are releasing factors produced in the hypothalamus; presumably higher centers in the central nervous system are responsive to external and internal stimuli that promote the liberation of the so-called releasing factors from the hypothalamus. The function of corticosteroids is to mitigate cellular damage,

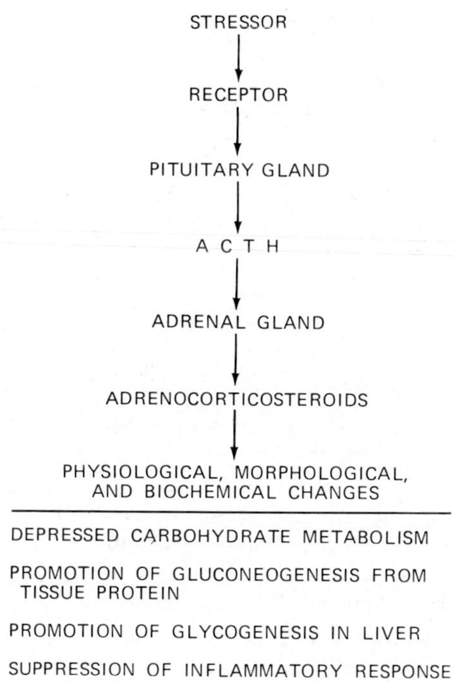

Fig. 2 Resistance stage of the general adaptation syndrome.

which could be induced as a consequence of long-term stressor input. It is also known that some corticosteroids cause significant changes in carbohydrate metabolism and suppress inflammatory reactions. According to Selye, these responses collectively provide protection against the stressor. We should note, however, that the action of corticosteroids in suppressing inflammation may actually be counterproductive since the organism simultaneously becomes more vulnerable to infection by pathogenic organisms. A significant body of literature exists which details the impact of increased corticosteroid output in terms of reducing both natural and acquired resistance (for details of the relationship between stress and parasitism, see Esch, Gibbons, and Bourque, 1975).

The third stage of the GAS, exhaustion, occurs when, after long and persistent stressor input, the cells of the adrenal cortex become exhausted. At this time there is functional and structural deterioration of the cortical cells, resulting in cessation of corticosteroid production. If this occurs, death of the stressed organism rapidly follows.

Thus, if stressor input continues over a long period of time or if it is of sufficient magnitude, there is a potential for mortality either from exhaustion of the adrenals or from side effects such as stroke, hypertension, bleeding ulcers, arthritis, and infection with pathogenic agents. We should emphasize that, although the concept of stress is accepted by most biomedical scientists, there is discussion about whether stress is necessarily related to some of the diseases mentioned.

Definitions of Stress

Stress as a process can be more or less adequately described at the individual level, but most definitions of stress appear to be less than acceptable. This is especially true since most of the definitions were developed to apply only at the individual level and only for animals. We know, however, that the stress concept can be extended to plants (Harper, 1967; Vadas et al., 1976) and to the population (George, 1977) and ecosystem (Cairns, 1976) levels of organization.

Let us consider, for example, the definition of stress offered by Selye (1956): the "sum of all physiological responses by which an animal attempts to maintain or re-establish a normal metabolism in the face of a chemical or physical force." In some ways this definition is acceptable, but we feel it is too restrictive because it excludes plants and it is not applicable at the ecosystem level. Homeostatic processes operating at the individual level and forces that tend to maintain stability or equilibrium at the ecosystem level are, we believe, analogous. Such an assertion is more meaningful if ecosystem stability is viewed as the capacity to maintain equilibrium or to return to equilibrium after the system has been perturbed. Certainly a perturbed system can be considered in terms of stress if ecosystem stability is perceived in this way.

Brett (1958) proposed a more acceptable definition of stress: "a state produced by any environmental or other factor which extends the adaptive response of an animal beyond normal range, or which disturbs the normal functioning process to such an extent that, in either case, the chances for survival are significantly reduced." The obvious utility of this definition rests with the implication that stress can be viewed in terms of chance or probability, which indicates an implicit assumption that stress can be quantified. Although this is clearly a positive aspect of Brett's definition, there are several objections similar to those raised for Selye's definition. Brett does not consider plants nor ecosystems, and, in addition, he implies that the outcome of stress must be viewed in negative terms. This is not always the case; although it is clearly negative for an individual organism if it dies as a result of exhaustion caused by stress, it is not necessarily negative for the success or survival of a population. For example, mortality of some individuals in a crowded population may actually serve to ensure species survival if space and/or nutrient resources are limiting. It is interesting to note that a similar mechanism may operate among some host—parasite systems, enhancing survival potential of both host and parasite species in an evolutionary sense. Pimentel (1961) and Pimentel and Bellotti (1976) noted that "co-evolution in a host and/or parasite toward a balanced supply—demand economy and regulation of parasite numbers is possible" through operation of a genetic feedback mechanism. Thus, if virulent parasites are eliminated by parasite-induced host mortality, the probability of propagating less-virulent parasites is enhanced, and survival and reproduction of more-resistant hosts may also be concomitantly increased. Another positive aspect of stress was suggested by Gibbons (1976) when he indicated that a process which may ultimately be negative at the population level can actually be preceded by an enhancement of the same process. Reviewing available literature, he noted, for example, that increases in temperature caused by release of thermal effluent in aquatic systems were shown initially to increase both primary and secondary productivity (Gibbons, 1970) and to alter a variety of species interactions (Saks et al., 1974).

All these objections to Selye's and Brett's definitions of stress were also discussed by Esch, Gibbons, and Bourque (1975), who, in an attempt to obviate the objections, suggested that stress be defined as "the *effect* of any force which tends to extend any homeostatic or stabilizing process beyond its normal limit, at any level of biological organization" such that the result will be either an enhancement or a diminishment in the probability of mortality, natality, or permanent change. This definition is broad enough in scope to permit application at individual, population, or ecosystem levels of organization but is explicit enough that the possibilities for quantification of stress are still maintained.

Model for Response to Stressor Input at the Individual Level

Possible response of an individual organism to stressor input is shown in Fig. 3. A resting or undisturbed individual operates within the constraints of a normal homeostatic range. In addition, there are wider limits within which the organism may continue to function, although not optimally. The dimension of these ranges varies in both space and time among individuals within a single population and among individuals in different populations.

If the organism is subjected to a stressor input of the proper kind and with an appropriate magnitude, the alarm stage of the GAS will be initiated. Generally, the duration of the alarm stage is brief, and the direction of the response is dictated by the nature of the specific biochemical or physiological process being measured. If the stressor input is withdrawn, the original steady state of the organism will return.

If the stressor input persists in time or grows in magnitude, then the respondent organism enters the resistance stage. The length of time an individual is able to resist stressor input and maintain itself within maximum homeostatic limits is variable. It depends to some extent, of course, on the inherent (or genotypic) potential of the individual, in combination with such intrinsic factors as its state of well-being, age, etc. Other extrinsic factors that modify the length of time the organism is able to operate within homeostatic limits include the season of the year and the nature, magnitude, and duration of the stressor input. As the model suggests, if the stressor is withdrawn during the resistance stage, then the original steady state returns.

If the stressor input continues long enough or if the magnitude of input increases beyond a critical point, the individual organism will no longer be able to cope homeostatically. The final stage, exhaustion, then ensues. If this end point is reached, the probability of an individual organism's surviving, reproducing, or returning to its original steady state is permanently altered.

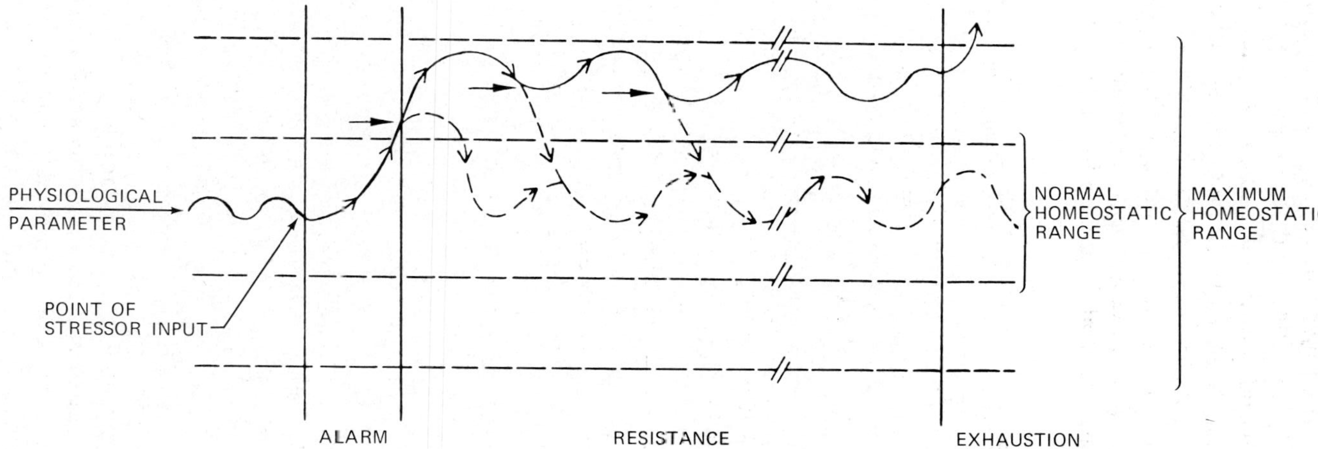

Fig. 3 Consequences of stressor input in an individual organism. Dashed lines with arrows indicate withdrawal of stressor input before point of no return; the system is able to readjust or reestablish original equilibrium.

This concept of stress was developed for vertebrate animals with autonomic nervous systems and endocrine glands capable of producing epinephrine and adrenocorticosteroids. Obviously many organisms, both plant and animal, have neither. The question that then occurs is, Can these organisms be considered within the framework of Fig. 3? The answer to the question is a tentative yes. We believe the schematic is general enough in scope to permit us to view the response of any organism, plant or animal, vertebrate or invertebrate, to the force of stressor input.

Model for Response to Stressor Input at the Ecosystem Level

Figure 3 represents what may occur when an individual organism is subjected to stressor input. But what happens when an ecosystem is perturbed? Clearly an ecosystem response to perturbation is far too complex to model or represent by a series of solid or dashed lines, especially if the perturbation is subtle. If, however, as previously noted, the forces operating to maintain homeostatic steady state in an individual are analogous to those maintaining equilibrium at the ecosystem level, models for stress at these two levels should be somewhat similar and, for the sake of this discussion, simple.

Before dealing with a conceptualization of stress at the ecosystem level, we should emphasize that ecosystem stability and complexity do not necessarily go hand-in-hand. Thus increased stability does not necessarily follow from increased complexity (May, 1976). Stability at the ecosystem level, as used here, refers only to the ability of an ecosystem to return to equilibrium following perturbation.

We have assumed that there is a clear parallel between stress at individual and ecosystem levels (Fig. 4). Initially, we perceive an equilibrium, or steady state, operating at more or less a constant level. If the ecosystem is perturbed, a new steady state, or equilibrium, will be established. If the stressor input is withdrawn, then either the original or the new equilibrium will be established, depending on an array of factors, including the nature, magnitude, and duration of the perturbing force and the initial fragility of the system. These factors will also be of importance in determining the length of time it would take the ecosystem to reach a stage of exhaustion. Cairns (1976) referred to the capacity of an ecosystem to resist insult in terms of inertia.

Exhaustion at the individual level is manifested as a permanent change in either mortality or natality probability. At the ecosystem level exhaustion leads to death (rarely) or an irreversible change in a

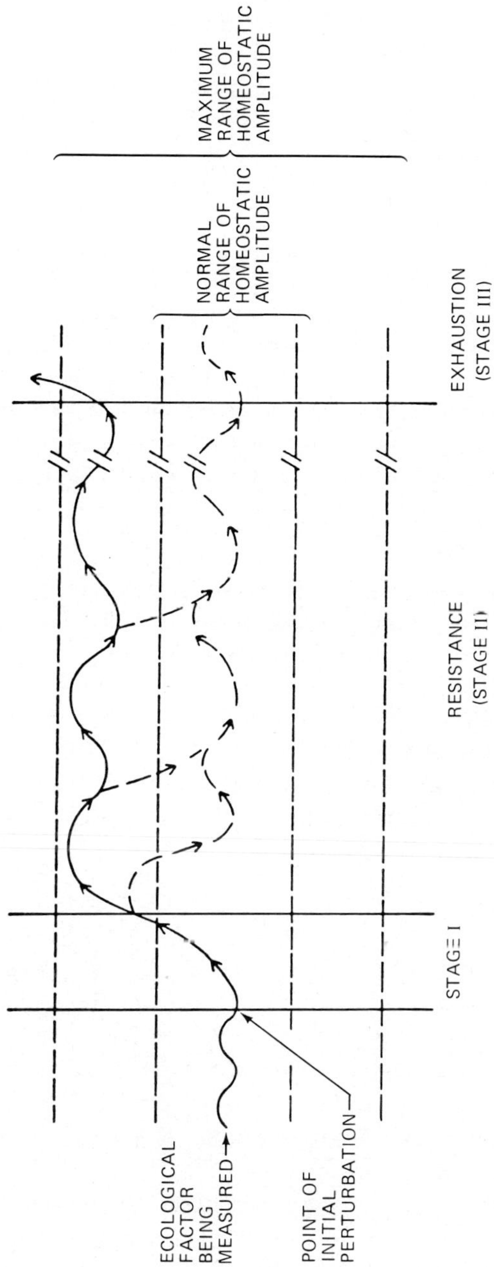

Fig. 4 Consequences of stressor input at the ecosystem level of organization.

given functional attribute (more commonly). In the latter course, the new steady state may be a close facsimile of the original, but again this depends on the nature, magnitude, and duration of the perturbing force. For example, let us consider the consequences of cultural eutrophication and its reversal. It is well known that enrichment causes an oligotrophic ecosystem to undergo eutrophication and that this process can be reversed if the source of enrichment is diverted or stopped, in which case the original "pristine" character of the system will be restored. Many of the functional characteristics of the original system may be restored, but the system still will not be the same. The explanation is simple when we consider that some species in the system will be unable to withstand conditions of stress associated with eutrophication and will become locally extinct. If this occurs, many of the species interactions that were characteristic of the original ecosystem are not reestablished in the new one. The result is a new steady state, or equilibrium, even though functionally the two oligotrophic systems are similar in many ways.

Thermal Ecology and Stress

Temperature is a universal influence on the normal physiological, metabolic, and behavioral processes of individual plants and animals, especially those living as heterotherms in aquatic systems. Most species of plants and animals have evolved a strategy not only of coping with normal environmental temperatures but also of exploiting temperature variations to such an extent that they are either partially or totally dependent on annual, seasonal, or diel fluctuations. With the construction of electricity-generating facilities producing heated effluent, many species of plants and animals that have evolved successful strategies for dealing with normal temperature cycles are now faced with levels of temperature that can be considered excessive. In responding to these elevated temperatures, each species has several alternative courses of action: (1) It can migrate away from or toward the high temperatures; (2) it can become locally extinct or flourish; or (3) it can cope until forced to follow one of the first two courses of action or until such time as it can evolve a new strategy to permit it to survive. Under any of these situations, the individual is subjected to stress and will behave in a manner consistent with Fig. 3.

Stress may be manifested in a number of insidious ways, none more so perhaps than increased susceptibility to disease of both organic and external origin. Our own specific interest over the years has been the relation between thermal effluent and parasitism and the various ways in which host–parasite relationships can be affected. Recently we have begun to deal with a disease in fish which we know to be related not only to thermal effluent but also to organic loading. Indeed, we now believe it to be basically a problem of stress, induced by a series of etiologies with which we can associate a number of environmental variables.

RED-SORE DISEASE, TEMPERATURE, AND STRESS

Description of Problem

Epistylis, a stalked, colonial ciliate, has been identified as the causative agent for red-sore disease, which affects several fish species in various aquatic systems throughout the southeastern United States (Rogers, 1971). The protozoan has a motile telotroch stage that attaches to a substrate, forms a stalk, and produces feeding bodies called zooids. When attached to the surface of a fish, it is said to cause scale erosion, producing pit-like lesions. The gram-negative bacterium *Aeromonas hydrophila* enters the tissues via these lesions and in time produces hemorrhagic septicemia and death.

Because of the heavy mortality from the disease, it has drawn considerable attention over the past few years. For example, more than 37,500 fish were killed in a few weeks in 1973 in Badin Lake, on the Yadkin River in North Carolina (Dean, 1974). During the fall of 1976, approximately 95% of the white perch population in Albemarle Sound, North Carolina, was killed by an epizootic of red-sore disease (Cook, 1976). During the same outbreak, 50% of the commercial catch of all species was discarded because of the presence of unsightly surface lesions associated with red-sore.

Despite the widespread nature of red-sore disease in the southeast, its epizootiology has not been extensively studied. Indeed, some literature on the problem is confusing, if not actually contradictory. For example, Rogers (1971) stated that *Epistylis* is the primary invader, with *A. hydrophila* then producing the secondary infection. On the other hand, Lom (1973) emphatically stated that *Epistylis* is incapable of producing lesions and that, as a bacterivore, it only secondarily associates with fish. Earlier Lom (1966) reported that heavy infections of fish with *Epistylis* were seasonal (occurring mostly in winter) and were independent of the amount of organic solids and the density of bacteria in the water. Bullock and McLaughlin (1970) and Meyer (1970) reported the most severe outbreaks to occur during summer when temperature is high and dissolved oxygen is low. During winter of 1973, Esch and Gibbons (unpublished observations), noting the presence of red-sore disease among several species of centrarchids in a South Carolina cooling reservoir, tentatively concluded that it could be related to thermal effluent from a nuclear production reactor.

From this brief overview, we see clearly that red-sore disease is a serious problem, that its epizootiology is not well understood, and that, because of its impact on both commercial and sport fisheries, it deserves rigorous study. An investigation of the problem was begun in the fall of 1974. The initial objectives were (1) to ascertain the identity of the causative agent, (e.g., is it *Epistylis* or *Aeromonas*); (2) to follow the course of the disease seasonally and relate it, if possible, to one or more water-quality parameters; (3) to determine whether thermal effluent affects the course of the disease; and (4) to develop a model for possible use in predicting when, where, and under what circumstances an epizootic may occur in a given body of water. During the first 2.5 years of the study, it became apparent that the disease could be related to the stress phenomenon. This notion has since been incorporated into the second and third objectives listed.

Description of the Study Site

Par Pond, an 1120-ha reservoir located at the Savannah River Plant near Aiken, South Carolina, serves as a cooling pond for a nuclear production reactor. Extensive descriptions of the temperature and biotic characteristics have been given by Holland et al. (1974), Lewis (1974), and Parker, Hirshfield, and Gibbons (1973). Figure 5 shows the primary collecting sites. Most of the fish from the thermally altered area were taken within 1 km of the point of entry of thermal effluent from Pond C. Temperatures in this area varied because of variability in reactor activity and season but generally averaged between 5 and 10°C above ambient, depending on season and distance from point of entry of thermal effluent into Par Pond.

From the fall of 1974 through the present time, temperature, dissolved oxygen, pH, redox potential, and conductivity have been measured weekly, in profile, at several selected sites in the reservoir. A depth profile for these parameters, measured in midwinter and midsummer at representative ambient and thermal locations (Fig. 6), clearly shows the monomictic nature of the impoundment. This is characteristic of most impoundments in the southeastern United States. Hazen (1978) provided a complete description of the water-quality parameters in Par Pond during the course of this study.

Identifying the Causative Agent for Red-Sore Disease

As previously indicated, the literature is contradictory as to whether *Epistylis* or *Aeromonas* is the etiological agent for red-sore

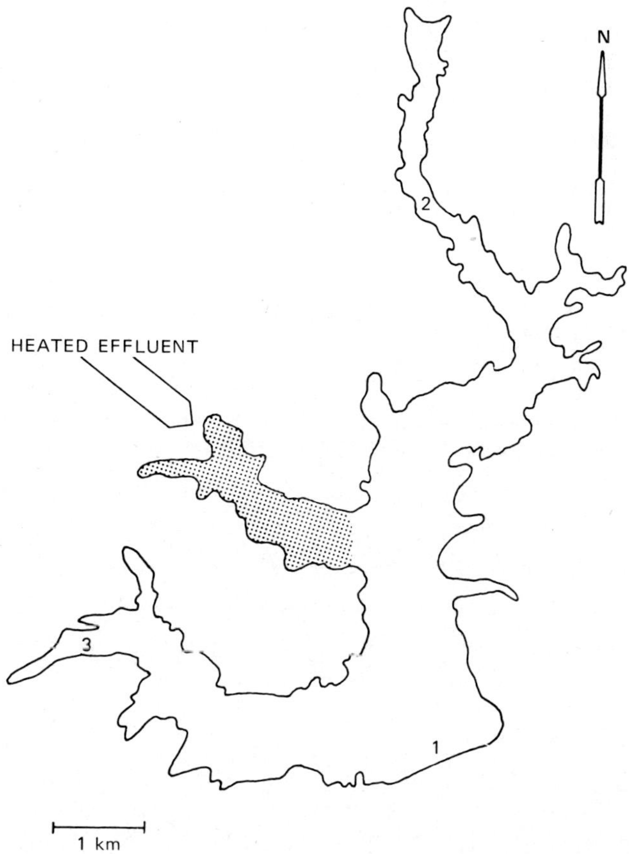

Fig. 5 **Map of Par Pond showing entry of thermal effluent. Shading represents the area with elevated temperature; all other locations in reservoir have ambient temperatures. Bass in thermal areas were mostly taken within 1 km of the point of entry of thermal effluent. Bass in ambient locations were mostly taken from sites marked 1, 2, and 3.**

disease. Rogers (1971) stated that *Epistylis* induces scale erosion, permitting secondary infection by *A. hydrophila*, but Lom (1973) indicated that there was no evidence to suggest that *Epistylis* could produce the histolytic enzymes required to cause scale erosion. In our study of the problem, we found three lines of evidence to suggest that *A. hydrophila* is the etiological agent for red-sore disease.

First, examination by scanning electron microscopy (Fig. 7) of the site of attachment by *Epistylis* to a lesion on the surface of a largemouth bass (*Micropterus salmoides*) does not suggest that the

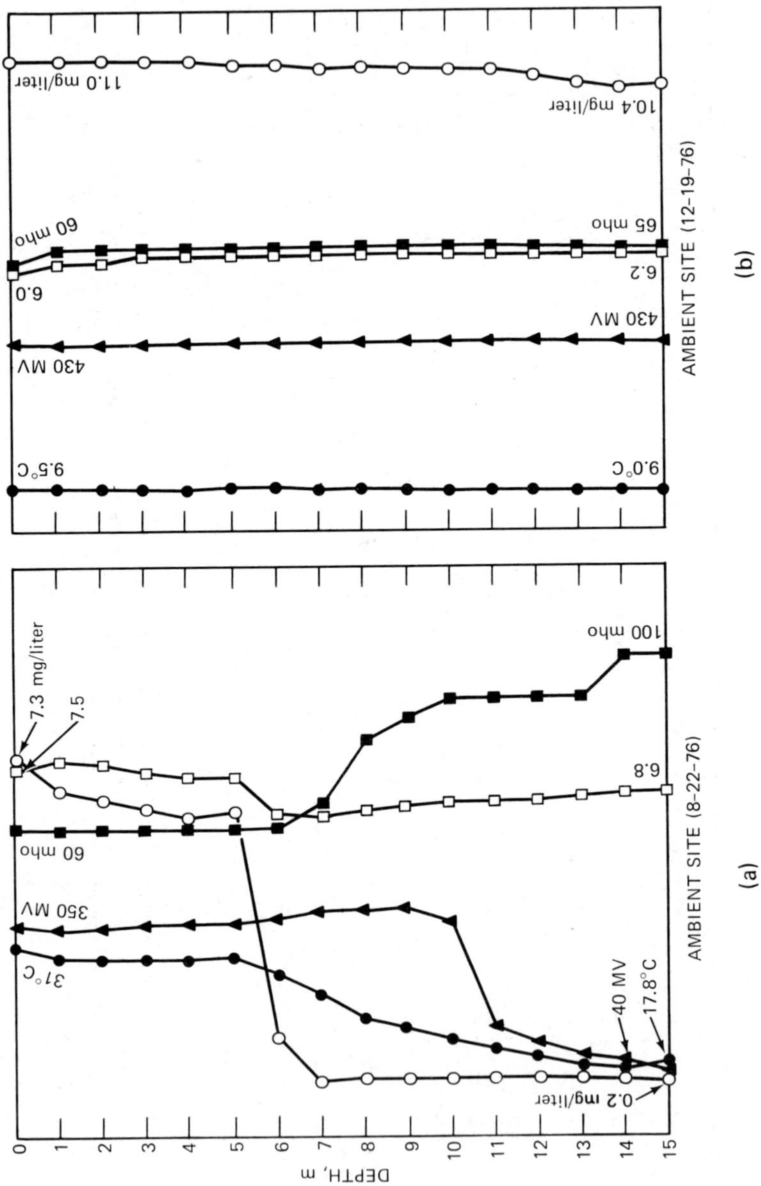
11.0 mg/liter
10.4 mg/liter
60 mho
65 mho
6.0
6.2
430 MV
430 MV
9.5°C
9.0°C
AMBIENT SITE (12-19-76)
(b)
7.3 mg/liter
7.5
100 mho
6.8
60 mho
350 MV
31°C
40 MV
17.8°C
0.2 mg/liter
0 1 2 3 4 5 6 7 8 9 10 11 12 13 14 15
DEPTH, m
AMBIENT SITE (8-22-76)
(a)

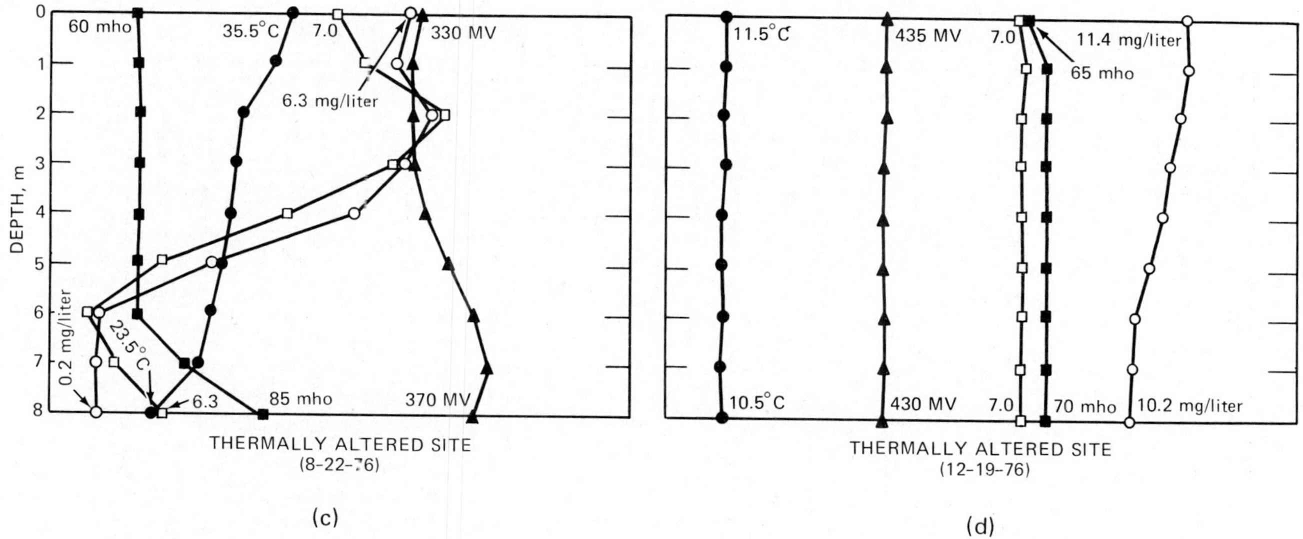

Fig. 6 Depth profiles for dissolved oxygen (○), temperature (●), pH (□), conductivity (■), and redox potential (▲) measured at a representative ambient location in (a) midsummer and (b) midwinter and a representative thermal location in (c) midsummer and (d) midwinter.

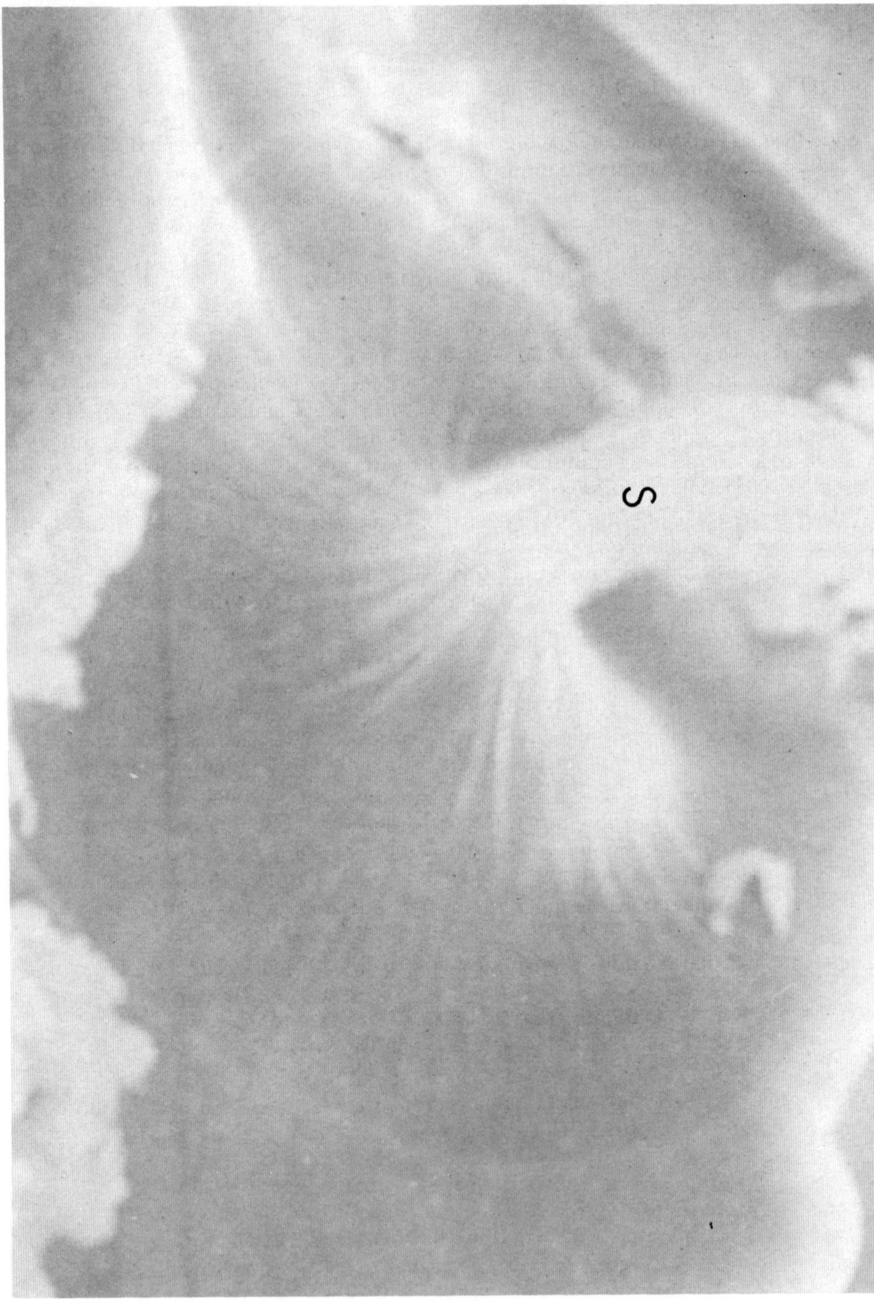

Fig. 7 Site of attachment by stalk of *Epistylis* sp. to surface of bass scale (S = stalk). 2000 ×.

protozoan produces erosion of the mucous-epithelium layer of the scale (Hazen et al., in press; Hazen, Raker, and Esch, 1977). Bacteria (presumably *A. hydrophila*) were observed adhering to the stalk of *Epistylis* but not in association with the zooids, or feeding bodies (Fig. 8). Histological sections of the surface lesions suggest a very loose association of the colonial ciliate with the scale surfaces and intensive inflammation in muscles beneath the scales. Within the

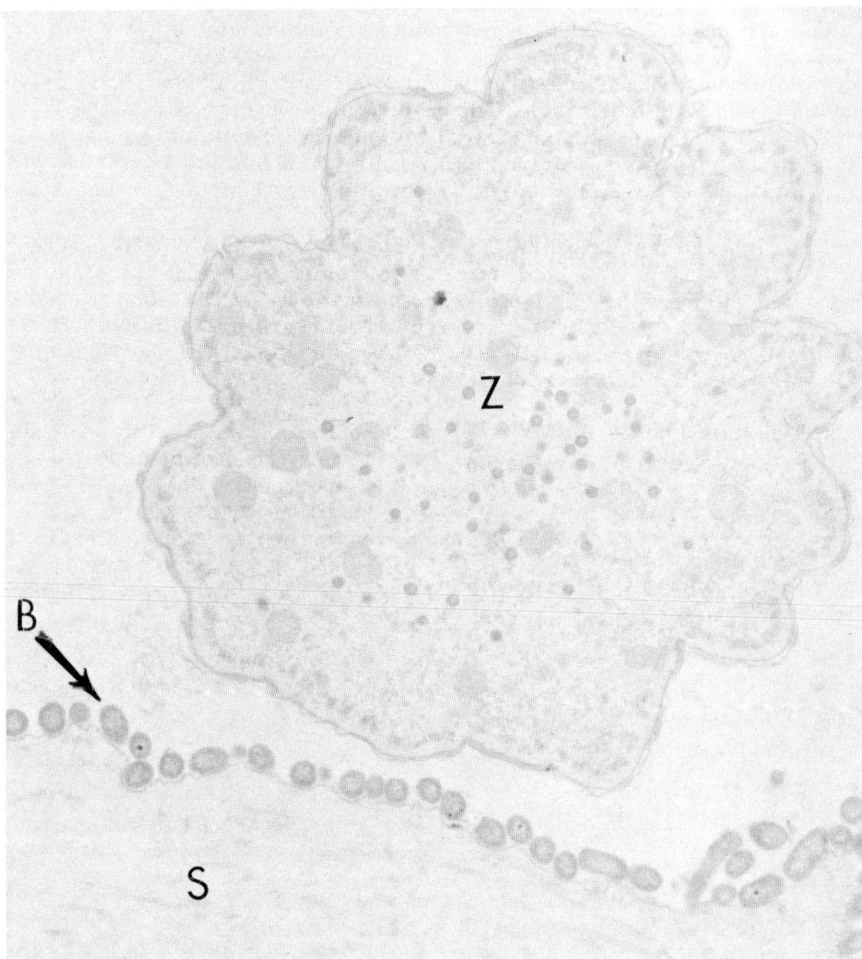

Fig. 8 Cross section of *Epistylis* sp. zooid and stalk viewed by transmission electron microscopy. Note presence of bacteria closely adhering to stalk, but not to the zooid (S = stalk, Z = zooid, B = bacteria). 4600 ×.

muscle, there is also substantial infiltration of bacterial cells, presumably *A. hydrophila* (Huizinga, Esch, and Hazen, manuscript in preparation; Hazen, Raker, and Esch, 1977).

A second line of evidence is provided by collaborative studies conducted by Robert Gorden and colleagues at the Savannah River Laboratory, Savannah River Ecology Laboratory, University of South Carolina Medical School, South Carolina Wildlife and Marine Resources Department, and Wake Forest University. During the spring and summer of 1975, a series of unexpected alligator mortalities occurred in Par Pond. In virtually all cases, *A. hydrophila* was cultured from lungs and other internal organs at the time of necropsy. This suggests that, in some manner, mortality may have been related to the presence of the bacteria. A review of available literature indicated that in 1971 red-sore disease induced by *A. hydrophila* was responsible for mortality of approximately 120,000 fish of several species in Lake Apopka, Florida (Shotts et al., 1972). In addition, 16 alligators died suddenly during the same period of time, all with symptoms of red-sore disease. During the fall and winter of 1976–1977, Gorden and his colleagues were able to generate substantial experimental evidence to show that *A. hydrophila* was capable of producing skin lesions on the surface of alligators and that *A. hydrophila* could be isolated from the tissues of dead and/or dying alligators. Furthermore, it was shown that surface lesions and mortality in alligators could be induced without a primary infection by *Epistylis*. Indeed, *Epistylis* was not present in 50 alligators with experimentally induced red-sore disease.

Finally, for *A. hydrophila* to be the primary invader, it is essential that the bacteria be capable of producing an extracellular toxin. Liu (1961) and others reported that *A. hydrophila* produces a number of potent exotoxins capable of inducing lesions such as those typically associated with red-sore disease.

On the basis of these observations, we feel confident in stating that the etiological agent for red-sore disease is *A. hydrophila*. Therefore, our further discussion is directed toward understanding the biology of the bacterium and its relation to the disease in fish.

Seasonal and Other Factors Associated with Red-Sore Disease

Previous studies have shown that red-sore disease differentially affects five species of centrarchid fish in Par Pond (Esch et al., 1976); infection percentages were consistently highest for largemouth bass (*Micropterus salmoides*). Since this trend has been the same since 1974, efforts have focused on this species of fish.

Red-sore disease among bass in Par Pond shows a very striking seasonal periodicity. The highest incidence of infection occurs during the spring months (March, April, and May), followed by lower levels in summer, a decline in fall, and the lowest incidence in winter (Fig. 9). Although there are differences in amplitude from year to year, these seasonal variations were consistent for 36 months, beginning in the fall of 1974.

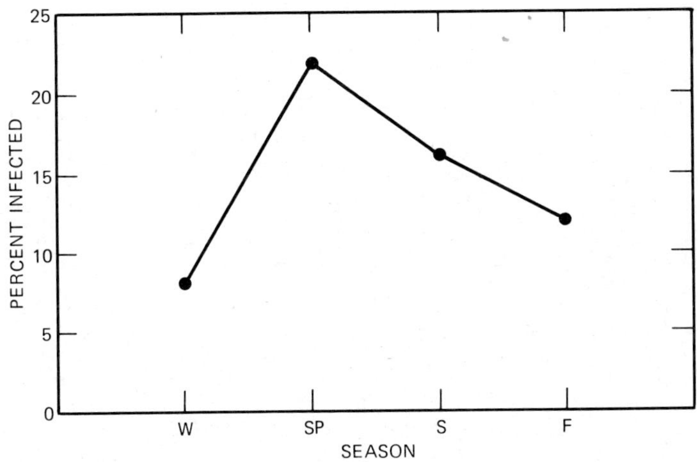

Fig. 9 Seasonal changes (composite for 36 consecutive months) in infection percentages among largemouth bass in Par Pond.

Since August 1975, the density of *A. hydrophila* has been determined in vertical profile at various locations within Par Pond. Temperature, dissolved oxygen, pH, redox potential, and conductivity were also recorded simultaneously (Hazen, manuscript in preparation). As might be expected, there were seasonal changes in each water-quality parameter, and *A. hydrophila* densities also varied seasonally. The seasonal changes in *A. hydrophila* densities showed a strong relation to the incidence of red-sore disease (Fig. 10); many of the significant increases and decreases in disease among bass were preceded by corresponding modulations in the density of *A. hydrophila* in water.

Meyer (1970) suggested that seasonal outbreaks of red-sore disease may be related to depressed levels of dissolved oxygen during summer months, which would lead to stress in certain fish species and then to increased vulnerability to infection. Since increased

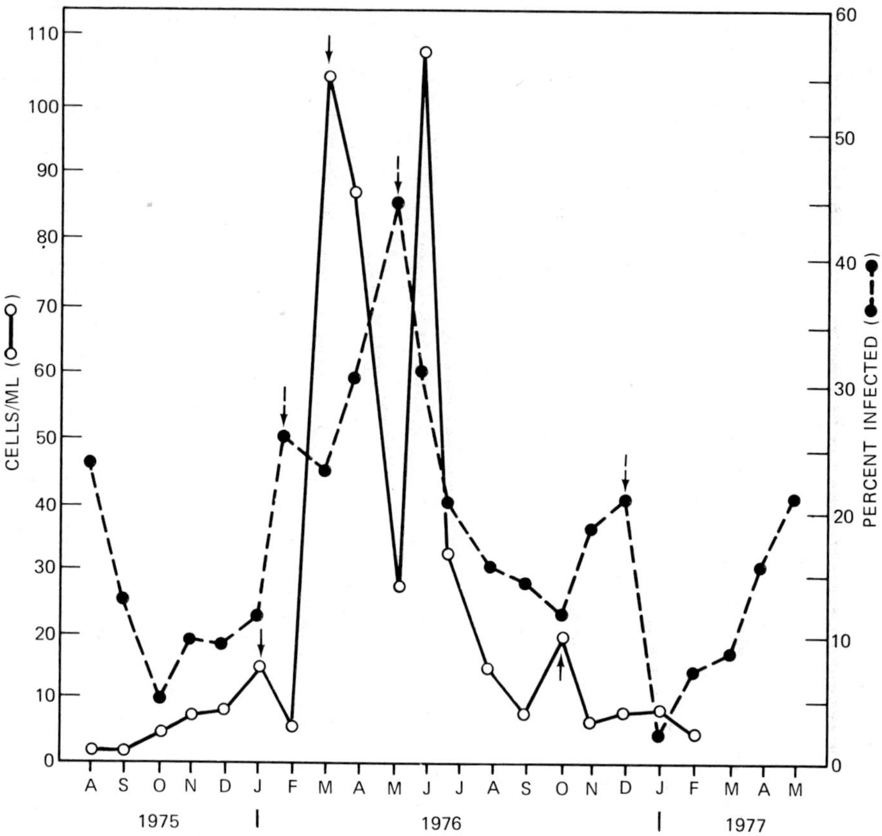

Fig. 10 Monthly changes in mean number of *A. hydrophila* cells per milliliter and in infection percentage of largemouth bass from Par Pond. Note that a rise in bacterial cell density (solid arrow) frequently precedes a rise in infection percentage among bass (dashed arrow). (CV ⩽ 35%.)

organic loading may lead to seasonal depression of dissolved oxygen in hypolimnetic water, efforts were made to measure total organic carbon at several sites in Par Pond during each of four consecutive seasons. These measurements were made simultaneously with the other five water-quality parameters and the density of *A. hydrophila*. As shown in Table 1, there was some seasonal variability, as might be expected. There was not, however, a relation between total organic carbon and the other water-quality parameters nor *A. hydrophila* density.

TABLE 1

SEASONAL DEPTH PROFILES FOR TOTAL ORGANIC CARBON* (mg/liter)

Depth,	Summer 1976		Fall 1976		Winter 1976		Spring 1977	
m	H	A	H	A	H	A	H	A
0	353.3	12.3	4.9	2.3	24.1	6.6	2.7	0.5
1	169.9	263.2	143.3	72.1	216.5	340.8	322.2	192.6
3	3.9	101.0	3.2	27.6	59.8	48.2	33.8	91.5
5	5.6	8.9	11.0	5.8	34.0	28.1	11.3	12.1
7	2.1	2.5	4.4	7.1	22.8	15.8	2.9	1.1
9		4.9		2.1		4.9		0.4
11		8.7		3.3		8.7		0.5
13		2.2		4.4		2.2		1.3
15		1.1		5.4		1.1		2.4
					SR water = 0.71		SR water = 1.0	

*Abbreviations are H, thermally altered location; A, ambient location; and SR, Savannah River.

Red-Sore Disease and Temperature

If the mean seasonal surface temperatures are compared with changes in incidence of red-sore disease among bass in Par Pond, then a parallel pattern emerges (Fig. 11). The lowest infection percentages are in winter, followed by peak infections in spring, and subsequent declines in summer and fall. The seasonal changes in infection percentages parallel seasonal temperature changes, except during summer.

The overall incidence of red-sore disease was only slightly higher among bass from thermal areas (N, 2956; infected, 19%) as compared with bass from ambient locations (N, 2431; infected, 16%). This pattern does not provide a realistic view of the differences in infection percentages among bass in ambient and thermally altered locations, however, because of other variables impacting on the bacteria and the bass. If, for example, the incidence of disease among bass in thermal and ambient locations is compared on a seasonal basis (Fig. 12), the influence of temperature can be seen more clearly. Thus the levels of infection were significantly higher in bass from thermal locations during the fall of 1974, the spring of 1975, and the winter months of all 3 years of study.

Since it appears that temperature is a highly significant variable, the question that must be considered is, In what way does temperature influence red-sore disease in Par Pond? At the present

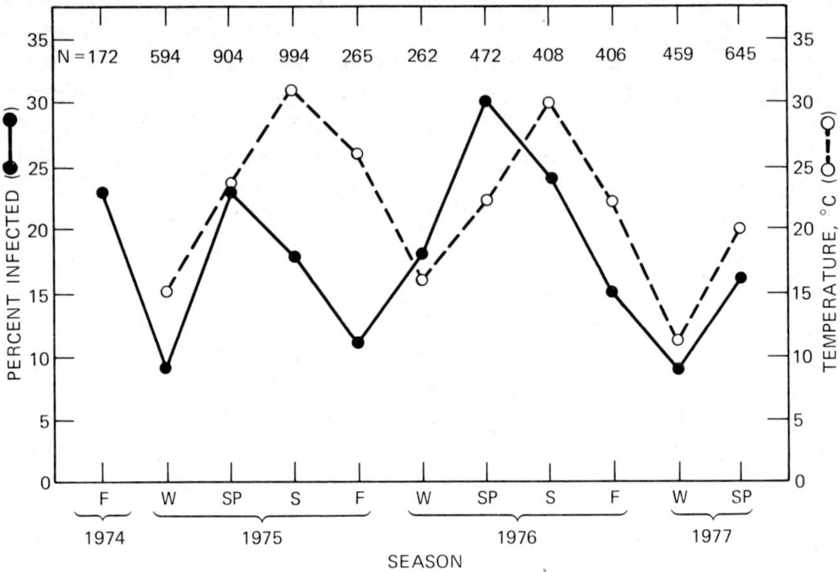

Fig. 11 Mean seasonal surface temperatures (composite for reservoir) and infection percentages among bass.

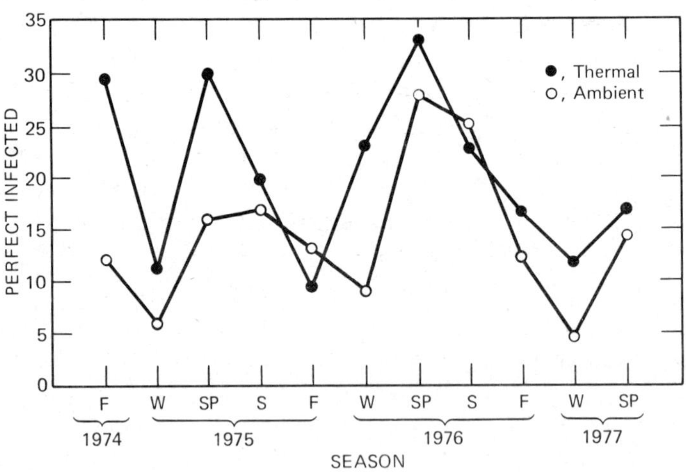

Fig. 12 Seasonal changes in infection percentages of bass from ambient and thermal locations.

time, two explanations appear to be plausible. First, it is conceivable that elevated temperature in thermally altered locations may act as a selection force for a more virulent strain of *A. hydrophila*. Although it is indirect and not unequivocal, there is some evidence to support this hypothesis. Hazen, Fliermans, and Esch (manuscript in preparation) have shown serological and immunological differences in certain strains of *A. hydrophila* isolated from fish, alligators, and Par Pond water. This line of study has promise and is being pursued. There appears to be a more sound explanation, however, for the relation between temperature and red-sore disease in the reservoir.

Since 1967, more than 10,000 bass have been captured in Par Pond (Gibbons et al., in press). The weight—length relationships of each of these fish were recorded, and the body condition, or K-factor, of each individual was determined. Body condition is a measure of individual fitness, or physical well-being (Carlander, 1969). Parenthetically, it is important to note that mark-recapture studies of many of the same 10,000 bass indicate that the vast majority (> 98%) appear to remain locally within discrete home ranges of the reservoir and, consequently, do not move long distances (Gibbons and Bennett, 1971; Quinn et. al., 1978; Hazen and Esch, 1978. When K-factors for all bass are shown seasonally (Fig. 13), a distinct pattern emerges. Generally, maximum body condition occurs in winter, with lowest conditions in summer. The exceptions during the fall of 1975 and in the fall of 1976 are due either to variations in reactor activity or to differences in sample sizes in thermally altered and ambient locations.

When body conditions of bass from ambient and thermally altered locations are compared (Fig. 14), individuals from thermally altered areas are, in general, less fit than those from ambient locations. Exceptions to this trend occur in the fall of 1975 and again in the winter of 1977. Even when these data are included with those from all other seasons, there are significant differences in body conditions of bass taken in thermally altered and ambient locations.

Because of the within-season variability in body condition and infection percentage, it was surmised that there could be a relationship between body condition and the probability of a bass being infected with *A. hydrophila*. Comparing the infection percentages for each 0.2-unit K-factor subclass between 1.0 and 3.0 (Fig. 15), we can see clearly that bass with the lowest body conditions are most likely to be infected. The decline in infection percentages begins between K-factors of 1.8 and 2.0 and continues to decrease as body conditions improve. It is interesting to note that virtually all bass with body conditions below 2.0 are without any

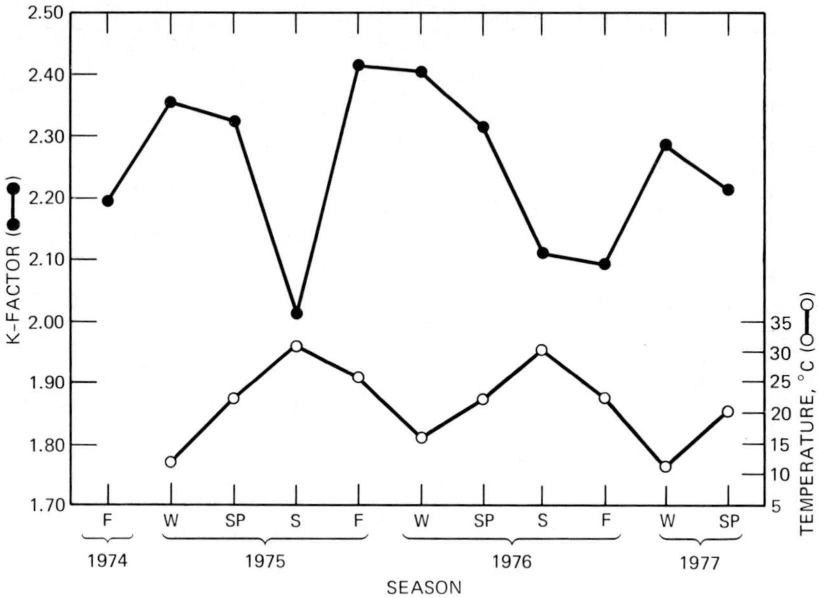

Fig. 13 Seasonal changes in body condition (K-factor) of bass in relation to changes in mean surface temperature.

dissectable body fat and that as the percent of body fat increases, body condition improves (Gibbons et al., in press).

Infection percentages were compared in each 0.2-unit K-factor subclass for bass from ambient and thermal locations (Fig. 16). For bass from thermally altered areas, infection percentages were high when body conditions were 1.8 or less. When body conditions were >1.8, infection percentages were lower, but variably so. The pattern for bass from ambient locations was clear; infection percentage declined beginning at 1.8 and continued to decrease as body conditions improved (the zero infection at 3.0 is believed to be artifact of the small sample size). Attempts to show these kinds of relationships seasonally are ineffective because of relatively small sample sizes in certain seasons in all 3 years. If bass are separated into two groups, however (>1.8 and <1.8), and the data for each season for all 3 years are pooled (e.g., spring, 1975—1977), the patterns show the impact of temperature on both infection percentage and body condition (Fig. 17).

Several conclusions can be drawn from these observations. First, among bass in both thermally altered and ambient locations in Par Pond, there is a clear relationship between body condition and the

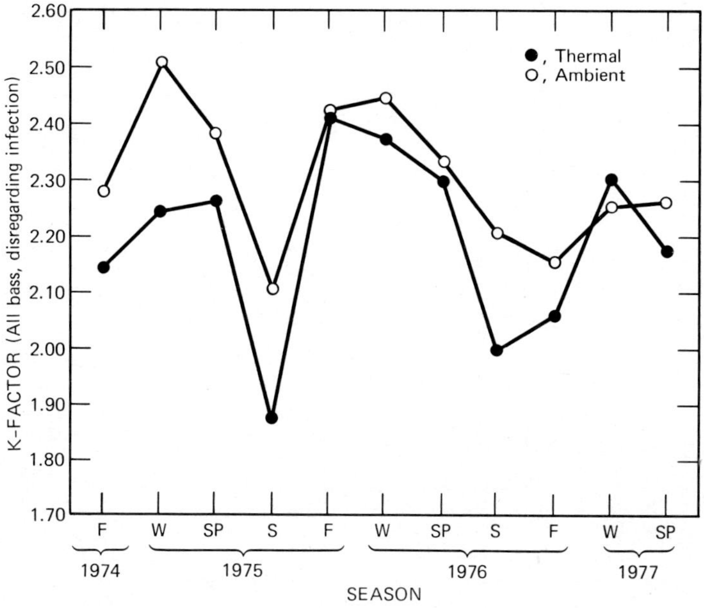

Fig. 14 Comparison of seasonal changes in body condition of bass from ambient and thermal locations.

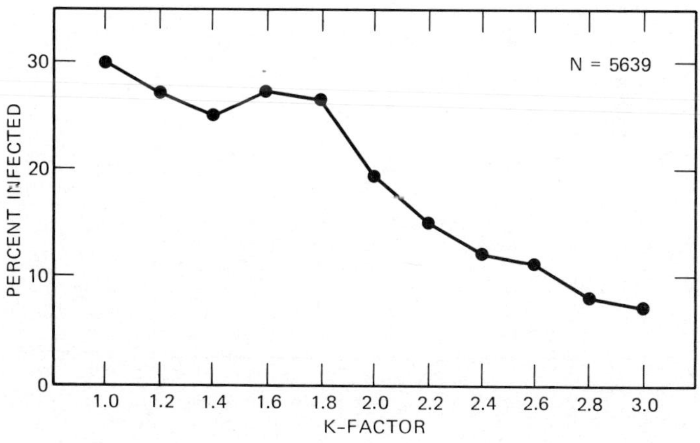

Fig. 15 Comparison of infection percentages among 0.2 unit body condition subclasses beginning at 1.0 and extending through 3.0.

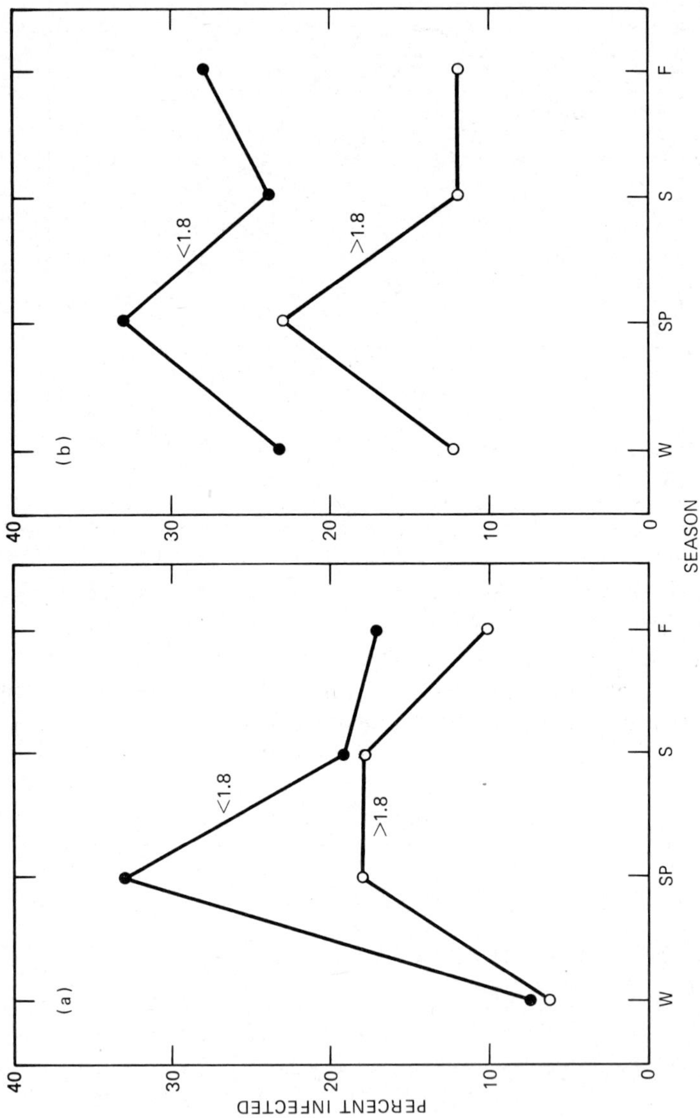

Fig. 16 Infection percentages among 0.2 unit body condition subclasses for bass from (a) ambient and (b) thermal locations.

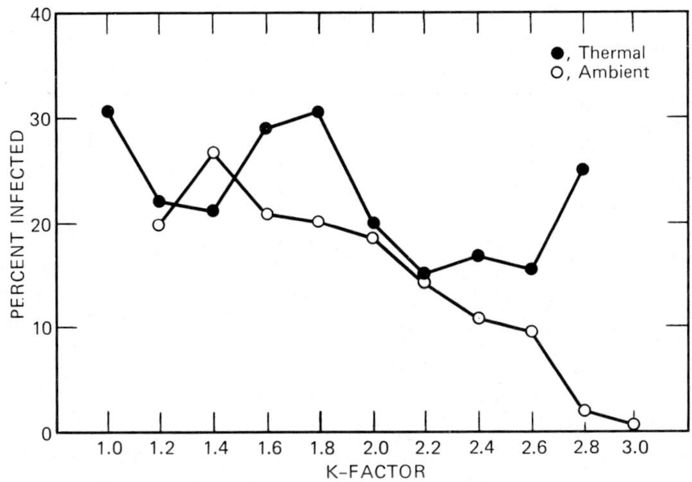

Fig. 17 Seasonal changes at ambient and thermally altered locations in infection percentages among bass with body conditions <1.8 and >1.8. Thermal samples at K-factors of 2.8 and 3.0 were pooled because of small sample sizes.

probability of a fish's being infected with red-sore disease. Second, body condition and infection probability appear to be dependent variables, except that the relationship is muted among bass in thermally altered locations. It is suggested that the muted effect is produced by the higher mean annual temperature in the heated areas of Par Pond and that the effect may operate via an immediate and direct stress on bass in these locations, which increases susceptibility to infection without necessarily having to induce an initial reduction in body condition.

Red-Sore Disease and Stress: A Hypothesis

On the basis of the evidence presented thus far, there seem to be relations among red-sore disease, bass body condition, and water temperature. Since these relationships are consistent from year to year, can a hypothesis be generated to provide an explanation? The answer to this question appears to be yes, and, furthermore, the hypothesis will incorporate the stress concept into our thinking on the epizootiology of red-sore disease. Indeed, since red-sore disease is known to occur in aquatic systems that are not affected by thermal effluent (e.g., Albemarle Sound and Badin Lake, North Carolina), it is conceivable that stress may be of greatest overall significance.

As water temperature increases in an aquatic system, the metabolic rates of bass will also increase (Fig. 18). With an increase in metabolism, there is a concomitant rise in catabolic processes, initially involving body fat but ultimately involving body protein as well (Gibbons et al., in press). The body condition, or K-factor, of bass affected by elevated temperature will, accordingly, decline in time (assuming, of course, that caloric intake is exceeded by metabolic demand). The exceedingly rapid and extensive growth of luxuriant stands of the submergent, rooted, vascular macrophyte, *Myriophyllum spicatum*, contributes to a fall in body condition in summer (aside from a normal post-spawn decline). Large masses of this plant provide excellent refuge for some species of forage fish. Reduced foraging success among bass, coupled with increased energy expenditure would exacerbate metabolic processes already conducive to reducing body condition. Lowered body conditions increase the probability of infection with red-sore disease. This explanation certainly describes the situation that occurs during part of the spring, throughout the summer, and into the fall months, before *Myriophyllum* dies back and disappears. It does not explain the high infection percentages that develop early in spring, however. During early spring, sexually mature bass are involved with activities associated with spawning and levels of circulating sex hormones (all steroids) are highest. These hormones function primarily to promote sexual behavior and develop secondary sexual characteristics, but they are also known to enhance the establishment, maintenance, and/or growth of numerous species of parasitic organisms (see Esch, Gibbons, and Bourque, 1975, for review). It is, thus, conceivable that elevated levels of circulating sex hormones also increase vulnerability to infection with *A. hydrophila*.

We believe that persistent, elevated temperature in thermally altered areas of the reservoir during most seasons of the year and during summer in ambient locations also promotes stress in bass. Stress, in the classical sense (Selye, 1950), necessarily implies the production and release of excess levels of adrenocorticosteroids, some of which have a striking anti-inflammatory action that promotes increased susceptibility to invasion by pathogenic organisms. We must point out that we have no evidence for increased production of corticosteroids during periods when there are high levels of red-sore disease among Par Pond bass, but this line of study is being pursued. We do know, however, that thyroxine levels are highest in bass during the summer months (Hazen et al., 1978) and that an increased level of circulating thyroxine is an indication of stress among mammals.

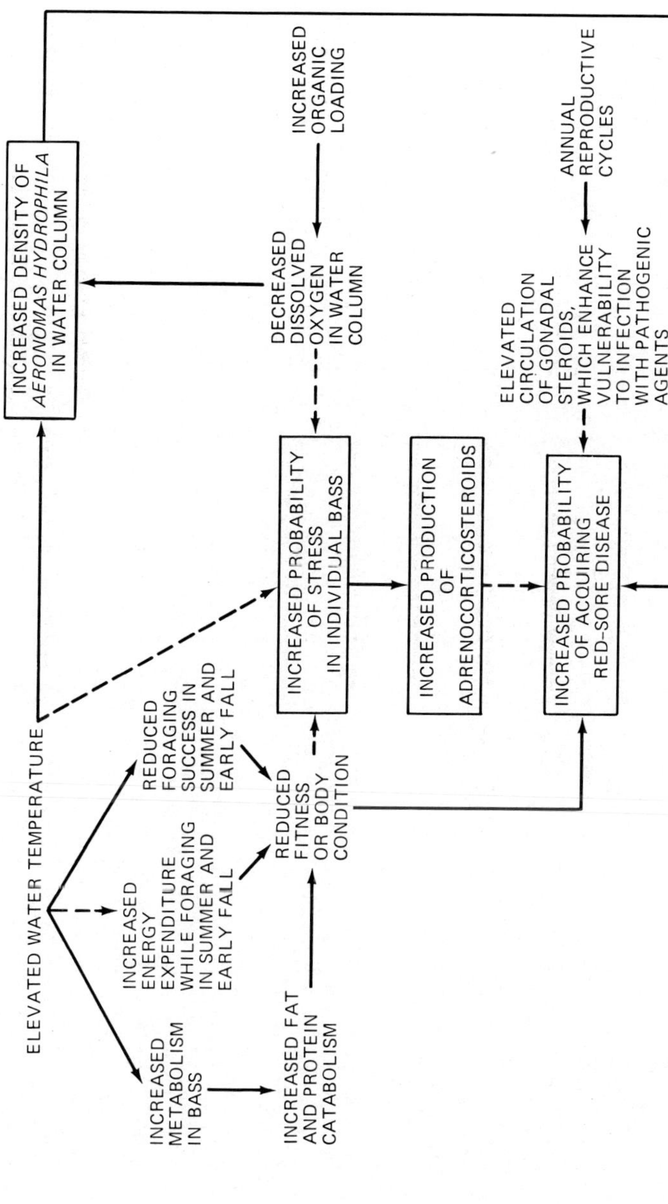

Fig. 18 Probable relationships among elevated temperature, increased organic loading, stress, and red-sore disease in largemouth bass. Known relationships are shown as solid lines (———); speculated relationships are shown as dashed lines (— — — —).

As previously noted, red-sore disease also occurs among fish species in aquatic systems that are not affected by thermal effluent. In these systems the disease is associated with such conditions as lowered dissolved oxygen and increased organic loading (Meyer, 1970; Rogers, 1971; Dean, 1974). Under these circumstances, it is quite possible that bass are stressed, and this leads to increased circulation of corticosteroids and then to increased susceptibility to infection with red-sore disease.

This scenario, which describes the relationships among red-sore disease, body condition, and temperature in Par Pond (see Fig. 18), includes the possibility that stress may be of significance in reducing innate or acquired resistance of bass to infection with *A. hydrophila.* As indicated, we have not yet generated data indicating that corticosteroids are higher in bass from thermally altered areas, but there is evidence in other fish species that epinephrine and corticosteroids vary in direct proportion to various types of stressors (Nakano and Tomlinson, 1967; Hill and Fromm, 1968).

CONCLUSIONS

The aim of this discourse has been twofold. First, an effort was made to describe the stress phenomenon in physiological terms and to illustrate how it has application at the individual and ecosystem levels of organization. Second, we attempted to represent these relationships by describing the case history of red-sore disease among largemouth bass in the southeastern United States.

Hazen (manuscript in preparation) has isolated *Aeromonas hydrophila* from lakes and streams in 34 states, from Maine in the northeast, to Montana in the west, to Texas in the southwest, and to Florida in the southeast. The largemouth bass, *Micropterus salmoides*, is present in virtually all the localities from which *A. hydrophila* has been isolated, yet red-sore disease has been reported only in the southeastern United States. On the basis of these observations and of our studies in Par Pond, it seems reasonable to conclude that only a unique assortment of physicochemical properties in a given aquatic system, an assemblage of variably susceptible hosts, and the presence of virulent *A. hydrophila* will promote an epizootic outbreak of disease. Stress and its impact at individual, population, and ecosystem levels of organization would, of course, temper the potential for outbreak. Thus, for red-sore disease to reach epizootic proportions, a wide range of interacting biotic and abiotic variables are clearly necessary. Perhaps, with additional effort, conditions conducive to such outbreaks can be identified. If so, we will have the means for predicting and perhaps minimizing the impact of the disease.

ACKNOWLEDGMENTS

We especially want to thank James Mathews, Mark Raker, and Melanie Trogdon for their excellent technical assistance and thank Jean Coleman for her usual excellence in preparing all the figures. The research reported here was supported by contract EY-76-S-09-0900 between the Energy Research and Development Administration (ERDA) and Wake Forest University and contract EY-76-C-09-0819 between ERDA and the University of Georgia. The work was also supported in part by grants from the North Carolina Board of Science and Technology and from the North Carolina Water Resources Research Institute.

REFERENCES

Brett, J. R., 1958, Implications and Assessments of Environmental Stress, in *The Investigation of Fish Problems*, H. R. MacMillan Lectures in Fisheries, pp. 69-97, University of British Columbia, Vancouver.

Bullock, G. L., and J. A. McLaughlin, 1970, Advances in Knowledge Concerning Bacteria Pathogenic to Fishes (1954—1968), in *A Symposium on Diseases of Fishes and Shellfishes*, S. F. Snieszko (Ed.), pp. 231-242, Special Publication No. 5, American Fisheries Society, Washington, D. C.

Cairns, J., Jr., 1976, Heated Waste-Water Effects on Aquatic Ecosystems, in *Thermal Ecology II*, ERDA Symposium Series, Augusta, Ga., Apr. 2—5, 1975, G. W. Esch and R. W. McFarlane (Eds.), pp. 32-38, CONF-750425, NTIS.

Carlander, K. D., 1969, *Handbook of Freshwater Fishery Biology*, Vol. 1, Iowa State University Press, Ames.

Cook, Grover, 1976, Division of Environmental Management, Department of Natural and Economic Resources, Raleigh, N.C., personal communication.

Dean, J., 1974, Fish Disease Outbreak, *Wildlife in North Carolina*, February 1974, pp. 18-19.

Esch, G. W., J. W. Gibbons, and J. E. Bourque, 1975, An Analysis of the Relationship Between Stress and Parasitism, *Am. Midl. Nat.*, 93: 339-353.

——, T. C. Hazen, R. V. Dimock, and J. W. Gibbons, 1976, Thermal Effluent and the Epizootiology of the Ciliate *Epistylis* and the Bacterium *Aeromonas* in Association with Centrarchid Fish, *Trans. Am. Microsc. Soc.*, 95: 687-693.

George, C. J., 1977, The Implication of Neuroendocrine Mechanisms in the Regulation of Population Character, *Fisheries*, 2: 14-19.

Gibbons, J. W., 1970, Reproductive Dynamics of a Turtle (*Pseudemys scripta*) Population in a Reservoir Receiving Heated Effluent from a Nuclear Reactor, *Can. J. Zool.*, 48: 881-885.

——, 1976, Thermal Alteration and the Enhancement of Species Populations, in *Thermal Ecology II*, ERDA Symposium Series, Augusta, Ga. Apr. 2—5,

1975, G. W. Esch and R. W. McFarlane (Eds.), pp. 27-31, CONF-750425, NTIS.

——, and D. H. Bennett, 1971, Abundance and Local Movement of Largemouth Bass (*Micropterus salmoides*) in a Reservoir Receiving Thermal Effluent from a Reactor, in *Radionuclides in Ecosystems,* Proceedings of the Third National Symposium on Radioecology, May 10—12, 1971, Oak Ridge, Tenn., D. J. Nelson (Ed.), CONF-710501-P1, pp. 524-527, NTIS.

——, D. H. Bennett, G. W. Esch, and T. C. Hazen, in press, Effects of a Heated Reservoir on Body Condition of Largemouth Bass, *Nature.*

Harper, J. L., 1967, Darwinian Approach to Plant Ecology, *J. Ecol.*, 55: 247-270.

Hazen, T. C., 1978, *The Ecology of* Aeromonas hydrophila *in a South Carolina Cooling Reservoir,* Ph.D. Thesis, Wake Forest University, Winston—Salem, N.C.

——, and G. W. Esch, 1978, Observations on the Ecology of *Clinostomum marginatum* in Largemouth Bass *(Micropterus salmoides), J. Fish. Biol.*, 12(5): 411-420.

——, G. W. Esch, A. B. Glassman, and J. W. Gibbons, 1978, Relationship of Season, Thermal Loading and Red-Sore Disease with Various Hematological Parameters in Largemouth Bass *(Micropterus salmoides), J. Fish. Biol.*, 12(5): 491-498.

——, M. L. Raker, and G. W. Esch, 1977, Light and Electron Microscope Studies on Lesions Associated with Red-Sore Disease in Largemouth Bass, *Bull. Assoc. Southeast. Biol.*, 24: 57.

——, M. L. Raker, G. W. Esch, and C. B. Fliermans, in press, Ultrastructure of Red-Sore Lesions on Largemouth Bass; Association of the Peritrich *Epistylis* sp. and the Bacterium *Aeromonas hydrophila, J. Protozool.*

Hill, C. W., and P. O. Fromm, 1968, Response of the Interrenal Gland of Rainbow Trout (*Salmo gairdneri*) to Stress, *Gen. Comp. Endocrinol.*, 11: 69-77.

Holland, W. E., M. H. Smith, J. W. Gibbons, and D. H. Brown, 1974, Thermal Tolerances of Fish from a Reservoir Receiving Heated Effluent from a Nuclear Reactor, *Physiol. Zool.*, 47: 110-118.

Lewis, W. M., Jr., 1974, Evaluation of Heat Distribution in a South Carolina Reservoir Receiving Heated Water, in *Thermal Ecology,* AEC Symposium Series, Augusta, Ga., May 3—5, 1973, J. W. Gibbons and R. R. Sharitz (Eds.), pp. 1-27, CONF-730505, NTIS.

Liu, P. V., 1961, Observations on the Specificities of Extracellular Antigens of the Genera *Aeromonas* and *Serratia, J. Gen. Microbiol.*, 24: 145-153.

Lom, J., 1966, Sessiline Peritrichs from the Surface of Some Freshwater Fishes, *Folia Parasitol.* (*Prague*), 1: 36-56.

——, 1973, The Mode of Attachment and the Relation to the Host in *Apiosoma Pissicola* (Blanchard) and *Epistylis lwoffi* (Faure-Fremiet), Ectocommensals of Freshwater Fish, *Folia Parasitol. (Prague)*, 20: 105-112.

May, R. M., 1976, Models for Two Interacting Populations, in *Theoretical Ecology, Principles and Applications,* R. M. May (Ed.), pp. 49-70, W. B. Saunders Company, Philadelphia.

Meyer, F. P., 1970, Seasonal Fluctuations in the Incidence of Disease on Fish Farms, in *A Symposium on Diseases of Fishes and Shellfishes*, S. L. Snieszko (Ed.), pp. 21-29, Special Publication No. 5, American Fisheries Society, Washington, D.C.

Nakano, T., and N. Tomlinson, 1967, Catecholamine and Carbohydrate Concentrations in Rainbow Trout *(Salmo gairdneri)* in Relation to Physical Disturbance, *J. Fish. Res. Board Can.*, 24: 1701-1715.

Odum, E. P., 1969, The Strategy of Ecosystem Development, *Science*, 164: 262-270.

Parker, E. D., M. F. Hirshfield, and J. W. Gibbons, 1973, Ecological Comparisons of Thermally Affected Aquatic Habitats, *J. Water Pollut. Control Fed.*, 45: 726-733.

Pimentel, D., 1961, Animal Population Regulation by the Genetic Feedback Mechanism, *Am. Nat.*, 95: 65-79.

——, and A. C. Bellotti, 1976, Parasite-Host Population Systems and Genetic Stability, *Am. Nat.*, 110: 877-888.

Quinn, T., G. W. Esch, T. C. Hazen, and J. W. Gibbons, 1978, Long Range Movement and Homing by Largemouth Bass *(Micropterus salmoides)* in a Thermally Altered Reservoir, *Copeia*, 1978(3): 542-545.

Rogers, W. A., 1971, Disease in Fish Due to a Protozoan *Epistylis* (Ciliata: Peritricha) in the Southeastern U. S., *Proc., Annu. Conf. Southeast. Assoc. Game Fish Comm.*, 25: 493-496.

Saks, N. M., J. J. Lee, W. A. Miller, and J. H. Tietjen, 1974, Growth of Salt-Marsh Microcosms Subjected to Thermal Stress, in *Thermal Ecology*, AEC Symposium Series, Augusta, Ga., May 3—5, 1973, J. W. Gibbons and R. R. Sharitz (Eds.), pp. 391-398, CONF-730505, NTIS.

Selye, H., 1950, Stress and the General Adaptation Syndrome, *Br. Med. J.*, 1: 1383-1392.

——, 1956, *The Stress of Life*, McGraw-Hill Book Company, Inc., New York.

Shotts, E. B., J. L. Gaines, C. Martin, and A. K. Prestwood, 1972, *Aeromonas*-Induced Deaths Among Fish and Reptiles in an Eutrophic Inland Lake, *J. Am. Vet. Med. Assoc.*, 161: 603-607.

Slobodkin, L., 1967, Toward a Predictive Theory of Evolution, in *Population Biology and Evolution*, R. C. Lewontin (Ed.), pp. 187-205, Syracuse University Press, Syracuse, N.Y.

Vadas, R. L., M. Kocor, P. C. Rusanowski, and B. R. Larson, 1976, The Effects of Thermal Loading on the Growth and Ecology of a Northern Population of *Spartina alterniflora*, in *Thermal Ecology II*, ERDA Symposium Series, Augusta, Ga., Apr. 2—5, 1975, G. W. Esch and R. W. McFarlane (Eds.), pp. 54-63, CONF-750425, NTIS.

SIZE-FRACTIONATED PRIMARY PRODUCTIVITY IN LAKE MICHIGAN NEAR THE KEWAUNEE NUCLEAR POWER PLANT

STEPHAN I. ZEEMAN* and RALPH GRUNEWALD
Department of Botany and the Center for Great Lakes Studies,
The University of Wisconsin, Milwaukee, Wisconsin

ABSTRACT

Primary productivity rates were measured at the site of the Kewaunee Nuclear Power Plant from Aug. 26, 1975, to July 23, 1976. Productivity was determined for three size fractions by sequential filtration through 64-, 10-, and 0.45-μm porosity filters. Total unfractionated productivity was also measured. Univariate analysis of variance showed no difference ($P > 0.05$) between pre- and postcondenser productivity rates of the unfractionated samples. Multivariate analysis of variance applied to size-fractionated samples detected lower rates only when there was a ΔT across the condenser. These differences were caused by changes in the <10- and >64-μm fractions. Average productivity rates for the year at the postcondenser station were within 13% of those at the precondenser station. Chlorophyll a values were within 4% of precondenser values. It was concluded that, although individual phytoplankton may be killed by passage through the cooling system, populations suffer no permanent damage. Both productivity rates and chlorophyll a concentrations for the net plankton (>64 μm) averaged over 55% of the sum of the three fractions. The 10- to 64-μm fraction averaged about 12% for productivity and 24% for chlorophyll, but the values for the <10-μm fraction were about 30 and 20% of the sum of the fractions for productivity and chlorophyll, respectively. Assimilation numbers (milligrams of carbon per milligram of chlorophyll a per cubic meter per hour) were generally low (<3) for the unfractionated plankton. For the size-fractionated samples, the two larger fractions were most often low, and the <10-μm fraction had numbers >3 on 44% of the sampling dates.

*Present address: Marine Science Program, University of South Carolina, Columbia, S. C.

This study assesses the effect on the rate of phytoplankton productivity of passage through the cooling system at the Kewaunee Nuclear Power Plant. An important facet of this research was the determination of photosynthetic rates for three size classes of phytoplankton, as well as for the total community.

Passing through the cooling system of a steam electric-power station subjects phytoplankton to a variety of stresses, including: (1) mechanical stress from physical abrasion, pressure changes, and turbulence; (2) thermal stress; and (3) chemical effects of antifouling agents, corrosion products, and concentrations of dissolved gases (Lauer, Walter, and Beck, 1972). The effects of these stresses are not at all certain. The severity of a stress depends on its intensity and duration and on the physiological condition of the organisms.

The effects of temperature on microalgae were reviewed by Hoogenhout and Amesz (1965). Patrick (1969) gave temperature tolerance ranges for freshwater algae and stated that diatoms preferred temperatures below 30°C and that condenser passage causes little or no harm to algae if the temperature remains below 34.5°C. Furthermore, if the volume of the entrained water is small relative to the volume of the receiving water body, any effect on the phytoplankton population as a whole will likely be negligible (Patrick, 1974).

Many studies have shown that photosynthetic rates may be affected by temperature changes and condenser passage. Warinner and Brehmer (1966) showed that enhancement or suppression of photosynthesis depends on initial temperature as well as the temperature increase. Similar results were reported by other investigators (e.g., Morgan and Stross, 1969; Brooks, 1972; Fox and Moyer, 1973).

In a preoperational study at the Kewaunee Nuclear Power Plant, Bremer and Redmond (1974) found that, in nine of ten sampling periods, analyses detected a significant difference between the precondenser and discharge productivity rates at Kewaunee. They showed that mechanical effects could reduce productivity rates from 1 to 34%. Operational studies on immediate effects (7 hr after sample collection) showed that chlorophyll was significantly lower in the discharge on only one occasion (Jones, Brown, and Redmond, 1975). Significant difference in productivity between the precondenser and discharge sites was not detected. Furthermore, mechanical effects were responsible for mean annual decreases of only 8% for productivity and 5% for chlorophyll. Delayed effects (after 24 and 48 hr), although more variable, were still minimal. These studies suggest that normal plant operations might have only limited effects on phytoplankton assemblages entrained in the cooling water.

Differential injury and recovery of specific size classes of entrained phytoplankton could be important to community structure in the receiving waters. The small nannoplankton are thought to be more important to primary production than the larger net plankton (Pomeroy, 1974). Rodhe, Vollenweider, and Nauwerck (1958) found that nannoplankton were responsible for more than 90% of production in spring and early summer and that net plankton became more important in summer and fall. Similar results were presented by Goldman and Wetzel (1963).

As far as we know, the application of multivariate analysis of variance (MANOVA) to a data set of this type is novel, but multivariate analyses in general have been successfully applied to algal communities (Allen and Skagen, 1973).

Since there were multiple measures for each sample (i.e., size-fractionated productivities), the chances for correlation among the variables was high (Van de Geer, 1971). Thus a separate analysis for each variable would not constitute independent tests. The MANOVA technique then provides an important analytical tool, and separate analyses should be avoided unless the MANOVA null hypothesis has been rejected (Cooley and Lohnes, 1971). The MANOVA procedure assumes a multivariate normal distribution for variables with equality of dispersions.

MATERIALS AND METHODS

Site Description

The Kewaunee Nuclear Power Plant is located on the shore of Lake Michigan, ~43.5 km east—southeast of Green Bay, Wisc. The facility has once-through cooling and can produce a net output of 540 MW(e). Ambient temperatures ranged from 0.5 to 18.3°C and ΔT's, from 0 to 14.5°C. The maximum discharge temperature recorded during this study was 28.8°C.

Sampling

Sampling was initiated on Aug. 26, 1975, and concluded on July 23, 1976. Samples were collected from a precondenser station in the forebay (FRB) and a postcondenser station in the discharge flume (KD) on eighteen dates. Two other stations were sampled on six dates from August to November, one at the intake structure (KWI) and one at a reference station (REF) about 4 km north of the plant. Figure 1 shows the location of the plant and the sampling sites.

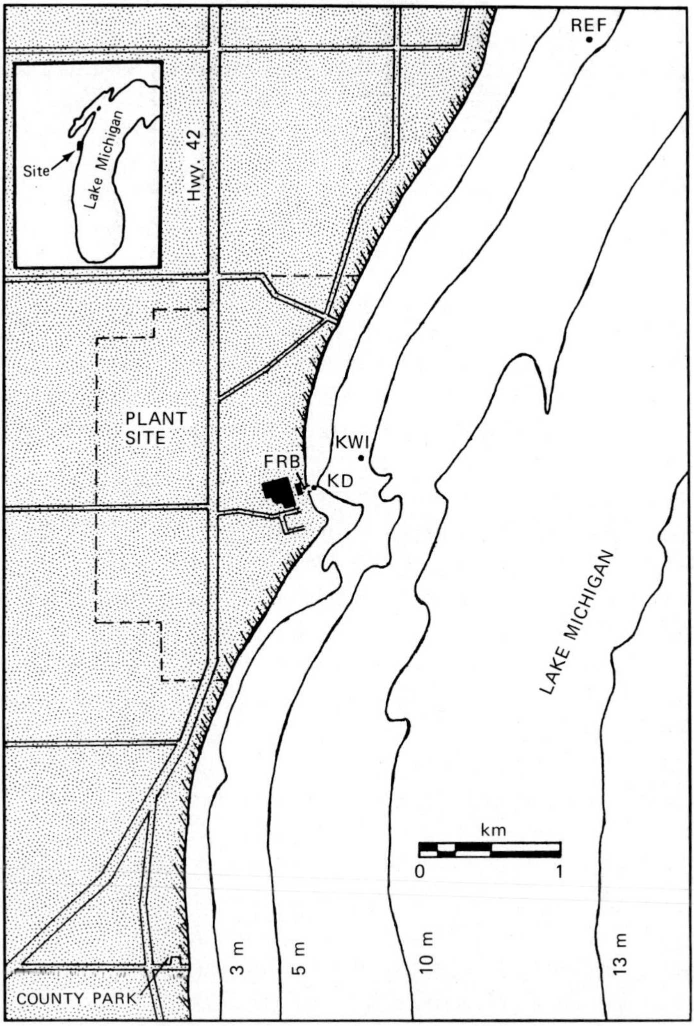

Fig. 1 Location of study area and sampling stations.

Water samples were obtained with an opaque, 2-liter, PVC Van Dorn sampler at a depth of 3 m at all sites except KD, where the depth of the samples varied because the current was too fast. Five separate casts were made at each site, and the water was stored in opaque, 2-liter plastic jars. Three samples were used for the carbon uptake experiments and two for chlorophyll extraction. At each site a 1-liter polyethylene bottle was also filled with lake water for inorganic carbon determination.

Carbon Uptake

Photosynthetic rates of phytoplankton were estimated by the radiocarbon method described by Steeman-Nielsen (1952). Individual samples were mixed well, and subsamples were siphoned into sets of two light and one dark bottle (125-ml glass-stoppered bottles). For each site, three sets were used in the size-fractionated procedure, one set from each of three different samples. In addition, one or two other sets were filled to measure the total unfractionated carbon uptake.

Each bottle was inoculated with 5 μCi of $Na_2H^{14}CO_3$ and incubated for 4 hr in a constant temperature (ambient) and light incubator. All incubations were started between 1000 and 1100 hr local time to minimize any effects of diurnal rhythms (Doty and Oguri, 1957).

Light was provided by 40-W cool-white fluorescent tubes. Intensities were checked before each experiment and adjusted to give not less than 150 microeinsteins m^{-2} sec^{-1} (~900 ft-c).

Samples were filtered immediately after the incubation period, at a vacuum of ⅙ atm.

The radiocarbon and chlorophyll samples were fractionated sequentially. Bothwell (1975) showed that this was superior to estimating from differential filtrations. The apparatus, based on a design by Schubel and Schiemer (1969), used filters of 10- and 64-μm pore size Nitex netting (Tetko, Inc.) and 0.45-μm HA Millipore filters.

After filtration, all filters were fumed with concentrated HCl and placed in vials containing 15 ml of Aquasol (New England Nuclear). Samples were later counted in a liquid scintillation counter.

Total inorganic carbon in the water samples was determined potentiometrically as described by Golterman and Clymo (1969).

Chlorophyll

Chlorophyll a and phaeopigments were measured fluorometrically (Strickland and Parsons, 1972). After filtration, the filters were placed in screw-cap vials containing 10 ml of 90% acetone, and the vials were wrapped in aluminum foil and kept refrigerated for 24 hr. No grinding was attempted since it was impractical with the Nitex filters. Thus there was a possibility of incomplete extraction (Yentsch and Menzel, 1963), but Bothwell (1975) indicated that this was not a serious problem.

Statistics

Univariate analysis of variance (ANOVA) computations were made with the program NWAY1 of the STATJOB series (Academic Computing Center, University of Wisconsin, Madison). The MANOVA computations were made with program BMD12V of the BMD series (Dixon, 1968).

RESULTS

Primary Productivity

The unfractionated productivity measurements ranged from <1 to about 30 mg C m^{-3} hr^{-1}, with the highest values occurring in late October. Interestingly, the <10- and the 10- to 64-μm fractions remained relatively low and constant throughout the year, usually contributing less than 5 mg C m^{-3} hr^{-1}. The >64-μm fraction, on the other hand, was responsible for the large seasonal variations observed in the unfractionated productivity measurements, reaching peak values of about 20 mg C m^{-3} hr^{-1} in the fall.

The results of the productivity measurements were tested by the MANOVA technique. The model used as a first step was a three-way partial hierarchical design. The three factors were date, station, and sample nested within station. The dependent variables were the three size fractions (the unfractionated totals and the sums of the fractions were tested separately). With the experimental design used, we hoped to estimate adequately the within-station variance and factor it out from the true between-station variance. Each sample was replicated (the two light bottles) to account for technique errors. The three samples taken at each station should estimate within-station variance.

The MANOVA results are shown in Table 1. The first analysis, which was for the six dates when all four stations were sampled, indicates that all main effects and their interactions were significant. When the analysis was repeated for the 18 dates with only samples from stations FRB (precondenser) and KD (postcondenser), again all effects were significant (Table 1).

The MANOVA was also applied to the results for the four dates when there was a ΔT of zero. The only significant factor at $\alpha = 0.05$ was date (Table 1). A test on the remaining dates with a ΔT again showed all effects to be significant. This indicates that productivity rates at KD were different from FRB because of plant operations. To make the tests more comparable, since only four dates had no ΔT, we selected four dates randomly (June 25, Sept. 20, July 12, and Jan. 15). The MANOVA test on these four dates again showed all effects to be significant. Mean productivity rates were slightly lower

TABLE 1

MANOVA RESULTS FOR PRIMARY PRODUCTIVITY MEASUREMENTS, WITH THE RATES OF THREE SIZE FRACTIONS AS DEPENDENT VARIABLES

Source	df	F	Sign
All Four Stations (6 dates)			
Dates	15, 193.64	113.0055	*
Stations	9, 170.51	40.8262	*
Samples within (stations)	24, 203.62	2.8252	*
Date X station	45, 208.73	7.6476	*
Date X sample (station)	120, 210.63	1.6923	*
Stations FRB and KD (18 dates)			
Dates	51, 316.38	62.3049	*
Stations	3, 106.0	9.4821	*
Samples (station)	12, 280.74	2.9821	*
Date X station	51, 316.38	2.8676	*
Date X sample (station)	204, 318.80	1.7987	*
Stations FRB and KD ($\Delta T = 0$)			
Dates	9, 53.63	76.0505	*
Stations	3, 22	1.5719	
Samples (station)	12, 58.5	0.9330	
Date X station	9, 53.69	0.6656	
Date X sample (station)	36, 65.73	1.0489	
Stations FRB and KD ($\Delta T \neq 0$, 4 random dates)			
Dates	9, 53.69	96.9242	*
Stations	3, 22	14.5135	*
Samples (station)	12, 58.5	4.3724	*
Date X station	9, 53.69	10.7098	*
Date X sample (station)	36, 65.73	4.0723	*

*Significant at $\alpha = 0.05$.

for all three fractions at station KD when a ΔT was present across the condenser. The difference is more noticeable for the < 10- and > 64-μm fractions. We should note that, on three of the four dates with no ΔT, only one pump was in operation. (Flow rates were 1.086×10^3 and 1.56×10^3 m^3/min with one and two pumps operating, respectively.) Thus mechanical effects may play an important role.

The results of a two-way MANOVA run on data from stations KD and FRB for individual dates, with station and sample as independent variables, are shown in Table 2. On 8 of the 18 dates, the stations were significantly different. Samples also differed significantly on 8 dates, but not always the same dates as those for station effects. The four dates with no ΔT showed no significant differences between stations or samples at $\alpha = 0.05$.

In an attempt to see which variables were contributing most to the observed MANOVA results, we carried out univariate three-way mixed-model ANOVA tests on each size fraction separately. The results of these tests, shown in Table 3, indicate that the 10- to 64-μm fraction is consistent over all stations and samples. The < 10- and > 64-μm fractions, however, form the basis of distinctions between stations. This is not wholly unexpected in view of the fact that the medium fraction, on the average, contributed only about 12% to the sum or productivity rates. Note also that sample variances are significant for the < 10-μm fraction but not for the other fractions.

Univariate ANOVA tests were also run on the sums of the fractions and the unfractionated totals of the productivity measurements. A three-way mixed-model ANOVA was again used for the sums of the fractions, but a two-way mixed model was required for the unfractionated totals because of unequal cell numbers. The results are shown in Table 4. Only the sample factor was not significant for the sums of the fractions. The ANOVA results for the unfractionated totals showed that stations were not significantly different, however.

To estimate relative variations within a station, we determined the coefficients of variation. These values ranged from 0.118 to 224.2%, with one exceptional value at 1217%. This considerable variability is not totally unexpected since patchiness of plankton is well documented (e.g., Platt, Dickie, and Trites, 1970; McAlice, 1970). Furthermore, methodological problems are indicated by the fact that the sums of the fractions and the unfractionated total are not equal for any given experiment. This disparity has been documented in other studies (e.g., Rodhe, Vollenweider, and Nauwerck, 1958; McCarthy, Taylor, and Loftus, 1974).

Chlorophyll a

The unfractionated chlorophyll a concentrations ranged from ~2 to 13 mg/m^3. Little difference could be seen among the stations for either unfractionated or fractionated concentrations. As with the productivity measurements, the < 10-μm fraction was relatively low and constant throughout the year (ranging between 0.5 and 3.5 mg/m^3). The > 64-μm fraction was responsible for the large seasonal fluctuations observed (ranging between 1 and 11 mg/m^3).

TABLE 2

MANOVA RESULTS FOR PRIMARY PRODUCTIVITY MEASUREMENTS AT STATIONS FRB AND KD, WITH THE RATES OF THE THREE SIZE FRACTIONS AS DEPENDENT VARIABLES

Date	ΔT	Number of pumps	Source†	Sign
8/26/75	9.5	2	A	*
			B	
9/6/75	6.1	2	A	*
			B	*
9/20/75	3.0	2	A	
			B	
10/4/75	9.5	2	A	
			B	*
10/24/75	9.5	2	A	
			B	
11/1/75	0	1	A	
			B	
11/15/75	6.0	2	A	
			B	
12/17/75	14.5	1	A	
			B	
1/15/76	6.3	1	A	*
			B	*
2/14/76	0	1	A	
			B	
3/27/76	0	1	A	
			B	
4/14/76	3.5	2	A	*
			B	
5/1/76	8.0	2	A	
			B	*
5/19/76	0	2	A	
			B	
6/9/76	9.5	2	A	*
			B	*
6/25/76	9.5	2	A	*
			B	*
7/12/76	9.7	2	A	*
			B	*
7/23/76	10.0	2	A	*
			B	*

*Significant at $\alpha = 0.05$
†A, station; B, sample.

TABLE 3

ANOVA RESULTS FOR PRIMARY PRODUCTIVITY MEASUREMENTS ANALYZING EACH FRACTION INDIVIDUALLY

Source†	df	MS	F	Sign
		Three-Way ANOVA for All Stations (6 dates)		
		0.45- to 10-μm Fraction		
A	5, 72	17.90529	217.88	*
B	3, 72	18.07965	220.00	*
C (B)	8, 72	0.49502	6.02	*
A × B	15, 72	3.15367	38.37	*
A × C (B)	40, 72	0.20671	2.52	*
		10- to 64-μm Fraction		
A	5, 72	20.77367	163.24	*
B	3, 72	0.18209	1.43	
C (B)	8, 72	0.14739	1.16	
A × B	15, 72	0.23573	1.85	*
A × C (B)	40, 72	0.19897	1.56	*
		>64-μm Fraction		
A	5, 72	430.87411	316.38	*
B	3, 72	11.16415	8.20	*
C (B)	8, 72	1.50015	1.10	
A × B	15, 72	7.57823	5.56	*
A × C (B)	40, 72	2.21260	1.62	*
		Three-Way ANOVA for Stations FRB and KD (all dates)		
		0.45- to 10-μm Fraction		
A	17, 108	9.86904	60.86	*
B	1, 108	1.37600	8.49	*
C (B)	4, 108	0.45492	5.11	*
A × B	17, 108	0.96713	5.96	*
A × C (B)	68, 108	0.16214	2.81	*
		10- to 64-μm Fraction		
A	17, 107	15.38948	105.38	*
B	1, 107	0.16729	1.15	
C (B)	4, 107	0.12321	1.84	
A × B	17, 107	0.24905	1.71	
A × C (B)	68, 107	0.17070	1.17	
		>64-μm Fraction		
A	17, 108	332.05368	261.38	*
B	1, 108	24.05937	18.94	*
C (B)	4, 108	3.51762	2.77	
A × B	17, 108	2.04306	1.61	*
A × C (B)	68, 108	2.31443	1.82	*

*Significant at $\alpha = 0.05$.

†A, date; B, station; and C (B), sample (station).

TABLE 4

ANOVA RESULTS FOR PRIMARY PRODUCTIVITY MEASUREMENTS ANALYZING THE SUMS OF FRACTIONS AND UNFRACTIONATED TOTALS

Source	df	MS	F	Sign
Three-Way ANOVA on Sums of Fractions for All Stations (6 dates)				
Date	5, 72	759.55246	387.69	*
Station	3, 72	26.03235	13.29	*
Sample (station)	8, 72	2.46270	1.26	
Date X station	15, 72	21.20638	10.82	*
Date X sample (station)	40, 72	1.95916	1.85	*
Three-Way ANOVA on Sums of Fractions for Stations FRB and KD (all dates)				
Date	17, 108	575.05704	298.14	*
Station	1, 108	32.13914	16.66	*
Sample (station)	4, 108	3.89665	2.02	
Date X station	17, 108	3.68309	1.91	*
Date X sample (station)	68, 108	1.92879	2.06	*
Two-Way ANOVA on Unfractionated Totals for All Stations (6 dates)				
Date	5, 24	289.55586	182.17	*
Station	3, 24	1.47682	0.93	
Date X station	15, 24	9.25700	5.82	*
Two-Way ANOVA on Unfractionated Totals for Stations FRB and KD (all dates)				
Date	17, 76	345.29883	305.87	*
Station	1, 76	0.81450	0.72	
Date X station	17, 76	3.19544	2.83	*

*Significant at $\alpha = 0.05$.

Evidence of patchiness was found on several occasions. For example, on Aug. 26 there was a relatively high concentration for the >64-μm fraction at station KWI which was not observed at any of the other stations. Similarly, high values were recorded for the >64-μm fraction on Sept. 20 at all stations except REF.

On the average, pre- and postcondenser chlorophyll a concentrations were comparable. The relative concentrations were 104, 98.5, 102, 99, and 101% of those found at FRB for the <10-, 10- to 64-, and >64-μm fractions, the unfractionated total, and the sum of the fractions, respectively.

Assimilation Ratios

The relationship of productivity per unit of chlorophyll a is known as the assimilation ratio. Current usage of assimilation ratio implies that carbon uptake should be measured at light saturation (Talling, 1974). During this study light intensity was kept constant, and no provision was made to determine saturating intensities, which are species dependent. For most of the year, the small fractions have the highest assimilation numbers. This is to be expected since small organisms have higher surface-to-volume ratios and, thus, are able to absorb nutrients more readily (Dugdale, 1967). The assimilation ratio has also been said to be a measure of physiological status (Thomas, 1970; Eppley, 1972). High ratios tend to indicate that factors such as temperature or nutrients are not limiting.

The medium fraction consistently had the lowest assimilation ratios. On the basis of size alone, the ratios in this fraction would be expected to be higher than the large fraction. Therefore, they may consist largely of detrital material, damaged cells, or broken filaments from the large fraction, or, alternatively, they may require different temperatures or light intensities.

The large fraction, although usually intermediate in terms of assimilation numbers, did reach a significant peak in spring (between 5.5 and 9.5).

The size-fractionated assimilation ratios showed some differences among the stations. Stations KWI and REF generally had higher ratios for the small fraction than did stations KD and FRB. There was also a spike in the small fraction at FRB on Sept. 20 which was not apparent at the other stations. The peaks in assimilation numbers were generally higher for station FRB than the corresponding peaks for KD. This could indicate lowered photosynthetic capacity after passage through the plant.

DISCUSSION

In this report we address power-plant operation only in relation to the physiological status of entrained phytoplankton. More general discussions on the ecology of the nearshore phytoplankton are given elsewhere (Zeeman, 1977).

Entrainment Effects

In this study, we attempted to distinguish between variances caused by sampling methods and true differences in primary productivity among stations by means of both multivariate and univariate analysis of variance. The models were designed to factor out effects of repetitive sampling to estimate true differences. The statistical results seem contradictory, however. When the three size fractions were analyzed simultaneously, power-plant operations appeared to be having an effect, however small. Differences among samples from the same station were also found to be significant when the plant was in operation. This indicates patchy distribution of phytoplankton. When the sums of the size fractions were subjected to ANOVA techniques, station differences were once again seen, but sample effects were not observed. The unfractionated productivity measurements, on the other hand, showed no differences among stations. Sample effects in the last analysis could not be tested.

One explanation for the observed results is that perhaps the size-fractionation method produces artifacts. The sums of the fractions and the unfractionated totals are not the same in any given experiment. It is, therefore, not unreasonable to assume that this could influence the outcome of statistical tests. Holmes and Anderson (1963) and Lasker and Holmes (1957) showed how variable retention can be on different filters. Evidence that size fractionation is responsible for much of the variance in productivity among stations comes from the ANOVA results. Analysis on the unfractionated totals showed no significant differences among stations, yet the ANOVA on the sums of the fractions showed that differences did exist. If the methodology were as precise as desired, the results would agree.

According to Strickland and Parsons (1972), the precision of the radiocarbon technique is ~7% for two replicates at the 1.5 mg C m^{-3} hr^{-1} level and ~8% at the 25 mg C m^{-3} hr^{-1} level. For six replicates the precision is increased to about 5 and 4% for the two levels, respectively. From the coefficients of variation in this study, it was evident that any attempt to characterize differences in productivities among the sites was going to be complicated by large variations, possibly caused by patchiness and analytical error.

Differences in productivity rates between the precondenser (FRB) and the postcondenser stations (KD) were small when averaged over the entire year. Percent differences (in comparison to FRB) were −9.08, +5.32, −12.12, −2.62, and −9.27% for the small-, medium-, and large-size fractions, the unfractionated total, and the sum of the fractions, respectively. Even though these values showed a general decrease at KD, we do not feel they are biologically significant in view of the methodologies involved in their determination. The slight increase observed for the medium-size fraction, if real, was possibly caused by breakage of filamentous or colonial algae into smaller fragments, which are apparently undamaged. The chlorophyll concentration was unaffected (98.5% of FRB), however. Therefore either the power plant stimulated photosynthesis in this fraction or the results are related to the method of fractionation.

Rousar (1973) reported a mean productivity rate of 9.6 mg C m^{-3} hr^{-1} at an inshore Wisconsin station. Our estimates were 9.3 and 8.6 mg C m^{-3} hr^{-1} at stations FRB and KD, respectively. This indicates that power-plant operation has not radically altered productivity rates in the surrounding region. Rather, deaths of individual phytoplankton caused by entrainment were balanced by natural processes of population growth. We also calculated that, at most, only 0.41% of the nearshore water (0 to 30 m in depth) could pass through the plant annually.

Natural variability poses a significant obstacle to determining small changes in productivity or chlorophyll a. Carpenter, Anderson, and Peck (1974) stated that inability to detect changes in photosynthetic rates could be caused by sampling error or patchiness or could be because there was no actual difference. They also concluded that for chlorophyll a 88 replicates would have to be taken to detect a ±5% change. Similarly, 22 replicates were required for a ±10% change and 6 replicates for a ±20% change.

The fact that the unfractionated productivities at the pre- and postcondenser stations were virtually identical and were not significantly different statistically was a very strong indication that the power plant was not affecting the phytoplankton. Any differences observed between the stations were probably negligible when the precision of sampling techniques and analytical procedures is taken into account. Statistical tests indicate that if certain components of the phytoplankton were affected, whether by stimulatory or inhibitory action, it would be those having dimensions < 10 and > 64 μm.

This discussion presents the pertinent facts regarding the effects of power-plant operation on primary productivity at the Kewaunee Power Plant. Some of the statistical results suggest that plant operation may be injurious to individual phytoplankton passing through the cooling system, but the effect on populations is negligible. Furthermore, any observed changes could be artifacts of the methods used in the study. If they are real, the effects are small and detectable only by statistical techniques. Indeed, they may not even be as significant as such natural perturbations as wave action during storms.

The lack of any obvious effects on the phytoplankton is probably a result of several factors. The ambient temperatures of

Lake Michigan are relatively low, and temperature rises caused by the power plant are not sufficiently great to exceed the thermal tolerance limits of the organisms, even during the summer. Also important is that this power plant does not have to use chlorine or other biocide to maintain efficient heat exchange in the condenser tubes.

Estimates of size-fractionated productivity are important to our understanding of aquatic ecosystems. Their utility will increase further when our knowledge of size-selective grazing is improved. Entrainment does not seem significant to phytoplankton population changes at the Kewaunee Nuclear Power Plant, but conditions may be different elsewhere. Changes in productivity of certain size classes could have far-reaching implications to trophic structure.

Multivariant analysis of variance, as demonstrated here, could prove to be a useful tool in analyzing structural changes in phytoplankton communities. The ability to analyze several variables simultaneously adds new dimensions to ecological investigations which have been neglected previously. These variables need not be limited to productivity rates. Indeed, it might prove enlightening to use other measurements (e.g., of adenosine triphosphate or chlorophyll a) in combination with productivity estimates.

ACKNOWLEDGMENTS

This research was supported in part by a grant to The University of Wisconsin, Milwaukee, Department of Botany, from the Wisconsin Public Service Corporation and by a research assistantship to Stephan Zeeman from the Center for Great Lakes Studies, UWM. This study was part of an M.S. thesis in botany by Zeeman.

REFERENCES

Allen, T. F. H., and S. Skagen, 1973, Multivariate Geometry as an Approach to Algal Community Analysis, *Br. Phycol. J.*, 8: 267-287.

Bothwell, M. L., 1975, *Studies on the Distribution of Phytoplankton Pigments and Nutrients in the Milwaukee Harbor Area and Factors Controlling Assimilation Numbers*, Ph.D. Thesis, University of Wisconsin, Madison.

Bremer, K. E., and D. G. Redmond, 1974, Phytoplankton Entrainment, in *Preoperational Thermal Monitoring Program of Lake Michigan Near the Kewaunee Nuclear Power Plant: January—December 1973*, Chap. 5, third annual report prepared by Industrial Bio-test Laboratories for Wisconsin Public Service Corporation, Green Bay.

Brooks, A. S., 1972, *The Influence of a Thermal Effluent on the Phytoplankton Ecology of the Indian River Estuary, Delaware*, Ph.D. Thesis, Johns Hopkins University, Baltimore, Md.

Carpenter, E. J., S. J. Anderson, and B. B. Peck, 1974, Copepod and Chlorophyll a Concentrations in Receiving Waters of a Nuclear Power Station and Problems Associated with Their Measurement, *Estuarine Coastal Mar. Sci.*, 2: 83-88.

Cooley, W. W., and P. R. Lohnes, 1971, *Multivariate Data Analysis*, John Wiley & Sons, Inc., New York.

Dixon, W. J. (Ed.), 1968, *BMD: Biomedical Computer Programs*, University of California Press, Berkeley.

Doty, M. S., and M. Oguri, 1957, Evidence for a Photosynthetic Daily Periodicity, *Limnol. Oceanogr.*, 2: 37-40.

Dugdale, R. C., 1967, Nutrient Limitation in the Sea: Dynamics, Identification, and Significance, *Limnol. Oceanogr.*, 12: 685-695.

Eppley, R. W., 1972, Temperature and Phytoplankton Growth in the Sea, *Fish. Bull.*, 70: 1063-1085.

Fox, J. L., and M. S. Moyer, 1973, Some Effects of a Power Plant on Marine Microbiota, *Chesapeake Sci.*, 14: 1-10.

Goldman, C. R., and R. G. Wetzel, 1963, A Study of the Primary Productivity of Clear Lake, Lake County, California, *Ecology*, 44: 283-294.

Golterman, H. L., and R. S. Clymo (Eds.), 1969, *Methods for Chemical Analysis of Fresh Waters*, IBP Handbook No. 8, Blackwell Scientific Publications, Oxford.

Holmes, R. W., and G. C. Anderson, 1963, Size Fractionation of C^{14}-labeled Natural Phytoplankton Communities, in *Symposium on Marine Microbiology*, C. H. Oppenheimer (Ed.), pp. 241-250, Charles C Thomas, Publishers, Springfield, Ill.

Hoogenhout, H., and J. Amesz, 1965, Growth Rates of Photosynthetic Microorganisms in Laboratory Cultures, *Arch. Mikrobiol.*, 50: 10-25.

Jones, P. A., C. L. Brown, and D. G. Redmond, 1975, Phytoplankton Entrainment, in *Operational Environmental Monitoring Program of Lake Michigan Near the Kewaunee Nuclear Power Plant: Chemical and Biological Studies, January—December 1974*, Chap. 5, fourth annual report prepared by Industrial Bio-test Laboratories for Wisconsin Public Service Corporation, Green Bay.

Lasker, R., and R. W. Holmes, 1957, Variability in Retention of Marine Phytoplankton by Membrane Filters, *Nature*, 180: 1295-1296.

Lauer, G. J., W. T. Walter, and A. P. Beck, 1972, Problems and Potential Solutions to Biological Entrainment and Chemical Discharges, paper presented at the Atomic Industrial Forum: Conference on Water Quality Considerations, Washington, D.C., Oct. 2, 1972.

McAlice, B. J., 1970, Observations on the Small Scale Distribution of Estuarine Phytoplankton, *Mar. Biol.*, 7: 100-111.

McCarthy, J. J., W. R. Taylor, and M. E. Loftus, 1974, Significance of Nanoplankton in the Chesapeake Bay Estuary and Problems Associated with the Measurement of Nanoplankton Productivity, *Mar. Biol.*, 24: 7-16.

Morgan, R. P., II, and R. G. Stross, 1969, Destruction of Phytoplankton in the Cooling Water Supply of a Steam Electric Station, *Chesapeake Sci.*, 10: 165-171.

Patrick, R., 1969, Some Effects of Temperature on Freshwater, in *Biological Aspects of Thermal Pollution*, P. A. Krenkel and F. C. Parker (Eds.), pp. 161-185, Vanderbilt University Press, Nashville, Tenn.

——, 1974, Effects of Abnormal Temperatures on Algal Communities, in *Thermal Ecology*, AEC Symposium Series, Augusta, Ga., May 3—5, 1973, J. W. Gibbons and R. R. Sharitz (Eds.), pp. 335-349, CONF-730505, NTIS.

Platt, T., L. M. Dickie, and R. W. Trites, 1970, Spatial Heterogeneity of Phytoplankton in a Near-Shore Environment, *J. Fish. Res. Board Can.*, 27: 1463-1473.

Pomeroy, L. R., 1974, The Ocean's Food Web, a Changing Paradigm, *Bioscience*, 24: 499-504.

Rodhe, W., R. A. Vollenweider, and A. Nauwerck, 1958, The Primary Production and Standing Crop of Phytoplankton, in *Perspectives in Marine Biology*, A. A. Buzzati-Traverso (Ed.), pp. 229-322, University of California Press, Berkeley.

Rousar, D. C., 1973, Seasonal and Spatial Changes in Primary Production and Nutrients in Lake Michigan, *Water, Air, Soil Pollut.*, 2: 497-514.

Schubel, J. R., and W. E. Schiemer, 1969, On the Microscopic Determination of the Volume Size Distribution of Fine-Grained Suspended Particles, *Limnol. Oceanogr.*, 14: 438-441.

Steeman-Nielsen, E., 1952, The Use of Radioactive Carbon (^{14}C) for Measuring Organic Production in the Sea, *J. Cons. Int. Explor. Mer*, 18: 117-140.

Strickland, J. D. H., and T. R. Parsons, 1972, *A Practical Handbook of Seawater Analysis*, Bulletin 167, 2nd ed., Fisheries Research Board of Canada, Ottawa, Ont.

Talling, J. F., 1974, Relations Between Primary Production and Population Density ("standing crop"), in *A Manual on Methods for Measuring Primary Production in Aquatic Environments*, IBP Handbook No. 12, R. A. Vollenweider (Ed.), pp. 150-152, Blackwell Scientific Publications, Oxford.

Thomas, W. H., 1970, On Nitrogen Deficiency in Tropical Pacific Oceanic Phytoplankton: Photosynthetic Parameters in Poor and Rich Water, *Limnol. Oceanogr.*, 15: 380-385.

Van de Geer, J. P., 1971, *Introduction to Multivariate Analysis for the Social Sciences*, W. H. Freeman & Co., San Francisco.

Warinner, J. E., and M. L. Brehmer, 1966, The Effects of Thermal Effluents on Marine Organisms, *Int. J. Air Water Poll.*, 10: 277-289.

Yentsch, C. M., and D. W. Menzel, 1963, A Method for Determination of Phytoplankton Chlorophyll and Phaeophytin by Fluorescence, *Deep-Sea Res.*, 10: 221-231.

Zeeman, S. I., 1977, *Size-Fractionated Primary Productivity in Lake Michigan near the Kewaunee Nuclear Power Plant*, M.S. Thesis, University of Wisconsin, Milwaukee.

PRIMARY PRODUCTIVITY: ANALYSIS OF VARIANCE IN A THERMALLY ENRICHED AQUATIC SYSTEM

MARGARET O. WELCH and C. H. WARD
Departments of Biology and Environmental Science and Engineering, Rice University, Houston, Texas

ABSTRACT

Primary production experiments were performed in a 400-ha cooling reservoir for a 530-MW electric generating plant in eastern Texas over a 15-month period. Annual primary productivity of the heated and ambient surface waters and the nutrient enrichment response were investigated in situ by ^{14}C methods. On the basis of analysis of variance techniques, primary productivity was significantly higher at the heated station on 18 of 26 sampling dates, but the magnitude of the difference was not correlated with the difference in water temperature at the stations. Nutrient enrichment had no effect on primary productivity, but temperature did. Regression analysis indicated that the temperature optimum for the natural population of phytoplankton was ~25°C. The temperature tolerance range for the phytoplankton community subjected to higher temperatures was higher than for the community at ambient temperatures.

An understanding of the biology of high-temperature aquatic systems is necessary for predicting and managing the effects of man-induced thermal additions. The extensive literature (e.g., Patrick, 1969; Fogg, 1975) documenting the thermal tolerance range and the thermal optimum of many species of freshwater algae enables us to generalize that more species of the Cyanophyta are thermal tolerant and grow well above 35°C, that the Chlorophyta tend to grow best up to 35°C, and that the diatoms, the Bacillariophyta, succeed best below 30°C. Most investigators agree that the thermal optimum for all species in the laboratory is generally higher than that in natural aquatic systems. Castenholz and Wickstrom (1975) and Patrick (1969) pointed out that most studies deal with the effects of thermal additions on species or on populations, but not a great deal is known about the effects at the community level. When the thermal optimum for existing species is exceeded in natural waters, the species are unable to compete for the available resources, and the community structure of the system changes. The seasonal variation of resources superimposes more complicating factors. The question is, Will system function, energy flow, and productivity also be altered?

The objectives of this study were to measure and evaluate the effect of increased temperature on a phytoplankton community as indicated by primary productivity. Annual primary productivity of the heated and unheated surface waters of a cooling reservoir were investigated in situ by ^{14}C methods.

LITERATURE

Warinner and Brehmer (1966), studying the effects of thermal effluents on a community of marine organisms in a riverine estuary, found an increase in primary production during the winter months and a decrease during the summer months. Patrick (1969) noted that, as long as nutrients and light are sufficient, productivity may increase with increases of temperature within the thermal tolerance range of the existing algae. Simmons and Armitage (1974) determined that heated power-plant effluent had no effect on algal blooms in the Potomac River. Foerster, Trainor, and Buck (1974), on finding that blooms correlated well with increased temperature, thought that the mechanism might be an increase in the rate of diffusion across the depletion zone surrounding the algal cell. Tilly (1973; 1974; 1975) and Marshall and Tilly (1971) reported on investigations of phytoplankton and periphyton in Par Pond in South Carolina. Maximum productivity and integral productivity were strongly correlated with temperature increases, but primary productivity per unit of chlorophyll was not. Tilly theorized that elevated temperatures enabled phytoplankton to use higher light intensities without photoinhibition, a phenomenon well documented in laboratory studies (Sorokin and Myers, 1953).

DESCRIPTION OF THE STUDY AREA

Lewis Creek Reservoir (Fig. 1) is a 404-ha reservoir constructed in 1970 by Gulf States Utilities as a semiclosed system to cool the condensers of a 530-MW electric generating plant. The reservoir,

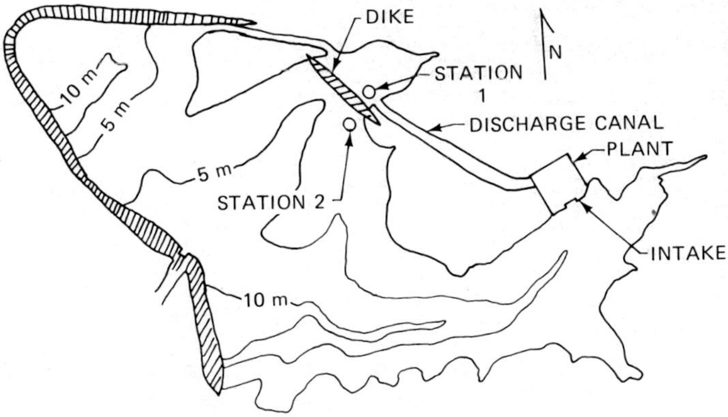

Fig. 1 Approximate bathometry of Lewis Creek Reservoir. Contour lines show depth.

formed by constructing an earthen dam over Lewis Creek, an intermittent stream, has a maximum depth of 5.34 m, and its elevation is 83 m. It is located in Montgomery County, Texas, at latitude 30°26′ and longitude 95°32′. The climate is subtropical, with mild winters, hot, humid summers, and a rainfall average of 120 cm/year. The drainage basin is ~1000 ha, and the land is used principally as pine timberland and livestock range. The area is sparsely populated.

Station 1 was located at the outfall of the canal, which discharges into a small preliminary cooling pond ~1 km from the generating plant. The temperature at Station 1 approximated the temperature of the effluent as it left the plant. Station 2 was in a cove of the lake ~2.5 km by water from Station 1. Because of the construction of a dike, these two stations were separated by land by about 15 m. This permitted almost simultaneous sampling and incubation. The temperature of Station 2 approximated the temperature of the water at the intake screen.

METHODS

Sampling

Sample collections were made from September 1976 to August 1977. To minimize the errors inherent in sampling the heterogeneously distributed phytoplankton, we took one large water sample from the surface to a depth of 30 cm at each station. Portions of the large sample were used for chemical analysis, primary productivity measurements, and phytoplankton cell counts.

Physicochemical Measurements

Measurements of temperature, pH, water transparency, solar radiation, and alkalinity were made at each station on each sampling date. Nitrate nitrogen, orthophosphate, and dissolved oxygen measurements were made monthly. Water temperature at the incubation depth (0.5 m) was measured with a mercury centigrade thermometer; hydrogen ion concentration was measured with a pH meter in the laboratory; water transparency was measured with a Secchi disk; and instantaneous solar radiation in foot-candles was measured with a portable light meter. The instantaneous light readings were converted to total langleys for the sampling date by correlating the data with data from the Texas A & M University Meteorological Center. Phenolphthalein and methyl orange alkalinity were measured by titration with 0.02*N* sulfuric acid (American Public Health Association, 1971). Duplicate dissolved oxygen samples were fixed by the azide modification of the Winkler method (American Public Health Association, 1971). Nitrate nitrogen was measured by the ultraviolet spectrophotometric method, and orthophosphate was measured by the stannous chloride method using a benzene—isobutanol extraction (American Public Health Association, 1971).

Biological Measurements

Carbon-14 techniques were used to estimate primary productivity, following the procedures of Goldman et al. (1974) and the liquid scintillation techniques of Schindler, Moore, and Vollenweider (1974). An ampul of 1 ml of $NaHCO_3$ with an activity of 5 μCi was added to each 125-ml bottle, and the bottles were incubated for 2 hr between 10:00 a.m. and 2:00 p.m. Twelve liters of the large initial sample were poured through a plankton net of No. 20 nylon-mesh bolting cloth with 68 threads to the centimeter. The concentrate was centrifuged at 1500 x *g* in the laboratory, resuspended in a known volume, and counted according to the methods of Edmondson (1974).

Ten nutrient-enrichment experiments were conducted to determine the effect of two levels of nitrogen ($NaNO_3$) and phosphorus (K_2HPO_4) at the two temperature levels by use of a factorial design (Ramm and Karn, 1976; Jordan and Bender, 1973). Levels of nitrogen were 0 and 0.025 mg/liter and of phosphorus, 0 and 0.005 mg/liter.

Eight temperature-effects experiments were conducted by reversing four bottles from each station and incubating the hot-water sample at Station 2 and the cool-water sample at Station 1.

Statistical Measurements

The data from each sampling date were treated with analysis of variance (ANOVA) to determine if there was a statistically significant effect of temperature or treatment (Snedecor and Cochran, 1967). The data for the year were graphed as a regression of primary productivity on temperature. The graphs of data from both stations exhibited a curvilinear relationship. Two linear components could be analyzed; however, by separating the data at 25°C and performing a least-squares linear regression on each component. In the equation for the regression line ($Y = \alpha + \beta X$), α is height, where $X = \overline{X}$, and β is slope (Cohen and Cohen, 1975; Snedecor and Cochran, 1967).

RESULTS

Physicochemical Measurements

The temperature range at Station 1 was 17 to 39°C, and the average temperature for the year was 27°C. At Station 2 the range was 10 to 34°C, and the average was 21.5°C.

The difference in temperature (ΔT) between Stations 1 and 2 was greater in the winter months. The largest ΔT (8°C) occurred in January, and the smallest (2°C) occurred three times, in October and November 1976 and June 1977.

The pH ranged from 6.5 to 7.0 at Station 1 and 6.4 to 7.0 at Station 2 and was frequently the same at both stations. The seasonal variation was slight.

The transparency of the water decreased during the fall and winter months and increased during the spring and summer, with the maximum occurring in August and the minimum in January. The range of the Secchi disk depth was 1.13 to 2.20 m. Although we attempted to sample on clear, cloudless days, daily variation in solar radiation was considerable, ranging from 530 to 120 langleys/day.

Alkalinity, as $CaCO_3$, varied from 89 to 120 mg/liter and was generally the same at both stations. Alkalinity was caused by bicarbonate since phenolphthalein alkalinity was never detected. Seasonal variation was slight.

Nitrate nitrogen ranged from undetectable to 0.45 mg/liter. Orthophosphate was never detected by the method used. Dissolved oxygen was always over 90% saturation and often over 100% saturation at both stations.

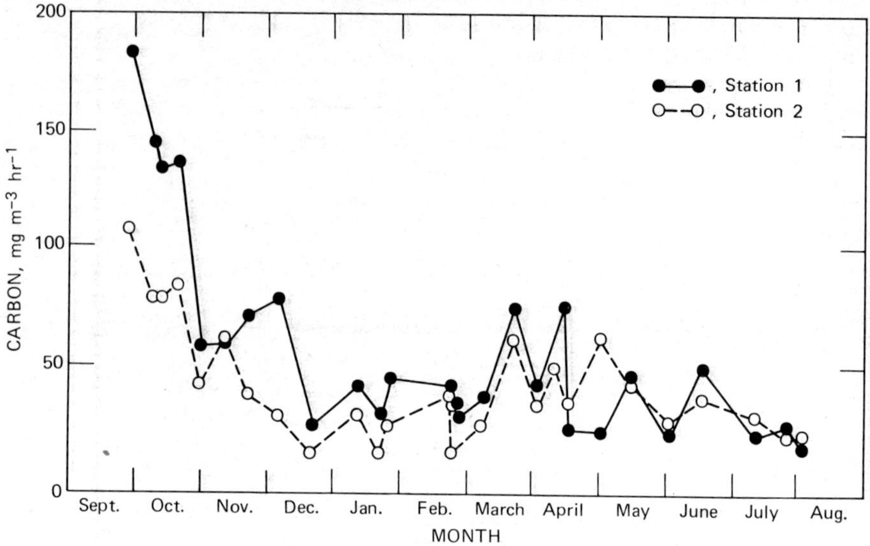

Fig. 2 Seasonal variation of the surface-water primary productivity.

Biological Measurements

On the basis of analysis of variance techniques, primary productivity was significantly higher at the discharge station on 18 of 26 sampling dates (Fig.2). On five sampling dates there were no significant differences between stations in the amount of carbon fixed. On three sampling dates productivity was significantly higher at the unheated station. There was no apparent relationship between higher productivity and biomass or community composition between the two stations. At all times, dark-bottle fixation was ~10% of light-bottle fixation at both stations.

The difference between the two stations in the amount of carbon fixed could be considered the difference in photosynthesis (ΔP). There was no correlation between the difference in temperature, ΔT, and the difference in carbon fixation, ΔP.

Analysis of variance showed nutrient enrichment had no effect on primary productivity in any of the experiments. Experiments incubating hot-water samples at the ambient station and ambient-water samples at the heated station showed statistically significant effects of temperature in five of eight experiments (Fig. 3). Ambient-water samples incubated at increased temperatures showed increased productivity rates, and, conversely, samples from the discharge

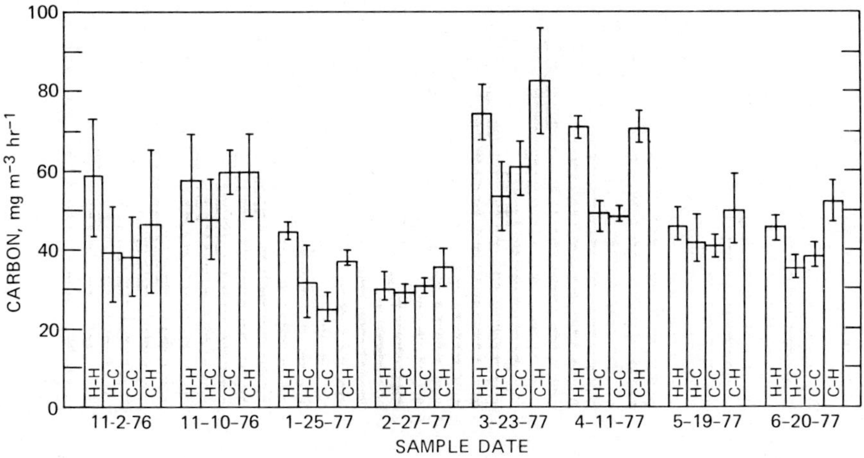

Fig. 3 Carbon-14 fixation rates during temperature effect experiments. H—H, hot samples incubated at Station 1. H—C, hot samples at Station 2. C—C, cold samples at Station 2. C—H, cold samples at station 1. [, 2 standard deviation.

stations incubated at cooler temperatures showed lower rates of productivity. On Nov. 10, Feb. 27, and May 19, there was no significant difference between the primary productivity at Stations 1 and 2 and no temperature effect when samples were reversed.

Figures 4 and 5 are plots of the productivity as a function of temperature for Stations 1 and 2, respectively. Regression lines and equations are given for each component; r^2 is a measure of variance in the data, and r is the Pearson product-moment correlation coefficient, a measure of how well the data fit the regression line or of the "scatter" in the data. On both figures the values at 25°C were included in the calculations for both components.

DISCUSSION

The rate of carbon fixation is affected by increased temperature, as evidenced by the ANOVA for 69% of the sampling dates. The maximum values in September and October for primary production rates could be associated with lake turnover and vertical mixing of nutrients into the epilimnion after disintegration of the thermocline, but this was not reflected in the chemistry of the water. This hypothesis is further supported by the concurrent decrease in transparency of the water during the fall of 1976. Texas lakes are generally monomictic, with mixing occurring during the cool months and stratification occurring during the warmer months.

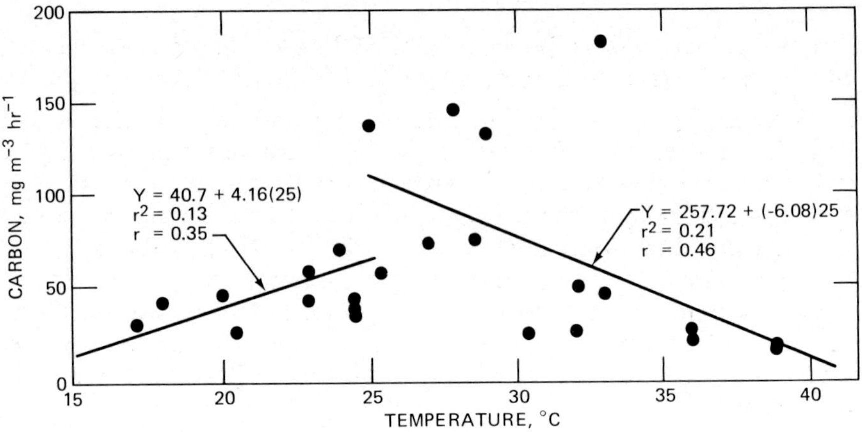

Fig. 4 Carbon fixation as a function of temperature at Station 1.

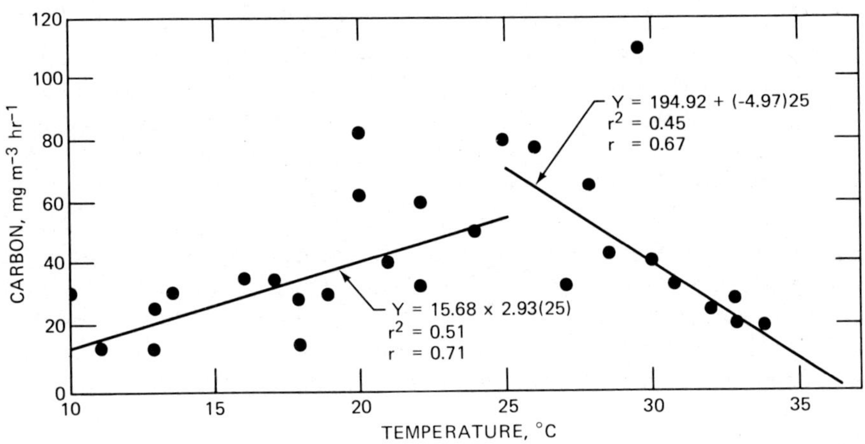

Fig. 5 Carbon fixation as a function of temperature at Station 2.

That temperature stimulates the rate of carbon fixation in Lewis Creek Reservoir is supported by the temperature-effect experiments. On dates when the data showed a significant effect of temperature on the rate of carbon fixation, it was significant to the 99% level, except on Nov. 2, when sampling error was such that the level of significance was only 90%.

The results of the nutrient experiments could be caused by an insufficient nutrient spike or could be associated with nutrient uptake lag time. It is possible that a 2-hr incubation time is not sufficient to detect the effects of additions of a limiting nutrient.

Biological processes are not often linear, and relationships are not generally continuous. Regression analysis assumes that the underlying relationships between variables are linear and additive. In attempting to analyze the data, however, we found it more expedient to divide the data from each station at 25°C and compute the linear regression equation and measures of correlation before using more complicated methods. The objectives were, not to develop a model for phytoplankton productivity, but to explain and predict the functioning of the phytoplankton community as it relates to temperature.

The rate of carbon fixation tends to increase β units with every degree of increase in temperature up to 25°C and to decrease $-\beta$ units with every degree of increase in temperature above 25°C; i.e., it appears that the temperature optimum for the natural phytoplankton community in Lewis Creek Reservoir lies in the vicinity of 25°C, and temperatures in excess of that tend to suppress the rate of carbon fixation.

By setting Y equal to 0, we can estimate a temperature tolerance range below and above which productivity is suppressed entirely. For Station 1 the estimated temperature tolerance range is 10 to 42°C and for Station 2, 5 to 39°C. This assumption excludes the possibility of a complete change in community structure. It is interesting to note that the estimated temperature tolerance ranges are not the same and that the estimate for the community subjected to consistently higher temperatures at Station 1 is higher. This suggests that organisms at Station 1 are adapted to higher temperatures.

The proportion of variance explained by temperature, r^2, is larger at Station 2; i.e., 51% of the variance at Station 2 between 10 and 25°C is explained by temperature and only 13% is explained at Station 1. Thus other factors at Station 1 contribute to the variability of rates of carbon fixation. One possibility is stress to the organism caused by passage through the condensers (Lanza and Cairns, 1972; Gurtz and Weiss, 1974).

Our work centered only on the phytoplankton of Lewis Creek Reservoir. It became obvious soon after this study commenced that, throughout much of the year, the system is dominated by a large population of benthic algae. Detailed studies on the effects of temperature on benthic productivity and biomass accumulation would be of considerable interest.

ACKNOWLEDGMENT

We want to express our appreciation to the Gulf States Utilities Company for its cooperation and support in the conduct of this study.

REFERENCES

American Public Health Association, 1971, *Standard Methods for the Examination of Water and Wastewater*, 13th ed., Washington, D. C.

Castenholz, R. W., and C. E. Wickstrom, 1975, Thermal Streams, in *River Ecology*, B. A. Whitton (Ed.), University of California Press, Berkeley.

Cohen, J., and P. Cohen, 1975, *Applied Multiple Regression: Correlation Analysis for the Behavioral Sciences*, John Wiley & Sons, Inc., New York.

Edmondson, W. T., 1974, A Simplified Method for Counting Phytoplankton, in *A Manual on Methods for Measuring Primary Production in Aquatic Environments*, R. A. Vollenweider (Ed.), International Biological Program Handbook 12, Blackwell Scientific Publications, Oxford.

Foerster, J. W., F. R. Trainor, and J. D. Buck, 1974, Thermal Effects on Connecticut River: Phycology and Chemistry, *J. Water Pollut. Control Fed.*, 46: 2138-2152.

Fogg, G. E., 1975, *Algal Cultures and Phytoplankton Ecology*, 2nd ed., University of Wisconsin Press, Madison.

Goldman, C. R., E. Steeman-Nielson, R. A. Vollenweider, and R. G. Wetzel, 1974, in *A Manual on Methods for Measuring Primary Production in Aquatic Environments*, R. A. Vollenweider (Ed.), International Biological Program Handbook 12, Blackwell Scientific Publications, Oxford.

Gurtz, M. E., and C. M. Weiss, 1974, Effect of Thermal Stress on Phytoplankton Productivity in Condenser Cooling Water, in *Thermal Ecology*, AEC Symposium Series, Augusta, Ga., May 3—5, 1973, J. W. Gibbons and R. R. Sharitz (Eds.), pp. 490-507, CONF-730505, NTIS.

Jordan, R. A., and M. E. Bender, 1973, *An In Situ Evaluation of Nutrient Effects in Lakes*, Report EPA-Re-73-018, Environmental Protection Agency, Washington, D. C.

Lanza, G. R., and J. Cairns, Jr., 1972, Physio-Morphological Effects of Abrupt Thermal Stress on Diatoms, *Trans. Am. Microsc. Soc.*, 91: 276-298.

Marshall, J. S., and L. J. Tilly, 1971, Temperature Effects on Phytoplankton Productivity in a Reactor Cooling Pond, in *Radionuclides in Ecosystems*, Proceedings of the Third National Symposium on Radioecology, Oak Ridge, Tenn., May 10—12, 1971, D. J. Nelson (Ed.), USAEC Report CONF-710501-P1, pp. 645-651, Oak Ridge National Laboratory, NTIS.

Patrick, R., 1969, Some Effects of Temperature in Freshwater Algae, in *Biological Aspects of Thermal Pollution*, P. A. Krenkel and F. L. Parker (Eds.), Proceedings of National Symposium on Thermal Pollution, Vanderbilt University Press, Nashville, Tenn.

Ramm, A. E., and B. P. Karn, 1976, A Note on the Design of Nutrient Enrichment Studies, *J. Water Pollut. Control Fed.*, 48: 2211-2212.

Schindler, D. W., J. Moore, and R. A. Vollenweider, 1974, Liquid Scintillation Techniques, in *A Manual on Methods for Measuring Primary Production in*

Aquatic Environments, R. A. Vollenweider (Ed.), International Biological Program Handbook 12, Blackwell Scientific Publications, Oxford.

Simmons, G. M., and B. J. Armitage, 1974, Evaluation of Heated Water Discharge on Phytoplankton Blooms in the Potomac River, *Hydrobiologia*, 45: 441-465.

Snedecor, G. W., and W. G. Cochran, 1967, *Statistical Methods*, 6th ed., Iowa State University Press, Ames.

Sorokin, C., and J. Myers, 1953, A High Temperature Strain of *Chlorella*, *Science*, 117: 330-331.

Tilly, L. J., 1973, Comparative Productivity of Four Carolina Lakes, *Am. Midl. Nat.*, 90: 356-365.

——, 1974, Periphyton Colonization and Productivity in the Reactor Cooling Reservoir—Par Pond, paper presented at the 35th Annual Meeting of the Association of Southeastern Biologists, Savannah, Ga., April 18—20, 1974.

——, 1975, Periphyton Crops and Productivity in a Reactor Thermal Effluent, paper presented at the 38th Annual Meeting of the American Society of Limnology and Oceanography, Halifax, Nova Scotia, June 23—26, 1975.

Warinner, J. E., and M. L. Brehmer, 1966, The Effects of Thermal Effluents on Marine Organisms, *Int. J. Air Water Pollut.*, 10: 277-289.

NITRATE REDUCTASE ACTIVITY AND PRIMARY PRODUCTIVITY OF PHYTOPLANKTON ENTRAINED THROUGH A NUCLEAR POWER STATION ON NORTHEASTERN LONG ISLAND SOUND

BRADFORD B. PECK* and R. SCOTT WARREN
Department of Botany, Connecticut College, New London, Connecticut

ABSTRACT

The effects of temperature and various concentrations of chlorine on nitrate reductase activity and primary productivity of phytoplankton were studied at the Millstone Nuclear Power Station on northeastern Long Island Sound. During August the ambient temperature at the cooling water intake ranged from 19.5 to 20°C. Power generation during this period resulted in temperature increases of 11 and 14°C at the discharge and depressed phytoplanktonic nitrate reductase activity by 88 to 89% and phytoplanktonic primary productivity by 42 to 52%. The decrease occurred during the 6- to 9-hr transit through the cooling pond. Nitrate reductase activity, maximally depressed after exposure to a mean increase of 13°C above ambient temperature on seven days in August, did not recover to intake control levels after 24 hr of incubation at ambient intake temperature. In March and April, when the ambient temperature of Long Island Sound water was 4.3 to 9.9°C, phytoplanktonic nitrate reductase activity was stimulated 25% above that of controls after 6 to 9 hr of exposure at 11.5 to 18.1°C above ambient temperature. The productivity results are similar to previous unpublished findings on temperature influence at this site. The nitrate reductase activity findings support the hypothesis that nitrate reductase is a heat-labile enzyme. Chlorine concentrations below and above those required to eliminate fouling organisms (0.50 ppm) produced large decreases in the photosynthetic rate of entrained phytoplankton. Previous work at this site reported similar findings. Nitrate reductase activity decreased 15% at 1.0 ppm and 1.2 ppm, the two highest chlorine dosages applied.

*Present address: Peaks Island, Maine.

The effects of various levels of temperature and chlorine in the cooling water entrained through the Millstone Nuclear Power Plant on nitrate reductase activity and primary productivity of the natural phytoplankton population were tested.

Nitrate reductase catalyzes the reduction of nitrate to nitrite:

$$NO_3^- + 2H^- + 2e^- \rightarrow NO_2^- + H_2O$$

The initial reduction of nitrate is the rate-limiting step in the assimilation of nitrate (Beevers and Hageman, 1969). The nitrite produced is further reduced to ammonia by nitrite reductase and is assimilated by glutamic dehydrogenase to produce glutamic acid, the nitrogen source for other amino acids.

Nitrate reductase is light and heat labile and is relatively unstable in vivo and in vitro (Beevers and Hageman, 1969; Schrader et al., 1968). Eppley, Packard, and MacIsaac (1970) and Packard et al. (1971) reported an optimum temperature of around 15°C for the nitrate reductase activity of a natural population of marine phytoplankton in the Peru current. There is evidence that a large and sudden rise in temperature lowers nitrate reductase activity of *Hordeum vulgare* L. (barley) (Travis, Jordan, and Huffaker, 1969).

Electric-power generating plants usually entrain large volumes of water for cooling purposes and discharge heated effluents that exceed maximum ambient temperatures. The nitrate reductase activity of entrained phytoplankton may be affected by thermal or chlorination stresses during passage through power-plant condensers.

Briand (1975) attributed large reductions in phytoplankton biomass to the magnitude of increases above ambient temperature at two southern California coastal power plants. Carpenter, Peck, and Anderson (1972), reported substantial phytoplankton productivity losses at the Millstone plant at all levels of continuous and intermittent chlorination, but the effects of temperature shock on nitrate reductase activity and productivity were not studied. If the level of phytoplanktonic nitrate reductase is influenced by photosynthetic rate, then chlorination could mask any temperature effects on this enzyme. To determine the effects of temperature increase resulting from power-station generation, we measured nitrate reductase activity and primary productivity over an annual cycle of temperature fluctuations from January 1973 through January 1974. Coordinating sampling with plant operations allowed us to monitor nitrate reductase and primary productivity over a range of temperatures and at chlorination levels of 0.0 to 1.2 ppm [the highest dosage studied by Carpenter, Peck, and Anderson (1972)].

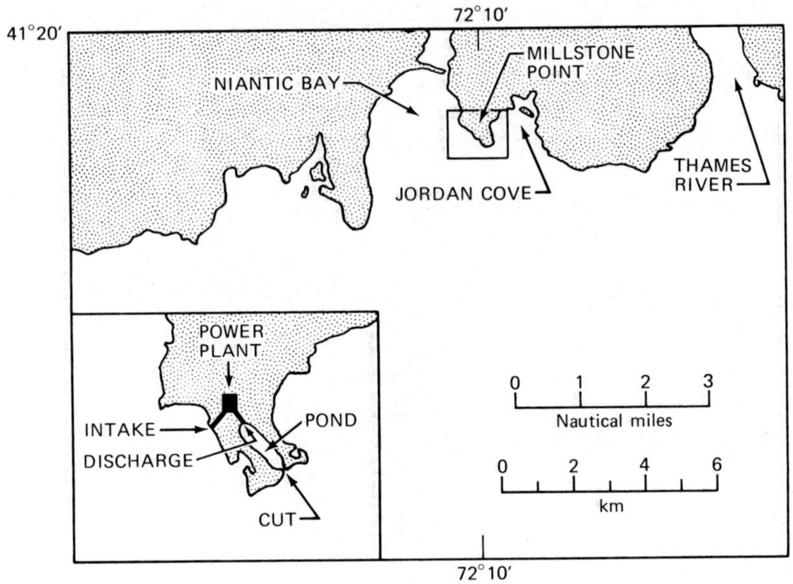

Fig. 1 Map of study area, Long Island Sound, showing nuclear power plant.

MATERIALS AND METHODS

Study Area

The Millstone Nuclear Power Station, Unit I, located on northeastern Long Island Sound (Fig. 1), entrains 9.5×10^4 m^3/hr (3.2×10^6 ft^3/hr) of seawater (salinity 28 to 30‰) for cooling from Niantic Bay at a depth of ~3 m. The ΔT from intake to discharge is 13°C when the plant is operating at a capacity of 650 MW(e). The water passes through the plant in about 2 min, is discharged into an effluent pond, and then enters Long Island Sound via a cut.

The fjord-shaped effluent pond, which was formerly a granite quarry, has a volume of 850,000 m^3, a maximum depth of ~30 m, and a minimum depth at the shallow sill, the cut where entrained water enters Long Island Sound, of ~3 m. When the plant is operating, water in the effluent pond is both isothermal and isohaline. Mean residence time of water in the pond is 6 to 9 hr. Measurements with a surface float show that surface flow from discharge to cut through the effluent pond occurs in ~1.5 hr to a depth of 1 m. The surface-to-volume ratio of the pond is small, and water cools by only about 0.5 to 1.0°C in passage through the pond to the end of the cut.

Sampling

Sampling was conducted to assure, as much as possible, that the same water masses (intake, discharge, and cut) were sampled from each station. All samples were collected around noon because maximum solar radiation occurred near noon, and maximum nitrate reductase activities have been reported about noon. Commencement of sampling deviated by a maximum of 30 min before or after noon during the entire study period. Twelve casts were made to a depth of 1 m with a 1-liter Kemmerer sampler at the intake, discharge, and cut (Fig. 1). Intake samples were the controls. Discharge samples were collected within 15 min after collection of intake samples. Cut samples were collected no more than 1.5 hr after collection of discharge samples.

All treatments, with and without temperature and chlorine additions, were randomized and replicated throughout the 1-year study, and samples discussed here represent trends observed. Samples were collected five times monthly from April through September 1973, three times in October 1973, twice monthly in February, March, and November 1973, once monthly in January and December 1973, and once in January 1974. On several occasions samples were collected from three depths in the middle of the effluent pond (1, 15, and 30 m) to determine nitrate reductase activity and primary productivity in the water column of the pond. Also at each station (intake, discharge, and cut) salinity and temperature were measured with an induction salinometer. Percent surface illumination at each depth was calculated according to Holmes (1970) from Secchi disk lowerings and was validated by a submarine photometer.

Chlorine

Chlorine was added at the plant intake at rates of 95 to 1100 g/hr to give 0.1 to 1.2 ppm chlorine. Free residual chlorine was measured with an *ortho*-tolidine chlorometric method at a 1-m depth at intake, discharge, and cut. This method is not sensitive enough to measure residual chlorine below 0.05 ppm.

Nitrate reductase activity and primary productivity were measured at various levels of continuous chlorine applications ranging from 0.0 to 1.2 ppm. Continuous chlorination was begun at least 12 hr before measurements were made. Samples without chlorination were taken no less than 12 hr after chlorine applications ceased. This assured flushing of the effluent pond so that only the chlorine concentration to be tested for a particular treatment was present.

Nitrate Reductase

Nitrate reductase was measured by a biochemical assay modified from that of Eppley, Coatsworth, and Solórzano (1969). Ten liters of seawater from each station were placed in a 5-gal green Nalgene (Nalge Co.) bottle and kept at ambient temperature in a seawater bath until the enzyme assay could be run, usually within 3 hr of sample collection. Nitrate reductase activity is expressed as microgram atoms of NO_2 formed per liter per hour and per cell per hour. Nitrite produced was determined on a colorimeter by the absorption at 550 nm after addition of sulfanilamide and *N*-(1-naphthyl)ethylenediamine dihydrochloride solutions (Strickland and Parsons, 1968). Nitrite standards ranging from 1 to 8 μg atoms/liter were determined in the same way.

Total Protein

Protein was determined by the method of Lowry et al. (1951).

Primary Productivity

Primary productivity was measured with the light–dark bottle ^{14}C method of Steeman-Nielson (1952). For each sample station 500 ml of the 12 liters of water collected was used to fill four (three light and one dark) 125-ml Pyrex bottles. To each bottle was added 0.7 μCi of $NaH^{14}CO_3$. At each station the four bottles were shaken, separately attached to curtain hooks, and lowered into the water to a depth of 1 m for 4 hr of incubation in situ. After incubation the contents of each bottle were filtered with suction onto glass-fiber filters, which were then suspended in 20-ml vials containing Bray's fluor and counted for 20 min on a liquid scintillation counter (Bray, 1960). Total carbon per cubic meter was determined from temperature–salinity data and the tables of Strickland and Parsons (1968). Percent productivity for discharge and cut, relative to intake productivity, was calculated from milligrams of carbon fixed per cubic meter per hour.

Phytoplankton Cell Concentration

Subsamples of phytoplankton (100 ml) from the pooled 12 liters of water collected at each station were preserved in Lugol's solution in 100-ml bottles (Lund, Kipling, and Lecren, 1958). After concentration by centrifugation, cells were counted by the Sedgewick–Rafter strip-count method (Jackson and Williams, 1962). Usually about 500 cells/ml were counted.

Nutrient Concentration in Seawater

Subsamples (100 ml) from the pooled 12 liters of water collected at each station were frozen for later nutrient analysis. The nitrite concentration of the water was determined by the diazotization–coupling reaction; the nitrate was reduced to nitrite and determined by passage through a copperized cadmium column (Strickland and Parsons, 1968); and ammonium was determined by the alkaline hypochlorite method of Solórzano (1969).

RESULTS

Phytoplankton Population

Phytoplankton cell concentrations did not vary greatly from intake to cut over the 1-year sampling period. Two blooms occurred, one in the spring, peaking at about 2×10^6 cells/liter, and one in midsummer, peaking at about 3.8×10^6 cells/liter (Fig. 2).

Of the fifteen different genera identified, the majority are diatoms, *Skeletonema costatum*, *Thalassiosira nordenskioldii*, *Asterionella japonica*, *Chaetoceros* sp., *Thalassionema nitzschoides*, and *Rhizosolenia fragillissima*. Other dominants include the flagellates *Rhodomonas minuta* and *Rhodomonas amphioxeia* and the dinoflagellates *Dinophysis* sp., *Peridinium* sp., and *Prorocentrum* sp. These genera were entrained through the plant from January 1973 to January 1974. The same species were found to be abundant in Long Island Sound by Conover (1956), in Block Island Sound by Riley (1952), and in the Niantic Bay estuary by Marshall and Wheeler (1965).

Chlorination, Primary Productivity, and Nitrate Reductase Activity

When there was no chlorine application during a study period, primary productivity was unaffected by entrainment (Table 1). Chlorination at every concentration used in this study reduced primary productivity. Continuous chlorine application at the highest dosage (1.2 ppm) reduced productivity by 89% at the discharge and 84% at the cut, as compared with intake water productivity. Free residual chlorine concentrations at the discharge ranged from 0.4 ppm at the highest chlorine application (1.2 ppm) to below measurable amounts at addition rates of 0.4 ppm and less. No

TABLE 1

PHYTOPLANKTONIC PRODUCTIVITY AND NITRATE REDUCTASE ACTIVITY AT VARIOUS CHLORINE APPLICATIONS*

Date and station	Temp., °C	ΔT, °C	Chlorine added at intake, ppm	Residual chlorine at discharge, ppm	Nitrogen in seawater, μg atoms/liter			10^6 cells/liter	Nitrate reductase activity†		NR activity of samples, % $cell^{-1}$ hr^{-1}	Carbon fixed, mg m^{-3} hr^{-1}	Productivity of samples, %
					NO_3^-	NO_2^-	NO_4^+		NR $liter^{-1}$ hr^{-1}	NR $cell^{-1}$ hr^{-1}			
Without Chlorination													
Feb. 21, 1973													
Intake	3.5		0.0		1.06	0.11	3.62	0.9	0.116	1.32×10^{-7}		5.58	
Discharge	3.4	+0.1		0.00	1.98	0.19	2.17	0.6	0.076	1.27×10^{-7}	−3.9	7.53	+26.0
Cut	3.2	−0.3		0.00	1.31	0.16	3.05	0.8	0.104	1.32×10^{-7}	0.0	4.86	−13.0
With Chlorination													
May 15, 1973													
Intake	10.7		0.1		1.63	0.42	0.21	0.6	0.102	1.63×10^{-7}		2.32	
Discharge	10.7	0.0		0.05	2.55	0.51	0.09	0.5	0.076	1.62×10^{-7}	−0.7	0.71	−65.1
Cut	10.7	0.0		0.00	1.99	0.42	0.22	0.4	0.072	1.66×10^{-7}	+1.8	0.36	−84.5
May 17, 1973													
Intake	10.5		0.2		2.45	0.29	0.12	0.3	0.044	1.29×10^{-7}		1.10	
Discharge	10.5	0.0		0.05	1.48	0.27	0.21	0.3	0.034	1.21×10^{-7}	−6.3	0.46	−58.2
Cut	10.9	+0.4		0.00	2.30	0.18	0.08	0.3	0.034	1.21×10^{-7}	−6.3	0.29	−73.6
Jul. 30, 1973													
Intake	19.0		0.4		1.16	0.03	2.98	0.2	0.964	4.38×10^{-7}		8.48	
Discharge	19.0	0.0		0.05	1.98	0.06	3.43	0.8	0.836	4.62×10^{-7}	+5.4	2.49	−71.0
Cut	19.0	0.0		0.00	2.12	0.08	3.41	0.9	0.930	4.68×10^{-7}	+6.8	2.05	−66.0

Table 1 (Continued)

Date and station	Temp., °C	ΔT, °C	Chlorine added at intake, ppm	Residual chlorine at discharge, ppm	Nitrogen in seawater, μg atoms/liter			10^6 cells/liter	Nitrate reductase activity†		NR activity of samples, % $cell^{-1}$ hr^{-1}	Carbon fixed, mg m^{-3} hr^{-1}	Productivity of samples, %
					NO_3^-	NO_2^-	NO_4^+		NR $liter^{-1}$ hr^{-1}	NR $cell^{-1}$ hr^{-1}			
					With Chlorination								
Jul. 26, 1973													
Intake	18.5		0.6		1.30	0.03	4.02	2.5	1.074	4.32×10^{-7}		10.52	
Discharge	18.5	0.0		0.05	1.81	0.11	3.17	1.9	0.812	4.34×10^{-7}	+0.5	2.37	−77.0
Cut	19.0	+1.0		0.00	0.47	0.11	3.61	1.9	0.914	4.86×10^{-7}	+12.5	2.32	−78.0
Jul. 21, 1973													
Intake	18.6		0.8		1.19	0.05	4.38	2.5	1.142	4.54×10^{-7}		9.00	
Discharge	18.6	0.0		0.17	0.34	0.09	5.34	2.3	0.990	4.38×10^{-7}	−4.6	3.48	−61.0
Cut	18.5	−0.1		0.00	0.35	0.06	4.44	1.9	0.890	4.62×10^{-7}	+1.7	2.55	−72.0
Jul. 12, 1973													
Intake	18.8		1.0		0.54	0.03	2.09	3.2	1.396	4.34×10^{-7}		12.19	
Discharge	18.8	0.0		0.25	0.85	0.12	1.35	2.1	0.666	3.22×10^{-7}	−26.0	1.47	−88.0
Cut	18.7	−0.1		0.00	1.15	0.07	2.09	2.5	0.908	3.70×10^{-7}	−15.0	2.75	−77.0
Jul. 3, 1973													
Intake	17.5		1.2		0.92	0.07	1.83	2.9	1.250	4.38×10^{-7}		11.80	
Discharge	17.5	0.0		0.40	0.56	0.19	2.12	2.2	0.508	2.31×10^{-7}	−47.3	1.25	−89.0
Cut	17.5	0.0		0.00	1.34	0.16	1.98	3.8	1.440	3.78×10^{-7}	−14.0	1.84	−84.0

*Intake values are used as controls for comparison. No measurable residual chlorine was detected at the cut for any application level.

†NR is nitrate reductase activity expressed as microgram atoms of NO_2^- formed per hour.

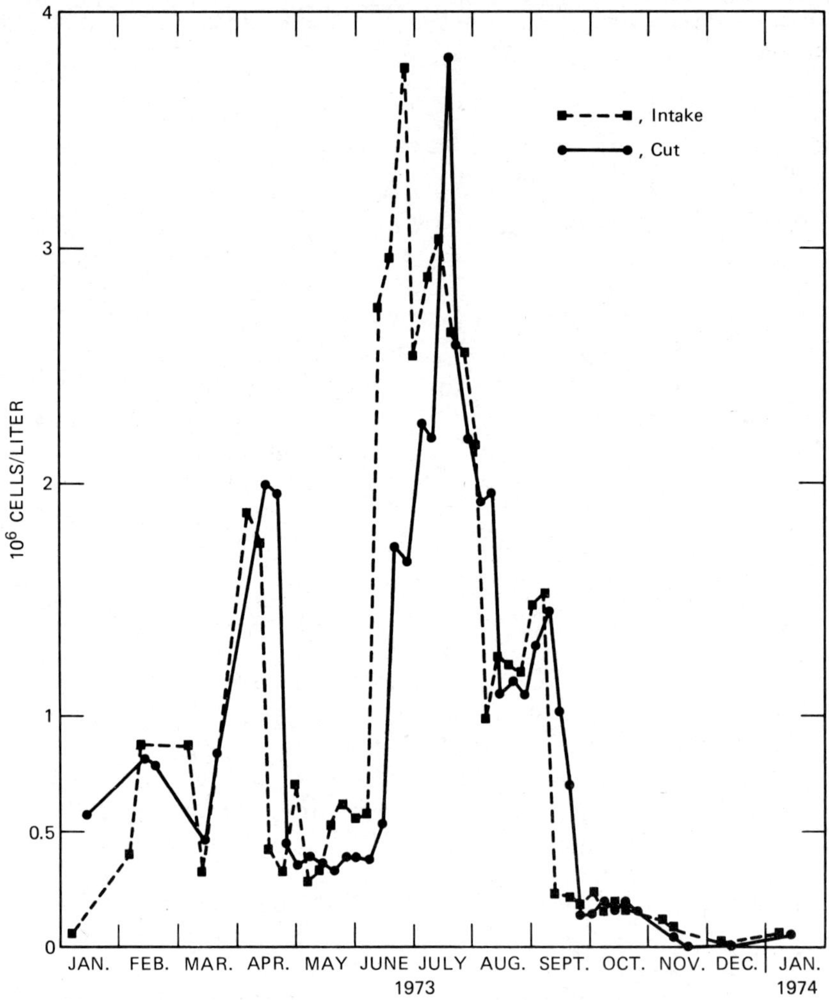

Fig. 2 Intake and cut phytoplankton cell concentrations from January 1973 to January 1974.

measurable residual chlorine was detected at the cut for any of the chlorine doses tested. At the lowest dosage (0.1 ppm), the productivity declined 65% at discharge and 84% at the cut. Thus a chlorine dosage smaller by over an order of magnitude yielded essentially the same effect as the highest dosage.

Nitrate reductase activity declined only at the two highest chlorine dosages. A 1.2-ppm chlorine application reduced nitrate reductase activity 47% in the discharge effluent and 14% at the cut, as compared with intake levels. A 1.0-ppm chlorine application reduced nitrate reductase activity 26% at the discharge and 15% at the cut.

Temperature, Ammonium, Primary Productivity, and Nitrate Reductase Activity

Generally, temperature increases above 9°C resulting from power-station generation depressed productivity at the discharge and cut relative to that at the intake (Table 2). Percent nitrate reductase activity per cell per hour at the cut was stimulated after 6 to 9 hr of exposure to temperatures increasing to 17°C through the pond and to the cut, but, when temperatures exceeded 25°C, activity was greatly depressed. The power plant was not chlorinating on any of these sampling dates. Ambient nitrate, nitrite, and ammonium concentrations remained essentially the same after passage through the power plant.

Figure 3 compares nitrate reductase activity at the plant intake with that at the cut for random sampling dates from January 1973 to January 1974. Figures 4 and 5 show changes in temperature and ammonium concentrations during this time. During March and April activity at the cut was stimulated above ambient nitrate reductase activity at the intake. Intake temperatures ranged from 4.3 to 9.9°C during this time, and cut temperatures were 11.5 to 18.1°C. From August through January 1974, nitrate reductase activity at the cut dropped sharply below that at the intake. Intake temperatures during this interval ranged from 22 to 6.4°C, and cut temperatures ranged from 34 to 17.9°C. From September through January 1974, ammonium concentrations at the cut were highest, ranging from 8 to 14 μg atoms/liter. Therefore, the depression of nitrate reductase activity at the cut in August 1973 appears to be caused by temperature increases resulting from station generation, and that from September through January 1974 is caused by increases in ammonium concentration.

On seven dates in August 1973, the recovery of nitrate reductase activity after heating was tested by use of 10 liters of seawater, sampled from a depth of 1 m at each station (intake, discharge, and cut), at the ambient intake temperature. The water was held in dialysis tubing for periods of 1, 4, and 24 hr, and a nitrate reductase assay was run after each incubation period. The nitrate reductase activity of these samples, depressed at the cut after transit through the quarry at a temperature range of 30 to 34°C, did not recover to intake levels after 24 hr of incubation at ambient intake temperature.

TABLE 2

PHYTOPLANKTON PRODUCTIVITY AND NITRATE REDUCTASE ACTIVITY AT VARIOUS TEMPERATURES WITH NO CHLORINATION*

Date and station	Temp., °C	ΔT, °C	Nitrogen µg atoms/liter			10^6 cells/liter	Nitrate reductase activity†		NR activity of samples,† % $cell^{-1}$ hr^{-1}	Carbon fixed, mg m^{-3} hr^{-1}	Productivity of samples, %
			NO_3^-	NO_2^-	NO_4^+		NR $liter^{-1}$ hr^{-1}	NR $cell^{-1}$ hr^{-1}			
Without Temperature Increase											
Feb. 21, 1973											
Intake	3.5		1.06	0.11	3.62	0.9	0.116	1.32×10^{-7}		5.58	
Discharge	3.4	+0.1	1.98	0.19	2.17	0.6	0.076	1.27×10^{-7}	−3.9	7.53	+26.0
Cut	3.2	−0.3	1.31	0.16	3.05	0.8	0.104	1.32×10^{-7}	0.0	4.86	−13.0
Jul. 24, 1973											
Intake	18.0		2.45	0.04	3.91	2.1	0.832	4.06×10^{-7}		8.39	
Discharge	17.9	−0.1	1.75	0.11	4.65	2.3	0.854	3.68×10^{-7}	−9.4	10.25	+22.2
Cut	17.9	−0.1	0.46	0.12	3.90	2.2	1.022	4.62×10^{-7}	+13.8	10.92	+30.2
With Temperature Increase											
Mar. 30, 1973											
Intake	7.7		2.39	0.03	0.05	0.3	0.042	1.31×10^{-7}		0.73	
Discharge	17.9	+10.2	2.16	0.13	0.08	0.8	0.106	1.33×10^{-7}	+1.5	0.75	+2.7
Cut	17.4	+9.7	2.19	0.13	0.06	0.9	0.142	1.65×10^{-7}	+25.9	0.69	−6.5

Table 2 (Continued)

Date and station	Temp., °C	ΔT, °C	Nitrogen µg atoms/liter			10^6 cells/ liter	Nitrate reductase activity†		NR activity of samples,† % $cell^{-1}$ hr^{-1}	Carbon fixed, mg m^{-3} hr^{-1}	Productivity of samples, %
			NO_3^-	NO_2^-	NO_4^+		NR $liter^{-1}$ hr^{-1}	NR $cell^{-1}$ hr^{-1}			
With Temperature Increase											
Apr. 3, 1973											
Intake	9.5		2.96	0.06	0.61	2.0	0.262	1.32×10^{-7}			
Discharge	17.9	+8.4	2.16	0.02	0.65	2.3	0.300	1.32×10^{-7}	0.0		
Cut	17.4	+7.9	2.28	0.05	0.63	2.4	0.392	1.65×10^{-7}	+25.0		
Aug. 2, 1973											
Intake	19.5		1.36	0.04	2.76	1.9	0.492	2.64×10^{-7}		4.99	
Discharge	31.5	+12.0	1.82	0.12	2.09	1.3	0.352	2.64×10^{-7}	0.0	0.80	−84.0
Cut	31.3	+11.8	2.29	0.21	1.96	1.7	0.056	3.25×10^{-8}	−88.0	2.38	−52.0
Aug. 6, 1973											
Intake	20.0		2.05	0.05	2.98	1.3	0.348	2.63×10^{-7}		4.59	
Discharge	34.0	+14.0	1.73	0.17	3.02	1.4	0.376	2.64×10^{-7}	+0.3	1.41	−69.0
Cut	34.0	+14.0	1.65	0.19	1.98	1.2	0.034	2.96×10^{-8}	−89.0	2.66	−42.0
Oct. 9, 1973											
Intake	17.5		2.83	0.23	6.25	0.2	0.050	2.59×10^{-7}		0.59	
Discharge	29.2	+11.7	3.43	0.21	2.85	0.2	0.050	2.69×10^{-7}	+3.9	0.23	−61.0
Cut	27.9	+10.4	2.17	0.29	13.25	0.3	0.000	0.00	−100.0	0.30	−49.0

*Intake values are used as controls for comparison.

†NR is nitrate reductase activity expressed as microgram atoms of NO_2^- formed per hour.

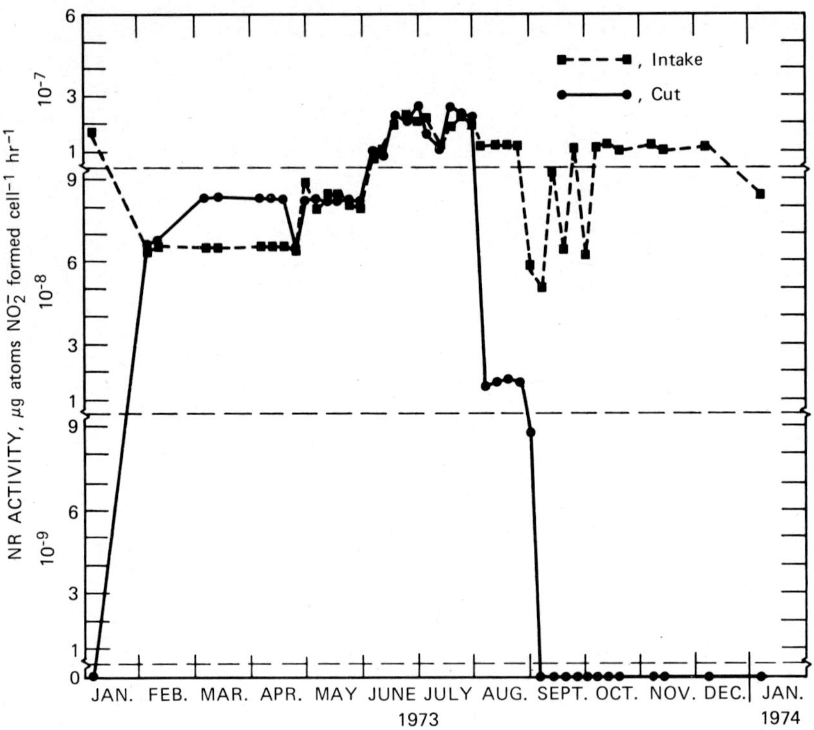

Fig. 3 Nitrate reductase activity at power-plant intake and at the cut on various sampling dates from January 1973 to January 1974. The power plant was not generating from the end of April to the beginning of August 1973.

DISCUSSION

Our data suggest that there is no immediate decrease in phytoplankton population as a result of mechanical stresses of entrainment through the power plant. There are indications that phytoplankton are destroyed during passage through power-plant cooling systems, however. Briand (1975) reported large reductions in phytoplankton numbers and volumes at two southern California coastal power plants. Morgan and Stross (1969) and Hamilton et al. (1970) noted large decreases in the chlorophyll a content of cells concomitant with decreased primary productivity after entrainment with chlorination. Residual chlorine concentrations of 1.5 to 2.3 ppm killed *Skeletonema costatum* in laboratory cultures, and sublethal chlorine concentrations adversely affected growth (Hirayama and Hirano, 1970).

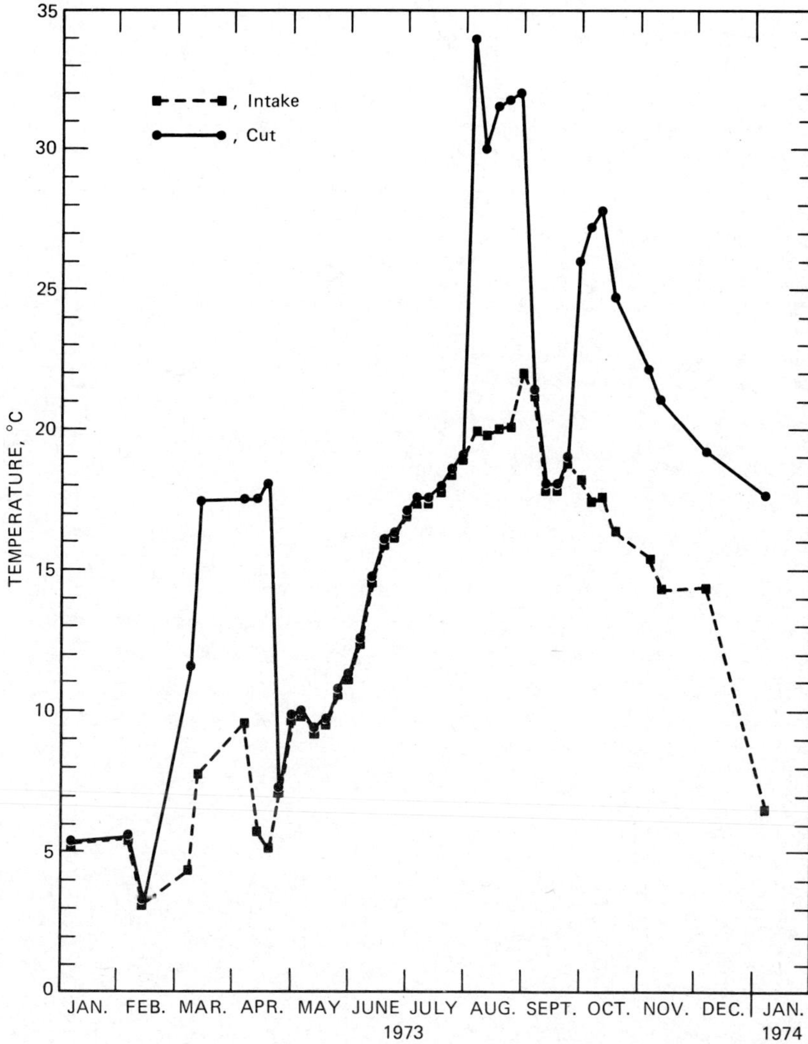

Fig. 4 Temperature at power-plant intake and at the cut on various sampling dates from January 1973 to January 1974. When the curves are the same, the plant was not operating.

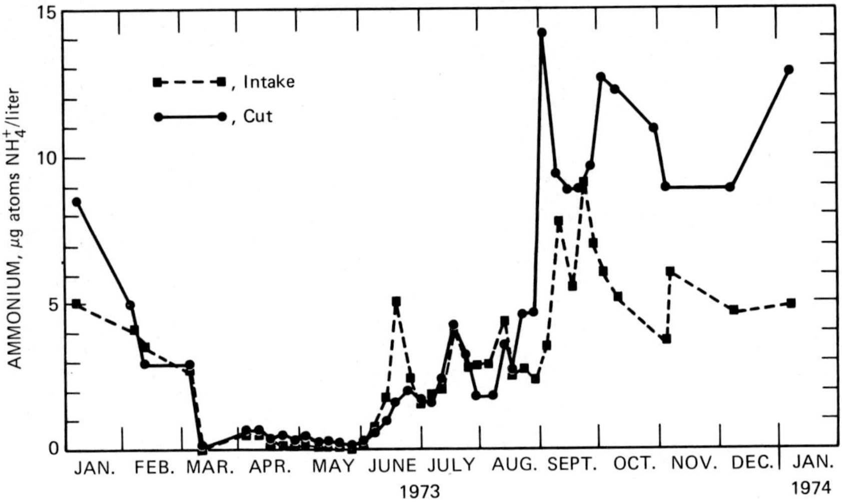

Fig. 5 Ammonium concentrations at power-plant intake and at the cut on various sampling dates from January 1973 to January 1974.

Our results indicate that photosynthesis is more sensitive to chlorination than is nitrate reductase activity. Either a real difference between photosynthesis and nitrogen assimilation occurs, or nitrate reductase activity measures an effect on the level of an enzyme, whereas productivity measures an activity. Recovery of the nitrate reductase enzyme at the cut after depression at the discharge by the two highest chlorine dosages is consistent with a fairly rapid turnover rate for the enzyme and may indicate that chlorine does not have any long-term effects on the ability of phytoplankton to synthesize nitrate reductase. Further recovery of nitrate reductase activity may take place after exposure to ocean water with no added chlorine.

It appears that nitrate reductase activity is stimulated in transit through the pond during 6 to 9 hr of exposure to temperatures increasing to ~18°C, such as occurred in March and April. At some temperature between 18°C, occurring at the cut in the spring, and 27°C, occurring at the cut in the fall, there is a shift from stimulation to inhibition of nitrate reductase. For example, on Mar. 30, when the temperature at the cut was 17.4°C, nitrate reductase activity at the cut increased 25.9% above intake activity, but on Aug. 2, when the temperature was 31.3°C, activity decreased 88%. Nitrate reductase activity never decreased after a 2-min exposure to temperature increases during passage through the plant but was affected after transit through the pond and exposure for 6 to 9 hr to a change in temperature. The assay used measures the amount of enzyme present not its in vivo activity. The temperature shock of August through October is, therefore, either inactivating the enzyme already present or slowing or stopping the synthesis of the enzyme. This could be either a generalized effect on all protein or a specific effect for nitrate reductase.

Nitrate reductase is an inducible enzyme, and, in an inducible enzyme system, temperature can influence enzyme synthesis (Langridge and McWilliam, 1967). In some instances this effect may be the result of a progressive loss of the response of the control system to the inducer. Thus the temperature-related decrease in phytoplankton nitrate reductase activity observed at the cut in August could be the result of high-temperature impairment of the system controlling production of the enzyme.

In December 1973 and January 1974, ammonium concentrations were high and nitrate reductase activities were undetectable at the cut (Figs. 3 and 5). Also, as can be seen in Figs. 3 to 5, high ammonium concentrations at the cut during two sampling dates in the middle of September were accompanied by low nitrate reductase activities (0 μg atoms NO_2^- cell^{-1} hr^{-1}) even though the intake and cut temperatures were the same (18°C) because the plant was not operating. Eppley, Coatsworth, and Solórzano (1969) found that ammonium concentrations in the range of 5 to 15 μg atoms/liter inhibited nitrate reductase activity in samples of tropical Pacific phytoplankton. Therefore, from September through January, ammonium concentration appears to be responsible for nitrate reductase depression at the cut, whereas in August depression is probably caused by temperature.

To determine the correlation among nitrate reductase activity, temperature, and ammonium statistically, we used a stepwise regression analysis (the May 2, 1966, version of the Health Sciences Computing Facility, UCLA at Connecticut College). Temperature and nitrate reductase activity and ammonium and nitrate reductase activity are significantly correlated.

Ammonium concentrations exceeding 8 μg atoms/liter observed at the cut from September through January could be the result of increased copepod excretion or decay in the depths of the effluent pond. Carpenter, Peck, and Anderson (1974), working at this same site, reported that live copepods sink relatively rapidly after passing through the power plant. They found a large number (54 to 63%) of dead copepods in the deep water of the effluent pond, in contrast to a few dead (6 to 13%) at the surface. Conditions in the effluent pond itself are not responsible for copepod mortality. Mortality is apparently related to the mechanical effects of passage through the power plant.

The failure of the heat-depressed cut samples to recover any nitrate reductase activity after 24 hr of incubation suggests that the ability of this phytoplankton population to synthesize nitrate reductase after heat shock was seriously altered. During incubation at intake temperatures, cells were contained within dialysis tubing, and it is possible that this containment influenced the nitrate reductase activity of the cells. Ammonium concentrations in August did not exceed 5 μg atoms/liter. However, the complicating effect of ammonium on changes in the nitrate reductase activity of contained cells must be recognized when interpreting any failure of nitrate reductase activity to recover. Vargo, Hargraves, and Johnson (1975) reported that detrital deposition and epiphytic growth on dialysis membranes at summer temperatures may decrease the membrane area available for nutrient flux in less than 3 days. This would tend to decrease the nitrate reductase present by limiting the nitrate entering the dialysis membrane. In our study this was not the case. Intake controls held at ambient intake temperature for 24 hr show no change in nitrate reductase activity.

ACKNOWLEDGMENTS

We thank E. J. Carpenter of the Marine Sciences Research Center, State University of New York, Stony Brook, Long Island, and Ian Morris and T. T. Packard of the Bigelow Laboratory for Ocean Sciences, West Boothbay Harbor, Me., for their comments on the original manuscript. Gratitude is also expressed to Jerry Lamb of the Naval Underwater Systems Center, New London, Conn., for statistical advice.

REFERENCES

Beevers, L., and R. H. Hageman, 1969, Nitrate Reduction in Higher Plants, *Annu. Rev. Plant Physiol.*, 20: 495-522.

Bray, G. A., 1960, A Simple Efficient Liquid Scintillator for Counting Aqueous Solutions in a Liquid Scintillation Counter, *Anal. Biochem.*, 1: 279-285.

Briand, F. J.-P., 1975, Effects of Power Plant Cooling Systems on Marine Phytoplankton, *Mar. Biol.*, 33: 135-146.

Carpenter, E. J., B. B. Peck, and S. J. Anderson, 1972, Cooling Water Chlorination and Productivity of Entrained Phytoplankton, *Mar. Biol.*, 16: 37-40.

——, B. B. Peck, and S. J. Anderson, 1974, Survival of Copepods Passing Through a Nuclear Power Station on Northeastern Long Island Sound, USA, *Mar. Biol.*, 24: 49-55.

Conover, S. A. M., 1956, Oceanography of Long Island Sound, 1952—1954. IV. Phytoplankton, *Bull. Bingham Oceanogr. Collect.*, 15: 62-111.

Eppley, R. W., J. L. Coatsworth, and L. Solórzano, 1969, Studies of Nitrate Reductase in Marine Phytoplankton, *Limnol. Oceanogr.*, 14(1): 194-205.

——, T. T. Packard, and J. J. MacIsaac, 1970, Nitrate Reductase in Peru Current Phytoplankton, *Mar. Biol.*, 6(3): 195-199.

Hamilton, D. H., Jr., D. A. Flemer, C. W. Keefe, and J. A. Mihursky, 1970, Power Plants: Effects of Chlorination on Estuarine Primary Production, *Science*, 169: 197-198.

Hirayama, K., and R. Hirano, 1970, Influences of High Temperature and Residual Chlorine on Marine Phytoplankton, *Mar. Biol.*, 7: 205-213.

Holmes, R. W., 1970, The Secchi Disk in Turbid Coastal Waters, *Limnol. Oceanogr.*, 15(2): 688-694.

Jackson, H. W., and L. G. Williams, 1962, Calibration and Use of Certain Plankton Counting Equipment, *Trans. Am. Microsc. Soc.*, 81(1): 96-103.

Langridge, J., and J. R. McWilliam, 1967, Heat Responses of Higher Plants, in *Thermobiology*, A. H. Rose (Ed.), pp. 232-284, Academic Press, Inc., New York.

Lowry, H. O., N. J. Rosebrough, A. L. Farr, and R. J. Randall, 1951, Protein Measurement with the Folin Phenol Reagent, *J. Biol. Chem.*, 193: 265-275.

Lund, J. W. G., C. Kipling, and E. D. Lecren, 1958, The Inverted Microscope Method of Estimating Algal Numbers and the Statistical Basis of Estimations by Counting, *Hydrobiology*, 11: 143-170.

Marshall, N., and B. M. Wheeler, 1965, Role of the Coastal and Upper Estuarine Waters Contributing Phytoplankton to the Shoals of the Niantic Estuary, *Ecology*, 46: 665-673.

Morgan, R. P., II, and R. G. Stross, 1969, Destruction of Phytoplankton in the Cooling Water Supply of a Steam Electric Station, *Chesapeake Sci.*, 10: 165-171.

Packard, T. T., D. Blasco, J. J. MacIsaac, and R. C. Dugdale, 1971, Variations of Nitrate Reductase Activity in Marine Phytoplankton, *Invest. Pesq.*, 35(1): 209-219.

Riley, G. A., 1952, Phytoplankton of Block Island Sound, 1949, *Bull. Bingham Oceanogr. Collect.*, 13: 40-64.

Schrader, L. E., G. L. Ritenour, G. L. Eilrich, and R. H. Hageman, 1968, Some Characteristics of Nitrate Reductase from Higher Plants, *Plant Physiol.*, 43: 930-940.

Solórzano, L., 1969, Determination of Ammonia in Natural Waters by the Phenolhypochlorite Method, *Limnol. Oceanogr.*, 14(5): 799-801.

Steeman-Nielson, E., 1952, Use of Radioactive (C^{14}) for Measuring Organic Production in the Sea, *J. Cons. Int. Explor. Mer*, 23: 178-198.

Strickland, J. D. A., and T. R. Parsons, 1968, *A Manual of Seawater Analysis*, Bulletin 167, Fisheries Research Board of Canada, Ottawa, Ont.

Travis, R. L., W. R. Jordan, and R. C. Huffaker, 1969, Evidence for an Inactivating System of Nitrate Reductase in *Hordeum vulgare* L. During Darkness That Requires Protein Synthesis, *Plant Physiol.*, 44: 1150-1156.

Vargo, G. A., P. E. Hargraves, and P. Johnson, 1975, Scanning Electron Microscopy of Dialysis Tubes Incubated in Flowing Seawater, *Mar. Biol.*, 31: 113-120.

GROWTH OF DUCKWEED UNDER CONSTANT AND VARIABLE TEMPERATURES

REBECCA R. SHARITZ and JEFFREY C. LUVALL
Savannah River Ecology Laboratory, Aiken, South Carolina

ABSTRACT

Effects of variable temperatures on the growth and vegetative reproduction of an aquatic plant were examined by exposing populations of *Spirodela oligorrhiza* to cyclic, acyclic, and constant temperature regimes. Growth rates under the cyclic and acyclic regimes (each of which ranged from 10 to 25°C) were not significantly different from those at a constant temperature equal to the mean of the range (17.5°C). Population growth was directly related to temperature in both constant and variable thermal regimes.

Laboratory studies of effects on organisms of increases in temperature have commonly been based on exposure to constant temperature regimes. Such studies do not reflect conditions generally existing in nature, and data on life-history phenomena of organisms at constant temperatures may be misleading if extrapolated to field conditions (Hagstrum and Hagstrum, 1970). Many freshwater organisms are confined to shallow-water habitats and exposed to both diurnal and seasonal fluctuations in temperature.

The significance of cyclic and recurring temperatures on growth and development in a variety of organisms has been demonstrated. For example, a diurnal thermoperiodicity with night temperatures 4 to 8°C lower than day temperatures enhances the growth of sporelings and fronds of *Porphyra* (Shimo, 1977). Insects frequently have been shown to develop more rapidly under fluctuating temperatures than constant temperatures if the range of fluctuation falls within the optimal range for the organisms' development (Hagstrum and Hagstrum, 1970). Heath (1963) reported that maximum tolerance to temperature extremes in the sea-run cutthroat trout *(Salmo clarki clarki)* occurred under a 24-hr temperature cycle rather than shorter or longer cycles. He suggested that most fish respond to the extremes rather than the mean of the temperature cycle. Heath also summarized several studies indicating that certain organisms have higher thermal tolerances under cyclic than under continuous exposure to high temperatures. Such other processes as larval development in the mud crab, *Rhithropanopeus harrisii* (Costlow and Bookhout, 1971; Christiansen and Costlow, 1975), and byssal thread formation by *Modiolus demissus* (Van Winkle, 1969) apparently respond to cyclic temperatures in a manner similar to that of a constant temperature equal to the mean of the cycle.

In addition to constant temperatures and cyclic fluctuations, many organisms must contend with randomly varying temperatures or with thermal regimes in which the temperature change is much less predictable. This circumstance is becoming more common as many industrial facilities and power plants release heated water into streams and lakes. Even though effluent temperatures may not exceed the tolerance levels of the organisms, the thermal regimes may change frequently in unpredictable ways.

A few studies of growth and development in aquatic animals under fluctuating temperatures are available (e.g., Grainger, 1959; Thorp and Wineriter, 1978) but the effects of irregularly changing temperatures have received little attention. This study compares the effects of three types of thermal regimes (constant, cyclic, and acyclic or randomly fluctuating) on survivorship and growth of an aquatic plant.

METHODS

A common species of duckweed, *Spirodela oligorrhiza* (Kurz) Hegelm, was chosen for study because of its abundance in southeastern aquatic habitats, because of its ease of manipulation in experimental systems (Clatworthy and Harper, 1962; Hodgson, 1970), and because of the large body of literature on growth, metabolism, and flowering in the Lemnaceae, as summarized by Hodgson (1970) and Hillman (1976). Populations of *S. oligorrhiza* were collected from five separate locations at the U. S. Department of Energy's Savannah River Plant in South Carolina. These were (1) a beaver pond along a creek that had received cooling water from nuclear reactor operations 9 years previously, (2) a sewage-effluent pond, (3) a roadside ditch, (4) a pond receiving chemical effluent, and (5) a ponded area along a natural stream. *Spirodela oligorrhiza* plants from each of these sites were grown under similar conditions in the laboratory (at a temperature range of 15 to 20°C) for 8 weeks before the initiation of the temperature study. Only fronds from actively growing clones were used in the experiments.

From each population one adult frond with a juvenile attached was placed in culture solution in each of 24 4.3- by 5.1-cm compartments in uncovered clear plastic trays. Each sample tray was treated to one of five thermal regimes in controlled environment chambers. This design provided 24 replicates of each temperature and population treatment. Frond division and growth of the clones were examined under constant temperatures of 10, 17.5, and 25°C; a daily square-wave cyclic regime of 10 to 25°C; and a daily square-wave acyclic regime varying within 10 to 25°C limits, with a mean temperature of 17.8°C. Cyclic temperatures changed every 12 hr, with the maximum (25°C) occurring in the day. Acyclic temperatures changed every 12 hr, with the maximum occurring either in the day or night so that the next temperature was unpredictable within the specified range. Temperatures within 0.5°C and 12-hr light—dark photoperiods with light intensities of approximately 800 ft-c (cool-white fluorescent light) were maintained in all chambers. These light conditions are within the range for optimal growth of the Lemnaceae (Ashby and Oxley, 1935). Each tray compartment contained 50 ml of half-strength Hoagland's culture medium (Hoagland and Arnon, 1950) plus 5 ppm ferric citrate. The pH was adjusted to 6.8 with 300 ml 0.1*M* KOH per 35 liters of solution. To reduce algal infection, all fronds were washed initially in distilled water, and the plants in each treatment were transferred to fresh culture solutions weekly to remove waste products and replenish nutrients.

Growth of *S. oligorrhiza* clones was determined at weekly intervals by adult frond count. The number of adult fronds was used as a measure of population growth because high correlations have been demonstrated between the rate of increase in frond number and the net assimilation rate, frond area, protein content, and depth of frond color in *Lemna* sp. (White, 1939). Frond weight was not used since Hicks (1934) demonstrated that in *L. minor* frond weight depends on the amount of stored starch and may not accurately reflect population growth. The experiment was continued for 4 weeks until the surface area in the culture chambers was covered by plants in the more rapidly growing populations.

RESULTS

The mean number of adult fronds of *S. oligorrhiza* at the end of the 4-week growing period under each set of experimental conditions

TABLE 1

NUMBER OF ADULT FRONDS OF *Spirodela oligorrhiza* GROWN IN 4 WEEKS FROM ONE ADULT FROND IN EACH OF FIVE TEMPERATURE REGIMES*

	Temperature regime†				
Population	Constant, 10°C	Cyclic, 10–25°C	Constant, 17.5°C	Acyclic, 10–25°C	Constant, 25°C
1	3.0 ± 0.2	44.2 ± 6.6[a]	62.5 ± 5.7[a,b]	79.8 ± 7.0[b]	235.1 ± 28.3
2	2.4 ± 0.5[b]	17.3 ± 2.7[a,b]	38.4 ± 5.1[a]	53.0 ± 8.6[a]	57.9 ± 9.5[a]
3	3.2 ± 0.2	72.8 ± 7.8[a]	63.3 ± 10.4[a]	66.9 ± 10.1[a]	236.6 ± 23.6
4	3.1 ± 0.2[b]	40.5 ± 4.8[a]	46.2 ± 5.7[a]	33.2 ± 7.3[a,b]	145.7 ± 24.2
5	3.3 ± 0.4	61.8 ± 7.6[a]	59.0 ± 9.1[a]	76.7 ± 16.6[a]	216.3 ± 41.8

*Population 1, beaver pond; 2, sewage-effluent pond; 3, roadside ditch; 4, chemical-effluent pond; and 5, natural stream. Values are $\overline{X} \pm$ standard error for N = 24 replicates for each temperature and population treatment.

†Values in each row with like superscripts do not differ significantly ($\alpha = 0.05$) (Tukey's HSD test; Kirk, 1968).

is given in Table 1. Significant differences (multiple significant difference; Morrison, 1967) ($\alpha = 0.05$) in population (clone) growth occurred at the extremes of the temperature treatments. The greatest number of fronds was produced by each population at a constant temperature of 25°C. Except in the sewage-effluent population, which showed reduced growth under all treatments (Table 1), the increased growth at high temperatures was significant and represented the production of a daughter frond by each adult every 4 days. There was a significant reduction in growth at 10°C constant temperature in three of the *S. oligorrhiza* populations, and all populations produced fewer fronds at the low temperature than under the other temperature regimes. Only one new generation was produced at this temperature. Cyclic and acyclic thermal regimes generally affected the growth rate of *S. oligorrhiza* in the same way as exposure to a constant temperature equal to the mean of the variable temperatures. In all populations, growth under cyclic and acyclic conditions was not statistically different ($\alpha = 0.05$) from that at 17.5°C, and in only one population did the two fluctuating thermal regimes produce growth responses that differed from each other (Table 1). A new frond was produced every 5 to 6 days under these three treatments.

In all populations rate of growth was highest at 25°C during the first 3 weeks (Fig. 1). Four populations showed a decline in the growth rate at this temperature during the final week of the study. Occasional differences in growth rates under the fluctuating and

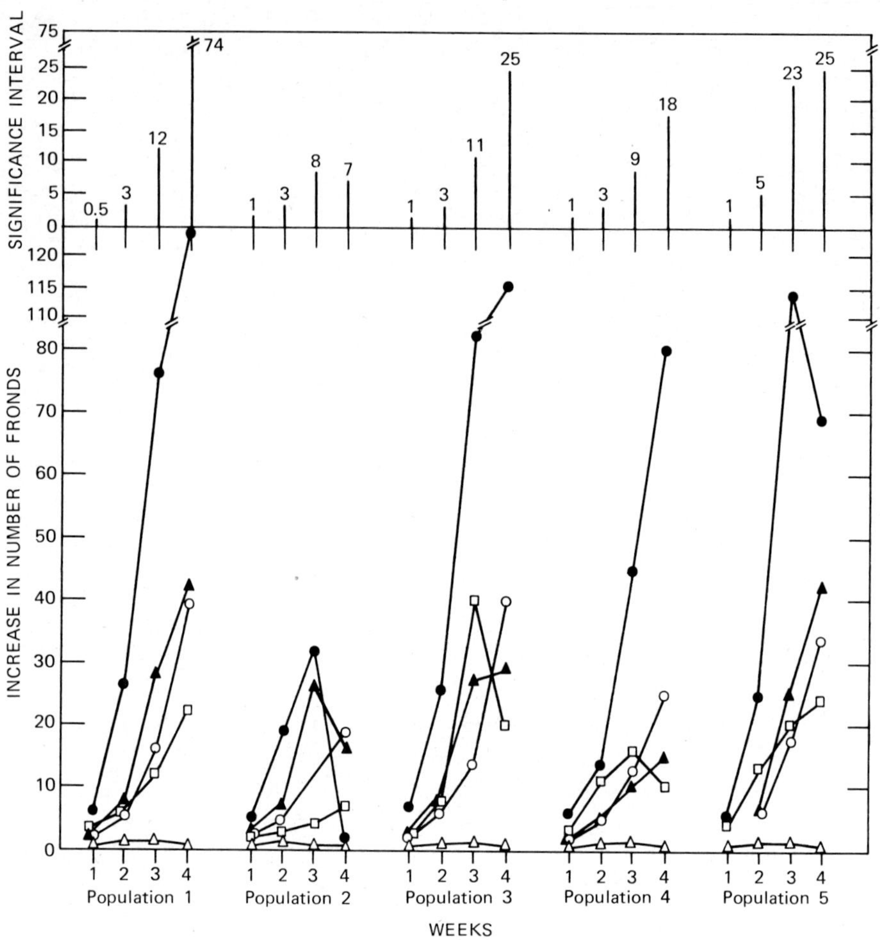

Fig. 1 Rate of frond increase in five populations of *Spirodela oligorrhiza* under five temperature regimes: constant exposures to 10°C (△—△), 17.5°C (○—○), and 25°C (●—●); cyclic exposure at 10 to 25°C (□—□); and acyclic exposure at 10 to 25°C (▲—▲) over a 4-week period. Differences between temperature treatments greater than the bars at the top are significant ($\alpha = 0.05$) (multiple significant difference; Morrison, 1967). N = 24. Populations are described in Table 1.

constant 17.5°C regimes during the early weeks disappeared in the later part of the study as growth rates were reduced.

Growth differences among the local populations of *S. oligorrhiza* were seen under the high-temperature treatments. Plants from the sewage-effluent pond grew less under all regimes, and those from the roadside ditch generally showed the highest rate of population growth. The significance of these differences cannot be determined without further investigation.

DISCUSSION

Differences in growth rates among populations of *S. oligorrhiza* were expected. Several investigators suggested that variations in growth rates of *L. minor* populations may be related to genetic or environmental differences (Wangermann and Ashby, 1951), age of the original parent plant (Claus, 1972), annual cycles (White, 1936), or other cyclic variations (Dickson, 1938). Because of the observed differences among the five *S. oligorrhiza* populations in this study, discussion is restricted to within-population comparisons across temperature treatments.

Only limited frond mortality was observed in the *S. oligorrhiza* plants grown during the study. Although little information is available on longevity in this species, Rejmánková (1973a) indicated that fronds of *Lemna* species in Czechoslovakian lakes usually live about 4 weeks. Frond length of life is related to environmental conditions, especially temperature (Ashby, Wangermann, and Winter, 1949; Ashby and Wangermann, 1951). This study was designed to be completed within the projected lifetime of a single plant.

Spirodela oligorrhiza demonstrated highest population growth rates at a constant temperature of 25°C. Only during the final week, as crowding in the culture chambers became apparent, was growth at this temperature reduced. Similar curves consisting of an exponential growth phase, a linear growth phase, and a later steady-state stage of population maintenance were described by Clatworthy and Harper (1962) in laboratory populations of *Lemna* species. At the lower extreme, the 10°C constant-temperature regime severely restricted *S. oligorrhiza* frond multiplication. This pattern is consistent with the report by Jacobs (1947) that optimum growth of *S. polyrhiza* occurred at 25°C and the population deteriorated at 7°C.

The relationship between temperature and growth in other members of the Lemnaceae has been well established. In an early study of *L. minor*, Hicks (1934) reported an exponential increase in frond number at temperatures between 15 and 30°C. Above 35°C, the growth rate declined rapidly. Similarly, Ashby and Oxley (1935) reported that the relative multiplication rate of *L. minor* fronds increased linearly with temperature up to 20°C and confirmed that temperatures above 35°C were deleterious to frond increase. At a light intensity of 500 ft-c, frond production in *L. minor* was twice as fast at 30°C as at 20°C (Ashby and Wangermann, 1951). More recently Hodgson (1970), after observing the growth of *L. minor* at constant temperatures of 12.5, 17.5, 22.5, and 27.5°C, reported that the population leaf area was highest at the 27.5°C exposure but the maximum net assimilation rate was achieved at 17.5°C, with a marked decline at higher temperatures. These results disagree with those of Ashby and Oxley (1935), who indicated that the assimilation rate in *L. minor* was independent of temperature between 18 and 29°C. Hodgson attributes the discrepancy to differences in experimental conditions, especially illumination. There appears to be general agreement, however, that population growth (frond multiplication) in the Lemnaceae under constant-temperature regimes increases with temperature to a maximum at approximately 30°C.

Combining the data from populations 1 through 5 in Table 1 yields mean frond densities of 3, 47, 54, 62, and 178 for the 10°C, cyclic, 17.5°C, acyclic, and 25°C regimes, respectively. Because the individual populations demonstrated different responses, these combined means cannot be taken as predictions of response for any given population, but they can be used to evaluate the relative importance of constant and cyclic regimes. According to Tukey's HSD test (Kirk, 1968), two of the combined means must differ by at least 12 to be significantly different at the 5% level. Thus the densities at constant temperatures of 10, 17.5, and 25°C are significantly different, but the densities under cyclic, 17.5°C, and acyclic regimes are not. The interval of 12 required for significance was computed from a mean-square error term that may have been inflated by inclusion of the highly variable replicates from the 25°C regime. An interval computed from a mean-square error term involving only the cyclic, 17.5°C, and acyclic populations would be smaller and might indicate significant differences among means. This would not alter the conclusion that cyclic and acyclic regimes had relatively little effect on the growth response of duckweed, however.

Although few efforts have been made to examine effects of fluctuating temperatures on growth in the Lemnaceae, Rejmánková (1973a; 1973b) demonstrated seasonal changes in growth of field populations of *L. minor* and *L. gibba* and suggested that frond multiplication is regulated by temperature during early and late periods of the growing season.

The effects of cyclic temperatures on growth of other organisms have been much more carefully examined. Shimo (1977) demonstrated the influence of diurnal thermoperiodicity on the growth of sporelings and fronds of *Porphyra* and reported greater elongation of plantlets grown at temperatures of 18 to 22°C when night temperatures were lowered 4 to 8°C below day temperatures. Heath

(1963) referred to a number of experimental studies demonstrating that organisms generally have higher thermal tolerance when temperature exposure is cyclic rather than continuous. His work demonstrated that certain species of fish respond to the upper extremes of the thermoperiod and have maximum thermal tolerance under a 24-hr temperature cycle. Van Winkle's (1969) study of *Modiolus demissus* showed that the net effect of a cyclic temperature regime (26 to 34°C) on byssal thread formation was similar to that of a constant 30°C exposure. Likewise, survival of the mud crab, *Rhithropanopeus harrisii*, exposed to 10°C cycles is similar to that at constant exposure to the mean temperature of the cycle (Costlow and Bookhout, 1971). Insects have frequently been shown to develop more rapidly under fluctuating temperatures than under constant temperature exposures (Hagstrum and Hagstrum, 1970). In a study paralleling this one, Thorp and Wineriter (1978) observed higher mortality in juvenile crayfish, *Procambarus acutus acutus*, under the cyclic regime (10 to 25°C) than under acyclic or constant 17.5°C exposures. This organism apparently suffered negative effects at 25°C, even for short periods of exposure.

Within the 10 to 25°C temperature range studied, frond multiplication of *S. oligorrhiza* appears to respond to the total temperature exposure. Cyclic and randomly fluctuating temperatures yielded the same growth response as constant-temperature treatments at the same mean temperature (17.5°C). The adaptability of duckweed species to a wide range of environmental conditions has been noted by other investigators (Hicks, 1934; Ashby and Oxley, 1935; Hodgson, 1970; Hillman, 1976). One explanation for duckweed's adaptability to a wide thermal range, even in natural populations, is that surface-dwelling aquatic species are more likely to be selected for tolerance to extreme temperatures than are bottom-dwelling forms, such as crayfish, which inhabit sites where temperatures normally remain more constant. This explanation is also applicable to the apparent selection for high thermal tolerance in bluegill, *Lepomis macrochirus*, living in a heated reservoir (Holland et al., 1974); whereas populations of the surface-dwelling mosquitofish, *Gambusia affinis*, from thermal and natural areas showed no differences in thermal tolerance (Smith, 1978). It is likely that this species, like the duckweeds, has a high thermal tolerance in the conditions of its natural habitat (Smoak, 1959).

The ability of *Spirodela* to develop turions (specially resistant vegetative fronds produced under adverse environmental conditions) also contributes to its wide distribution and its effectiveness as a colonizing species in disturbed aquatic habitats. Since many man-induced environmental disturbances occur intermittently, plant and animal species that have already undergone natural selection to withstand fluctuating conditions with low predictability may be of major importance in maintaining ecosystem structure.

ACKNOWLEDGMENTS

We thank J. H. Thorp and D. H. Nelson for their ideas and development of related studies, S. A. Wineriter for aid in maintaining the temperature chambers, K. A. VandenBosch for assistance in counting the plants, J. E. Pinder III for statistical advice and analysis, and J. B. Coleman for drafting the figure. The manuscript was reviewed by E. H. Liu and J. W. Gibbons. Research was performed under contract EY-76-C-09-0819 between the U. S. Department of Energy and the University of Georgia.

REFERENCES

Ashby, E., and T. A. Oxley, 1935, The Interaction of Factors in the Growth of *Lemna*. VI. An Analysis of the Influence of Light Intensity and Temperature on the Assimilation Rate and the Rate of Frond Multiplication, *Ann. Bot. (London)*, XLIX: 309-336.

——, E. Wangermann, and E. J. Winter, 1949, Studies in the Morphogenesis of Leaves. III. Preliminary Observations on Vegetative Growth in *Lemna minor*, *New Phytol.*, 48: 374-381.

——, and E. Wangermann, 1951, Studies in the Morphogenesis of Leaves. VII. Part II. Correlative Effects of Fronds in *Lemna minor*, *New Phytol.*, 50: 200-209.

Christiansen, M. E., and J. D. Costlow, Jr., 1975, The Effect of Salinity and Cyclic Temperature on Larval Development of the Mud-Crab *Rhithropanopeus harrisii* (Brachyura: Xanthidae) Reared in the Laboratory, *Mar. Biol.*, 32: 215-221.

Clatworthy, J. N., and J. L. Harper, 1962, The Comparative Biology of Closely Related Species Living in the Same Area. V. Inter- and Intraspecific Interference Within Cultures of *Lemna* spp. and *Salvinia natans*, *J. Exp. Bot.*, 13: 307-324.

Claus, W. D., 1972, Lifespan and Budding Potential of *Lemna* as a Function of Age of the Parent, *New Phytol.*, 71: 1081-1095.

Costlow, J. D., and C. G. Bookhout, 1971, The Effect of Cyclic Temperatures on Larval Development in the Mud-Crab *Rhithropanopeus harrisii*, in *Fourth European Marine Biology Symposium*, D. J. Crisp (Ed.), pp. 211-220, Cambridge University Press, London.

Dickson, H., 1938, The Occurrence of Long and Short Cycles in Growth Measurements of *Lemna minor*, *Ann. Bot. (London)*, II: 97-106.

Grainger, J. N. R., 1959, The Effect of Constant and Varying Temperatures on the Developing Eggs of *Rana temporaria* L., *Zool. Anz.*, 163: 267-277.

Hagstrum, D. W., and W. R. Hagstrum, 1970, A Simple Device for Producing Fluctuating Temperatures, with an Evaluation of the Ecological Significance of Fluctuating Temperatures, *Ann. Entomol. Soc. Am.*, 63: 1385-1389.

Heath, W. G., 1963, Thermoperiodism in Sea-Run Cutthroat Trout *(Salmo clarki clarki), Science*, 142: 486-488.

Hicks, P. A., 1934, Interaction of Factors in the Growth of *Lemna.* V. Some Preliminary Observations upon the Interaction of Temperature and Light on the Growth of *Lemna, Ann. Bot. (London)*, XLVIII: 515-525.

Hillman, W. S., 1976, Calibrating Duckweeds: Light, Clocks, Metabolism, Flowering, *Science*, 193: 453-458.

Hoagland, D. R., and D. I. Arnon, 1950, The Water-Culture Method for Growing Plants Without Soil, Circular 347, pp. 1-32, California Agricultural Experiment Station, College of Agriculture, University of California, Berkeley.

Hodgson, G. L., 1970, Effects of Temperature on the Growth and Development of *Lemna minor* Under Conditions of Natural Daylight, *Ann. Bot. (London)*, 34: 365-381.

Holland, W. E., M. H. Smith, J. W. Gibbons, and D. H. Brown, 1974, Thermal Tolerances of Fish from a Reservoir Receiving Heated Effluent from a Nuclear Reactor, *Physiol. Zool.*, 47: 110-118.

Jacobs, D. L., 1947, An Ecological Life History of *Spirodela polyrhiza* (Greater Duckweed) with Emphasis on the Turion Phase, *Ecol. Monogr.*, 17: 437-469.

Kirk, R. E., 1968, *Experimental Design: Procedures for the Behavioral Sciences*, Brooks/Cole Publishing Company, Belmont, Calif.

Morrison, D. F., 1967, *Multivariate Statistical Methods*, McGraw-Hill Book Company, Inc., New York.

Rejmánková, E., 1973a, *Biomass, Production and Growth Rate of Duckweeds* (*Lemna gibba* and *L. minor*), PT-PP Report No. 3, pp. 101-106, International Biological Program, Trebon, Czechoslovakia.

——, 1973b, Seasonal Changes in the Growth Rate of Duckweed Community *(Lemnetum gibbae), Folia Geobot. Phytotaxon.*, 8: 1-13.

Shimo, S., 1977, *The Thermal Effects on the Marine Organisms: The Status of the Arts*, Bio-environment Laboratory, Central Research Institute of Electric Power Industry, Tokyo.

Smith, M. H., 1978, Savannah River Ecology Laboratory, personal communication.

Smoak, G. H., Jr., 1959, Some Considerations in Upper Thermal Tolerance on *Gambusia affinis holbrooki* (Girard) in the Savannah River Plant Area, *S. C. Acad. Sci.*, 1959: 44-53.

Thorp, J. H., and S. A. Wineriter, 1978, Growth in a Stochastic Environment: Acyclic Temperature Regime, in preparation.

Van Winkle, W., Jr., 1969, Physiological Effects of Short-Term, Cyclic Environmental Changes, *Am. Zool.*, 9: 1100.

Wangermann, E., and E. Ashby, 1951, Studies in the Morphogenesis of Leaves. VII. Part I. Effects of Light Intensity and Temperature on the Cycle of Ageing and Rejuvenation in the Vegetative Life History of *Lemna minor, New Phytol.*, 50: 186-199.

White, H. L., 1936, The Interaction of Factors in the Growth of *Lemna.* IX. Further Observations on the Effect of Light Intensity on Growth and Multiplication, *Ann. Bot. (London)*, L: 827-848.

——, 1939, The Interaction of Factors in the Growth of *Lemna.* XIV. The Interaction of Potassium and Light Intensity in Relation to Growth and Assimilation, *Ann. Bot. (London)*, III: 619-648.

GROWTH AND ECOLOGY OF *Spartina alterniflora* IN MAINE AFTER A REDUCTION IN THERMAL STRESS

M. KESER,* B. R. LARSON,* R. L. VADAS,† and W. McCARTHY*
*Department of Botany and Plant Pathology, and †Department of Botany and Plant Pathology and Oceanography and Zoology, University of Maine, Orono, Maine

ABSTRACT

When a surface thermal discharge was relocated through a diffuser system in May 1975, the thermal impact on a previously stressed and moribund population of *Spartina alterniflora* in Montsweag Bay, Maine, was terminated. Post-stress growth and recovery were examined over three growing seasons. Plant biomass and density were lowest in 1975 because of continued surface discharge into Bailey Cove during the winter and early growing season of 1975. In 1976 and 1977, however, biomass and density increased slightly and shoot growth returned to prestress levels, as indicated by size-frequency distributions. The partial recovery of *Spartina* resulted from viable patches of plants and rhizomes in ~40% of the study area. In relation to previous values, biomass declined in the control marsh during this 3-year period. The decline was attributed to increased salinity, greater tidal amplitude, and decreased summer temperatures resulting from the removal of a causeway between the mainland and Westport Island in the fall of 1974.

Spartina alterniflora, which dominates extensive coastal areas of eastern North America (Teal, 1962; Redfield, 1972), is generally considered eurytopic and relatively insensitive to aquatic stresses (Young, 1974; Adams, 1963; Anderson, 1969). Populations near their northern limit of distribution are not as tolerant as others, however, and, when they were exposed to thermal stress in Maine, they were affected adversely and collapsed within 2.5 years (Vadas, Keser, Rusanowski, and Larson, 1976). The disappearance of plants in 60% of the marsh was related to the destruction of the rhizome system, which had metabolized its reserves and was unable to produce viable shoots in the spring of 1974 and 1975.

In May 1975 the thermal discharge into Bailey Cove from the Maine Yankee Atomic Power Company (MYAPCO) was discontinued, and 2 months later the discharge was diverted into a multiport diffuser system in the main channel of Montsweag Bay. This paper analyzes the growth and potential recovery of this previously stressed population of *Spartina.*

SITE AND METHODS

We studied two populations of *Spartina* in Montsweag Bay from spring 1975 to August 1977. Populations were located in Bailey Cove, which was previously stressed by a surface thermal discharge, and in Causeway Marsh, a control area located 3 km north of the power plant (Fig. 1). Detailed descriptions of Bailey and Causeway marshes are available elsewhere (Vadas et al., 1976).

Ten randomly selected quadrats ($\frac{1}{16}$ m^2) at each site were sampled for aboveground biomass in July and August 1975 and 1977 and monthly from May to October 1976. Because of the disruptive nature of biomass collections, limited sampling was performed in 1975 and 1977 to minimize damage to previously stressed plants. In 1975 and 1977 samples were taken during previously established peak biomass periods (Vadas et al., 1976). Sampling in Bailey Marsh was stratified since only 40% of the marsh contained *Spartina.* Four samples were taken at random in areas containing plants, and six were taken in areas devoid of plants. Plants were cut with shears at the surface of the mud and were transported to the laboratory, where they were washed, counted, measured, and dried to a constant weight at 70°C. Salinities and temperatures were measured monthly over mud flats adjacent to each marsh during high tide at depths of 0.15 m (surface) and 3.0 m (bottom) with a salinometer–thermometer. Light extinction coefficients were determined with a Secchi disc (Holmes, 1970). Incident solar radiation was measured with a pyreheliometer at MYAPCO.

After removal of the causeway in the fall of 1974, the average tidal range increased by 0.34 m, reflecting greater water flow in and out of Montsweag Bay. The average high tide was 0.06 m higher, and the average low tide 0.27 m lower. The resultant average tidal range, 2.88 m, exposed an additional 109 to 158 ha of mud flat during each tide (Anonymous, 1975).

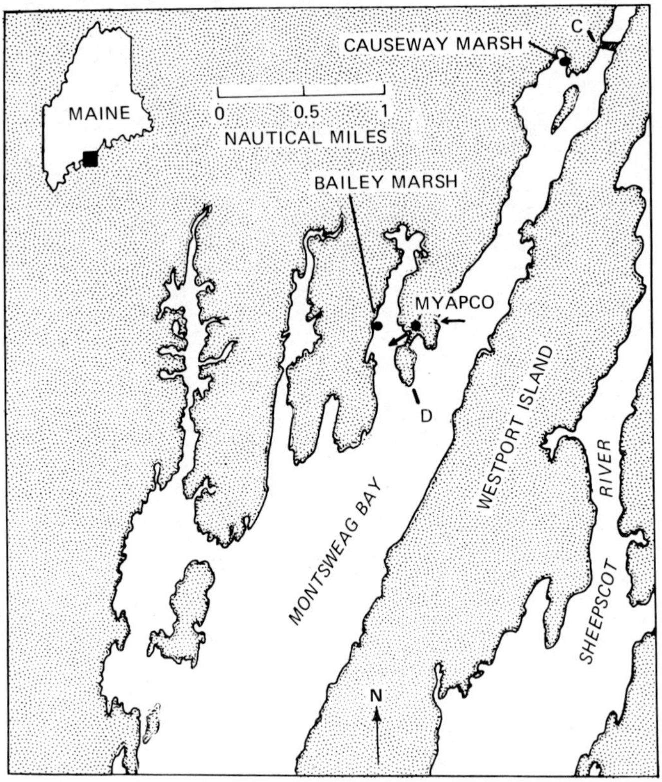

Fig. 1 Maine Yankee Atomic Power Company (MYAPCO) at Wiscasset, Me., showing the stressed (Bailey) and control (Causeway) marshes; Cowseagan Narrows causeway, C; and diffuser discharge, D. Water intake and former surface discharge are indicated by arrows.

RESULTS AND DISCUSSION

Hydrographic Data

Average salinities at Causeway Marsh during 1975 to 1977 increased by 4.2‰ at the surface and by 3.4‰ at a depth of 3 m (Fig. 2) compared with values from 1972 to 1974 (Vadas et al., 1976). Variabilities among monthly measurements declined, however. In contrast, salinity patterns at Bailey Marsh from 1975 to 1977 were similar to those from 1972 to 1974.

Lower temperatures at Causeway Marsh during the early growing seasons of 1975 to 1977 (April—July) resulted in lower yearly

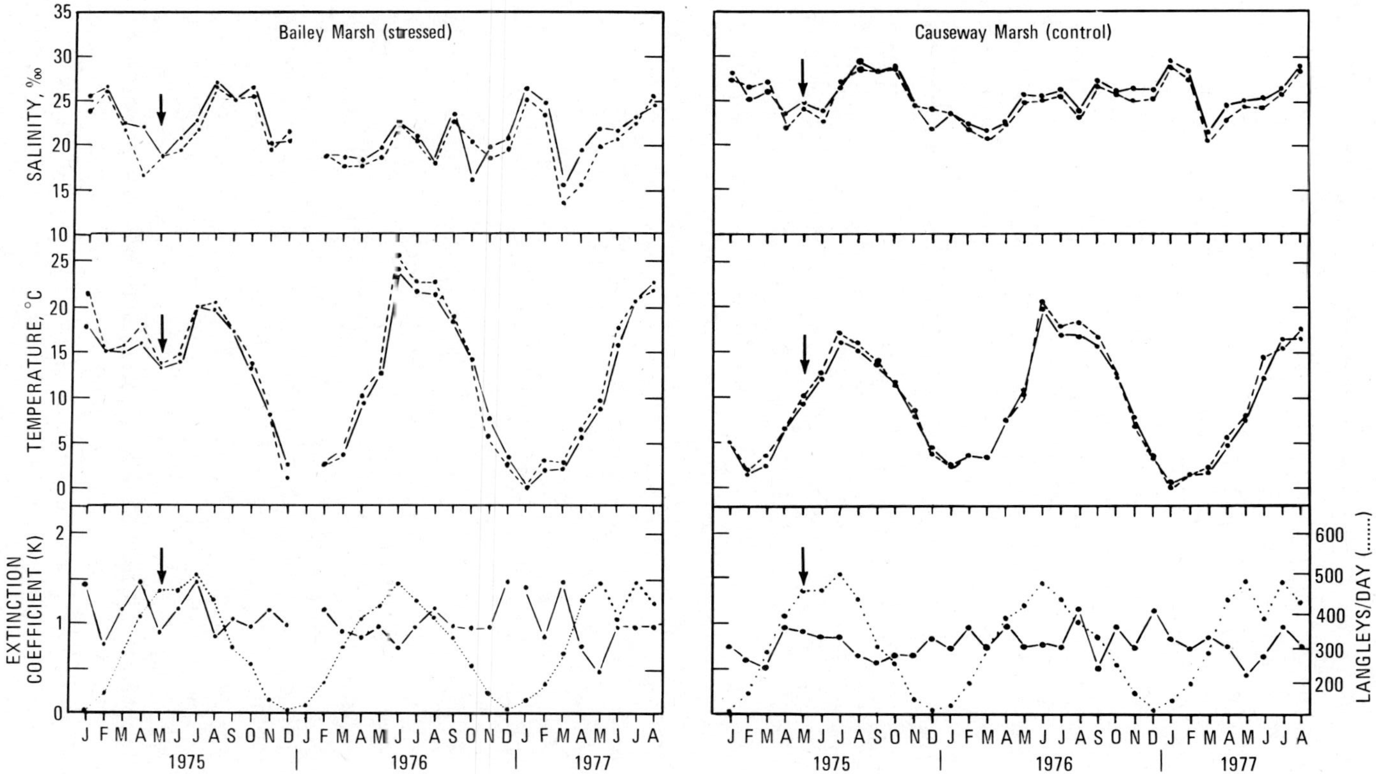

Fig. 2 Hydrographic data for Bailey and Causeway marshes since diversion of thermal effluent through a diffuser system. ·- - -·, surface (0.15-m) values. ·—·, bottom (3-m) values. Arrows indicate termination of surface discharge.

average temperatures (1.3°C at 0.15 m and 2.1°C at 3 m) than in 1972 to 1974. These were caused by increased exchange with the colder waters when the causeway was removed. Temperatures at Bailey Marsh declined significantly after relocation of the thermal discharge. Surface and bottom temperatures were higher at Bailey Marsh (2.7°C and 3.6°C, respectively) in spring and early summer (April–July) than at Causeway Marsh during this post-stress phase. The differential was caused by increased insolation and heat conduction to the water column from newly exposed mud flats (Dean and Officer, 1977). Average yearly temperatures were lower in Bailey Marsh, however, because of the entrapment of freshwater and the presence of ice. Turbidity decreased slightly after completion of the diffuser. Incoming solar radiation patterns were similar throughout stress and post-stress phases.

Plant Density

The density of *Spartina* at Causeway Marsh exhibited two distinct peaks, one in late spring and one in early fall (Fig. 3). Densities were highest from 1972 to 1974, averaging 704 and 1118 shoots/m^2 in July and August, respectively (Vadas et al., 1976). Mean densities from 1975 to 1977 averaged 434 and 686 shoots/m^2 in July and August, respectively. The decline in density during this period was attributed to removal of the causeway, which allowed a greater influx of colder and more saline water.

At Bailey Marsh the late spring peak did not develop in 1974, and its absence was attributed to the thermal discharge. This peak was reestablished in 1976, however, 1 year after the discharge into Bailey Cove was discontinued (Fig. 3). The density of *Spartina* in July and August 1972 was 560 and 900 shoots/m^2, respectively, whereas in stressed years (1973 and 1974), densities in July and August averaged 625 and 750 shoots/m^2, respectively. Average shoot density for the marsh was lowest in 1975 but increased in 1976 and 1977 to 222 and 476 shoots/m^2 for July and August, respectively. In 40% of the marsh, however, where healthy plants and rhizomes were present, plant densities in poststressed years (1976, 1977) averaged 672 and 1188 shoots/m^2 in July and August. These values were higher than those in 1972 and higher than densities at Causeway Marsh for the same periods. The plants at Bailey Marsh had thin shoots and leaves, however, and were not as robust as those at Causeway Marsh. These data suggest that recovery in the overall marsh is sparse and that, where plants survived, they recovered only partially.

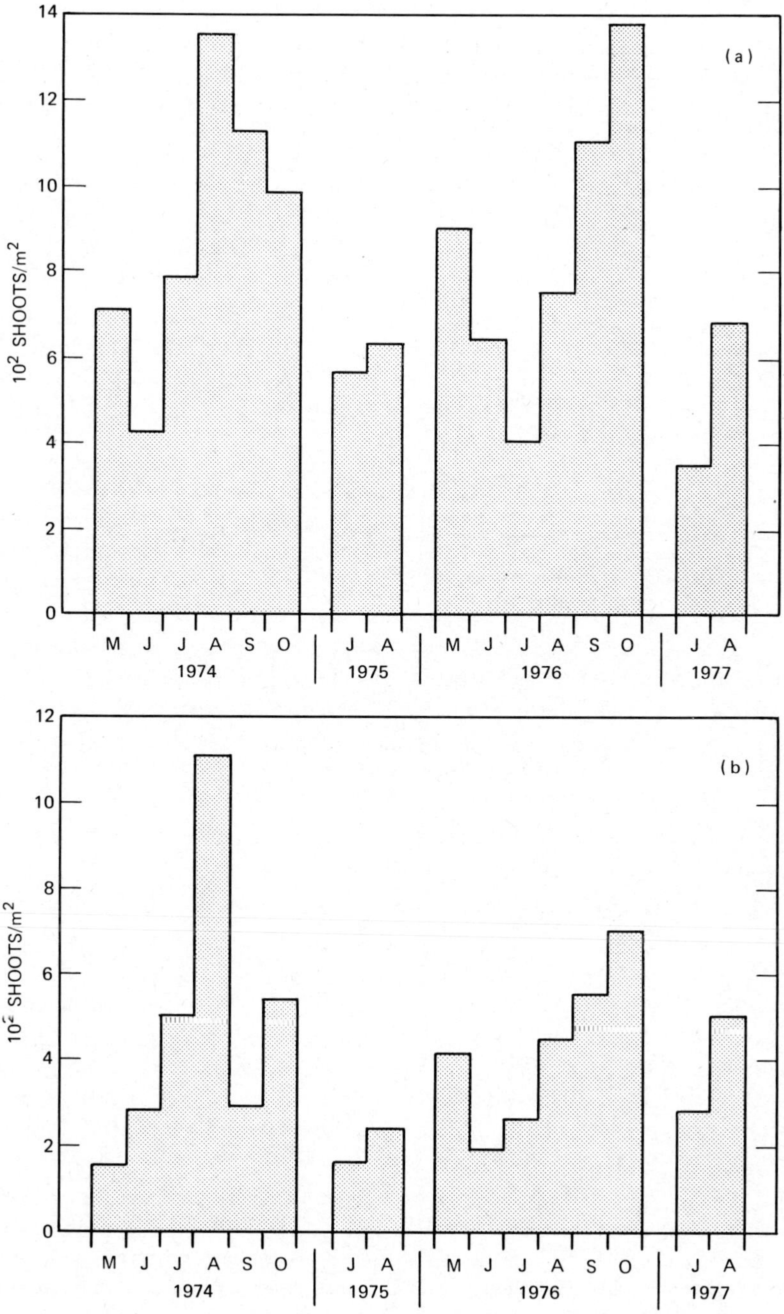

Fig. 3 **Density of** ***Spartina alterniflora*** **during the last year of direct thermal discharge (1974) and since relocation of the thermal discharge through a diffuser (1975–1977) at (a) Causeway Marsh (control) and (b) Bailey Marsh (stressed).**

Growth

At Bailey Marsh the number of new shoots increased in August, forming the typical bimodal pattern that was evident before 1972 but was lacking during the second year of thermal stress. The size and numbers of *Spartina* continued to decline in 1975 because the surface discharge continued until May of that year. The bimodal growth pattern was reestablished in 1976 and 1977, however. The recovery of *Spartina* was limited to 40% of the marsh, where rhizomes had survived, and the remaining portion has not been recolonized by seeds or rhizomes. Growth patterns for *Spartina* at Causeway Marsh have not changed over the course of these studies (Fig. 4). Shoot growth began in early May, and by July the majority of plants were in the larger size ranges (50 to 125 cm in length).

These results suggest that rhizome systems of northern populations of *Spartina* are intolerant of thermal stress and that revegetation of a marsh is a slow process in northern areas. The sensitivity of rhizomes to stress is not unique. The destruction of marshes through compaction and damage to rhizome systems by walking was described by Teal and Teal (1969). Similarly, Thomas (1973) reported that the mortality of *S. alterniflora* was most severe the second year after an oil spill because of the demise of the rhizome system.

Flowering

Flowering in *Spartina* at Bailey Marsh during stressed years began in July and continued through September (Vadas et al., 1976). When the stress was ameliorated (in 1975 to 1977), flowering was initiated later and lasted through October (Table 1). The onset of flowering in Causeway Marsh (control) occurred in August throughout both phases of these studies. Plants in the 75- to 100- and 100- to 200-cm size classes flowered at Bailey Marsh, but only plants taller than 100 cm flowered at Causeway Marsh. The average height of reproductive plants in the Control Marsh during 1975 to 1977 ranged from 135 to 149 cm, whereas average shoot heights at Bailey Marsh ranged from 98 to 116 cm. These size ranges were similar to those recorded in 1973 and 1974 (Vadas et al., 1976). Plants in the 50- to 75-cm range at Causeway Marsh did not flower in 1975 to 1977 although that size class had flowered previously. Since marsh plants are less fertile at high salinities (Adams, 1963), it is possible that the reduction in flowering at this site was caused by increased salinities.

Biomass

The increase in biomass of *Spartina* in Maine follows an exponential growth curve that usually peaks in August. Biomass estimates at Causeway Marsh for 1972 to 1974 averaged 1.04 kg/m^2 for July and August (Vadas et al., 1976). Average biomass values for July and August 1975 to 1977 declined by 42% over the average of the previous three summers. These changes resulted from natural variability (Vadas et al., 1976) and likely from physical changes (especially salinity) in Montsweag Bay. Negative correlations between biomass and salinity were observed for other marsh grasses, e.g., *S. alterniflora* (Adams, 1963) and *S. foliose* (Phleger, 1971). However,

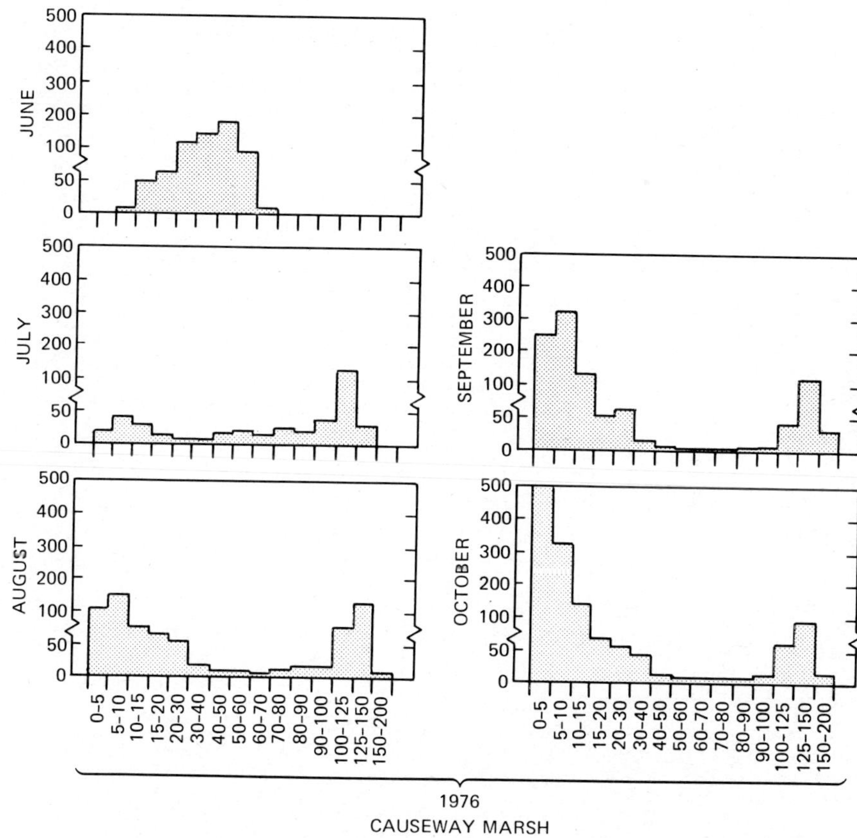

Fig. 4 Monthly size distributions for *Spartina alterniflora* at Causeway Marsh (control) (1976 only) and at Bailey Marsh (stressed) during the last full year of surface discharge (1974) and after diversion of effluent through the diffuser (1975—1977).

(Figure continues on following page.)

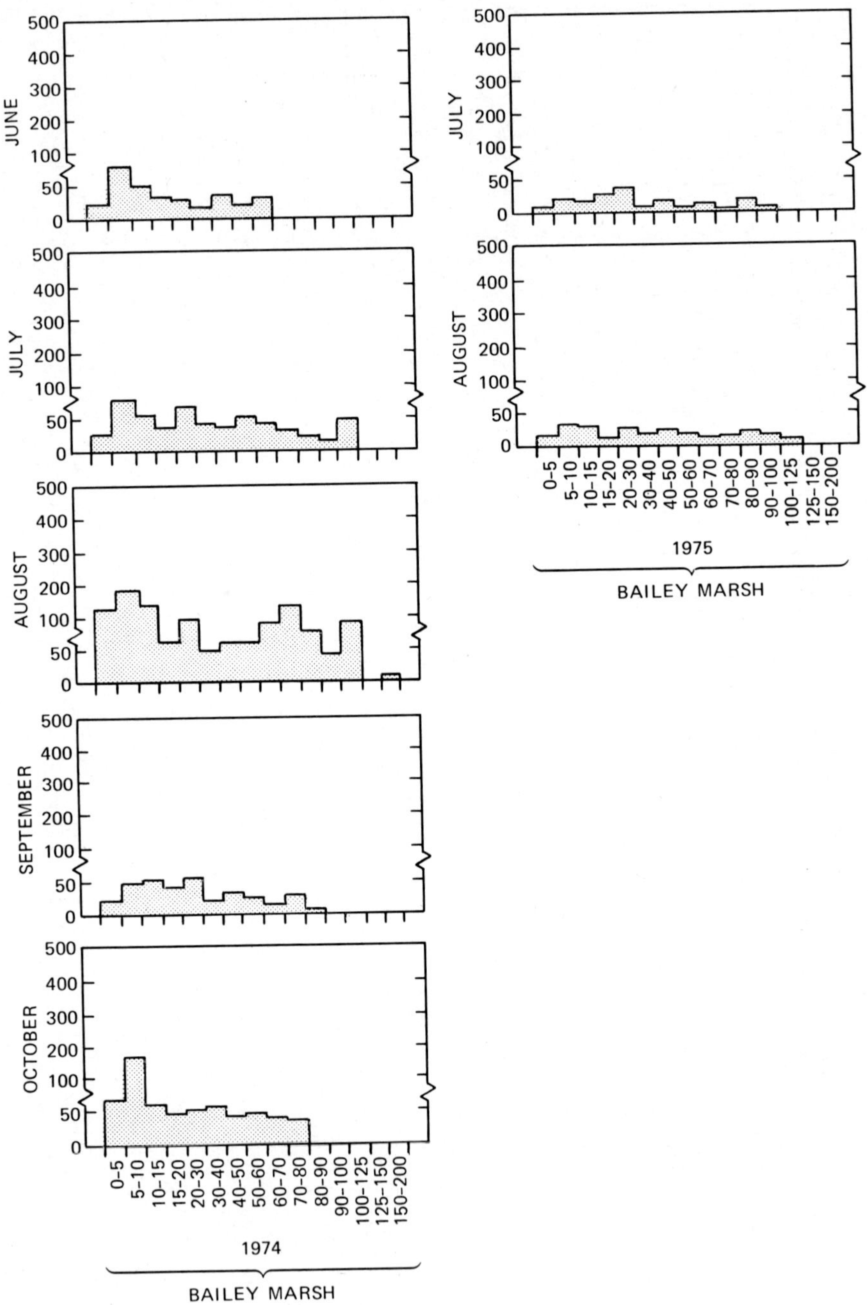

Fig. 4 (Continued)

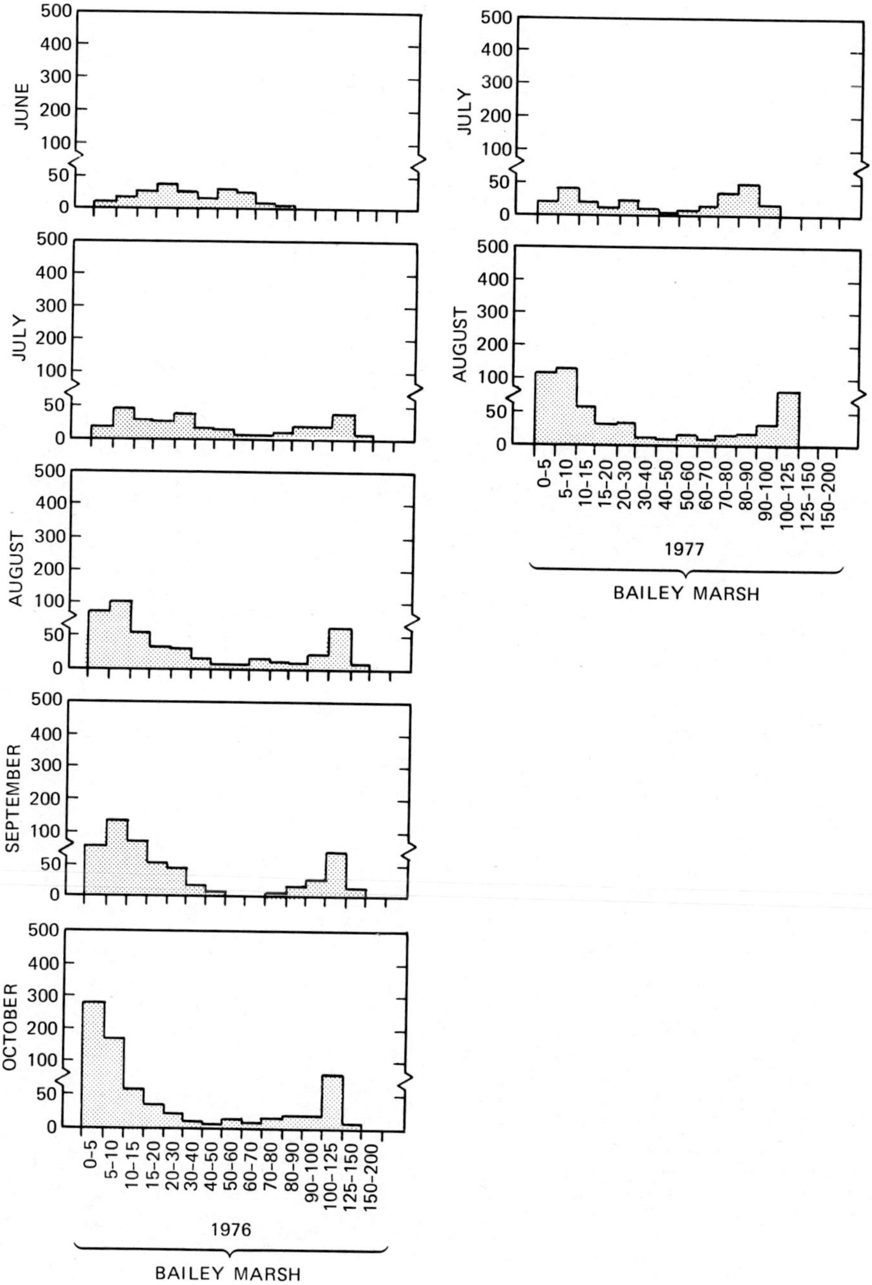

Fig. 4 (Continued)

(See p. 427 for legend.)

TABLE 1

SIZE FREQUENCY OF REPRODUCTIVELY MATURE *Spartina alterniflora*

Date	Size class, cm						Mean height* of reproductive plants, cm
	50–75		75–100		100–200		
	Shoots/m^2	% Flowering	Shoots/m^2	% Flowering	Shoots/m^2	% Flowering	
Causeway Marsh (control)							
Aug. 1975	10	0	50	0	283	1.00	149.9 ± 14.99
Aug. 1976	15	0	32	0	213	18.80	135.47 ± 9.74
Sept. 1976	5	0	13	0	214	11.19	149.15 ± 12.26
Oct. 1976	14	0	18	0	203	15.75	146.48 ± 12.37
Aug. 1977	14	0	51	0	203	6.90	137.64 ± 9.25
Bailey Marsh (stressed)							
Aug. 1975	38	0	38	8.00	8	60.00	98.86 ± 7.48
July 1976	0	0	35	4.76	38	54.17	113.73 ± 8.88
Aug. 1976	24	0	34	19.04	67	83.00	111.11 ± 9.35
Sept. 1976	16	0	59	10.81	117	80.82	116.48 ± 10.70
Oct. 1976	27	0	56	0	107	65.67	113.46 ± 7.05
Aug. 1977	18	0	53	24.52	77	62.50	106.9 ± 7.19

*± Standard deviation.

similar findings by Nixon and Oviatt (1973) in Rhode Island were confounded additionally by negative correlations between salinity and nutrient levels which masked the causes of variability. High soil—sediment salinities have also been implicated in reduced growth in *Spartina* (Mooring, Cooper, and Seneca, 1971; Broome, Woodhouse, and Seneca, 1973).

The idea that the decrease in biomass at Causeway was caused by increased salinity is supported by the lack of correlation between biomass and temperature. Despite the apparent correlation between increasing temperatures and biomass from 1972 to 1974 (Vadas et al., 1976), warmer temperatures in 1976 did not result in higher than normal biomass estimates. This suggests that the growth of *Spartina* in Maine is not wholly temperature dependent, but rather that temperature and salinity, and perhaps other factors, act in concert to control growth.

The decline in biomass at Bailey Marsh during the thermal-discharge phase, however, was attributed directly to increased temperatures. Direct surface discharge continued until May 2, 1975, and resulted in the lowest recorded biomass estimates for August (Fig. 5). Biomass increased slightly in 1976 and 1977, suggesting that the thermal stress and apparent shock has been alleviated and that the marsh may be returning to more stable (prestress) conditions.

SUMMARY

During the first year of diffuser operation, the biomass and shoot density of *Spartina* at Bailey Marsh continued to decline. In 1976 and 1977 the decline, which was associated with the previous thermal stress on the rhizome system, was arrested, and recovery was evident.

The biomass, plant density, and flowering of *Spartina* at Causeway Marsh (control) decreased after the Cowseagan Narrows causeway was removed in the fall of 1974. These declines were attributed to increased tidal amplitude and salinities and decreased temperatures during the growing season. Light was not considered important since changes in incoming solar radiation were not evident and since turbidity had decreased from 1975 to 1977.

The most noticeable effect of the thermal effluent was the reduction in coverage of *Spartina* in Bailey Marsh. Approximately 60% of the study area was devoid of plants. Colonization by seeds or rhizomes had not occurred as of 1977. Examination of rhizomes in the area showed them to be badly decomposed. Redfield (1972)

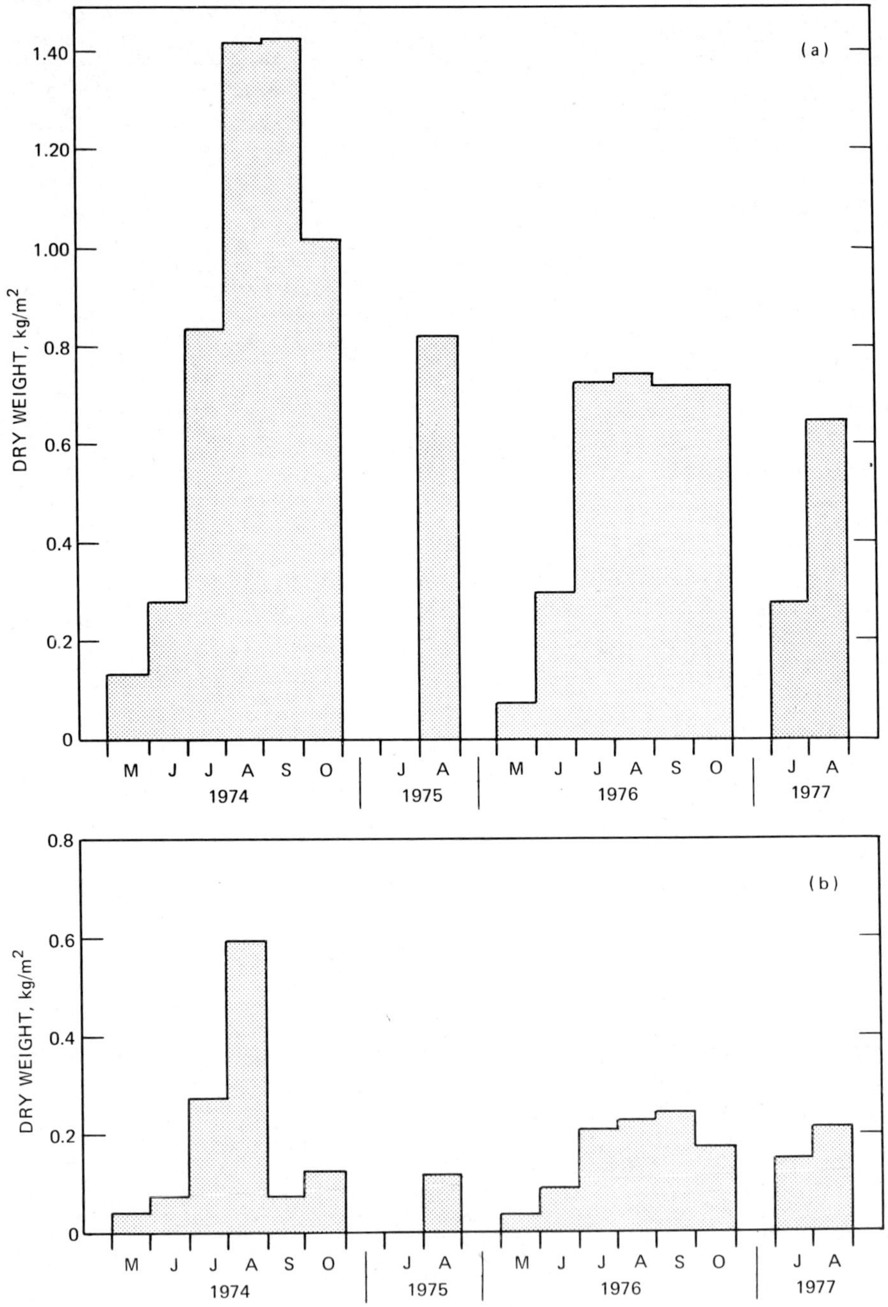

Fig. 5 Biomass of *Spartina alterniflora* at (a) Causeway Marsh (control) and (b) Bailey Marsh (stressed) during the last full year of surface thermal discharge (1974) and after diversion of the effluent through the diffuser (1975—1977).

found that lateral extension of *Spartina* by rhizomes and colonization via seeds was a slow process in Massachusetts marshes. If plants in Bailey Marsh respond similarly, it will be several years before complete recovery occurs.

REFERENCES

Adams, D. A., 1963, Factors Influencing Vascular Plant Zonation in North Carolina Salt Marshes, *Ecology*, 44: 445-456.

Anderson, R. R., 1969, Temperature and Rooted Aquatic Plants, *Chesapeake Sci.*, 10: 157-164.

Anonymous, 1975, *Hydrography*, Semi-Annual Report No. 6, pp. 1.7-3—1.7-11, Maine Yankee Atomic Power Company, Wiscasset.

Broome, S. W., W. W. Woodhouse, and E. D. Seneca, 1973, *An Investigation of Propagation and Mineral Nutrition of* Spartina alterniflora, Sea Grant Publication UNC-SG-73-14, University of North Carolina, Chapel Hill.

Dean, D., and C. B. Officer, 1977, *Development Document in Support of Alternative Effluent Limitations Pursuant to Section 316(a) of the Federal Water Pollution Control Act for Maine Yankee Nuclear Generating Station*, Maine Yankee Atomic Power Company, Wiscasset.

Holmes, R. W., 1970, The Secchi Disk in Turbid Coastal Waters, *Limnol. Oceanogr.*, 15: 688-694.

Mooring, M. T., A. W. Cooper, and E. D. Seneca, 1971, Seed Germination Response and Evidence for Height Ecophenes in *Spartina alterniflora* from North Carolina, *Am. J. Bot.*, 58: 48-55.

Nixon, S. W., and C. A. Oviatt, 1973, Analysis of Local Variation in the Standing Crop of *Spartina alterniflora, Bot. Mar.*, 16: 103-109.

Phleger, C. F., 1971, Effect of Salinity on Growth of Salt Marsh Grass, *Ecology*, 52: 908-911.

Redfield, A. C., 1972, Development of a New England Salt Marsh, *Ecol. Monogr.*, 42: 201-237.

Teal, J. M., 1962, Energy Flow in the Salt Marsh Ecosystem of Georgia, *Ecology*, 43: 614-624.

——, and M. Teal, 1969, *Life and Death of the Salt Marsh*, Atlantic Monthly Press, Boston.

Thomas, M. L. H., 1973, Effects of Bunker C Oil on Intertidal and Lagoonal Biota in Chedabucto Bay, Nova Scotia, *J. Fish. Res. Board Can.*, 30: 83-90.

Vadas, R. L., M. Keser, P. C. Rusanowski, and B. R. Larson, 1976, The Effects of Thermal Loading on the Growth and Ecology of a Northern Population of *Spartina alterniflora*, in *Thermal Ecology II*, ERDA Symposium Series, Augusta, Ga., Apr. 2—5, 1975, G. W. Esch and R. W. McFarlane (Eds.), pp. 54-63, CONF-750425, NTIS.

Young, D. L., 1974, Studies of Florida Gulf Coast Salt Marshes Receiving Thermal Discharges, in *Thermal Ecology*, AEC Symposium Series, Augusta, Ga., May 3—5, 1973, J. W. Gibbons and R. R. Sharitz (Eds.), pp. 532-550, CONF-730505, NTIS.

EFFECTS OF REDUCED TEMPERATURES ON PREVIOUSLY STRESSED POPULATIONS OF AN INTERTIDAL ALGA

R. L. VADAS,* M. KESER,† and B. LARSON†
*Departments of Botany and Plant Pathology and Oceanography and Zoology, and †Department of Botany and Plant Pathology, University of Maine, Orono, Maine

ABSTRACT

Relocation of a surface thermal discharge through a multiport diffuser and removal of a causeway substantially reduced temperatures in an estuary surrounding a nuclear power plant. Although stressed and moribund during thermal discharge, basal portions of adult thalli of *Ascophyllum nodosum* showed considerable resilience to thermal stress and potential competition from other species and survived. Previously stressed populations recovered fully; all measures of vigor being indistinguishable from prestress years. Growth in apical tips was sensitive to small shifts in ambient water temperature, suggesting that *Ascophyllum* might be a good indicator for thermal and perhaps other stresses in marine ecosystems. Growth data indicate that thermal enhancement occurs in *Ascophyllum*, artificially from thermal effluent and naturally from increased insolation to newly exposed mud flats.

Altering physical, chemical, or biological components of ecological systems often provides insight into the structure and role of individual components. Manipulative experiments, for example, are powerful tools in analyzing causal relationships in marine communities (Connell, 1961; Paine, 1966; Dayton, 1971). Gradients of stress, whether natural or man-induced, also provide the opportunity for gaining insight into the responses of species to such stresses and potentially to their roles in community organization.

The Maine Yankee Atomic Power Company (MYAPCO) produced a sequence of varying thermal and physical stresses on rocky intertidal communities in Montsweag Bay, Maine. The sequence included, chronologically, a surface thermal discharge into a shallow cove, the removal of a causeway, and the installation of a diffuser discharge.

The stresses produced by the thermal discharge and, to some extent, the causeway removal were documented previously (Vadas, Keser, and Rusanowski, 1976). Briefly, the effects of the thermal discharge were confined to the vicinity of the discharge channel. There the dominant intertidal algae, *Ascophyllum nodosum* (L.) Le Jol. and *Fucus vesiculosus* L., were affected adversely. The cover of *Fucus* decreased with the onset of the thermal effluent and was never reestablished. The growth and survivorship of *Ascophyllum* increased initially but subsequently declined. Increased plant and apical tip mortality and the absence of new growth contributed to the gradual but incomplete demise of *Ascophyllum* populations in the discharge channel. The relocation of the discharge and power-plant shutdowns during the initial two summers of operation ameliorated the total impact and inevitable destruction of these populations. Although significant declines occurred in percent cover, biomass, growth, and survival of *Ascophyllum*, basal sections of many plants survived. This paper is concerned with the recovery of previously stressed plants and the effects of the present diffuser discharge on these algae.

STUDY SITES AND METHODS

Since detailed information was presented previously (Vadas, Keser, and Rusanowski, 1976), only aspects unique to this study or necessary for interpretation are included here and with the results. The MYAPCO diffuser consists of two 152-m-long fiber-glass pipes with 42 ports, each 43 cm in diameter. The multiport diffuser is located on the southwest side of Foxbird Island (Fig. 1) at a depth of 13.4 m below mean low water (MLW). Six sites were monitored in Montsweag Bay, three of which are experimental sites (Bailey Point, Young Point, and Foxbird Island). These sites are near the diffuser (potentially stressed) and are discussed here. Three sites served as regional controls, but only one of these, Hodgson Island, is considered here. Hydrographic samples were taken monthly at each site during high tide at depths of 0.15 and 3.0 m with a salinometer—thermometer. Light extinction coefficients were determined with a Secchi disk. Incident solar radiation was measured with a pyrheliometer at MYAPCO.

Algal coverage was determined seasonally (spring, summer, and fall) in three permanent, randomly established, vertical-belt transects per site. Data were analyzed by tidal zones. Percentages were

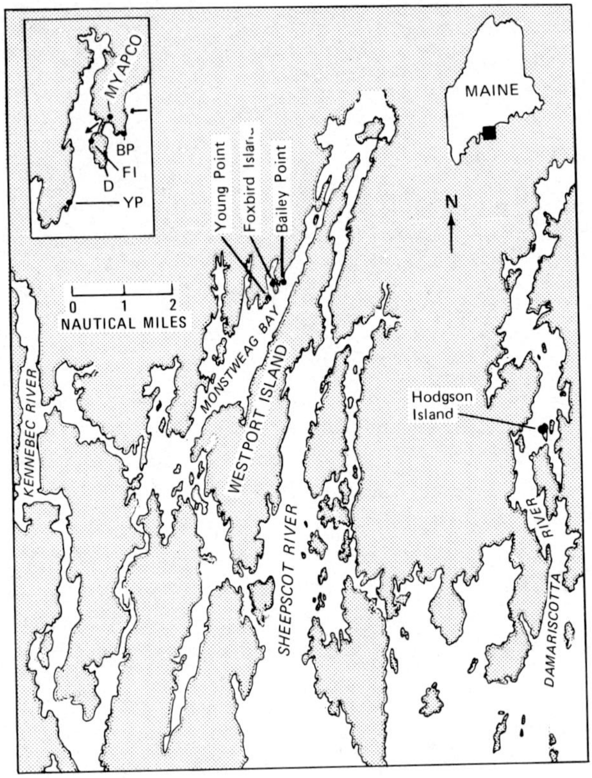

Fig. 1 Location of Maine Yankee Atomic Power Company (MYAPCO) and the control, Hodgson Island (HI), and experimental, Bailey Point (BP), Young Point (YP), and Foxbird Island (FI), sites. Arrows on left inset indicate sea water intake and the former surface discharge, and D denotes the position of the diffuser system.

transformed with an arc sine function and averaged and compared with an analysis of variance and Duncan's multiple range test.

The standing crop of *Ascophyllum* was determined by scraping the holdfasts of plants from rock surfaces in 25- by 25-cm quadrats. Samples were collected only once at Foxbird Island since the discharge was relocated because of the stressed nature of this population and the destructiveness of harvest sampling, especially to the reestablishment of *Ascophyllum* (Virville, 1953; Printz, 1959).

Apical tip elongation was shown to be a very rapid and sensitive indicator of thermal stress (Vadas, Keser, and Rusanowski, 1976) and was used to characterize growth. As in our previous study, two groups of 20 plants each were tagged in zone 2. Five apical tips per plant were further delineated and coded with colored plastic tape. Length measurements in June and through most of the growing season, therefore, represent the average of 200 tips. Mortality gradually reduced the number of plants and tips by the following June, but there were rarely fewer than 20 tips per site. Data were averaged for the last two stressed years (1973–1974 and 1974–1975) and were presented individually for subsequent years (see Fig. 3). The data for Foxbird Island during the stressed years represent only the lengths for 1973–1974 since viable apexes were not formed in 1974. The 95% confidence limits were used to distinguish differences among years at each site. The survivorship of plants and apical tips was determined concurrently with growth measurements.

RESULTS

Environmental Parameters

Two significant changes have occurred since the initial period of surface discharge into Bailey Cove. First, the Cowseagan Narrows causeway, a major impediment to water flow in and out of Montsweag Bay, was removed during the fall of 1974. Second, the cooling-water discharge was relocated and released through a multiport diffuser in May 1975. Both alterations contributed to changes in environmental conditions in Montsweag Bay. Extinction coefficient patterns were excellent indicators of the construction activities involved in these changes (Fig. 2). Throughout construction phases sediments were continually resuspended into the water column. The surface discharge into Bailey Cove additionally increased turbidity at the Foxbird Island and Young Point sites. Except after storms, turbidity values have decreased since the diffuser became operational but are slightly higher than preoperation values (Vadas, Keser, and Rusanowski, 1976; Vadas, Keser, and Larson, 1977).

Salinities increased and generally, with the exception of Foxbird Island and Young Point, became less variable throughout the bay, e.g., Bailey Point (Fig. 2) and (Vadas, Keser, and Larson, 1977). Higher salinities were caused by increased flushing and exchange with higher salinity waters. The relatively damped amplitude in salinity fluctuations in Montsweag Bay, however, does not approach that of most control sites, e.g., Hodgson Island (Fig. 2). Reduced temperatures at experimental sites were the direct result of cooler waters entering the bay and of rapid heat dissipation with the diffuser.

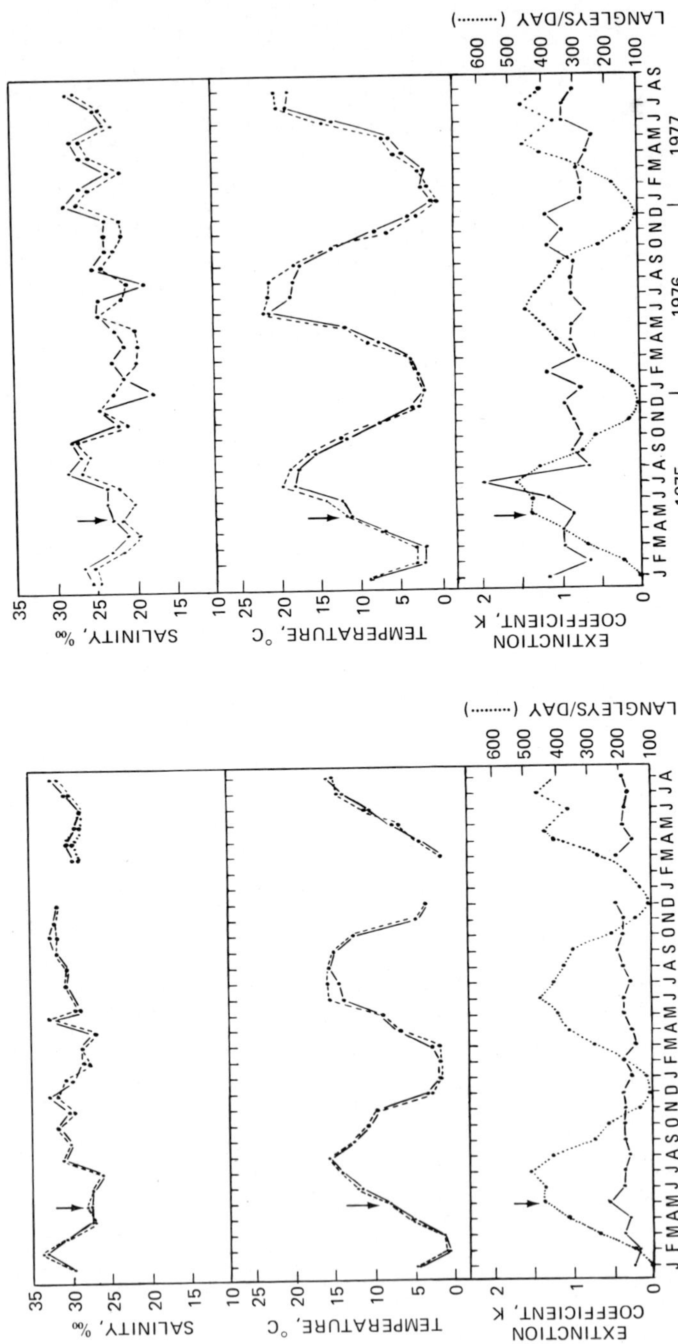

(Figure continues on following page.)

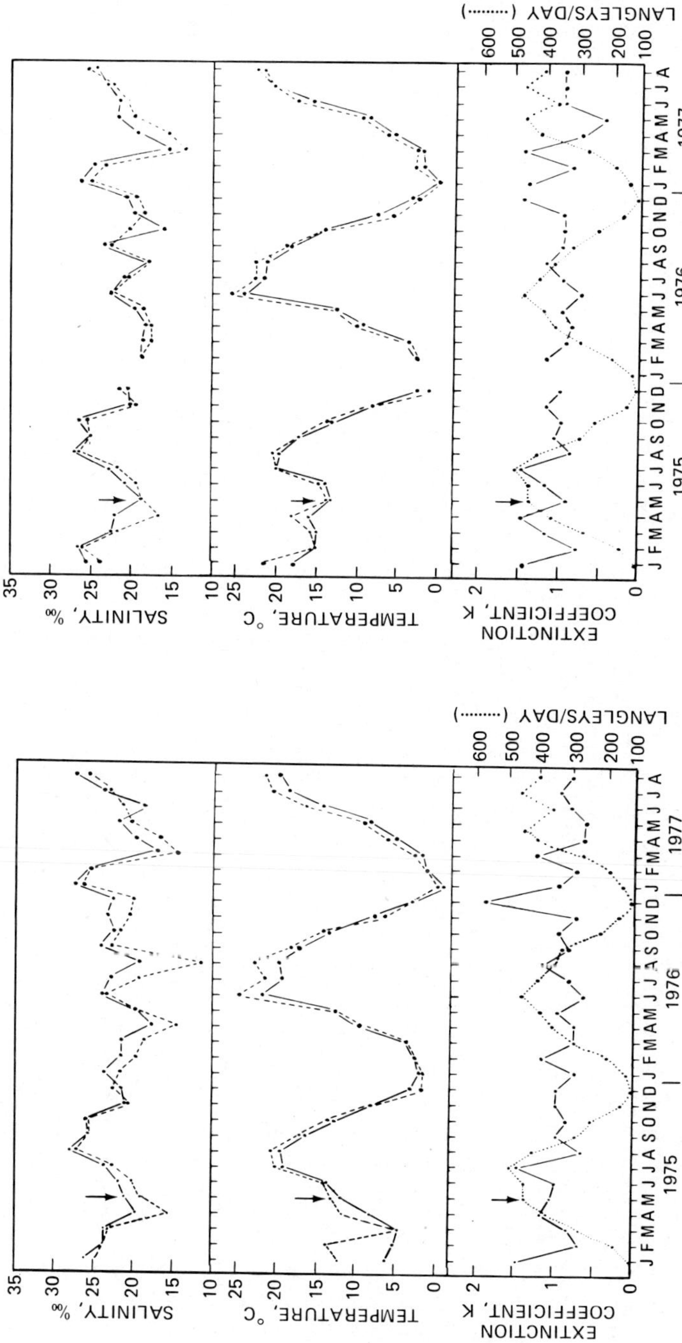

Fig. 2 Hydrographic data for control, Hodgson Island (a), and experimental, Bailey Point (b), Young Point (c), and Foxbird Island (d), sites. — — —, surface (0.15 m). ——, bottom (3 m). Arrows indicate termination of the surface discharge.

Regional and local temperature variability complicates these patterns, however. Annual mean water temperatures in Maine in 1976 were the highest recorded since the 1950s (Welch, 1977). Additionally, the removal of the causeway increased the depth of high tides by 0.06 m and lowered that of low tides by 0.27 m (Maine Yankee Atomic Power Company, 1975). This change increased significantly the amount of mud flat exposed during low tide and affected water temperatures in Montsweag Bay (Dean and Officer, 1977).

Percent Cover

The surface area covered by *Ascophyllum*, the dominant species, varied within normal limits (Vadas, 1972) at the four sites. Variability was lowest at Hodgson Island (control) and highest at Foxbird Island (previously stressed site). Generally, variability rarely exceeded ±6% between years at any site, and most values were not significant ($P > 0.05$). Coverage in the upper intertidal, zone 1, at Foxbird Island, except for 1975, was significantly below prestress periods, however. Conversely, plants in the mid- to low-intertidal, zone 2, recovered gradually but completely; differences between 1977 and the first stressed year (1973) were not significant ($P > 0.05$).

The cover of *Fucus* was highly variable at all sites. Significant differences among years occurred periodically at all sites. This was caused primarily by the relatively short life span (3 to 5 years) of *Fucus*, by ice damage, and by rapid recolonization patterns rather than by predictable environmental influences. At Foxbird Island the thermal discharge destroyed most adult *Fucus* populations. Normally sporelings of *Ascophyllum* do not readily colonize areas denuded naturally or artificially (Keser and Vadas, unpublished), but both *Ascophyllum* and *Fucus* reestablished in large numbers in 1975 after relocation of the effluent. *Fucus* reestablished at a slow rate in zone 2 but since 1976 has been significantly higher ($P < 0.05$) than during the stressed years. This increase in *Fucus* was caused in part by the availability of bare substrate that previously was occupied by *Ascophyllum* and *Fucus*.

To permit comparisons with earlier studies (Vadas, Keser, and Rusanowski, 1976), we listed the combined and transformed percentages of both species in Table 1. The actual area covered by these two species increased significantly ($P < 0.05$), by approximately 29% between 1974 and 1977.

TABLE 1

PERCENT COVER OF DOMINANT ALGAE, *Ascophyllum nodosum* AND *Fucus vesiculosus*, DURING AND AFTER THERMAL STRESS*

Year	Zone 1				Zone 2			
	Ascophyllum	*Fucus*	N†	Total	*Ascophyllum*	*Fucus*	N†	Total
				Hodgson Island (Control)				
1973	64.10 ± 19.11a	12.54 ± 18.61a	27	72.38 ± 14.35a	79.75 ± 11.19a	1.58 ± 4.68a	27	80.63 ± 10.81a
1974	68.10 ± 19.20a	22.17 ± 18.71a	27	74.38 ± 12.55a	83.93 ± 7.54a	1.32 ± 4.95a	27	85.07 ± 6.80a
1975	71.13 ± 15.48a	5.10 ± 7.52a	27	73.12 ± 14.44a	80.86 ± 7.00a	1.31 ± 4.13a	26	81.46 ± 6.35a
1976	69.93 ± 12.87a	2.68 ± 5.47a	27	70.76 ± 11.99a	81.16 ± 8.69a	0.00	27	81.16 ± 8.69a
1977	60.96 ± 24.26a	2.53 ± 7.36a	18	62.87 ± 20.05a	73.74 ± 9.98a	1.32 ± 3.13a	18	74.15 ± 10.07a
				Bailey Point				
1973	76.48 ± 10.47a	7.53 ± 11.14a	23	81.74 ± 6.21a	76.85 ± 10.22a	5.06 ± 9.44a	23	79.88 ± 8.65a
1974	74.08 ± 14.21a	2.86 ± 4.79a	21	75.72 ± 14.87a	77.07 ± 13.79a	1.81 ± 6.69a	21	78.15 ± 12.14a
1975	71.91 ± 18.32a	7.20 ± 13.25a	21	76.76 ± 14.85a	74.27 ± 11.14a	3.45 ± 7.00a	21	75.92 ± 10.53a
1976	62.65 ± 14.90a	13.46 ± 16.05a	21	72.41 ± 13.85a	64.28 ± 16.89b	0.61 ± 2.81a	21	64.42 ± 16.69b
1977	80.66 ± 6.78a	0.00	14	80.66 ± 6.78a	58.90 ± 14.36b	8.69 ± 11.32a	14	62.71 ± 14.14b

(Table continues on following page.)

Table 1 (Continued)

Year	Zone 1				Zone 2			
	Ascophyllum	*Fuctus*	N†	Total	*Ascophyllum*	*Fucus*	N†	Total
Young Point								
1973	77.30 ± 13.16a	1.49 ± 4.79a	21	78.25 ± 13.04a	80.54 ± 10.60a	0.56 ± 2.69a	23	80.70 ± 10.34a
1974	72.93 ± 17.77a	0.00	21	72.93 ± 17.77a	77.77 ± 12.11a	0.00	27	77.77 ± 12.11a
1975	74.57 ± 15.28a	2.65 ± 4.92a	21	75.60 ± 15.35a	79.07 ± 12.38a	0.00	25	79.07 ± 12.38a
1976	65.10 ± 19.82a	1.74 ± 5.26a	21	65.64 ± 19.66a	76.39 ± 13.27a	0.23 ± 1.17a	24	76.44 ± 13.26a
1977	69.81 ± 21.71a	0.58 ± 2.17a	14	70.39 ± 19.72a	64.49 ± 24.12a	0.00	16	64.49 ± 24.12a
Foxbird Island								
1973	69.96 ± 14.51a	13.56 ± 14.50b	25	79.41 ± 8.85a	71.91 ± 14.75a	0.91 ± 4.55b	25	72.45 ± 14.67a
1974	52.59 ± 10.70c	2.85 ± 5.24c	27	53.24 ± 10.57b	48.30 ± 14.53d	1.01 ± 2.74b	24	48.43 ± 14.31c
1975	62.49 ± 12.57ab	15.39 ± 14.41b	26	72.07 ± 10.78a	56.52 ± 23.18cd	2.65 ± 4.80b	24	57.27 ± 22.88bc
1976	53.90 ± 19.48bc	25.12 ± 18.15a	27	71.10 ± 14.90a	60.30 ± 17.71bc	7.50 ± 9.78a	24	63.40 ± 16.40ab
1977	56.05 ± 17.78bc	24.56 ± 19.82a	18	72.62 ± 10.10a	67.37 ± 14.00ab	11.32 ± 14.04a	16	70.18 ± 12.60a

*Original data were transformed with an arc sine function and analyzed with an analysis of variance and Duncan's multiple range test. Values given are means $(\overline{X}) \pm$ standard deviation, and those having a common letter (a—d) are not significantly different. $P \leqslant 0.05$ in all cases.

†N is the number of 0.25-m^2 quadrats.

TABLE 2

STANDING CROP* OF *Ascophyllum nodosum* DURING SURFACE DISCHARGE (1973—1974) AND AFTER 2 YEARS OF DIFFUSER DISCHARGE AT FOXBIRD ISLAND (1977)

	Standing crop, kg/m² dry weight					
	Zone 1			Zone 2		
Year	N	$\overline{X}$	SD	N	$\overline{X}$	SD
1973	19	2.647a	4.599	27	1.895a	2.690
1974	20	0.492b	0.984	27	0.589b	1.059
1977	10	2.489a	4.047	10	3.188a	3.363

*Values given are mean ($\overline{X}$) and standard deviation (SD), and those having a common letter (a,b) are not significantly different ($P \leqslant 0.05$ Student's t-test). N is the number of quadrats.

Biomass

Standing-crop estimates for Foxbird Island are given in Table 2. Means and standard deviations for zones 1 and 2 are given for two stressed years (1973 and 1974) and a poststress year (1977). Biomass estimates in 1977 for zones 1 and 2 were 83 and 82% higher, respectively, than in 1974, the year of maximum stress. Both estimates are significantly higher ($P < 0.05$) than values for 1974 but are not different from 1973 nor from earlier predischarge periods (Vadas, Keser, and Rusanowski, 1976).

Growth

Apical tips in August and November at Hodgson Island, the regional control, were significantly shorter in poststress than in stressed years because of extremely high growth rates (20% above normal) in the spring of 1973 and 1974 (Vadas, Keser, and Larson, 1977). Increased growth was anticipated at Hodgson Island during 1976 because of the presence of warmer waters regionally, but it never materialized.

Growth patterns in Montsweag Bay during poststress periods closely reflect local differences in water temperature. During stressed years at Foxbird Island, growth, when present at all, was significantly lower than it is today. Growth at Young Point during 1973—1974 was marginally stressed because of the proximity of the discharge, and, although apical tips were longer in spring, they were normal later in the year (Vadas, Keser, and Rusanowski, 1976). During poststress years growth at both these sites was significantly increased over stressed years. This was caused by: (1) increased temperatures from insolation to newly exposed mud flats and the subsequent conduction of heat to the water column, (2) the regional warming effect in 1976, and (3) possibly the diffuser discharge.

The effectiveness of the diffuser was evident at Bailey Point, the site nearest to it. With the exception of 1976–1977, growth there was significantly lower than in stressed years. To fully understand the growth patterns at Bailey Point, we must understand the effects of the surface discharge. Previously, a flow of warm water moved northward from the southern end of Foxbird Island, past Bailey Point, on the ebb tide. The periodically elevated temperatures increased the growth of *Ascophyllum* from spring to fall, as shown in Fig. 3. Thus reduced growth in 1975–1976 and the early part of 1977–1978 indicates that discharge through the diffuser has less effect at this site than the surface discharge had. These data suggest that seasonal or local variability alone (e.g., in 1976–1977), may be more important to the growth of *Ascophyllum* in Montsweag Bay than thermal stress from the diffuser discharge.

Mortality

Except in 1976, plant mortality was consistently lowest (10 to 20%) at Hodgson Island (Fig. 4). This site provided an excellent control on tag losses, which rarely exceeded 1%. The reasons for the unusually low survivorship in 1976 are not known. It is possible that seaweed harvesters cut plants at this site. The sustained mortality throughout the rest of the year may have been induced by weakening of holdfasts from the initial raking and pulling process, which accelerated mortality in all periods. Yearly plant survivorship at Bailey Point was similar for stressed and poststress years, ranging from 40 to 60%. Survivorship at Young Point and Foxbird Island, with the possible exception of 1977–1978 at Young Point, increased substantially with the relocation of the thermal discharge. It is possible that plants at Young Point were harvested in 1977.

Survival of apical tips was considerably lower and less variable than plant survival, even at Hodgson Island. This supports the view that plant holdfasts were weakened at Hodgson Island from harvesting or other causes. Tip survival at the three Montsweag Bay sites was quite similar during stressed and poststress phases, with the exception of Foxbird Island in 1974–1975 and possibly Young Point in 1977–1978.

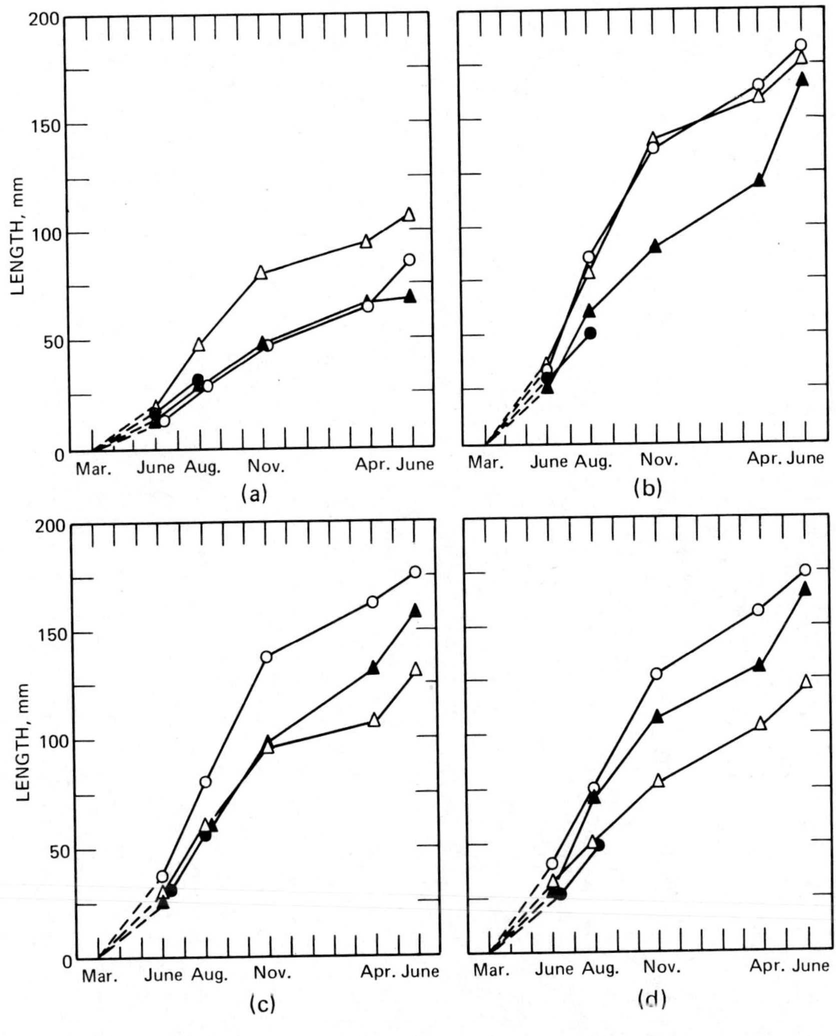

Fig. 3 Growth of *Ascophyllum nodosum* during surface discharge and after relocation of the discharge through a diffuser at control, Hodgson Island (a), and experimental, Bailey Point (b), Young Point (c), and Foxbird Island (d), sites. △, the average of 1973–1974 and 1974–1975. ▲, 1975–1976. ○, 1976–1977. ●, 1977–1978.

DISCUSSION

All measurable parameters of vigor indicate that previously stressed populations of *Ascophyllum* on Foxbird Island have recovered. With the relocation of the thermal discharge, apical tips

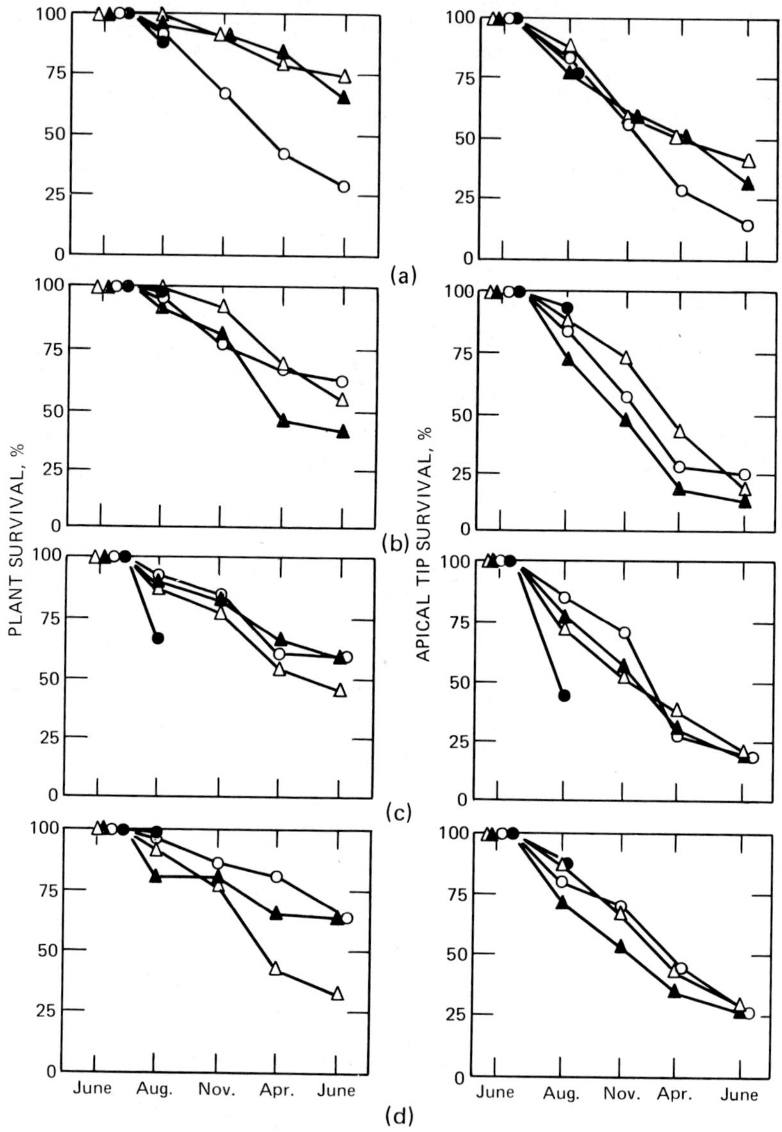

Fig. 4 Thallus (plant) and apical tip survival as percent of plants initially tagged during surface discharge and after relocation of the discharge through a diffuser at control, Hodgson Island (a), and experimental, Bailey Point (b), Young Point (c), and Foxbird Island (d), sites. △, the average of 1973–1974 and 1974–1975. ▲, 1975–1976. ○, 1976–1977. ●, 1977–1978.

regenerated from stunted and moribund thalli. Growth during poststress periods in this population was equivalent to prestress growth and to growth at control sites. With few exceptions, apical tip mortality neither increased nor decreased as a result of amelioration of the thermal stress. Plant survival increased markedly at previously stressed sites, however. Increased plant survival and increased apical growth at Foxbird Island brought about a significant increase in plant biomass by 1977.

Several aspects of our data suggest that thermal enhancement (Gibbons, 1976) occurred in *Ascophyllum*. These effects were best seen in the growth studies at experimental sites (Fig. 3 and Vadas, Keser, and Rusanowski, 1976, Fig. 3). First, as we reported previously, growth was initiated earlier in the spring at sites influenced by warmer waters. Second, higher temperatures during the summer accelerated growth processes, as predicted by Kanwisher (1966). Thermal enhancement in Montsweag Bay also occurred naturally and was best seen in 1976, when growth at all sites except Hodgson Island was significantly higher than normal. Summer temperatures in the Gulf of Maine in 1976 were the highest recorded for the last 20 years (Welch, 1977). Averaging Welch's temperature data (taken daily) over the maximum growth periods of *Ascophyllum* (April through July) yields means of 11.4, 11.9, and 11.4°C for 1955, 1957, and 1976, respectively. A wide shoulder on all Montsweag Bay and regional control temperature graphs for 1976 (Fig. 2) and the data of Vadas, Keser, and Larson (1977, Fig. 89) support this trend for our study sites even though these temperatures were measured monthly. Growth at Bailey Point illustrates both types of thermal enhancement, artificial in 1973 and 1974 and natural in 1976. The enhanced growth rates during poststress periods at Foxbird Island and Young Point are mainly the result of natural warming processes.

These data provide strong support for the potential use of *Ascophyllum* as an indicator of environmental stress in the North Atlantic (Vadas, Keser, and Rusanowski, 1976). Borowitzka (1972) suggested that benthic organisms, because of their attached habit, would be expected to integrate and reveal the cumulative effects of long-term exposure to stress. In areas of high industrial and sewage pollution, the diversity of benthic algae declined (Grenager, 1957; Borowitzka, 1972; Edwards, 1975), whereas stimulation of particular species has also been observed (Cotton, 1911; Hartog, 1959; Klavestad, 1967; Edwards, 1972).

The reversal in *Ascophyllum* growth after elimination of thermal stress and the appearance in our relatively long-term measurements of changes in growth caused by thermal alterations suggest that this species might be an ideal candidate for monitoring stress in marine ecosystems. The use of *Ascophyllum* has several advantages, including its wide distribution, the similarity in phenology throughout its range (MacFarlane, 1932; Printz, 1956), and the simplicity with which plants can be tagged and monitored. Additionally, we have shown that small-to-moderate temperature changes (natural or artificial) are detectable in the field. Perhaps short-term field measurements (monthly or weekly) or short-term laboratory growth (Stromgren, 1977), combined with respirometry (Kanwisher, 1966), might provide a more rapid assay for thermal stress. The potential of *Ascophyllum* as an indicator organism for other stresses (pollutants) remains to be tested. Rueness (1973), studying adult *Ascophyllum* transplanted to the polluted inner Oslofjord, observed no differences in growth in transplanted vs. control plants. His sample size may have been too small to detect differences, however (Rueness, 1973, p. 452).

Studying the sequence of stresses placed on the populations of *Ascophyllum* and *Fucus* in Montsweag Bay provides insight into the response of individual organisms to stress but little into the impact of stress on community organization. Although the dominant species, *Ascophyllum*, was disturbed significantly during surface discharge, the population showed considerable resilience. This was aided by power-plant shutdowns during the summers of stressed years when plants were most vulnerable. Irrespective of the initial thermal impact, enough individuals survived to prevent *Ascophyllum* from being overgrown or outcompeted. Moreover, the most likely competitor, *Fucus*, was also severely stressed. A few species colonized the recently exposed surfaces (Vadas, Keser, and Rusanowski, 1976), but none were able to displace the dominant algae. In a sense the basal system of *Ascophyllum* is analogous to the underground perennial systems of higher plants that undergo regular physical or biological disturbance. Regeneration also occurs normally in wave-exposed populations of *Ascophyllum* (Baardseth, 1970). This resilience may, in fact, contribute to its continued and long-recognized dominance in sheltered and moderately exposed coasts of the North Atlantic.

The effects of thermal stress on *Ascophyllum* and the possible ecological consequences of this stress are given in Fig. 5. In one sense the graphic model is a simplification of Shelford's Law of tolerance, where organisms respond differentially over a gradient of stress. In this case, however, the stress declined exponentially with distance from the source. Theoretically, the resultant zones of stress should be concentric, but, because of the restriction of macroalgae to rocky substrates, the actual zones are linear along the nearest shore.

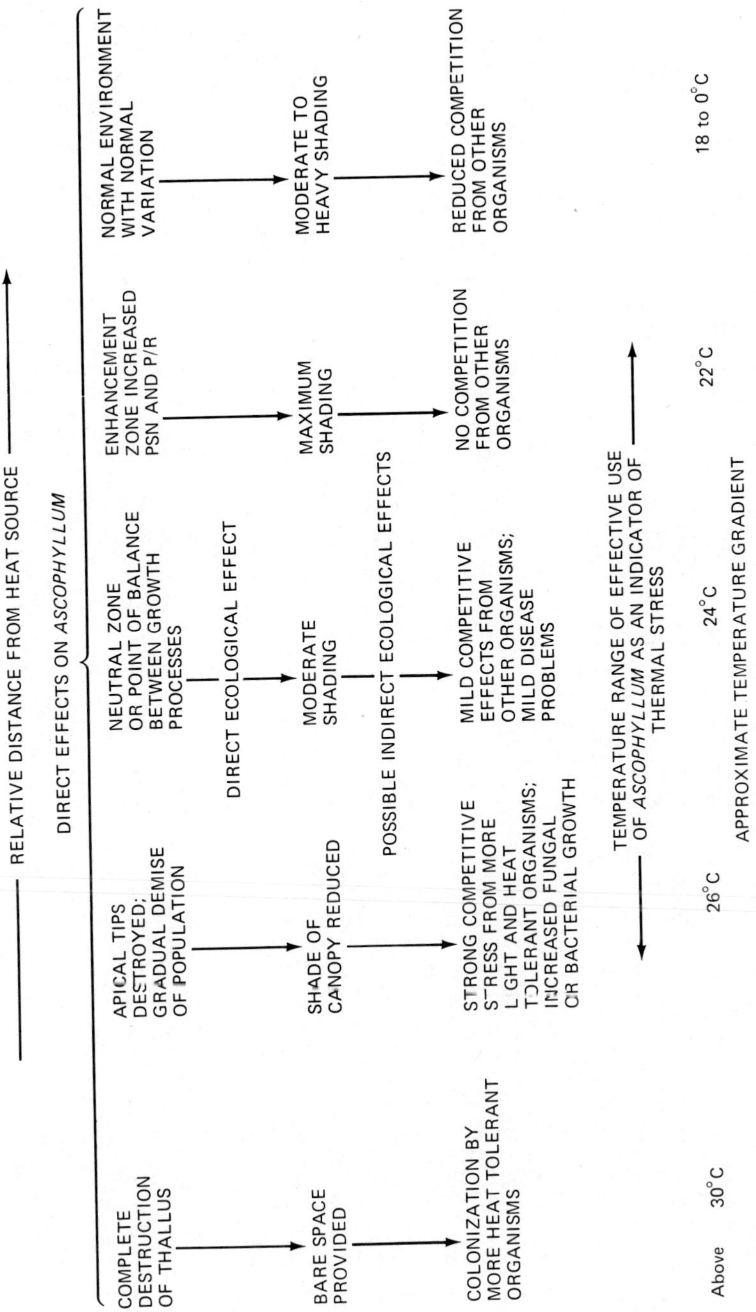

Fig. 5 Graphic model of potential impact of elevated temperatures on populations of *Ascophyllum nodosum*.

All but one of the direct effects on *Ascophyllum* were observed in Montsweag Bay. The proposed neutral zone (the region between gradual demise and enhancement) was not identifiable in our growth data, if it exists at all. No attempt was made to identify all the direct ecological effects in the field. Given the large variation in field data, it is unlikely that moderate shading in the neutral zone could be distinguished from normal variation, and it is questionable whether increased shading could be detected in the enhancement zone. Although we made no attempt to investigate indirect ecological effects fully, the absence of potential competitors and colonization by one warm-water alga were noted. We saw no evidence of competitive displacement of the dominant organism, even though it was severely stressed. Finally, it is questionable, given the generally reduced number of species occurring in these and similar estuaries (Mathieson and Fralick, 1973), whether any species besides *Fucus* and a few small annuals could colonize the intertidal ledges of these estuaries.

ACKNOWLEDGMENTS

Research reported here was supported by a grant from the Maine Yankee Atomic Power Company. We acknowledge the assistance of Steve Evans and Carroll Morse and the environmental studies group at MYAPCO and the following students: Dennis Healy, Robin Burrill, Heather Saunders, and William McCarthy. (Contribution 111, Ira C. Darling Center, University of Maine, Walpole.)

REFERENCES

Baardseth, E., 1970, Synopsis of Biological Data on Knobbed Wrack *Ascophyllum nodosum* (Linnaeus) Le Jolis, Synopsis No. 38, Rev. 1, U. N. Food and Agriculture Organization, Fisheries, New York.

Borowitzka, M. A., 1972, Intertidal Algal Species Diversity and the Effect of Pollution, *Aust. J. Mar. Freshwater Res.*, 23: 73-84.

Connell, J., 1961, The Influence of Interspecific Competition and Other Factors on the Distribution of the Barnacle *Chthamalus stellatus*, *Ecology*, 42: 710-723.

Cotton, A. D., 1911, On the Growth of *Ulva latissima* L. in Water Polluted by Sewage, *Kew Bull.*, 15-19.

Dayton, P. K., 1971, Competition, Disturbance and Community Organization: The Provision and Subsequent Utilization of Space in a Rocky Intertidal Community, *Ecol. Monogr.*, 41: 351-389.

Dean, D., and C. B. Officer, 1977, *Development Document in Support of Alternative Effluent Limitations Pursuant to Section 316(a) of the Federal Water Pollution Control Act for Maine Yankee Nuclear Generating Station*, Maine Yankee Atomic Power Company, Wiscasset.

Edwards, P., 1972, Benthic Algae in Polluted Estuaries, *Mar. Pollut. Bull.*, 3: 55-60.

——, 1975, An Assessment of Possible Pollution Effects Over a Century on the Benthic Marine Algae of Co. Durham, England, *Bot. J. Linnean Soc.*, 70: 269-305.

Gibbons, J. W., 1976, Thermal Alteration and the Enhancement of Species Populations, in *Thermal Ecology II*, ERDA Symposium Series, Augusta, Ga., Apr. 2—5, 1975, G. W. Esch and R. W. McFarlane (Eds.), pp. 27-31, CONF-750425, NTIS.

Grenager, B., 1957, Algological Observations from the Polluted Area of the Oslofjord, *Nytt. Mag. Bot.*, 5: 41-60.

Hartog, C. den, 1959, The Epilithic Algal Communities Occurring Along the Coast of the Netherlands, *Wentia*, 1: 1-241.

Kanwisher, J. W., 1966, Photosynthesis and Respiration in Some Seawoods, in *Some Contemporary Studies in Marine Science*, H. Barnes (Ed.), pp. 407-420, George Allen & Unwin Ltd., London.

Klavestad, N., 1967, Undersokelser over Benthos-algevegtasjoneh i indre Oslofjord i 1962—1965, in *Oslofjorden og dens forurensningsproblemer.* 1. Undersokelsen 1962—1965. *Norweg. Inst. Water Res.*, 9: 1-119. (Cited in Rueness, 1973.)

MacFarlane, C., 1932, Observations on the Annual Growth of *Ascophyllum nodosum, Proc. Trans. N. S. Inst. Sci.*, 18: 27-33.

Maine Yankee Atomic Power Company, 1975, *Hydrography*, Semiannual Report No. 6, pp. 1.7-3 to 1.7-11, Wiscasset.

Mathieson, A. C., and R. A. Fralick, 1973, Benthic Algae and Vascular Plants of the Lower Merrimack River and Adjacent Shoreline, *Rhodora*, 75: 52-64.

Paine, R. T., 1966, Food Web Complexity and Species Diversity, *Am. Nat.*, 100: 65-75.

Printz, H., 1956, Phenology of *Ascophyllum nodosum, Proc. Int. Seaweed Symp.*, 2: 198-202.

——, 1959, Investigations of the Failure of Recuperation and Repopulation in Cropped *Ascophyllum* Areas, AVH. *Norske Vidensk. Akad., Mat. Nat. Kl.*, 3: 1-15.

Rueness, J., 1973, Pollution Effects on Littoral Algal Communities in the Inner Oslofjord, With Special Reference to *Ascophyllum nodosum, Helgol. Wiss. Meeresunters.*, 24: 446-54.

Stromgren, Tor, 1977, Short-Term Effects of Temperature upon the Growth of Intertidal Fucales, *J. Exp. Mar. Biol. Ecol.*, 29: 181-195.

Vadas, R. L., 1972, Marine Algae, in *Fourth Annual Report on Environmental Studies*, Vol. 1, pp. 195-261, Maine Yankee Atomic Power Company, Wiscasset.

——, M. Keser, and B. Larson, 1977, Marine Algae, in *Semiannual Environmental Surveillance Report*, No. 9, pp. 290-318, Maine Yankee Atomic Power Company, Wiscasset.

——, M. Keser, and P. C. Rusanowski, 1976, Influence of Thermal Loading on the Ecology of Intertidal Algae, in *Thermal Ecology II*, ERDA Symposium Series, Augusta, Ga., Apr. 2—5, 1975, G. W. Esch and R. W. McFarlane (Eds.), pp. 202-212, CONF-750425, NTIS.

Virville, D. de, 1953, Dépeulement de la Flora Marine sur les Côtes Occidentales du Cotentin, *Proc. Int. Seaweed Symp.*, 1: 26-28.

Welch, W. R., 1977, Monthly and Annual Means of Surface Seawater Temperature at Boothbay Harbor, Maine, 1905—1976, unpublished report, Maine Department of Sea and Shore Fisheries, Augusta.

GENETIC AND PHYSIOLOGICAL FLEXIBILITY OF A CALANOID COPEPOD IN THERMAL STRESS

BRIAN P. BRADLEY
Department of Biological Sciences, University of Maryland, Catonsville, Maryland

ABSTRACT

The copepod *Eurytemora affinis* has considerable potential for both physiological and genetic adaptation to temperature stress. Physiological variance was expressed more in females, and genetic variance was expressed more in males. Estimates of both variances were obtained by several methods, all giving essentially the same results. The assay used to measure temperature tolerance depended on time to enter a coma after a temperature shock. Information was obtained on individuals that could later be used for breeding or retested. The assay was short term (30 min maximum) but was closely related to long-term survival in high temperatures. On the basis of field observations and laboratory experiments in cycling temperatures, the copepod populations seem to be at or near their limit of survival in summer temperatures in the Chesapeake Bay. This raises the question of whether all the genetic and physiological flexibility observed is usable in further adaptation or whether reproductive capacity, for example, becomes limiting.

The copepod *Eurytemora affinis* is present throughout the year in the Chesapeake Bay and tributaries but is most abundant in winter and early spring. Since no resting stages have been found, the free-swimming adults and preadults are adapted to the entire range of environmental variables, such as temperature and salinity.

Temperatures in the bay range between 0 and 30°C. Above 26°C, the species seems to be under considerable stress, judging by the very low densities found. On the other hand, there is considerable flexibility in temperature tolerance at both individual (physiological differences within individuals) and population (genetic differences between individuals) levels which ought to allow further adaptation when necessary.

This paper presents and discusses data on genetic and physiological flexibility in temperature tolerance and on the range of adaptation of individuals, in the context of widening the range of temperatures by adding waste heat from electric-power generating plants. Practical measurements of (1) the temperature tolerance of individuals, (2) the genetic variance in temperature tolerance, and (3) the physiological variance in temperature tolerance, at least in organisms that can be cultured in the laboratory, are also described.

METHODS

Eurytemora affinis has a short life cycle (12 days at 20°C), always reproduces sexually, has clearly distinguishable sexes, and can be cultured in the laboratory in filtered bay water at 0 to 7‰ salinity with a mixture of several species of algae for food. Stock cultures were routinely kept in 3-liter flasks; matings were set up in shell vials; and individual broods were reared in 125-ml flasks. The animals used were collected from Bear Creek near Baltimore Harbor with a No. 10 plankton net. Each experiment included only animals from a single collection, so that seasonal differences did not contribute to measured genetic variability.

Temperature tolerances of individual copepods were measured as the time to inactivity or to succumb (TS) to a temperature shock. Individual animals were placed in 2 ml of water in vials in racks of 12 and partially immersed into an aquarium initially at 32°C. In the experiments measuring genetic variance, the temperature was raised 0.5°C each 5 min and held constant between adjustments of the thermostatically controlled heating—stirring unit. Vials containing an animal that had fallen to the bottom were removed, and the animals were allowed to recover. The tests took a maximum of 0.5 hr, and different broods were equally represented, as much as possible, in each test. The few animals not inactivated after 30 min were assigned the maximum score (30). In the experiments measuring physiological variance by acclimation, the temperature was held constant (34.5°C) throughout the 30-min test; the animals remained in the aquarium; and the time to recover (TR) was also noted. Tolerance in this case was measured as an index (30 + TS — TR) having a maximum value of 60 for animals that did not succumb at all. Although they obviously have different variances, the two measures of temperature tolerance are closely related since TR actually adds little information because of its high negative correlation with TS. Both the index and

TS were found to be closely related to survival time at high temperatures (Bradley, 1976). This means that the assays are not only convenient (short term) and practical (unambiguous) but also realistic in measuring the resistance of the animals to thermal stress.

Population and individual flexibilities were defined in terms of components of phenotypic variance. Observed phenotypic variance in temperature tolerance can be partitioned into genetic and nongenetic components:

$$\sigma_P^2 = \sigma_A^2 + \sigma_{NA}^2 + \sigma_{E_g}^2 + \sigma_{E_s}^2$$

where σ_A^2 = additive genetic variance (caused by genes themselves), a measure of population flexibility

σ_{NA}^2 = nonadditive genetic variance (caused by difference between genes at the same or different loci)

$\sigma_{E_g}^2$ = general environmental variance (nongenetic variance between individuals)

$\sigma_{E_s}^2$ = general environmental variance (within individuals), a measure of individual flexibility

The additive genetic variance, σ_A^2, indicates the degree to which a population can adapt or change genetically in temperature tolerance. The rate of change is directly proportional to the heritability of the characteristic, σ_A^2/σ_P^2. This ratio varies from 0, when none of the variations between animals is caused by additive effects of the genes themselves, to 1, when all the variation is additive and genetic. The additive portion rather than the total genetic variance is the critical measure because combinations of genes at a locus are not passed on, nor are most combinations of genes at different loci.

Therefore, the heritability measure indicates the potential of the population to change genetically the characteristic concerned. In the present case the heritability of temperature tolerance indicates the potential of a local population of copepods to increase in average temperature tolerance, assuming other factors (e.g., fertility) do not change. I should emphasize that heritability itself will change as selection for increased tolerance proceeds since additive genetic variability generally decreases because of the restricted range of genotypes present after selection.

Tolerance to cold temperature was measured in a fashion similar to that described, with some differences. Obviously the shock temperature was different (1.5°C), and animals were removed after 10 min (at which time most had succumbed) and allowed to recover at room temperature. Again TS and TR were observed and included in the index (30 + TS — TR), but the maximum tolerance for animals not succumbing was arbitrarily set at 50 rather than 60 since exposure to the cold temperature was for only 10 min.

Special environmental variance ($\sigma^2_{E_s}$) was considered to be a measure of individual flexibility since it is a measure of the range of tolerance within individuals, especially when individuals are exposed to a change in external temperature and the change in their average tolerance (acclimation) can be measured.

The measurements of additive genetic variance are all based on known genetic relationships, the principle being that any corresponding phenotypic correlation (excluding a common environment) is caused by the common genes. Hence similarities and differences among individuals are in part genetic, and the characteristic concerned varies genetically. In this study genetic variance was measured in three ways (Appendix A) by use of the relationships between parents and offspring, between half-siblings, and between full-siblings. As can be seen, the estimate of heritability obtained from the ratio of 2 COV_{FS} to σ^2_P is inflated if any nonadditive (dominant and epistatic) variances or common environmental effects are present. The standard errors for heritability estimates were calculated according to Falconer (1963).

Physiological flexibility can be measured as random variation in temperature tolerance within individuals, directly from test—retest correlations and indirectly by subtracting genetic variance from total phenotypic variance (Appendix B). In the context of adaptation to thermal stress, a more appropriate measure is the variance in tolerance between sets of animals exposed to different temperatures before testing. The details of estimating this variance from the analysis of variance are also shown in Appendix B. The variance between exposure temperatures can also be expressed as a fraction of the total variance. Because the variance between exposure temperatures and the total variance are both estimates, in one case the former is actually larger than the latter, as can be seen from the data in Table 3. In addition to its expression as a variance, physiological flexibility can also be shown as the difference in average tolerances (Figs. 1 and 2) or the percentage change in tolerance when two random groups of copepods are exposed to different temperatures before testing for temperature tolerance.

The range of temperatures over which reproduction occurred was tested by exposing hatched eggs to 4°C and raising temperature gradually to 27 and 29°C, when reproduction was observed at the adult stage. Ten flasks, each containing 20 ovigerous females, were held at 4°C until the egg sacs were dropped and then at 10°C for 2 days and 20°C for 2 days. They were then placed in a water bath in

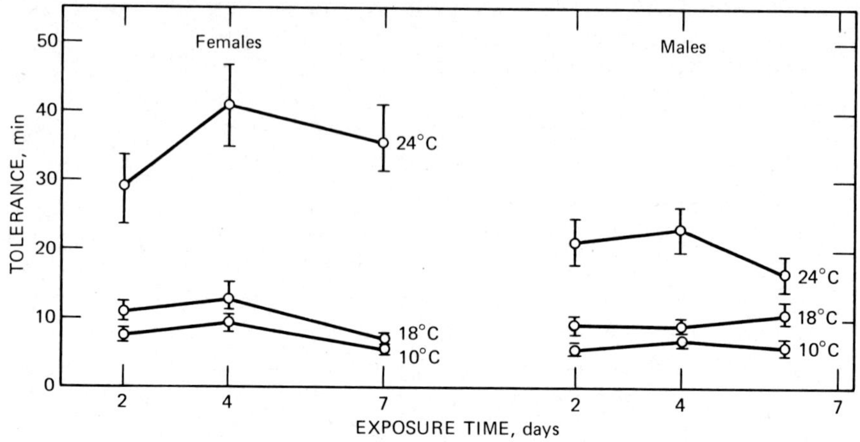

Fig. 1 Means ± standard errors (I) of tolerances to high temperatures of *E. affinis* exposed to 10, 18, and 24°C.

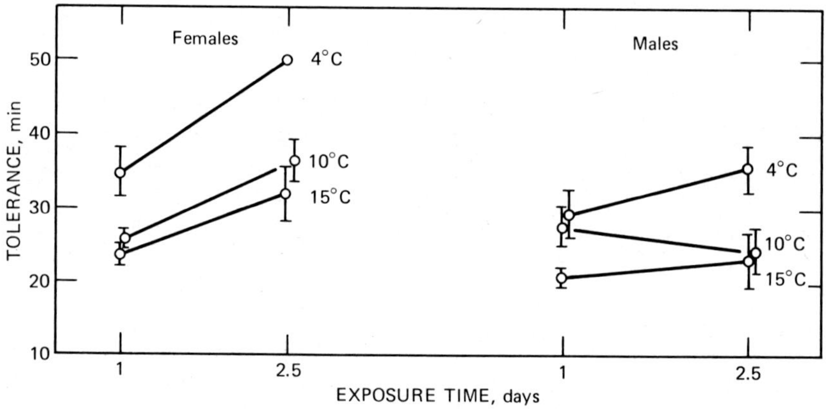

Fig. 2 Means ± standard errors (I) of tolerances to low temperatures of *E. affinis* exposed to 4, 10, and 15°C. Females exposed at 4°C for 2.5 days all had the maximum score (50).

which the temperature was raised 1°C every day. Five of the flasks were removed when the temperature reached 27°C and kept at that temperature. The other five remained in the water bath until the temperature reached 29°C. The number of ovigerous females among the progeny in each flask was counted over a 3-day period.

In another experiment on reproductive success, females and males collected at 0°C were exposed to 27, 28, and 29°C and

TABLE 1

ESTIMATES OF HERITABILITY* (±SE) AND GENETIC VARIANCE IN TEMPERATURE TOLERANCE OBTAINED FROM FOUR EXPERIMENTS

	Half-sibs (Experiment 1)	Full-sibs (Experiment 1)	Full-sibs (Experiment 2)	Full-sibs (Experiment 3)	Offspring—parent (Experiment 4)
Heritability					
Female	0.40 ± 0.18	0.20 ± 0.09	0.11 ± 0.10	0.28 ± 0.18	0.11 ± 0.44
Male	0.84 ± 0.35	0.79 ± 0.24	0.89 ± 0.45	0.78 ± 0.29	0.76 ± 0.26
Genetic variance					
Female	4.8	2.4	3.4	20.8	19.4
Male	9.6	9.0	29.6	36.4	42.3
Number of families	19	68	23	36	57

*Heritability estimates from full-sibs are biased upward (see Appendix A and discussion in text). Standard error of heritability was calculated according to Falconer (1963).

reproduction observed. Ten pairs of males and females in each of two flasks were placed at each temperature for 3 days, and any ovigerous females were noted.

RESULTS

Table 1 summarizes data presented more extensively elsewhere (Bradley, 1978b). Several conclusions can be drawn. First, *Eurytemora* populations vary considerably in genetic influence on temperature tolerance. Second, the genes express themselves more clearly in males than in females, perhaps because the greater physiological variances in females may tend to obscure genetic differences. Third, on the basis of the similarity of estimates of heritability from full-sib covariances, on the one hand, and half-sib covariances and offspring—parent regressions, on the other, most of the genetic variance seems to be additive (see Appendix A). We should note that the lower heritabilities in female progeny are not the result of higher phenotypic variances; the genetic variances are also lower.

There were two other sources of evidence for genetic variance in temperature tolerance: (1) Significant and persistent differences in tolerance were found between two separate populations of

TABLE 2

ESTIMATES OF PHYSIOLOGICAL VARIANCE IN MALES AND FEMALES*

Method				
1, Test—retest correlation				
Female	38.4			
Male	14.0			
Experiment number	1	2	3	4
2, Subtraction of genetic variance				
Female	9.5	22.2	52.8	27.5
Male	2.2	3.5	9.2	11.0

*Estimates were made using methods 1 and 2 in Appendix A.

Eurytemora collected at different seasons (Bradley, 1975). Since they were treated identically, the consistent difference implied a genetic basis. (2) Genetic variance in temperature tolerance was found when the progeny of animals collected from the intake canal were compared with the progeny of animals collected from the discharge canal of a steam electric-power plant on the Patuxent River in Maryland. Progeny rather than parents were used to avoid effects of exposure to water of different temperatures. The average tolerances of the male progeny collected at the intake and discharge canals were 20.3 and 32.5°C, respectively, and of the female progeny, 17.5 and 27.8°C, respectively. Both these differences were significant ($P < 0.01$); this indicates that more-tolerant parent animals had been selected by the power plant. The differences among the progeny indicated genetic differences among the parents since all the progeny were treated exactly alike.

Table 2 shows the estimates of physiological variance in temperature tolerance in males and females, measured by the first two methods in Appendix A. The pair of estimates from test—retest correlations and all estimates obtained from the four experiments on genetic variance indicate that physiological variance within the female group is much higher than that within the male. These variances were observed in the absence of systematic changes in external temperature. The physiological variances of females are comparable to the genetic variances of males (Table 1) and vice versa. Taken together, the estimates in Tables 1 and 2 indicate that *Eurytemora* has considerable potential for adaptation to thermal stress, both genetically and physiologically. Of course, physiological potential may well depend on the animal's genotype.

As suggested earlier, in the context of adaptation to thermal stress, a more appropriate measurement of physiological variability may be the variance in temperature tolerance between different temperature treatments (the third method in Appendix B).

Some of the data on which the latter estimates of physiological variance are based were reported elsewhere (Bradley, 1978a). Variance components are shown separately between 10°C and the 18 and 24°C classes combined, between the 18 and 24°C classes, and among all three classes. In every case the variance in mean tolerances of female progeny in the different classes is greater than that in the means of male progeny. The critical range of temperature seems to be 18 to 24°C; every variance between 18 and 24°C is greater than that between 10 and 18 + 24°C, except that for male progeny at 7 days. Acclimation, or physiological adjustment to an extreme temperature, seems to be virtually complete by 4 days or earlier. This conclusion can also be drawn from the quite different measurement of acclimation to high temperatures shown in Fig. 1. Thus, on the basis of both variances among exposure temperatures and mean tolerances after exposure to different temperatures, *Eurytemora*, and especially *Eurytemora* females, can adapt to thermal stress to a significant extent at the individual level.

Analogous experiments were done on tolerance to cold temperature, but the emphasis here is on high-temperature tolerance since *Eurytemora affinis*, a north-temperate species, is probably under greater stress at the warm end of the temperature range. Similar methods were used to measure physiological flexibility in cold tolerance; the results are shown in Table 3 (variances) and Fig. 2 (mean tolerances). Females seem to have greater physiological flexibility in cold tolerance as well as heat tolerance. The sizes of the components of variance between the 10 and 15°C classes indicate that little acclimation to low temperature (1.5°C) occurs in this range.

The discussion of adaptation to increased temperatures from thermal addition raises the further question of the relationship between heat tolerance and cold tolerance. In one set of experiments, 24 copepods were tested for heat tolerance and then for cold tolerance, or vice versa, with a lapse of 3 hr to allow recovery. Each animal was assayed twice, once for heat and once for cold, each day. The three correlations among heat and cold tolerances averaged 0.5, which was comparable to the cold—cold retest correlation (0.6) and the heat—heat retest correlation (0.5). However, in other experi-

TABLE 3

ACCLIMATION OF *Eurytemora affinis* TO HIGH AND LOW TEMPERATURES, EXPRESSED AS VARIANCES AMONG EXPOSURE TEMPERATURES*

A. Heat Tolerance

Variance	Exposure time, days: 2 Female	2 Male	4 Female	4 Male	7 Female	7 Male
10 and 18 + 24°C (σ_1^2)	67.5	39.7	137.9	38.2	115.4	31.2
18 and 24°C (σ_2^2)	145.5	66.5	367.1	96.9	406.6	15.8
Overall (σ_B^2)	116.6	59.1	273.2	73.2	278.2	28.2
Total variance†	248.3	102.6	472.3	84.1	477.0	34.1

B. Cold Tolerance

Variance	Exposure time, days: 1 Female	1 Male	2.5 Female	2.5 Male
4 and 10 + 15°C (σ_1^2)	42.0	7.1	124.2	69.1
10 and 15°C (σ_2^2)	0	12.0	2.8	0
Overall (σ_B^2)	25.8	10.6	82.1	41.8
Total variance†	95.8	102.6	233.6	173.6

*Method 3, Appendix B.
†Total variance was calculated as $\sigma_B^2 + \sigma_W^2$.

ments where there was a delay of 1 day in measuring for heat or cold tolerance, the correlations were always around zero. Hence, the earlier positive correlations between heat and cold tolerances may be artifacts of retesting animals within a few hours instead of allowing them a day to recover. In no case was there any detectable effect of testing for heat or cold tolerance first. The conclusion must be that heat and cold tolerances are not negatively related and may not be related at all. Hence, I assume that adaptation to heat stress will not necessarily compromise adaptation to cold stress, but I do not know whether the genetic correlation (the proportion of genes affecting both characteristics) is positive, negative, or zero.

This discussion has been focused on temperature tolerance in terms of individual resistance to temperature stress, in particular heat stress. To mean anything to the species, consideration of temperature tolerance should also include reproductive success. As a test of the effective range of temperatures over which reproduction occurs in an individual copepod, 20 ovigerous females and 10 males collected at 0°C were placed in each of ten flasks, at 4°C, and the resulting progeny were allowed to mature at temperatures increasing to 27 and 29°C (five flasks at each temperature). At 27°C, three of the five flasks had survivors, but only one flask contained ovigerous females. At 29°C, four of the five flasks had survivors, but only one ovigerous female was seen. No progeny were obtained at either 27 or 29°C. The outer limit of tolerance of animals collected at 0°C and grown in a slowly increasing temperature seems to be below 30°C, at least in the laboratory.

In the second experiment, a test of whether females and males collected at 0°C would reproduce at the other end of the temperature range, six of ten females in one flask produced egg sacs at 27°C. In the other flask, however, all animals except one male died. At 28 and 29°C, all animals were dead at 3 days except one female in each flask. These experiments indicate that the reproductive range of one phenotype is less than the total range of environmental temperatures. Information is needed on the relationship between temperature tolerance and reproductive capacity to determine if adaptation to thermal stress may actually be influencing the reproductive capacity of the species.

DISCUSSION

The method used to measure temperature tolerance in individuals was not conventional. Some of the existing assays of temperature tolerance were unsuitable because data on individuals could not be obtained. These assays included time to death of 50% of the individuals (Mattice and Dye, 1976), temperature causing 50% mortality after 24-, 48-, or 96-hr intervals (many investigators), and proportion surviving after constant time at high temperature (Carlson, 1974). I tried using the temperature at which activity ceased as a criterion, in a manner similar to the widely used critical thermal maximum (CTM) assay (Gonzalez, 1974; Wallis, 1975; Hassan and Spotila, 1976; McFarlane, Moore, and Williams, 1976; Murphy et al., 1976). Although Gonzalez used CTM successfully with copepods, I found very little variation in the temperature at which they became inactive; it was almost always around 35°C. Furthermore, I needed to be sure they survived the assay, at least in the genetics experiments in which they were to be bred.

The time to enter a coma after a temperature shock, the assay I used to measure temperature tolerance, proved to be repeatable and, equally important, to be closely related to survival at high temperatures (Bradley, 1976). In this context, it is interesting to note that Hamby (1975), working with the marine snail *Littorina littorea*, found that the temperature at which an animal entered a coma was significantly shifted by acclimation, but the lethal temperature was affected very little. He concluded that the nervous system of *Littorina* was most vulnerable to thermal extremes, as is the case with other poikilotherms (Prosser, 1973).

The great genetic variance in temperature tolerance implies seasonal selection caused by changing water temperatures. There have been some laboratory demonstrations of genetic variance associated with environmental variability (McDonald and Ayala, 1974; Beardmore, 1961; Gibson and Bradley, 1974). Battaglia (1967; 1970) reported visible differences in nature in copepods controlled by single loci and influenced by temperature variation. There are many other instances of genetic variance in other species associated with temperature variability (see Parsons, 1973, for a review). A complete discussion of the relevant theory on genetic and environmental variability is inappropriate here, but the reader is referred to Levins (1968) and to the work of Gillespie (e.g., Gillespie, 1974).

There are relatively few examples of demonstrated genetic variance in temperature tolerance as such. Banta and Wood (1939) reported that work they did in 1928 showed much higher tolerance in one clone of *Daphnia magna*, obviously caused by a genetic difference. Fatt and Dougherty (1963) found a single locus effect on temperature tolerance in the nematode *Caenorhabditis elegans*. In a sense, the whole class of temperature-sensitive mutants also demonstrates genetic variance in temperature tolerance.

One of the earliest demonstrations of genetic variance in temperature tolerance was described by Dallinger (1887) in his presidential address to the Royal Microscopical Society. He collected three species of flagellated bacteria at 15.6°C, gradually raised the temperature, and observed (under the microscope) whether or not they seemed uncomfortable. When they became adjusted, he again raised the temperature, repeating the process until the temperature reached 70°C. Wild specimens did not survive at this temperature, and bacteria living at 70°C did not survive the original 15.6°C optimum. His experiment was ended "with such pain as I presume is natural" when "the accident" happened; this was presumably a malfunction of his thermostat after 7 years of painstaking observation and a remarkable demonstration of adaptation to thermal stress.

There are far more precedents for the finding of physiological than of genetic flexibility, especially as demonstrated by acclimation.

As Heinle (1978) pointed out, however, not much work has been done on zooplankton. There are several comprehensive reviews that include discussions of acclimation, e.g., Kinne (1970), Vernberg and Vernberg (1972), and Gilles (1975). Acclimation has been demonstrated in several ways, at the molecular level (Baldwin and Hochachka, 1970; Bowler, 1963), as change in respiration rate (Fitzpatrick and Atebara, 1974), as change in respiration and heart rates (Markel, 1974), as change in other rate functions or development times (Landry, 1975; Bernard, 1970), and simply as survival (McLeese, 1956). As mentioned earlier, the temperature at which the animal entered a heat coma was used by Hamby (1975) as a criterion to show acclimation.

Not much work has been done on acclimation to cold temperatures. McLaren (1966) found that the constant describing a "biological zero" in Belehradek's equation (which is used to describe the relationship between the rate of a process and temperature) varied less than 1°C among populations of copepod species over a wide range of temperature environments. From this and other studies on acclimation to high temperatures, Heinle (1978) speculated that adaptation to extreme low temperatures may be primarily genetic and adaptation to high temperatures may be both genetic and nongenetic. Variation between populations may have little to do with variations within populations, however. The work reported here (and some not reported) suggests that variation in heat and cold tolerance within populations of *Eurytemora*, and hence adaptation to both extremes of temperature, can be both genetic and nongenetic, the relative importance depending on the sex of the animals.

The sexual dimorphism in genetic variance in high-temperature tolerance (Table 1) has some precedence in other copepod traits studied by McLaren (1976). He found that females had significantly higher heritability of age at maturity (at 15°C) than males and that males had a higher heritability of adult size (also at 15°C) than females. Heritabilities of these characteristics were not significant in either sex at 10 and 12.5°C. McLaren attributed this to natural selection occurring at these "normal" temperatures. Enfield, Hartung, and Hefeneider (1975) reported differential gene expression in pupal weight in male and female *Tribolium* and cited other examples of sexual dimorphism.

The importance of sexual dimorphism in both genetic variance and physiological flexibility (Tables 2 and 3) in the adaptation of *Eurytemora* to variable temperatures is not obvious. It suggests that males are more vulnerable in the short term. If females stored sperm and produced multiple egg sacs per mating, males might be more expendable, but this is not the case (Heinle, 1970). Perhaps the question of sex dimorphism cannot be answered until the mechanism of sex determination in copepods is known. There have been some suggestions of a multifactorial basis (e.g., Battaglia, 1967; McLaren, 1976), but, aside from evidence of the lability of the sex ratio to density and perhaps temperature and salinity (Heinle, 1970; Katona, 1970), little is known about the control of sex in copepods.

The relationship between genetic and physiological variance in the adaptation of organisms was discussed by Slobodkin (1968) and Slobodkin and Rapoport (1974). Slobodkin's view is that environmental stresses that can be dealt with physiologically cause a minimum of genetic effects in later recurrences. Hence, genetic variance in a particular characteristic would not even be maintained in species with a high physiological flexibility in that characteristic since genotypic differences would be obscured and the resistance of the individual to the stress would be less related to its genotype. This theory is supported by the results of Levins (1969) and Marshall and Jain (1968). Levins found that *Drosophila* species with broad niches had larger physiological tolerances to dissication and smaller genetic differences than narrow-niched species, but he did not measure genetic variance directly. Marshall and Jain found that one species of wild oat (*Avena barbata*) was developmentally more flexible than another (*A. fatua*) but was less variable genetically.

The data presented do not give definitive answers to some practical questions. On the basis of experiments on the range of temperatures over which an individual can reproduce and of informal observations of animals in cycling temperatures in the laboratory (and the very low population densities in the Chesapeake Bay in summer), *Eurytemora affinis* seems to be at the limits of adaptation even in normal temperatures. A large potential for adaptation to temperature exists, however, both genetically and physiologically. This begs the question, What does this genetic and physiological flexibility mean if it cannot be used to allow the population to adjust to thermal stress? I must conclude that the measures of flexibility used are not sufficient by themselves to predict the survival of *Eurytemora* in an even wider range of temperatures.

To answer the question of the apparently unusable flexibility, we need to know (1) whether the most tolerant animals are also the most flexible physiologically, (2) whether genetic variance can be exhausted, and (3) whether reproductive capacity and temperature tolerance are negatively related.

If selection for temperature tolerance, which increases the average tolerance in the population, results in lowered physiological flexibility (acclimation), then the effective tolerance of the population might not increase much since acclimation increases the range of tolerance. If genetic variance in temperature tolerance cannot, in fact, be exhausted, there may well be a negative correlation between reproduction and tolerance so that the most tolerant genotypes reproduce less than average. I should point out that the two most obvious models explaining the large residual genetic variance are both unlikely. Heterozygous superiority, which results in large variance caused by genic segregation, is unlikely in view of the apparently small average deviation of heterozygotes from the means of homozygotes (dominance variance) (Table 1). The other model, seasonal selection, also seems unlikely because, since heat and cold tolerance do not seem to be opposite characteristics, selection would not be reversed seasonally and genetic variance would not be maintained.

Finally, if genetic variance for heat tolerance can be exhausted, the question of reverse adaptation arises. Is adaptation to more "normal" conditions compromised by adaptation to extreme stress conditions? Adaptation of species or communities to novel stresses might be tolerable if there was a guarantee that further stresses were not on the way.

ACKNOWLEDGMENTS

This work could not have been done without the assistance of Richard Imbach, Kenneth Keeling, Denise Markoff, Richard Muths, and Gregory Taylor and the financial support of the Office of Water Research and Technology, U. S. Department of the Interior (Grant B-021-MD) and the National Science Foundation (Grant BMS-75-20282).

Finally, I thank Carol Ferrigno for her patience in typing the manuscript and Tim Ford for preparing the figures.

REFERENCES

Baldwin, J., and P. W. Hochachka, 1970, Functional Significance of Isoenzymes in Thermal Acclimation: Acetylcholinesterase from Trout Brain, *Biochem. J.*, 116: 883-887.

Banta, A. M., and T. R. Wood, 1939, Genetical Studies in Sexual Reproduction, in *Studies on the Physiology, Genetics, and Evolution of Some Cladocera*, A. M. Banta (Ed.), Paper No. 39, Department of Genetics, Carnegie Institute of Washington, Washington, D. C.

Battaglia, B., 1967, Genetic Aspects of Benthic Ecology in Brackish Waters, in *Estuaries*, H. W. Lauff (Ed.), pp. 574-577, Publication No. 83, American Association for the Advancement of Science, Washington, D. C.

——, 1970, Cultivation of Marine Copepods for Genetic and Evolutionary Research, *Helgol. Wiss. Meeresunters.*, 20: 385-392.

Beardmore, J. A., 1961, Diurnal Temperature Fluctuation and Genetic Variance in *Drosophila* Populations, *Nature*, 189: 162-163.

Bernard, M., 1970, Quelques aspects de la biologie du copepode pelagique *Temora stylifera* en Mediterranee, *Pelagos*, 11: 1-96.

Bowler, K., 1963, A Study of Factors Involved in Acclimatization to Temperature and Death at High Temperatures in *Astacus pallipes*. 2. Experiments at the Tissue Level, *J. Cell. Comp. Physiol.*, 12: 133-146.

Bradley, B. P., 1975, The Anomalous Influence of Salinity on Temperature Tolerances of Summer and Winter Populations of the Copepod *Eurytemora affinis*, *Biol. Bull.*, 148: 26-34.

——, 1976, The Measurement of Temperature Tolerance: Verification of an Index, *Limnol. Oceanogr.*, 21: 596-599.

——, 1978a, Increase in Range of Temperature Tolerance by Acclimation in the Copepod *Eurytemora affinis*, *Biol. Bull.*, 154: 177-187.

——, 1978b, Genetic and Physiological Adaptation of the Copepod *Eurytemora affinis* to Seasonal Temperatures, *Genetics*, 90: in press.

Carlson, D. M., 1974, Responses of Planktonic Cladocerans to Heated Waters, in *Thermal Ecology*, AEC Symposium Series, Augusta, Ga., May 3—5, 1973, J. W. Gibbons and R. R. Sharitz (Eds.), pp. 186-206, CONF-730505, NTIS.

Dallinger, W. H., 1887, Transactions of the Society. V. The President's Address, *J. R. Microsc. Soc.*, April 1887, pp. 185-199.

Enfield, F. D., N. Hartung, and S. H. Hefeneider, 1975, Sex Differences in Gene Expression for Pupa Weight in Long Term Selected Lines of *Tribolium*, *Can. J. Genet. Cytol.*, 17: 9-13.

Falconer, D. S., 1960, *Introduction to Quantitative Genetics*, Oliver and Boyd, Ltd., Edinburgh.

——, 1963, Quantitative Inheritance, in *Methodology in Mammalian Genetics*, W. J. Burdette (Ed.), pp. 193-216, Holden—Day, Inc., San Francisco.

Fatt, H. V., and E. C. Dougherty, 1963, Genetic Control of Differential Heat Tolerance in Two Strains of the Nematode *Caenorhabditis elegans*, *Science*, 141: 266-272.

Fitzpatrick, L. C., and M. Y. Atebara, 1974, Effects of Acclimation to Seasonal Temperatures on Energy Metabolism in the Toad *Bufo woodhousei*, *Physiol. Zool.*, 47: 119-129.

Gibson, J. B., and B. P. Bradley, 1974, Stabilising Selection in Constant and Fluctuating Environments, *Heredity (London)*, 33: 293-301.

Gilles, R., 1975, Mechanisms of Thermoregulation, in *Marine Ecology*, O. Kinne (Ed.), Vol. II, Part 1, p. 251, John Wiley & Sons, Inc., New York.

Gillespie, J. H., 1974, The Role of Environmental Grain in the Maintenance of Genetic Variation, *Am. Nat.*, 108: 831-836.

Gonzalez, J. G., 1974, Critical Thermal Maxima and Upper Lethal Temperatures for the Calanoid Copepods *Acartia tonsa* and *A. clausi*, *Mar. Biol.*, 27: 219-223.

Hamby, R. J., 1975, Heat Effects on a Marine Snail, *Biol. Bull.*, 149: 331-347.

Hassan, K. C., and J. R. Spotila, 1976, The Effect of Acclimation on the Temperature Tolerance of Young Muskellunge Fry, in *Thermal Ecology II*, ERDA Symposium Series, Augusta, Ga., Apr. 2—5, 1975, G. W. Esch and R. W. McFarlane (Eds.), pp. 136-140, CONF-750425, NTIS.

Heinle, D. R., 1970, Population Dynamics of Exploited Cultures of Calanoid Copepods, *Helgol. Wiss. Meeresunters.*, 20: 360-392.

——, 1978, Zooplankton, in *Functional Adaptations of Marine Organisms*, F. John Vernberg and Winona B. Vernberg (Eds.), Chap. 4, Academic Press, Inc., New York, in press.

Katona, S. K., 1970, Growth Characteristics of the Copepods *Eurytemora affinis* and *E. herdmani* in Laboratory Cultures, *Helgol. Wiss. Meeresunters.*, 20: 373-384.

Kinne, O., 1970, Temperature, Animals, Invertebrates, in *Marine Ecology*, O. Kinne (Ed.), Vol. I, Part I, p. 407, John Wiley & Sons, Inc., New York.

Landry, M. R., 1975, Seasonal Temperature Effects and Predicting Development Rates of Marine Copepod Eggs, *Limnol. Oceanogr.*, 20: 434-440.

Levins, R., 1968, *Evolution in Changing Environments*, Princeton University Press, Princeton, N. J.

——, 1969, Thermal Acclimation and Heat Resistance in *Drosophila* Species, *Am. Nat.*, 103: 483-499.

McDonald, J. F., and F. J. Ayala, 1974, Genetic Response to Environmental Heterogeneity, *Nature*, 250: 572-574.

McFarlane, R. W., B. C. Moore, and S. E. Williams, 1976, Thermal Tolerance of Stream Cyprinid Minnows, in *Thermal Ecology II*, ERDA Symposium Series, Augusta, Ga., Apr. 2–5, 1975, G. W. Esch and R. W. McFarlane (Eds.), pp. 141-144, CONF-750425, NTIS.

McLaren, I. A., 1966, Predicting Development Rate of Copepod Eggs, *Biol. Bull.*, 131: 457-469.

——, 1976, Inheritance of Demographic Characters in Copepods, *Biol. Bull.*, 151: 200-213.

McLeese, D. W., 1956, Effects of Salinity, Temperature, and Oxygen on the Survival of the American Lobster, *J. Fish. Res. Board Can.*, 13: 247-272.

Markel, R. P., 1974, Aspects of the Physiology of Temperature Acclimation in the Limpet *Acmaea limatula* Carpenter (1964): An Integrated Field and Laboratory Study, *Physiol. Zool.*, 47: 99-109.

Marshall, D. R., and S. K. Jain, 1968, Phenotypic Plasticity of *Avena fatua* and *A. barbata*, *Am. Nat.*, 102: 457-467.

Mattice, J. S., and L. L. Dye, 1976, Thermal Tolerance of the Adult Asiatic Clam, in *Thermal Ecology II*, ERDA Symposium Series, Augusta, Ga., Apr. 2–5, 1975, G. W. Esch and R. W. McFarlane (Eds.), pp. 130-135, CONF-750425, NTIS.

Murphy, J. C., C. T. Garten, Jr., M. H. Smith, and E. A. Standora, 1976, Thermal Tolerance and Respiratory Movement of Bluegill from Two Populations Tested at Different Levels of Acclimation Temperature and Water Hardness, in *Thermal Ecology II*, ERDA Symposium Series, Augusta, Ga., Apr. 2–5, 1975, G. W. Esch and R. W. McFarlane (Eds.), pp. 145-147, CONF-750425, NTIS.

Parsons, P. A., 1973, Genetics of Resistance to Environmental Stresses in *Drosophila* Populations, *Annu. Rev. Ecol. Systemat.*, 4: 239-265.

Prosser, C. L., 1973, *Comparative Animal Physiology*, 3rd ed., W. B. Saunders Co., Philadelphia.

Roberts, R. C., 1967, Quantitative Genetic Analysis, in *Behavior—Genetic Analysis*, J. Hirsch (Ed.), pp. 212-257, McGraw-Hill Book Company, Inc., New York.

Slobodkin, L. B., 1968, Toward a Predictive Theory of Evolution, in *Population Biology and Evolution*, R. C. Lewontin (Ed.), pp. 187-205, Syracuse University Press, Syracuse, N. Y.

——, and A. Rapoport, 1974, An Optimal Strategy of Evolution, *Q. Rev. Biol.*, 49: 181-199.
Vernberg, W. B., and F. J. Vernberg, 1972, *Environmental Physiology of Marine Animals*, Springer-Verlag, New York.
Wallis, R. L., 1975, Thermal Tolerance of *Mytilus edulis* of Eastern Australia, *Mar. Biol.*, 30: 183-191.

APPENDIX A: GENETIC FLEXIBILITY OF A POPULATION

Genetic flexibility is estimated as σ^2_A (additive genetic variance) or as a ratio of total phenotypic variance (heritability, h^2):

$$h^2 = \frac{\sigma^2_A}{\sigma^2_P}$$

Estimates of h^2 or σ^2_A are obtained from (1) offspring—midparent regression and from covariances (2) between half-sibs and (3) between full-sibs:

$$(1)\ b_{O\bar{P}} = h^2$$

$$(2)\ COV_{HS} = 0.25\sigma^2_A$$

$$(3)\ COV_{FS} = 0.5\sigma^2_A$$ + nonadditive genetic and environmental components common to broods

COV_{HS} and COV_{FS} are obtained from analysis of variance of half-sib and full-sib families:

	df	Expected components of mean squares
Between male parents	m − 1	$\sigma^2_w + w\sigma^2_f + fw\sigma^2_m$
Between female and male parents	m(f − 1)	$\sigma^2_w + w\sigma^2_f$
Within broods	mf(w − 1)	σ^2_w

The symbols σ^2_w, σ^2_f, and σ^2_m are components of variance within broods, between female parents, and between male parents, respectively, and m, f, and w are the number of male parents, the number of females bred to each male, and the number of offspring per brood, respectively. Clearly, the design will not always be perfectly balanced. The appropriate coefficients must then be estimated (Falconer, 1960; 1963; Roberts, 1967). σ^2_m is an estimate of COV_{HS}, and $\sigma^2_m + \sigma^2_f$ is an estimate of COV_{FS}.

APPENDIX B: MEASURING INDIVIDUAL (PHYSIOLOGICAL) FLEXIBILITY

Physiological flexibility, $\sigma^2_{E_s}$ can be estimated (1) by test—retest correlation, (2) indirectly by subtracting genetic variance, and (3) by acclimation tests:

$$(1)\ r_{tt} = \frac{\sigma^2_p - \sigma^2_{E_s}}{\sigma^2_p} = 1 - \frac{\sigma^2_{E_s}}{\sigma^2_p} \quad \text{(from Falconer, 1960)}$$

$$(2)\ \sigma^2_{E_s} = \sigma^2_p - (\sigma^2_A + \sigma^2_{NA}) \quad \text{if general environmental variance is not large}$$

$$(3)\ \sigma^2_{E_s} = \sigma^2_B$$

where σ^2_B is the variance in tolerance between prior exposure temperatures calculated from the analysis of variance of tolerances following exposure to various temperatures:

	df	SS	MS	Exp. MS
Between temperatures	2	SS_B	$SS_B/2$	$\sigma^2_w + k_1\sigma^2_B$
10 vs. 18 + 24	1	SS_1	SS_1	$\sigma^2_w + k_2\sigma^2_1$
18 vs. 24	1	SS_2	SS_2	$\sigma^2_w + k_1\sigma^2_2$
Within temperatures	$n - 3$	SS_W	$SS_{W/n-3}$	σ^2_w
Total	$n - 1$	SS_T		

The symbols σ^2_B, σ^2_1, and σ^2_2 are obtained from the mean squares and give estimates of the variances between all three classes (overall), between 10°C and 18 + 24°C and between 18 and 24°C. Note that k_1 is the number of animals per class, and k_2 is the harmonic mean of k_1 and $2k_1$, thus $k_2 = 1.3k_1$.

The third measure is most appropriate in the context of individual adaptation to temperature stress.

EFFECTS OF THERMAL EFFLUENTS ON REPRODUCTION IN A SEA ANEMONE

BRIAN L. JENNISON*
Department of Zoology, University of California, Berkeley, California

ABSTRACT

Reproduction of the sea anemone *Anthopleura elegantissima* was studied in the thermal-outfall canal of a large (1350-MW) Pacific Gas and Electric Company power plant and at an adjacent control site at Morro Bay, San Luis Obispo County, California. Female reproductive cycles, determined by oocyte diameters, were similar at both sites for 3 years; oocytes appeared in the fall, and spawning occurred in the late summer. Outfall females spawned as much as 1 month before control females, however. Spermatogenesis of outfall males was delayed and compressed compared with control populations, but outfall males developed quickly and spawned before control males. Anemones transplanted from the control to the outfall in November 1975 spawned by the end of March 1976; undisturbed animals from both sites did not spawn until the end of the summer. It is suggested that this species of sea anemone spawns during the summer period of highest annual temperature and that anemones in the outfall have become acclimated to the temperatures there, which are often 10°C above ambient.

The sea anemone *Anthopleura elegantissima* is a conspicuous and abundant intertidal animal along the central California coast (Hand, 1955). It occurs in large numbers in the thermal-outfall canal of the Pacific Gas and Electric Company's large (1350-MW) power plant at Morro Bay, San Luis Obispo County, California.

Ford (1964) showed that *A. elegantissima* exhibits an annual reproductive cycle in central California, with gametogenesis beginning in the winter, oocytes increasing in size through the spring and

*Present Address: Harbor Branch Foundation, Inc., Fort Pierce, Florida.

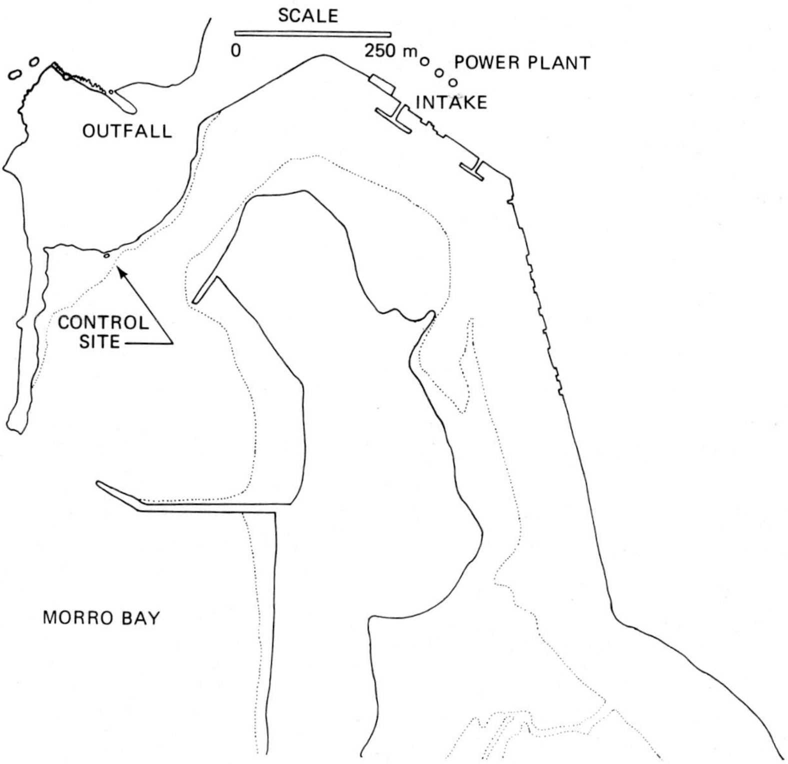

Fig. 1 Collection sites near a 1350-MW Pacific Gas and Electric Company power plant at Morro Bay, San Luis Obispo County, California.

summer, and spawning occurring in the early fall. He speculated that reproduction might somehow be cued by the increasing seawater temperature in summer.

My purpose in this study was to determine whether the timing of reproductive events in this species was altered by subjection to the elevated temperature of a power-plant thermal-outfall canal.

MATERIALS AND METHODS

Twelve to 24 anemones were collected at both control and outfall sites (Fig. 1) at 4- to 8-week intervals between Feb. 4, 1974, and Sept. 27, 1976. Animals were relaxed in 50 : 50 seawater and isotonic $MgCl_2$, cleaned, fixed in Bouin's fluid, dehydrated, embedded in paraffin, and sectioned at thicknesses of 8 to 10 μm on a rotary microtome. Sections were made transversely through the column at the level of the gonads, which run longitudinally from the basal disk to the bottom of the actinopharynx (Stephenson, 1928). The sections were mounted on glass slides, stained with picro-indigo carmine and basic fuchsin, and examined microscopically.

Oocytes from each identifiable female were measured with a calibrated ocular micrometer, after the method of Dunn (1975), to assess female maturation over time at both sites. At least 50 gametogenic cells were measured from one section of each female. As a correction for differences in the diameters of oocytes due to the plane of the section, only cells having a distinct nucleolus were included in the total. If fewer than 50 cells were present in the section, all gametogenic cells were measured regardless of position. Males were classified into arbitrary stages based on an assessment of their gonadal maturity (Table 1).

Reciprocal transplants of anemones were made between the outfall and control sites on Nov. 17, 1975. Eight cages with 1.25-cm-mesh stainless-steel screen fronts, sides, and tops; masonite floors; and neoprene mesh backs held together by nylon fishing line were built. Sixty anemones collected from one clone in the control area were distributed evenly among four cages. Two cages were hung at zero tidal height at the power-plant intake structure (caged experimental control group), and two were hung in the outfall canal (transplants). Similar procedures were followed for four cages of outfall animals. The cages were collected on Mar. 25, 1976, and the anemones were analyzed histologically.

RESULTS

Anthopleura elegantissima at the Morro Bay control site reproduced sexually on an annual cycle (Fig. 2). Although the January 1974 control collections were anomalous and there were no females in the samples again until July, it is clear that oocyte diameter reached a peak in summer, followed by spawning between August and September. No oocytes were visible again in prepared sections until the end of December 1974, but they increased in size through the spring and summer, and spawning occurred again in August 1975. The third year followed this pattern, with oocytes first visible in October and a gradual increase in diameter leading to spawning between August and September 1976.

In the power-plant thermal outfall at Morro Bay, distinct annual cycles of gametogenesis and spawning were obtained (Fig. 3). No gonads were found in the first samples collected (Feb. 4, 1974), but

TABLE 1

MATURATION STAGES OF MALES

Stage	Description
Stage 1	First appearance of spermatogonia
Stage 2	Spermatogonial proliferation; swelling of spermatogonial vesicles to occupy one-fourth of mesenterial width
Stage 3	First appearance of spermatocytes; further vesicular swelling (up to one-half of mesenterial width)
Stage 4	First appearance of tailed sperm; vesicles occupy up to three-fourths of mesenterial width
Stage 5	Gonads ripe, occupy entire width of mesentery
Spawning	Sperm vesicles breaking down; many tailed sperm present
Residual	Only a few unspawned sperm remain

this may be an artifact of early collecting technique (i.e., all small anemones from a single clone). Beginning in March, oocytes appeared and grew through the summer. An examination of live material indicated that some females were spawning on July 21, 1974. Subsequent analysis of the histologically prepared material supported this finding. Spawning continued for more than a month; samples taken Aug. 18 showed two females ripe and one having spawned. By Sept. 17, only residual oocytes remained. Small oocytes appeared again at the end of October and developed through the winter and spring, reaching peak size by July 11, 1975. Spawning was not observed until Aug. 8, by which time it was complete, except for a few residual gametes that persisted until early October. The third year showed the same course of development.

Spermatogonia were present in samples from the Morro Bay control site in February, and the first few tailed sperm had appeared by March 1974 (Fig. 4). Ripe gonads occurred by April, and spawning had begun by Aug. 18. Although spawning was apparently complete by Sept. 17, residual gametes remained through the end of October. The 1975 pattern was quite similar to that found in 1974, with ripe sperm appearing as early as April and spawning beginning in August. A few late-spawning individuals were collected on Nov. 17 although the third year's cycle had already begun. The cycle was repeated in 1976.

In the outfall, only 14 of 27 samples taken over the course of this project had male gonads present; all these occurred between late spring and early fall. Males ranging from early spermatogonial proliferation to first presence of tailed sperm were found in April 1974, and the population was ripe by May. Although the June

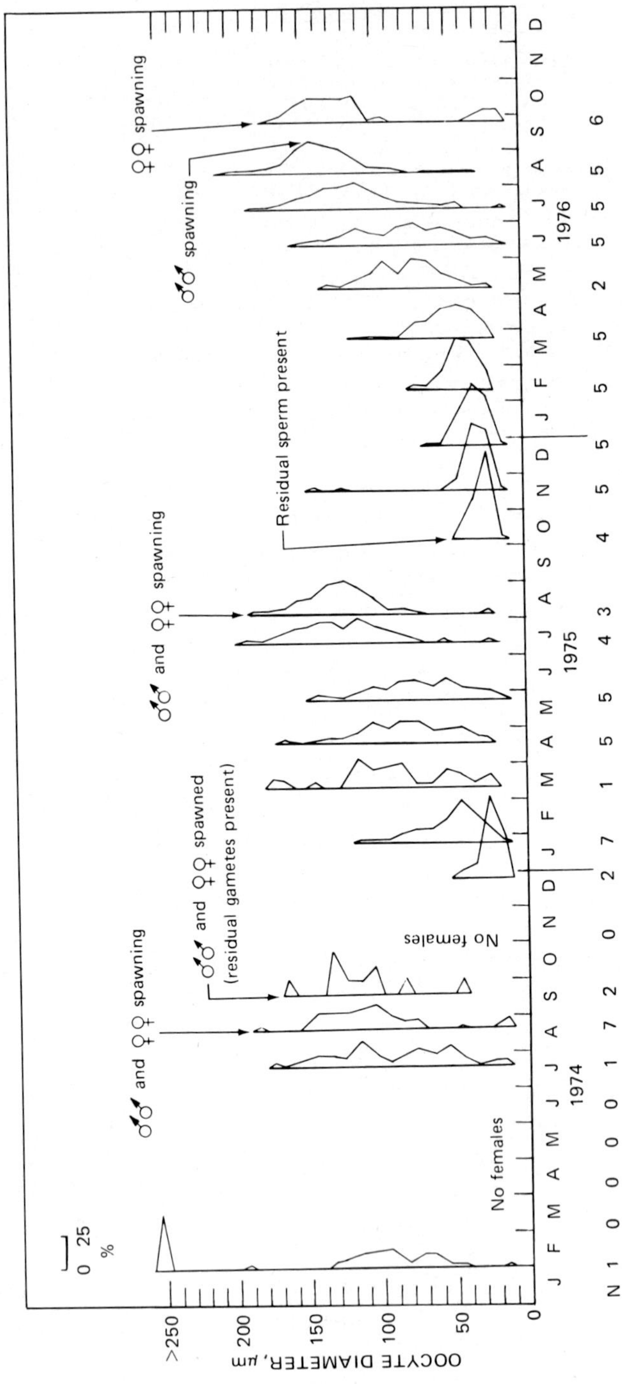

Fig. 2 Oocyte growth cycles of *A. elegantissima* at the Morro Bay control site.

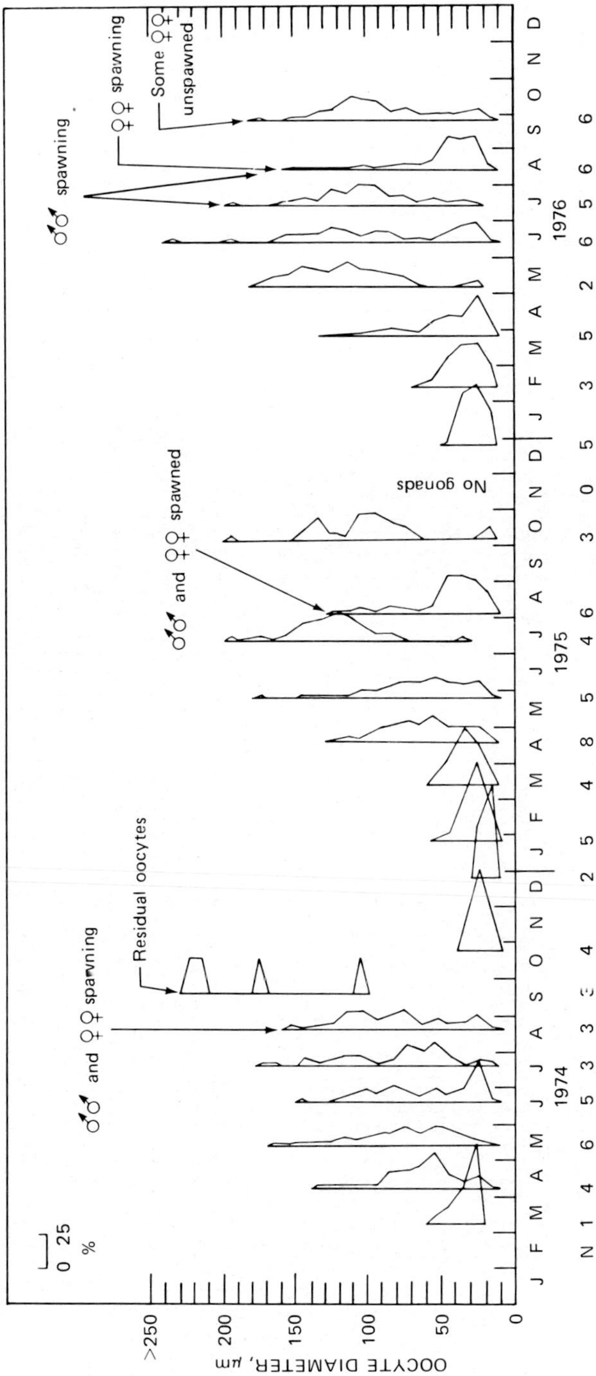

Fig. 3 Oocyte growth in the power-plant outfall.

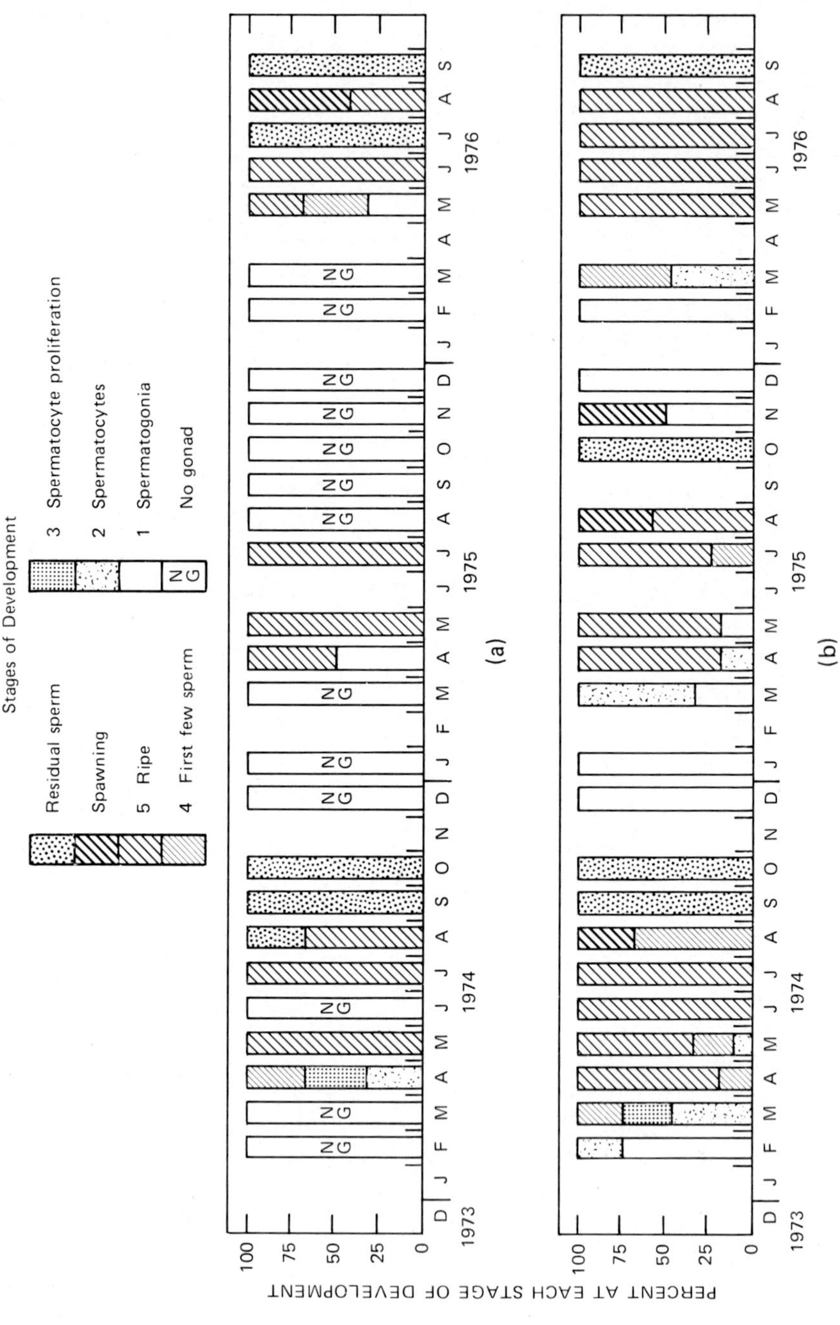

Fig. 4 Percentages of male *A. elegantissima* at each maturation stage. (a) At the power-plant outfall. (b) Control.

samples contained no males (possibly a sampling artifact), males with mature sperm were present in July, and by August some of the population had spawned. Residual gametes persisted until October. In 1975, no gonads could be found until April, when anemones with only early spermatogonia occurred along with those with ripe gonads. Mature sperm remained through July, but by Aug. 8 there were no identifiable males. This situation remained until May 1976, when again a variety of stages of maturity appeared at once. Although only residual gametes were found on July 13 (this indicated that some of the population had spawned), the August sample again contained ripe anemones, as well as some which were spawning. By September, only residual sperm were found.

Outfall temperatures were significantly higher than those of the control area throughout the project, except during April and May 1975 (Fig. 5). Control temperatures showed a rise through the spring to a summer plateau in all 3 years and a subsequent drop-off each winter. Outfall temperatures showed a similar pattern, but at a considerably higher level. Fluctuations in outfall temperatures were caused by variations in power-plant activity and occasionally by high spring tides, which backflushed the outfall canal to some extent.

Of the anemones collected on Mar. 25, 1976, from the transplant cages, the size-frequency histograms (Fig. 6) of outfall (where five of the six anemones examined had gonads) and outfall caged but not transplanted (three of seven had gonads) animals were very similar to those of unmanipulated anemones from the Morro Bay control site (where five of six had gonads). Of the control-site anemones transplanted to the outfall, however, two showed a much larger average oocyte diameter, and three more had spawned. Unfortunately, the transplant cages hung in the intake area became silted up, and the anemones were lost.

DISCUSSION

Numerous studies have correlated seasonally changing sea temperatures with the timing of reproductive events in marine invertebrates (see Kinne, 1970; Giese and Pearse, 1974, for reviews). Hedgpeth and Gonor (1969) suggested that temperature is the major environmental factor controlling reproductive cycles of benthic marine invertebrates, but they added that photoperiod and other factors are also important. Thorson (1946), Giese (1959), and others cautioned that a distinction must be made between the effect of gradual temperature changes on gametogenesis and that of a sharp temperature change on spawning.

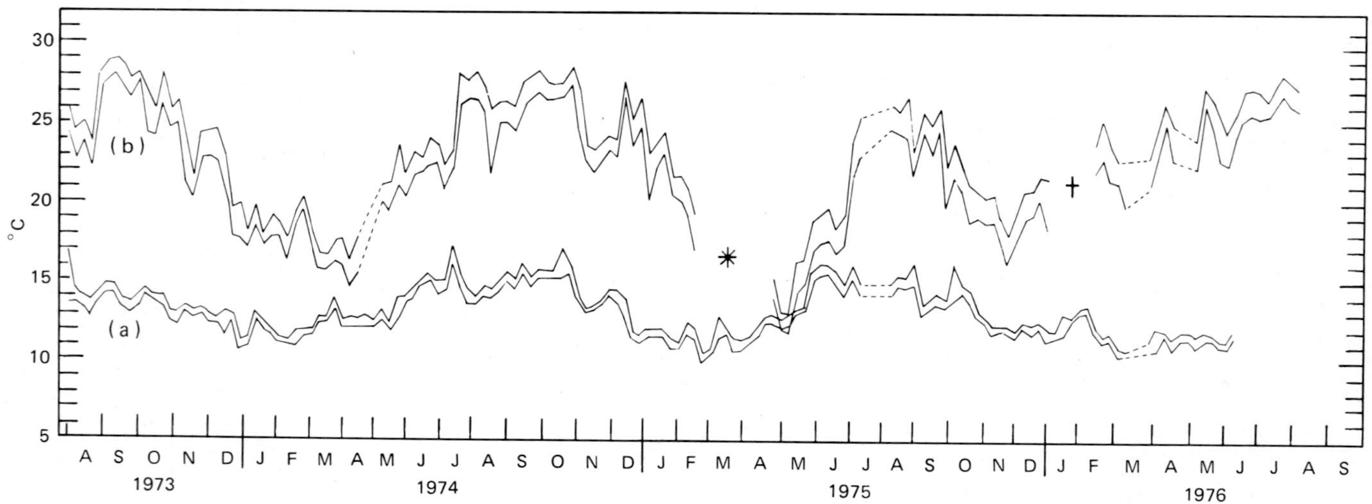

Fig. 5 Comparison of seawater temperatures of Morro Bay control (a) and outfall (b) during this study. Temperatures were measured with Ryan 45-day temperature recorders and are ±2 standard errors of weekly means (28 6-hr means estimated from continuous traces). Breaks in the outfall curves (b) in 1975 (*) occurred because a recorder was lost and in 1976 (†), because of recorder malfunction.

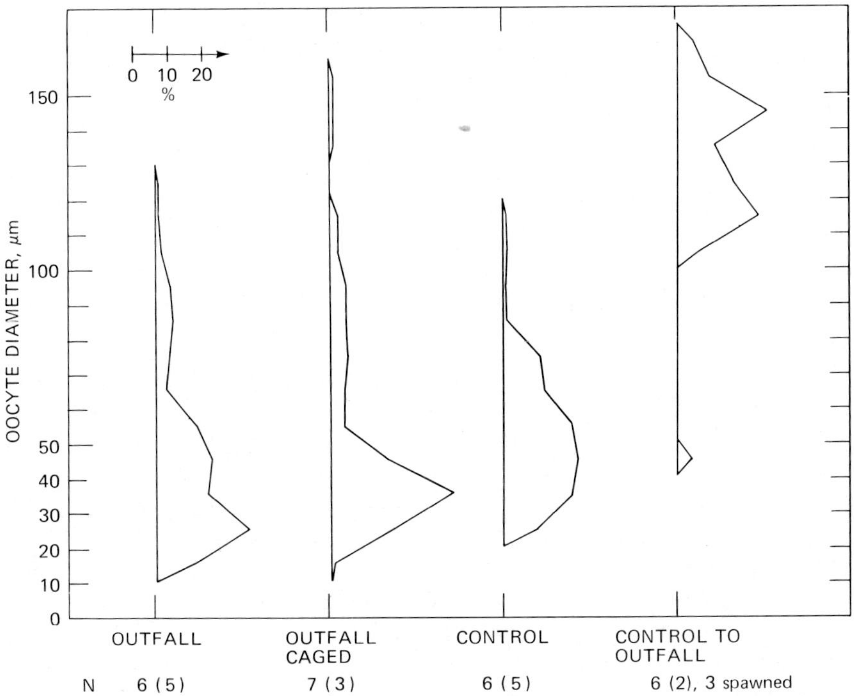

Fig. 6 Oocyte size-frequency histograms from the Mar. 25, 1976, transplant collections. N is the number of animals examined, and the numbers in parentheses are those having gonads.

Although temperatures are significantly higher in the Morro Bay outfall than at the control site (Fig. 5), the reproductive cycles of females appear to parallel each other well, except for the timing of spawning. The observed early spawning of outfall animals was consistent from one year to the next and suggests a possible effect of the warm-water regime on this critical aspect of reproduction. Comparing the timing of spawning of outfall anemones (Fig. 3) with the temperature data shows that, in 1974 and 1975, anemones spawned within 2 weeks of the time that mean temperatures first exceeded 25°C. In both years this temperature was reached abruptly. In 1976, spawning did not occur until the mean temperature had stabilized at 26°C; earlier brief fluctuations above this point did not bring about spawning. It is possible that anemones are unable to retain and nourish oocytes at sustained temperatures above this threshold level because of the metabolic demands placed upon them by these high temperatures.

In the Morro Bay control populations, female spawning also occurs during periods of maximum temperature, but not with the same timing exhibited by anemones in the outfall. Spawning took place in the control samples during periods of mean temperatures between 14 and 16°C, but it did not occur until a month after these temperatures had been reached. In 1974 and 1975, these temperatures were reached in June, but spawning did not occur until July. In 1976, these temperatures were reached in July, and the population did not spawn until September. A similar spawning pattern was seen by Ford (1964) in populations from Marin County and by Jennison (1977) in Sonoma County, California.

Despite the significant temperature differences between outfall and control areas, we see the same general pattern of a winter temperature decline and a summer increase at both sites. If anemones time their reproductive events by temperature cues, it is likely that a predictable pattern of temperature change would be more important than the absolute temperature level (Giese and Pearse, 1974). Galtsoff (1961) observed that different physiological races of the same species may spawn at different temperatures. Outfall anemones, although not necessarily of a different race than those in the control, may, nevertheless, be acclimated to the warm-water regime in which they are found. Possibly they are responding at 25°C to the same cues that affect the control animals at 15°C. The difference in response time (2 weeks in the outfall to 1 month in the control) may be simply a function of the more rapid metabolic rate of outfall anemones caused by the higher absolute temperatures in which they occur. The delay in spawning of outfall females in 1976 might have resulted from the gradual rise in temperature, which permitted a period of metabolic accommodation. Thus the precipitous spawning that was observed in 1974 and 1975, when the temperature threshold was reached abruptly, did not occur.

Males of *A. elegantissima* at the Morro Bay control site develop over an annual time course similar to that of the females, with early gametogenic stages proliferating in February and March, ripe sperm apparent by April, and spawning occurring between August and September in each of the 3 years studied (Fig. 4). Males in the outfall, however, showed an abbreviated reproductive season, with no early stages visible before April of each year. This appears to be caused by the high temperatures in the outfall. Male development is suppressed until late spring, possibly for metabolic reasons, but sperm mature in outfall populations by May each year. It is not unreasonable to assert that males of *A. elegantissima* can develop quickly. Using autoradiography, Beeman (1970) determined that it would take a minimum of 10 days to develop free spermatozoa from primary spermatocytes in the opisthobranch *Phyllaplysia taylori*. Only 9 days were required in the sea urchin *Strongylocentrotus purpuratus* (Holland and Giese, 1965).

Despite the delayed and compressed cycle exhibited by outfall males, these anemones spawn at about the same time as do their control counterparts. Anemones in the outfall began spawning sometime after July 21 and had finished by Sept. 17, 1974, and control animals spawned during the same period. In 1975, no spawning was observed in the outfall, but, since ripe gonads were found on July 11 and no gonads could be seen by Aug. 8, we can infer that spawning occurred between these two dates. In the control, anemones were spawning on Aug. 8. Not until 1976 was there an appreciable difference in the timing of spawning of the two populations. All outfall anemones collected on July 13, 1976, had spawned; others in the same clone were still ripe or spawning on Aug. 11; but all had spawned by the end of September. In the control, complete spawning took place between Aug. 11 and Sept. 27.

In the transplant experiment, outfall, control, and outfall caged but not transplanted anemones all showed very similar oocyte size-frequency histograms (Fig. 6). Control anemones that were transplanted to the outfall underwent a rapid maturation, however. Of six anemones sectioned, two were ripe, and three more had recently spawned. This occurred during a period of increasing temperature in the outfall (the temperature at the time of the March collection was around 22°C), and it appears that increasing temperature accelerated vitellogenesis and induced spawning in anemones from the cold-water control area. Anemones acclimated to the outfall do not normally spawn until summer, when the outfall reaches its average maximum temperature of 25°C, but by March 1976 the control anemones that were transplanted into the outfall were already 7°C past the 15°C maximum that the control area reaches in summer.

CONCLUSIONS

The sea anemone *Anthopleura elegantissima* reproduces on an annual cycle at Morro Bay, with gametogenic cells being visible first in late fall and winter and spawning occurring in late summer in natural populations.

Males of *A. elegantissima* from a thermal-outfall canal exhibit a delayed onset of gametogenesis but spawn as much as 1 month before those in natural populations. Females in the outfall begin gametogenesis at the same time as control animals but spawn earlier.

Control females transplanted to the outfall in November 1975 underwent accelerated vitellogenesis and spawned by the end of March 1976, 4 to 5 months ahead of either unmanipulated control or acclimated outfall anemones.

Control anemones spawn during the period of peak summer temperatures, which may reach 15°C. Outfall anemones spawn after temperatures in the thermal canal reach 25°C. It is suggested that outfall anemones have become acclimated to the warmer water (which is regularly 10°C higher in temperature than that in the control area) and are responding at 25°C to the same temperature cues that influence spawning in control populations at 15°C.

ACKNOWLEDGMENTS

This paper is part of a thesis submitted for the Ph.D. degree in zoology at the University of California, Berkeley. I thank my major advisor, Cadet Hand, and the members of my committee. The cooperation and assistance of the Pacific Gas and Electric Company are gratefully acknowledged. This work was performed while I was supported by a traineeship associated with Project #R/E-10, Grant #04-6-158-44021, awarded to Cadet Hand and Ralph I. Smith by the University of California Sea Grant College Program, National Oceanic and Atmospheric Administration, U. S. Department of Commerce. The Harbor Branch Foundation, Inc., provided support during manuscript preparation. This is contribution number 91, Harbor Branch Foundation, Inc.

REFERENCES

Beeman, R. D., 1970, An Autoradiographic and Phase Contrast Study of Spermatogenesis in the Anaspidean Opisthobranch *Phyllaplysia taylori* Dall, 1900, *Arch. Zool. Exp. Gen.*, III(I): 5-22.

Dunn, D. F., 1975, Reproduction of the Externally Brooding Sea Anemone *Epiactis prolifera* Verrill, 1869, *Biol. Bull.*, 148: 199-218.

Ford, C. E., Jr., 1964, Reproduction in the Aggregating Sea Anemone, *Anthopleura elegantissima*, *Pac. Sci.*, 18(2): 138-145.

Galtsoff, P. S., 1961, Physiology of Reproduction in Molluscs, *Am. Zool.*, 1: 273-289.

Giese, A. C., 1959, Comparative Physiology: Annual Reproductive Cycles of Marine Invertebrates, *Annu. Rev. Physiol.*, 21: 547-576.

——, and J. S. Pearse (Eds.), 1974, *Reproduction of Marine Invertebrates*, Vol. I, Academic Press, Inc., New York.

Hand, C., 1955, The Sea Anemones of Central California. Part II. The Endomyarian and Mesomyarian Anemones, *Wasmann J. Biol.*, 13: 37-99.

Hedgpeth, J. W., and J. J. Gonor, 1969, Aspects of the Potential Effect of Thermal Alteration on Marine and Estuarine Benthos, in *Biological Aspects of Thermal Pollution,* P. A. Krenkel and F. L. Parker (Eds.), pp. 80-118, Vanderbilt University Press, Nashville, Tenn.

Holland, N. D., and A. C. Giese, 1965, An Autoradiographic Investigation of the Gonads of the Purple Sea Urchin (*Strongylocentrotus purpuratus*), *Biol. Bull.*, 128: 241-258.

Jennison, B. L., 1977, *An Analysis of the Environmental Factors Which Influence Gametogenesis, Spawning and Nutrient Storage in the Sea Anemone* Anthopleura elegantissima, Ph.D. Thesis, University of California, Berkeley.

Kinne, O., 1970, Temperature, Animals, Invertebrates, in *Marine Ecology,* O. Kinne (Ed.), Vol. 1, part 1, pp. 407-514, John Wiley & Sons, New York.

Stephenson, T. A., 1928, *The British Sea Anemones,* Vol. 1, Ray Society, London.

Thorson, G., 1946, Reproduction and Larval Development of Danish Marine Bottom Invertebrates, *Meddr. Kommu. Danm. Fisk.-og Havunders., Serie: Plankton,* 4: 1-523.

A COMPARISON OF MORPHOMETRIC, BIOCHEMICAL, AND PHYSIOLOGICAL INDEXES OF CONDITION IN MARINE BIVALVE MOLLUSCS

ROGER MANN
Woods Hole Oceanographic Institution, Woods Hole, Massachusetts

ABSTRACT

A number of morphometric, biochemical and physiological parameters are compared and evaluated for their potential use as indexes of condition in marine bivalve molluscs. Condition is defined as "the ability of an animal to withstand an adverse environmental stress, be this physical, chemical or biological." Laboratory and field data are compared to determine the effect of one environmental stressor, temperature, on gravimetric and volumetric meat/shell condition indexes; percentages of organic content, carbohydrate, carbon, and nitrogen; C/N and carbohydrate/N ratios; weight-specific ammonia excretion rates; and oxygen-consumption/ammonia-excretion ratios in a number of commercially valuable species of oyster, clam, and mussel.

With the increasing use of the sessile, adult bivalve mollusc as an indicator of environmental quality (Goldberg, 1975) and the considerable interest in culturing commercially valuable bivalve species in thermal effluents, there is a pressing need to determine which of many available morphometric, biochemical, and physiological parameters can be used as indexes of condition in bivalves. Condition is defined as "the ability of an animal to withstand an adverse environmental stress, be this physical, chemical or biological." Stress is defined as "a measurable alteration of a physiological, or behavioral, or biochemical, or cytological steady-state which is induced by environmental change, and which renders the individual (or the population, or the community) more vulnerable to further environmental change" (Bayne, 1975).

Although many stressors, such as disease, starvation, salinity, and exposure, affect the condition of bivalve molluscs, this paper examines the variation in a number of previously used indexes of condition in relation to only one environmental stressor, temperature. Furthermore, discussion is limited to temperature ranges comparable to those experienced by the species in their natural habitats. The natural occurrence of a species in one particular seawater temperature range should not be taken to indicate that the species is not under temperature stress for at least some portion of the year. For example, the southern geographical limit of *Mytilus edulis* corresponds approximately to a mean summer water temperature of 26.7°C (Wells and Gray, 1960). When water temperature is held constant, however, this species is unable to acclimate its metabolic and feeding rates to temperatures greater than 20 to 22°C (Bayne, Widdows, and Worrall, 1977). Gonzalez and Yevich (1976) made similar observations for populations of *M. edulis* in the effluent canal of a steam generating plant.

INDEXES OF CONDITION

Volumetric and Gravimetric Meat-to-Shell Ratios

Grave (1912) suggested expressing the "fatness" of an oyster as the percentage of the shell-cavity volume occupied by meat. This can be directly related to the ratio used by Baird (1958) to measure the condition of mussels and oysters, i.e.,

$$\frac{\text{Meat volume}}{\text{Shell-cavity volume}} \times 100 = \text{Index of condition} \qquad (1)$$

Baird reported that for oysters this index is not affected by the size of the specimen under examination but is size related for mussels. Thus for mussels the index is valid only for comparing animals of the same size (i.e., length). The index should not be used in cases where spawning is in any degree possible, however. In natural populations the spawning period may last for up to 6 months, and in populations exposed to abiotic thermal stimulation it can be even longer. Therefore, the index appears to be of limited value.

Hopkins (in Higgins, 1938) modified the approach of Grave (1912):

$$\frac{\text{Dry weight of meat (g)} \times 100}{\text{Shell-cavity volume (ml)}} = \text{Index of condition} \qquad (2)$$

Similar derivations subsequently made by Medcof and Needler (1941) and Korringa (1955) have been used by a large number of investigators since that date. This index is an improvement over the volume-to-volume ratio index in that it accommodates the increase in water content of the soft tissues and the decrease in dry-meat weight which occur on spawning (Masumoto, Masumoto, and Hibino, 1934; Quayle, 1970) and, thus, remains a valid test during periods when spawning is possible. However, Galtsoff (1964) stated that the comparison is valid (for oysters at least) only if specimens of the same species and age are used because oysters from over-crowded natural reefs and young oysters are usually flat, with very little inner space between the valves. Consequently, this index of condition will be relatively high because the soft tissues occupy almost the entire shell cavity.

Recently an alternative gravimetric index was used by Walne and Mann (1975) and Mann and Ryther (1977):

$$\frac{\text{Dry weight of meat (g)} \times 1000}{\text{Dry weight of shell (g)}} = \text{Index of condition} \qquad (3)$$

This allows both entities used in the calculation to be assayed by the same method and, thus, facilitates the processing of a large number of samples. The method does not account for possible variations in shell-cavity volume caused by changes in overall shape of an individual animal or in shell thickness, but it does incorporate absolute shell size per se.

Changes in shape of bivalve molluscs have been extensively documented in the scientific literature. Ansell (1965), describing the interrelationship of a number of parameters of growth in mature *Venus (Mercenaria) mercenaria*, concluded that, if only the length is known for a given group of animals, the only transformations that can be accurately made are estimates of total weight, total volume, and shell-cavity volume. Thus, although his data indicate isodiametric shell growth estimates of wet and dry flesh weights, total body nitrogen and carbohydrate levels could not be estimated from length data because of changes related to the gametogenic cycle. Ohba (1959) suggested that growth in mature *Tapes japonica* was not isodiametric since year classes could be separated by significant differences in their breadth-to-length and breadth-to-height ratios. This conclusion was not supported by the data of Nosho and Chew (1972), however, who worked with the same species in Hood Canal, Washington. Seed (1968) documented a number of factors influencing shell shape in *M. edulis.*

A fourth index of condition was used recently by de Wilde (1975) in temperature and growth studies of *Macoma balthica*:

$$\frac{\text{Ash-free dry-meat weight} \times 1000}{\text{Shell length}^3} = \text{Index of condition} \qquad (4)$$

Using the weight of ash-free dry meat rather than that of dry meat negates the effects of fluctuations in the ash content of the meat, and using the cube of shell length again assumes isodiametric growth but neglects any potential changes in shell thickness.

Indexes of condition (1 to 4) are all basically similar in that they attempt to compare present meat content with its potential maximum size using the shell as a reference material. This comparison is valid because, although meat content can either increase or decrease, the shell can only increase or remain constant in weight and volume (apart from some very small losses associated with abrasion). All these indexes can be somewhat misleading with respect to meat, however, because only the dry-meat indexes account for fluctuations in water content. None of these indexes give data on the proportionate quantities of available metabolic reserves, structural components, and gonadal material in the soft tissues.

Biochemical Indexes

Within the seawater temperature range experienced annually by a bivalve mollusc, there is generally a close relationship between the absolute temperature and the cycles of storage and gametogenesis in the animal (Masumoto, Masumoto, and Hibino, 1934; Loosanoff, 1936; Chipperfield, 1953; Walne, 1958; Millar, 1963; Porter, 1964; Shaw, 1964; 1965; Holland and Chew, 1974; Dare and Edwards, 1975). A brief overview of this relationship is relevant here before we discuss biochemical indexes per se. Obviously species differ in their overall strategies of storage and gametogenesis, as do different geographical populations of the same species (Seed, 1975). The four phases of the complete gametogenic cycle which can usually be identified are: (1) low metabolic activity, often coincident with cold water temperatures in the winter months; (2) high metabolic activity, often coincident with rising water temperatures in spring and with phytoplankton blooms; (3) rapid gametogenesis or completion of gametogenesis, coincident with depletion of carbohydrate reserves and accumulation of lipid in yolky gametogenic products; and (4) low levels of stored reserves. Spawning follows phase 3, and the subsequent period of low reserve levels (phase 4) is a result of the reserve's having been used in gonad proliferation and lost on spawning and the high metabolic demands associated with high water temperatures. Decreasing water temperatures in the fall months are often accompanied by some replenishing of reserves and, in some species (e.g., *M. edulis*), by the initiation of gametogenic processes before low metabolic activity is again evident in relation to low water temperatures.

During the late stages of gametogenesis, glycogen reserves are rapidly depleted (Masumoto, Masumoto, and Hibino, 1934; Mann, unpublished data). Thus a statement like that of Galtsoff (1964): "Oysters of good quality have a relatively large amount of meat in relation to their total volume. Their glycogen content is high, and the meat has a creamy color . . . ." can be misleading. The creamy color, caused by developing gonadal material, is produced at the expense of the glycogen reserves. Once gonad development is initiated and a temperature stimulus is maintained, development continues at the expense of stored reserves whatever the physiological state of the animal. This was well demonstrated for *M. edulis* by Bayne, Gabbott, and Widdows (1975). Also, Mann (unpublished data) observed gonad proliferation in juvenile *C. gigas* starved for a period of 3 weeks at 14.5°C.

A simple gross biochemical analysis for major components (carbohydrate, protein, lipid, and ash) or calorie content during late gametogenesis can also be misleading because the possibility exists that a considerable proportion of the energy-rich lipid component is stored in gametogenic products and is not generally available for energy metabolism by the animal.

Larval forms of oysters (and probably other bivalves) have a lipid—protein-based respiratory metabolism, but there is a gradual change to a carbohydrate—protein-based respiratory metabolism after settlement (Holland and Hannant, 1974). In an ecological sense this is a useful strategy since a high lipid content provides bouyancy in a pelagic larval form. This is unnecessary in a sessile adult in which the requirement for a readily mobilized reserve material is adequately served by glycogen. If the initial effect of a temperature stress is a drain of such reserve material, then, by definition, the quantity of available reserves can be used as an index of condition:

$$\%\text{ Carbohydrate content} = \text{Biochemical index} \qquad (5)$$

As for the gravimetric and volumetric indexes (1 to 4), some problems are evident when we apply this index to a wide size range of animals. Metabolic functions, which will inevitably influence the rate of utilization of reserves, are generally related allometrically to the absolute size of the animal. Thus, when Holland and Hannant (1976) examined the summer and winter glycogen contents of five age groups of *Ostrea edulis* L., they found that, on a percentage basis, summer levels were similar in all groups but winter levels increased with increasing age.

Since carbohydrate and protein are the major biochemical components of adult bivalve soft tissues (Masumoto, Masumoto, and Hibino, 1934; Dare and Edwards, 1975; Walne and Mann, 1975), an increase in the percentage content of one is usually reflected in a decrease in percentage content of the other. Since the depletion of carbohydrate is accompanied by an increased use of protein as a respiratory substrate, the ratio of these two components can also be used as an index of condition:

$$\frac{\text{Carbohydrate}}{\text{N}} = \text{Biochemical index} \qquad (6)$$

The recent development of analytical instrumentation to both carbon and nitrogen contents simultaneously stimulated Ansell and Sivadas (1973) to suggest that the carbon-to-nitrogen ratio may be useful as an index of condition in the bivalve *Donax vittatus* (da Costa). This ratio gives a measure of the total organic content of the soft tissues in relation to protein content and thus, by inference, a measure of nonprotein organic material available for use as a respiratory substrate. Thus

$$\frac{\text{C}}{\text{N}} = \text{Biochemical index} \qquad (7)$$

As in other gross biochemical components, there is a seasonal temperature-related cycle in the ash (inorganic) content of soft tissues (ash is defined as the material remaining after ignition at 500°C or higher for a period of 12 hr). In the oysters *Crassostrea gigas* and *Ostrea edulis*, percentage ash content is usually highest in the winter months and lowest in summer (Walne and Mann, 1975). The decrease in relative organic content in winter partially reflects the utilization of stored reserves during this period and, thus, can be considered a further index of condition:

$$\%\ \text{Organic (100-ash) content} = \text{Biochemical index} \qquad (8)$$

Physiological Indexes: Ammonia Excretion and the Oxygen-to-Ammonia Ratio

All the indexes discussed previously are based on quantitative measurements of a potentially utilizable reserve material, and all suffer from inadequacy because they do not account for rates of accumulation and depletion of such material. In this context certain physiological parameters can be of value. Oxygen consumption rate can be usefully employed as an index of physiological rate. In bivalve molluscs, however, the respiratory response to temperature stress is complex; both short-term acute and long-term acclimation responses may be involved. These have been extensively described for *M. edulis* by Bayne and co-workers (Bayne, 1973; Widdows, 1973; see also a comprehensive review in Bayne, 1976). Furthermore, these responses can be superimposed on short-term fluctuations caused by feeding activity (Bayne, 1973; Bayne and Scullard, 1977b). The nature of such responses may vary somewhat between species, however; e.g., Ansell and Sivadas (1973) stated that *Donax vittatus* does not exhibit thermal acclimation of its respiratory processes.

It is evident that the use of respiration data alone can provide some distinct interpretive problems. Its use in conjunction with other physiological parameters, however, can alleviate some of these problems. The cycles of accumulation and depletion of carbohydrate reserves in bivalves and the increasing use of protein as a respiratory substrate after depletion have been briefly discussed earlier. Increased use of protein as a respiratory substrate inevitably results in an increased rate of ammonia excretion. A combination of oxygen consumption rate (O) and ammonia excretion rate (NH_3) can be used in the form of an O/NH_3 ratio to assess what proportion of the respiratory substrate is protein and, thus, to infer the status of an animal's reserves:

$$\frac{O}{NH_3} = \text{Physiological index} \qquad (9)$$

The O/NH_3 ratio was used previously in studies of isopods (Wieser, 1972) and the bivalves *M. edulis* (Bayne, 1973; Bayne and Scullard, 1977a) and *Donax vittatus* (Ansell and Sivadas, 1973). Note that the ammonia excretion rate, not total nitrogen elimination, should be used to evaluate the nature of the substrate being oxidized (see Ansell and Sivadas, 1973). Total nitrogen includes a variable amount of amino acids and urea (Potts, 1967; Hammen, 1968; Allen and Garrett, 1971; Bayne, 1976) which may, in part, be connected with osmoregulatory rather than excretory functions. The ratio of oxygen consumed to total nitrogen excreted has been used in studies of copepods (Conover and Corner, 1968; Corner and Cowey, 1968), prawns (Snow and Williams, 1971), and *M. edulis* (Bayne and Thompson, 1970).

A seasonal variation in the O/NH_3 ratio occurs in conjunction with the natural gametogenic cycle, but using a ratio rather than the absolute physiological rates can partially eliminate some of the problems involved in comparing animals of differing sizes. Physiological functions can be used to compare animals of differing sizes if we know the allometric relationship of that function to size. Thus, if

$$P = aW^b$$

where P is the specific rate function, W is the weight, and a and b are constants, animals of differing sizes can be compared by expressing the rate function using the "physiological weight" (W^b) rather than the absolute weight (W). Thus comparisons of weight-specific ammonia excretion rates change from

$$\mu g\ NH_3 - N\ g\ dry\ W^{-1}\ hr^{-1}$$

to

$$\mu g\ NH_3 - N\ g\ dry\ (W^b)^{-1}\ hr^{-1}$$

The nature of the allometric relationship may also change with the seasonal gametogenic cycle. Bayne and Scullard (1977a) described this change for ammonia excretion in *M. edulis*. Ryther and Mann (1977) described a similar increase in weight-specific ammonia excretion rates and a change in the allometric relationship of this rate to dry-meat weight after spawning in the oysters *Crassostrea gigas* and *Ostrea edulis*.

COMPARING GRAVIMETRIC AND BIOCHEMICAL INDEXES: MATERIALS AND METHODS

Literature Data

Data from Masumoto, Masumoto, and Hibino (1934) for *C. gigas* (Kusatsu stock grown at Kusatsu), from Dare and Edwards (1975) for 65-mm-long *M. edulis*, and from Walne and Mann (1975) for *O. edulis* were recalculated to obtain values for percentage carbohydrate, carbohydrate/N ratios, percentage organic content, and dry-meat/dry-shell ratios for the shellfish stocks examined. Data for each biochemical index for each species were compared to the gravimetric condition index by linear regression analysis.

Laboratory Studies

An experiment was performed to delineate the effect of temperature on growth, biochemical storage cycles, and gametogenesis in three species of bivalves when all other environmental factors (salinity, water flow, and phytoplankton concentration) were constant. Four populations of each of three species *(Crassostrea gigas, Ostrea edulis,* and *Tapes japonica),* matched by numbers of animals and mean individual live weight, were selected from common parental stocks. Populations of each species were maintained at sustained temperatures of 12, 15, 18, and 21°C for 19 weeks. Samples of 15 to 20 animals of each species were sacrificed at biweekly intervals for assay of live weight, freeze-dried meat weight, dry-shell weight (100°C for 24 hr), and the following biochemical components:

1. Carbon and nitrogen assayed by a Perkin—Elmer CHN analyzer
2. Carbohydrate assayed by the anthrone method of Strickland and Parsons (1968) after extraction in cold 5% wt/vol trichloroacetic acid
3. Ash, defined as the material remaining after ignition for 12 hr at 500°C

Within each species—temperature combination, data for each of the biochemical indexes (% carbohydrate, carbohydrate/N ratio, C/N ratio, and % organic content) were compared to the gravimetric dry-meat/dry-shell condition index by linear regression analysis. Since no significant differences in values of both slope and intercept were found between temperatures within each species, data for all temperatures were pooled and recalculated to obtain one regression line for each biochemical—condition index combination for each species.

RESULTS AND DISCUSSION

Linear regressions of biochemical and gravimetric indexes for natural and laboratory populations are summarized in Tables 1 and 2, respectively. A good correlation is evident in all regressions for all species tested. It is notable, however, that, although slope and intercept values are similar for *C. gigas* in both natural and laboratory populations for percentage carbohydrate and carbohydrate/N ratio computations, only slope values are similar for natural and laboratory populations of *O. edulis*. This indicates a consistently higher carbohydrate content in laboratory animals. A complete explanation of this phenomenon must await further study, however.

Previously a number of investigators examined the use of glycogen content as an index of condition in bivalve molluscs, usually by direct comparison of dry-meat/shell-cavity ratios (index 2) with percentages of glycogen. Ingle (1949), comparing 34 samples of

TABLE 1

LINEAR REGRESSIONS OF BIOCHEMICAL INDEXES FOR NATURAL POPULATIONS*

Biochemical index	Species		r	df	P
% Carbohydrate	*C. gigas*	Y = +0.28X − 1.597	+0.805	24	<0.001
	O. edulis	Y = +0.665X − 5.087	+0.812	10	<0.001
	M. edulis	Y = +0.093X + 2.277	+0.610	23	<0.001
Carbohydrate/N ratio	*C. gigas*	Y = +0.043X − 0.595	+0.833	24	<0.001
	O. edulis	Y = +0.075X − 0.857	+0.831	10	<0.001
	M. edulis	Y = +0.0115X − 0.025	+0.541	23	<0.02
% Organic content	*C. gigas*	Y = +0.059X + 89.3	+0.783	24	<0.001
	O. edulis	Y = +0.526X + 64.01	+0.728	10	<0.01
	M. edulis	Y = +0.035X + 80.62	+0.784	23	<0.001

*All data take the form Y = MX + b. Y = % carbohydrate, carbohydrate/N ratio, and % organic (100-ash) content and X = dry-meat (mg)/dry-shell (g) ratio in natural populations of *C. gigas* (Masumoto, Masumoto, and Hibino, 1934), *O. edulis* (Walne and Mann, 1975), and *M. edulis* (Dare and Edwards, 1975).

TABLE 2

LINEAR REGRESSIONS OF BIOCHEMICAL INDEXES FOR LABORATORY POPULATIONS*

Biochemical index	Species		r	df	P
% Carbohydrate	*C. gigas*	Y = +0.192X − 1.832	+0.811	38	<0.001
	O. edulis	Y = +0.576X + 10.37	+0.701	38	<0.001
	T. japonica	Y = +0.097X + 2.71	+0.564	38	<0.001
Carbohydrate/N ratio	*C. gigas*	Y = +0.032X − 0.791	+0.802	38	<0.001
	O. edulis	Y = +0.092X + 1.784	+0.725	38	<0.001
	T. japonica	Y = +0.014X + 0.062	+0.558	38	<0.001
C/N ratio	*C. gigas*	Y = +0.021X + 3.308	+0.804	38	<0.001
	O. edulis	Y = +0.054X + 2.73	+0.803	38	<0.001
	T. japonica	Y = +0.008X + 3.612	+0.688	38	<0.001
% Organic content	*C. gigas*	Y = +0.201X + 64.12	+0.528	38	<0.001
	O. edulis	Y = +0.306X + 70.66	+0.844	38	<0.001
	T. japonica	Y = +0.059X + 80.07	+0.855	38	<0.001

*Y = % carbohydrate, carbohydrate/N ratio, C/N ratio, and % organic (100-ash) content and X = dry-meat (mg)/dry-shell (g) ratio in populations of *C. gigas*, *O. edulis*, and *T. japonica* maintained at 12, 15, 18, and 21°C for 19 weeks. Data, pooled for all temperatures, take the form Y = MX + b.

oysters, found a positive correlation between the two parameters ($r = 0.69$), but omitting the three highest values for glycogen content reduced the r value to 0.38. Ingle's general conclusion was that the relationship between these two parameters was poor, but it is notable that the glycogen contents of the oysters used in the study are very low (1 to 2% in most cases), and this probably is not a representative sample. By contrast, Walne (1970) and Gabbott and Stephenson (1974) found a good correlation in *O. edulis* between the dry-weight/shell-cavity ratio (index 2) and a glycogen condition index calculated as glycogen (g)/internal shell volume (ml) ($P < 0.01$ and $P < 0.001$, respectively).

Despite the constraints involved in the use of each regressed parameter, the present data suggest that any of these parameters can be used as equally valuable indexes of condition in the bivalves described here. Only further studies of other species of bivalves can confirm or eliminate these general relationships, but it is indeed reassuring to suggest that an easily affected volumetric or gravimetric assay of condition could be equally as meaningful as a more complex biochemical one. The use of physiological indexes, such as the O/NH_3 ratio, must await further investigation for a complete assessment of value.

In a more general context, the use of any index of condition to predict the bivalve's ability to withstand stress must rest upon an understanding of basic life processes in the animals. Despite many decades of intensive research, there are still areas markedly depauperate in documentation. For example, Bayne et al. (1976) recently noted changes in the taurine/glycine ratio in mussel tissue as a response to temperature stress. A complete physiological explanation of this phenomenon is not yet forthcoming, however, let alone an explanation that incorporates the effect of environmental stresses. In studies of the response of bivalves to thermal perturbation, further research is required at both the laboratory and field levels. In this respect the recent studies of Gonzalez and Yevich (1976) and Bayne, Widdows, and Worrall (1977) serve as excellent examples of the use of modern techniques to provide evidence of thermal stress at the biochemical, physiological, and cytological levels and will undoubtedly stimulate similar studies in the near future.

ACKNOWLEDGMENTS

The research reported here was completed under grants from the NOAA Office of the Sea Grant 04-6-158-44106 and the Jessie Smith Noyes Foundation.

I would like to thank J. M. Capuzzo and J. H. Ryther for their critical reading of the manuscript.

REFERENCES

Allen, J. A., and M. R. Garrett, 1971, The Excretion of Ammonia and Urea by *Mya arenaria* L. (Mollusca: Bivalvia), *Comp. Biochem. Physiol.* 39A: 633-642.

Ansell, A. D., 1965, Some Parameters of Growth of Mature *Venus mercenaria* L., *J. Cons. Int. Explor. Mer,* 29: 214-220.

——, and P. Sivadas, 1973, Some Effects of Temperature and Starvation on the Bivalve *Donax vittatus* (da Costa) in Experimental Laboratory Populations, *J. Exp. Mar. Biol. Ecol.*, 13: 229-262.

Baird, R. H., 1958, Measurement of Condition in Mussels and Oysters, *J. Cons. Int. Explor. Mer,* 23: 249-257.

Bayne, B. L., 1973, Physiological Changes in *Mytilus edulis* L. Induced by Temperature and Nutritive Stress, *J. Mar. Biol. Assoc. U. K.*, 53: 39-58.

——, 1975, Aspects of Physiological Condition in *Mytilus edulis* (L.), with Special Reference to the Effects of Oxygen Tension and Salinity, in *Proceedings of the Ninth European Marine Biology Symposium*, H. Barnes (Ed.), pp. 213-238, Aberdeen University Press, Aberdeen, Scotland.

——, 1976 (Ed.), *Marine Mussels: Their Ecology and Physiology*, Cambridge University Press, New York.

——, P. A. Gabbott, and J. Widdows, 1975, Some Effects of Stress in the Adult on the Eggs and Larvae of *Mytilus edulis* L., *J. Mar. Biol. Assoc. U. K.*, 55: 675-689.

——, D. R. Livingstone, M. N. Moore, and J. Widdows, 1976, A Cytochemical and a Biochemical Index of Stress in *Mytilus edulis* L., *Mar. Pollut. Bull.*, 7(12): 221-224.

——, and C. Scullard, 1977a, Rates of Nitrogen Excretion by Species of *Mytilus* (Bivalvia: Mollusca), *J. Mar. Biol. Assoc. U. K.*, 57: 355-369.

——, and C. Scullard, 1977b, An Apparent Specific Dynamic Action in *Mytilus edulis* L., *J. Mar. Biol. Assoc. U. K.*, 57: 371-378.

——, and R. J. Thompson, 1970, Some Physiological Consequences of Keeping *Mytilus edulis* in the Laboratory, *Helgol. Wiss. Meeresunters.*, 20: 526-552.

——, J. Widdows, and C. Worrall, 1977, Some Temperature Relationships in the Physiology of Two Ecologically Distinct Bivalve Populations, in *Physiological Responses of Marine Biota to Pollutants,* F. J. Vernberg, A. Calabrese, F. P. Thurberg, and W. V. Vernberg (Eds.), pp. 379-400, Academic Press, Inc., New York.

Chipperfield, P. N. J., 1953, Observations on the Breeding and Settlement of *Mytilus edulis* (L) in British Waters, *J. Mar. Biol. Assoc. U. K.*, 32: 449-476.

Conover, R. J., and E. D. S. Corner, 1968, Respiration and Nitrogen Excretion by Some Marine Zooplankton in Relation to Their Life Cycles, *J. Mar. Biol. Assoc. U. K.*, 48: 49-75.

Corner, E. D. S., and C. B. Cowey, 1968, Biochemical Studies on the Production of Marine Zooplankton, *Biol. Rev.*, 43: 393-426.

Dare, P. J., and D. B. Edwards, 1975, Seasonal Changes in Flesh Weight and Biochemical Composition of Mussels (*Mytilus edulis* L.) in the Conway Estuary, North Wales, *J. Exp. Mar. Biol. Ecol.*, 18: 89-97.

de Wilde, P. A. W. J., 1975, Influences of Temperature on Behavior, Energy Metabolism, and Growth of *Macoma balthica* (L.), in *Proceedings of the Ninth European Marine Biology Symposium*, H. Barnes (Ed.), pp. 239-256, Aberdeen University Press, Aberdeen, Scotland.

Gabbott, P. A., 1975, Storage Cycles in Marine Bivalve Molluscs: A Hypothesis Concerning the Relationship Between Glycogen Metabolism and Gametogenesis, in *Proceedings of the Ninth European Marine Biology Symposium*, H. Barnes (Ed.), pp. 191-211, Aberdeen University Press, Aberdeen, Scotland.

——, and R. R. Stephenson, 1974, A Note on the Relationship Between the Dry Weight Condition Index and the Glycogen Content of Adult Oysters (*Ostrea edulis* L.) Kept in the Laboratory, *J. Cons. Int. Explor. Mer*, 35: 359-361.

Galtsoff, P. S., 1964, *The American Oyster*, Crassostrea virginica *Gmelin*, Fishery Bulletin 64, U. S. Fish and Wildlife Service, Department of the Interior, Washington, D. C.

Goldberg, E. D., 1975, The Mussel Watch. A First Step in Global Marine Monitoring, *Mar. Pollut. Bull.*, 6: 111.

Gonzalez, J. G., and P. Yevich, 1976, Responses of an Estuarine Population of the Blue Mussel, *Mytilus edulis*, to Heated Water from a Steam Generating Plant, *Mar. Biol.*, 34: 177-189.

Grave, Caswell, 1912 (Ed.), *A Manual of Oyster Culture in Maryland*, Fourth Report of the Shell Fish Commission of Maryland, 1912, pp. 279-348, reprinted in 1944 by King Brothers, Inc., Baltimore.

Hammen, C. S., 1968, Amino Transference Activities and Amino Acid Excretion of Bivalve Molluscs and Brachiopods, *Comp. Biochem. Physiol.*, 26: 697-705.

Higgins, E., 1938, *Progress in Biological Inquiries, 1937*, Report of the Commissioner of Fisheries for the Fiscal Year 1938, Appendix 1, Administrative Report No. 30, pp. 1-70, U. S. Bureau of Fisheries, Washington, D. C.

Holland, D. A., and K. K. Chew, 1974, Reproductive Cycle of the Manila Clam *(Venerupis japonica)*, from Hood Canal, Washington, *Proc. Natl. Shellfish. Assoc.*, 64: 53-58.

Holland, D. L., and P. J. Hannant, 1974, Biochemical Changes During Growth of the Spat of the Oyster, *Ostrea edulis* L., *J. Mar. Biol. Assoc. U. K.*, 54: 1004-1016.

——, and P. J. Hannant, 1976, The Glycogen Content in Winter and Summer of Oysters, *Ostrea edulis* L., of Different Ages, *J. Cons. Int. Explor. Mer*, 36(3): 240-242.

Ingle, R. M., 1949, A Comparative Study of Oyster Condition, *Science*, 109: 593.

Korringa, P., 1955, Quality Estimation of Mussels and Oysters, *Arch. Fischereiwiss.*, 46: 189-193.

Loosanoff, V. L., 1936, Sexual Phases in Quahog, *Science*, 83: 287-288.

Mann, R., and J. H. Ryther, 1977, Growth of Six Species of Bivalve Molluscs in a Waste Recycling—Aquaculture System, *Aquaculture*, 11: 231-245.

Masumoto, B., M. Masumoto, and M. Hibino, 1934, The Biochemical Studies of Magaki (*Ostrea gigas* Thunberg). II. The Seasonal Variation in the Biochemical Composition of *Ostrea gigas* Thunberg, *J. Sci. Hiroshima Univ. Ser. A*, 4: 47-56.

Medcof, J. C., and A. W. H. Needler, 1941, The Influence of Temperature and Salinity on the Condition of Oysters *(Ostrea virginica)*, *J. Fish. Res. Board Can.*, 5(3): 253-257.

Millar, R. H., 1963, Breeding and Gonadal Cycle of Oysters in Loch Ryan, Scotland, *J. Cons. Int. Explor. Mer*, 28: 432-439.

Nosho, T., and K. K. Chew, 1972, The Setting and Growth of the Manila Clam, *Venerupis japonica* (Deshayes), in Hood Canal, Washington, *Proc. Natl. Shellfish. Assoc.*, 62: 50-58.

Ohba, S., 1959, Ecological Studies in the Natural Population of a Clam, *Tapes japonica,* with Special Reference to Seasonal Variation in the Size and Structure of Population and to Individual Growth, *Biol. J. Okayama Univ.*, 5: 13-42.

Porter, H. J., 1964, Seasonal Gonadal Changes in Adult Clams, *Mercenaria mercenaria* (L.), in North Carolina, *Proc. Natl. Shellfish. Assoc.*, 55: 35-52.

Potts, W. T. W., 1967, Excretion in the Molluscs, *Biol. Rev.*, 42: 1-41.

Quayle, D. B., 1970, *Pacific Oyster Culture in British Columbia,* Bulletin 169, Fisheries Research Board of Canada, Ottawa.

Ryther, J. H., and R. Mann, 1977, *Bivalve Mollusc Culture in a Waste Recycling Aquaculture System,* Technical Report 77-79, Woods Hole Oceanographic Institution, Woods Hole, Mass.

Seed, R., 1968, Factors Influencing Shell Shape in *Mytilus edulis* L., *J. Mar. Biol. Assoc. U. K.*, 48: 561-584.

——, 1975, Reproduction in *Mytilus* (Mollusca: Bivalvia) in European Waters, *Pubbl. Stn. Zool. Napoli,* 39(Suppl.): 317-334.

Shaw, W. N., 1964, Seasonal Gonadal Changes in Female Soft-Shell Clams, *Mya arenaria,* in the Tred Avan River, Maryland, *Proc. Natl. Shellfish. Assoc.*, 53: 121-132.

——, 1965, *Seasonal Gonadal Cycle of the Male Soft-Shell Clam,* Mya arenaria, *in Maryland,* Special Scientific Report, Fisheries Series No. 508, U. S. Fish and Wildlife Service, Department of the Interior, Washington, D. C.

Snow, N. B., and P. J. Le B. Williams, 1971, A Simple Method to Determine the O:N Ratio of Small Marine Animals, *J. Mar. Biol. Assoc. U. K.*, 51: 105-109.

Strickland, J. D. H., and T. R. Parsons, 1968, *A Practical Handbook of Sea Water Analysis,* Bulletin 167, Fisheries Research Board of Canada, Ottawa.

Walne, P. R., 1958, Growth of Oysters (*Ostrea edulis* L.), *J. Mar. Biol. Assoc. U. K.*, 37: 591-602.

——, 1970, The Seasonal Variation of Meat and Glycogen Content of Seven Populations of *Ostrea edulis* L. and a Review of the Literature, *Fish. Invest., London, Ser. 2,* 26(3): 1-35.

——, and R. Mann, 1975, Growth and Biochemical Composition in *Ostrea edulis* and *Crassostrea gigas,* in *Proceedings of the Ninth European Marine Biology Symposium,* H. Barnes (Ed.), pp. 587-607, Aberdeen University Press, Aberdeen, Scotland.

Wells, W. H., and E. E. Gray, 1960, The Seasonal Occurrence of *Mytilus edulis* on the Carolina Coast as a Result of Transport Around Cape Hatteras, *Biol. Bull. (Woods Hole, Mass.),* 119: 550-559.

Widdows, J., 1973, Effect of Temperature and Food on the Heart Beat, Ventilation Rate and Oxygen Uptake of *Mytilus edulis, Mar. Biol.*, 20: 269-276.

Wieser, W., 1972, Oxygen Consumption and Ammonia Excretion in *Ligia beaudiana* M. E., *Comp. Biochem. Physiol.*, 43A: 869-876.

RESPONSE OF MOSQUITOFISH TO THERMAL EFFLUENT

DAVID H. BENNETT* and C. PHILLIP GOODYEAR†
Savannah River Ecology Laboratory, Aiken, South Carolina

ABSTRACT

Reproductive activity, sex ratios, size at sexual maturity, percentage of fat, and size structure were examined in five populations of mosquitofish (*Gambusia affinis)* from lentic environments experiencing different thermal conditions at the Savannah River Plant near Aiken, S. C. Populations were sampled during February, June, October, and December 1970. Mosquitofish from three thermally affected stations inhabited waters from 28 to 40°C, whereas mosquitofish at the unaffected stations were found in waters from 12 to 29°C. Periods of reproductive activity were different for thermally affected and unaffected populations. Females with eyed embryos were found in mosquitofish populations from unaffected waters in June; eyed embryos were found in the affected populations throughout the study period. Mean brood size of populations from thermally affected areas generally varied inversely with water temperature. Size at sexual maturity and sex ratios were generally similar in the five populations. Body fat content varied seasonally for each population. Percentages of fat were significantly higher in male and juvenile fish from thermally affected waters than in those from unaffected waters, whereas fat content was similar in females. Mean size of females from unaffected waters was directly related to water temperature, whereas the size structure of mosquitofish populations from affected waters varied inversely with water temperature. This study indicates that *G. affinis* is a eurythermal species that can adapt to conditions of severe thermal stress.

Recent increases in electrical power demands and concomitant increases in release of waste heat into natural waters have stimulated an intensive evaluation of the effects of abnormally high tempera-

*Present address: College of Forestry, Wildlife and Range Sciences, University of Idaho, Moscow, Idaho.
†Present address: U. S. Fish and Wildlife Service, Ann Arbor, Michigan.

tures on aquatic life. Early thermal studies concentrated on establishing upper tolerance limits for various species, and more recently ecological impacts of thermal effluents have been evaluated on-site. As more information on the effects of thermal effluents becomes available, it is apparent that some fishes can adapt to relatively high temperature conditions better than others.

Mosquitofish, *Gambusia affinis*, is one species that has been reported to have a relatively high thermal tolerance (Hart, 1952). Since *Gambusia* sp. are indigenous to areas that have warm climates (Blair et al., 1968) and generally inhabit shallow, nonmoving waters throughout their life cycle, they are naturally exposed to warmer waters than are deepwater forms of aquatic life (Falke and Smith, 1974). Data on the life history of *G. affinis* from areas unaffected by thermal discharge are available (Krumholz, 1948; Rosen and Bailey, 1963), but little is known about mosquitofish in thermally affected waters.

The U. S. Department of Energy's Savannah River Plant (SRP) is an area of diverse thermal research. Specifically, the Par Pond system has been the site of extensive thermal ecological evaluations on turtles, largemouth bass, alligators, and aquatic plants (Gibbons et al., 1974). The Par Pond system was constructed in the late 1950s to provide cooling water for nuclear production reactors. Water is pumped to the reactor, heated, and discharged into a canal system where it gravitates through several cooling ponds before entering a cove in Par Pond about 5 km from the pumping site. One cooling pond in the system is Pond C, where water temperatures commonly exceed 40°C (Ferens and Murphy, 1974; Clugston, 1973). This provides an ideal site for field studies to evaluate the effects of thermal discharges on biota in receiving waters.

The presence of mosquitofish inhabiting waters in Pond C at temperatures near their reported upper tolerance limit prompted comparative field and laboratory studies to determine their mechanisms for survival in an abnormally warm and widely fluctuating thermal environment. The objectives of this study were to compare size structure, percentage of fat, reproductive activity, sex ratios, and size at sexual maturity of mosquitofish populations inhabiting thermally stressed areas with those from unaffected waters.

STUDY AREAS

Mosquitofish were sampled from five locations at the Savannah River Plant. Three populations (at stations PC1, PC2, and PC3) were sampled in Pond C, a primary cooling pond in the Par Pond system.

Station PC1 was the nearest to the effluent canal, followed by PC2 and PC3. Station PC1 was separated by about 300 m of open water from PC2, and PC3 was approximately 650 m from PC1 and PC2. Mosquitofish populations at PC1 and PC2 appeared to be inhabiting waters at similar thermal regimes, whereas the population at PC3 had access to a small inlet stream at temperatures "natural" for the area. Two control or unaffected populations were also sampled, one from an unheated section of Par Pond, a 1120-ha cooling pond, and the other from Risher Pond, a 0.1-ha farm pond (stations PP and RP, respectively). Also during the February sampling period a second Par Pond population was sampled, but, because of its similarity to that at the other Par Pond station, it was not sampled for the remainder of the study period. Station PP was approximately 2 km from the Pond C populations. Mosquitofish populations in RP were about 20 km southwest of the Par Pond and Pond C populations.

METHODS

Mosquitofish populations were sampled with a 2-m seine and a dip net in February, June, October, and December 1970. Fish were transported alive to the laboratory in Styrofoam coolers and were processed within 24 hr. All fish were measured (total length to 0.1 mm), weighed (nearest milligram), and sexed by the absence or presence of a gonapodium, and females were examined internally for embryos in the oviduct. The stage of development of the embryos was classified as: stage 1, no enlargement—no development; stage 2, enlarged embryo without development; stage 3, embryo with eyespots; stage 4, embryo with well-developed eyes; stage 5, embryo at late yolk sac; stage 6, embryo mobile and capable of surviving externally from the female. For comparative purposes, stages 1 and 2, 3 and 4, and 5 and 6 were combined as early, intermediate, and advanced stages, respectively, of embryo development.

Samples from February, June, and October were separated by sex, quickly frozen, and lyophilized in a freeze dryer. Samples were considered dry when consecutive daily weighings were identical. Dried samples were homogenized in a small blender and placed in dessicators until fat determinations were made. Fat content was determined by hot ether extraction for 3 hr. Ether was evaporated from samples in drying ovens (30°C), and the weight of the extract was recorded. Results were expressed as percent of dry body weight. Data were log transformed and tested by analysis of variance, and class means were compared using the Student—Neuman—Keuls test ($P \leqslant 0.05$) (Sokal and Rohlf, 1969).

TABLE 1

WATER TEMPERATURES (°C) AT THE SITES OF COLLECTION OF SAMPLE *Gambusia affinis**

Sample time	Sample station: Heated			Sample station: Unheated	
	PC1	PC2	PC3	PP	RP
2/19/70	32.2	32.8	28.8	16.4	12.0
6/14/70	33.5	38.0	35.7	28.1	31.0
10/1/70	40.0	37.2	28.1	29.1	26.2
12/12/70	32.8	36.4	31.0	16.4	17.0
Mean	34.6	36.1	30.9	22.2	20.0
Range	32–40	33–38	28–36	16–29	12–31

*Stations PC1, PC2, and PC3 are in areas affected by thermal effluent; RP and PP are at temperatures natural for the area.

Temperatures and observations of behavior were recorded at the sampling stations when samples were collected and at other times throughout the year.

RESULTS

Water-temperature regimes at the five sampling stations differed among and within stations during the study period (Table 1). Water temperatures at PC1 and PC2 were significantly higher than those at the unaffected stations (PP and RP); temperatures at PC3 were intermediate. Water temperatures at PC1, PC2, and PC3 often changed 20°C during a 24-hr period with changes in reactor power levels. Nearshore areas inhabited by mosquitofish were generally 5 to 10°C lower than waters 1 to 2 m from shore.

At stations PC1 and PC2 mosquitofish survived by inhabiting the shallow, cooler waters near shore which were slightly sheltered from the high temperatures. Fish at these stations usually inhabited the upper few centimeters of water immediately adjacent to shore, being restricted to these areas by high temperatures that precluded lateral, shoreline dispersal. When lethal temperatures prevailed at PC1 and PC2, individuals who were physically forced to swim from the shore to deeper waters immediately lost equilibrium and died within a few seconds. Shoreline observations made at PC2 revealed that mosquitofish seemed to "select" temperatures at 39.5°C when higher and lower temperature waters were available.

TABLE 2

MEAN WEIGHT (mg) FROM FIVE SAMPLE POPULATIONS* OF *Gambusia affinis*

Sex and sample time	Sample station PC1	PC2	PC3	PP	RP
Male					
February	198	173	175	152	218
June	130	103	202	104	189
October	187	204	139	197	249
December	169	174	212	200	191
$\overline{X}$	171	164	182	159	216
Female					
February	580	485	277	202	467
June	471	307	369	558	950
October	318	405	238	357	522
December	680	628	298	269	150
$\overline{X}$	536	473	289	314	523

*Stations PC1, PC2, and PC3 are in thermally affected areas; PP and RP are in areas at natural temperatures.

Mosquitofish behavior at other stations was apparently correlated with predation. Mosquitofish populations at PC3 and PP were confined to the shoreline by continual predation by juvenile largemouth bass (*Micropterus salmoides*). *Gambusia* at Risher Pond, however, were found to inhabit deeper waters, probably because of relatively low predation pressure by sunfishes (*Lepomis* sp.). Largemouth bass are not found in Risher Pond, and sunfishes did not seem to be effective predators on mosquitofish.

The size distribution of mosquitofish collected during February, June, October, and December from the various populations was generally similar (Fig. 1). Populations at PP, PC2, and PC3 had a higher percentage of small individuals than those at RP and PC1. The modal length class for each population sampled was RP, 33 to 38 mm; PP, PC3, and PC2, 21 to 26 mm; and PC1, 27 to 32 mm. During the various collecting periods, large males were sampled from populations at RP and PC3, whereas smaller average-sized males were collected at PC2 and PP. Large females were collected at PC1 and RP, however. Although the size distributions among populations were generally similar for the combined samples, larger individuals were captured at different times of year (Table 2). Mean sizes of

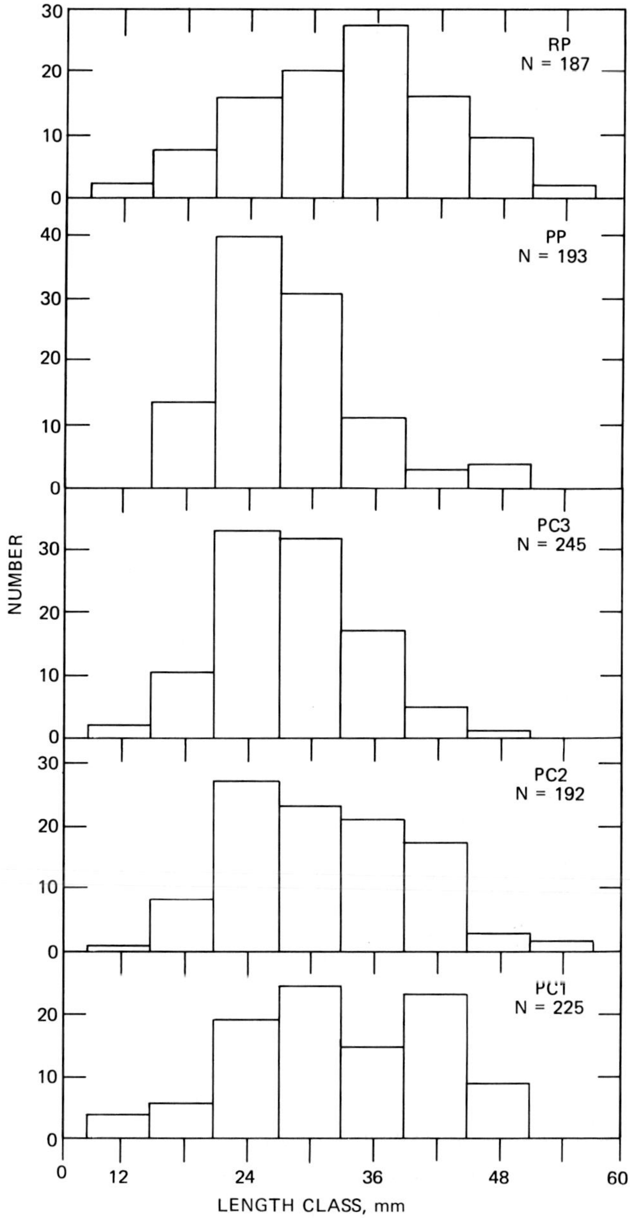

Fig. 1 Frequency distribution of *Gambusia affinis* collected at thermally affected (PC1, PC2, and PC3) and unaffected (RP and PP) stations. Collections from February, June, October, and December 1970 were combined.

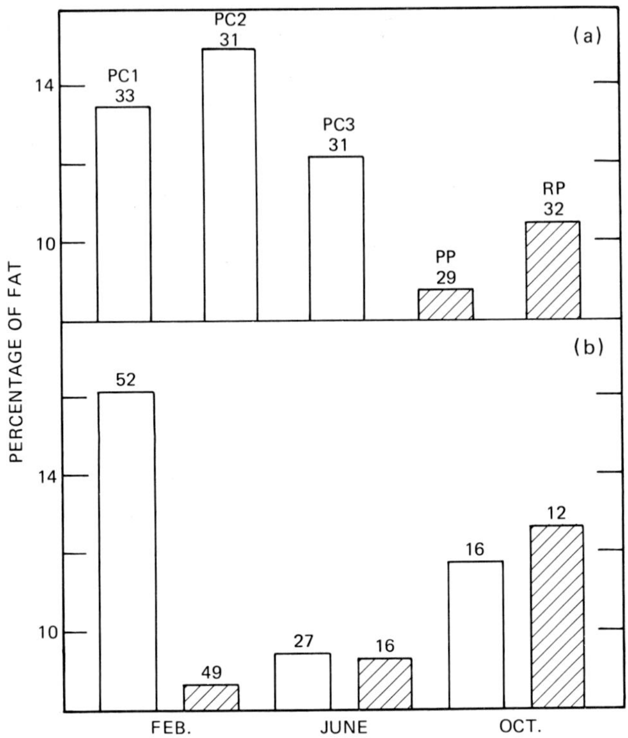

Fig. 2 Mean percentage of fat for *Gambusia affinis* collected from thermally affected areas (PC1, PC2, and PC3), open bars (□), and unaffected areas (PP and RP), striped bars (▨). (a) Samples from February, June, and October 1970 combined within stations; means for PC1, PC2, PC3, and RP were significantly higher than the mean for PP ($\alpha = 0.05$). (b) Mean values based on pooled samples from thermally affected and unaffected areas for February, June, and October; means from the thermally affected areas were significantly higher than those from unaffected areas ($\alpha = 0.05$) in February.

male and female mosquitofish at PC1 and PC2 were usually larger in February and December when periods of low temperatures occurred at these stations. On the average, male and female mosquitofish from the unaffected areas were larger in the summer and fall.

Mean percentage of body fat of fish from heated areas (PC1, PC2, and PC3) was significantly higher ($\alpha = 0.05$) than that of fish from unaffected station PP, whereas fish at RP had body fat reserves similar to those of fish from affected stations (Fig. 2a). Male and juvenile mosquitofish from thermally affected areas had a significantly higher ($\alpha = 0.05$) percentage of fat than those from unaffected

TABLE 3

MEAN NUMBERS OF EMBRYOS PER FEMALE *Gambusia affinis* FROM FIVE SAMPLE POPULATIONS*

Sample station	Sample time†			
	Feb.	June	Oct.	Dec.
PC1	11.8 (70)	5.0 (10)	13.8 (11)	27.9 (12)
PC2	9.2 (68)	12.2 (12)	23.2 (8)	27.4 (12)
PC3	5.6 (78)	8.8 (21)	10.0 (18)	0.0 (12)
PP	0.0 (45)	23.0 (21)	0.0 (11)	0.0 (11)
RP	12.0 (65)	24.5 (12)	6.0 (12)	0.0 (5)

*Stations PC1, PC2, and PC3 are in thermally affected areas; PP and RP are in areas at natural temperatures.

†Numbers in parentheses are numbers of females examined.

areas. Female mosquitofish from all stations, however, contained similar amounts of fat. Juvenile mosquitofish from thermally affected areas had the highest mean percentage of fat (14 to 18% dry weight).

Seasonal variation in percentage of fat was found in fish from thermally affected and unaffected areas (Fig. 2b). Percentages of fat were significantly higher ($\alpha = 0.05$) in fish from affected areas than in those from unaffected areas in February and were similar in June and October. Body fat content was lower in June and higher in October.

The period of reproductive activity was altered in Pond C as a result of the thermal effluent (Table 3). Embryos were found in female mosquitofish from PC1 and PC2 in February, June, October, and December. This indicates that fish were reproductively active throughout the year. Mean brood size generally increased at PC1 and PC2 as water temperatures decreased. Embryos were found in female mosquitofish at PC3 in all sampling periods except December. Several embryos in the early development stage were found in one female at Risher Pond in February, but the stage of development and the small proportion of gravid females indicated that the reproductive period had not commenced at the time the February sample was taken. Although water temperatures were suitable for reproduction in October (Table 1), few embryos were found in female mosquitofish from PP and RP (Table 3). The maximum number of embryos per female (61) was found in a female from Par Pond.

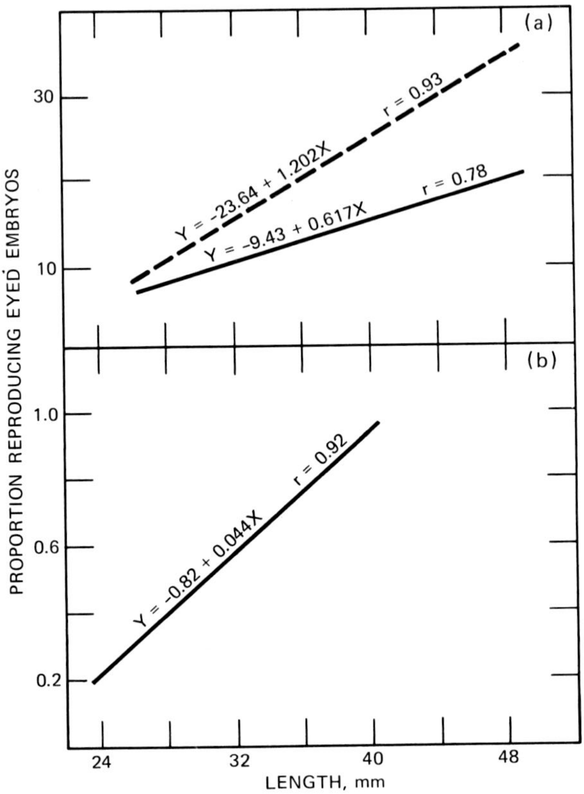

Fig. 3 (a) Relationship between number of eyed embryos and length of *Gambusia affinis*. ———, thermally affected samples. — — —, unaffected samples. (b) Proportion of females reproducing as a function of length. Curve in part b is based on data from thermally affected populations.

The relationship between number of eyed embryos and body length was not significantly different ($t = 1.74$, 103 df, $\alpha = 0.05$) in thermally affected and unaffected areas (Fig. 3a). These data indicate that female mosquitofish from both areas had similar numbers of eyed embryos per unit body length.

The proportion of females reproducing in thermally affected areas increased linearly ($Y = -0.82 + 0.04X$, $r = 0.92$) from 23 to 40 mm (Fig. 3b). An average of 65% of the females longer than 40 mm were reproductively active in the thermally affected areas. All females collected from unaffected areas in June were reproductively active.

Size at sexual maturity and sex ratios were generally similar among all populations of mosquitofish. Few females 25 mm in length contained embryos at PC2, PC3, and PP. One 23-mm female at PC1 was pregnant. The smallest reproductively active female from Risher Pond was 33 mm, but the smallest male was 24 mm. Males from other populations sampled were sexually developed at 21 mm. Sex ratios for the study period were also similar; female-to-male ratios were approximately 1.6 at PC2, PC3, RP, and PP, and the sex ratio at PC1 was 2.

DISCUSSION

Temperature is reportedly the primary exogenous factor regulating reproduction of mosquitofish (Medlen, 1951). Medlen indicated that the minimum critical temperature for reproduction is about 16°C. On the basis of our results, however, there does not appear to be an upper critical temperature threshold after which embryo formation is curtailed. We found that mosquitofish collected from water at 40°C were reproductively active, and Ferens and Murphy (1974) reported that mosquitofish collected from 32 to 44°C water were also reproductively active. Our data indicated, however, that mean number of embryos was generally inversely related to water temperature in areas subjected to extremely high water temperatures (PC1 and PC2) (Table 3).

Several investigators reported a linear relationship between the number of eyed embryos and body length (Krumholz, 1948; Ferens and Murphy, 1974). Our data support these studies; graphically, the relationship (Fig. 3a) suggests that females from thermally unaffected areas have a higher number of embryos per length than those from affected areas. The regression coefficients were not statistically different between affected and unaffected populations, however, because of the variability in number of embryos per female. Ferens and Murphy also found no significant difference in the number of eyed embryos as a function of length between thermally affected and unaffected mosquitofish populations.

Brood size of mosquitofish varies widely among populations (Krumholz, 1948). The highest mean number of embryos found in this study was 27.9 at PC1. Self (1940) found an average of 56.7 embryos per female in Oklahoma, and Krumholz reported a range in average brood size of 5.3 to 68.3 in Illinois. A mosquitofish population sampled in Tennessee averaged 33.2 embryos per female (Blaylock, 1969). We do not feel that differences in fecundity were directly related to temperature since other factors can also affect brood size in mosquitofish. Krumholz indicated that body size, time of year, previous brood size, and fertility can affect the number of young per brood. We feel that although brood size is comparable to that reported in the literature the number of generations per year is probably increased in mosquitofish from thermally affected areas. Discussion and data presented by Ferens and Murphy (1974) support this hypothesis.

Size at sexual maturity and sex ratios are parameters that vary greatly among populations. Krumholz (1948) found that females reproduced at 31 mm in one population and at 24 mm in another. The variation in size at sexual maturity (23 to 33 mm) that we found agrees with these data. The sex ratios found in this study (1.6 : 1 and 2 : 1) were also within the range reported by Krumholz. He found that females doubled males in "standing" populations, whereas new populations were generally 1 : 1. It is interesting to note that the lowest proportion of males was found at PC2, the station subjected to the highest temperatures (Table 1). Krumholz also found that males were generally less resistant to stressful conditions than females.

Percentages of body fat were generally similar for the populations studied, with the exception of the Par Pond samples, which had significantly lower amounts of body fat (Fig. 2a). The results of our analyses of body fat content were generally similar to those of Falke and Smith (1974), except that we found lesser amounts of body fat in females than in males. Falke and Smith reported higher percentages of fat in females. Since their fish were sampled in August, time differences may be responsible for the disparity.

Fish exhibit changes in body fat content generally associated with changes in metabolism and reproductive activity (Nikolsky, 1963). Peaks in fat content are usually found in late summer (Maki, 1967) or fall (Dahlberg, 1969) and are probably associated with high quality and quantity of food ingested and assimilated (Love, 1957) and with reduced channelization of energy into reproduction. Although fish are actively reproducing in Pond C and are subjected to abnormally high temperatures, which should increase metabolic demands, percentages of body fat were similar to those for fish from unaffected areas. One possible explanation is that elevated-temperature regimes increase productivity at lower trophic levels, which increases the available food supply. Thus fish are able to ingest and assimilate enough food to store fat even though temperatures are abnormally high and reproduction is generally occurring throughout the year. Our data support the conclusion of Falke and Smith (1974) that environmental temperature cannot account directly for differences in body composition.

Mosquitofish in Pond C may be living near their maximum thermal tolerance (Falke and Smith, 1974). On the basis of our observations, mosquitofish at PC1 and PC2 were inshore inhabitants when the reactor was operating. Fish had to be forced to swim to the deeper, warmer waters (45°C), and this resulted in nearly instantaneous loss of equilibrium and mortality. Shoreline dispersal was prevented by extreme temperatures. When temperatures decreased, however, bass and sunfish preyed on the mosquitofish.

Observations on temperature selection indicated that mosquitofish selected 39.5°C water when cooler waters were available for habitation. This temperature was 2°C higher than the reported upper tolerance limit for *Gambusia* acclimated to 20 and 35°C (Hart, 1952) and 12°C higher than temperatures selected by mosquitofish in a thermal gradient tank (Bacon, Neill, and Kilambi, 1967).

Our results show that *G. affinis* is a eurythermal species that can adapt to conditions of severe thermal stress. Mosquitofish from thermally affected areas were found to have similar population size structure, percentage of body fat, proportion of females reproducing, size at sexual maturity, sex ratios, and numbers of eyed embryos per unit length of female. Reproductive activity was increased in the thermally affected areas, however. Because of the eurythermal nature of *Gambusia*, management agencies should consider introducing the mosquitofish into reactor cooling ponds because of its economic importance as a predator on mosquito larvae and its biological value as a forage species.

ACKNOWLEDGMENTS

The research reported here was supported by contract EY-76C-09-0819 between the U.S. Department of Energy and the University of Georgia. Manuscript preparation was aided by the College of Forestry, Wildlife and Range Sciences, University of Idaho.

REFERENCES

Bacon, E. J., Jr., W. H. Neill, Jr., and R. V. Kilambi, 1967, Temperature Selection and Heat Resistance of the Mosquitofish, *Gambusia affinis*, *Proc., Annu. Conf. Southeast. Assoc. Game Fish Comm.*, 21: 411-415.

Blair, W. F., A. P. Blair, P. Brodkorb, F. R. Cagle, and G. A. Moore, 1968, *Vertebrates of the United States*, 2nd ed., pp. 117-118, McGraw-Hill Book Company, New York.

Blaylock, B. G., 1969, The Fecundity of a *Gambusia affinis affinis* Population Exposed to Chronic Environmental Radiation, *Radiat. Res.*, 37: 108-117.

Clugston, J. P., 1973, *The Effects of Heated Effluents from a Nuclear Reactor on Species Diversity, Abundance, Reproduction, and Movement of Fish*, Ph.D. Thesis, University of Georgia, Athens.

Dahlberg, M. D., 1969, Fat Cycles and Condition Factors of Two Species of Menhaden, Brevoortia (Clupeidae), and Natural Hybrids from the Indian River of Florida, *Am. Midl. Nat.*, 82(1): 117-126.

Falke, J. D., and M. H. Smith, 1974, Effects of Thermal Effluent on the Fat Content of Mosquitofish, in *Thermal Ecology*, AEC Symposium Series, Augusta, Ga., May 3—5, 1973, J. W. Gibbons and R. R. Sharitz (Eds.), pp. 100-108, CONF-730505, NTIS.

Ferens, M. C., and T. M. Murphy, Jr., 1974, Effects of Thermal Effluents on Populations of Mosquitofish, in *Thermal Ecology*, AEC Symposium Series, Augusta, Ga., May 3—5, 1973, J. W. Gibbons and R. R. Sharitz (Eds.), pp. 237-245, CONF-730505, NTIS.

Gibbons, J. W., R. R. Sharitz, F. G. Howell, and M. H. Smith, 1974, The Ecology of Artificially Heated Streams, Swamps, and Reservoirs on the Savannah River Plant, in *Environmental Effects of Cooling Systems at Nuclear Power Plants,* Symposium Proceedings, Oslo, 1974, STI/PUB/378, International Atomic Energy Agency, Vienna.

Hart, J. S., 1952, *Geographic Variations of Some Physiological and Morphological Characters in Certain Freshwater Fish,* University of Toronto Studies Biological Series No. 60, Ontario Fisheries Research Laboratory Publication 72.

Krumholz, L. A., 1948, Reproduction in the Western Mosquitofish *Gambusia affinis affinis* (Baird and Girard), and Its Use in Mosquitofish Control, *Ecol. Monogr.*, 18: 1-43.

Love, R. M., 1957, Biochemical Composition of Fishes, in *The Physiology of Fishes,* M. E. Brown (Ed.), Vol. 1, pp. 401-418, Academic Press, Inc., New York.

Maki, I., 1967, Population Studies of Honmoroko, *Gnathopogon caerulescens,* Pisces in Lake Biwa. Japan III. On the Seasonal Fluctuation of the Fat Content in the Fish Body of the Under-Yearling and the "Over-Wintering" of Them, *Jpn. J. Ecol.*, 17(5): 199-205.

Medlen, A. B., 1951, Preliminary Observations on the Effects of Temperature and Light upon Reproduction in *Gambusia affinis, Copeia,* 1951(2): 148-152.

Nikolsky, G. V., 1963, *The Ecology of Fishes,* Academic Press, Inc., New York.

Rosen, D. E., and R. M. Bailey, 1963, The Poeciliid Fishes (*Cyprinodontiformes*), Their Structure, Zoogeography, and Systematics, *Bull. Am. Mus. Nat. Hist.*, 126: 1-176.

Self, J. T., 1940, Notes on the Sex Cycle of *Gambusia affinis affinis,* and on Its Habits and Relation to Mosquitofish Control, *Am. Midl. Nat.*, 23(2): 393-398.

Sokal, R. R., and F. J. Rohlf, 1969, *Biometry,* W. H. Freeman and Company, San Francisco.

RESPONSE OF A MOBILE INVERTEBRATE TO HETEROTHERMAL CONDITIONS

S. J. LOZANO and J. F. KITCHELL
Laboratory of Limnology, University of Wisconsin, Madison, Wisconsin

ABSTRACT

Distribution and abundance of *Hyalella azteca,* a freshwater amphipod, were studied to determine responses to temperature of an organism that spends its entire life cycle in a spatially heterothermal system. The study site, a 190-ha power-plant cooling lake, has a stable 15°C horizontal temperature gradient throughout the year. Seasonal temperatures range from 4°C at the cool end of the lake in winter to 45°C at the warm end in summer. High water temperatures restricted the distribution and abundance of *H. azteca.* At other seasons, lower temperatures allowed wider distribution of amphipods in the lake. Reproduction and growth were also influenced by the temporal and spatial heterothermal conditions. Reproduction occurred only during short time intervals in spring and fall and differed from the continuous summer reproduction evident in surrounding waters. Field data on distribution, reproduction, and growth, plus laboratory studies, show evidence for a modified life history and temperature-directed movement in *H. azteca.*

Temperature is one of the most important environmental factors influencing the life history and distribution of aquatic invertebrates (Jensen et al., 1969; Hutchison, 1976). Growth rate, fecundity, and survivorship are temperature-dependent population features that change seasonally in natural environments. Changes in distribution are mediated through behavioral and/or population responses. Motile organisms are capable of moving to more-favorable habitats in response to changes in water temperature.

Cooper (1965) and Mathias (1971), analyzing the population and production biology of *Hyalella azteca,* a freshwater amphipod, in lakes with typical annual temperature cycles, found that growth, onset of reproduction, and fecundity in this organism are temperature dependent. Strong (1972; 1973) showed that there are inheritable differences among populations of *H. azteca* from localities with different thermal and predatory characteristics. In our study, we detected changes in distribution, growth, and abundance of *H. azteca* which were attributable to the thermal characteristics of a heterothermal system. This system has a temporal and spatial thermal gradient different from that of nearby natural waters.

STUDY SITE AND TEMPERATURE CHARACTERISTICS OF LAKE COLUMBIA

Lake Columbia, Columbia County, Wisconsin, is a 190-ha closed-cycle cooling lake owned by the Wisconsin Power and Light Company. The lake is contained by dikes, which were erected on a sedge meadow. A central dike separates the east and west arms of the lake (Fig. 1). Water depth is kept between 1.9 and 2.4 m, at an average of 2.1 m. Water losses from seepage and evaporation are made up from the nearby Wisconsin River; makeup water is pumped into a settling basin before it enters the main body of the lake.

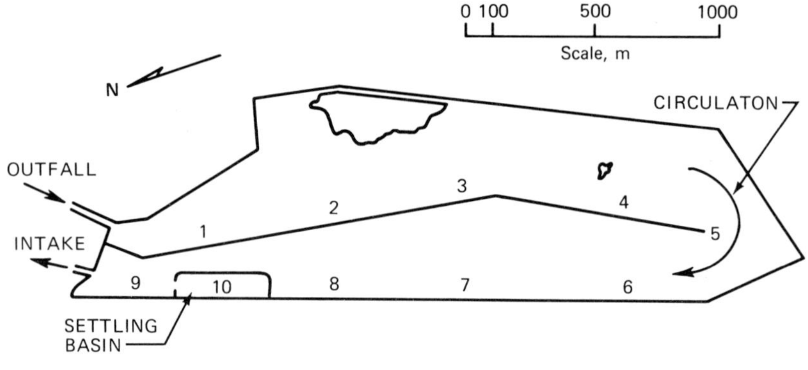

Fig. 1 Map of invertebrate sampling stations (1 to 10) on Lake Columbia, Wisconsin.

Cooling water is removed from the lake at the intake channel and pumped through the power-plant condensers, where the temperature is raised 15°C. Heated effluent flows from the outfall channel and circulates around the center dike back to the intake channel in approximately 5 days. Heat is dissipated to the atmosphere at an exponential rate from outfall to intake channel. Because of the lake's configuration and shallow depth, vertical mixing is easily accomplished with very little thermal and chemical stratification. Except for temperature, there are no consistent horizontal gradients of physical and chemical parameters (Andren et al., 1976). Seasonal and spatial variation in water temperature between outfall and intake are shown in Fig. 2. During periods of power-plant shutdown (October and June), homothermal conditions existed in Lake Columbia. Water temperatures measured in the settling basin were 5 to 6°C cooler than intake temperatures.

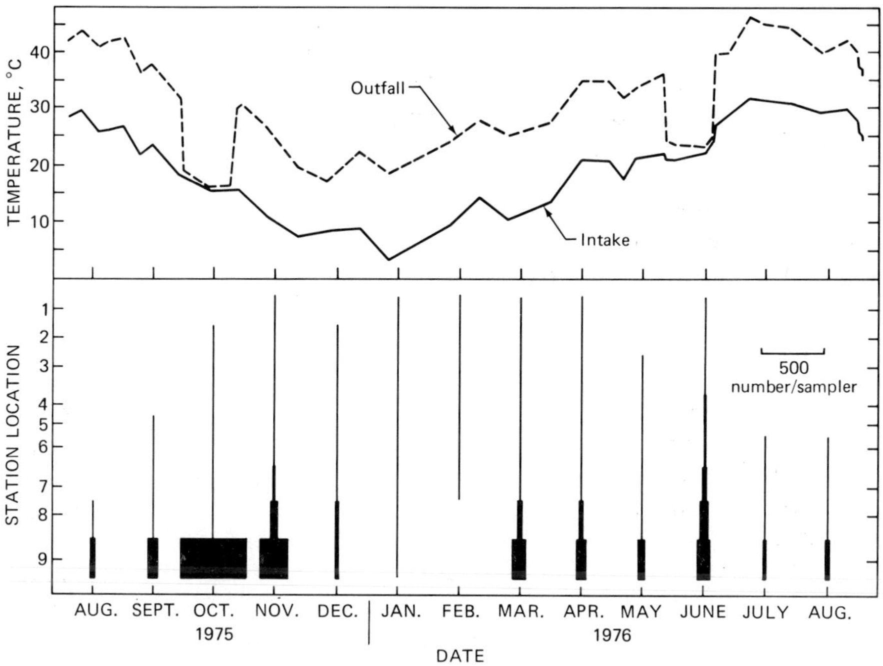

Fig. 2 **Seasonal changes in distribution and abundance of *H. azteca* compared with the annual temperature regime in Lake Columbia.**

METHODS AND MATERIALS

Field Techniques

Ten sampling sites were chosen at distances 300 to 700 m apart along the west and center dikes of Lake Columbia (Fig. 1). Three artificial substrate samplers, modified from Hester and Dendy (1962), were suspended 3 m offshore and 0.5 m below the surface at each sampling site. The modified Dendy sampler consisted of six squares (64 cm^2) of fiberboard alternating with five layers of conservation webbing. Samplers were presoaked before use and were suspended in the cooling lake for a 2-week sampling period each month. Preliminary test runs showed that numbers of *H. azteca* in samplers reach a constant level within 2 days.

Upon collection, samplers were placed in separate plastic bags filled with 90% ethyl alcohol (EtOH). Organisms were removed from the sampler, washed into a 0.28-mm Nitex mesh concentrator (Tetko, Inc.), and preserved in 70% EtOH. Amphipods were separated from other invertebrates, cleared in xylene, and mounted in Hoyer's solution on glass slides (Cooper, 1965).

Amphipod abundance was recorded as number of *Hyalella* per sampler. Head length was measured with a microprojector and digitizer (Hewlett–Packard Co., 9197A) along the dorsal margin from the posterior edge to the base of the first antennal segment. Antennal segment number was determined as the total number of segments from the first and second antennae (Cooper, 1965).

Laboratory Techniques

We conducted laboratory experiments in November 1976 to show the relative importance of density and temperature on the colonization of samplers by *H. azteca*. Amphipods collected from the field were acclimated for 1 week at 13, 21, and 23 ± 0.5°C in 60-liter aquariums. *Myriophyllum* sp. and *Cladophora* sp. were added for substrate and food. In one experiment 25, 50, and 100 amphipods were placed in separate 4-liter vessels maintained at their acclimation temperature. Small samplers (15 cm^2) were suspended in the vessels for 12 hr. In a second experiment, groups of 35 amphipods were placed in 4-liter vessels at their acclimation temperature, with six replicates for each temperature. The small samplers suspended in the vessels were removed after 3 hr. The number of amphipods per sampler was recorded for both experiments. We measured the size of the amphipods from the second experiment to test for differential colonization of substrate samplers by large amphipods. Amphipod head lengths from the second experiment were measured according to the methods described. Nonparametric statistical tests, the Mann–Whitney U test and Kruskal–Wallis H test, were used to determine whether differences existed between median scores (Gibbons, 1976).

RESULTS

Field Results

Seasonal distribution and abundance of *H. azteca*, presented in Fig. 2, are related to seasonal and spatial differences of water temperature measured in the outfall and intake channels. Only in July 1976 were amphipods found at sampling sites that had water temperatures above 33°C, which is the upper lethal temperature, as determined in laboratory studies (Sprague, 1963; Bovee, 1949). We found no amphipods at 50% of the sampling sites during the warmest months of the year. As temperatures declined, either from seasonal variations or power-plant shutdowns, amphipods began to colonize a greater proportion of the cooling lake. We concluded that amphipods moved from two sites nearest the intake to station 2, which is near the outfall. We assumed that any amphipods in water pumped over the power-plant condensers would die because of the large rise in temperature. Since entrainment can be discounted as a mechanism of dispersal, the population of amphipods in the cooling lake recolonized over 3000 m in 2 months.

Amphipod movement was found to be related to size; larger *H. azteca* moved more rapidly during the fall migration. There was a significant difference between median head lengths of animals collected at stations 1 to 5 and 6 to 9 in the fall of 1975 (Mann–Whitney U test, Table 1). Analyses of collections from October, a period of homothermal conditions, support the observa-

TABLE 1

MEDIAN HEAD LENGTHS OF AMPHIPODS, *H. azteca*, IN NEWLY COLONIZED, OUTFALL AREAS (STATIONS 1 TO 5) OF LAKE COLUMBIA AS COMPARED TO INTAKE AREAS (STATIONS 6 TO 9)*

	Median head length,† mm	
Date	Stations 1 to 5	Stations 6 to 9
September	0.53	0.45
October	0.55	0.47
November	0.58	0.50

*Intake water temperatures are equivalent to or slightly greater than ambient for nearby lakes, and temperatures near the outfall are 5 to 10°C above ambient.

†$P < 0.05$ for all size differences between the two groups of stations.

tion that differences in size are a result of differential movement of larger amphipods and are independent of temperature.

Growth is also temperature dependent. Amphipods grew larger per instar during the cooler months in fall (Table 2); in November each instar was at least 19% larger than the corresponding instar collected in August. Sizes of instars at different dates were found to be significantly different for all age groups (Kruskal—Wallis H test).

We compared the seasonal abundance of amphipods at station 9, near the intake, with that at station 10, in the settling basin and, thus, not influenced by the thermal effluent from power-plant operations (Fig. 3). Both populations of amphipods exhibited a bimodal shift in seasonal abundance, but with different seasonal maxima. Young instars (amphipods with 16 or fewer antennal segments) were found in the cooling lake in the fall (September through October) and spring (March through May) but were absent from samples collected during the warm summer months. In the settling basin young instars were present throughout the summer and early fall but were proportionately most abundant in June.

TABLE 2

SEASONAL SIZE DIFFERENCES OF *H. azteca* IN LAKE COLUMBIA FROM ALL STATIONS (1 TO 9)

	Median head length,* mm		
Instar	August (28°C)	September (22°C)	November (15°C)
V	0.35	0.37	0.43
VI	0.38	0.41	0.46
VII	0.43	0.49	0.51
Adult	0.50	0.61	0.67

*$P < 0.05$ for differences among all dates.

Laboratory Results

Amphipods rapidly colonized samplers in the laboratory during 12-hr tests (Table 3). Although neither temperature nor density influenced colonization rate during the 12-hr experiments, we did find a significant difference ($P < 0.05$) in colonization rate at two temperatures (13 and 21°C) in the 3-hr experiment (Mann—Whitney U test, Table 4). Amphipods that colonized the samplers were larger than those remaining outside the samplers ($P < 0.05$).

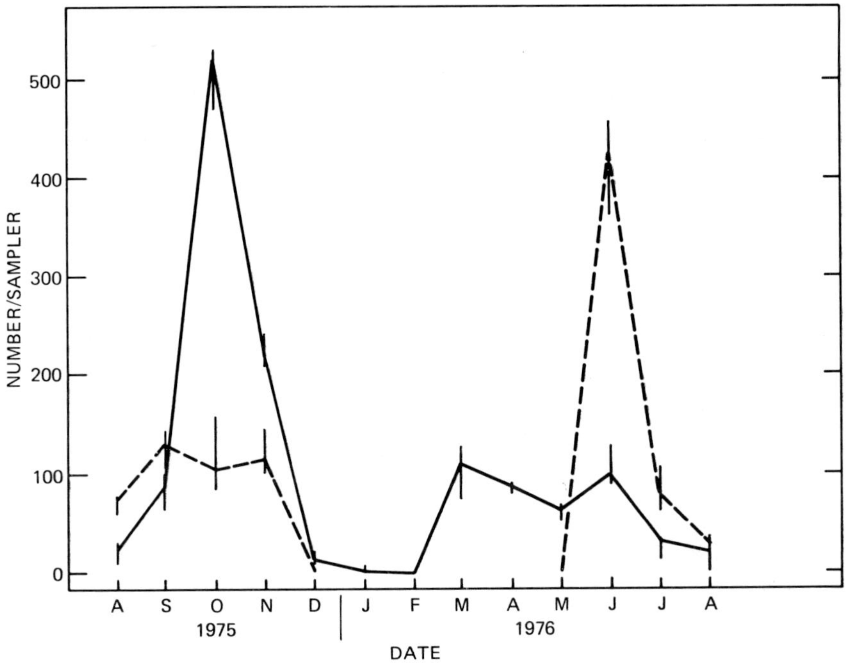

Fig. 3 Seasonal abundance (medians and ranges) of *H. azteca* collected from Lake Columbia (———), at station 9, and from the settling basin (— — —), at station 10. Ranges are shown as vertical bars.

TABLE 3

TEMPERATURE AND DENSITY EFFECTS ON COLONIZATION OF SAMPLERS BY *H. azteca* IN 12-hr LABORATORY EXPERIMENTS

Temperature, °C	Number of amphipods	% of amphipods in sampler after 12 hr*
13	25	96
	50	98
	100	97
23	25	100
	50	94
	100	98

*Percentage differences between the two temperatures were not significant.

TABLE 4

DIFFERENTIAL MOVEMENT OF *H. azteca* BY TEMPERATURE AND SIZE DURING 3-hr EXPERIMENTS*

Characteristic	13°C		21°C
Mean colonization rate, number/hour	2.22	$P < 0.05$ ↔	3.20
Mean head length of amphipods inside sampler, mm	0.49	NS ↔	0.50
	↑ $P < 0.05$ ↓		↑ $P < 0.05$ ↓
Mean head length of amphipods outside sampler, mm	0.44	NS ↔	0.46

*N = 35 amphipods per experiment; 6 replicates at each temperature.

DISCUSSION

In Lake Columbia the distribution and growth of *H. azteca* are restricted by the thermal characteristics of the cooling lake. High summer temperatures exclude amphipods from 50% or more of the lake; large numbers of amphipods were found only at the coolest sampling site from March until November. Although growth rate and duration of instar are accelerated by increased temperature (Cooper, 1965; Strong, 1973), length frequency distributions of samples show greater growth at each instar at cooler temperatures in Lake Columbia. This suggests that optimal temperatures for growth are exceeded during summer months in Lake Columbia.

There is also evidence from field and laboratory results that cool water temperatures influence the activity of *H. azteca*. The marked decline in amphipod abundance in the cooling lake in winter, when water temperatures were below 7°C, could be a result of decreased activity and could, therefore, result in slow colonization of samplers. Winter mortality for amphipods is low (Cooper, 1965) and can be discounted. Cool water temperatures were associated with lower colonization rates in laboratory experiments. The colonization rate at 13°C was still very high, however, with a median of 2.6 amphipods per hour. This suggests that virtually all colonization of samplers must cease at temperatures less than 7°C since samplers located near the intake, where temperatures were less than 10°C, yielded an average of one or less amphipod per station in January and February.

Another feature of *H. azteca*'s distributional response to temperature was the fall migration of amphipods into previously unavailable habitats. Within two months, amphipods were found in samplers 3 km away from their original site of dispersal. We observed that amphipods colonized samplers sequentially from the cool to the warm end of the lake; this observation suggests that migration is influenced by temperature. Kromrey et al. (1976) reached similar conclusions in a study of weekly changes in the distribution and abundance of *H. azteca* in Lake Columbia. Although *H. azteca* are known to make diel migrations from the littoral zone to areas offshore (Mullen, 1969), extensive migrations similar to those in Lake Columbia have not been reported in the literature. Laboratory experiments and field data gave corroborating evidence of size-dependent activity.

Size-selective predation by fish must be considered in interpreting data comparing amphipod sizes with their locations in the lake (Strong, 1973). Larger amphipods are presumably preferred prey. However, concomitant studies of fish distribution and abundance revealed that fish densities are greatest at stations where temperatures are higher (> 28°C) than those tolerated by amphipods (Lozano, Rondorf, and Kitchell, 1978). If predation were important, we would expect an inverse relationship between predator density and prey size. In fact, the reverse is true. The largest amphipods were proportionately more abundant in samples taken from stations where fish were most abundant. We are unable to determine whether selective predation on small amphipods was occurring, but the more probable explanation is simply that vagility is size dependent. Larger amphipods are probably more mobile and, hence, represent a greater proportion of the population of a newly colonized habitat (Table 1).

Seasonal changes in life-history data for amphipods collected in the settling basin (Fig. 3) were similar to those described by other investigators (Cooper, 1965; Mathias, 1971; Strong, 1973). Reproduction of amphipods in the settling basin began in late May or early June and continued through September. Population abundance increased very rapidly in June and then declined throughout the summer. High summer mortality is common for *H. azteca* (Cooper, 1965). There is also evidence of seasonal bimodality in abundance during the summer and fall. The high abundance of amphipods from September through November could have resulted from lower mortality rates during the cool months for amphipods born in late August and September.

Amphipod life-history features in the cooling lake differ from the typical annual pattern seen in aquatic invertebrate populations with summer reproductive periods. Amphipods in the cooling lake began reproducing during the spring and fall, 3 months earlier, and continued 2 months longer than did amphipods in the settling basin. Although reproduction of *H. azteca* continued year-round in a geyser hot spring with water temperatures between 12 and 40°C (Strong, 1972), reproduction in the cooling lake ceased at all stations when temperatures rose above 28°C. Using Dendy samplers on a monthly schedule may not be sufficient to measure low summer reproduction rates. Hargrave (1970) found a high correlation between algal production and density of *H. azteca*. Although it seems unlikely that food could be limiting in eutrophic Lake Columbia, food quality may be important. Benthic production measurements taken in the summer of 1975 show an abundance of periphyton and other invertebrates at all locations in Lake Columbia (Lozano, Rondorf, and Kitchell, 1978). Even though food was abundant, the high metabolic cost of maintenance during the warm summer months could have inhibited reproduction.

In conclusion, amphipods respond to the unique thermal conditions in Lake Columbia by altering their distribution through temperature-induced migrations. Reproduction and growth are inhibited by high temperatures in some regions of the lake and are generally reduced during the warm summer period.

ACKNOWLEDGMENTS

The research reported here was supported in part by research grants from the Office of Water Resources Technology (A-070-WIS) and the Environmental Protection Agency (R803971020).

We thank Anne Pilli and Ron Hall for advice and technical assistance and Betsy Colburn for reviewing early drafts of the manuscript.

REFERENCES

Andren, A., M. Anderson, N. Loux, and R. Talbot, 1976, Aquatic Chemistry, in *Documentation of Environmental Change Related to the Columbia Electrical Generating Station*, 9th Semi-Annual Report, IES Report 69, University of Wisconsin, Madison.

Bovee, E. C., 1949, Studies on the Thermal Death of *Hyalella azteca* (Saussure), *Biol. Bull.*, 96: 123-128.

Cooper, W. E., 1965, Dynamics and Production of a Natural Population of a Freshwater Amphipod, *Hyalella azteca*, *Ecol. Mongr.*, 35: 377-394.

Gibbons, J. D., 1976, *Nonparametric Methods for Quantitative Analysis*, Holt, Rinehart & Winston, Inc., New York.

Hargrave, B. T., 1970, Distribution, Growth, and Seasonal Abundance of *Hyalella azteca* (Amphipoda) in Relation to Sediment Microflora, *J. Fish. Res. Board Can.*, 27: 685-699.

Hester, F. E., and J. S. Dendy, 1962, A Multiple-Plate Sampler for Aquatic Macroinvertebrates, *Trans. Am. Fish. Soc.*, 91: 420-421.

Hutchison, V. H., 1976, Factors Influencing Thermal Tolerances of Individual Organisms, in *Thermal Ecology II*, ERDA Symposium Series, Augusta, Ga., Apr. 2—5, 1975, G. W. Esch and R. W. McFarlane (Eds.), pp. 10-26, CONF-750425, NTIS.

Jensen, L. D., R. M. Davies, A. S. Brooks, and C. D. Meyrs, 1969, *The Effects of Elevated Temperatures upon Aquatic Invertebrates*, RP-49, Report 4, Edison Electric Institute, New York.

Kromrey, G., J. Boxrucker, D. Mills, B. Morgan, E. Nebrija, and J. Young, 1976, The Distribution of Macroinvertebrates Along a Thermal Gradient in a Power Plant Cooling Lake, paper presented at the 24th Annual Meeting, North American Benthological Society, March 1976, La Crosse, Wisc.

Lozano, S. J., D. W. Rondorf, and J. F. Kitchell, 1978, *Assessment of a Cooling Lake Ecosystem*, Technical Report, Water Resource Center, University of Wisconsin, Madison.

Mathias, J. A., 1971, Energy Flow and Secondary Production of the Amphipods *Hyalella azteca* and *Crangonyx richmondensis occidentalis* in Marion Lake, British Columbia, *J. Fish. Res. Board Can.*, 28: 711-726.

Mullen, R. E., 1969, *An Investigation of the Diel Swimming Activity of* Hyalella azteca *(Saussure) in Lake Mendota, Wisconsin*, M.A. Thesis, University of Wisconsin, Madison.

Sprague, J. B., 1963, Resistance of Four Freshwater Crustaceans to Lethal High Temperatures and Low Oxygen, *J. Fish. Res. Board Can.*, 20: 387-415.

Strong, D. R., Jr., 1972, Life History Variations Among Populations of an Amphipod *(Hyalella azteca)*, *Ecology*, 53: 1103-1111.

——, 1973, Amphipod Amplexus, the Significance of Ecotypic Variation, *Ecology*, 54: 1383-1388.

TEMPERATURE SELECTION BY YOUNG TOPSMELT: LABORATORY AND FIELD INVESTIGATIONS

KARL F. EHRLICH,* G. E. McGOWEN, and G. MUSZYNSKI*
Biology Department, Occidental College, Los Angeles, California

ABSTRACT

Topsmelt *(Atherinops affinis)* were reared from eggs for 6.5 months. Their laboratory preferred-temperature range throughout this period (3 to 198 days post-hatching) was 19 to 23°C, and they avoided water temperatures warmer than 27°C. Selected temperatures were independent of acclimation temperatures, which were 10 to 21°C. Temperature did not act as a strong directing force for these fish (which have a wide thermal preference) until their lethal temperature was approached. Their behavior resulted in a frequency distribution significantly skewed toward the lower temperatures. Our ichthyoplankton survey found that the peak occurrence of young atherines was associated with their laboratory-determined preferred-temperature range.

Topsmelt [*Atherinops affinis* (Ayres)] occur from the Gulf of California to Vancouver Island, British Columbia, and are common in bays, sloughs, and kelp beds (Miller and Lea, 1972). These inshore areas are often heavily impacted by man; e.g., warm effluents from power plants are one source of environmental change. Little is known of the thermal ecology of topsmelt; however, Doudoroff (1938) observed a wide temperature preference (20 to 28°C) for a group of unidentified atherines and determined (Doudoroff, 1945) the 96-hr LT_{50} to be ~31°C for juvenile topsmelt. Embryonic development of topsmelt occurs from 12.8 to 26.8°C (Hubbs, 1965). Stephens (1977) reported that atherines (unidentified species) showed a positive orientation to warm water in King Harbor,

*Present address: Lockheed Center for Marine Research, Carlsbad, California.

Redondo Beach, Calif. This harbor is a receiving body for thermal effluent and contains many horizontal and vertical temperature gradients.

Nearshore areas are often nursery grounds for many fishes. Alteration of the physical environment could result in these areas' being less desirable or suitable to many species and, thereby, in reduced recruitment. With this in mind, we initiated our investigation, combining a laboratory investigation with field surveys to provide further understanding of topsmelt's behavioral responses to temperature.

MATERIALS AND METHODS

Laboratory Procedures

We collected ripe adult topsmelt *(Atherinops affinis)* in June 1975 with gill nets placed in King Harbor, Redondo Beach, Calif. Eggs were stripped, fertilized, and incubated according to a modified version of the methods of Ehrlich and Farris (1971) for grunion *(Leuresthes tenuis).* The incubating eggs remained totally submerged, and gravel was not used since topsmelt do not spawn in sand as do grunion. The incubation temperature was 18°C, and that for rearing was 21°C. We cultured the fish in round, blue, 100-liter polyethylene containers through which water flowed at a rate of 1.5 liters/min from an open seawater system. The salinity was generally in the range of 32 to 34‰, with rare maximum excursions of 1.5‰. Young larvae were fed brine shrimp *(Artemia salina)* nauplii supplemented with natural plankton seived through a 750-μm screen and collected on a 20-μm seive. As the fish grew, we increased the size of the brine shrimp fed to them, until adult *Artemia* was the sole food source. The brine shrimp were fed cultured green algae (*Brachyamonas* sp.) so that they would be of a more uniform nutritive value to the fish.

The temperature preference of topsmelt larvae was tested 3, 11, and 18 days post-hatching. We also tested juvenile fish 120 days post-hatching and, at this time, transferred three groups of 30 individuals each to new containers for acclimation to additional temperatures of 10, 15, and 20°C. The temperature was changed at the rate of 1°C/day. The fish were held approximately 60 days at the acclimation temperatures before their thermal preference was tested again.

We used the two horizontal temperature gradients and the methodology of Ehrlich et al. (1978) to delineate behavioral responses. The gradients allowed experimentation with groups and permitted these pelagic fish to swim freely. This apparatus combined some of the best attributes of spatial (e.g., Norris, 1963) and temporal (e.g., Beitinger, 1976) gradients. Larvae up to 18 days post-hatching were tested in a 1.5-m-long round-bottom gradient 5 cm in diameter and filled to a depth of 1.5 cm. This apparatus contains 11 evenly spaced thermistor probes and operates on a countercurrent principle, with hot and cold water entering opposite ends of the outer temperature-controlling trough and the inner experimental one.

Potentially harmful changes in water quality were avoided by using only six larvae at once and having aerated seawater flowing slowly through the gradient. Seawater velocities of 0.1 to 0.9 mm/sec were selected on the basis of known larval swimming speeds (Blaxter, 1969; Rosenthal and Hempel, 1970; Hunter, 1972). Velocities in this range did not present the larvae with any significant difficulty in maintaining a chosen position. Wilson (1974), studying behavioral responses to pollutants, successfully used velocities of up to 10.8 mm/sec with pelagic marine fish larvae. Our observations in the present study showed that the flow rates were low enough to avoid any potential rheotactic interference.

The larger gradient (which is 300 cm long by 15 cm wide by 10 cm deep) is static and is controlled by heating and cooling the opposite ends. Surface and bottom baffles, in conjunction with gentle aeration between them, disrupt convection currents. This aeration also maintains saturation levels of oxygen and helps to remove any supersaturated gases, which Gift (1977) reported as potential problems in temperature gradients. Eleven thermistor probes are evenly spaced along the test chamber.

Both experimental chambers are enclosed in lightproof cabinets with viewing ports. Observations were made from a darkened room. A temperature differential of at least 10°C was established before data were collected for determining thermal preference.

Low levels of lighting, based on minimum intensities for larval feeding (10 to 15 ft-c) and schooling (60 to 70 ft-c) (Blaxter, 1970), were used during the experiments on larvae and juveniles, respectively. These levels of illumination did not appear to disturb the fish as brighter light sometimes did. Additionally, Sullivan and Fisher (1954) reported that temperature selection was more precise at low light intensities. Natural day lengths were used during preexperimental holding, as well as during testing. Hasler (1956) reported that fishes could align themselves with small deformities in a tank's structure. Isotherms were shifted during an experiment, and hot and cold ends were reversed between experiments in an attempt to partition position from temperature preferences.

Topsmelt are gregarious, and we tested them in groups of six to nine fish to assess the effects of temperature on populations rather than on unnaturally isolated individuals. The fish were placed in the experimental apparatus when the water was uniformly at their acclimation temperature. A gradient was established about the fish after they were habituated to the new surroundings. Larvae were observed every 5 min for 4 to 5 hr and juveniles every 15 min for 6 to 8 hr or until a constant response was attained. Ehrlich et al. (1978) suggested that the time required to reach the final preferendum is related to the difference between it and the acclimation temperature. In all 16 species of temperate marine fishes tested, the final preferendum was attained within the experimental times used (Ehrlich et al., 1978).

The number of fish observations was defined as the total number of observations of the experimental group times the number of fish in the group. The frequency of occurrence of all experimental temperatures was not uniform because of the shifting of the gradient. This caused a bias in the number of fish observed at a particular temperature when numbers were summed over an entire experiment. To compensate for this, we adjusted the data before calculating the mean and modal selected temperatures by using the number of fish per total occurrence of a particular temperature rather than the actual number of fish at each temperature (see Ehrlich et al., 1978).

Field Investigation

Plankton samples were collected at the southern end of San Diego Bay, San Diego, Calif., on approximately a 3-week basis from June 1972 through June 1973, to relate the distribution of atherine larvae to field temperatures. A 0.5-m-diameter plankton net with 0.505-mm mesh openings and a centrally located flowmeter was used for collections, which were made during the day. The tows were made just below the surface for 2 min at a speed of ~4 knots. Specimens were preserved in 4% Formalin.

RESULTS AND DISCUSSION

Juvenile topsmelt (*Atherinops affinis*) swam freely along the length of the experimental chambers before the temperature gradient was established but showed some attraction to both ends. A total of 53 fish were observed during seven experiments; they were distributed along the length of the test chamber in the following pattern: 21 on the left, 12 in the center, and 20 on the right. The water at these times was at the fish's acclimation temperatures. A similar attraction for tank ends was noted by Doudoroff (1938) for atherines of undesignated species. As soon as a thermal gradient was established, the fish moved toward the warm end of the tank. They were then distributed in the following pattern: 1 in the cold portion, 10 in the middle, and 42 in the hot portion of the gradient. The differences in fish distribution before and after establishment of a temperature differential were observed just before the change in temperatures began and 15 min later.

Although the fish moved freely throughout the experimental chamber when it was at a uniform temperature, after the formation of a thermal gradient, the topsmelt became positioned in the compartment they initially selected; this was generally in the temperature range of 19 to 23°C. Shifting isotherm positions did not cause the fish to move until they contacted temperatures of 26 to 27°C, at which time they started tracking this temperature range and avoiding warmer waters. This behavioral pattern was shown by 18-day-old larvae and by all juveniles tested (Figs. 1 and 2). Younger fish showed less discrimination with respect to thermal avoidance (Fig. 1). Doudoroff (1945) found juvenile topsmelt to be eurythermal (96-hr viable temperature range, ~10 to 31°C). Hubbs (1965) found a slightly narrower range during the embryonic stages. It may be that, for species with a wide thermal tolerance, as Brett (1956) suggested for salmonids, temperature is not a strong directing force until the fish's lethal range is approached. The upper viable limit for hatching of topsmelt eggs is 26.8°C (Hubbs, 1965). Assessment of temperatures allowing development may give a more sensitive indication of the viable range than simply placement of juveniles in water baths at different temperatures for 4 days. Apparently, in the terminology of Reynolds (1977), temperature for this species should be considered a proximate rather than an ultimate factor in habitat selection. Topsmelt show a stronger avoidance than preference response. Beitinger (1977) showed that little variation in avoidance of water near lethal temperatures was tolerated by bluegill sunfish *(Lepomis macrochirus).* The relationship between selected temperature and time in the gradient is shown in Figs. 1 and 2 for fish at 11, 18, and 192 days post-hatching. The two younger groups were larvae. Comparison of the results from these two groups shows that by 18 days the larvae started to show the behavioral patterns of juveniles, i.e., avoidance of water warmer than 27°C and aggregation of individuals within the gradient. The tendency of older larvae and

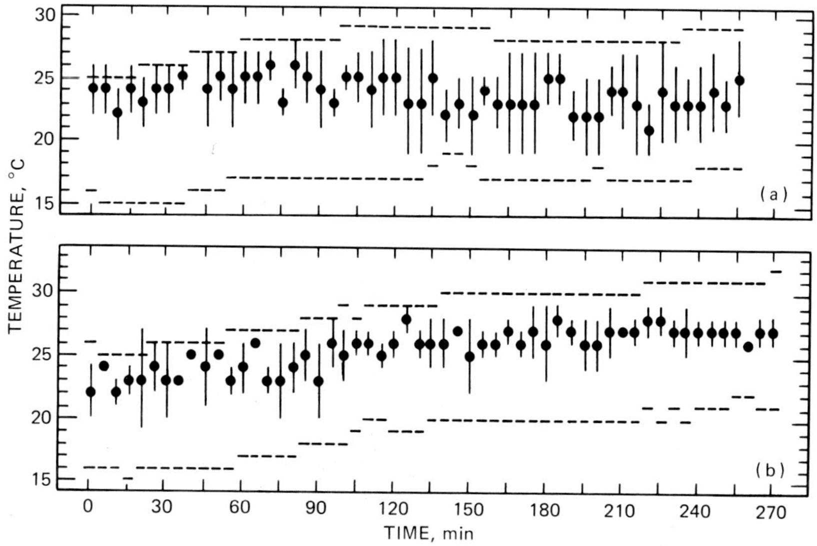

Fig. 1 Changes in temperature selected by (a) 11- and (b) 18-day-old larvae during an experiment. There were 312 fish observations at 11 days and 330 at 18 days. Solid circles represent the mean selected temperature at each observation; vertical lines show one standard deviation above and below the mean selected temperature; and the horizontal bars demarcate the experimental gradient.

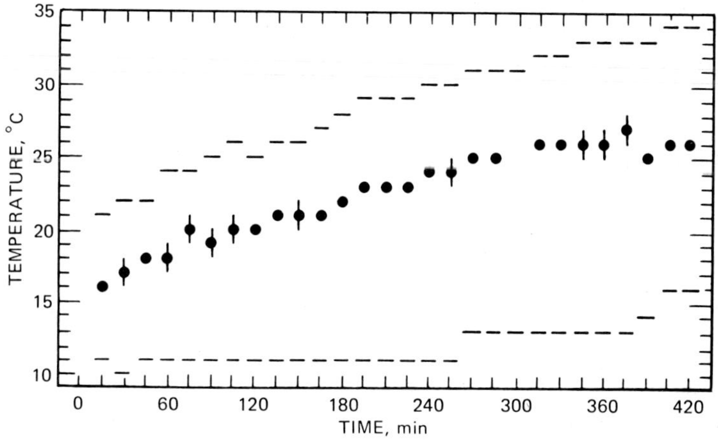

Fig. 2 Changes in temperature selected by 192-day-old juveniles during an experiment. There were 252 fish observations. (See Fig. 1 for explanation of symbols.)

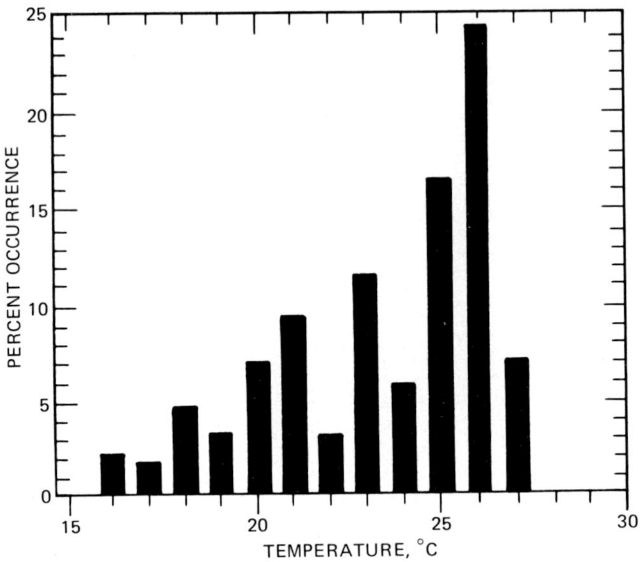

Fig. 3 Temperature-specific occurrences. The skewed frequency distribution of 192-day-old fish is shown.

juveniles to aggregate and school is shown by the smaller variation about the temperature selected at each observation than was demonstrated by 11-day-old and younger larvae. The older fish also showed greater precision of temperature selection despite their increasing activity with age.

The habit of topsmelt of remaining in a given position independent of temperature until conditions became extreme presents difficulties with respect to data interpretation. This response, in conjunction with the experimental technique of moving isotherms, gives little value to mean and modal preferred temperatures since these are a function of the rate of shifting of the gradient. Furthermore, once the fish started following an isotherm of 26 to 27°C, these temperatures became heavily weighted with respect to fish occurrence; this resulted in a skewed frequency distribution ($g_1 = -0.72$, $P < 0.005$) (Fig. 3). Apparently, selection of temperatures near lethal limits produces skewed frequency distributions, which vary with the duration of the experiment. Because of this, the common practice of reporting mean and modal preferred temperatures may not be adequate. Various investigators have reported that the final preferendum is often near the upper lethal limit (Lowe and Heath, 1969; Reynolds and Thomson, 1974; Reynolds and Casterlin, 1976). Beitinger (1977) stated that bluegill *(Lepomis macrochirus)* show sharper avoidance of water near their upper than near their lower lethal temperature. This more precise avoidance of undesirably warm water is probably caused by their higher level of activity in warm temperatures.

The temperatures selected by topsmelt 15 min after the formation of a temperature gradient may provide a more meaningful insight into their thermal preference because of their tendency to become positioned within the gradient. In nine out of ten experiments, this initial selected temperature (IST) was in the range of 19.2 to 23.1°C (Table 1) and was independent of acclimation temperatures ($r = 0.28$, $n = 10$, $P > 0.05$). Additional evidence that the IST is the best indicator of the topsmelt's thermal preference was obtained from the field investigation.

Atherine larvae were collected in every sampling period from January through October, with the highest densities obtained from March through June, but the fish were not identified below the family level. It appears, however, that this group of specimens consisted of topsmelt and grunion *(Leuresthes tenuis)* larvae and possibly a few individuals of the third atherine species, jacksmelt *(Atherinopsis californiensis).* Topsmelt and grunion have been found to occur together during their larval and juvenile stages (Ehrlich and McGowen, unpublished data). These determinations were based on the field collection and subsequent rearing of larvae to a size when they could be speciated. The temperature-specific frequency distributions of the larval atherines are shown in Table 2. Temperature occurrences for the fish density calculations were considered only for samples containing atherines. Although the range of 22 to 23°C occurred only 15.4% of the time, 40.3% of the fish were collected at these temperatures (Table 2). Similarly, 19 to 23°C occurred 40.4% of the time, but 59.3% of the total fish population sampled was found at these temperatures. Furthermore, temperatures 27°C and higher accounted for 23.1% of all occurrences, but only 9% of the larvae were found in this range. This indicates a strong avoidance response of these temperatures. Ehrlich (1977) demonstrated that the laboratory-determined preferred-temperature range of juvenile grunion was 22 to 23°C; this was also the temperature range that maximized larval yolk utilization efficiency. The secondary mode at 16°C (Table 2) may indicate that larvae were inadequately feeding; Ehrlich (1977) showed that even 1 day of food deprivation results in a decrease in the preferred temperatures of larval grunion and many other species. Thus the laboratory and field data corroborate each other and demonstrate that topsmelt have a eurythermal preference; they showed little discrimination in the range from 19 to 23°C, and

TABLE 1

THERMAL PREFERENCE OF LARVAL JUVENILE TOPSMELT IN A HORIZONTAL TEMPERATURE GRADIENT

Date	Age, days post-hatching	Number of fish tested	Number of fish observations	Mean standard length, mm	Acclimation temperature, °C	Selected temperature, °C			
						Mean	Standard deviation	Modal	Initial
7/16/75	3	6	372	7.3	21.1	21.2	4.4	21	22.4
7/24/75	11	6	312	11.0	21.1	23.4	3.2	22	23.1
7/31/75	18	6	330	14.4	21.1	25.2	2.9	27	21.9
11/03/75	120	9	243	51	17.7	25.8	2.6	27	19.8
1/12/76	190	9	315	67	20.2	21.6	3.1	19	20.2
1/13/76	191	9	180	67	20.2	17.9	2.7	18	21.1
1/14/76	192	9	252	60	15.0	23.3	3.2	26	16.4
1/15/76	193	9	306	60	15.0	21.0	1.0	21	21.0
1/19/76	197	8	184	55	10.0	24.2	1.9	25	19.2
1/20/76	198	7	245	55	10.0	25.0	1.2	24	23.0

TABLE 2

OCCURRENCE OF LARVAL ATHERINE FISHES IN THE FIELD AT DIFFERENT TEMPERATURES DURING A 1-YEAR SAMPLING PERIOD

Temperature		Abundance	
°C	Occurrence	Mean number/50 m^3	%
13	5	0.8	0.7
14	0		
15	3	2.3	2.0
16	3	12.0	10.7
17	1	1.0	0.9
18	0		
19	5	7.3	6.5
20	5	2.9	2.6
21	3	11.2	9.9
22	4	22.6	20.1
23	4	22.7	20.2
24	3	2.7	2.4
25	3	11.6	10.3
26	1	5.4	4.8
27	2	0.4	0.4
28	1	1.2	1.1
29	1	4.6	4.1
30	1	0.3	0.3
31	1	1.6	1.4
32	1	0.7	0.6
33	4	0.7	0.6
34	1	0.6	0.5

possibly to 26°C, but they generally avoided water warmer than 27°C.

The temperatures preferred by topsmelt, particularly the higher temperatures found in San Diego Bay, are uncommon or do not occur at all throughout most of the geographical range of this fish. The question then arises, Of what value is a preferred temperature that does not naturally occur? It was suggested by Brett (1971) that the final temperature preferendum may be synonymous with the physiological optimum. If fish prefer water warmer than they will ever naturally encounter, does this imply that they have evolved to function below their physiological optimum? There are several potential advantages of selecting a wide range of warm temperatures, such as 19 to 26°C, which will act as a mechanism to direct the fish inshore. First, the warmer nearshore water will reduce the duration of the highly vulnerable larval stages. Second, movement inshore may direct the fish to areas of reduced predation. Norris (1963) showed that preference of warm water by young opaleye *(Girella nigricans)* directs them from offshore into intertidal areas. Third, food levels may be higher inshore or in bays and harbors. In King Harbor many atherine larvae are found in the back basins that are warmest and contain the highest concentration of food organisms (McGowen, unpublished data). Beitinger (1977) suggested that seeking the warmest water available, below temperatures that are detrimental, may be beneficial to fishes.

The thermal behavioral responses of topsmelt suggest that they may be attracted to warmed water, such as power-plant effluents, as has been observed by Stephens (1977). This does not imply, however, that these conditions are necessarily beneficial to them. Before any conclusion can be reached, such factors as the fish's physiological thermal optimum, susceptibility to predation and disease, and ability to compete must also be considered.

REFERENCES

Beitinger, T. L., 1976, Behavioral Thermoregulation by Bluegill Exposed to Various Rates of Temperature Change, in *Thermal Ecology II*, ERDA Symposium Series, Augusta, Ga., Apr. 2—5, 1975, G. W. Esch and R. W. McFarlane (Eds.), pp. 176-179, CONF-750425, NTIS.

——, 1977, Thermopreference Behavior of Bluegill *(Lepomis macrochirus)* Subjected to Restrictions in Available Temperature Range, *Copeia*, 1977: 536-541.

Blaxter, J. H. S., 1969, Development: Eggs and Larvae, in *Fish Physiology*, W. S. Hoar and D. J. Randall (Eds.), Vol. 3, pp. 177-252, Academic Press, Inc., New York.

——, 1970, Light-Fishes, in *Marine Ecology*, O. Kinne (Ed.), Vol. I, pp. 213-285, John Wiley & Sons, Inc., New York.

Brett, J. R., 1956, Some Principles in the Thermal Requirements of Fishes, *Q. Rev. Biol.*, 31: 75-87.

——, 1971, Energetic Responses of Salmon to Temperature. A Study of Some Thermal Relations in the Physiology and Freshwater Ecology of Sockeye Salmon *(Oncorhynchus nerka)*, *Am. Zool.*, 11: 99-113.

Doudoroff, P., 1938, Reactions of Marine Fishes to Temperature Gradients, *Biol. Bull.*, 75: 494-509.

——, 1945, The Resistance and Acclimation of Marine Fishes to Temperature Changes. II. Experiments with *Fundulus* and *Atherinops*, *Biol. Bull.*, 88(2): 194-206.

Ehrlich, K. F., 1977, Laboratory Thermal Effects Investigations on Marine Fishes, in *Effects of Thermal Effluent from Southern California Edison's Redondo Beach Steam Generating Plant on the Warm Temperate Fish Fauna of King Harbor Marina*, Annual Report for Phase II, Mar. 1, 1975, to Feb. 29, 1976, Southern California Edison, Redondo Beach.

——, and D. A. Farris, 1971, Some Influences of Temperature on the Development of the Grunion *Leuresthes tenuis* (Ayres), *Calif. Fish Game*, 57: 58-68.

——, J. M. Hood, G. Muszynski, and G. E. McGowen, 1978, Thermal Behavioral Responses of Selected California Littoral Fishes, *Fish. Bull.*, 76: in press.

Gift, J. J., 1977, Application of Temperature Preference Studies to Environmental Impact Assessment, *J. Fish. Res. Board Can.*, 34: 746-749.

Hasler, A. D., 1956, Influence of Environment Reference Points on Learned Orientation in Fish *(Phoxinus)*, *Z. Vgl. Physiol.*, 38: 303-310.

Hubbs, C., 1965, Developmental Temperature Tolerances and Rates of Four Southern California Fishes, *Fundulus parvipinnis*, *Atherinops affinis*, *Leuresthes tenuis*, and *Hypsoblennius* sp., *Calif. Fish Game*, 51: 113-122.

Hunter, J. R., 1972, Swimming and Feeding Behavior of Larval Anchovy, *Engraulis mordax*, *Fish. Bull.*, 70: 821-838.

Lowe, C. H., and W. G. Heath, 1969, Behavioral and Physiological Responses to Temperature in the Desert Pupfish *Cyprinodon macularis*, *Physiol. Zool.*, 42: 53-59.

Miller, D. J., and R. N. Lea, 1972, *Guide to the Coastal Marine Fishes of California*, Fisheries Bulletin 157, California Department of Fish and Game, Sacramento.

Norris, K. S., 1963, The Functions of Temperature in the Ecology of the Pericoid Fish *Girella nigricans* (Ayres), *Ecol. Monogr.*, 33: 23-62.

Reynolds, W. W., 1977, Temperature as a Proximate Factor in Orientation Behavior, *J. Fish. Res. Board Can.*, 34: 734-739.

——, and M. E. Casterlin, 1976, Thermal Preferenda and Behavioral Thermoregulation in Three Centrachid Fishes, in *Thermal Ecology II*, ERDA Symposium Series, Augusta, Ga., Apr. 2—5, 1975, G. W. Esch and R. W. McFarlane (Eds.), pp. 185-190, CONF-750425, NTIS.

——, and D. A. Thomson, 1974, Responses of Young Gulf Grunion, *Leuresthes sardina*, to Gradients of Temperature, Light, Turbulence and Oxygen, *Copeia*, 1974: 747-758.

Rosenthal, H., and G. Hempel, 1970, Experimental Studies in Feeding and Food Requirements of Herring Larvae (*Clupea harengus* L.), in *Marine Food Chains*, J. H. Steele (Ed.), pp. 344-364, Oliver and Boyd, Edinburgh.

Stephens, J. S., 1977, Field Studies, in *Effects of Thermal Effluent from Southern California Edison's Redondo Beach Steam Generating Plant on the Warm Temperate Fish Fauna of King Harbor Marina*, Annual Report for Phase II, Mar. 1, 1975, to Feb. 29, 1976, Southern California Edison, Redondo Beach.

Sullivan, C. M., and K. C. Fisher, 1954, The Effects of Light on Temperature Selection in Speckled Trout *(Salvelinus fontinalis)*, *Biol. Bull.*, 107: 278-288.

Wilson, K. W., 1974, The Ability of Herring and Plaice Larvae to Avoid Concentrations of Oil Dispersants, in *The Early Life History of Fish*, J. H. S. Blaxter (Ed.), pp. 589-602, Springer-Verlag, New York.

MOVEMENT OF THREE SPECIES OF FISHES PAST A THERMALLY INFLUENCED AREA IN THE COOSA RIVER, ALABAMA

JERRY L. MOSS,* SITDHI BOONYARATPALIN, and WILLIAM L. SHELTON
Alabama Cooperative Fishery Research Unit, Auburn University,
Auburn, Alabama

ABSTRACT

Movement of ultrasonic-tagged largemouth bass and flathead and channel catfish passing the thermally influenced area of the Coosa River, Alabama, adjacent to the Ernest C. Gaston Steam Plant was studied during the warm part of the year. The area of thermal influence changed with the seasons, and movement of largemouth bass and catfish past the thermal zone was affected but was not prevented. Movement patterns suggested that the fishes avoided passing directly through the area of maximum temperature influence by traveling under or around it; thus the heated water did not form an impassable barrier.

Concern has increased in recent years regarding the effects of heated discharges on the aquatic environment. The release of heated water into a riverine habitat is a potential barrier to the passage of fishes. Although many changes unquestionably occur when heated effluents are released into an aquatic system, there is considerable disagreement as to whether these changes are harmful or beneficial (Alabaster, 1962; Cairns, 1969; Davies, 1966; Langford, 1971; Nakatani, 1969; Proffit, 1969). Inasmuch as most power generating stations must dispose of 50 to 65% of the energy they produce as heat (Federal Power Commission, 1969, cited in Langford, 1972), the disposal of heated water is expected to become an increasingly serious burden on our aquatic resources.

Since recent technological advancements in biotelemetry have been especially useful in studies of fish movement in relation to

*Present address: Game and Fish Division, Montgomery, Alabama.

thermal effluents (Coutant, 1969; Merriman, 1970; Ross, 1970; Kelso, 1974; Wrenn, 1976; Leggett, 1976) we chose this approach for our study. Major objections to sonic tagging include the possible alteration of normal behavior, the small number of individuals that can be studied because of overlap in signals, and the expense of the transmitters.

The objective of our study was not to describe normal movement, however, but to determine whether the cooling-water discharge from a coal-powered steam generating plant on the Coosa River, Alabama, was an impassable barrier during warm months when thermal influence was maximum. Sonic tagging was meant to demonstrate only the feasibility of passage through a thermal effluent and not the percentage of a species which might do so.

MATERIALS AND METHODS

Three species of catfish and the largemouth bass were studied. All bass were collected by electrofishing within the study area. Catfish came from various sources. Two channel catfish (*Ictalurus punctatus*) were seined from ponds at the Auburn University Fisheries Research Unit. Two of the three flathead catfish (*Pylodictis olivaris*) were captured in gill nets in the study area, and one flathead and two blue catfish (*I. furcatus*) were caught in Lay Reservoir about 13 km downstream from the study area. We felt justified in using nonresident fishes because a description of normal movement was not the objective.

Several techniques have been used successfully to attach transmitters to fishes. We used the surgical implantation techniques described by Hart and Summerfelt (1975). The fishes were anesthetized, and a 3-cm medial incision was made along the abdomen parallel to the ventral midline and anterior to the pelvic girdle. We inserted the tag in a lateral position and closed the incision with 00 Ethicon silk suture (Johnson & Johnson), using 5 to 8 interrupted surgeons' stitches. The closed incision was painted with a solution of malachite green to inhibit fungal growth. The average time necessary to complete the surgical procedure was 9 min. Each fish was retained for a 24-hr postoperative period to monitor recovery and permit the fish to become adjusted to the mass of the transmitter.

The tracking vessel was a 5-m flathull aluminum boat with a 40-hp outboard motor. A Smith—Root, Inc., TA-60 receiver and SR-70H hydrophone were used in tracking. The hydrophone had an 8° cone of reception and could be rotated through 360°. Two types of transmitters were used. The Smith—Root, Inc., SR-69-A had a maximum advertised transmitting time of 8 weeks, a range of 2 to 5 km, and different pulse rates to permit identification of individual fish on a common carrier frequency of 74 kHz. The transmitter weighed 49 g in air and 30 g in water.

Bayshore Systems T-3T series tags incorporate a thermistor in which the pulse rate varies with the water temperature. An analog temperature decoder was used in conjunction with the receiver to display the pulse rate, which was converted to temperature on a calibration curve. The T-3T transmitter had a maximum advertised transmission time of 6 to 8 weeks and a transmission range of about 825 m. It weighed 24 g in air and 15 g in water. Different transmitting frequencies were used to identify individuals.

STUDY AREA

The Gaston Steam Plant, operated by Alabama Power Company and located on the Coosa River about 3 km northeast of Wilsonville, Ala., is one of the largest steam plants in Alabama, with a total capacity of 1880 MW from five operational units. One unit is equipped with cooling towers and, therefore, does not discharge heated water. Cooling water is drawn from Yellowleaf Creek and discharged into the Coosa River.

Our study was limited to a 10.6-km section of the Coosa River and a 4.5-km portion of Yellowleaf Creek, a large tributary adjacent to the steam plant. The Coosa River in this region averages about 270 m in width and 6 m in depth. The maximum channel depth varies from 1.5 to 30 m. The plant is approximately 27.2 km downstream from Logan Martin Dam and 36.8 km upriver from Lay Dam. The mainstream currents were generally moderate (33 to 36 m/min) but fluctuated according to the relative discharges from Logan Martin and Lay reservoirs. Water level fluctuations were small because discharge from the downstream reservoir was coordinated with release from the upstream one.

Water released from Logan Martin Dam altered the temperature pattern that had been established around the plant during the preceding period of no hydroelectric discharge. Temperatures in the area affected by the discharge from the plant frequently exceeded those in the upstream river channel by as much as 10°C, but the increased flow of cooler water from upstream decreased the water temperature as much as 6 to 10°C. The heated effluent dissipated rapidly during cool months but noticeably affected ambient river temperatures during the warmer months. The thermal effluent was discharged at a depth of about 6 to 9 m but rose rapidly to the surface and spread laterally. During the summer, when flow was low, a well-defined thermally stratified zone spread across the entire width of the river and sometimes extended as much as 2 to 3 km upriver and 3 to 6 km downriver. Cooler water flowed under the thermal zone throughout the year, however, and composed about 50 to 60% of the cross-sectional area during the maximum thermal development. This distribution changed when water was released from the upstream reservoir; consequently, the thermal zone was extremely variable even at its maximum development.

The assimilation of waste heat from the generating facility is a function of river water temperature, which varies with the season and the flow rate of the river. A diurnal variation is expected as a result of production demands, but this was complicated by the variation in flow caused by discharge from Logan Martin Reservoir. A general seasonal pattern was evident, however. During cool months water temperatures were little affected except in the vicinity of the outfall. As the season progressed, the temperature influence spread (Fig. 1) until the maximum extent was evident in July to September, when river water temperatures were highest and flow was lowest (Fig. 2).

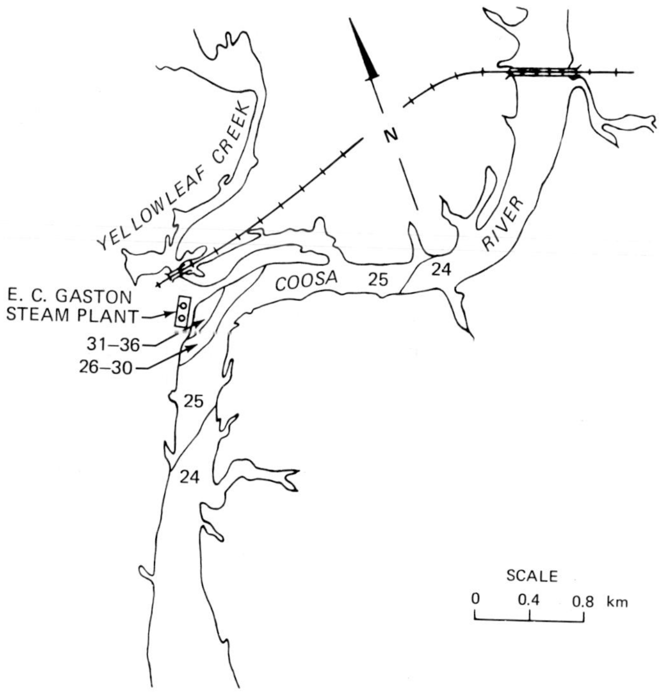

Fig. 1 Surface temperature pattern, June 7, 1974. Temperatures given in °C.

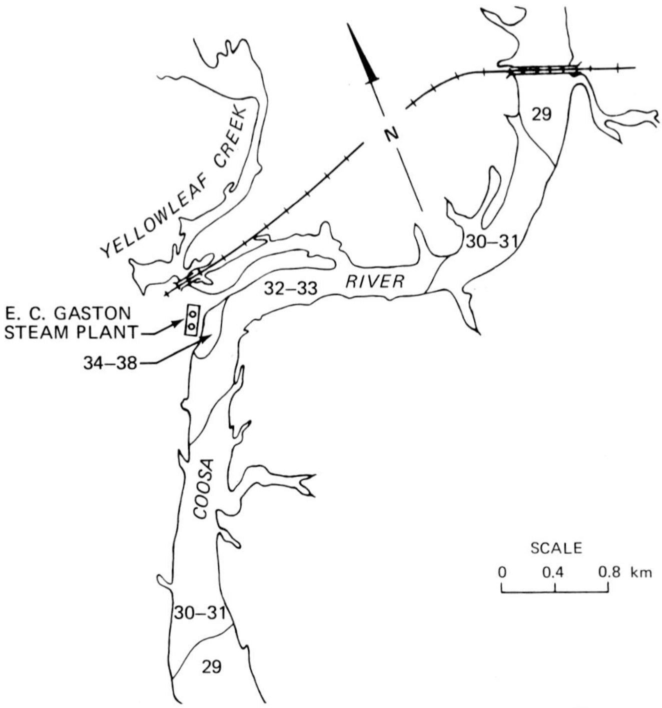

Fig. 2 Surface temperature pattern, Sept. 3, 1974. Temperatures given in °C.

RESULTS

Seven catfish (two channel, two blue, and three flathead) and eight largemouth bass were tagged and tracked during the study. Little information was collected from the two blue catfish tagged because both died within 36 hr after surgery. They seemed more sensitive to the stress of handling and transport than the other species. Since two of the flathead catfish and three largemouth bass moved little, they are omitted from the following discussion. Five largemouth bass, two channel catfish, and one flathead catfish moved considerably, and three individuals tagged with thermistor transmitters were selected to demonstrate successful passage past the thermal zone during the warmest months.

Channel Catfish No. 2

Channel catfish no. 2, a 3.3-kg male, was seined from an experimental pond at the Department of Fisheries, Auburn Univer-

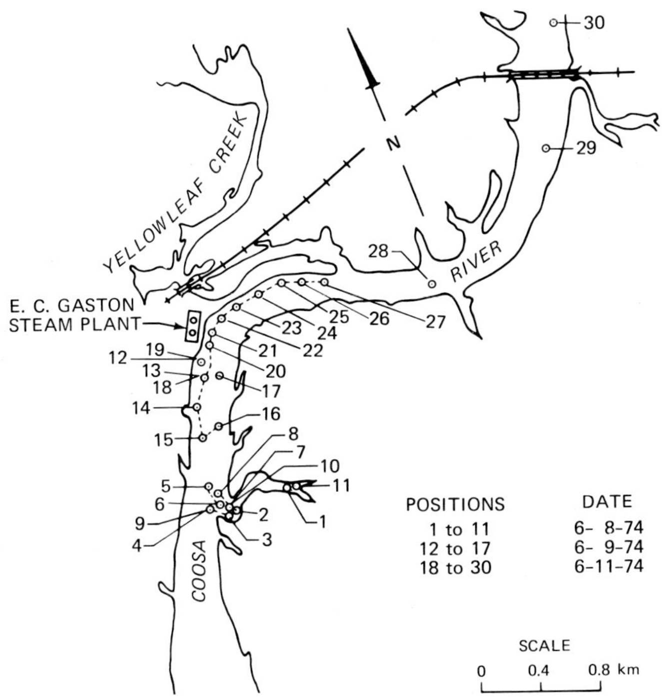

Fig. 3 Plot of location of channel catfish no. 2, June 8 to 11, 1974. Dashed line connecting positions (○ — — — ○) indicates continuous tracking.

sity, implanted with a calibrated thermistor-type transmitter, transported, and released June 8, 1974, into a large cove about 900 m below the steam plant. The position plots for this fish (Fig. 3) represent only 3 days of tracking although many unsuccessful searches were made later. The fish moved considerably during the period of tracking. Positions 1 to 11 were recorded during a 10-hr period on the day of release, and 12 to 17 were recorded during a 1-hr tracking period the next morning. The greatest movement was recorded on June 11 (positions 18 to 30), when the fish moved directly past the generating plant and continued more than 3 km upstream, going beyond the study area.

During the monitoring of channel catfish no. 2, periodic surface temperatures and profiles were taken for comparison with the transmitted temperature. The surface temperatures at the fish's position were comparable with the general surface-temperature-distribution patterns depicted in Fig. 1. During the tracking days of June 9 and 11, the fish maintained his position in water of 22 to

23°C. Surface temperatures corresponding to the fish's location varied between 27 and 32°C. The fish appears to have spent several days downstream from the thermal discharge zone before beginning a steady upstream movement on June 11, passing directly under the thermal plume. The temperature profiles taken along the path of movement indicated that his swimming depth was greater than 5 m. Five hours and 20 minutes passed between the recordings of positions 19 and 30. Horizontal movement of about 12 m/min against an 18- to 36-m/min current was continuous between positions 20 and 27. The fish continued swimming steadily upstream until he passed out of the study area (positions 28 to 30). The path and the transmitted temperature suggest that the fish moved in the cooler underflow of water in the river channel.

Flathead Catfish

A 3.8-kg male flathead catfish collected from Lay Reservoir was implanted with a temperature transmitter on Aug. 26, 1974, and released the next day about 550 meters downstream from the plant in the river (Fig. 4). His movement was tracked for 7 days before contact was lost. The surface temperature distribution depicted in Fig. 2 was typical for the period. Positions 1 to 20 were recorded on 3 consecutive days after release. Transmitted temperatures were between 28 and 30°C, and corresponding surface temperatures were 31 to 33°C. The relation between transmitted temperatures and river temperature profiles indicated that the fish was maintaining a depth of 1.5 to 3.0 m. Movement did not appear to have a particular pattern but was rather general in a 0.8- to 1.2-km portion of the river. However, 5 days later the fish was located 0.8 km upriver from the plant, and the next day he was 1.5 km downstream, near the area of release. Six days later (Sept. 10) the fish was again upstream from the plant, and the next day he had moved more than 1.5 km downstream. Despite the lack of continuous observations, we know the fish moved past the thermal zone during the period of maximum development at least four times in the 2-week tracking period. The depth of movement suggests that the fish probably passed under the thermal area, as did channel catfish no. 2.

Largemouth Bass No. 8

Largemouth bass no. 8 was captured Sept. 3 in Yellowleaf Creek about 4 km upstream from the Coosa River. The fish, a 1.7-kg female, was kept overnight in cooled water (18°C) after implantation of a Bayshore Systems thermistor transmitter. The ambient water

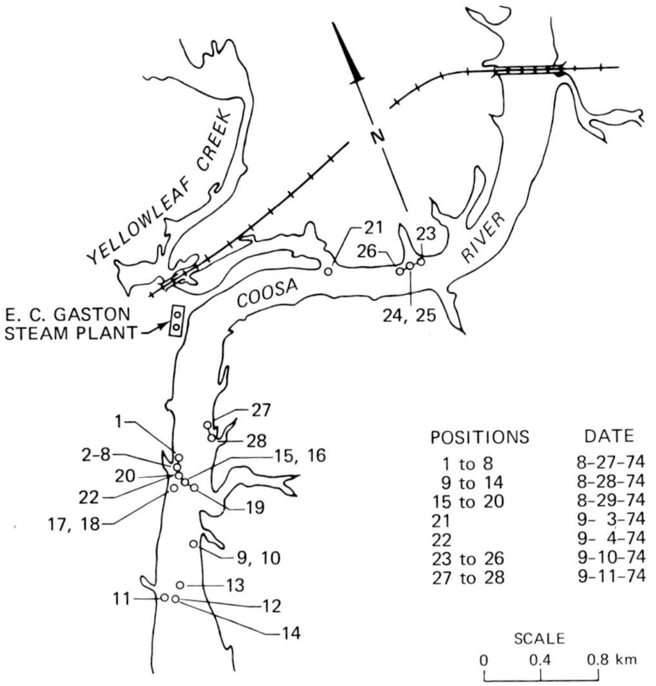

Fig. 4 Plot of location of flathead catfish, Aug. 27 to Sept. 11, 1974. Dashed line connecting positions (○ — — — ○) indicates continuous tracking.

temperature was about 31°C. After a brief reacclimation period, the fish was released in the Coosa River about 400 m below the steam plant at 1110 hours on Sept. 4. Position plots represent 8 days of tracking. This fish was not located after Sept. 19. Positions 2 to 10 represent 4 consecutive tracking days. Positions 11 to 17 were recorded at intermittent periods between Sept. 10 and 19 (Fig. 5).

After her release, the fish moved rapidly toward the east side (positions 1 and 2) and then upstream at a depth of 3 to 5 m (positions 3 and 4). Her movement was not continuous but included apparent stationary periods. She moved upstream past the thermal zone within 3 hr (positions 5 and 6). As the fish moved past the plant, transmitted temperatures were between 28.9 and 29.4°C, and corresponding surface temperatures were 29 to 33°C. Swimming depth was estimated to be about 3 m on the basis of vertical temperature profiles. Upstream from the plant the fish appeared to be deeper, about 5 to 7 m. Upstream movement stopped opposite the mouth of the Yellowleaf Creek (position 7), and the direction

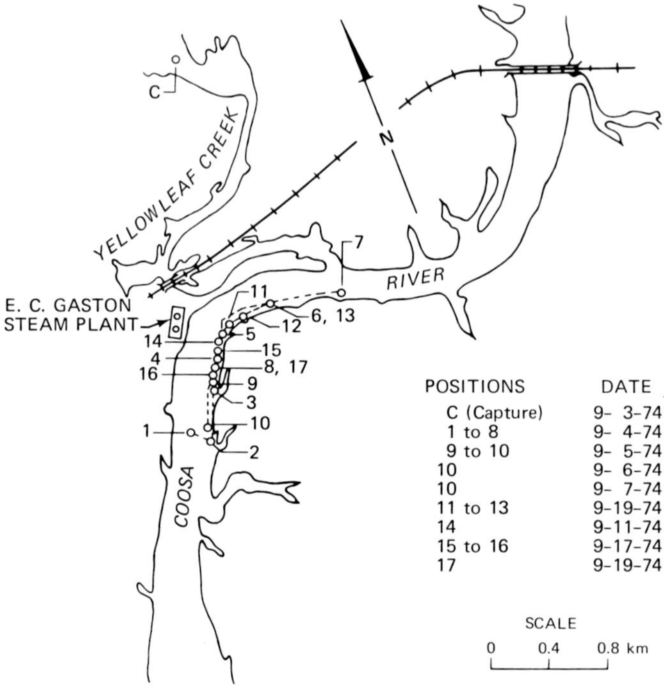

Fig. 5 Plot of location of largemouth bass no. 8, Sept. 3 to 19, 1974. Dashed line connecting positions (○ — — — ○) indicates continuous tracking.

was reversed. Within 1 hr she had passed through the thermal zone again to a position 100 m downstream from the plant (position 8). The next morning, Sept. 5, she was located a short distance further downstream (position 9), and 4 to 5 hr later was located near the point of release, where she apparently remained for the next 5 days (position 10). On Sept. 10, the fish was located opposite the plant (position 11), and the transmitted temperature indicated a water temperature of 24.4°C. Movement past the plant was again recorded, but she did not proceed as far upriver as previously observed before returning (positions 12 and 13). The next morning the fish was again located opposite the plant (position 14). Subsequently, movement was only slightly downstream along the east shore. Surface temperatures averaged 30 to 32°C, and transmitted temperatures were 25 to 27°C. The fish was not located on Sept. 19.

The movements of this fish indicate that passage was possible in the period of maximum thermal development. The path of movement and the estimated depths suggest that the fish avoided maximum temperature areas horizontally and vertically by moving in the cooler water.

DISCUSSION

The seasonal temperature patterns associated with the thermal discharge followed trends typical for the ambient temperature. The diurnal flow fluctuations imposed by discharge from upstream altered temperature distributions, however. During the cooler months heated water was rapidly diluted, and there was little effect, even immediately across the river from the plant. During the summer and early fall, however, when river flow was lowest and ambient river temperature was highest, the heated effluent had a marked effect on the area within several kilometers of the plant. The fluctuating thermal zone occasionally spread completely across the river, creating a potential intermittent surface barrier with a continuous underflow of cooler water.

During the critical months of late summer, channel catfish no. 2 passed the thermal zone by avoiding the high surface temperatures and moving instead in the cooler, deeper water. The flathead catfish moved past the plant at least four times between Aug. 27 and Sept. 11. No observations were recorded as the fish moved directly through the thermal zone, but these passages indicated that the thermal plume did not prevent these two species from moving upriver or downriver past the plant.

Based on the observed movements of largemouth bass no. 8 in September, we can state that this species seems capable of passing the thermally influenced area during most critical times of the year. The bass passed the thermal zone during the season of maximum temperature development (September) at least four times, each time moving past on the side of the river opposite the plant at a depth of 3 to 7 m, apparently in the cooler water.

Thermal discharge with the flow pattern described did not present an impassable barrier to the fishes studied during the warmest part of the year. Little can be inferred as to the proportion of a population that might pass, but since individuals of three species were not impeded we can conclude that passage was possible for those species.

ACKNOWLEDGMENT

The research reported here was supported by the U. S. Fish and Wildlife Service; the Game and Fish Division, Alabama Department of Conservation and Natural Resources; and the Department of Fisheries and Allied Aquaculture, Auburn University. Financial assistance was provided by Alabama Power Company.

Reference to companies or trade name does not imply endorsement of commercial products.

REFERENCES

Alabaster, J. S., 1962, The Effects of Heated Effluent on Fish, *Int. J. Air Water Pollut.*, 7: 541-563.

Cairns, J., Jr., 1969, The Response of Freshwater Protozoan Communities to Heated Waste Waters, *Chesapeake Sci.*, 10(3,4): 177-185.

Coutant, C. C., 1969, Behavior of Sonic-Tagged Chinook Salmon and Steelhead Trout Migrating Past Hanford Thermal Discharges, in *Biological Effects of Thermal Discharges*, Annual Progress Report for 1968, USAEC Report BNWL-1531, pp. 21-26, Battelle-Northwest, Richland, Washington, Pacific Northwest Lab., NTIS.

Davies, I., 1966, Chemical Changes in Cooling Water Towers, *Int. J. Air Water Pollut.*, 10: 853-863.

Federal Power Commission, 1969, *Problems in Disposal of Waste Heat from Steam Electrical Plants*, Staff Report, Washington, D. C.

Hart, L. G., and R. C. Summerfelt, 1975, Surgical Procedures for Implanting Ultrasonic Transmitters into Flathead Catfish (*Pylodictis olivaris*), *Trans Am. Fish. Soc.*, 104(1): 56-59.

Kelso, J. R. M., 1974, Influence of a Thermal Effluent on Movement of Brown Bullhead (*Ictalurus nebulosus*) as Determined by Ultrasonic Tracking, *J. Fish. Res. Board Can.*, 31: 1507-1513.

Langford, T. E., 1971, The Distribution, Abundance, and Life Histories of Stoneflies (Plecoptera) and Mayflies (Ephemeroptera) in a British River Warmed by Cooling-Water from a Power Station, *Hydrobiologia*, 38(2): 300-377.

——, 1972, A Comparative Assessment of Thermal Effects in Some British and North American Rivers, in *River Ecology and Man*, R. T. Oglesby, C. A. Carlson, and J. A. McCann (Eds.), pp. 319-352, Academic Press, Inc., New York.

Leggett, W. C., 1976, The American Shad (*Alosa sapidissima*), with Special Reference to Its Migration and Population Dynamics in the Connecticut River, in *The Connecticut River Ecological Study, The Impact of a Nuclear Power Plant*, D. M. Merriman and L. M. Thorpe (Eds.), pp. 169-225, Monograph No. 1, American Fisheries Society, Bethesda, Md.

Merriman, D., 1970, The Calefaction of the Connecticut River, *Sci. Am.*, 22(5): 42-52.

Nakatani, R. E., 1969, Effects of Heated Discharges on Anadromous Fishes, in *Biological Aspects of Thermal Pollution*, P. A. Krenkel and F. L. Parker (Eds.), pp. 294-353, Vanderbilt University Press, Nashville, Tenn.

Proffit, M. A., 1969, *Effects of Heated Discharges upon Aquatic Resources of White River at Petersburg, Indiana*, Report 3, Indiana Water Research Center.

Ross, F. F., 1970, Warm Water Discharges into Rivers and the Sea, in *Proceedings of the Annual Conference of the Institute of Water Pollution Control*, pp. 12-18, Kent, England.

Wrenn, W. B., 1976, Temperature Preference and Movement of Fish in Relation to a Long, Heated Discharge Channel, in *Thermal Ecology II*, ERDA Symposium Series, Augusta, Ga., Apr. 2—5, 1975, G. W. Esch and R. W. McFarlane (Eds.), pp. 191-194, CONF-750425, NTIS.

EFFECTS OF THERMAL EFFLUENT ON BENTHIC MARINE INVERTEBRATES DETERMINED FROM LONG-TERM SIMULATION STUDIES

RICHARD F. FORD, DALLAS G. FOREMAN, KENNETH J. GRUBBS, CHERYL D. KROLL, and DONALD G. WATTS
Center for Marine Studies, San Diego State University, San Diego, California

ABSTRACT

Laboratory experiments of 14-weeks duration were conducted to determine the effects of simulated thermal-effluent regimes on survival, growth, and body condition of seven motile invertebrate species from shallow, rocky habitats. Treatment conditions reproduced dynamic temperature regimes measured in the thermal plume of a large generating station in southern California 84 and 355 m from the discharge structure. Cam patterns of field-temperature records were used with time-cycle controllers and pneumatic mixing valves supplied with effluent and ambient-source water to control temperatures in groups of replicate 1100-liter tanks. High survival and normal growth and condition of control animals indicated that laboratory conditions were satisfactory. Two sea urchin species (*Strongylocentrotus purpuratus* and *S. franciscanus*) exhibited relatively low survival and impaired growth and condition in the warmer effluent regime, the effects being most pronounced among large adults. Growth and condition of two asteroid starfish species (*Pisaster ochraceus* and *P. giganteus*) were adversely affected by both effluent regimes. Two species of gastropod molluscs (*Kelletia kelletii* and *Roperia poulsoni*), both warm—temperate forms, were relatively unaffected by either of the effluent regimes. Their survival, growth, and condition were not significantly different from controls in most cases. Growth and condition of a third gastropod (*Astraea undosa*) were adversely affected by effluent conditions, but survival was not.

Little specific information is available on temperature tolerances and related responses of benthic marine animals affected by thermal effluent on the Pacific Coast of the United States. A number of studies have addressed problems of temperature tolerances, preferences, and effects in fishes and invertebrates (e.g., Norris, 1963; Des Voigne, 1970; Hair, 1971; Vlaming, 1971; Wolcott, 1973;

Ulbricht, 1973; Wolfson, 1974; Adams and Price, 1974; Jennison, 1975; Brewer, 1976; Lockheed, 1977). Much can be learned about the general biological effects of thermal effluent from well-designed field sampling and experimentation, but laboratory simulation studies are often necessary to provide detailed information. This is particularly true for motile invertebrate species, which cannot be followed easily in the field by sequential sampling to determine survival, growth, and condition of individuals.

This paper describes long-term simulation experiments to determine the effects of thermal effluent from a large electric generating station on seven ecologically important species of motile, benthic invertebrates. Combined effects of temperature and other water-quality characteristics of the effluent were evaluated. The species studied are the echinoid echinoderms *Strongylocentrotus purpuratus* and *S. franciscanus* (grazers), the asteroid echinoderms *Pisaster ochraceus* and *P. giganteus* (carnivores), and the gastropod molluscs *Astraea undosa* (grazer), *Kelletia kelletii* (carnivore), and *Roperia poulsoni* (carnivore). Representative juvenile and adult size groups of each species were maintained in large tanks provided with flowing seawater. The survival, growth, and condition of control groups maintained both in generating-station intake water and ocean water at ambient temperatures were used as standards against which to compare characteristics of two treatment groups held in thermal-effluent regimes. The treatment conditions simulated temperatures and chemical conditions experienced by animals at two representative locations in the thermal plume.

MATERIALS AND METHODS

A detailed description of the primary laboratory at which these studies were conducted is given by Ford et al. (1976). The laboratory is located at a large, fossil-fueled generating station on the coast near Los Angeles, Calif. Characteristics of operation and of the thermal-effluent seawater discharged from this plant are similar to those of other electrical generating stations along the southern California coast (Adams, 1969).

Separate laboratory supply systems provided thermal-effluent seawater and seawater of ambient temperature from a cooling-water intake of the generating station. Filtered water from these sources was supplied to all parts of the laboratory through polyvinyl chloride lines.

One group of controls was maintained at ambient ocean temperature in a laboratory at the Scripps Institution of Oceanography, La Jolla, Calif. Filtered seawater supplied from depths of 3 to 5 m at the end of the Scripps pier was used.

Field-Temperature Records

Water temperatures at two locations (stations 2 and 3) influenced by the thermal plume from the largest generating units of the station were monitored continuously. Station 2 was located 84 m (275 ft) and station 3, 335 m (1100 ft) from the discharge structure. Spring-driven Powers model 101 two-pen circular-chart temperature recorders (Powers Regulator Co.) were installed at each location to provide a 7-day temperature record accurate to ±0.5°C.

One of the temperature sensors at each station was placed on the rock substrate at a depth of 0.6 m (2 ft) below mean lower low water (MLLW), and the second was placed at a depth of 3.7 m (12 ft) below MLLW. Temperature records from the 0.6-m depth were used as treatment conditions for the simulation experiments because the primary influence of the thermal plume on the shoreline was near the surface.

Laboratory System for Simulation Experiments

Circular, fiber glass holding tanks 150 cm in diameter, 76 cm deep, and with a capacity of 1100 liters were used for all size groups of test species. Each tank was fitted with surface and bottom drains to assure effective removal of detritus and uniform, rapid water circulation by vortex action. The experimental tanks were arranged in interconnected groups of 20 replicates, with common recirculation through two heated sumps, each serving 10 tanks. Supplementary electrical heating of the sump water was necessary to maintain experimental temperature conditions because temperatures produced by the generating units were sometimes quite variable and did not correspond to those required in the experiments.

Water temperatures in different groups of 20 replicate treatment tanks were controlled to duplicate the varying temperatures measured continuously at stations 2 and 3 in the thermal plume. Weekly chart records of field temperatures were used as patterns to cut aluminum cams, and the cams, in turn, were used to drive separate time-cycle controllers associated with each group of tanks. A Powers model VP657 three-way mixing valve operated pneumatically by a Powers model 101 time-cycle controller was used to blend effluent water from the heated sumps with ambient source water in changing proportions to duplicate the temperature regime measured at a field station 2 weeks previously. The 2-week lag was required to allow time for the field record to be made, retrieved, and converted to a cam pattern. One time-cycle controller and mixing valve controlled the seawater flow to each group of 20 experimental tanks. The mixing valve was supplied with warm effluent water through a common line from two centrifugal pumps, each drawing from a separate effluent sump unit. Each of these sumps, in turn, received part of the drain flow from 10 experimental tanks, and the remainder of the drain water was discharged to waste. Thus the system recycled part of the warm water to conserve heat but had a substantial and continuous supply of makeup water (10 to 30 liters/min) from both the effluent and ambient-temperature laboratory supply lines. The flow rate of mixed seawater to the individual tanks averaged 8 liters/min (range, 6 to 10 liters/min).

Continuous temperature records were made in the experimental tanks and were compared with temperatures called for by the cam patterns. Under normal operating conditions the accuracy of the controller system in duplicating field-temperature records was within ±0.5°C. An air stone operating continuously at a high flow rate in each tank maintained dissolved oxygen concentrations at normal levels and increased vertical and horizontal mixing.

Twenty control tanks, identical with those used for the treatment groups, were maintained at the generating-station laboratory and were supplied with water at varying ambient temperatures from the intake screenwell of the station. Twelve control tanks at the Scripps Institution of Oceanography were supplied with water of varying ambient ocean temperatures. Flow rates, oxygenation, and other conditions were the same as those in the treatment tanks.

Banks of fluorescent lights providing natural and uniform illumination were controlled by standard clock timers, which were reset weekly to follow sunrise and sunset times. Sunlight entered the laboratories through translucent windows, providing supplementary light and reasonably natural twilight conditions.

Estimates of the relative amounts of thermal-effluent and ambient-source water entering the treatment tanks, which were made from reading in-line flow meters, indicated that the test animals were held in varying effluent mixtures similar to those at the two field locations. This suggests that the animals were exposed to water conditions simulating the field locations fairly closely in temperature and chemical composition. Significant biological effects observed in the treatment tanks were assumed to be the combined results of temperature and chemical composition of the water.

Because of the location of the laboratory, the ambient-intake and thermal-effluent water used in the experiments came from the cooling-water system of different generating units than those simulated in the experiments. Data on the chemical composition of the water in these two systems were obtained seasonally and at representative times of the tidal cycle in studies conducted concurrently. Comparative measurements were made of the trace metals Cu, Zn, Cd, Pb, Co, Cr, and As (Ford et al., 1976; Dorband et al., 1977); dissolved oxygen, phosphate, ammonia, nitrate, and nitrite (Ford et al., 1976); and Cu, Cr, pH, oil and grease, and chlorine (unpublished monitoring data from generating-station records).

On the basis of the results of t-tests comparing sample means, there were no significant differences ($P > 0.05$) in the concentrations of these chemical constituents in the two cooling-water systems. This indicates that the chemical composition of the effluent and intake water supplied to the laboratory was essentially the same as that of the cooling-water system simulated.

Collection, Stocking, and Maintenance of Animals

Animals used in experiments at the generating-station laboratory were collected in the vicinity of the station at locations outside the limits of the thermal plume. The Scripps control animals were collected in shallow-water areas off Point La Jolla, Calif., which had temperature regimes similar to those of the collecting sites near the generating station.

Animals were placed in the 1100-liter tanks and held for 3 to 5 weeks before stocking. This quarantine procedure served several functions. It allowed accumulation of large numbers of individuals so that representative size groups could be selected and sorted before the tanks were stocked, provided adequate time for individuals to acclimate to laboratory holding conditions at ambient temperatures, and enabled investigators to observe and eliminate individuals injured by collecting and handling.

In each unit of 20 tanks used for a given experimental treatment or control group at the generating-station laboratory, five tanks stocked with the same combinations of species, size groups, and densities of experimental animals were used as replicates. The Scripps controls used only three replicates because space limitations allowed use of only 12 tanks.

Animals of each test species used in the experiments were separated into three size categories representing the size ranges of juveniles and adults commonly encountered in the study area. Five individuals of each size category were randomly assigned to each of the five (or three) replicate tanks in a series.

Natural foods were introduced daily and were maintained in the tanks at all times in slight excess to promote optimum growth. Both

Strongylocentrotus species were fed multicellular algae (95%), supplemented by frozen, whole fish of various species (5%). All the algal species used are known to be natural foods of these sea urchins (Lees, 1970; Baker, 1973). Both *Pisaster* species were fed approximately 85% live *Mytilus* and 15% whole fish. *Mytilus* is a major dietary component of these starfish in nature. *Kelletia kelletii* and *Roperia poulsoni* were fed 50% live, crushed *Mytilus* sp. and other small organisms associated with the mussels and 50% whole fish. *Astraea undosa* was fed 99% algae and 1% whole fish.

Methods for Determining Biological Characteristics

Detailed data on size-specific mortality and survival were used to develop cohort life tables. The data were based on twice-daily censuses of mortality in the experimental and control tanks.

Size-specific growth rates were determined by monthly measurements of changes in linear size (nearest 0.1 mm), live wet weight (nearest 0.1 g), and displacement volume (nearest 5 cc). External water was shaken from the animals before their weights were determined. Maximum test diameter and body weight were determined for *Strongylocentrotus.* Body weight and displacement volume of the body were measured for *Pisaster.* Body weight and linear measurements of the shell were determined for the three gastropod mollusc species. For *K. kelletii* and *R. poulsoni,* maximum shell length, from the apex of the longest portion of the spire to the anterior end of the siphonal canal, was used as the linear measurement. For *A. undosa* maximum shell diameter between opposite ribs or nodes on the lowest (most ventral) shell whorl was used.

RESULTS

Effects of Thermal Effluent on Survival

Comparative life table summaries (Tables 1 to 4) provide a concise means of describing and evaluating the effects of different, varying thermal-effluent conditions on survival of the test species. These data were summarized separately for each experiment and for each size group. Complete life table summaries are given by Ford et al. (1976). Values of l_x, the proportion of individuals surviving at the beginning of each 1-week time interval (x) of the experiment, and d_x, the proportion of individuals dying within each 1-week time interval, are expressed as relative frequencies. The values are based on an initial cohort of animals; the number of individuals in the cohort (N) is indicated in the table.

TABLE 1

COMPARATIVE LIFE TABLES FOR COHORTS OF *Strongylocentrotus purpuratus** IN STATION 2 AND 3 EXPERIMENTS AND IN CONTROLS†

	Scripps control group (N = 10)				Generating-station control group (N = 25)				Station 2 (N = 25)				Station 3 (N = 25)			
	Temperature, °C				Temperature, °C				Temperature, °C				Temperature, °C			
Week	Mean	Range	d_x	l_x	Mean	Range	d_x	l_x	Mean	Range	d_x	l_x	Mean	Range	d_x	l_x
1 (Aug. 1)	18.8	18—21	0.00	1.00	20.4	19—22	0.00	1.00	20.4	19—22	0.00	1.00	20.4	19—22	0.00	1.00
2 (Aug. 8)	19.0	18—21	0.10	1.00	21.1	19—22	0.00	1.00	21.1	19—22	0.00	1.00	21.1	19—22	0.00	1.00
3 (Aug. 15)	20.6	19—21	0.00	0.90	21.4	19—22	0.00	1.00	20.7	17—24	0.00	1.00	20.5	18—24	0.00	1.00
4 (Aug. 22)	17.5	18—21	0.00	0.90	19.8	18—21	0.00	1.00	22.8	19—23	0.04	1.00	20.3	18—24	0.00	1.00
5 (Aug. 29)	18.5	18—19	0.00	0.90	18.4	17—21	0.00	1.00	21.4	19—23	0.00	0.96	20.6	18—22	0.00	1.00
6 (Sept. 5)	18.4	17—19	0.00	0.90	19.8	17—23	0.04	1.00	21.4	18—24	0.52	0.96	20.0	18—23	0.00	1.00
7 (Sept. 12)	18.2	17—19	0.00	0.90	20.4	19—22	0.08	0.96	18.7	16—21	0.00	0.44	20.1	18—22	0.04	1.00
8 (Sept. 19)	18.9	17—20	0.00	0.90	20.7	19—22	0.00	0.88	18.9	17—21	0.00	0.44	20.3	18—23	0.00	0.96
9 (Sept. 26)	19.8	19—20	0.00	0.90	17.4	18—21	0.00	0.88	19.4	17—21	0.00	0.44	20.5	18—26	0.00	0.96
10 (Oct. 3)	18.6	17—20	0.00	0.90	19.8	18—21	0.00	0.88	20.9	18—23	0.00	0.44	19.7	18—22	0.00	0.96
11 (Oct. 10)	18.2	18—19	0.00	0.90	20.4	19—22	0.00	0.88	20.6	18—23	0.04	0.44	20.8	19—22	0.00	0.96
12 (Oct. 17)	18.1	17—19	0.00	0.90	18.6	17—20	0.00	0.88	19.4	18—21	0.00	0.40	18.5	17—20	0.00	0.96
13 (Oct. 24)	17.1	17—18	0.00	0.90	18.1	17—19	0.00	0.88	18.0	17—19	0.00	0.40	18.3	17—20	0.00	0.96
14 (Oct. 31)	16.4	16—17	0.00	0.90	16.6	15—18	0.00	0.88	17.6	14—21	0.00	0.40	17.8	16—20	0.00	0.96
15 (Nov. 7)				0.90				0.88				0.40				0.96

*Test animals were 8 to 19 mm in diameter.

†See text for explanation of symbols d_x and l_x.

TABLE 2

COMPARATIVE LIFE TABLES FOR COHORTS OF *Strongylocentrotus purpuratus** IN STATION 2 AND 3 EXPERIMENTS AND IN CONTROLS†

	Scripps control group (N = 10)				Generating-station control group (N = 25)				Station 2 (N = 25)				Station 3 (N = 25)			
	Temperature, °C				Temperature, °C				Temperature, °C				Temperature, °C			
Week	Mean	Range	d_x	l_x	Mean	Range	d_x	l_x	Mean	Range	d_x	l_x	Mean	Range	d_x	l_x
1 (Aug. 1)	18.8	18–21	0.00	1.00	20.4	19–22	0.00	1.00	20.4	19–22	0.00	1.00	20.4	19–22	0.00	1.00
2 (Aug. 8)	19.0	18–21	0.00	1.00	21.1	19–22	0.00	1.00	21.1	19–22	0.00	1.00	21.1	19–22	0.00	1.00
3 (Aug. 15)	20.6	19–21	0.00	1.00	21.4	19–22	0.00	1.00	20.7	17–24	0.00	1.00	20.5	18–24	0.00	1.00
4 (Aug. 22)	17.5	18–21	0.00	1.00	19.8	18–21	0.04	1.00	22.8	19–23	0.04	1.00	20.3	18–24	0.00	1.00
5 (Aug. 29)	18.5	18–19	0.00	1.00	18.4	17–21	0.00	0.96	21.4	19–23	0.40	0.96	20.6	18–22	0.00	1.00
6 (Sept. 5)	18.4	17–19	0.00	1.00	19.8	17–23	0.04	0.96	21.4	18–24	0.00	0.56	20.0	18–23	0.00	1.00
7 (Sept. 12)	18.2	17–19	0.00	1.00	20.4	19–22	0.08	0.92	18.7	16–21	0.12	0.56	20.1	18–22	0.08	1.00
8 (Sept. 19)	18.9	17–20	0.00	1.00	20.7	19–22	0.00	0.88	18.9	17–21	0.00	0.44	20.3	18–23	0.00	0.92
9 (Sept. 26)	19.8	19–20	0.00	1.00	17.4	18–21	0.00	0.88	19.4	17–21	0.00	0.44	20.5	18–26	0.00	0.92
10 (Oct. 3)	18.6	17–20	0.00	1.00	19.8	18–21	0.00	0.88	20.9	18–23	0.04	0.44	19.7	18–22	0.00	0.92
11 (Oct. 10)	18.2	18–19	0.00	1.00	20.4	19–22	0.04	0.88	20.6	18–23	0.00	0.40	20.8	19–22	0.00	0.92
12 (Oct. 17)	18.1	17–19	0.00	1.00	18.6	17–20	0.04	0.84	19.4	18–21	0.00	0.40	18.5	17–20	0.00	0.92
13 (Oct. 24)	17.1	17–18	0.00	1.00	18.1	17–19	0.04	0.80	18.0	17–19	0.08	0.40	18.3	17–20	0.00	0.92
14 (Oct. 31)	16.4	16–17	0.00	1.00	16.6	15–18	0.00	0.76	17.6	14–21	0.00	0.40	17.8	16–20	0.00	0.92
15 (Nov. 7)				1.00				0.76				0.32				0.92

*Test animals were 40 to 70 mm in diameter.
†See text for explanation of symbols d_x and l_x.

TABLE 3

COMPARATIVE LIFE TABLES FOR COHORTS OF *Strongylocentrotus franciscanus** IN STATION 2 AND 3 EXPERIMENTS AND IN CONTROLS†

	Scripps control group (N = 10)				Generating-station control group (N = 25)				Station 2 (N = 25)				Station 3 (N = 25)			
	Temperature, °C				Temperature, °C				Temperature, °C				Temperature, °C			
Week	Mean	Range	d_x	l_x	Mean	Range	d_x	l_x	Mean	Range	d_x	l_x	Mean	Range	d_x	l_x
1 (Aug. 1)	18.8	18—21	0.10	1.00	20.4	19—22	0.00	1.00	20.4	19—22	0.00	1.00	20.4	19—22	0.00	1.00
2 (Aug. 8)	19.0	18—21	0.00	0.90	21.1	19—22	0.00	1.00	21.1	19—22	0.00	1.00	21.1	19—22	0.00	1.00
3 (Aug. 15)	20.6	19—21	0.00	0.90	21.4	19—22	0.00	1.00	20.7	17—24	0.00	1.00	21.4	19—22	0.00	1.00
4 (Aug. 22)	17.5	18—21	0.00	0.90	19.8	18—21	0.00	1.00	22.8	19—23	0.00	1.00	19.8	18—21	0.00	1.00
5 (Aug. 29)	18.5	18—19	0.00	0.90	18.4	17—21	0.04	1.00	21.4	19—23	0.32	1.00	18.4	17—21	0.00	1.00
6 (Sept. 5)	18.4	17—19	0.00	0.90	19.8	17—23	0.00	0.96	21.4	18—24	0.04	0.68	19.8	17—23	0.00	1.00
7 (Sept. 12)	18.2	17—19	0.00	0.90	20.4	19—22	0.00	0.96	18.7	16—21	0.04	0.64	20.4	19—22	0.00	1.00
8 (Sept. 19)	18.9	17—20	0.00	0.90	20.7	19—22	0.00	0.96	18.9	17—21	0.00	0.60	20.7	19—22	0.00	1.00
9 (Sept. 26)	19.8	19—20	0.10	0.90	17.4	18—21	0.00	0.96	19.4	17—21	0.04	0.60	17.4	18—21	0.04	1.00
10 (Oct. 3)	18.6	17—20	0.0	0.80	19.8	18—21	0.00	0.96	20.9	18—23	0.04	0.54	19.8	18—21	0.00	0.96
11 (Oct. 10)	18.2	18—19	0.0	0.80	20.4	19—22	0.04	0.96	20.6	18—23	0.20	0.50	20.4	19—22	0.00	0.96
12 (Oct. 17)	18.1	17—19	0.0	0.80	18.6	17—20	0.00	0.92	19.4	18—21	0.08	0.30	18.6	17—20	0.00	0.96
13 (Oct. 24)	17.1	17—18	0.0	0.80	18.1	17—19	0.00	0.92	18.0	17—19	0.08	0.28	18.1	17—19	0.00	0.96
14 (Oct. 31)	16.4	16—17	0.00	0.80	16.6	15—18	0.00	0.92	17.6	14—21	0.00	0.20	16.6	15—18	0.12	0.96
15 (Nov. 7)				0.80				0.92				0.20				0.84

*Test animals were 90 to 125 mm in diameter.

†See text for explanation of symbols d_x and l_x.

TABLE 4

COMPARATIVE LIFE TABLES FOR COHORTS OF *Pisaster ochraceus** IN STATION 2 AND 3 EXPERIMENTS AND IN CONTROLS†

Week	Scripps control group (N = 15)				Generating-station control group (N = 25)				Station 2 (N = 25)				Station 3 (N = 25)			
	Temperature, °C				Temperature, °C				Temperature, °C				Temperature, °C			
	Mean	Range	d_x	l_x	Mean	Range	d_x	l_x	Mean	Range	d_x	l_x	Mean	Range	d_x	l_x
1 (Aug. 1)	18.8	18–21	0.00	1.00	20.4	19–22	0.00	1.00	20.4	19–22	0.00	1.00	20.4	19–22	0.00	1.00
2 (Aug. 8)	19.0	18–21	0.00	1.00	21.1	19–22	0.00	1.00	21.1	19–22	0.00	1.00	21.1	19–22	0.00	1.00
3 (Aug. 15)	20.6	19–21	0.00	1.00	21.4	19–22	0.00	1.00	20.7	17–24	0.00	1.00	20.5	18–24	0.00	1.00
4 (Aug. 22)	17.5	18–21	0.00	1.00	19.8	18–21	0.00	1.00	22.8	19–23	0.00	1.00	20.3	18–24	0.08	1.00
5 (Aug. 29)	18.5	18–19	0.00	1.00	18.4	17–21	0.00	1.00	21.4	19–23	0.04	1.00	20.6	18–22	0.00	0.92
6 (Sept. 5)	18.4	17–19	0.00	1.00	19.8	17–23	0.00	1.00	21.4	18–24	0.00	0.96	20.0	18–23	0.00	0.92
7 (Sept. 12)	18.2	17–19	0.00	1.00	20.4	19–22	0.00	1.00	18.7	16–21	0.04	0.96	20.1	18–22	0.04	0.92
8 (Sept. 19)	18.9	17–20	0.00	1.00	20.7	19–22	0.00	1.00	18.9	17–21	0.00	0.92	20.3	18–23	0.00	0.88
9 (Sept. 26)	19.8	19–20	0.00	1.00	17.4	18–21	0.00	1.00	19.4	17–21	0.04	0.92	20.5	18–26	0.04	0.88
10 (Oct. 3)	18.6	17–20	0.00	1.00	19.8	18–21	0.00	1.00	20.9	18–23	0.00	0.88	19.7	18–22	0.00	0.84
11 (Oct. 10)	18.2	18–19	0.00	1.00	20.4	19–22	0.00	1.00	20.6	18–23	0.00	0.88	20.8	19–22	0.00	0.84
12 (Oct. 17)	18.1	17–19	0.00	1.00	18.6	17–20	0.00	1.00	19.4	18–21	0.00	0.88	18.5	17–20	0.00	0.84
13 (Oct. 24)	17.1	17–18	0.00	1.00	18.1	17–19	0.00	1.00	18.0	17–19	0.00	0.88	18.3	17–20	0.00	0.84
14 (Oct. 31)	16.4	16–17	0.00	1.00	16.6	15–18	0.00	1.00	17.6	14–21	0.00	0.88	17.8	16–20	0.00	0.84
15 (Nov. 7)				1.00				1.00				0.88				0.84

*Test animals were 100 to 300 g in wet body weight.

†See text for explanation of symbols d_x and l_x.

Weekly means and ranges of water temperatures in the tank systems, from continuous measurements read hourly, are shown for each control and treatment group. These allow direct comparisons between the temperature regime to which the test animals were exposed and their mortality and survival. Temperature conditions of the station 3 experiment were less severe than those of the station 2 experiment, particularly in the degree and range of temperature fluctuations (Table 1). The temperature conditions reflect differences in thermal plume characteristics and distances from the discharge structure at stations 2 and 3.

Statistical Evaluations

Chi-square tests for homogeneity among control and treatment groups were used as the standard method of evaluation. These tests contrasted the total numbers of individuals of each species and size group surviving in the two control systems with the numbers surviving in each of the two effluent treatments at the start of a given time interval (x). Inspection of the life table data and the process of elimination were used to avoid multiple tests, and in almost all cases only one test was applied to a given set of data.

Separate series of chi-square tests were applied to survival data at the midpoint of the station 2 and station 3 experiments on Sept. 19 and at the end of the 14-week experiments on Nov. 7, 1975. The results of these evaluations are discussed in the following sections.

Evaluation of Control Conditions

Data shown in the life tables for the control groups suggest that the controls were adequate and provided good conditions for survival of all test species. In the Scripps control groups, percentages of individuals surviving to the end of the experiments ranged from 58% for the smallest size group of *K. kelletii* to 80 to 100% for all the other species and size groups. The chi-square test comparisons and the process of elimination were used to assess possible differences in the numbers of individuals of each species and size group surviving at the end of the experiments between the generating-station and Scripps control groups. There was a significant difference in survival ($P < 0.05$) between the two controls in only one case. Survival of the smallest size group of *K. kelletii* was markedly lower (58%) in the Scripps controls as a result of failure of a seawater system in the Scripps tanks. Final survival values for most of the other species were at least slightly higher for the Scripps controls than for controls in higher temperature water at the generating-station laboratory (Tables 1 to 4).

Echinoderms

Results of the chi-square tests for all size groups of *S. purpuratus* indicated that survival was significantly lower ($P < 0.05$) in the station 2 treatment group than in either of the controls (Tables 1 and 2). Uniform results were obtained for both the midpoint and the end of the experiment. In contrast, the results of chi-square evaluations for *S. purpuratus* of all sizes in the station 3 experiment (Tables 1 and 2) showed no significant differences in survival between treatment and control groups ($P > 0.05$) at either the midpoint or the end of the experiment. These differences in survival are reflected clearly in the life tables. Apparently temperatures in the station 3 experiment were mild enough to have little effect on the survival of any of the three size groups; there was 92 to 96% survival at the end of the experiment. In contrast, survival of the largest size group of *S. purpuratus* at the end of the station 2 experiment was only 30% and of the smallest and intermediate size groups, 40% and 60%, respectively.

Strongylocentrotus franciscanus in the station 2 experiment showed even poorer survival, ranging from 20% for the largest (Table 3) to 44% for the smallest size group at the end of the experiment. Results of chi-square tests for *S. franciscanus* of all sizes in this experiment indicated that survival was significantly lower ($P < 0.05$) in the treatment group than in either of the controls. There were no significant differences in survival between the control and treatment groups in the station 3 experiment ($P > 0.05$).

Pisaster ochraceus and *P. giganteus* of all sizes showed relatively high survival in both effluent-treatment groups. Results of all chi-square tests comparing both species in both experiments showed no significant differences ($P > 0.05$) among the effluent-treatment groups and either of the control groups. These results were the same for the midpoint and the end of both experiments. The largest size group of *P. ochraceus* (525 to 825 g body weight) was omitted from the station 3 experiment because few individuals of that size could be found in the study area during the period of stocking.

Survival of *P. ochraceus* (Table 4) was only slightly lower in the station 2 treatment groups (80 to 88% at the end of the experiment) than in station 3 treatment groups (84 to 92%). Survival of *P. giganteus* in the station 2 treatment groups ranged from 92 to 100%, and that of the station 3 treatment groups ranged from 96 to 100%. This suggests that survival of all size categories of both species was at most only slightly affected, with *P. ochraceus* showing somewhat greater evidence of possible adverse effects.

Gastropod Molluscs

Conditions of the station 2 and station 3 experiments had no evident effects on survival of *K. kelletii.* Results of chi-square tests indicated there were no significant differences in survival ($P > 0.05$) between control and treatment groups for any size group at the midpoint or at the end of the experiment. Survival of the smallest treatment size group was 84% at the end of the station 2 experiment, and for all other treatment groups survival was 96 to 100%.

Thermal effluent had no apparent effect on survival of *R. poulsoni* in the station 3 experiment; final survival of the treatment groups ranged from 92 to 100%. Results of chi-square tests on these data indicated no significant differences ($P > 0.05$). In the station 2 experiment, however, the largest size group had markedly lower survival (52%) than did the controls at the end of the experiment. Results of chi-square tests indicated that survival of this group was significantly lower ($P < 0.05$) than that of either corresponding control group at the end of the experiment but not at the midpoint ($P > 0.05$). Survival of the smaller two size groups was also moderately low (72 and 76%) in thermal effluent at the end of the experiment, but the results of chi-square tests and the process of elimination indicated that survival of these treatment groups was not significantly different from that of either corresponding control group ($P > 0.05$).

Astraea undosa of all sizes exhibited high survival in both the station 2 and station 3 experiments, with final values ranging from 88 to 100%. Results of chi-square tests at both the midpoint and the end of these experiments indicated that there were no significant differences in survival between the effluent-treatment and the control groups ($P > 0.05$).

Effects of Thermal Effluent on Growth and Condition

Mean values of linear size, weight, volume, and a measure of body condition for Scripps control and effluent-treatment groups at the start and end of both simulation experiments are shown in Table 5. The change between each of the initial and final mean values (Δ) and the numbers of individuals on which the means were based (N) are indicated. The ratio of wet body weight to linear size (W/L) was used as a somewhat arbitrary but standard indication of body condition, like the relationship W/L^3 used in fisheries studies. For *Pisaster*, separate weight and volume determinations were used to assess condition because no accurate, linear measurements could be made.

Statistical Evaluations

Data on growth and condition of both the Scripps and generating-station control groups were evaluated by standard t-tests for paired observations. In each case the initial and final size, weight, volume, or W/L values for an individual test animal were treated as paired observations to reduce effects of unwanted variability caused by differences in size and other characteristics. Differences in growth rates among the two control and the two treatment groups were evaluated by comparing the slopes of regression lines for the growth data by analysis of covariance and the Newman–Keuls multiple range test.

Evaluation of Control Conditions

The control data in Table 5 suggest that conditions at the Scripps laboratory were adequate for normal growth for most of the size groups and species considered. The results of t tests for paired observations indicated significant increases in linear size among all size categories of *S. purpuratus* and for the smaller two size groups of *S. franciscanus* in both the Scripps and generating-station controls ($P < 0.05$). The largest size group of *S. franciscanus*, on the other hand, showed no significant changes in size or W/L in either control group ($P > 0.05$), possibly because of the slow growth of these individuals and the short duration of the experiment. The results of t tests for *P. ochraceus* and *P. giganteus* in both control groups indicated significant increases in weight and volume among all size categories ($P < 0.05$). On the basis of the limited information available in the literature (Boolootian, 1966), the growth rates exhibited by control animals of both species appear to be normal.

The smallest size group of *K. kelletii* showed significant increases in weight and W/L in both control groups; the largest size group exhibited a significant increase in weight ($P < 0.05$). Weight and W/L for all size categories of *R. poulsoni* showed no significant changes in either control group ($P > 0.05$). Laboratory growth rates of *R. poulsoni* controls were similar to those estimated for a natural population by Fotheringham (1971). With one exception, the results of t tests for *A. undosa* showed significant increases in body weight and W/L ratios ($P < 0.05$) for all size groups in both Scripps and generating-station controls. The largest size category showed no significant change in the W/L ratio in either control group; their growth was comparable to that reported by Schwalm (1973) for a natural population. These results suggest that the control conditions at both the Scripps and generating-station laboratories were adequate

TABLE 5

MEAN VALUES AND CHANGES (Δ) FOR LINEAR SIZE, WEIGHT, VOLUME, AND CONDITION (W/L) OF CONTROL GROUPS AND STATIONS 2 AND 3 SIMULATION EXPERIMENTS, JULY TO NOVEMBER 1975

Species, size group, and growth or condition measurement	Scripps control group				Station 2				Station 3			
	N	Initial mean	Final mean	Δ	N	Initial mean	Final mean	Δ	N	Initial mean	Final mean	Δ
Strongylocentrotus purpuratus												
8- to 19-mm size group	10				23				21			
Test diameter, mm		16.8	26.4	9.6		18.8	21.9	2.1		15	18.9	3.9
W/L		0.14	0.35	0.21		0.2	0.27	0.07		0.12	0.17	0.05
20- to 30-mm size group	10				23				15			
Test diameter, mm		25.5	35.8	10.3		28.2	29.7	0.5		24	25.5	1.5
W/L		0.34	0.6	0.26		0.42	0.43	0.01		0.29	0.32	0.03
40- to 70-mm size group	10				23				13			
Test diameter, mm		56.4	59.8	3.4		41.8	42	0.04		45.8	45.4	−0.4
W/L		1.40	1.43	0.03		0.83	0.83	0.0		0.92	0.8	−0.12
Strongylocentrotus franciscanus												
20- to 40-mm size group	9				25				25			
Test diameter, mm		28.2	35.1	6.9		39.8	40.3	0.5		35.2	36	0.8
W/L		0.39	0.54	0.15		0.77	0.77	0.0		0.61	0.59	−0.02
55- to 85-mm size group	10				23				25			
Test diameter, mm		72.8	75.5	2.7		67.4	67.3	−0.3		72.5	71.9	−0.6
W/L		2.42	2.45	0.03		2.18	2.01	−0.17		2.5	2.2	−0.3
90- to 125-mm size group	8				24				20			
Test diameter, mm		98.9	100.6	1.7		91.1	91.0	−0.01		93.4	92.4	−1.0
W/L		4.07	4.23	0.16		3.91	3.71	−0.2		4.0	3.7	−0.3
Pisaster ochraceus												
100- to 300-g size group	15				21				21			
Weight, g		230.4	316.2	85.8		243.7	269.6	26		267.7	292	24.3
Volume, cc		195.7	287.3	91.6		185.9	237.1	51.2		233.5	260.2	26.7
318- to 500-g size group	14				25				20			
Weight, g		417.8	534.4	116.6		405.6	420.5	14.9		398.4	412.0	13.6
Volume, cc		364.1	481.9	117.8		348.4	369.6	21.2		343.6	365.5	21.9
525- to 825-g size group	7				24				23			
Weight, g		607	637.4	30.4		684	657	−27		735	705.9	−29.1
Volume, cc		538.4	577.6	39.2		584.8	585.8	1		666.2	626.2	−40.0

Pisaster giganteus												
200- to 425-g size group	14				25				14			
Weight, g		350.1	510.7	160.6		261.9	318.1	56.2		324.3	398.7	74.4
Volume, cc		306.4	476.2	169.8		229.1	278	48.9		276	360.4	84.4
450- to 650-g size group	11				25				12			
Weight, g		582.7	735.9	153.2		538.4	600	61.6		558.8	649.2	90.4
Volume, cc		514.8	666.8	152.0		482.4	534	51.6		479.8	587.8	108
675- to 1175-g size group	6				25				8			
Weight, g		758.2	953.6	195.4		810.6	839	28.4		820.3	892.6	72.3
Volume, cc		681	866.2	185.2		716.6	773.2	56.6		730	805.4	75.4
Kelletia kelletii												
50- to 70-mm size group	8				25				23			
Weight, g		32	34.1	2.1		22.2	26	3.8		20.9	24.2	3.3
W/L		0.47	0.50	0.03		0.35	0.40	0.05		0.34	0.38	0.04
75- to 95-mm size group	14				25				23			
Weight, g		74.2	74.1	−0.1		68	73.8	5.8		55.5	59	3.5
W/L		0.85	0.85	0.00		0.75	0.81	0.06		0.65	0.68	0.03
100- to 127-mm size group	15				24				21			
Weight, g		158.5	156.9	−1.6		122.8	128.3	5.5		116.8	121.5	4.7
W/L		1.4	1.4	0.0		1.14	1.19	0.05		1.04	1.08	0.04
Roperia poulsoni												
27- to 33-mm size group	4				20				18			
Weight, g		3.5	3.5	0.0		3.0	3.3	0.3		3.04	3.05	0.01
W/L		0.1	0.1	0.0		0.10	0.10	0.00		0.10	0.10	0.00
35- to 40-mm size group	14				25				22			
Weight, g		5.1	5.2	0.1		5.4	5.6	0.2		5.79	5.77	−0.02
W/L		0.1	0.1	0.0		0.2	0.2	0.0		0.15	0.19	0.04
40- to 48-mm size group	5				24				20			
Weight, g		5.8	6.0	0.2		9.4	9.7	0.3		8.12	8.09	−0.03
W/L		0.2	0.2	0.0		0.19	0.19	0.0		0.19	0.19	0.00
Astraea undosa												
30- to 58-mm size group	14				25				22			
Weight, g		15.6	19.8	4.20		23.8	22.8	−1.0		24.3	24.0	−0.39
W/L		0.4	0.4	0.0		0.48	0.45	−0.03		0.48	0.47	−0.1
60- to 81-mm size group	14				25				24			
Weight, g		69.4	73	3.6		80.5	74.1	−6.4		77.5	70.0	−7.5
W/L		1.0	1.01	0.01		1.11	1.01	−0.10		1.07	0.97	−0.10
90- to 125-mm size group	15				25				24			
Weight, g		343	350	7		221.4	195.5	−25.9		239.8	215.1	−24.7
W/L		3.2	3.3	0.1		2.28	2.01	−0.27		2.39	2.15	−0.24

for normal growth of the three gastropod species. The short experimental period probably accounted for the relatively small increases in weight and length observed, particularly among large individuals.

Echinoderms

That growth and condition of both *Strongylocentrotus* species were adversely affected by conditions in both simulation experiments is indicated by comparisons of growth increments of control and treatment groups (Table 5). Effects were evident for all size groups of *S. franciscanus,* which exhibited marked decreases in mean test diameter in both thermal-effluent experiments. Such "negative" growth has been observed in natural populations of sea urchins (Ebert, 1967). Results of analyses of covariance and multiple range tests indicated that growth rates for all size groups of the two *Strongylocentrotus* species were significantly lower in both effluent regimes than in the Scripps and generating-station controls ($P < 0.05$). The depressive effect on growth of the smallest size group of *S. purpuratus* in the station 2 experiment is indicated clearly by the growth plots in Fig. 1. Loss of spines and failure to hold the spines erect were common among individuals of the larger size groups of both species exposed to effluent. These effects were also observed among large individuals of both species near station 2 in the field during the period of warmest water in late summer and early fall. None of the individuals in the Scripps and generating-station controls exhibited these effects.

There was similar evidence of effects on growth and condition of the two *Pisaster* species. Mean weights and volumes of the largest size group of *P. ochraceus* declined under both station 2 and station 3 effluent conditions (Table 5). In contrast, weights and volumes for all three size categories of *P. giganteus* increased under both treatment conditions. Results of analyses of covariance and multiple range tests indicated that increases in weight for all *Pisaster* size groups were significantly lower in both simulated effluent regimes than in Scripps and generating-station controls ($P < 0.05$). The adverse effects appeared to be somewhat more severe for *P. ochraceus* than for *P. giganteus.*

Gastropod Molluscs

Data for all size groups of *K. kelletii* indicate that their growth and condition were not adversely affected by either simulation experiment (Table 5). Weight and W/L values for all three size groups

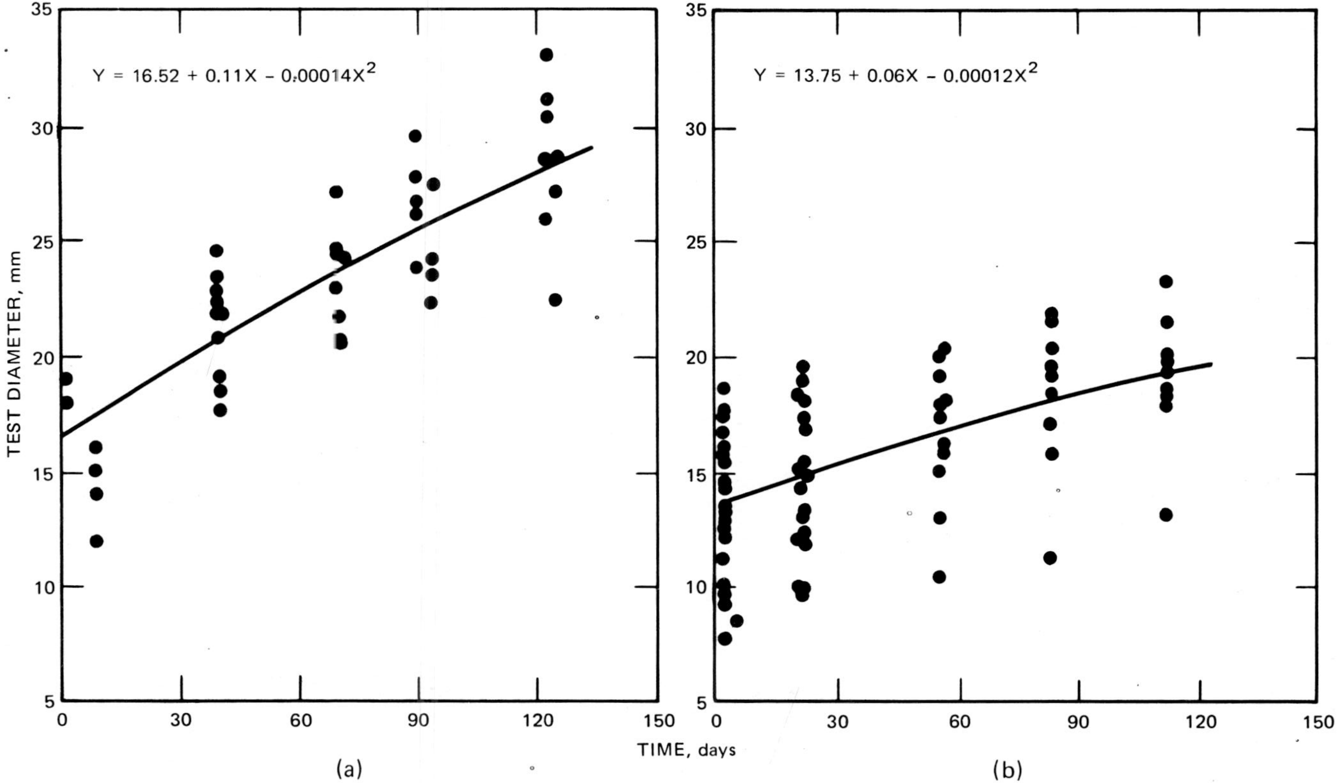

Fig. 1 Growth of juvenile *Strongylocentrotus purpuratus*, 8 to 19 mm in initial diameter, in (a) Scripps Institution control groups and (b) station 2 simulation experiment, July to November 1975. Equations for the curves are indicated.

increased under both treatment conditions. Results of analyses of covariance and multiple range tests indicated that growth in weight of the two larger size groups was significantly greater in both effluent regimes than in Scripps and generating-station controls ($P < 0.05$). *Roperia poulsoni* of all three size categories exhibited only slight changes in weight and W/L ratios both in the control and effluent regimes. Results of analyses of covariance indicated that growth in weight for all size groups was not significantly different among the control and effluent-treatment conditions ($P > 0.05$).

In contrast, all size groups of *A. undosa* exposed to both effluent conditions showed evident decreases in both weight and W/L ratios (Table 5). Results of analyses of covariance and multiple range tests indicated that growth in weight for the two smaller size groups of this species was significantly lower in both simulated effluent regimes than in the two controls ($P < 0.05$).

Chemical Conditions

Analyses were conducted simultaneously to evaluate major chemical constituents of the laboratory water sources which might be affected by operation of the generating station. The trace metals Cu, Zn, Cd, Pb, Co, Cr, and As were found to be present in very low concentrations, and there were no significant differences in concentrations in the control and thermal-effluent water supplies used at the Scripps and generating-station laboratories (Ford et al., 1976; Dorband et al., 1977). Monitoring data for the period from July to November 1975, obtained from generating-station records, were used to compare levels of oil and grease and chlorine in control and thermal-effluent water supplies at the generating-station laboratory. Results of a t-test comparing sample means indicated no significant difference ($P > 0.05$) in concentrations of oil and grease between the intake water supplied to the controls ($\overline{x} = 0.79$ mg/liter) and the thermal effluent ($\overline{x} = 0.72$ mg/liter). Concentrations of total residual chlorine in both intake and effluent water supplies at the points where they entered the laboratory were below the detection limit for the amperometric method of analysis used (< 0.01 mg/liter) during and after release of chlorine into the cooling-water system.

Weekly analyses indicated that dissolved oxygen concentrations were maintained near air-saturation levels uniformly in all control and treatment tanks. This was accomplished by using air stones and a high flow rate of seawater. Salinity and pH also remained very constant and were uniform among the control and effluent tanks.

These results indicate that there were no evident differences in any of the chemical constituents among the control and effluent water sources used in the experiments. It is important to note that the chemical measurements were not made continuously during the experiments.

DISCUSSION

The simulation experiments were designed to assess combined effects of temperature and other water-quality characteristics of two representative thermal-effluent regimes. Absence of significant differences in chemical constituents among control and thermal-effluent water sources and the low concentrations of potentially toxic chemicals suggest that the significant effects observed in the treatment groups were caused primarily by temperature conditions. The results do not rule out the possibility of direct or synergistic effects involving chemical conditions, however.

Survival of both *Strongylocentrotus* species was adversely affected by the simulated effluent regime of station 2 but apparently was not affected by the less severe effluent conditions of station 3. Similarly, growth and body condition of both species were adversely affected by the simulated effluent regime of station 2 and affected to a lesser extent by that of station 3.

Results of the station 2 experiment show that survival and growth of large adult *S. purpuratus* and *S. franciscanus* generally were affected to a greater extent than those of the smaller size groups. This probably indicates more severe metabolic limitations for large sea urchins at high temperatures (Johansen and Vadas, 1967; Ulbricht and Pritchard, 1972; Ulbricht, 1973). The results also indicate that all size groups of both species could not tolerate temperatures consistently above 22°C. Loss of spines and failure to hold the spines erect by both species, which were very common in the treatment tanks following prolonged exposure to temperatures above 22°C, usually preceded death of the individuals. This did not occur in the control tanks. Ebert (1968) and Baker (1973) showed that spine abrasion and loss affect growth markedly, and presumably both are indications of poor condition.

In a given effluent regime, effects on survival and growth generally were more severe for *S. franciscanus* than for *S. purpuratus*. *Strongylocentrotus purpuratus* occurs in shallow subtidal and intertidal areas, and *S. franciscanus* inhabits only shallow to deep subtidal areas in southern California. Thus *S. purpuratus* might be expected to show somewhat greater tolerance to high temperatures than *S. franciscanus* because of its natural exposure.

This information suggests that both species would be good thermal-indicator organisms for use in field studies. The fact that they are both generally abundant and easily sampled in nearshore areas also enhances their value as indicators.

Survival of *P. ochraceus* and *P. giganteus* was affected very little in simulations of the two thermal-effluent regimes. In contrast, there were evident effects of both effluent regimes on growth and condition of these species.

Effects on the two sea urchin species were most pronounced among large, adult individuals; this suggests metabolic limitations for asteroid starfish of large size at high temperatures. Effects on survival, growth, and condition also appeared to be somewhat more severe for *P. ochraceus* than for *P. giganteus*.

Survival of the three species of gastropod molluscs apparently was not affected by the station 3 simulation experiment, and there were no effects on *A. undosa* and *K. kelletii* of all sizes exposed to the station 2 simulation. Survival of large *R. poulsoni* and, to a lesser extent, of the two smaller size groups apparently was affected in the effluent regime at station 2.

Growth and condition of *K. kelletii* and *R. poulsoni* were not affected by either thermal-effluent regime, but those of *A. undosa* apparently were depressed in both simulations. The relatively slow natural growth rates of these species and the short duration of the experiments makes interpretation of these results difficult.

All three gastropod species are warm—temperate forms abundant in shallow water in southern California and Baja California, Mexico. *Roperia poulsoni* and *A. undosa* also occur intertidally. Thus, it is not surprising that they are relatively tolerant of the water temperature conditions to which they were exposed in these experiments.

Laboratory studies of this kind are always open to question because confinement stress and other artifacts of the experimental system can produce results not representative of natural occurrences. We attempted to duplicate important field conditions, such as water quality, water circulation, space, and natural food supply. The high survival and relatively normal growth and condition exhibited by control animals suggest these aspects of the simulation experiments were successful. The controller system duplicated effluent temperatures accurately, exposing the animals to varying temperatures and probably other water-quality conditions similar to those in shallow water at stations 2 and 3. For these reasons, we believe that most of the information obtained on survival, growth, and condition is representative of what would occur at the two field locations we simulated. It is particularly important to note that the more severe temperature conditions measured in the field for the station 2 simulation experiment influenced only a very small section of the shoreline immediately adjacent to the point of discharge.

Although all the benthic invertebrates considered in the study are relatively slow-moving forms, it is extremely difficult to mark, follow, and measure individuals for extended periods in the field. Long-term simulation experiments provide a convenient, alternate means of obtaining detailed data on survival, growth, and condition for such motile invertebrates. The fact that such species do move slowly suggests they may be vulnerable to changes in temperature and other characteristics of the effluent as the structure of the thermal plume changes.

ACKNOWLEDGMENTS

The research reported here was supported by research contracts to the San Diego State University Foundation. We thank David B. Innis, Stephen V. Orr, David C. Sommerville, Frederic L. Steinert, and William R. Weber, who provided technical assistance. We also thank Jon C. Van Olst of San Diego State University for reviewing the manuscript.

Contribution No. 32 from the San Diego State University Center for Marine Studies.

REFERENCES

Adams, J. R., 1969, Thermal Power, Aquatic Life, and Kilowatts on the Pacific Coast, paper presented at American Power Conference Annual Meeting, Chicago, Illinois, April 1967.

——, and D. G. Price, 1974, Thermal Shock Tolerances of Larval Red Abalone (*Haliotis rufescens*), in *Environmental Investigations at Diablo Canyon, 1972—1973*, J. R. Adams and J. F. Hurley (Eds.), pp. 155-174, Pacific Gas and Electric Company, San Ramon, Calif.

Baker, S. L., 1973, *Growth of the Red Sea Urchin* Strongylocentrotus franciscanus *(Agassiz) in Two Natural Habitats*, M.S. Thesis, San Diego State University, San Diego, Calif.

Boolootian, R. A. (Ed.), 1966, *Physiology of Echinodermata*, Wiley—Interscience Publishers, Inc., New York.

Brewer, G. D., 1976, Thermal Tolerance and Resistance of the Northern Anchovy *Engraulis mordax*, *Fish. Bull.*, 74: 433-455.

Des Voigne, D. M., 1970, Preliminary Observations of the Thermal Tolerances of Selected Fauna of Sequim Bay, Washington, *Northwest Sci.*, 44: 113-121.

Dorband, W. R., J. C. Van Olst, J. M. Carlberg, and R. F. Ford, 1977, Effects of Chemicals in Thermal Effluent on *Homarus americanus* Maintained in

Aquaculture Systems, *Proceedings, World Mariculture Society Seventh Annual Meeting*, San Diego, Calif., January 1976, pp. 391-411.

Ebert, T. A., 1967, Negative Growth and Longevity in the Purple Sea Urchin *Strongylocentrotus purpuratus* (Stimpson), *Science*, 157(3788): 557-558.

——, 1968, Growth Rates of the Sea Urchin *Strongylocentrotus purpuratus* (Stimpson) Related to Food Availability and Spine Abrasion, *Ecology*, 49: 1075-1091.

Ford, R. F., D. G. Foreman, K. J. Grubbs, D. B. Innis, D. C. Sommerville, and F. L. Steinert, 1976, *Effects of Thermal Effluent on Marine Organisms*, Technical Report, Center for Marine Studies, San Diego State University, San Diego, Calif.

Fotheringham, N., 1971, Life History Patterns of the Littoral Gastropods *Shaskyus festivus* (Hinds) and *Ocenebra poulsoni* Carpenter (Prosobranchia: Muricidae), *Ecology*, 52(5): 742-757.

Hair, J. R., 1971, Upper Lethal Temperature and Thermal Shock Tolerances of the Opossum Shrimp *Neomysis awatschenis*, from the Sacramento—San Joaquin Estuary, California, *Calif. Fish Game*, 57: 17-20.

Jennison, B. L., 1975, The Effect of Increased Temperature on Reproduction in the Sea Anemone *Anthopleura elegantissima*, *Am. Zool.*, 15: 787.

Johansen, K., and R. L. Vadas, 1967, Oxygen Uptake and Responses to Respiratory Stress in Sea Urchins, *Biol. Bull.*, 132: 16-22.

Lees, D. C., 1970, *The Relationship Between Movement and Available Food in the Sea Urchins* Strongylocentrotus franciscanus *and* Strongylocentrotus purpuratus, M.S. Thesis, San Diego State University, San Diego, Calif.

Lockheed Center for Marine Research, 1977, *Marine Biofouling Control Studies, San Onofre Nuclear Generating Station.* Phase II Thermal Tolerance Studies, 1975—1977, Carlsbad, Calif.

Norris, K. S., 1963, The Functions of Temperature in the Ecology of the Percoid Fish *Girella nigricans* (Ayers), *Ecol. Monogr.*, 33: 23-62.

Schwalm, C. C., 1973, *Population Dynamics and Energetics of* Astraea undosa, M.S. Thesis, San Diego State University, San Diego, Calif.

Ulbricht, R. J., 1973, The Influence of Temperature Acclimation upon the Metabolic Rate of the Purple Sea Urchin, *Strongylocentrotus purpuratus:* Alternate Interpretations, *Comp. Biochem. Physiol.*, 45: 667.

——, and A. W. Pritchard, 1972, Effect of Temperature on the Metabolic Rate of Sea Urchins, *Biol. Bull.*, 142: 178-185.

Vlaming, V. L. de, 1971, Thermal Selection Behavior in the Estuarine Goby *Gillichthys mirabilis* Cooper, *J. Fish. Biol.*, 3: 277.

Wolcott, T. C., 1973, Physiological Ecology and Intertidal Zonation in Limpets (*Acmaea*): A Critical Look at "Limiting Factors," *Biol. Bull.*, 145: 389.

Wolfson, A. A., 1974, *Some Effects of Increased Temperature on the Settlement and Development of a Marine Community in the Laboratory*, IMR Reference No. 74-13, Institute of Marine Resources, University of California, La Jolla.

EFFECTS OF THERMAL ALTERATION ON MACROINVERTEBRATE FAUNA IN THREE ARTIFICIAL CHANNELS

DALLAS E. ALSTON, J. M. LAWRENCE, DAVID R. BAYNE, and FRANCES F. CAMPBELL
Department of Fisheries and Allied Aquacultures, Agricultural Experiment Station, Auburn University, Auburn, Alabama

ABSTRACT

Three earthen channels constructed at the Greene County Steam-Electric Generating Plant in Alabama were used in a 2-year study to evaluate the effects of heated effluents on aquatic biota. Macroinvertebrates were collected with modified Hester—Dendy multiple-plate samplers and an Ekman dredge at 4- to 6-week intervals throughout the period. Channel 1, a control habitat, received unheated river water; channel 2, a thermally stratified habitat, received both unheated water and heated effluent waters from the steam plant; and channel 3 received only heated water. The highest average weekly water temperature in channel 1 was 30.8°C and in channel 3, 35.5°C. During most periods oligochaetes and chironomid larvae (Diptera) were the dominant organisms found in both plate and dredge samples. Total fish-food organisms collected from plate and dredge samples were in channel 1, 132,000, and in channels 2 and 3, 160,000. The overall mean diversities for dredge samples indicated that channels 1 and 2 were not significantly different, but channel 3 had a lower diversity. The plate samplers indicated that channel 1 had the highest diversity; channel 2, intermediate; and channel 3, lowest. Channels 1 and 2 had more mayflies (Ephemeroptera) than did channel 3.

Much work has been done to determine the effects of thermal alteration on aquatic ecosystems (Federal Water Pollution Control Administration, 1967; Raney and Menzel, 1969; Coutant and Goodyear, 1972; Coutant and Pfuderer, 1973; 1974; Coutant and Talmage, 1975; Gibbons and Sharitz, 1974; Esch and McFarlane, 1976; Lehmann, 1976). Some of these studies were conducted in the vicinity of heated-water effluents with little or no control over variables, and controlled studies used extremely artificial in-vitro experiments. Our study attempted to compromise the two extremes by controlling the amount of flow and water temperature in earthen channels. The channels were constructed in 1974 by the Alabama Power Company to evaluate the environmental effects of heated effluents on aquatic biota. The experiments were designed to gather specific information to be used in an overall effort to determine the environmental impact of heated water released from steam electric-generating plants into Alabama streams and reservoirs. Specifically, the investigative emphasis was on water quality, production of fish-food organisms, and health, reproduction, and growth of confined fishes. In this paper the only objectives considered are determining the effects of heated water on fish-food organisms in earthen channels and investigating seasonal changes in fish-food organisms which occur over a 2-year period (October 1974 through October 1976).

MATERIALS AND METHODS

Channels

Three earthen channels, 6 by 160 m, with a maximum depth of 2 m (Fig. 1), were constructed in Greene County, Alabama, on the site of the 500-MW Greene County Steam-Electric Generating Plant. Water entered the southern end of each channel to provide a

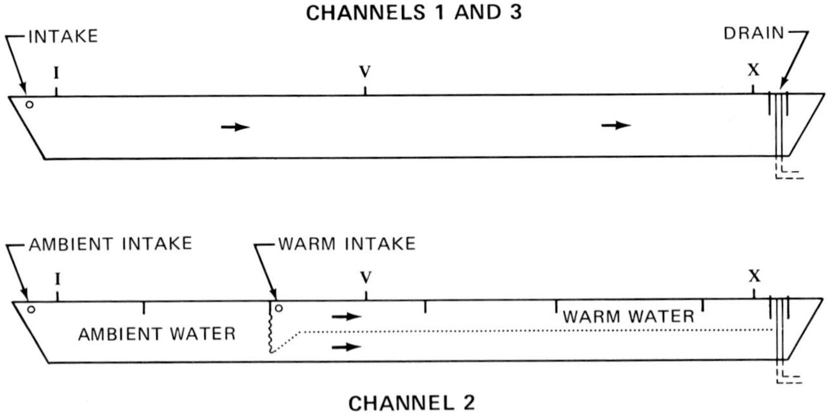

Fig. 1 Longitudinal section of the channels. Roman numerals are stations; warm water is the heated-water density current, and arrows indicate direction of flow.

constant, unidirectional water flow. At the northern end of each channel a standpipe was installed to accommodate overflow. Concrete collars were poured to a depth of 61 cm around the periphery of each channel to prevent waterline erosion.

Water Supply

Unheated and heated water was pumped directly from the steam plant. Channel 1 received only unheated water; channel 2 received both unheated and heated water; and channel 3 received only heated water. Unheated water was pumped from the steam-plant intake canal leading from the Black Warrior River. Heated water was pumped from the head end of the steam-plant discharge canal to the channels. Invertebrates could enter each channel through the supply lines; thus organisms entering each channel depended to some extent on the fauna of either the intake or discharge canal. The total design capacity for each water supply was 20 liters/sec. A series of wooden baffles was used to reduce surface currents, and heated water was introduced at the surface of channel 2 to create a warm layer over about two-thirds of this channel (Fig. 1). Motile organisms could move into or out of the warm-water layer.

Macroinvertebrate Sampling

Ten permanent, equally spaced stations were established along the length of each channel (Fig. 1). Macroinvertebrates were sampled with modified Hester–Dendy multiple-plate samplers (Hester and Dendy, 1962) and an Ekman dredge. The dredge samples were taken at each end of the three channels between stations I and III and between VIII and the drain. Each dredge sample consisted of three random grabs along the middle of the channel to give a composite sample of 697 cm^2. Collections were made at 4- to 6-week intervals. All mud samples were screened through a U. S. Standard No. 30 sieve in the field, and the residue was preserved with 5% formalin and stained with rose bengal to aid in the sorting of organisms. The samples were then transported to the fisheries laboratory at Auburn University, where the organisms were floated from the detritus by use of a saturated salt solution. The residue left in the salt solution was checked for bivalves, caddis-fly cases, and other organisms. The organisms in the samples were sorted into various taxa and counted.

The plate samplers were constructed of three circular plexiglass plates with two spacers between each plate to allow a total available surface area of 562.3 cm^2. These samplers were suspended vertically in the water column at depths of 0.6 and 1.2 m. Duplicate plate samplers were placed at stations I and IX in each of the three channels and allowed to collect organisms. Samples were obtained at monthly intervals throughout the study period.

Subsamples were often taken to estimate the number of organisms in each sample, and all organisms were keyed to the lowest taxa possible within the limitations of time and available expertise.

Diversity Indexes

Weber (1973) recommended the Shannon–Weaver function to achieve greater uniformity of diversity ($\bar{d}$) calculations. For simplified calculations, the machine formula presented by Lloyd, Zar, and Karr (1968) was used:

$$\bar{d} = \frac{C}{N} (N \log N - \Sigma n_i \log n_i) \qquad (1)$$

In this formula C = 3.3219 (to convert base 10 log to base 2), N is the total number of individuals, and n_i is the total number of individuals in the ith species.

RESULTS

Channel Habitat

The highest average weekly water temperatures in channel 1 in 1975 and 1976 were 30.4 and 30.8°C, respectively, and in channel 3, 34 and 35.5°C, respectively. The highest average weekly temperatures of the unheated and heated portions of channel 2 were between those given for channels 1 and 3. Throughout the study period the temperature differential between channels 1 and 3 ranged from 3 to 10°C. The highest average daily water temperature for channel 3 was 36.7°C on Aug. 1, 1976.

Water-quality data indicated that most of the parameters measured were similar. Channel 1 had the greatest turbidity and suspended-matter concentration. Dissolved oxygen concentration was never less than 3.7 ppm in any channel. Complete listings of the water-quality data are given in the Alabama Power Company's final report (Lawrence and Bayne, 1977).

The channels had no macrophytes but did have both attached and unattached algae populations as enumerated by Bright (1977). Fish populations in the channels, which were stocked as closely as possible to the natural fish populations in the Black Warrior River, are listed by Turner, Lawrence, and Bayne (in press).

Macroinvertebrate Dominance

To aid in processing the data, we divided the 27-month sampling interval into 9 equal periods. During most periods oligochaetes and chironomids (Diptera) were the dominant organisms found in both the plate and dredge samples. Other major groups were the caddis flies (Trichoptera), ceratopogonids (Diptera), gastropods (Mollusca), mayflies (Ephemeroptera), and water boatmen (Hemiptera).

Macroinvertebrate Totals

The estimated total numbers of larger fish-food organisms and taxa collected by the plates and dredge samples are indicated for each period in Tables 1 and 2. Plate samplers in channel 3 collected the greatest total number of organisms (130,000) but had fewer taxa in each period than did the other channels. Dredge samples collected in channel 2 yielded the greatest number of organisms (55,000), with channel 3 again having the fewest taxa. Totaling the organisms found for all the periods for each channel for both the dredge and plate samplers, we find that channels 2 and 3 each had ~160,000 organisms, and channel 1 had ~130,000.

Macroinvertebrate Diversity Index

Diversity indexes ($\bar{d}$) were calculated for the larger fish-food organisms (oligochaetes, amphipods, snails, clams, and insects) for each station in each channel, and the values were averaged to obtain a mean diversity index ($\bar{D}$) for each period for each channel. Means were compared using Duncan's new multiple range test ($P = 0.05$) (Tables 1 and 2).

Winter

Even though it was not statistically significant, channel 1 had the highest $\bar{D}$ for both winter periods (1975 and 1976) for the plate samplers. Dredge samples from channel 1 ($\bar{D} = 1.78$) were not significantly different from either channel 2 or 3 even though the heated-water channel (3) had a $\bar{D}$ of only 0.67 during the winter of 1976.

Spring

The plate-sampler data indicated no significant differences during the spring periods even though channel 1 maintained the highest $\bar{D}$ for both periods. During the first spring (1975), the $\bar{D}$ for channel 1 was significantly higher than that for channel 3 for the dredge samples. A year later (spring 1976), the $\bar{D}$'s had dropped in all three

TABLE 1

MEAN DIVERSITY ($\bar{D}$), AVERAGE NUMBER OF TAXA ($\bar{T}$), AND TOTAL NUMBER OF LARGER FISH-FOOD ORGANISMS (N) FOR NINE PERIODS FOR THE CHANNEL PLATE SAMPLERS*

Period	Plate-sampler data		
Fall 1974			
Channel	2	3	1
$\bar{D}$	1.93 ± 0.84	2.06 ± 0.92	2.09 ± 0.58
$\bar{T}$	8.8	9.8	8.8
N	4200	5500	12,000
Winter 1975			
Channel	3	2	1
$\bar{D}$	1.37 ± 0.32	1.95 ± 0.50	2.15 ± 0.32
$\bar{T}$	14.3	22.0	16.2
N	7400	4300	5700
Spring 1975			
Channel	3	2	1
$\bar{D}$	1.83 ± 0.72	1.96 ± 0.23	2.48 ± 0.42
$\bar{T}$	11.5	15.0	17.2
N	8400	6200	6300
Summer 1975			
Channel	3	2	1
$\bar{D}$	1.11 ± 0.28	2.00 ± 0.86	2.21 ± 0.24
$\bar{T}$	15.3	21.3	18.5
N	45,000	26,000	23,000
Fall 1975			
Channel	3	2	1
$\bar{D}$	1.77 ± 0.86	2.30 ± 0.48	2.46 ± 0.46
$\bar{T}$	8.8	13.5	13.5
N	44,000	33,000	28,000
Winter 1976			
Channel	3	2	1
$\bar{D}$	1.66 ± 0.54	2.06 ± 0.53	2.44 ± 0.24
$\bar{T}$	8.8	15.5	18.5
N	14,000	28,000	10,000
Spring 1976			
Channel	2	3	1
$\bar{D}$	2.10 ± 0.64	2.48 ± 0.31	2.94 ± 0.49
$\bar{T}$	16.8	15.3	20.8
N	3000	2300	2900
Summer 1976			
Channel	3	1	2
$\bar{D}$	2.10 ± 0.97	2.40 ± 0.34	2.93 ± 0.15
$\bar{T}$	13.5	22.0	19.5
N	3700	6200	3400
Fall 1976			
Channel	2	3	1
$\bar{D}$	1.00 ± 0.54	1.12 ± 0.66	2.28 ± 0.50
$\bar{T}$	14.5	10.0	15.3
N	5300	2000	5100

*Mean diversity ($\bar{D}$) is calculated by averaging station diversities during the period and given ± standard deviation; $\bar{D}$'s compared with Duncan's new multiple range test are not significantly different ($P > 0.05$) if underlined by a common line.

TABLE 2

MEAN DIVERSITY ($\bar{D}$), AVERAGE NUMBER OF TAXA ($\bar{T}$), AND TOTAL NUMBER OF LARGER FISH-FOOD ORGANISMS (N) FOR NINE PERIODS FOR THE CHANNEL DREDGE SAMPLES*

Period	Dredge data		
Fall 1974			
Channel	1	3	2
$\bar{D}$	2.49 ± 0.98	2.68 ± 0.47	2.81 ± 0.93
$\bar{T}$	20.5	14.5	20.0
N	12,000	5500	4200
Winter 1975			
Channel	3	1	2
$\bar{D}$	0.94 ± 0.14	1.01 ± 0.61	1.92 ± 0.12
$\bar{T}$	14.0	16.0	21.0
N	7400	5700	4300
Spring 1975			
Channel	3	2	1
$\bar{D}$	1.45 ± 1.59	2.57 ± 0.38	3.29 ± 0.01
$\bar{T}$	8.0	14.0	16.0
N	8400	6200	6300
Summer 1975			
Channel	3	1	2
$\bar{D}$	1.14 ± 0.73	1.85 ± 0.30	2.17 ± 2.16
$\bar{T}$	16.0	18.0	14.0
N	45,000	23,000	26,000
Fall 1975			
Channel	2	3	1
$\bar{D}$	2.47 ± 0.48	2.67 ± 0.33	3.00 ± 0.17
$\bar{T}$	15.5	14.5	16.5
N	33,000	44,000	28,000
Winter 1976			
Channel	3	1	2
$\bar{D}$	0.67 ± 0.10	1.78 ± 0.98	2.06 ± 0.22
$\bar{T}$	6.0	8.5	7.0
N	14,000	10,000	28,000
Spring 1976			
Channel	3	2	1
$\bar{D}$	0.41 ± 0.05	0.65 ± 0.28	0.78 ± 0.80
$\bar{T}$	9.0	11.0	0.0
N	2300	3000	2900
Summer 1976			
Channel	3	2	1
$\bar{D}$	0.76 ± 0.77	0.98 ± 0.28	2.40 ± 0.09
$\bar{T}$	10.5	11.0	13.0
N	3700	3400	6200
Fall 1976			
Channel	3	1	2
$\bar{D}$	1.89 ± 0.78	2.20 ± 0.38	3.10 ± 0.15
$\bar{T}$	16.0	19.0	14.5
N	2000	5100	5300

*Mean diversity ($\bar{D}$) is calculated by averaging station diversities during the period and given ± standard deviation; $\bar{D}$'s compared with Duncan's new multiple range test are not significantly different ($P > 0.05$) if underlined by a common line.

channels (channel 1, 0.78; channel 2, 0.65; channel 3, 0.41), and there were no significant differences.

Summer

The plate-sampler data indicated that $\bar{D}$ in 1975 for channel 1 was significantly greater than that for channel 3. Dredge-sample data for this same time did not indicate that channel 1 was significantly different from the other channels. In 1976, however, $\bar{D}$ (2.40) for channel 1 was significantly higher than for either channel 2 (0.98) or 3 (0.76).

Fall

The plate-sampler data indicated that $\bar{D}$ for channel 1 in 1976 was significantly greater than that for channel 2 or 3. The $\bar{D}$ for channel 2 was highest, 3.10, compared with 2.20 for channel 1 [illegible] 1.89 for channel 3.

All Seasons

The following overall mean diversities were computed for each channel for all of the periods and years and were compared using Duncan's new multiple range test [means underlined with a common line were not significantly different, (P = 0.05)].

Sample	Channel 3	Channel 2	Channel 1
Ekman dredge	1.40	2.08	2.09
Plate sampler	1.72	2.07	2.34

Selected Fish-Food Organisms

Oligochaetes

In general, there were more oligochaetes in channel 3 than channel 1. One of the species of oligochaetes which was prevalent in channel 3 was *Branchiura sowerbyi* Beddard (family Tubificidae), an exotic species that was introduced into Europe and North America from Southeast Asia on tropical and subtropical plants (Aston, 1968; Brinkhurst and Cook, 1974).

Snails

During the summer of 1975, channel 2 maintained the highest population of snails, with an average of 960 individuals/m^2. The next summer the snail population was smaller, but channel 2 again had the highest average, 400 snails/m^2, compared with 200 for channel 1 and 280 for channel 3. Throughout the 27-month study, channel 3 generally had the lowest population of snails of the three channels. *Physa* was the most prevalent snail, but *Ferrissia*, *Gyraulus*, and *Lymnaea* were also found.

Bivalves

Corbicula manilensis (Philippi), the Asiatic clam, dominated the clam population in all channels. Channel 2 had many more *Corbicula* than did channel 1 or 3.

Mayflies

The highest population of mayflies for the plate samplers was in channel 1, with an intermediate population in channel 2 and the smallest in channel 3. The dredge data indicated similar values, except that during the winter of 1976 channel 2 had more mayflies than channel 1.

Chironomids

Dredge samples from channel 3 had fewer chironomid larvae than did either of the other channels. Except for the initial periods when the chironomids were becoming established in both channels 1 and 3, there was only one period, spring of 1976, when the population of chironomids in channel 1 fell below that of channel 3. The predaceous chironomids were more numerous in channel 3 than channel 1. Predaceous chironomid larvae include the genera *Clinotanypus*, *Procladius*, *Psectrotanypus*, and *Ablabesmyia* (Darby, 1962).

Ceratopogonids

Ceratopogonids (Diptera) were more prevalent in the plate samplers during summer months in channels 2 and 3 than in channel 1. No significant trends for the nine periods could be established for the dredge samples.

DISCUSSION

The decrease in water temperature the winter of 1976 probably caused the lower $\bar{D}$'s found in all channels for the spring of 1976. The high temperatures in the summer of 1976 seemed to affect the

$\bar{D}$'s of the heated-water channels. The overall $\bar{D}$'s for the Ekman dredge samples indicated that channels 1 and 2 were not significantly different but that $\bar{D}$ in channel 3 was significantly lower. This indicates that the floating heated-water plume had no effect on bottom organisms. The mean $\bar{D}$ for plate samplers indicated that each channel had a different $\bar{D}$, 1 being highest; 2, intermediate; and 3, lowest. The floating plume lowered the $\bar{D}$ of large fish-food organisms in channel 2 but was not as severe as the effects measured in channel 3.

The prevalence of *Branchiura sowerbyi* in channel 3 agrees with results reported by Brinkhurst and Cook (1974), who found that they attained large numbers and reached sexual maturity in heated effluents. Our findings on mayfly populations corroborate results reported by Howell and Gentry (1974), who stated that mayflies increase in numbers of both individuals and species in a progression from thermally affected to unheated waters. The floating plume in channel 2 did not affect mayfly abundance. Large populations of mayflies were found in the dredge samples. Channel 3 contained numerous predaceous chironomids. Fish prey heavily upon larvae of all chironomids, and, since these predaceous forms are not protected by a silken case, they are more vulnerable than the case-building chironomid larvae. Oligochaetes and ceratopogonids, which are also prime fish food, had substantial populations in channel 3.

Even though channel 3 was maintained as a heated-water habitat, it appeared to have supported a substantial population of desirable fish-food organisms. Channel 2 diversity was not affected as much as that of channel 3 since mayflies were present. This was probably because of the two separate environments in channel 2, a heated-water layer and unheated-water layer. It appears that heated water caused shifts in populations and changed the diversity of channels 2 and 3, but enough organisms were present in both channels to support a healthy fish population.

ACKNOWLEDGMENT

Funding and field facilities for the research reported here were provided by the Alabama Power Company.

REFERENCES

Aston, R. J., 1968, The Effect of Temperature on the Life Cycle, Growth and Fecundity of *Branchiura sowerbyi* (Oligochaeta: Tubificidae), *J. Zool.*, 154: 29-40.

Bright, D. J., 1977, *Effects of Heated Water on Periphytic Communities of Natural and Artificial Streams*, M.S. Thesis, Auburn University, Auburn, Ala.

Brinkhurst, R. O., and D. G. Cook, 1974, Aquatic Earthworms (Annelida: Oligochaeta), in *Pollution Ecology of Freshwater Invertebrates*, C. W. Hart, Jr., and Samuel L. H. Fuller, (Eds.), pp. 143-156, Academic Press, Inc., New York.

Coutant, C. C., and C. P. Goodyear, 1972, Thermal Pollution—Biological Effects, *J. Water Pollut. Control Fed.*, 44: 1250-1294.

———, and H. A. Pfuderer, 1973, Thermal Pollution—Biological Effects, *J. Water Pollut. Control Fed.*, 45: 1331-1369.

———, and H. A. Pfuderer, 1974, Thermal Effects, *J. Water Pollut. Control Fed.*, 46: 1476-1540.

———, and S. S. Talmage, 1975, Thermal Pollution—Biological Effects, *J. Water Pollut. Control Fed.*, 47: 1656-1710.

Darby, R. E., 1962, Midges Associated with California Rice Fields, with Special Reference to Their Ecology (Diptera: Chironomidae), *Hilgardia*, 32: 1-206.

Esch, G. W., and R. W. McFarlane (Eds.), 1976, *Thermal Ecology II*, ERDA Symposium Series, Augusta, Ga., April 2—5, 1975, CONF-750425, NTIS.

Federal Water Pollution Control Administration, 1967, *Temperature and Aquatic Life*, Division of Pollution Surveillance, Laboratory Investigations No. 6, Cincinnati, Ohio.

Gibbons, J. W., and R. R. Sharitz (Eds.), 1974, *Thermal Ecology* AEC Symposium Series, Augusta, Ga., May 3—5, 1973, CONF-730505, NTIS.

Hester, F. E., and J. S. Dendy, 1962, A Multiple-Plate Sampler for Aquatic Macroinvertebrates, *Trans. Am. Fish. Soc.*, 91(4): 420-421.

Howell, F. G., and J. B. Gentry, 1974, Effect of Thermal Effluents from Nuclear Reactors on Species Diversity of Aquatic Insects, in *Thermal Ecology*, AEC Symposium Series, Augusta, Ga., May 3—5, 1973, J. W. Gibbons and R. R. Sharitz (Eds.), pp. 562-571, CONF-730505, NTIS.

Lawrence, J. M., and D. R. Bayne, 1977, *Final Report on Impact of Heated Water on Aquatic Communities of Experimental Fish Channels, 1974—1976*, Alabama Power Company, Birmingham, Ala.

Lehmann, E. J., 1976, *Thermal Pollution*. Part 2. Biological Effects, U. S. Department of Commerce, NTIS.

Lloyd, M., J. H. Zar, and J. R. Karr, 1968, On the Calculation of Information-Theoretical Measures of Diversity, *Am. Mid. Nat.*, 79(2): 257-272.

Raney, E. C., and R. W. Menzel, 1969, *Heated Effluents and Effects on Aquatic Life with Emphasis on Fishes: A Bibliography*, U.S. Department of the Interior, Office of Water Resources Research, Washington, D.C.

Turner, C. J., J. M. Lawrence, and D. R. Bayne, in press, The Effects of Heated Water on Warmwater Fish in Earthen Raceways, *Proc., Annu. Conf. Southeast. Assoc. Game Fish Comm.*

Weber, D. I. (Ed.), 1973, *Biological Field and Laboratory Methods for Measuring the Quality of Surface Waters and Effluents*, National Environmental Research Center, U.S. Environmental Protection Agency, Cincinnati, Ohio.

EFFECTS OF POWER-PLANT OPERATION ON THE LITTORAL BENTHOS OF BELEWS LAKE, NORTH CAROLINA

DAVID R. LENAT
Department of Environmental Sciences and Engineering,
University of North Carolina, Chapel Hill, North Carolina

ABSTRACT

The littoral benthos of Belews Lake in North Carolina, a 1500-ha cooling reservoir for a 2200-MW coal-fired power plant, was studied over a 3-year period, one preoperational year and two postoperational years. Comparing control areas with heated areas allowed separation of normal successional changes from changes associated with power-plant operation. Twenty-four taxa were analyzed in detail, and ten taxa appeared to be directly or indirectly affected by power-plant operation. The densities of *Polypedilum digitifer*, *Nais communis*, and *Hyalella azteca* decreased with increases in temperature. Densities of two other taxa, *Hexagenia munda* and *Limnodrilus hoffmeisteri*, increased with slight increases in temperature but decreased after further increases. Six other taxa appeared to have been indirectly affected by power-plant operation through changes in littoral macrophytes, periphyton, or other factors. Use of percent similarity coefficients showed that the imposition of a thermal gradient resulted in a corresponding "biological" gradient.

The effects of power-plant operation on aquatic environments have been frequently studied during the last decade. The impact of thermal effluents is still in question, however. Many studies suffer from a lack of preoperational data and/or inadequate controls, and others deal with environments having multiple pollutants; this makes it difficult to study only the impact of power-plant operation. The study design for this investigation attempted to minimize these problems.

Studies at Belews Lake, North Carolina, have dealt with chemical and physical parameters, phytoplankton, zooplankton, and both littoral and profundal benthos (C. M. Weiss, T. P. Anderson, Lenat, and P. H. Campbell, manuscript in preparation).

Benthic macroinvertebrates are well suited to assessment of environmental stress. The benthos are a less mobile and less variable group than plankton or fish and, therefore, serve well to describe the quality of the environment at a particular site (Garton and Harkins, 1970). Chemical and physical assessments provide a statement of water quality at the time of sampling, but the benthic community acts as a monitor or integrator of water quality over extended periods. It responds to maximum and minimum conditions of the environment rather than average values and thus functions as a detector of even occasional changes. Studies of the benthos also assess the overall conditions of the environment rather than testing only for specific parameters.

Since heated discharges tend to "float," we might expect that the greatest potential for thermal stress would occur in littoral areas. Littoral organisms, however, must adapt to a wide temperature range to exist under natural conditions. North Carolina surface waters naturally reach temperatures up to 33°C. This investigation attempts to assess the effects of artificial temperature increments superimposed over the natural temperature fluctuations.

STUDY SITE

Belews Lake is meso-oligotrophic (Weiss and Kuenzler, 1976) and has a surface area of 1500 ha. The maximum depth is 40 m, and the retention time is approximately 1000 days. The geographic and chemical characteristics of the lake act to maximize the effects of thermal stress. The North Carolina Piedmont is located near the transition from a temperate to a subtropical climate. Therefore, temperate species are near their southern limits, encountering temperatures near their tolerance limit (Cairns, 1975). The North Carolina Piedmont generally has very soft water. Alkalinities in Belews Lake, for example, are usually less than 25 mg/liter. According to Cairns, the capacity to adjust to thermal additions is greatest in hard water because of natural buffering processes.

Belews Lake has been investigated for 7 years. Year 1 (1970–1971) was a preimpoundment study, and years 2 and 3 were periods of lake filling. The first power-plant unit started operation in year 5, and the second unit started in year 6. This paper compares data from year 4, the best "control" data, with that from years 5 and 6.

The substrate at each benthic station is heterogeneous. The most common type of substrate is a firm clay—sand, but patches of rubble or detritus occur frequently. Macrophytes invaded Belews Lake in year 5 and occurred at all stations (except one) in year 6. *Potamageton diversifolias* and *Najas* sp. were the most common species. The density of macrophytes in the discharge area was much less than that observed in the main lake area. At discharge stations macrophytes usually occurred only as scattered patches in depths of 0 to 0.5 m.

The most complete temperature data are for years 5 and 6 and are presented in Table 1. Temperatures as high as 37°C occurred during the periods sampled. Dissolved oxygen (not shown) was usually near saturation and was not considered to be limiting.

METHODS

The 14 littoral benthic sampling stations at Belews Lake are shown in Fig. 1. Stations are located to form two longitudinal series, one in the main lake basin (65M, 8, 4, 1907, 2, 1906, and 1116) and one in the discharge basin (7, 5, 1904B, 1904A, and 1904). Two additional stations were located in the main lake basin, near the cooling water intake (3) and the ash basin outfall (1). Samples were taken quarterly in August, October, January, and April.

A box-type sampler was used which sampled approximately 0.2 m^2 at depths of 0.3 to 0.7 m. In all sample processing, screens with a mesh aperture of 300 μ were used. Samples were preserved in 95% ethyl alcohol and stored in quart mason jars.

Samples were stained with rose bengal and sorted under a 7× dissecting microscope. For the Oligochaeta and Chironomidae, mounting with CMC-10 or polyvinyl lactophenol and identification under a compound microscope were required.

Analysis of benthic collections includes computation of Shannon—Wiener diversity index (Wilhm, 1967) and percent similarity coefficients (Johnson and Brinkhurst, 1971).

Four samples were taken at each station, and an effort was made to sample all substrate types. Since this procedure generates very high variability among replicates, statistical analysis treats the four replicates as a composite sample. Comparisons are made, not among stations, but between two station groups.

The basic statistical design was a complete randomized-block procedure. All analyses were run on log-transformed data. Interaction terms were used as a divisor in computing F values. This procedure was facilitated by use of the program FACTOR of the

TABLE 1

LITTORAL TEMPERATURE (°C)

Sample time	Sample station													
	65M	8	4	1907	2	1906	1	3	1116	7	5	1904B	1904A	1904
							Year 5							
August	28.0	30.0	30.0	32.8	30.0	27.8	29.0	28.2	27.0	31.0	33.0	*	*	34.8
October	16.5	18.2	19.0	19.5	19.0	18.0	18.3	18.0	18.0	20.7	23.1	23.5	23.5	23.5
January	6.5	10.5	11.1	12.1	12.2	11.0	10.9	10.4	10.5	13.8	15.0	15.7	15.6	15.0
April†	19.5	18.5	18.5	17.5	17.5	17.5	17.8	18.0	17.5	18.5	18.3	17.5	17.5	17.5
							Year 6							
August	29.8	32.3	32.4	32.8	33.1	33.0	33.0	33.5	32.2	34.6	36.0	36.0	35.5	36.7
October	19.7	21.6	22.0	22.3	21.2	21.2	20.6	20.5	21.2	21.2	23.3	24.2	24.0	24.5
January	6.5	13.2	13.4	15.0	14.1	12.0	11.0	11.4	11.7	13.6	16.5	17.5	16.9	17.7
April	22.1	23.7	24.4	25.7	24.5	23.5	23.3	23.0	22.7	24.0	26.2	27.3	27.2	27.6

*No data.

†Power plant was shut down.

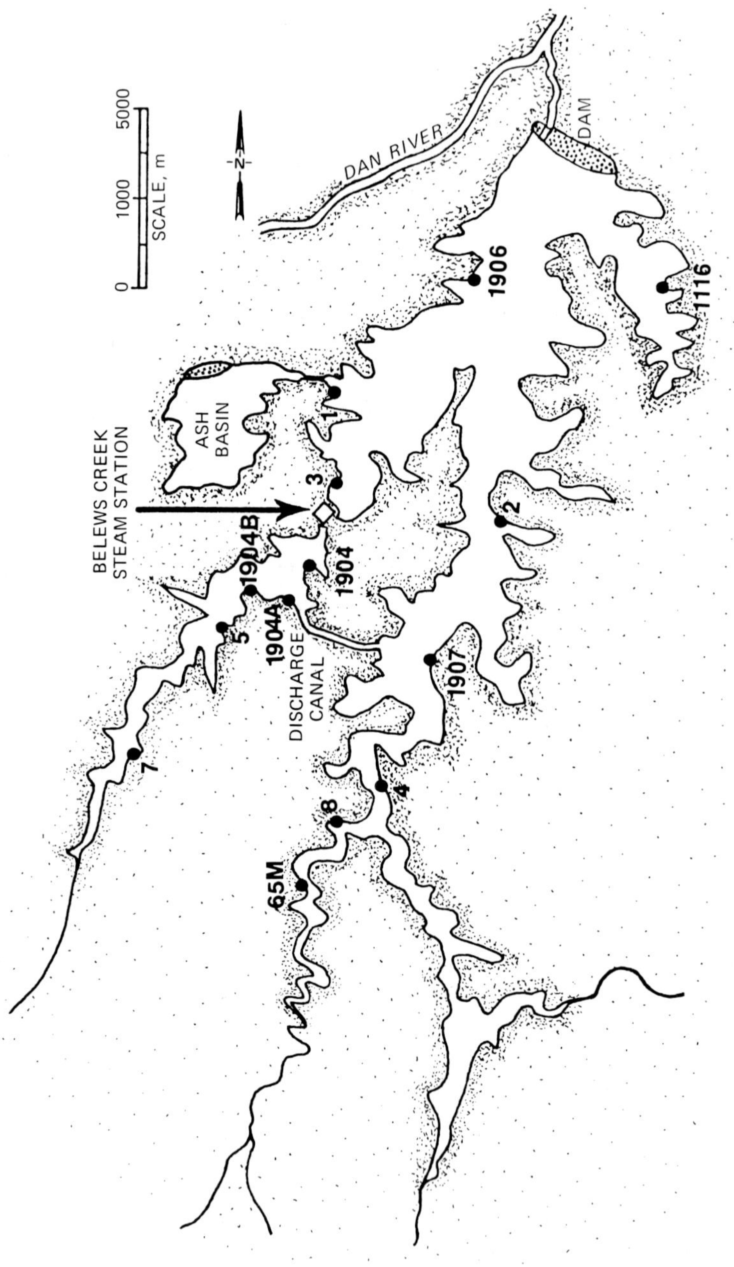

Fig. 1 Belews Lake sampling stations.

Department of Biostatistics time-sharing library, University of North Carolina. The basic analysis makes comparisons among years and between stations groups. The two groups were main lake (M) and discharge (D) stations. Occasionally a subset of the M group, termed upper main lake (U), is used (stations 65M, 8, 4, 1907, and 2). Because the station—date interactions frequently confounded analysis of between-year differences, only the yearly averages for stations could be compared.

For each dominant organism a 3 by 2 matrix was set up, showing the yearly standing-crop averages for both M and D stations over years 4 to 6, and F tests were run to check significant differences among years and between station groups. Years were compared by a simple two-way analysis of variance that used both year and station as variables. Comparing station groups was more difficult since an arbitrary pairing of stations was not possible. Therefore, the values within each station group were rank-ordered and station serves as a dummy variable. A minimum of seven F tests were performed for each 3 by 2 matrix.

For all dominant organisms, F tests were performed to assess both the differences between discharge (D) stations and main lake (M) stations and the differences among years. It was necessary to define a series of analysis criteria to facilitate comparisons among years. These criteria are:

1. Stability. No significant difference exists among years for either station group, or the significant difference is slight ($< 20/m^2$).

2. Convergence. A significant difference exists between station groups for the first year but not for the second.

3. Colonization or succession. A significant difference exists among years for both station groups. If there is a consistent difference between M and D stations, the change is proportionally equal (parallel change).

4. Power-plant effect.

a. No significant difference exists among years for M stations, but there is a significant difference among years for D stations.

b. A significant difference exists between basins for both years. The increase or decrease among years for the two basins is not proportionally equal (nonparallel change).

RESULTS AND DISCUSSION

Approximately 150 taxa were recorded in littoral samples from Belews Lake in year 4 to year 6. Twenty-three of the most abundant taxa were subjected to statistical analysis and were evaluated according to the criteria established. For ten of these organisms, changes in density were found to be associated with increases in temperature (i.e., criterion 4). Data for these organisms are presented in Tables 2 and 3.

Five organisms appeared to be directly affected by increases in temperature. Decreases in densities of three of these organisms *(Polypedilum digitifer, Nais communis,* and *Hyalella azteca)* were associated with increased temperature. Densities of the other two organisms *(Limnodrilus hoffmeisteri* and *Hexagenia munda)* increased with slight increases in temperature but declined at higher temperatures.

Polypedilum digitifer (Chironomidae) was a major component of the littoral benthos over the entire course of the study. During the course of this study, the density of *P. digitifer* remained remarkably stable at M stations. At D stations, however, an order of magnitude decline was noted from year 4 to year 5. The density of *P. digitifer* was significantly greater at D stations than at M stations in year 4, but this situation reversed in year 5. No significant changes occurred after year 5. This pattern suggests a change associated with power-plant operation.

Temperature has been shown to be one of the most important factors regulating the growth rate of Chironomidae (Konstantinov, 1958). Temperatures in the range of 30 to 37°C have been shown to be unfavorable or fatal (Curry, 1968; Nebeker, 1973; Gallup, Hickman, and Rasmussen, 1975). The maximum critical temperature shows considerable geographic variation, however. Temperatures reported as fatal in cold temperate areas are frequently encountered by organisms in subtropical regions. The only data available for subtropical areas are from a study of a thermal lake in New Zealand by Forsyth and McColl (1974), who found that 34°C was the critical temperature for Chironomidae.

Nais communis (Oligochaeta) was found to have maximum densities in winter and spring. Between years 4 and 5, a parallel increase in density was noted at both M and D stations. In year 6, density continued to increase at M stations but declined at D stations. Neither change was statistically significant, but a significant difference between M and D stations was noted for the first time in year 6. This is attributed to an increase in winter temperatures. *Nais communis* was the only member of the family Naididae which appeared to be negatively affected by high temperatures and also seemed to be the only naidid occurring during cold winter months.

The density of *Hyalella azteca* (Amphipoda) changed quickly with the start of power-plant operation in year 5; *H. azteca*

TABLE 2

SUMMARY OF STATISTICAL TESTS*

Organism	Years 4 to 5				Years 5 to 6			
	Stable	Convergence	Coloni-zation or succession	Power-plant effect	Stable	Convergence	Coloni-zation or succession	Power-plant effect
Diptera								
Cryptotendipes		X			X			
Cryptochironomus sp.			X(+)		X			
Dicrotendipes		X					X(+)	
Endochironomus nigricans				X(−)				X(−)
Polypedilum digitifer				X(−)	X			
Pseudochironomus sp.	X				X			
Cladotanytarsus sp.			X(−)				X(−)	
Tanytarsus sp.	X						X(−)	
Ablabesmyia sp.	X				X			
Procladius sp.	X						X(−)	
Nannocladius/Psectrocladius + *Parakiefferiella*		X	X(+)				X(−)	
Eukiefferiella sp.	X							X(−)
Palpomyia (complex)			X(−)					X(+)
Ephemeroptera								
Centroptilum sp.			X(+)				X(+)	
Hexagenia munda			X(+)					X(+,−)
Trichoptera								
Cyrnellus marginalis				X(−)	X			
Oligochaeta								
Limnodrilus spp.				X(+)		X		
Aulodrilus limnobius	X				X			
Dero/Nais sp.			X(+)					X(+)
Nais communis			X(+)					X(−)
Stylaria spp.			X(+)				X(−)	
Amphipoda								
Hyalella azteca				X(−)			X(−)	
Turbellaria								
Dugesia tigrina			X(−)		X			

*See text for definition of analysis criteria. (+) indicates an increase and (−) a decrease.

TABLE 3

DENSITY (NUMBER/m^2) AND STATISTICAL ANALYSIS FOR LITTORAL BENTHOS POSSIBLY AFFECTED BY POWER-PLANT OPERATION

	Station group		
Organism	**M**	**D**	**F value**
Polypedilum digitifer (Chironomidae)			
Year 4	125	543	4.8+
Year 5	130	60	9.0*
Year 6	118	59	66.7†
F, years 4–5	0.1(NS)	8.9*	Criterion 4: power plant
F, years 5–6	1.7(NS)	0.1(NS)	Criterion 1: stability
Hyalella azteca (Amphipoda)			
Year 4	202	322	0.1(NS)
Year 5	180	11	21.2†
Year 6	47	0	2.7(NS)
F, years 4–5	0.1(NS)	3.8	Criterion 4: power plant
F, years 5–6	5.5*		Criterion 3: succession
Limnodrilus hoffmeisteri (Tubificidae)			
Year 4	41	36	0.1(NS)
Year 5	293	576	8.9*
Year 6	263	321	1.3(NS)
F, years 4–5	22.9†	81.1†	Criterion 4: power plant
F, years 5–6	0.6(NS)	0.6(NS)	Criterion 2: convergence
Nais communis (Naididae)			
Year 4	47	66	2.1(NS)
Year 5	175	196	0.1(NS)
Year 6	281	109	13.6*
F, years 4–5	12.8*	5.2(NS)	Criterion 3: succession
F, years 5–6	1.7(NS)	0.9(NS)	Criterion 4: power plant
Hexagenia munda (Ephemeroptera)			
Year 4	1	2	2.6(NS)
Year 5	48	76	165.5‡
Year 6	54	493	32.3†
F, years 4–5	7.6*	15.1†	Criterion 3: succession
F, years 5–6	1.4(NS)	5.2(NS)	Criterion 4: power plant

(Table continues on following page.)

TABLE 3 (continued)

Organism	Station group M	Station group D	F value
Dero/Nais sp. (Naididae)			
Year 4	199	66	4.7+
Year 5	465	196	9.2*
Year 6	199	2106	137.5‡
F, years 4–5	14.0†	7.7(NS)	Criterion 3: succession
F, years 5–6	5.1(NS)	6.0(NS)	Criterion 4: power plant
Palpomyia complex (Ceratopogonidae)			
Year 4	325	495	1.0(NS)
Year 5	169	255	<0.1(NS)
Year 6	55	353	0.7(NS)
F, years 4–5	7.9*	2.1(NS)	Criterion 3: succession
F, years 5–6	12.6†	0.9(NS)	Criterion 4: power plant
Eukiefferiella sp. (Chironomidae)			
Year 4	4	2	1.9(NS)
Year 5	17	5	7.3*
Year 6	115	10	3.7(NS)
F, years 4–5	1.2(NS)	32.6†	Criterion 1: stability
F, years 5–6	5.9*	4.4(NS)	Criterion 4: power plant
Cyrnellus marginalis (Trichoptera)			
Year 4	5	1	4.8(NS)
Year 5	100	4	36.4†
Year 6	57	3	113.9‡
F, years 4–5	38.1‡	6.4(NS)	Criterion 4: power plant
F, years 5–6	1.7(NS)	1.4(NS)	Criterion 1: stability
Endochironomus nigricans (Chironomidae)			
Year 4	42	27	31.0†
Year 5	52	6	187.5‡
Year 6	383	98	87.0‡
F, years 4–5	0.5(NS)	3.5(NS)	Criterion 1: stability
F, years 5–6	34.9‡	3.5(NS)	Criterion 4: power plant

*Significant at the 5% level.
†Significant at the 1% level.
‡Significant at the 0.1% level.

disappeared almost completely from the discharge basin. At U stations, however, *H. azteca* showed no significant differences between years 4 and 5. In year 6, *H. azteca* was not collected at D stations, and a significant decline occurred at U stations. This distribution pattern appears to be related to temperature.

Data from studies of the effects of temperature on *H. azteca* (Bovee, 1949; Sprague, 1963) indicate that 32°C is the critical temperature for *Hyalella* irrespective of acclimation temperature. Gallup, Hickman, and Rasmussen (1975), studying the effects of a power plant operating on Lake Wabamun, Alberta, Canada, found that the density of *Hyalella* at control stations was 19 times greater than that at heated stations.

The results of these studies are consistent with the data from Belews Lake. Belews Lake data tend to confirm that 32°C is the critical temperature for *Hyalella. Hyalella* reached high levels during year 4 (preoperation) when natural lake temperatures rarely exceeded 32°C. In year 5, when temperatures greater than 32°C were found only in the discharge basin, *Hyalella* promptly disappeared from this area. In year 6, temperatures of 32 to 33°C were recorded both in the main lake basin and in the discharge basin. This seems to account for the sharp decline observed in *Hyalella* standing crop despite the favorable habitat. The decline of *Hyalella* observed in year 6 is probably to be expected in this geographic area. In the Piedmont of North Carolina, *Hyalella* is usually confined to rivers (Lenat, unpublished data), possibly because of the high temperatures recorded in lakes of this area.

Hexagenia munda is an important component of the littoral community because of its large size. Ephemeroptera (mostly *Hexagenia*) make up approximately 20% of the biomass collected in years 5 and 6.

Temperature appeared to exert a major influence on the distribution of *Hexagenia* in Belews Lake. Initial inspection of *Hexagenia* standing crop values (Table 3) suggested that the density of *Hexagenia* was increased by heat addition up to a certain critical temperature. *Hexagenia* increased at stations 5, 7, and 1907, where temperatures averaged 1.5 to 3°C above ambient (years 5 and 6). At stations with higher temperatures (1904B, 1904A, and 1904), however, *Hexagenia* occurred in relatively low densities. The depression of *Hexagenia* at high temperatures ($>$35 to 37°C) agrees with data from Lake Hyco, North Carolina (Lenat, 1975).

These data led to a hypothesis that *Hexagenia* standing crop might be severely affected by further increases in the heat load. It was thought that higher temperatures or an extension of the period of maximum temperature in year 7 might well cause a drop in the standing crop of *Hexagenia* in the discharge basin. Additional samples were taken at stations 7, 5, and 1904 on Sept. 22, 1976, to test this hypothesis (part of year 7 work is in progress). Data from years 5 and 6 indicated that maximum densities should occur during this period, but no specimens of *Hexagenia* could be found at any of these stations. Samples taken from stations 1907 and 8, however, indicated that *Hexagenia* was still present in the main lake basin in considerable numbers ($\sim 100/m^2$).

Limnodrilus spp., principally *L. hoffmeisteri* (Oligochaeta), showed a distinct maximum at discharge stations in year 5. This was the only period in which a significant difference could be observed between M and D stations. These data suggest that the year-5 temperature regime at D stations was optimal.

Tubificidae, especially *Limnodrilus*, have frequently been shown to have a positive response to increases in temperature (Boltovskoi, 1970; Weiderholm, 1972; Aston, 1973; Gallup, Hickman, and Rasmussen, 1975). This was the pattern observed at Belews Lake between years 4 and 5. Aston (1973) demonstrated a more complex pattern in laboratory cultures. He found that egg production increased up to 25°C but declined at 30°C. Egg production at 30°C was approximately equal to that measured at 15 to 20°C. These temperature data are not directly comparable to Belews Lake since constant temperatures were used, but the general pattern is analogous to the overall distribution seen for years 4 to 6.

Several taxa appear to have been indirectly affected by increases in temperature. Increases in the densities of *Dero/Nais* sp. (Oligochaeta) and *Palpomyia* (complex) (Diptera) were associated with growths of filamentous algae (probably *Anabaena*) at D stations in the end of year 5 and the beginning of year 6. Thermal effluents have previously been shown to stimulate periphyton growth, especially for blue-green algae (Foerster, Trainor, and Buck, 1974; Hickman, 1974; Bush, Welch, and Mar, 1974).

Three organisms developed maximum densities at M stations during the postoperational periods, *Cyrnellus marginalis* (Trichoptera), *Endochironomus nigricans* (Chironomidae), and *Eukiefferiella* sp. (Chironomidae). These organisms had low densities at all stations during year 4 and remained at low densities at D stations through year 6. This distribution parallels the development of macrophytes in the lake.

Endochironomus nigricans has frequently been associated with macrophyte growths (Berg, 1950; Buckley and Sublette, 1964; Lindeman, 1942). Macan (1975) indicated that the optimal habitat for polycentropid larvae (which includes *Cyrnellus*) is at the periphery of macrophyte beds. Sandberg (1969) found that most Orthocladinae (a group including *Eukiefferella*) are associated with vegetation zones. It is, therefore, reasonable to suggest that the distribution of these organisms is controlled by the distribution of macrophyte growths. The low densities of the organisms in the discharge basin may result from inhibition of macrophytes by power-plant effluents. However, increases in temperature may have been equally important. Macrophytes *(Potamogeton diversifolias)* became abundant at discharge stations in year 7, but the densities of *E. nigricans* and *C. marginalis* remained very low (Lenat, manuscript in preparation).

Changes in community structure were assessed by the use of percent similarity coefficients (PSC), which vary from 0 to 100%. The PSC values were computed for all possible station pairs with each quarterly sampling. This process generated 12 matrixes, which were reduced to three matrixes by the computation of yearly averages. These data were further reduced by the computation of within-group and between-group averages for M and D stations (Table 4).

TABLE 4

AVERAGE BETWEEN-GROUP AND WITHIN-GROUP SIMILARITY COEFFICIENTS

Station group	Year 4	5	6
M vs. M	38.9	39.3	44.2
D vs. D	48.0	46.4	42.4
M vs. D	36.2	36.7	32.5

Within-group comparisons for M stations show a slight increase from year 4 to year 6, indicating an increase in similarity at control stations. This is probably caused by the spread of macrophytes in this area. At D stations, however, a decline in similarity was noted during the same period. The imposition of a thermal gradient in this area appears to have induced a greater biological gradient. Note, also, that between-group similarities declined slightly from year 4 to year 6. Since it is not possible to evaluate these data statistically, only the general trends are discussed here.

Other summary statistics are presented in Table 5—total numbers of organisms collected, taxa richness, Shannon–Weiner diver-

TABLE 5

TOTAL NUMBER OF ORGANISMS, TAXA RICHNESS, DIVERSITY, AND EVENNESS

Year	Main lake stations										Discharge stations					
	65M	8	4	1907	2	1906	1	3	1116	Mean	7	5	1904B	1904A	1904	Mean
						Number of organisms* (N), N/m^2										
4	4521	2895	1787	2616	2650	1442	1496	2196	2605	2468	7707	5027	5188	3132	721	4355
5	3466	2411	5382	4101	6351	2960	2390	3907	5888	4095	6039	5468	3423	4316	4112	4672
6	2002	2282	2906	3520	3035	2207	4209	6469	4747	3486	5048	6049	6910	8191	4855	6211
						Taxa richness (S)										
4	41	29	28	28	30	28	30	30	34	30.9	37	31	27	29	21	29.2
5	32	30	30	32	28	24	24	35	34	29.7	31	26	29	27	23	27.4
6	27	28	30	32	28	26	31	34	29	29.2	30	28	29	26	25	27.4
						Diversity index ($\overline{H}$)										
4	3.42	3.07	3.26	3.30	3.36	2.94	3.44	3.07	3.53	3.27	3.02	3.41	2.98	3.15	2.63	3.07
5	3.26	3.70	3.25	3.67	3.22	2.99	3.36	3.62	3.81	3.44	3.67	3.18	3.31	3.21	2.40	3.15
6	3.07	3.70	3.59	3.68	3.25	3.50	3.80	3.67	3.66	3.55	3.86	3.23	3.12	2.46	2.69	3.07
						Evenness (E)										
4	0.62	0.63	0.67	0.68	0.66	0.62	0.69	0.69	0.69	0.65	0.58	0.69	0.63	0.65	0.58	0.63
5	0.68	0.76	0.67	0.73	0.67	0.65	0.76	0.71	0.75	0.71	0.74	0.68	0.68	0.67	0.53	0.66
6	0.65	0.79	0.74	0.79	0.73	0.75	0.77	0.73	0.76	0.74	0.79	0.67	0.65	0.54	0.60	0.65

*Rounded to nearest integer.

sity, and evenness. The total number of organisms collected at M stations was maximal during year 4, but numbers at D stations continued to increase through year 6. The high numbers of benthic organisms collected at this time were due largely to very high densities of Naididae collected at D stations in the summer and fall of year 6.

Average taxa richness varied little during the course of the study. Average values for both D and M stations were in the range of 27 to 30 taxa. Discharge stations were slightly lower than M stations during all years. Maximum numbers of taxa (up to 55) occurred at headwater areas where both pond and stream organisms were occasionally collected. One particular value appeared to be unusually low, that at station 1 in January of year 5. Only nine taxa were collected, but at nearby stations 25 to 35 taxa were collected. Only one chironomid species *(Psectrocladius)* was present at this time. The type of substrate present (the macrophyte *Potamogeton*) was not sufficient to explain this anomalous value. The low number of taxa suggested some toxic effect of ash basin discharge, possibly as a "slug," which occurred as a one-time event.

Diversity and evenness are always greater at M stations. This difference increased in years 5 and 6 but was not great enough to suggest a highly significant difference.

Approximately 40% of the dominant littoral taxa showed changes in density that were associated with increases in temperature. Many of these changes could be detected only because of the large number of stations sampled, however. The small changes in diversity and between-group similarity suggest that the changes had a relatively minor impact on the total community.

ACKNOWLEDGMENTS

The research reported here was supported by a grant from Duke Power Company, Charlotte, N. C. Charles M. Weiss and Terry P. Anderson contributed to the development of the ideas presented in this manuscript.

REFERENCES

Aston, R. J., 1973, Field and Experimental Studies on the Effects of a Power Station Effluent on Tubificidae (Oligochaeta: Annelida), *Hydrobiologia*, 42: 225-242.

Berg, C. I., 1950, Biology of Certain Chironomidae Reared from *Potamogeton*, *Ecol. Monogr.*, 20: 84-101.

Boltovskoi, M., 1970, The Effect of Heated Water, Discharged from the Cooling System of the Konakovskaya Thermal Power Station, on the Hydrology and Biology of Ivankovshoe Reservoir, in Preliminary papers for UNESCO—IBP Symposium on Productivity Problems of Freshwaters, Warsaw, Poland, pp. 291-295.

Bovee, E. C., 1949, Studies on the Thermal Death of *Hyallela azteca* Sanssure, *Biol. Bull.*, 96: 123-125.

Buckley, B. R., and J. E. Sublette, 1964, Chironomidae (Diptera) of Louisiana. II. The Limnology of the Upper Part of the Cane River Lake, Natchitoches Parish, Louisiana, with Particular Reference to the Emergence of Chironomidae, *Tulane Stud. Zool.*, 11: 151-166.

Bush, R. M., E. B. Welch, and B. W. Mar, 1974, Potential Effects of Thermal Discharge on Aquatic Systems, *Environ. Sci. Technol.*, 8: 561-568.

Cairns, J., Jr., 1975, Heated Waste-Water Effects on Aquatic Ecosystems, in *Thermal Ecology II*, ERDA Symposium Series, Augusta, Ga., Apr. 2—5, 1975, G. W. Esch and R. W. McFarlane (Eds.), pp. 32-38, CONF-750425, NTIS.

Collardeau, C., 1961, Influence de la température sur la consummation d'oxygen de quelques larvaes de Trichoptères, *Hydrobiologia*, 18: 252-264.

Curry, L. L., 1968, A Survey of Environmental Requirements for the Midge (Diptera:Tendipidae), in *Biological Problems in Water Pollution*, Third Seminar, C. M. Tarzwell (Ed.), Publication No. 999-WP-25, pp. 127-140, U. S. Public Health Service, Washington, D.C.

Foerster, J. W., F. R. Trainor, and J. D. Buck, 1974, Thermal Effects on the Connecticut River: Phycology and Chemistry, *J. Water Pollut. Control Fed.*, 46: 2138-2152.

Forsyth, D. J., and R. H. S. McColl, 1974, The Limnology of a Thermal Lake: Lake Rotowhero, New Zealand. II. General Biology with Emphasis on the Benthic Fauna of Chironomids, *Hydrobiologia*, 44: 91-104.

Gallup, D. N., M. Hickman, and J. Rasmussen, 1975, Effects of Thermal Effluents and Macrophyte Harvesting on the Benthos of an Alberta Lake, *Verh. Int. Verein. Limnol.*, 19: 552-561.

Garton, R. R., and R. D. Harkins, 1970, *Guidelines: Biological Surveys at Proposed Heat Discharge Sites*, U. S. Environmental Protection Agency, Pacific Northwest Water Laboratory, Corvallis, Ore.

Hickman, M., 1974, Effect of the Discharge of Thermal Effluent from a Power Plant on Lake Wabamun, Alberta, Canada. The Epipelic and Epipsammic Algal Communities, *Hydrobiologia*, 45: 199-215.

Johnson, M. G., and R. O. Brinkhurst, 1971, Benthic Community Metabolism in Bay of Quinte and Lake Ontario, *J. Fish. Res. Board Can.*, 28: 1715-1725.

Konstantinov, A. S., 1958, The Effect of Temperature on Growth Rate and Development of Chironomid Larvae, *Akad. Nauk. SSSR Doklady Biol. Sci. Sect.*, 120: 1362-1365.

Lenat, D. R., 1975, The Benthos, in C. M. Weiss et al. (Eds.), *An Assessment of the Environmental Stabilization of Belews Lake, Year IV, and Comparisons with Lake Hyco, North Carolina*, Department of Environmental Science and Engineering Publication No. 416, University of North Carolina, Chapel Hill.

Lindeman, R. L., 1942, Seasonal Distribution of Midge Larvae in a Senescent Lake, *Am. Midl. Nat.*, 27: 428-444.

Macan, T. T., 1975, Structure of the Community in the Vegetation of a Moorland Fishpond, *Verh. Int. Verein. Limnol.*, 19: 2298-2304.

Nebeker, A. V., 1973, Temperature Requirements and Life Cycle of the Midge *Tanytarsus dissimilis*, *J. Kansas Entomol. Soc.*, 46: 160-165.

Sandberg, G., 1969, A Quantitative Study of Chironomid Distribution and Emergence in Lake Erken, *Arch. Hydrobiol. Suppl.*, 35: 119-201.

Sprague, J. B., 1963, Resistance of Four Freshwater Crustaceans to Lethal High Temperature and Low Oxygen, *J. Fish. Res. Board Can.*, 20: 387-415.

Weiderholm, T., 1972, Bottom Fauna and Cooling Water Discharges in a Basin of Lake Mälaren, *Rep. Inst. Freshwat. Res., Drottningholm*, 51: 195-214.

Weiss, C. M., and E. J. Kuenzler, 1976, *The Trophic State of North Carolina Lakes*, Water Resources Research Institute, University of North Carolina, UNC-WRRI-76-119, North Carolina State University, Raleigh.

Wilhm, J. L., 1967 Comparison of Some Diversity Indices Applied to Populations of Benthic Macroinvertebrates in a Stream Receiving Organic Wastes, *J. Water Pollut. Control Fed.*, 39: 1673-1683.

EFFECTS OF POWER-PLANT OPERATION ON THE PHYTOPLANKTON COMMUNITY OF BELEWS LAKE, NORTH CAROLINA

PETER H. CAMPBELL
Department of Environmental Sciences and Engineering,
University of North Carolina, Chapel Hill, North Carolina

ABSTRACT

The phytoplankton community of Belews Lake, North Carolina, a 1500-ha cooling reservoir, was studied over a 4-year period. Two preoperational years served as a base line for 2 years of operation by the Belews Creek Steam Station, a 2200-MW coal-fired power plant. Circulation of surface waters by the power plant resulted in greater seasonal and spatial homogeneity of total standing-crop values and of distribution of dominant phytoplankton classes and taxa. No major changes in standing crop were found, but general successional changes in the phytoplankton community appeared to be accelerated after power-plant operation. Chlorophyceae dominance, measured by cell density, and Dinophyceae dominance, measured by biovolume, increased, but Bacillariophyceae dominance greatly decreased. Density and biovolume levels of all three classes exhibited no marked changes. No negative thermal influence on phytoplankton populations was found. No selection was observed for blue-green algae by the elevated temperatures of the thermal plume. Mean total cell densities and enrichment indexes exhibited steady decreases through all 4 years, and mean total biovolumes, numbers of taxa, and Shannon–Weaver diversity indexes exhibited steady increases. The values of these parameters all remained within ranges indicative of an oligomesotrophic lake.

Belews Lake was constructed by Duke Power Company as a cooling reservoir for their coal-fired Belews Creek Steam Station. Creation of the lake provided an opportunity to measure physical, chemical, and biological properties of a newly formed impoundment, to follow it through several years of natural development to obtain a base-line evaluation, and to determine the effects of heated-water effluent on the aquatic ecosystem after the completed power plant began operation.

The lake, which lies in the Piedmont section of North Carolina, has a surface area of 1500 ha, a retention time of ~1000 days, and a mean depth of 15 m (maximum depth is ≈40 m). Belews Lake has been the subject of an environmental study for 7 years. Year 1 (August 1970 to April 1971) was a preimpoundment study. Dam closure marked the beginning of year 2 (May 1971 to June 1972). Lake filling continued to completion in the spring of year 3 (July 1972 to June 1973), and in this year the phytoplankton studies began. Year 4 (July 1973 to June 1974) was the first 12-month period in which the lake existed at full-pool level. The first unit of the Belews Creek Steam Station, with a capacity of 1145 MW, commenced operation at the beginning of year 5 (July 1974 to June 1975). The second unit went on-line in the middle of year 6 (July 1975 to June 1976); this doubled the generating capacity and pumping rate. Results of studies in the lake over this period have been the subject of several reports (Weiss et al., 1974; 1975; 1978a; Weiss, Anderson, and Lenat, 1972).

Comparatively few studies have been made of the effects of thermal discharge on algae in lakes (Patrick, 1974). The possible effects of such conditions on phytoplankton are various. If the temperature tolerances of algae are exceeded because of excess thermal input, species composition will change, and, if the increase is great enough, shifts in the flora from diatoms to mainly green algae or blue-green algal flora may occur (Patrick, 1969). A heat-enhanced shift to a grazing-resistant blue-green community may eliminate many consumer species, and, once the blue-greens become sufficiently dominant, the noxious character of their abundance has an adverse effect on water quality (Foerster, Trainor, and Buck, 1974; Patrick, 1974). The heat shock to phytoplankton passing through power-plant condensers can inhibit productivity (Gurtz and Weiss, 1974). Heat addition can result in lower phytoplankton numbers and diversity (Warinner and Brehmer, 1966; Kullberg, 1968) and increased dominance by tolerant species at the expense of less hardy taxa (Knight, 1973). Under certain conditions the heat shock may cause synchrony in reproduction by creating incipient cell divisions (Foerster, Trainor, and Buck, 1974). Heat shock for extremely short periods of time may have only temporary adverse effects on many species of algae, however, and moderate temperature increases from the low end of the tolerance range toward optimum temperature can increase algal biomass and diversity (Patrick, 1969; 1974). Examples of such increased phytoplankton productivity have been found in waters warmed by thermal power-plant effluents (Pannel, Johnson, and Raymont, 1962; Hockley, 1963; Morgan and Stross, 1969; Foerster, Trainor, and Buck, 1974).

The objectives of this study were to focus on the phytoplankton community and to determine cyclical and successional changes in response to the evolving aquatic environment and the added thermal stress. The 4-year study (July 1972 to June 1976) encompassed two preoperational and two postoperational years.

MATERIALS AND METHODS

Three sampling stations were located on the discharge arm of the lake, four on the upper main lake arm connected to the discharge arm by a canal, two on the intake arm, and two in the lower lake near the dam (Fig. 1). Data were collected monthly for a complete year from five stations in year 3, ten in year 4, eleven in year 5, and eight in year 6. Four stations were studied through all 4 years, and eight stations were examined for the complete 3-year period after establishment of full-pool level.

A wide spectrum of water-quality parameters were measured to characterize the physical and chemical environment of the lake. Those referred to in this paper include (1) temperature, measured with a multiprobe surveyor, (2) water transparency, measured with a Secchi disk, (3) soluble reactive phosphorus, measured by the automated stannous chloride method (Environmental Protection Agency, 1974) with an autoanalyzer, and (4) inorganic nitrogen (NO_2—NO_3—NH_3—N), also measured with an autoanalyzer after procedures similar to Environmental Protection Agency methods.

For the phytoplankton investigation, monthly water samples were drawn from the Secchi depth, except in year 6, when composite samples representing the euphotic zone were constituted. Because there is presently no known fixative that can adequately preserve all classes of phytoplankton without disturbing their morphology or pigmentation (Reynolds, 1973), live samples were collected for examination to ensure accurate identification and classification. Samples were kept on ice without preservation and returned to the laboratory for prompt study.

Samples were concentrated by centrifugation and examined live. After a 10-ml aliquot was centrifuged and the centrifugate drawn off, the final 0.5 ml containing the concentrated algae was mixed and transferred to a microscope slide, where the drop was spread, by adding a cover slip, to the exact area of the cover slip. The edges were ringed with a paraffin—petroleum jelly mixture to prevent

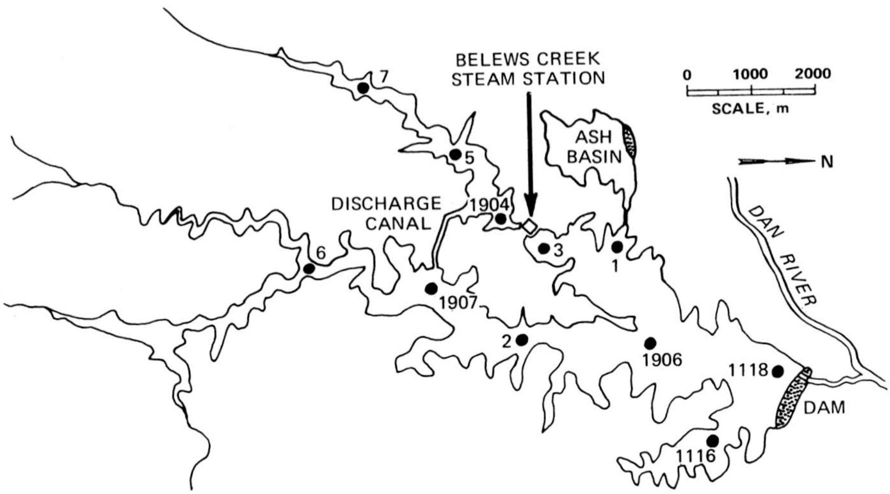

Fig. 1 Belews Lake sampling stations.

drying, and the phytoplankton in the preparation were enumerated in two transects of the cover slip at 450x magnification with a standard light microscope. Analyses revealed counts that approximated a Poisson distribution. Counts of at least 100 cells or units were made and were found to have a counting error not greater than ±20%. The entire cover-slip area was scanned for presence of scarcer taxa. Because oil-immersion objectives can be used in identifying taxa, this simple method of live phytoplankton analysis has the advantage over use of a hemocytometer, Sedgewick—Rafter, or Palmer—Maloney counting chamber. Further description and discussion of this technique are presented by Campbell (1973) and Weiss et al. (1974; 1975; 1978a). All taxa were identified by species where possible. Phytoplankton abundance was measured by density (number of cells or units/ml) and biovolume (mm^3/m^3). Biovolumes were calculated with a volume factor for each taxon determined by water displacement of plasticine clay models constructed to scale from observations and average measurements. For colonial or filamentous forms, the average number of cells per unit was taken into account. For diatoms, which have large central vacuoles, only the percentage of the total cell volume representing the plasma volume (Strathmann, 1967) was used. The enrichment index used is modified from Palmer (1969) to account for density. The points Palmer assigned to each of the pollution-tolerant species in a sample were multiplied by the number of units per milliliter for that species, and the accumulative total for the sample was then divided by the total number of taxa found (Weiss et al., 1974). Diversity was measured by the Shannon–Weaver index (Shannon and Weaver, 1949; Lloyd and Ghelardi, 1964).

RESULTS

Physical and Chemical Parameters

Mean Secchi disk values exhibited a general trend toward increasing depth over the 4 years, from 1.6 m in year 3 to 2.6 m in year 6. Nutrient levels in the lake were low. Yearly mean inorganic nitrogen levels decreased steadily from 127 μg/liter in year 3 to 67 μg/liter in year 6. Soluble reactive phosphorus levels were very low, with maximum values never exceeding 10 μg/liter, and mean values exhibited a steady decrease from 8.5 μg/liter to 2.3 μg/liter through the 4 years of study. This trend was at least partially due to refinements made in the analytical technique during this period. Nutrient inputs to the lake were primarily associated with periods of high runoff.

The mean yearly surface-water temperatures measured during sampling increased for each of the 4 years of study, from 17.5 and 18.9°C in the base-line period to 21.0 and 23.6°C in the postoperational years. Although this increase paralleled a general warming trend in the weather over this period, greater increases in mean water temperature occurred during the 2 years of power-plant operation. The mean water temperature of the warmest month was between 30.0 and 31.5°C in all years. During the 2 years of power-plant operation, the highest water temperatures encountered occurred in summer at the discharge station (1904). Temperatures at the time of sampling approached but never exceeded 35.0°C. Temperature differences between the intake station (3) and the discharge station (1904) during the 19 samplings when the power plant was in operation averaged 6.7°C, with no difference as great as 10.0°C.

The pumps that circulate lake water through the power-plant condensers began operation in June 1974, and in July the currents began carrying the thermal discharge from the plant. The main circulation pattern generated by the power plant ran from the intake near station 3 through the condensers to station 1904, then through the connecting discharge canal to 1907, down this main arm of the lake to 1906, and up the intake arm back to station 3 (Fig. 1). Except for September 1974 and the period from April to June 1975, all samplings for the two postoperational years occurred during periods of thermal discharge. The major influence of the surface thermal plume was at stations 1904 and 5 in the discharge arm. The influence extended, to a lesser degree, to both stations 7, farther up the discharge arm, and 1907, on the main lake adjacent to the discharge canal connecting the two arms. From station 1907 influence of the surface plume was sometimes found to reach upstream as far as station 6 and downstream in the circulation current to 1906. Calculations based on the average monthly pumping rates indicated that in the first year of operation the pumps had the capacity to circulate a volume of water equivalent to the epilimnion (7.6 m depth) in 5.5 weeks. This circulation time was reduced to 3 weeks after addition of the second generating unit in December 1975.

Total Phytoplankton Standing Crop

The means for five parameters used to assess various aspects of total phytoplankton standing crop are presented in Table 1. Mean total cell densities and enrichment indexes each decreased through the 4 years, and numbers of taxa per sample and Shannon—Weaver diversity indexes steadily increased.

The seasonal distributions of mean total density and biovolume values for Belews Lake are presented in Fig. 2. A biovolume peak in June 1974 stands out sharply here because of exceptional abundances of the filamentous pennate diatom *Fragilaria crotonensis*. If this month's values are not included in the averaging, mean total biovolume values show a slight but steady increase over the 4 years. The maximum mean density value occurred in July 1972, as a result of a great abundance of the minute colonial blue-green alga *Agmenellum quadriduplicatum*. Abundances of this blue-green alga alone were responsible for the very large enrichment indexes of year 3.

Minimum temporal mean values for total density, total biovolume, and total number of taxa per sample all occurred in the coldest period of each year, between December and March. Even during power-plant operation, there was no significant winter increase in total phytoplankton density or biovolume at stations influenced by the heated water. Spatially the maximum mean values for these three parameters occurred predominantly at the upstream stations nearest the nutrient supply provided by inflowing streams. The largest range in values for total density, biovolume, enrichment indexes, and Shannon—Weaver diversity indexes all occurred in year 3, when the lake was still filling.

TABLE 1

TOTAL PHYTOPLANKTON MEANS IN BELEWS LAKE, 1972–1976

Parameter and year	Yearly mean	Range
Total density, number/ml		
Year 3	2875	256–29334
Year 4	1620	220–6643
Year 5	1544	247–5299
Year 6	1474	364–4197
Total biovolume, mm^3/m^3		
Year 3	681	79–7116
Year 4	1058	87–6422
Year 5	870	101–4212
Year 6	991	97–6077
Total number of taxa		
Year 3	27	16–44
Year 4	32	14–54
Year 5	37	23–54
Year 6	40	28–63
Enrichment index		
Year 3	1172	0–17441
Year 4	128	0–2105
Year 5	26	0–142
Year 6	21	0–328
Shannon–Weaver diversity index		
Year 3	3.185	0.689–4.275
Year 4	3.455	1.838–4.519
Year 5	3.586	2.197–4.601
Year 6	3.846	2.930–4.617

Power-Plant Influence on Total Standing Crop

There were 19 sampling periods when thermal effluent was being discharged into the lake, four months when the plant was shut down, and 21 months in the preoperational period when the lake was free from thermal effects. Samples taken at the four stations in the circulation pattern (3, intake; 1904, discharge; 1907; and 1906) during periods of thermal discharge and during periods when there was no thermal effect were compared to determine how often there were significant differences in total phytoplankton density at adjacent stations. A fifth station (1118, near the dam and least affected by the circulation pattern) was included as a control. Differences were considered significant if there was no overlap in the 95% confidence limits for the density counts at adjacent stations

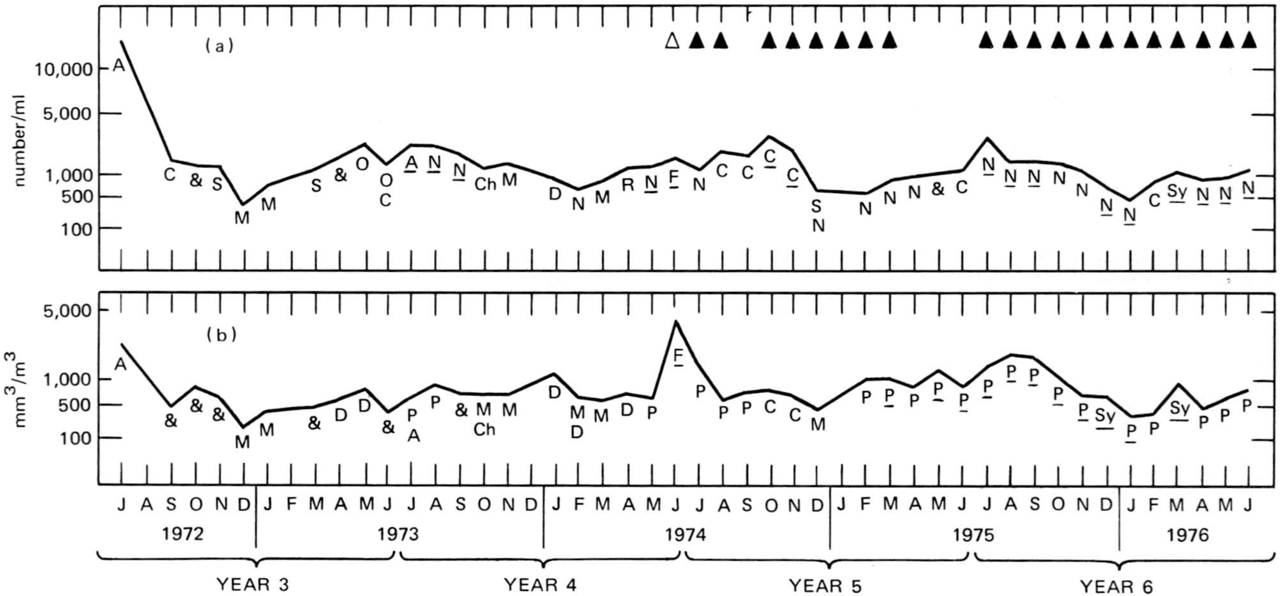

Fig. 2 Seasonal distribution of (a) mean total density and (b) biovolume for Belews Lake, 1972–1976. Most important dominant genera during each monthly sampling are A, *Agmenellum* (Cyanophyceae); C, *Chlorella* (Chlorophyceae); Ch, *Chrysochormulina* (Prymnesiophyceae); D, *Dinobryon* (Chrysophyceae); F, *Fragilaria* (Bacillariophyceae); M, *Melosira* (Bacillariophyceae); N, *Nannochloris* (Chlorophyceae); O, *Oocystis* (Chlorophyceae); P, *Peridinium* (Dinophyceae); R, *Rhizosolenia* (Bacillariophyceae); S, *Selenastrum* (Chlorophyceae); Sy, *Synedra* (Bacillariophyceae); &, several dominants. Genera underscored were dominant in all samples. ▲, thermal effluent. △, pump operation only.

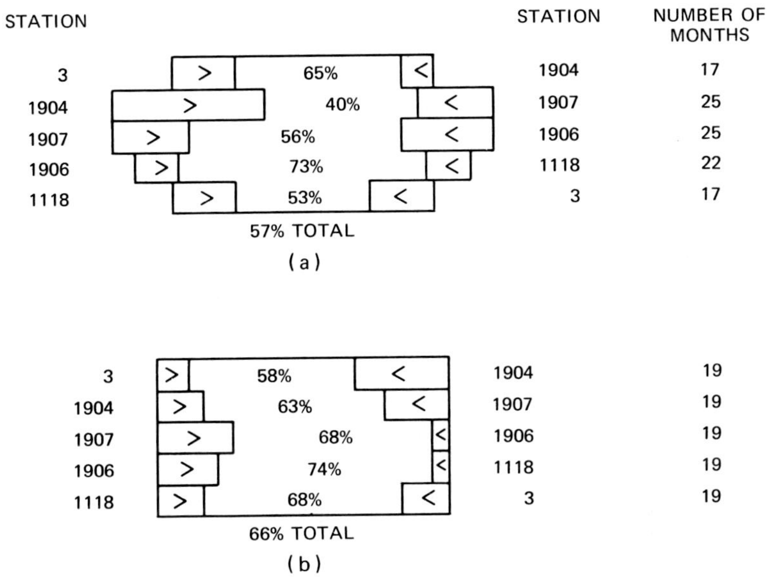

Fig. 3 Comparison of differences in total phytoplankton density during (a) nonstressed and (b) stressed (power plant in operation) periods at adjacent stations. The small blocks at the left and right indicate significant difference in counts. >, station to left greater; <, station to right greater. Percentages are samples with no significant difference (95% confidence limits for counts overlap).

during a particular sampling (Lund, Kipling, and LeCren, 1958). Results are summarized in Fig. 3.

In the station comparisons for months without thermal stress, 57% exhibited no significant differences in density. Where there were significant differences, the most frequent in all station comparisons was a significantly greater density at station 1904 than at main lake station 1907. All other station comparisons showed approximately the same low number of samples with significantly greater or lesser densities.

In the station comparisons for the months with thermal stress, 66% exhibited no significant difference in density. Where there were significant differences, the greatest were the number of times the discharge station, 1904, showed higher densities than station 3, near the intake, and the number of times 1907 densities were greater than 1906 densities. All station-pair relationships but one are in contrast to those found for the nonstressed period. In both periods, station 3 and the control station, 1118, were about equal in their distribution of greater densities.

Similar adjacent-station comparisons were made for total biovolume abundances and total numbers of taxa per sample at the five pairs of stations. The change in biovolume with power-plant operation was similar to and even more marked than that in total density. Biovolume differences of less than 300 mm^3/m^3 occurred in only 48% of the adjacent-station samples in the nonstressed periods, but, when the power plant began operation, 78% fell below this level. Similar comparisons of the total numbers of taxa per sample, however, showed no noticeable changes with power-plant operation.

Phytoplankton Classes

All the 11 algal classes that contain phytoplankton representatives [according to the classification of Christensen, as modified by Hommersand (1972)] were found in Belews Lake. By far the most important class in each of the 4 years by density was the Chlorophyceae (the green algae). A number of classes attained important dominance frequencies by biovolume. Most prominent of these was the Dinophyceae (the dinoflagellates), followed by the Bacillariophyceae (the diatoms), the Chlorophyceae, and the Chrysophyceae (the golden-brown algae). The frequency of dominance of the Cyanophyceae (blue-green algae) was extremely low by both density and biovolume determinations.

The Chlorophyceae was by far the most frequent dominant by density during all seasons of all 4 years; it was dominant in 76 to 78% of all samples in the base-line years, increasing to 100% in the postoperational years. The Dinophyceae was the most frequent summer dominant by biovolume of years 4 to 6. In years 5 and 6 it was also the most frequent winter and spring dominant, and by year 6 it had become the most frequent dominant in autumn. The Bacillariophyceae, with dominance in 16 to 22% of the samples in the preoperational years, was the most frequent dominant in the autumn and winter of years 3 and 4 and in the spring of year 4. During the 2 years of power-plant operation, frequencies of diatom dominance in these seasons became secondary to the dinoflagellates. During the 4 years of study, the Dinophyceae steadily increased in frequency of dominance, from 18 to 34 to 61 to 72%, respectively. The classes with the greatest frequencies of dominance all tended to exhibit similar degrees of importance throughout the lake. The few examples of regional variability in dominance distribution by density were found only in the preoperational years.

There were some changes in the pattern of class dominance over the 4 years, especially for the Chlorophyceae by density, but there were no obvious changes in the patterns of fluctuation in density or biovolume abundances by the classes after power-plant operation commenced. There was, however, somewhat less variability in the distribution of class abundances among the stations in the postoperational years when power-plant circulation connected the intake, discharge, and main-lake sections of the impoundment. The most notable example was the marked abundance of *Fragilaria crotonensis* Kitton, which coincided with the first month of pump operation (June 1974, one month before steam generation commenced) and resulted in Bacillariophyceae biovolume peaks at every station on the lake. Two other patterns that appear to have coincided with the postoperational years were the development of low but consistent abundances of the Cyanophyceae and the decrease in occurrence of notable abundances of Chrysophyceae.

Phytoplankton Taxa

The Chlorophyceae contained 15 taxa that attained dominance in one or more samples by density and 8 by biovolume. The Dinophyceae were second in importance by biovolume, with 7 taxa, and the Bacillariophyceae contained 6 dominant taxa by density and biovolume. In this study the principal dominant taxa were considered to be those exhibiting a notable frequency of dominance and subdominance by either density or biovolume. These taxa are listed in Table 2.

Succession in the principal taxa contributing to the yearly phytoplankton cycle has occurred over the 4 years, as Table 2 and Fig. 2 further illustrate. The tiny colonial blue-green alga *Agmenellum quadriduplicatum* (Menegh.) Breb. was a principal summer dominant by density in years 3 and 4 but made no further notable appearances. In these same 2 years, the small filamentous centric diatom *Melosira italica* var. *alpigena* (Grun.) A. Cleve was a principal autumn and winter dominant. In year 5 its importance was greatly decreased, however, and by year 6 it had become a minor taxon. The major importance of *Selenastrum minutum* (Naeg.) Collins, a small lunate-celled green alga, was in the autumn and winter of year 3. Although it was also important in autumn and winter of the following 2 years, by year 6 it too was no longer dominant. The minute spherical-celled green alga, *Nannochloris* sp., on the other hand, went from a dominant in only a few spring samples in year 3 to become the most important principal taxon of summer and winter in year 4. It continued this distribution of

TABLE 2

PRINCIPAL PHYTOPLANKTON TAXA* OF BELEWS LAKE, YEARS 3 TO 6

Year 3 (5 stations)		Year 4 (10 stations)		Year 5 (11 stations)		Year 6 (8 stations)	
By Density							
Selenastrum minutum	12.0	*Nannochloris* sp.	46.5	*Chlorella* sp.	61.0	*Nannochloris* sp.	90.0
Chlorella sp.	11.5	*Melosira italica* var. *alpigena*	25.0	*Nannochloris* sp.	48.0	*Chlorella* sp.	35.5
Oocystis pusilla	10.5	*Selenastrum minutum*	17.0	*Selenastrum minutum*	19.5	*Synedra filiformia*	8.0
Melosira italica var. *alpigena*	7.0	*Agmenellum quadriduplicatum*	14.0	*Nannochloris bacillaris*	18.5	*Nannochloris bacillaris*	6.5
Nannochloris sp.	6.0	*Fragilaria crotonensis*	10.0	*Melosira italica* var. *alpigena*	9.0		
Chrysococcus rufescens	5.0	*Nannochloris bacillaris*	9.5				
Agmenellum quadriduplicatum	4.5	*Rhizosolenia longiseta*	9.0				
By Biovolume							
Melosira italica var. *alpigena*	14.5	*Melosira italica* var. *alpigena*	37.0	*Peridinium inconspicuum*	37.0	*Peridinium inconspicuum*	70.0
Dinobryon divergens	11.5	*Dinobryon divergens*	24.0	*P. volzii*	25.0	*Synedra filiformis*	14.5
Asterionella formosa	6.0	*Fragilaria crotonensis*	17.5	*Melosira italica* var. *alpigena*	24.5	*Peridinium wisconsinense*	9.5
Ceratium hirundinella	4.5	*Peridinium gatunense*	12.5	*Chlorella* sp.	21.0	*P. intermedium*	8.5
		P. volzii	12.5	*Peridinium wisconsinense*	18.0	*Synedra delicatissima*	8.5
				P. gatunense	11.5		

*Taxa with a sum value of frequency of dominance plus one-half the frequency of subdominance in any month greater than 80% of the number of stations examined that month. The value given is the sum of total frequency of dominance plus one-half the total frequency of subdominance.

seasonal importance in year 5 and was joined by *Nannochloris bacillaris* Naumann in the winter of 1975. In year 6 it became the most important taxon in all four seasons. The only other taxon to approach this status as a perennial dominant was *Chlorella* sp., another small-celled green alga, which in year 5 was a principal dominant of all seasons except winter and in year 6 was the most frequent subdominant. It was also important in the summer and spring of year 3 but in year 4 was only a minor dominant.

Of the six principal taxa by density in years 5 and 6, only one, the needle-shaped pennate diatom *Synedra filiformis* Grun., was not also a principal taxon in the first 2 years. Of the ten principal taxa of the two base-line years, however, half did not occur again as principal taxa after the power plant began operation. Taxa with moderate frequencies of dominance by density were dominant or subdominant in all sections of the lake each year in which they were present. The few cases of localization exhibited in distribution of dominance were limited to the period before power-plant operation began.

By biovolume, only one principal taxon was common to more than two consecutive years. This was *Melosira italica* var. *alpigena*, which exhibited the same autumn and winter distribution of importance in years 3 and 4 and to a lesser degree in year 5 but became of minor importance in year 6. The dendritic colonial chrysophyte, *Dinobryon divergens* Imhof., was the only other principal taxon common to both years 3 (dominance in spring) and 4 (dominance in winter and spring). In the postoperational years, there were only occasional biovolume dominances of this taxon. *Peridinium gatunense* Nygaard and *P. volzii* Lemm., two moderately large armored dinoflagellates, were principal taxa common to years 4 and 5, with *P. gatunense* being the primary dominant in summer in year 4, and *P. volzii* an important dominant in the spring of year 4 and in both winter and spring of year 5. Neither taxon developed notable biovolumes in year 6, however. Years 5 and 6 shared two principal taxa, *P. wisconsinense* Eddy, another moderately large dinoflagellate that was most important as a spring dominant in year 5, and the smaller *P. inconspicuum* Lemm., which was the most frequent summer and spring dominant of year 5 and became the most important taxon of all four seasons in year 6. Thus, by the fourth year of this study, *P. inconspicuum* had become the perennial dominant by biovolume, just as *Nannochloris* sp. had become by density.

Three of the seven principal taxa in the base-line period were dinoflagellates, and five of the nine postoperational-period taxa were in this class, specifically in the armored genus *Peridinium*. The shift

TABLE 3

NUMBER OF TAXA IDENTIFIED FROM BELEWS LAKE AND THEIR DISTRIBUTION BY ALGAL CLASS, 1972–1976

Class	Study years				4-year total	% of total
	3	4	5	6		
Cyanophyceae	7	6	15	18	24	5.5
Cryptophyceae	3	5	9	11	13	3.0
Dinophyceae	10	16	17	20	25	5.7
Prymnesiophyceae		1	1	1	1	0.2
Chrysophyceae	24	26	30	26	41	9.4
Bacillariophyceae	25	36	44	41	61	14.0
(Centrales)	(6)	(16)	(14)	(13)	(21)	(4.8)
(Pennales)	(19)	(20)	(30)	(28)	(40)	(9.2)
Xanthophyceae	2	5	1	1	6	1.4
Rhaphidophyceae	3	1	1	1	3	0.7
Euglenophyceae	36	43	26	40	66	15.1
Prasinophyceae		1	2	2	3	0.7
Chlorophyceae	66	103	122	131	193	44.3
(Volvocales)	(7)	(11)	(14)	(15)	(23)	(5.3)
(Tetrasporales)	(3)	(4)	(4)	(4)	(4)	(0.9)
(Ulotrichales)	(1)	(1)	(1)	(1)	(1)	(0.2)
(Chlorococcales)	(45)	(68)	(81)	(88)	(116)	(26.6)
(Zygnematales)	(10)	(19)	(22)	(31)	(49)	(11.2)
Total number of taxa	176	243	268	293	436	100.0%
Total number of samples	(72)	(112)	(121)	(104)	(409)	

in biovolume dominance to representatives of this genus was a steady and continuing process since year 4, before the power plant began operation.

In several cases taxa exhibited frequencies of dominance by biovolume which were not distributed throughout all main regions of the lake. Although most of these were limited to the base-line period, a few minor examples occurred during the operational years. These distributions were confined to either the most upstream stations or the lower main lake, at the greatest distance from the influence of the power plant.

With each year of study there was an increase in the total number of taxa identified from the lake (Table 3). The Chlorophyceae contained by far the largest number of taxa and showed an increase each year. The Chlorococcales, the order of the green algae containing the majority of phytoplankton species, on its own embodied more taxa than any class. It also showed steady increases each year, as did the desmid-containing order Zygnematales, the flagellate-celled order Volvocales, and the classes Dinophyceae and Cryptophyceae. Other classes fluctuated in numbers, but none exhibited any notable decrease over more than a 1-year period. The only class to exhibit a sharp change exclusively between the preoperational and postoperational periods was the Cyanophyceae, which more than doubled in number of taxa with this change.

DISCUSSION

In comparison with other North Carolina lakes, the water of Belews Lake is very clear, with a green tint similar to that found in the state's mountain reservoirs. Phytoplankton abundances were too low to have any important effect on light penetration. Seasonally, water turbidity was lowest in the warmer months, the period when phytoplankton abundance was normally greatest.

Inorganic nitrogen and soluble phosphorus levels were low. Soluble phosphorus values were always less than 10 μg/liter. In Belews Lake, phosphorus was determined to be the critical limiting element responsible for inhibiting phytoplankton productivity (Weiss et al., 1978a). Mean inorganic nitrogen and soluble phosphorus levels both decreased gradually through the 4 years of study, as did mean total phytoplankton density and enrichment index levels. These tendencies were all indicative of a trend toward increasing oligotrophy in the lake. Shannon—Weaver diversity indexes exhibited a general increase in values throughout the study period, a further reflection of this trend. A low nutrient flux would lead to higher diversity because of the greater overlap between successive species populations caused by low growth rates (Moss, 1973). Also contributing to the increasingly higher diversity values was the steady yearly increase in the number of taxa present; this suggests a gradual maturing of the system.

The lowest individual Shannon—Weaver diversity index and the largest density, biovolume, and enrichment index values occurred in the first month of the study, while the lake was still filling and receiving nutrients washed and leached from newly flooded topsoils. Filling of the lake was completed in the following spring, and by the second summer only reduced levels of the blue-green alga were responsible for the first summer's bloom. Enrichment indexes at all other times were very low and indicated a lake free of any organic enrichment problems. Mean values for Secchi depth, total phosphorus, total phytoplankton density, and biovolume and enrichment indexes were together within the range indicative of an oligomesotrophic lake, according to the trophic levels set forth by Weiss et al. (1978b).

Mean yearly temperatures increased in the lake over the 4 years. Although there was a greater increase in mean water temperature during the two years of power-plant operation, at no sampling in the 4 years did the mean temperature exceed 32°C, and no measurement of the maximum water temperature at the discharge station ever exceeded 35°C. The optimum temperature for most green algae seems to be between 30 and 35°C, and that for many blue-greens is >35°C, whereas most diatom species prefer lower temperatures (Cairns, 1956; Patrick, 1969; Canale and Vogel, 1974). Thus, since the temperatures in Belews Lake, even during power-plant operation, were not great enough to actively select for blue-green algae, green algae would be expected to be the dominant class in the warmer seasons. In fact, the Chlorophyceae was the most frequent class by density during all seasons in all 4 years. Abundant blue-green algal densities were limited to July of the two base-line years. Blue-greens are favored by increases in phosphorus (Welch, Hendrey, and Stoll, 1975); thus the absence of any blue-green dominance in summer during power-plant operation was presumably caused by the very low nutrient supply.

Lowest seasonal water temperatures occurred in winter. These values increased with succeeding years, because of both a general climatic warming trend and the thermal input during the postoperational years. Temperature is considered to control phytoplankton populations in winter and early spring (Yentsch et al., 1974), and increases of ~8°C in temperature were reported to stimulate phytoplankton production when natural water temperatures were 16°C or cooler (Morgan and Stross, 1969). In Belews Lake, however, the discharge-receiving stations exhibited no apparent total phytoplankton density or biovolume increases over the intake stations in the cooler seasons.

Comparing the circulation patterns at the four stations revealed an increase in spatial homogeneity of the phytoplankton after power-plant operation commenced. In the postoperational period there was a 9% increase in the number of adjacent-station pairs with no significant difference in density over the number in the base-line period and a 30% increase in pairs with a difference of less than 300 mm^3/m^3 in biovolume. With the circulation and mixing effects by the currents generated, phytoplankton patchiness decreased, and abundances became more uniformly distributed throughout the lake.

Where there were significant density differences in the adjacent-station comparisons, there was a tendency toward greater density at the discharge station than at the intake station in the stressed period. This may have been caused by temperature effects on the phytoplankton since this trend is consistent with results from other studies of phytoplankton exposed to elevated temperatures. It was found that a heat shock of limited duration within certain temperature-tolerance limits of the algae can create incipient cell division and, thus, result in larger densities (Foerster, Trainor, and Buck, 1974; Goldman and Carpenter, 1974; Patrick, 1974; Rankin, Buck, and Foerster, 1974). The absence of any similar trend toward an increase in biovolume at the discharge station (1904) as compared with the intake station (3) further supports this interpretation. The proportion of samples exhibiting this trend was small, however, compared with the approximately two-thirds of the samples that showed no significant difference.

An increase in both spatial and seasonal homogeneity was shown in the distribution of dominance by phytoplankton classes and taxa for the operational years. In years 5 and 6 the Chlorophyceae, primarily *Nannochloris* and *Chlorella*, were dominant by density in all samples. By biovolume, the Dinophyceae, represented by *Peridinium*, became the major dominant of all seasons except winter in year 5 and of all seasons in year 6. Almost all samplings showing localized distribution of class or taxon dominance were limited to the base-line years. Because of power-plant operation, the mixing of surface waters in the different regions of the lake created more uniform conditions and a more even distribution of populations. Thus high frequencies of dominance would be expected from the best-adapted taxon. This occurred dramatically during the first month of pump operation when a one-time bloom occurrence of *Fragilaria* was spread to every station in the lake. Thereafter there was a trend toward fewer taxa attaining the status of a principal dominant.

The frequency-of-dominance differences occurring between base-line and postoperational years were not similarly reflected in any notable density or biovolume changes. Although autumn-to-spring dominance by Bacillariophyceae, represented primarily by *Melosira*, decreased markedly after year 4, there was no correspondingly great change in diatom abundance. The warmer winter water temperatures, caused by both the warming climatic trend and the added thermal discharges, may have favored algae more adapted to warmer temperatures than the diatom taxa. (In fact, preliminary data from year 7 indicate that the cold winter of 1977 was followed by a return of *Melosira* as an important dominant.)

The shift in increasing dominance to the principal genera *Nannochloris* by density and *Peridinium* by biovolume began during the base-line period and continued through the postoperational years. *Nannochloris* has been observed to increase in dominance at the expense of less-tolerant taxa after condenser passage (Knight, 1973). The similar perennial dominance of *Peridinium inconspicuum* in year 6 suggests that it, too, may be a taxon more tolerant of these conditions.

Increases in the number of taxa occurred each year for the important orders in the Chlorophyceae and for the class as a whole, as well as for the Dinophyceae. These trends were established before the postoperational period, as were the decreases in total density and enrichment indexes and the increases in total biovolume and diversity. The effects of power-plant operation, primarily that of increased homogeneity, have been mainly superimposed on these changes. Only the blue-green algae, which are noted for their tolerance of high temperatures, exhibited a sharp rise in the number of taxa exclusively between preoperational and postoperational periods, but their abundances remained uniformly low. The general long-term successional changes in the phytoplankton community appear to have been only somewhat accelerated after power-plant operation.

In each of the 4 years, the two most important taxa by density were small-celled organisms whose largest dimensions were under 10 μm. Such "ultraplankton" cells have high surface-to-volume ratios, high respiration-to-weight ratios, and, consequently, rapid turnover rates. This makes them potentially highly productive taxa capable of rapidly taking advantage of increases in limiting nutrients. By biovolume, the trend was toward dominance by larger-celled dinoflagellates. Members of this class produce algal toxins that encourage rejection of the cells by grazers, and their larger size has adaptive value both where there is pressure by filter-feeding zooplankton capable of grazing only small algae and where the lower respiration per unit weight of larger cells allows survival under high-temperature, low-light, and low-nutrient conditions, although with consequently slower turnover rates and lower productivity (Porter, 1977). Their motility would also enable them to actively migrate vertically throughout the thermocline to take advantage of the greater nutrient concentrations available in the hypolimnion (Talling, 1971; Kamykowski and Zentara, 1977). Thus the increased dominance of the dinoflagellates by biovolume is more likely to reflect their greater resistance to grazing pressure and the stresses of condenser passage and their greater capacity to deal with low nutrient levels than to indicate their importance to the productivity of the aquatic food web.

The phytoplankton community of Belews Lake appears to be proceeding through a series of long-term successional changes in response to the adjusting and equilibrating environment of this new impoundment and to local climatic changes. Power-plant operation has superimposed a greater homogeneity on the phytoplankton community and has not appeared to cause any detrimental effects within the period of this investigation.

ACKNOWLEDGMENTS

The research reported here was part of a project funded by the Duke Power Company, Charlotte, N. C., through a research grant to Charles M. Weiss, Project Director, Department of Environmental Sciences and Engineering, School of Public Health, University of North Carolina, Chapel Hill.

REFERENCES

Cairns, J., Jr., 1956, Effects of Increased Temperatures on Aquatic Organisms, *Ind. Wastes (Chicago)*, 1: 150-152.

Campbell, P. H., 1973, *Studies on Brackish Water Phytoplankton*, Sea Grant Publication, UNC-SG-73-07, University of North Carolina, Chapel Hill, NTIS.

Canale, R. P., and A. H. Vogel, 1974, Effects of Temperature on Phytoplankton Growth, *J. Environ. Eng. Div., Am. Soc. Civ. Eng.*, 100(EEI): 231-241.

Environmental Protection Agency, 1974, *Methods for Chemical Analysis of Water and Wastes*, Report EPA-625/6-74-003, EPA Water Quality Office, Analytical Quality Control Laboratory, Cincinnati.

Foerster, J. W., F. R. Trainor, and J. D. Buck, 1974, Thermal Effects on the Connecticut River: Phycology and Chemistry, *J. Water Pollut. Control Fed.*, 46: 2138-2152.

Goldman, J. C., and E. J. Carpenter, 1974, A Kinetic Approach to the Effect of Temperature on Algal Growth, *Limnol. Oceanogr.*, 19: 756-766.

Gurtz, M. E., and C. M. Weiss, 1974, Effect of Thermal Stress on Phytoplankton Productivity in Condenser Cooling Water, in *Thermal Ecology*, AEC Symposium Series, Augusta, Ga., May 3—5, 1973, J. W. Gibbons and R. R. Sharitz (Eds.), pp. 490-507, CONF-730505, NTIS.

Hockley, A. R., 1963, Some Effects of Warm Water Effluents in Southampton Water, *Annu. Rep. Challenger Soc.*, 3: 37-38.

Hommersand, M. H., 1972, Algae, in *Yearbook of Science and Technology*, 1971, pp. 106-112, McGraw-Hill Book Company, New York.

Kamykowski, D., and S. J. Zentara, 1977, The Diurnal Vertical Migration of Motile Phytoplankton Through Temperature Gradients, *Limnol. Oceanogr.*, 22: 148-151.

Knight, R. L., 1973, *Entrainment and Thermal Shock Effects on Phytoplankton Numbers and Diversity*, Department of Environmental Sciences and Engineering, Publication Number 336, University of North Carolina, Chapel Hill.

Kullberg, R. G., 1968, Algal Diversity in Several Thermal Spring Effluents, *Ecology*, 49: 751-755.

Lloyd, M., and R. J. Ghelardi, 1964, A Table for Calculating the "Equitability" Component of Species Diversity, *J. Ani. Ecol.*, 33: 217-225.

Lund, J. W. G., C. Kipling, and E. D. LeCren, 1958, The Inverted Microscope Method of Estimating Algal Numbers and the Statistical Basis of Estimations by Counting, *Hydrobiologia*, 11: 143-170.

Morgan, R. P., and R. G. Stross, 1969, Destruction of Phytoplankton in the Cooling Water Supply of a Steam Electric Station, *Chesapeake Sci.*, 10: 165-171.

Moss, B., 1973, Diversity in Freshwater Phytoplankton, *Am. Midl. Nat.*, 90: 341-355.

Palmer, C. M., 1969, A Composite Rating of Algae Tolerating Organic Pollution, *J. Phycol.*, 5: 78-82.

Pannel, J. P. M., A. E. Johnson, and J. E. G. Raymont, 1962, An Investigation into the Effects of Warmed Water from Marchwood Power Station into Southampton Water, *Proc. Inst. Civ. Eng. (London)*, 23: 35-62.

Patrick, R., 1969, Some Effects of Temperature on Freshwater Algae, in *Biological Aspects of Thermal Pollution*, P. A. Krenkel and F. L. Parker (Eds.), pp. 161-185, Vanderbilt University Press, Nashville, Tenn.

——, 1974, Effects of Abnormal Temperatures on Algal Communities, in *Thermal Ecology*, AEC Symposium Series, Augusta, Ga., May 3—5, 1973, J. W. Gibbons and R. R. Sharitz (Eds.), pp. 335-349, CONF-730505, NTIS.

Porter, K. G., 1977, The Plant—Animal Interface in Freshwater Ecosystems, *Am. Sci.*, 65: 159-170.

Rankin, J. S., J. D. Buck, and J. W. Foerster, 1974, Thermal Effects on the Microbiology and Chemistry of the Connecticut River—A Summary, in *Thermal Ecology*, AEC Symposium Series, Augusta, Ga., May 3—5, 1973, J. W. Gibbons and R. R. Sharitz (Eds.), pp. 350-355, CONF-730505, NTIS.

Reynolds, N., 1973, The Estimation of the Abundance of Ultraplankton, *Br. Phycol. J.*, 8: 135-146.

Shannon, C. E., and W. Weaver, 1949, *The Mathematical Theory of Communication*, University of Illinois Press, Urbana.

Strathmann, R. R., 1967, Estimating the Organic Carbon Content of Phytoplankton Cell Volume or Plasma Volume, *Limnol. Oceanogr.*, 12: 411-418.

Talling, J. F., 1971, The Underwater Light Climate as a Controlling Factor in the Production Ecology of Freshwater Phytoplankton, *Mitt. Int. Verein. Limnol.*, 19: 214-243.

Warinner, J. E., and M. C. Brehmer, 1966, The Effects of Thermal Effluents on Marine Organisms, *Int. J. Air Water Pollut.*, 10: 277-289.

Weiss, C. M., T. P. Anderson, P. H. Campbell, and D. R. Lenat, 1978a, *Environmental Effects of Power Plant Operation: Belews Lake V and VI, North Carolina, July 1974—June 1976*, Department of Environmental Sciences and Engineering, Publication Number 475, University of North Carolina, Chapel Hill, in preparation.

——, T. P. Anderson, P. H. Campbell, D. R. Lenat, J. H. Moore, and S. L. Pfaender, 1974, *Environmental Comparison, Belews Lake—Year III and Lake Hyco, North Carolina, July 1972—June 1973*, Department of Environmental Sciences and Engineering, Publication Number 370, University of North Carolina, Chapel Hill.

——, T. P. Anderson, and D. R. Lenat, 1972, *Environmental Assessment, Belews Creek—Belews Lake, North Carolina, Year II*, Department of Environmental

Sciences and Engineering, Publication Number 317, University of North Carolina, Chapel Hill.

——, P. H. Campbell, D. Y. Conlin, J. H. Moore, and S. L. Pfaender, 1978b, *The Limnology of John H. Kerr Reservoir*, Department of Environmental Sciences and Engineering, Publication Number 474, University of North Carolina, Chapel Hill.

——, et al., 1975, *An Assessment of the Environmental Stabilization of Belews Lake—Year IV, and Comparisons with Lake Hyco, North Carolina, July 1973—June 1974,* Department of Environmental Sciences and Engineering, Publication Number 416, University of North Carolina, Chapel Hill.

Welch, E. B., G. R. Hendry, and R. K. Stoll, 1975, Nutrient Supply and the Production and Biomass of Algae in Four Washington Lakes, *Oikos*, 26: 47-54.

Yentsch, C. S., C. M. Yentsch, L. R. Strube, and I. Morris, 1974, Influence of Temperature on the Photosynthetic Efficiency in Natural Populations of Marine Phytoplankton, in *Thermal Ecology*, AEC Symposium Series, Augusta, Ga., May 3—5, 1973, J. W. Gibbons and R. R. Sharitz (Eds.), pp. 508-517, CONF-730505, NTIS.

EFFECTS OF POWER-PLANT OPERATION ON THE ZOOPLANKTON COMMUNITY OF BELEWS LAKE, NORTH CAROLINA

TERRY P. ANDERSON and DAVID R. LENAT
Department of Environmental Sciences and Engineering, University of North Carolina, Chapel Hill, North Carolina

ABSTRACT

The euphotic-zone zooplankton community of Belews Lake, North Carolina, was studied over a 3-year period. Belews Lake is a 1500-ha cooling reservoir for a 2200-MW coal-fired power plant. This study included 1 year of preoperational data and 2 years of postoperational data. A detailed examination of 19 taxa showed that three species may have responded to power-plant operation. Two warm stenotherms *(Hexarthra* and *Ptygura)* increased in density, and an increase in the winter density of *Polyarthra* was also noted. An analysis of variance indicated that circulation of surface water by the power plant resulted in greater spatial and seasonal homogeneity for most taxa. Other changes in the zooplankton community appeared related to natural causes.

A lake ecosystem is temporally heterogenous in terms of a large number of physical and biological factors. Seasonal patterns in the zooplankton community have evolved to take advantage of these temporal changes, therefore reducing competitive interactions between species. Because of the marked seasonality of the zooplankton, comparison studies are best made between periods of at least 1 year. We studied the zooplankton community of Belews Lake, North Carolina, on a monthly basis for a period of 3 years and compared the resulting data on a yearly basis.

Considerable differences in the abundance of zooplankton are to be expected when making between-year comparisons (Ricker, 1938; Carlin, 1943; Edmondson, 1946; Hutchinson, 1967). Multivariant analysis of zooplankton densities vs. environmental variables has indicated that temperature and food concentrations are the most important factors controlling zooplankton abundance (Edmondson, 1965; Hayward and Gallup, 1976).

Temperature has been shown to affect the reproductive rates and seasonal cycles of both Crustacea and Rotifera. Selgeby (1975) compared the seasonal cycles of Crustacea in cold Lake Superior with those in warmer Lake Michigan and Lake Erie. He found that differences in temperature can advance breeding 1 to 3 months. Williams (1964) also has found differences in the seasonal patterns of rotifers when comparing areas with different temperature regimes. Another important variable affecting zooplankton densities is the presence of planktivorous fish (Brooks and Dodson, 1965; Stenson, 1973).

Between-year changes in the zooplankton community are most probably due to changes in these three factors. We, therefore, assessed zooplankton dynamics in Belews Lake in light of temperature, food, and planktivorous fish populations. Power-plant operation may have superimposed a fourth factor. Our main objective has been to distinguish power-plant effects from those due to natural causes. We have placed our emphasis on lake-wide effects instead of concentrating on local areas of impact, such as the discharge cove area. We believe that this is the proper focus of concern since localized studies inevitably lead to questions about what the results mean in terms of the functioning of the system as a whole.

STUDY AREA

Belews Lake (Fig. 1) was constructed by Duke Power Company as a cooling reservoir for the coal-fired Belews Creek Steam Station. The lake lies in the Piedmont section of North Carolina. It has a surface area of 1500 ha, a retention time of approximately 1000 days, and a mean depth of 15 m (maximum depth $\simeq$40 m). Belews Lake has been classified as oligomesotrophic (Weiss and Kuenzler, 1976).

The Belews Creek Steam Station has two operating units, each with a capacity of 1145 MW. Operational data for years 5 and 6 are given in Fig. 2. The mean pumping rate of 52 m^3/sec in year 6 is theoretically capable of circulating the entire epilimnion of the lake (average depth of 7.5 m) in 3 weeks. Note that several periods of shutdown occurred, especially in year 5.

Belews Lake has been the subject of an environmental study for 7 years, designated year 1 to year 7. Year 1 (August 1970 to April 1971) was a preimpoundment study with dam closure at the beginning of year 2 (May 1971 to June 1972). Lake filling continued through the end of year 3 (July 1972 to June 1973). Year 4 (July

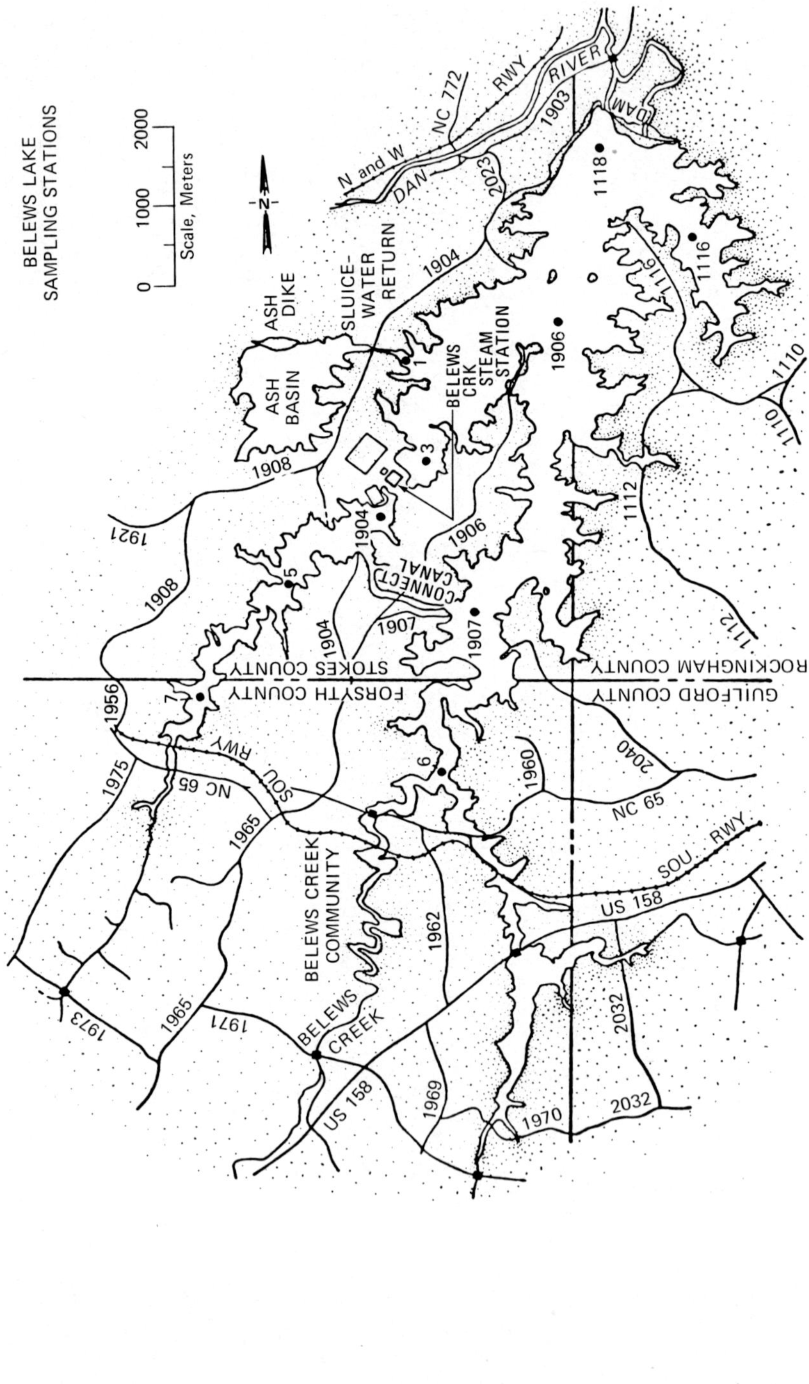

Fig. 1 Map of Belews Lake, North Carolina, showing the location of the Belews Creek Steam Station and the zooplankton sampling stations (●).

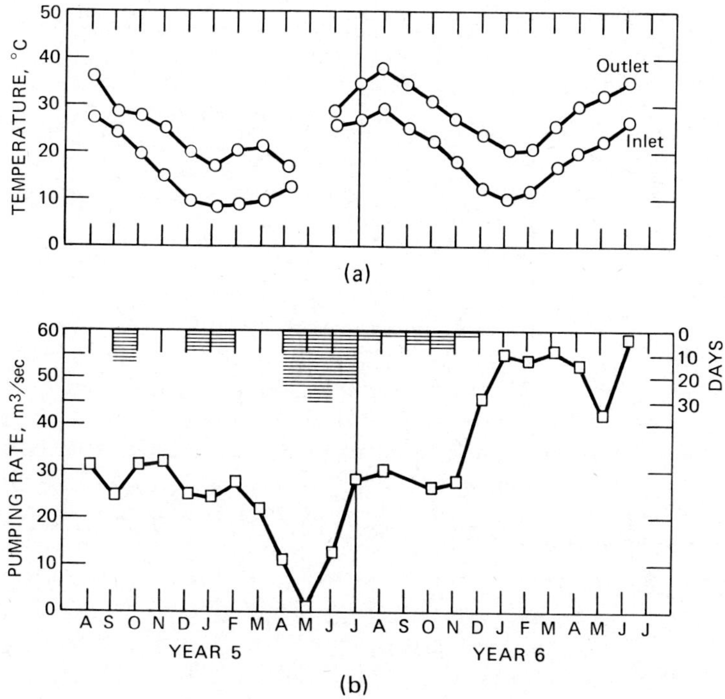

Fig. 2 (a) Monthly mean inlet and outlet temperatures, and (b) monthly mean pumping rate and days of each month that power plant was not generating electricity (bar graph), for the Belews Creek Steam Station during the period of study.

1973 to June 1974) was the first 12-month period that the lake existed at full-pool level. The first unit of the Belews Creek Steam Station commenced operation in year 5 (July 1974 to June 1975), and the second unit went on line in year 6 (July 1975 to June 1976). Zooplankton studies began in year 4. This study compares year 4 (control) data with data from years 5 and 6. The year 4 to year 6 comparisons give the strongest indication of power-plant effects.

The discussion in this paper focuses primarily on euphotic-zone data. These data were the most complete and encompass that zone of the lake which is the most important in terms of limnetic trophic level interactions and heat effects. Collections were also taken from the entire water column or the depths below the euphotic zone to ensure that an important component of the zooplankton was not missed because of migration or depth preferences that were deeper than the euphotic zone. This occurred for one species and is mentioned in the text.

MATERIALS AND METHODS

Collection Techniques

Zooplankton were collected through September of year 5 with a 12.6-cm diameter 20-mesh Birge closing net. Monthly duplicate samples were collected at the 10 stations shown in Fig. 1 by taking vertical hauls through the euphotic zone (defined as the depth to which 1% of the surface illumination penetrated as measured by a selenium cell underwater photometer). The depth of the euphotic-zone collections encompassed and usually surpassed the depth of the epilimnion (7 to 8 m) at all stations except 1 and 7. The entire water column was sampled at these shallower stations. A larger (24-cm diameter) 20-mesh net was used in October and November, and, beginning in December and continuing throughout the rest of the study, a 30-cm diameter 20-mesh net was used. The change in net diameter was made to increase the chances of collecting zooplankton that might have avoided the smaller nets.

The zooplankton collected in each year were preserved in the field with a 4% Formalin solution. A sample from one representative station had a 1% Neo-Synephrine solution added before preservation to narcotize the zooplankton. This sample was used to compile a preliminary species list for the month and to aid in identifying organisms that would contract when preserved with Formalin.

Statistical Techniques

We normalized the euphotic-zone zooplankton data first using a $\log(n + 1)$ transformation of the averaged duplicate counts from each monthly collection. This procedure aids in stabilizing variance. Two separate sets of statistical analyses were performed. The first set of tests examined significant differences between years for 19 taxa. The second set of tests examined spatial and seasonal homogeneity for these same taxa. All tests were conducted with data from eight stations that were consistently sampled over the period of years 4 to 6.

Differences between years were first tested for each taxon by treating each station separately with a two-way analysis of variance (ANOVA) using year and month as variables. This was followed by a three-way ANOVA, using stations, year, and month as variables, to assess lake-wide differences between years for each taxon tested. The year × month interaction term was used as the error term in these tests. The replicate data were tested for most of the species and did not add significantly to the source of variation. Therefore the averages of the replicate counts were transformed and used. The year × month interaction term was very conservative owing to the large and variable seasonal differences of the densities within a year. As a result we thought that using this as the error term would not increase the probability of a Type I error.

Tests were also conducted to assess seasonal and spatial homogeneity. Spatial homogeneity refers to a similarity of taxon density between stations in any given month within a year. Using a data set of 19 taxa, 8 stations, and 3 years and using the station × month interaction term of each year as the error term, we tested the significance of station as a variable within each year. Seasonal homogeneity refers to taxon density being similar between stations during the same seasons of the year. The station × month interaction term in ANOVA is a measure of the degree to which changing densities at stations parallel one another over the year. We tested for a significant decline in the station × month interaction by using the year 6 station × month interaction as the error term and compared it to the same term in year 4 and year 5. We partitioned the variance appropriately within each year and then tested for a significant decrease. Each taxon was tested separately.

RESULTS AND DISCUSSION

Changes in Environmental Parameters

Temperature

At Belews Lake the normal temperature pattern has been altered by the release of heated effluents from the Belews Creek Steam Station. Station 1904 did not stratify after power-plant operation and was vertically isothermal or nearly so because of the mixing effect of the discharged heated effluent. The heated effluent is discharged near station 1907 from the connecting canal as a surface plume about 5 m deep and dissipates its heat rapidly. A gradient of less-heated water extends from station 1907 down lake toward the dam station, 1118, and the intake station, 3. In the summer this gradient (using surface temperatures from August, year 6) ranged from 32.2 to 30.0 and 29.3°C, respectively. The average temperatures that the zooplankton experience are less than this since the heated water overlays cooler water deeper within the euphotic zone. The lake is stratified by April and usually remains so until late November. Under these conditions the average euphotic-zone temperatures are some 4°C less than surface temperatures. The temperature vs. depth profiles under stratified conditions are characterized by a warm layer (near surface temperatures) 4 to 6 m deep overlaying progressively cooler water which may include a portion of the metalimnion. From November to March the average main lake euphotic-zone temperatures (except for station 1904) vary less than 1 to 2°C from the surface temperatures owing to destratifying forces. The surface temperatures discussed in the following paragraphs are meant to give an idea of the general impact of power-plant operation on the temperature regime of the lake.

Average surface temperatures for 11 stations are given in Tables 1 and 2. Average surface temperatures increased from year 4 to year 6 (from 18.9 to 23.6°C), but this increase was not uniformly distributed over all months. Power-plant operation affected the temperature regime in two ways. First, winter temperatures increased. Winter temperatures in year 6 were 5 to 12°C greater than in year 4. There was a steady increase in the minimum temperature recorded from 7.5°C in year 4 to 13.0°C in year 6. A further effect of temperature increases has been a decrease in the annual temperature fluctuation, as reflected in the decreasing standard deviation of yearly average temperatures. Second, the period of maximum temperature increased, especially in year 6, although the maximum temperature remained approximately the same. Average surface temperatures greater than 30°C were recorded in only 1 month in year 4 but were observed in 3 months in year 6. The variation in mean air temperatures recorded at the Greensboro, N.C., airport (National Oceanic and Atmospheric Administration records) over the natural warming period (January through August) between years was less than 1°C. The contribution of natural heating to surface temperature was therefore thought negligible compared to that of the power plant.

It is important to note periods when average surface temperatures did not change appreciably. In early fall and late spring of year 5, the power plant was shut down. Comparisons between years 4 and 5 during these periods indicate that there was little systematic change in temperature; changes in the seasonal pattern of fall–spring species therefore should be observed mainly in year 6, although some shifts might be seen in the early spring of year 5.

Phytoplankton

Changes in food concentrations and food quality in Belews Lake were assessed by examining algal standing crop. Data on algal densities are given in Fig. 3. The line for total density primarily reflects the distribution of nannoplankton *(Nannochloris, Chlorella,*

TABLE 1

MEAN* MONTHLY SURFACE TEMPERATURES, STANDARD DEVIATION (SD), AND MINIMUM–MAXIMUM TEMPERATURE

	Month															
	7	8	9	10	11	12	1	2	3	4	5	6	Mean	SD	Min	Max
Year 4	31.3	29.1	27.4	19.4	12.0	†	7.5	10.9	10.2	14.7	20.4	25.3	18.9	8.4	7.5	31.3
Year 5	29.6	30.1	24.8‡	21.8	19.3	13.5	†	12.1	14.5	14.1‡	24.2‡	27.0‡	21.0	6.7	12.1	30.1
Year 6	31.0	31.4	26.7	25.9	23.7	14.7§	13.0	16.1	17.9	25.4	25.7	30.3	23.6	6.3	13.0	31.4

*Mean of 11 stations, °C.
†No data.
‡Power-plant shutdown.
§December 1975, second unit in operation.

TABLE 2

AVERAGE* MONTHLY SURFACE TEMPERATURE CHANGES† BETWEEN YEARS

	Month											
	7	8	9	10	11	12	1	2	3	4	5	6
Δ1, year 5 to 4	−1.7	+1.0	−2.6	+2.4	+7.3	+6.0‡	+6.0‡	+1.2	+4.3	−0.6	+3.8	+1.7
Δ2, year 6 to 5	+1.4	+1.3	+1.9	+4.1	+4.4	+2.2	+2.0‡	+4.0	+3.4	+11.3	+1.5	+3.3
Δ3, year 6 to 4	−0.3	+2.3	−0.7	+6.5	+11.7	+6.0‡	+5.5	+5.2	+7.7	+10.7	+5.3	+5.0

*Average of 11 stations, °C.
†Changes greater than 3°C are underlined.
‡Estimated.

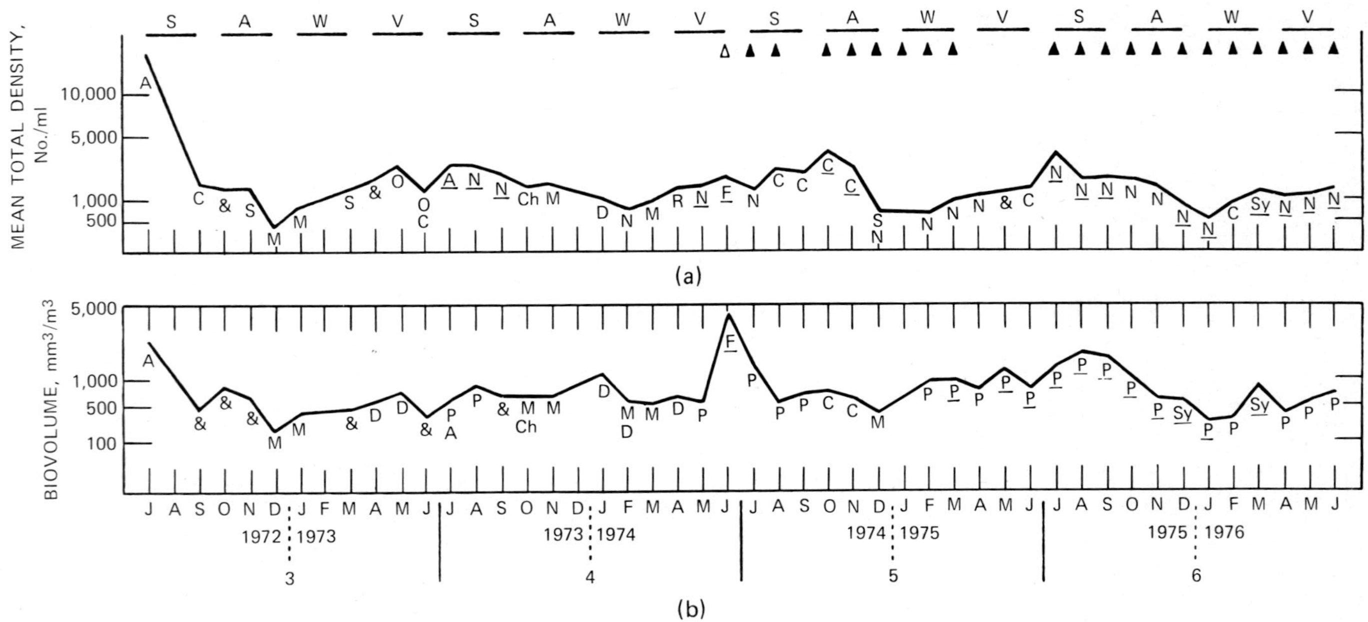

Fig. 3 Seasonal distribution of (a) mean total density and (b) biovolume for Belews Lake, 1972–1976. Most important dominant genera during each monthly sampling are indicated as follows: A, *Agmenellum;* C, *Chlorella;* Ch, *Chrysochromulina;* D, *Dinobryon;* F, *Fragilaria;* M, *Melosira;* &, several dominants; N, *Nannochloris;* O, *Oocystis;* P, *Peridinium;* R, *Rhizosolenia;* S, *Selanastrum;* Sy, *Synedra.* Those underlined were dominant in all samples.

and *Selenastrum*), which constitute a prime source of food available to microconsumer zooplankton. Few differences can be seen between years, except for an increase in the fall of year 5.

Figure 3 also presents data on changes in phytoplankton biovolume. This line primarily reflects changes in the large, or net, algal species. Biovolume declined slightly in year 6 with the appearance of *Synedra*, a long needle-shaped diatom, as an important component of the net phytoplankton. No important decline in biovolume was noted in year 5, but a species change occurred, with *Melosira italica* var. alpigena being replaced by the larger *Peridinium* spp. This successional sequence from years 4 to 6 may have caused a decline in the availability and food value of the phytoplankton because *Peridinium* and *Synedra* are larger in size and thus would be more difficult to consume.

Planktivores

One major change occurred in the planktivore population at Belews Lake during this study. Threadfin shad (*Dorosoma petenense*) were introduced into the lake by local sportsmen. Data supplied by the North Carolina Wildlife Resources Commission show that threadfin shad increased from a density of 1.3/ha (average of eight coves totaling 3.24 ha) in September of year 5 to 6463.4/ha in September of year 6 (Scott Van Horn, fishery biologist, personal communication). They declined to year-5 levels in September of year 7. Changes in the zooplankton community did not reflect the appearance and disappearance of this planktivore. For instance, crustacean populations that would be fed upon by young-of-the-year shad in the spring and summer (Applegate and Mullan, 1969) did not show declines in their spring—summer peaks in year 5 (Fig. 4).

Seasonal Abundance of the Major Zooplankton Taxa

Figure 5 shows the seasonal distribution, years 4 to 6, for Rotifera, Cladocera, and Copepoda. Several major trends are evident. Rotifers were numerically dominant in Belews Lake, comprising 81.2% of the total zooplankton in year 4, 72.1% in year 5, and 77.0% in year 6. Cladocera comprised only 7 to 9% of the total zooplankton during this period, and Copepoda (including nauplii) made up 11 to 19%. Rotifera generally exhibit a marked spring density peak. This peak was markedly lower in year 6 than in years 4 and 5. Winter densities of Rotifera appear to be increasing over the period of years 4 to 6. However, no significant differences (5% level) were noted between years when all months were considered together.

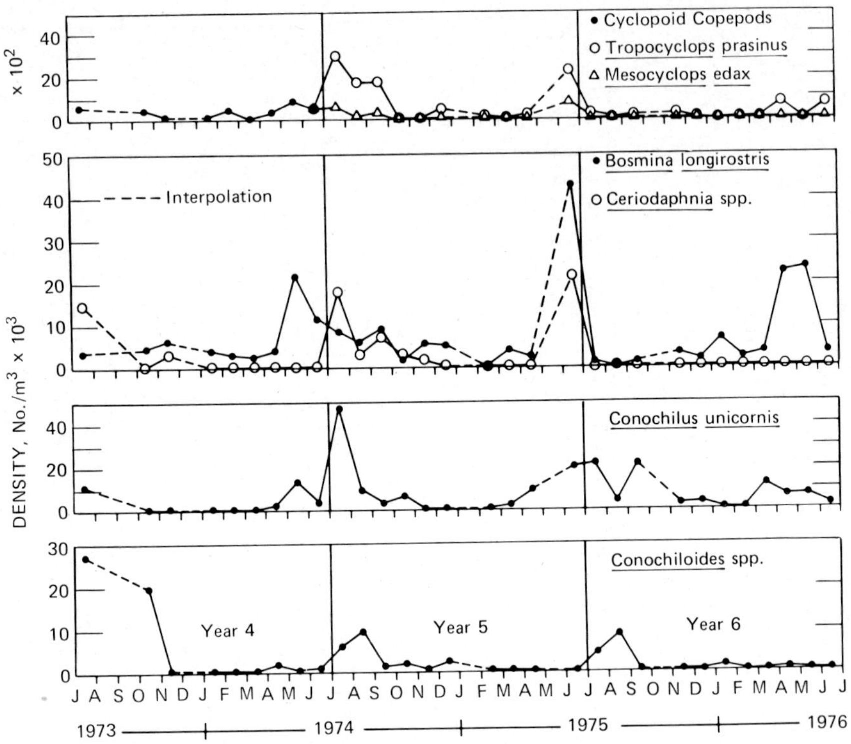

Fig. 4 Seasonal distribution of seven species of zooplankton from Belews Lake. Data plotted from mean euphotic-zone density, all stations.

Seasonal curves for the major zooplankton taxa are given in Figs. 4 to 7. Statistical analyses and relative abundances of these taxa are given in Table 3, and they are discussed in detail in the following paragraphs.

The taxa may be divided into three groups on the basis of changes between years 4 and 6: stable, declining, and increasing.

Stable Taxa

Six taxa showed no significant difference between years 4 and 6 (at a 5% level) and were considered stable: *Keratella cochlearis*, *K. earlinae*, *Bosmina longirostris*, *Polyarthra* spp., nauplii, and cyclopoid copepodids. We think that *Tropocyclops prasinus* was also stable over this period, although it was not identified to a species level until year 5. It exhibited a warm season peak in July and June of year 5.

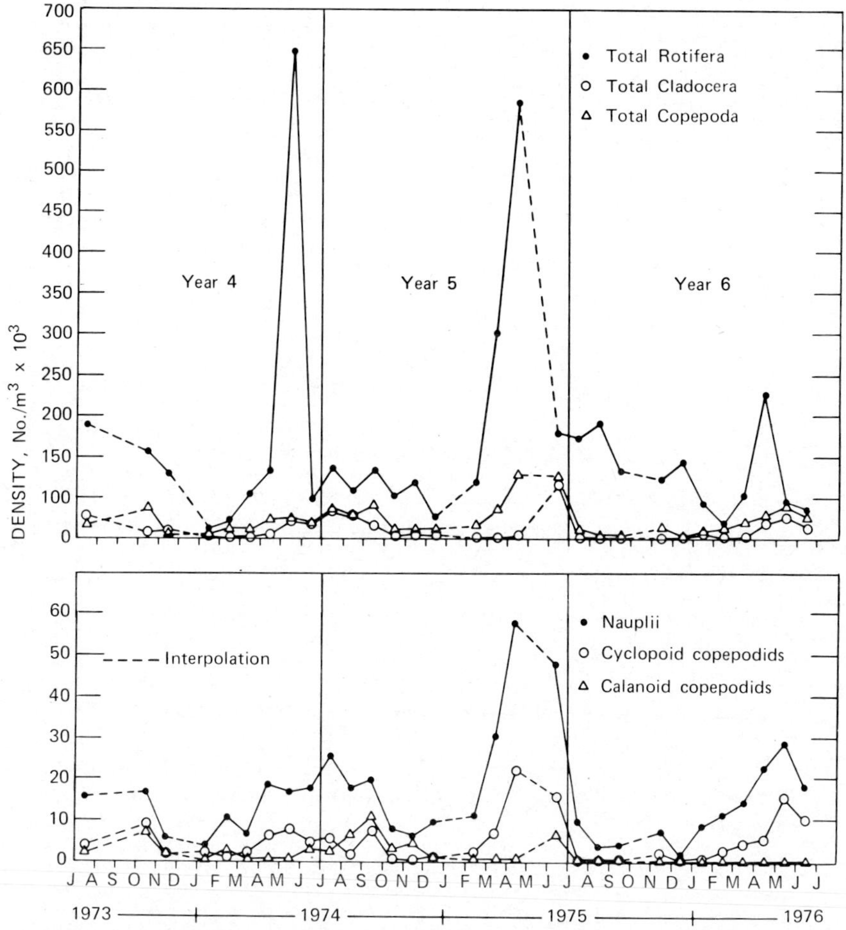

Fig. 5 Seasonal distribution of major zooplankton groups (Rotifera, Copepoda, and Cladocera) from Belews Lake. Data plotted from mean monthly euphotic-zone density, all stations.

This inflated the year-5 densities and accounted for the apparent decline in year 6. Most of these taxa had a year-6 maximum. Note that the spring peak of *K. cochlearis* was missed in year 5 (Fig. 6).

The average density of *Polyarthra* spp. increased greatly in postoperational years (Fig. 6). Comparisons between years, however, showed a significant increase at only one station (1) in the year 6 to year 4 comparison. Three stations in the year 5 to year 4 comparison showed significant increases, while no significant differences were

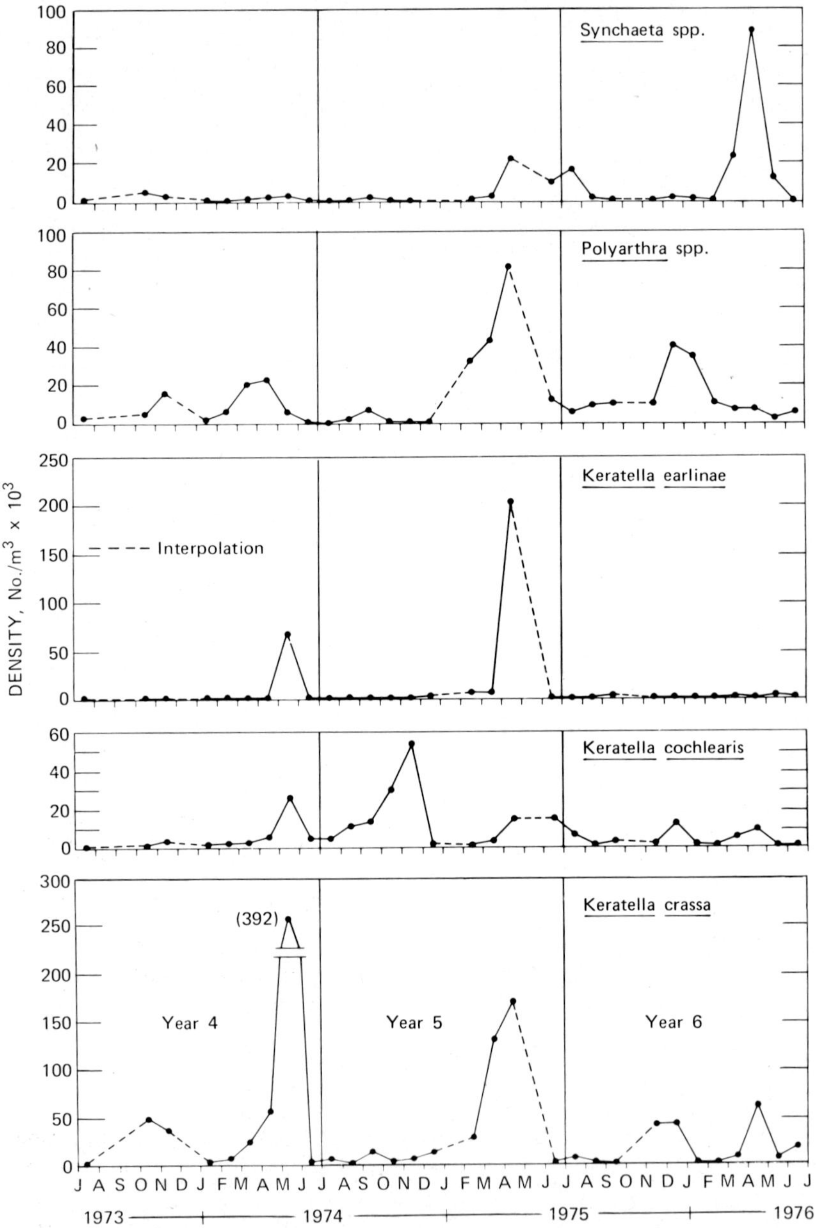

Fig. 6 Seasonal distribution of five major rotifer species from Belews Lake. Data plotted from mean monthly euphotic-zone density, all stations.

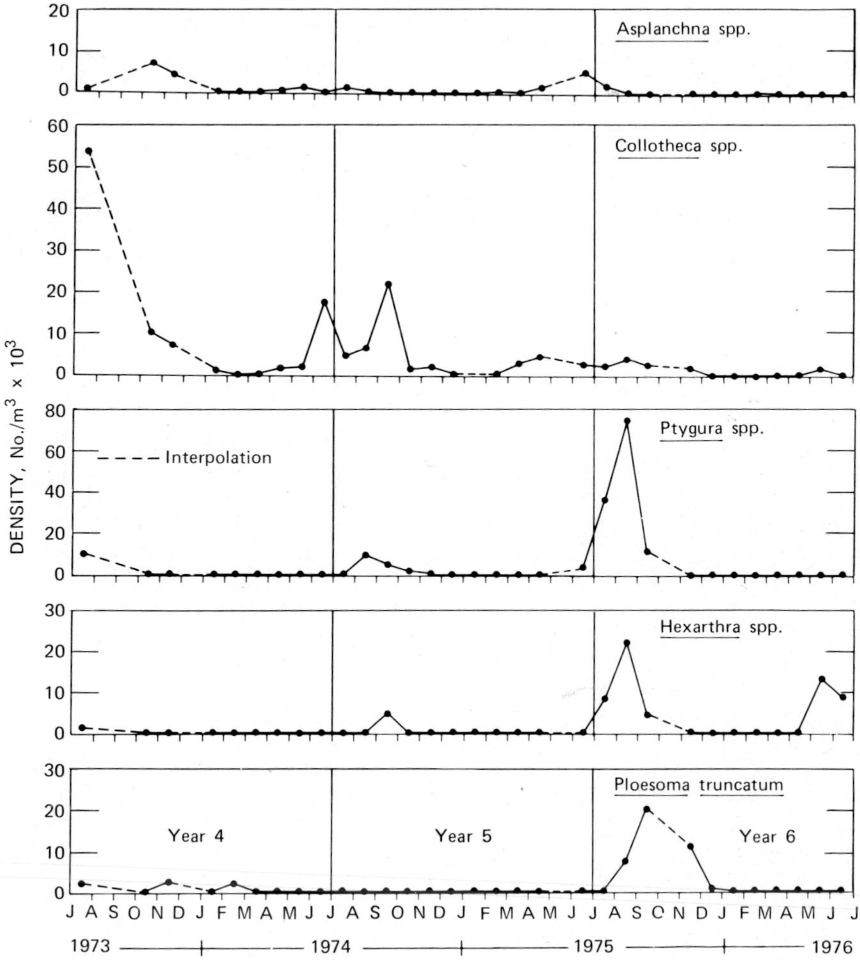

Fig. 7 Seasonal distribution of five rotifer species from Belews Lake. Data plotted from mean monthly euphotic-zone density, all stations.

noted when year 6 was compared to year 5. When all stations were pooled and years compared, no significant differences occurred. This was due to high year—month interaction. Variation between years was nonparallel.

Monthly sampling of *Polyarthra* spp. yielded a seasonal pattern with one to two peaks. Peaks were compared with average surface temperatures and two optimums were found, one at 25 to 27°C and one at 12 to 16°C. This suggests that at least two species were important.

TABLE 3

ABUNDANCE AND ANALYSIS OF VARIANCE OF MAJOR ZOOPLANKTON SPECIES: EUPHOTIC-ZONE HAULS

	All stations, years 4—6					
	Yearly averages, No./m^3			Analysis of variance*		
	4	5	6	4—5	4—6	5—6
Rotifera						
Keratella crassa	64,000	39,800	17,000	NS	S5↓	S5↓
K. cochlearis	5,400	15,200	4,000	S5↑	NS	S5↓
K. earlinae	8,100	27,900	1,300	S1↑	NS	S1↓
Polyarthra spp.	9,200	18,400	13,800	NS	NS	NS
Synchaeta spp.	1,200	3,400	13,000	NS	S5↑	NS
Conochilus unicornis	3,900	10,100	7,200	S1↑	S1↑	NS
Collotheca spp.	10,500	5,000	1,300	S5↓	S5↓	S5↓
Hexarthra spp.	200	500	5,900	NS	S1↑	S1↑
Ptygura spp.	1,000	2,300	11,400	S1↑	S1↑	NS
Conochiloides spp.	5,600	2,200	1,500	NS	S5↓	S1↓
Ploesoma truncatum	900	0	3,800	A(5)	S5↑	A(5)
Asplanchna spp.	1,800	1,000	200	NS	S5↓	S5↓
Cladocera						
Bosmina longirostris	6,900	8,800	6,600	NS	NS	NS
Ceriodaphnia spp.	2,300	6,000	100	S1↑	S1↓	S1↓
Copepoda						
Nauplii	12,500	23,300	12,000	S1↑	NS	S1↓
Calanoid copepodids	2,100	4,000	100	S1↑	S1↓	S1↓
Cyclopoid copepodids	4,400	6,500	4,000	NS	NS	S5↓
Mesocyclops edax	NI	200	2	NI	NI	S1↓
Tropocyclops prasinus	NI	1,000	200	NI	NI	S1↓

*Explanation of symbols:
NS, not significant.
S5, significant at the 5% level.
S1, significant at the 1% level.
NI, not identified in year 4.
↓↑, direction of significant change, i.e., decrease or increase.
A( ), organism absent in indicated year.

Power-plant operation may have had a marked positive effect on the average density of *Polyarthra* through an increase in winter temperatures. In year 4, low densities were recorded during January. By year 6, however, this was the period of maximum density. Since no significant differences were noted between years when all stations were compared, the implication is that power-plant operation did not affect yearly abundances but may have affected its seasonal distribution.

Conochilus unicornis showed a significant increase from year 4 to year 6. However, missing data in year 6 invalidates this comparison, and this species is considered stable. We observed, however, that the period of minimum densities decreased from 5 to 4 to 2 months as winter surface temperatures increased (Fig. 4). *Conochiloides* spp. is another taxon considered to be stable in spite of a statistically significant decline between years 4 and 6. Two species of *Conochiloides* were recorded from Belews Lake, *C. dossuarius* and *C. coenobasis.* The seasonal distribution of *Conochiloides* (Fig. 4) was apparently that of a warm stenotherm, but we think that this organism was present in the metalimnion at much lower temperatures. Makarewicz and Likens (1975) found that *Conochiloides* was present only in the hypolimnion of Mirror Lake, New York, and Makarewicz (1974) implied that it feeds mainly on bacteria and detritus, as it occurred below the zone of phytoplankton production. Bottom water of Mirror Lake was not anaerobic, and *Conochiloides* was present throughout the hypolimnion with a maximum near the bottom. In situations where the hypolimnion becomes anaerobic, however, rotifers tend to concentrate at the area of oxygen depletion, i.e., the metalimnion (Campbell, 1941). *Conochiloides*, therefore, was probably found in or near the metalimnion in Belews Lake because its hypolimnion is anaerobic during stratification. The apparent decline in *Conochiloides* from year 4 to year 6 was probably due to a deepening of the epilimnion, which caused *Conochiloides* to be missed in euphotic-zone collections, rather than to an actual drop in the density of this organism. This supposition is supported by data from deep samples (bottom-to-surface vertical hauls) in which the density of *Conochiloides* did not decline from year 4 to year 6. Moreover, the peak observed in euphotic-zone samples (August) was extended into the fall in deep samples for all 3 years.

Declining Taxa

Six taxa showed statistically significant declines in year 6 compared to year 4 and/or year 5: *Keratella crassa*, *Collotheca* spp., *Asplanchna* spp., Calanoid copepodids (=*Diaptomus* spp.), *Mesocyclops edax*, and *Ceriodaphnia* spp. *Keratella crassa* was the most abundant rotifer in Belews Lake during all years of this study. Little is known about its ecology. *Keratella crassa* is usually found to coexist with *K. cochlearis.* Makarewicz and Likens (1975) found that the maximum density for *K. crassa* in Mirror Lake, New York, was near the surface. *Keratella cochlearis* in this lake had a more uniform vertical distribution, with a maximum density at 3 m. A study by

Loeffelman (1976) at Smith Mountain Lake, Virginia, compared the densities of *K. crassa* and *K. cochlearis* in different parts of the lake. He found that *K. crassa* was favored in areas where pump storage operations increased vertical mixing.

In Belews Lake a significant decline was noted when year 6 was compared with the other years (5% level). This decline occurred primarily at main lake stations and not at the stations receiving direct heated discharge (stations 1904 and 1907). This indicates that factors other than temperature caused its decline.

The average yearly density of *Collotheca* spp. showed a steady decline from year 4 to year 6 (Fig. 7). *Collotheca* appears to be a warm stenotherm since peaks do not occur below 26°C. Pejler (1957) showed that *Collotheca libera* was a warm stenotherm and that *C. mutabilis* also was restricted to high temperatures in most lakes. Year 4 samples showed very high between-station variance. Maximum densities of *Collotheca* were found only at main lake stations, which implied that optimal conditions were confined to this area. Power-plant circulation in years 5 and 6, however, tended to reduce between-station variance (see later statistics) and exposed organisms in surface waters to a common range of environmental conditions. This change in environmental conditions may have been unfavorable for *Collotheca*, although other factors, such as interspecific competition, could also account for its decline.

Asplanchna spp. showed a significant decline (5% level) in year 6 when compared to year 4 and year 5. Year 4 and year 5 densities were not significantly different. The distribution of *Asplanchna* spp. was too irregular to deduce the cause of its decline. A possible explanation is competition with *Ploesoma truncatum*, a predaceous rotifer that increased in numbers at the same time.

Calanoid copepodids (*Diaptomus pallidus* and *D. reighardi*) declined sharply in the spring of year 5 and were largely absent in year 6. This phenomenon was probably not related to the increase in temperature. *Diaptomus* was found to be abundant in Lake Hyco, North Carolina, where surface temperatures up to 40°C were recorded (data reworked from Weiss et al., 1975a).

Mesocyclops edax appeared to have disappeared in year 6, however, it reappeared in July and August of year 7 in higher densities than previously recorded (Lenat and Anderson, in preparation). The seasonal distribution for this species seems to be somewhat irregular, and this peak shift in the warm season from year 6 to year 7 accounted for the decline and indicates that it was not detrimentally affected by power-plant operation.

Several species of *Ceriodaphnia* were found in Belews Lake. The most important were *C. lacustris* and *C. quadrangula.* The seasonal distribution of *Ceriodaphnia* was much more limited than that of *Bosmina* (Fig. 4). One peak generally was found to occur during the warm period of the year. In year 4 this peak occurred in July. However, in year 5 peaks were noted in July followed by another in June. This resulted in two peaks in year 5 and inflated the yearly average for this year. The June peak in year 5 was thought to be the annual warm-season peak and accounts for the significant decline in year 6. Smyly (1974) has presented 12 years of data on the seasonal distribution of *Ceriodaphnia quadrangula* in two English lakes. The maximum density encountered in any one year appeared to be similar, although the timing of peaks varied considerably between years.

Ceriodaphnia was relatively absent from Belews Lake during year 6. This decline does not appear to be due to temperature. Carlson (1974) showed that *Ceriodaphnia* was resistant to temperatures up to 42°C. In addition, it was the second most abundant cladoceran found in Lake Hyco, North Carolina, a cooling reservoir for a coal-fired power plant which had surface temperatures of 40°C (Weiss et al., 1975a). The *Ceriodaphnia* peak reappeared in the spring of year 7 in Belews Lake. We will analyze its distribution more critically in a future report. This reappearance shows that it was not permanently affected by whatever factors caused its decline.

Increasing Taxa

Four taxa showed a significant increase from year 4 to year 6 (5% level): *Hexarthra* spp., *Ptygura* spp., *Ploesoma truncatum*, and *Synchaeta* spp. It is interesting to note that two of the taxa *(Hexarthra* and *Ptygura)* are warm stenotherms.

Hexarthra spp.

Figure 7 had a temperature optimum in Belews Lake of 24 to 34°C. Maximum densities were recorded in August of year 6 at station 7 (32.3°C). Similar summer peaks have been recorded at Lake Hyco and the lower Catawba lakes, North Carolina (Weiss et al., 1975a; 1975b), Lake Anna, Virginia (Saunders, 1975), and Smith Mountain Lake, Virginia (Loeffelman, 1976). Edmondson (1960) found that *Hexarthra fennica* population densities were highly temperature dependent, and it was favored by high temperatures. Hutchinson (1967) reviewed this information and stated that "it is not unlikely that opportunistic species appearing at the height of the summer would exhibit marked temperature dependence."

Hexarthra may have been favored by power-plant operation, perhaps through an extension of the period of maximum temperature. Densities in year 6 were significantly greater (1% level) than in years 4 or 5.

Ptygura, like *Hexarthra*, exhibited a seasonal distribution typical of a warm stenotherm (Fig. 7). *Ptygura* peaked at the highest temperatures of all species studied: 30 to 35°C. Maximum densities were recorded in August, year 6, at temperatures of 32 to 34°C. The extension of the period of maximum summer temperatures may have favored an increase in the standing crop of *Ptygura*. The average density of *Ptygura* increased from $1000/m^3$ in year 4 to $11{,}400/m^3$ in year 6. *Ptygura* is rarely mentioned in the literature as a significant component of the zooplankton. However, collections from Belews Lake do not constitute the only record of *Ptygura* in Piedmont lakes. Very similar seasonal distributions have been found in Lake Anna, Virginia (Loeffelman, 1976) and in Lake Keowee, South Carolina (Duke Power Company, 1977). These records suggest that *Ptygura* may commonly occur in lakes of the southeastern United States, although only in the warmest months of the year.

The distribution of *P. truncatum* appeared to be unrelated to power-plant temperature effects (Fig. 7). It occurred irregularly in year 4, was absent in year 5, and exhibited a sharp late-summer peak in year 6. *Ptygura truncatum* is a predacious rotifer (Hutchinson, 1967), and its increase in September of year 6 may be related to the increase of such potential prey organisms as *Hexarthra* and *Ptygura*.

The average density of *Synchaeta* (Fig. 6) increased significantly (5% level) in year 6 when compared to year 4. Several authors have indicated that *Synchaeta* may be predacious on smaller rotifers, especially *Keratella* (Hutchinson, 1967). However, such feeding is probably only incidental (Pejler, 1964). *Synchaeta* peaks usually occur in early spring, preceding peaks by *Keratella* and other rotifers (Ahlstrom, 1934; Beach, 1960; Pejler, 1962). Pejler (1962) indicated that *Synchaeta* fed on live algae only, therefore, being favored at the start of the spring pulse. The spring peak of *Synchaeta* increased from year 4 to year 6 in Belews Lake. *Synchaeta* feeds on large algae (Pejler, 1957), and its increase may be related to the increase in *Peridinium inconspicua* (length approximately 15 to 25 μm).

Tests for Homogeneity

We hypothesized that the high power-plant pumping rates might tend to increase the homogeneity of the zooplankton community. This hypothesis was tested in two ways using the analysis of variance techniques discussed earlier. Table 4 illustrates such a test for one of

TABLE 4

ANALYSIS OF VARIANCE TESTS FOR HOMOGENEITY OF *PTYGURA* SPP.

Source	df	Sum of squares	Mean square	F*
Stations, year 4	7	29.634	4.233	4.562 S5
Stations, year 5	7	8.156	1.165	1.663 NS
Stations, year 6	7	3.322	0.474	0.990 NS
Station × month, year 4	77	71.447	0.928	1.937 S1
Station × month, year 5	77	53.944	0.700	1.462 S5
Station × month, year 6	77	36.889	0.479	0.0

*Explanation of symbols:
S5, significant at the 5% level.
S1, significant at the 1% level.
NS, not significant.

the 19 taxa. Table 5 lists the results of all tests and their interpretation.

An increase in spatial homogeneity was indicated if the significance of station as a variable declined from year 4 to year 6. This occurred for 13 of the 19 taxa tested and indicated that power-plant circulation may have brought about an increase in spatial homogeneity (station similarity).

An increase in seasonal homogeneity was indicated if the station × month interaction term was significantly less in year 6 than in year 4 or year 5 (year 6 was the year in which the power plant was fully operational with both units on line). This occurred for 12 of the 19 taxa tested. An increase in seasonal homogeneity for this many taxa indicated that power-plant circulation may have brought about an increase in seasonal homogeneity among stations. Tests among the Cladocera, Copepoda, and Rotifera were not as conclusive. The sensitivity of the analysis was lost owing to the high degree of variation between the species that were pooled within each group.

SUMMARY

The densities of 19 zooplankton taxa at 8 stations were examined separately by analysis-of-variance techniques to determine significant differences in two ways. The yearly density of each species at each station was compared among the 3 years of study. All stations were then pooled within a year, and the density of each

TABLE 5

STATION AND STATION x MONTH INTERACTION AS INDICATORS OF HOMOGENEITY BY ANALYSIS OF VARIANCE

Species/group	Station*†			Indication of spatial homogeneity	Station × month interaction‡		Indication of seasonal homogeneity
	4	5	6		4–6	5–6	
Bosmina longirostris	NS	NS	S5	No	NS	NS	No
Ceriodaphnia spp.	NS	S5	NS	No	S1	S1	Yes
Diaptomus spp.	S1	NS	NS	Yes	S1	S1	Yes
Mesocyclops edax	NI	S1	NS	Yes	NI	S1	Yes
Tropocyclops prasinus	NI	S5	NS	Yes	NI	NS	No
Calanoid copepodids	S1	S1	NS	Yes	S1	NS	Yes
Cyclopoid copepodids	S1	S1	S5	Yes	NS	NS	No
Nauplii	S1	S1	S5	Yes	NS	NS	No
Conochiloides spp.	S1	NS	NS	Yes	S1	S1	Yes
Conochilus spp.	S5	S5	NS	Yes	S1	NS	Yes
Keratella cochlearis	S1	S1	NS	Yes	NS	NS	No
K. crassa	NS	S1	S1	No	NS	NS	No
K. earlinae	NS	NS	NS	No	S5	NS	Yes
Polyarthra spp.	NS	NS	NS	No	S5	NS	Yes
Hexarthra spp.	S1	NS	NS	Yes	S1	S1	Yes
Synchaeta spp.	S1	NS	NS	Yes	S1	NS	Yes
Ploesoma truncatum	?	A	NS	?	?	?	?
Ptygura spp.	S1	NS	NS	Yes	S1	S5	Yes
Asplanchna spp.	S1	NS	NS	Yes	S1	S1	Yes
Cladocera	S5	S5	S1	No	NS	S5	Yes
Copepoda	S1	S1	S1	No	NS	NS	No
Rotifera	S5	S1	S1	No	NS	NS	No

*Explanation of symbols:
NS, not significant.
S5, significant at 5% level.
S1, significant at 1% level.
A, species absent in collection.
NI, not identified in year 4.
?, insufficient data for test.

†Station column indicates whether or not the monthly densities of each species/group varied significantly among stations for the year listed.

‡Station × month interaction column indicates whether or not the interaction term for each species/group was significantly less in year 6 than in the year being compared.

species was then tested between years. The taxa were then separated into three groups. The first group was composed of nine taxa that were considered stable over the 3-year period and therefore were not detrimentally affected by power-plant operation: *Keratella cochlearis*, *K. earlinae*, *Conochilus unicornis*, *Conochiloides* spp., *Polyarthra* spp., *Tropocyclops prasinus*, *Bosmina longirostris*, nauplii, and cyclopoid copepodids. An increase in winter water-column temperatures brought about by power-plant operation may have been influential in causing the shift of *Polyarthra* spp. abundance from the fall and spring of previous years to the winter in year 6.

The second group was composed of six taxa that declined over the 3-year period: *Keratella crassa*, *Collotheca* spp., *Asplanchna* spp., *Ceriodaphnia* spp., *Mesocyclops edax* and calanoid copepodids (*Diaptomus* spp.). None of these declines could be directly attributed to power-plant operation.

The third group was composed of four taxa that showed an increase in density over the 3-year period: *Hexarthra* spp., *Ptygura* spp., *Ploesoma truncatum* and *Synchaeta* spp. Both *Ptygura* spp. and *Hexarthra* spp. are warm stenotherms and may have responded favorably to the increase in the duration of maximum summer temperatures brought about by the power plant. The increase in the other two species was associated with natural causes.

Significant changes in density were not observed at higher taxonomic levels (Rotifera, Cladocera, Copepoda). This indicated that the power plant did not have a detrimental impact on the total standing crop of the zooplankton community of the lake. Changes among the majority of species within the major groups were most often associated with natural causes.

Circulation of surface water by the power plant appeared to reduce station density differences and bring about a more similar seasonal density distribution for most taxa. This was probably the most notable influence of the power-plant operation on the zooplankton community of the lake.

ACKNOWLEDGMENTS

This investigation was funded by the Duke Power Company, Charlotte, N. C., through a research grant to Charles M. Weiss, Project Director, Department of Environmental Sciences and Engineering, School of Public Health, University of North Carolina at Chapel Hill, Chapel Hill, N. C. We are indebted to Sheila L. Pfaender, Sarah R. Gnilka, and Arthur L. Buikema for performing the zooplankton taxonomy over the course of this study. Our colleague Peter H. Campbell provided the data on phytoplankton densities and biovolumes. Thomas G. Donnelly wrote the programs for the statistical tests.

REFERENCES

Ahlstrom, E. H., 1934, A Quantitative Study of Rotatoria in Terwilliger's Pond, Put-in-Bay, Ohio, *Bull. Ohio Biol. Survey*, 30(6): 3-38.

Applegate, R. L., and J. W. Mullan, 1969, *Ecology of* Daphnia *in Bull Shales Reservoir*, U. S. Bureau of Sport Fish. and Wildlife, Research Report 74.

Beach, N. W., 1960, A Study of the Planktonic Rotifers of the Ocqueoc River System, Presque Isle County, Michigan, *Ecol. Monogr.*, 30: 339-357.

——, 1962, The Zooplankton of Osbysjoni, Djursholm. II. Further Ecological Aspects, *Oikos*, 13: 216-231.

Brooks, J. L., and S. I. Dodson, 1965, Predation, Body Size and Composition of Plankton, *Science*, 150: 28-35.

Campbell, R. S., 1941, Vertical Distribution of the Plankton Rotifera in Douglas Lake, Michigan, with Special Reference to Submerged Depression Individuality, *Ecol. Monogr.*, 11(1): 3-19.

Carlin, B., 1943, Die planktonrotatorien des Motalastrom, *Meddn. Lunds Univ. Limnol. Instn.*, 5: 1-255.

Carlson, D. M., 1974, Responses of Planktonic Cladocerans to Heated Water, in *Thermal Ecology*, AEC Symposium Series, Augusta, Ga., May 3—5, 1973, J. W. Gibbons and R. R. Sharitz (Eds.), pp. 186-205, CONF-730505, NTIS.

Comita, G. W., 1972, The Seasonal Zooplankton Cycles, Production, and Transformations of Energy in Sverson Lake, Minnesota, *Arch. Hydrobiol.*, 70: 14-66.

Duke Power Company, 1977, *Oconee Nuclear Station Environmental Summary Report, 1971—1976*, Vol. 12, Charlotte, N. C.

Edmondson, W. T., 1946, Factors in the Dynamics of Rotifer Populations, *Ecol. Monogr.*, 16: 358-372.

——, 1960, Reproductive Rates of Rotifers in Natural Population, *Mem. Ist. Ital. Idrobiol. Dott. Marco de Marchi*, 12: 21-77.

——, 1965, Reproductive Rate of Planktonic Rotifers as Related to Food and Temperature in Nature, *Ecol. Monogr.*, 35: 61-110.

Hayward, R. S., and D. N. Gallup, 1976, Feeding and Assimilation in *Daphnia schoedleri* Sars as Affected by Environmental Conditions, *Arch. Hydrobiol.*, 77: 139-163.

Hutchinson, G., 1967, *A Treatise on Limnology*, Vol. II. *Introduction to Lake Biology and the Limnoplankton*, John Wiley & Sons, Inc., New York.

Lenat, D. R., and T. P. Anderson, in Environmental Changes as a Consequence of Power Plant Operation, Belews Lake, Year VII, July 1976—June 1977, Department of Environmental Sciences and Engineering, University of North Carolina, Chapel Hill, in preparation.

Loeffelman, P. H., 1976, *Effects of Hydroelectric Pumpback Storage Process on Rotifer and Cladoceran Populations in Smith Mountain Lake, Virginia*, M.S. Thesis, Virginia Polytechnic Institute and State University, Blacksburg, Va.

Makarewicz, J. C., 1974, *The Community of Zooplankton and Its Production in Mirror Lake*, Ph.D. Thesis, Cornell University, Ithaca, N. Y.

——, and G. E. Likens, 1975, Niche Analysis of a Zooplankton Community, *Science*, 190: 1000-1003.

Pejler, B., 1957, Taxonomic and Ecological Studies of Planktonic Rotatoria from Northern Swedish Lapland, *Kungl. Svenska Vetenskap. Handlingar*, Ser. 4, Band 6, No. 5.

——, 1964, Regional Ecological Studies of Swedish Freshwater Zooplankton, *Zool. Bidrag Uppsala*, 36: 407-515.

Ricker, W. E., 1938, Seasonal and Annual Variation in Quantity of Pelagic Net Plankton, Cultus Lake, British Columbia, *J. Fish. Res. Board. Can.*, 4: 33-47.

Saunders, J. F., 1975, *A Study of Zooplankton Communities of Lake Anna, 1973—1974*, M.S. Thesis, Virginia Polytechnic Institute and State University, Blackburg, Va.

Selgeby, J. H., 1975, Life Histories and Abundances of Crustacean Zooplankton in the Outlet of Lake Superior, 1971—1972, *J. Fish. Res. Board Can.*, 32: 461-470.

Smyly, W. J. P., 1974, Vertical Distribution and Abundance of *Ceriodaphnia quadrangula* (O. F. Muller) (Crustacea, Cladocera), *Freshwater Biol.*, 4: 257-266.

Stenson, J. A. E., 1973, On Predation and *Holopedium gibberum* (Zaddach) Distribution, *Limnol. Oceanogr.*, 18: 1005-1010.

Weiss, C. M., T. P. Anderson, P. H. Campbell, D. R. Lenat, J. H. Moore, S. L. Pfaender, and T. G. Donnelly, 1975a, *An Assessment of the Environmental Stabilization of Belews Lake—Year IV and Comparisons with Lake Hyco, North Carolina, July 1973—June 1974,* Department of Environmental Science and Engineering, Publication No. 416, University of North Carolina, Chapel Hill.

——, P. H. Campbell, T. P. Anderson, and S. L. Pfaender, 1975b, *Catawba Lakes, Characterization of Phyto- and Zooplankton Communities and Their Relationships to Environmental Factors,* Department of Environmental Science and Engineering, Publication No. 389, University of North Carolina, Chapel Hill.

——, and E. J. Kuenzler, 1976, *The Trophic State of North Carolina Lakes,* Report No. 119, Water Resources Research Institute of the University of North Carolina, Raleigh, N. C.

Williams, L. G., 1964, Dominant Planktonic Rotifers of Major Waterways of the United States, *Limnol. Oceanogr.*, 11: 83-91.

STOCHASTIC APPROACH TO PREDICT SURVIVAL OF ESTUARINE ANIMALS EXPOSED TO HOT DISCHARGE EFFLUENT

K. S. CHUNG* and KIRK STRAWN
Department of Wildlife and Fisheries Sciences, Texas A. & M. University, College Station, Texas

ABSTRACT

The stochastic model was used to predict survival probabilities for aquatic animals in intake water exposed to hot discharge effluent for 180 min (30 min afferent and 150 min efferent to the cooling towers) on the basis of water-passage times. The least-squares method was used to estimate seasonal mortality rates of 8 species of crustaceans and 46 fishes at several test temperatures for each of three seasons. During the hot season (June—September), mortality rates (fraction/min) were 23% or greater for most organisms at 40°C (discharge-canal temperature afferent to the cooling towers) and zero for most animals at 35°C (discharge temperature efferent to the towers). Survival probabilities for macrocrustaceans and fishes after a 180-min passage through the discharge-canal system, in terms of average seasonal temperature, ranged from 0 to 96% from June—September and from 0 to 61% for the 56 species tested. Almost all mortality occurred during the 30-min passage before the cooling towers. From October to May, predicted survival for all animals except the bay anchovy was 100% after a 180-min passage through the discharge-canal system (survival for the anchovy was 88% in the cool season, May and October).

Much research has been focused on industrial use of water for cooling purposes and the effects on aquatic organisms of heated discharge effluent (Naylor, 1965; Adams, 1969; Churchill and Wojtalik, 1969; Vernberg and Vernberg, 1969; Cairns, 1972; Gibbons and Sharitz, 1974; Esch and McFarlane, 1976). Few studies provide

*Present address: Instituto Oceanografico, Universidad de Oriente, Cumana, Venezuela.

data on the heat-resistance times of organisms acclimated in natural environments, however. These data are needed to predict survival of animal populations during passage through a power plant and its cooling-water system.

Previous studies at the P. H. Robinson Generating Station, Bacliff, Tex. (McCullough, 1971; French, 1973; Reimer and Strawn, 1973; Landry and Strawn, 1974), indicated that numerous entrained and entrapped organisms were killed by heat in the discharge canal. Cooling towers, which have been added since those studies, reduced summer temperatures in the lower half of the discharge canal. This should increase survival of entrained organisms, but they may die of heat shock in the discharge canal between the plant and the cooling towers.

Our study was designed to estimate seasonal survival times of crustaceans and fishes at test temperatures similar to those of the discharge canal and to predict, at mean seasonal temperatures, seasonal survival probabilities of aquatic animals entrained and entrapped from intake water and exposed to heated effluent during passage through the discharge canal. This type of data can be used in assessing the effect of power-plant design on the mortality of entrained organisms.

MATERIALS AND METHODS

Study Area

The study location was the Houston Lighting & Power Company's P. H. Robinson Generating Station, on state highway 146, ~1 mile south of Bacliff, Tex. Two 450,000-KW generating units, one 565,000-KW unit, and one 750,000-KW unit are capable of pumping cooling water through the condensers at a total rate of 265,056 m^3/hr. The plant uses cooling water drawn, via a 3.7-km intake canal, from Dickinson Bay, a subsystem of Galveston Bay. The water is heated as it passes through the plant's condenser-tube network and returns to Galveston Bay via a 3.2-km discharge canal. Organisms reach the discharge canal by passing through the revolving screens and/or by being impinged on the screens and sluiced into the discharge canal. A drop structure at the edge of Galveston Bay prevents water from entering the discharge canal downstream except during storm tides.

The cooling towers, which operate from June through September, are located on the discharge canal between the plant and Galveston Bay. Cooled water from the towers, which is mixed with

~40°C discharge-canal water, reduces canal temperatures to 35°C or less between the cooling towers and Galveston Bay.

Experiments were conducted over a 16-month period, June 1974 through September 1975. All-glass 38-liter tanks filled with water from the intake canal were aerated to provide circulation and were thermally controlled within 0.1°C of predetermined levels. Survival from June to September was determined at temperatures increasing from 32 to 41°C in 1°C increments. Test temperatures above 35°C represented those between the plant and the cooling towers, and temperatures of 35°C or lower represented those between the cooling towers and Galveston Bay. Average test temperatures were lower from October to May when discharge-canal temperatures were lower. A 1.5-mm mesh screen was used to confine small animals to the front of the tank for easier observation. We measured survival for 180-min exposure periods, which we felt were representative of exposure at the plant. Drift-bottle studies indicated that surface water passed from the plant discharge to the first cooling-tower intake in ~30 min. Temperature, taken by a hand thermometer, and salinity, measured by a refractometer (American Optical Instrument Company), in the intake and discharge canals at the water surface were taken in conjunction with each experiment. All organisms tested were captured from the intake canal by dip net, cast net, lift net, hoop net, and various traps on a daily basis during the hot season (June—September) and once a month from October to May. During the hot season, otter trawl and log and revolving screens were also used occasionally. Hook and line and seine were used once a week during the hot season and once a month during the remainder of the year. After capture, animals were placed in a 45-liter ice chest to minimize temperature change, aerated, and transported from the intake canal to the laboratory in less than 5 min. In the laboratory they were divided among experimental tanks according to species and previous results at various temperatures. The number of animals of one species in a tank never exceeded 30 and rarely exceeded 10, except for small common species; availability was usually limited to 1 or 2 for most species (Chung, 1977). Species were mixed in the test tanks, except highly predaceous species, such as the blue crab (*Callinectes sapidus*), which was tested separately. Care was taken to avoid overcrowding and to use only animals that appeared to be in good health. Preliminary tests indicated that animals very sensitive to handling, such as the bay anchovy (*Anchoa mitchilli*) and silversides (*Membras martinica* and *Menidia beryllina*), survived well beyond 180 min at intake-canal temperatures. For a given capture temperature, 100% of a sample frequently survived at reduced test temperatures; these, thus, served as controls. This was true even during hot weather for the easily injured Gulf menhaden (*Brevoortia patronus*). Animals captured in the afternoon and evening were kept in cages in the intake canal for later experimentation. The cylindrical cages, 61 cm in diameter and 63.5 cm deep, were constructed of plastic netting material of 0.325-cm or 1.3-cm mesh.

Cessation of opercular motion was the criterion of death for the fishes. Crustaceans were considered dead when appendages no longer moved. Survival times were measured to 0.1 min with an elapsed-time meter. Survivors were discarded.

Treatment of Data

After determining survival times of individual organisms at each test temperature, we calculated seasonal survival percentages for 180 min by 10-min intervals for each species. The least-squares method (Steel and Torrie, 1960) was used to estimate seasonal mortality rates of 8 species of crustaceans and 46 fishes by season at each test temperature.

To estimate survival probabilities and expected survival times at each test temperature, we used the stochastic approach (Matis and Hartely, 1971; Gross and Clark, 1975):

$$\rho_i(t) = e^{-\mu_i t} \qquad \text{for } 0 \text{ min} \leq t \leq 180 \text{ min}$$

$$\tau_i = (\mu_i)^{-1}$$

where t = time exposed
$\rho_i(t)$ = probability of survival to time t at test temperature i
μ_i = constant death rate at test temperature i
τ_i = expected survival time at test temperature i

To estimate survival probabilities for the population of a species exposed in the discharge canal afferent to the cooling towers, we used the stochastic model:

$$P_1(t) = e^{-\lambda_1 t} \qquad \text{for } 0 \text{ min} \leq t \leq 30 \text{ min}$$

where t is time exposed, $P_1(t)$ is probability of survival in the discharge canal afferent to the cooling towers, and λ_1 is death rate in the discharge canal afferent to the cooling towers.

To estimate expected population survival after exposure in the discharge canal efferent to the cooling towers, we used the stochastic model:

$$P_2(t) = (e^{-\lambda_1 30})[e^{-\lambda_2 (t-30)}] \qquad \text{for } 30 \text{ min} \leq t \leq 180 \text{ min}$$

where t = time exposed

$P_2(t)$ = probability of survival in the discharge canal efferent to the cooling towers

λ_1 = death rate in the discharge canal afferent to the cooling towers

λ_2 = death rate in the discharge canal efferent to the cooling towers

These two effects were assumed to be independent. On the basis of water flow, we assumed that animals took 30 min to reach the cooling towers and estimated that as much as 150 min could be required to travel from the cooling towers to the area of reduced temperature in Galveston Bay.

In these three analyses the year was divided into three seasons, on the basis of water temperature, hot (June—September), cool (May and October), and cold (November—April). Each season was analyzed separately. Death rates in the discharge canal afferent and efferent to the cooling towers were estimated from laboratory experiments using experimental temperatures similar to seasonal averages in the discharge canal. Average discharge temperatures were 27°C in the cold season, 32°C in the cool season, and, during the hot season, 40°C afferent and 35°C efferent to the cooling towers.

RESULTS AND DISCUSSION

Temperature and Salinity

Temperature ranged from 7.2 to 30.5°C in the intake canal, from 19.4 to 41.6°C in the discharge canal above cooling towers, and from 14.4 to 37.8°C below the towers. Average monthly temperatures in the intake canal and in the discharge canal afferent and efferent to the cooling towers ranged from 26 to 30, 36 to 40, and 33 to 35°C, respectively, in the hot season; 23 to 26, 32 to 35, and 31 to 34°C, respectively, in the cool season; and 13 to 19, 26 to 30, and 23 to 28°C, respectively, in the cold season (Fig. 1). From October to May, effluent temperatures were under 35°C and the cooling towers were not operated. The difference between average water temperatures in the intake canal and in the upper part of the discharge canal was about 10°C during all seasons. Average water-temperature reductions from the upper part of the discharge canal to below the cooling towers were approximately 4.5 and 5.5°C in the hot seasons of 1974 and 1975, respectively. The difference between average water temperatures in the two summers was caused by operation of an additional cooling tower in 1975.

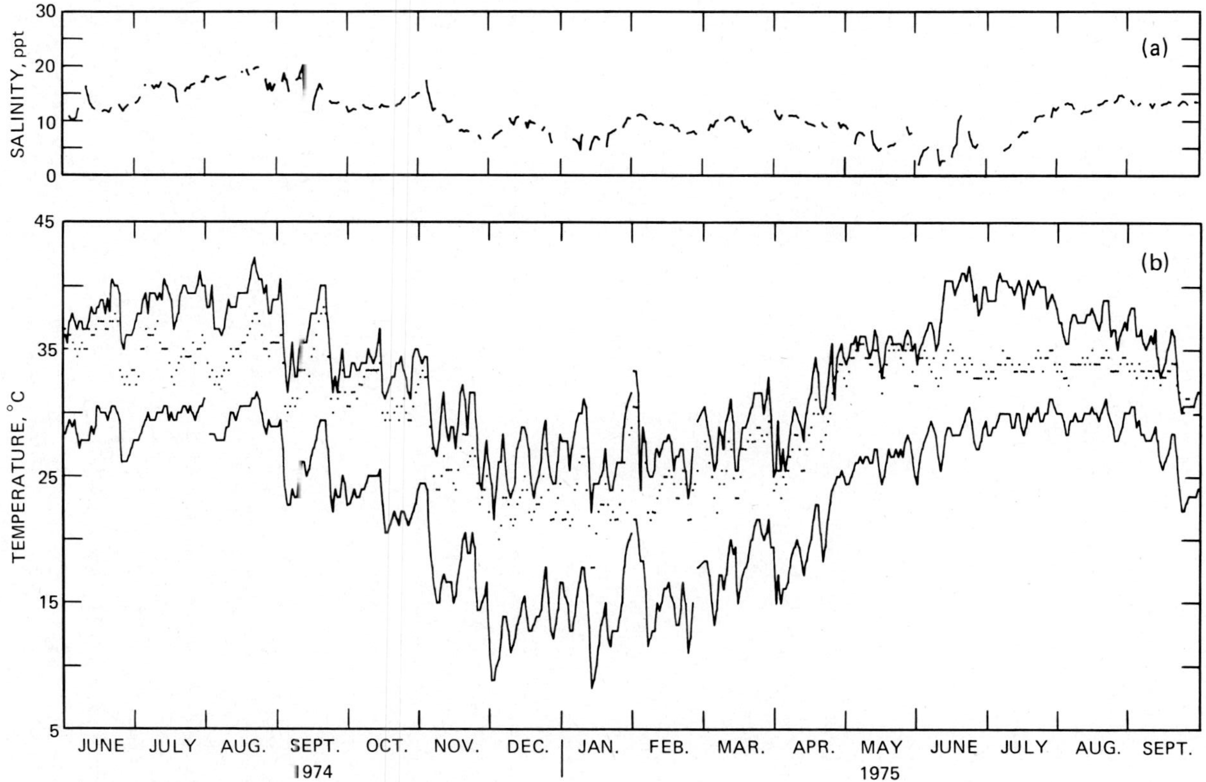

Fig. 1 (a) Average daily salinity in the intake canal. (b) Average daily temperature in the discharge canal afferent (upper solid line) and efferent (dotted line) to the cooling towers and in the intake canal (lower solid line) from June 1974 to September 1975 at the P. H. Robinson Generating Station, Bacliff, Texas.

Since recordings of salinity from the intake and discharge canals were similar, only intake-canal salinity is presented. Salinity ranged from 1 to 24.8 ppt in the intake canal (Fig. 1), and there was a striking difference between the 2 years, with values ranging from 6.8 to 24.8 ppt in 1974 and from 1 to 12.4 ppt in 1975. This was especially true in early June and early July; in 1975, salinities were 5.3 ppt or less and, in 1974, 10.3 ppt or more. This was caused by heavy rains in early June and early July 1975.

Multiple regression analyses indicated that intake-canal temperature was the most significant factor influencing thermal tolerance of aquatic animals and that salinity was less significant (Chung, 1977).

Predicted Survival in Discharge-Canal System

In the hot season (June—September), mortality rates of crustaceans and fishes tested at 35°C (discharge-canal temperature efferent to the cooling towers in the hot season) were 0 for all species except one crustacean, white shrimp (*Penaeus setiferus*), and 17 fishes (Table 1), and rates were very low for the white shrimp and nine of the fishes (0.0001 to 0.0024/min).

In the hot season most aquatic animals transported by the power plant from intake water and exposed to the elevated temperatures of the effluent die of thermal shock between the plant and the cooling towers (Table 1). Except for the rough silversides during the hot season, no mortalities were predicted for any species in any season downstream of the cooling towers because all individuals of most species died before reaching the towers and any surviving individuals of resistant species were able to withstand the greatly reduced temperatures between the towers and the ambient water in Galveston Bay. At the average cool-season temperature of the discharge canal, 100% of all species except the bay anchovy survived, and, at the average cold-season temperature, all animals tested were predicted to survive. Expected survival probabilities after direct passage through the discharge-canal system (30 min in the discharge canal afferent to the cooling towers and 150 min efferent to the towers) ranged from 0 to 96% for the species studied. All individuals from intake water, except for those of 6 species of crustaceans and 15 fishes (Table 1), would be expected to die during the summer (June—September) at the average discharge-canal temperature. The exceptions are grass shrimp, *Palaemonetes pugio* (48% survival) and *P. vulgaris* (53%); river shrimp, *Macrobrachium ohione* (3%); white shrimp, *Penaeus setiferus* (10%); brown shrimp, *Penaeus aztecus* (46%); blue crab, *Callinectes sapidus* (76%); ladyfish, *Elops saurus* (38%); sea catfish, *Arius felis* (18%); Atlantic toadfish, *Opsanus beta* (39%); sheepshead

(Text continues on page 658.)

TABLE 1

SEASONAL DIFFERENCES IN PREDICTED SURVIVAL OF ANIMALS EXPOSED 180 min* IN THE DISCHARGE-CANAL SYSTEM OF THE P. H. ROBINSON GENERATING STATION, BACLIFF, TEX., DETERMINED BY THE STOCHASTIC METHOD (JUNE 1974—SEPTEMBER 1975)

Species and season	Predicted survival,† %	Test temperature, °C	Mortality rate, number/min	Expected survival time,‡ min
Palaemonetes pugio				
Cold	100.00	34	0.0031	322.58
		35	0.0044	227.27
		36	0.0110	90.91
		38	0.1151	8.69
		39—41	0.2302	4.34
Cool	100.00	35	0.0000	NLT
		36	0.0025	400.00
		38	0.0055	181.82
		40	0.0767	13.04
Hot	47.66	35—36	0.0000	NLT
		38	0.0012	833.33
		40	0.0247	40.49
Palaemonetes vulgaris				
Cold	100.00	34	0.0011	909.09
		35	0.0054	185.19
		36	0.0069	144.93
		38	0.0575	17.39
		40—41	0.2302	4.34
Cool	—	36	0.0041	243.90
		38	0.0276	36.23
		40—41	0.2302	4.34
Hot	52.78	36	0.0000	NLT
		38	0.0051	196.08
		40	0.0213	46.95
Macrobrachium ohione				
Hot	3.16	36	0.0000	NLT
		38	0.0364	27.47
		40	0.1151	8.69
Penaeus setiferus				
Cold	—	34	0.0028	357.14
		35	0.0767	13.04
		38—39	0.2302	4.34
Cool	100.00	34	0.0000	NLT
		35	0.0007	1428.57
		36	0.0026	384.62
		38	0.0767	13.04
		39—41	0.2302	4.34
Hot	10.01	35	0.0001	10000.00
		36	0.0005	2000.00

(Table 1 continues on following page.)

Table 1 (Continued)

Species and season	Predicted survival,† %	Test temperature, °C	Mortality rate, number/min	Expected survival time,‡ min
Panaeus setiferus (continued)				
		38	0.0061	163.93
		40	0.0767	13.04
Penaeus aztecus				
Cold	100.00	35	0.0000	NLT
		36	0.0000	NLT
		38	0.0767	13.04
Cool	100.00	35	0.0004	2500.00
		36	0.0041	243.90
		38	0.0063	158.73
		40—41	0.2302	4.34
Hot	46.25	35	0.0000	NLT
		36	0.0005	2000.00
		38	0.0071	140.85
		40	0.0257	38.91
Alpheus heterochaelis				
Hot	0.00	38	0.0004	2500.00
Callinectes sapidus				
Cold	100.00	36	0.0018	555.56
		38	0.0050	200.00
		40	0.2302	4.34
Cool	100.00	38	0.0040	250.00
		40	0.0575	17.39
Hot	76.11	36	0.0000	NLT
		38	0.0004	2500.00
		40	0.0091	109.89
Panopeus herbstii				
Hot	0.00	38	0.0004	2500.00
Elops saurus				
Hot	38.28	36	0.0000	NLT
		38	0.0036	277.78
		40	0.0320	31.25
Brevoortia patronus				
Cold	100.00	25—28	0.0000	NLT
		32	0.0069	144.93
		34	0.0575	17.39
		35—39	0.2302	4.34
Cool	100.00	34	0.0003	3333.33
		35	0.0032	312.50
		36	0.0310	32.26
		38	0.1151	8.69
Hot	0.00	34	0.0000	NLT
		35	0.0002	5000.00
		36	0.0051	196.08
		38	0.0575	17.39

Table 1 (Continued)

Species and season	Predicted survival,† %	Test temperature, °C	Mortality rate, number/min	Expected survival time,‡ min
Brevoortia patronus (continued)		40	0.2302	4.34
Dorosoma petenense				
Cold	—	32	0.0013	769.23
		34	0.0084	119.05
		35	0.0767	13.04
		36—38	0.2302	4.34
Cool	100.00	34	0.0016	625.00
		35	0.0352	28.41
		36—38	0.2302	4.34
Hot	0.00	34	0.0011	909.09
		35	0.0315	31.75
		36	0.0356	28.09
		38	0.1151	8.69
Harengula pensacolae				
Cold	—	32	0.0315	31.75
Hot	0.00	34	0.0000	NLT
		35	0.0000	NLT
		36	0.0034	294.12
		38	0.0575	17.39
		39—40	0.2302	4.34
Anchoa hepsetus				
Hot	0.00	34	0.0289	34.60
		35—38	0.2302	4.34
Anchoa mitchilli				
Cold	100.00	32	0.0077	129.87
		34—38	0.2302	4.34
Cool	88.16	32	0.0007	1428.57
		34	0.0085	117.65
		35	0.0171	58.48
Hot	0.00	36	0.1151	8.69
		37—39	0.2302	4.34
		32	0.0014	714.29
		34	0.0046	217.39
		35	0.0064	156.25
		36	0.0257	38.91
		38	0.1151	8.69
		39—40	0.2302	4.34
Synodus foetens				
Hot	0.00	35	0.0000	NLT
		36	0.0356	28.09
		38—39	0.2302	4.34
Bagre marinus				
Hot	0.00	36	0.0000	NLT
		38	0.0259	38.61

(Table 1 continues on following page.)

Table 1 (Continued)

Species and season	Predicted survival,† %	Test temperature, °C	Mortality rate, number/min	Expected survival time,‡ min
Arius felis				
Cold	—	36	0.0050	200.00
		38	0.0767	13.04
		39—40	0.2302	4.34
Cool	—	35	0.0024	416.67
		36	0.0039	256.41
		38	0.0305	32.79
		40—41	0.2302	4.34
Hot	17.81	36	0.0000	NLT
		38	0.0128	78.13
		40	0.0575	17.39
Opsanus beta				
Cold	—	36	0.0072	138.89
		37—38	0.2302	4.34
Cool	—	38	0.0356	28.09
		39—40	0.2302	4.34
Hot	39.21	35—37	0.0000	NLT
		38	0.0035	285.71
		40	0.0312	32.05
Strongylura marina				
Hot	0.00	34	0.0000	NLT
		35	0.0024	416.67
		36	0.0026	384.62
		38	0.0289	34.60
Cyprinodon variegatus				
Cold	—	38	0.0364	27.47
Hot	95.31	40	0.0016	625.00
Fundulus grandis				
Cold	—	38—40	0.2302	4.34
Hot	95.88	40	0.0014	714.29
Gambusia affinis				
Cold	100.00	35—37	0.0000	NLT
		38—40	0.2302	4.34
Cool	100.00	36	0.0000	NLT
		38	0.0002	5000.00
		40	0.0313	31.95
Hot	62.43	35—37	0.0000	NLT
		38	0.0009	1111.11
		40	0.0157	63.69
Poecilia latipinna				
Cool	100.00	39	0.0000	NLT
Hot	91.94	35—38	0.0000	NLT
		40	0.0028	357.14
Membras martinica				
Cold	—	35	0.1151	8.69

Table 1 (Continued)

Species and season	Predicted survival,† %	Test temperature, °C	Mortality rate, number/min	Expected survival time,‡ min
Membras martinica (continued)				
Cool	100.00	36–37	0.2302	4.34
		34	0.0022	454.55
		35	0.0124	80.65
		36	0.0132	75.76
Hot	0.92	34	0.0001	10000.00
		35	0.0016	625.00
		36	0.0064	156.25
		38	0.0211	47.39
		40	0.1465	6.83
Menidia beryllina				
Cold	—	34–38	0.2302	4.34
Cool	—	35	0.0028	357.14
		36	0.0031	322.58
		38–39	0.2302	4.34
Hot	0.00	34–35	0.0000	NLT
		36	0.0016	625.00
		38	0.1151	8.69
		39–40	0.2302	4.34
Syngnathus louisianae				
Hot	0.00	35	0.0000	NLT
		36	0.0242	41.32
		38	0.1151	8.69
Trichiurus lepturus				
Cold	100.00	29	0.0000	NLT
Cool	—	32	0.0767	13.04
		34	0.2302	4.34
Hot	0.00	32	0.0330	30.30
		35	0.0767	13.04
		36	0.1151	8.69
Chaetodipterus faber				
Cold	100.00	36	0.0000	NLT
		38	0.1151	8.69
		39–40	0.2302	4.34
Cool	100.00	35	0.0000	NLT
		36	0.0021	476.19
		38	0.1151	8.69
		39–40	0.2302	4.34
Hot	3.16	35–36	0.0000	NLT
		38	0.0098	102.04
		40	0.1151	8.69
Caranx hippos				
Hot	0.00	35–36	0.0000	NLT
		38	0.0271	36.90

(Table 1 continues on following page.)

Table 1 (Continued)

Species and season	Predicted survival,† %	Test temperature, °C	Mortality rate, number/min	Expected survival time,‡ min
Chloroscumberus chrysurus				
Hot	0.00	33–34	0.0000	NLT
		35	0.0053	188.68
		36	0.0098	102.04
		38–39	0.2302	4.34
Oligoplites saurus				
Hot	3.16	35	0.0000	NLT
		36	0.0007	1428.57
		38	0.0090	111.11
		40	0.1151	8.69
Caranx bartholomaei				
Hot	0.00	38	0.0575	17.39
Eucinostomus sp.				
Hot	10.01	36	0.0000	NLT
		38	0.0021	476.19
		40	0.0767	13.04
Lagodon rhomboides				
Cold	100.00	25	0.0000	NLT
		32	0.0010	1000.00
		34	0.0127	78.74
		35–36	0.1151	8.69
		37–39	0.2302	4.34
Cool	100.00	35	0.0032	312.50
		36	0.0280	35.71
		38	0.1151	8.69
Hot	0.00	34	0.0005	2000.00
		35	0.0006	1666.67
		36	0.0009	1111.11
		38	0.0295	33.90
		40–41	0.2302	4.34
Archosargus probatocephalus				
Cold	—	36	0.1151	8.69
Cool	—	35	0.0289	34.60
		36	0.2302	4.34
Hot	0.00	35	0.0000	NLT
		36	0.0005	2000.00
		38	0.0323	30.96
		40	0.2302	4.34
Bairdiella chrysura				
Cool	100.00	34	0.0000	NLT
		35	0.0048	208.33
		36	0.0757	13.21
		38	0.2302	4.34

Table 1 (Continued)

Species and season	Predicted survival,† %	Test temperature, °C	Mortality rate, number/min	Expected survival time,‡ min
Bairdiella chrysura (continued)				
Hot	0.00	34	0.0000	NLT
		35	0.0003	3333.33
		36	0.0007	1428.57
		38	0.0243	41.15
		40	0.2302	4.34
Cynoscion arenarius				
Cold	100.00	32—33	0.0000	NLT
		34	0.0339	29.50
		35	0.1151	8.69
		36—38	0.2302	4.34
Cool	100.00	32—33	0.0000	NLT
		34	0.0031	322.58
		35	0.0311	32.15
		38	0.2302	4.34
Hot	0.00	35	0.0002	5000.00
		36	0.0038	263.16
		38	0.0575	17.39
		40	0.2302	4.34
Cynoscion nebulosus				
Cold	—	35	0.0283	35.34
Hot	0.00	34	0.0000	NLT
		35	0.0016	625.00
		36	0.0018	555.56
		38	0.0247	40.49
Leiostomus xanthurus				
Cold	100.00	32	0.0000	NLT
		34	0.1151	8.69
		35—38	0.2302	4.34
Hot	0.00	35	0.0039	256.41
		36	0.0053	188.68
		38	0.0767	13.04
		39—40	0.2302	4.34
Menticirrhus americanus				
Cold	—	36	0.1151	8.69
Cool	100.00	34	0.0000	NLT
		35	0.0331	30.21
		36	0.0767	13.04
Hot	0.00	35—36	0.0000	NLT
Micropogon undulatus				
Cold	100.00	27	0.0000	NLT
		32	0.0011	909.09
		34	0.1151	8.69
		35—38	0.2302	4.34

(Table 1 continues on following page.)

Table 1 (Continued)

Species and season	Predicted survival,† %	Test temperature, °C	Mortality rate, number/min	Expected survival time,‡ min
Micropogon undulatus (continued)				
Cool	—	34	0.0016	625.00
		35	0.0050	200.00
		36	0.0246	40.65
		38	0.2302	4.34
Hot	0.00	34	0.0000	NLT
		35	0.0039	256.41
		36	0.0073	136.99
		38	0.0260	38.46
		40	0.2302	4.34
Pogonias cromis				
Cold	100.00	34	0.0000	NLT
		35	0.0050	200.00
		36	0.0248	40.32
		38	0.1151	8.69
Hot	0.00	36	0.0000	NLT
		38	0.0353	28.33
Sciaenops ocellata				
Cold	—	36	0.2302	4.34
Cool	100.00	35	0.0000	NLT
Hot	0.00	35–36	0.0000	NLT
		38	0.0253	39.53
Stellifer lanceolatus				
Hot	0.00	35	0.0000	NLT
Mugil cephalus				
Cold	100.00	38	0.0073	136.99
		40–41	0.2302	4.34
Cool	100.00	38	0.0076	131.58
		40–41	0.2302	4.34
Hot	43.82	35	0.0000	NLT
		36	0.0004	2500.00
		38	0.0013	769.23
		40	0.0275	36.36
Mugil curema				
Cold	100.00	34	0.0000	NLT
		35	0.0015	666.67
		36	0.0022	454.55
Cool	100.00	36	0.0000	NLT
		38	0.0005	2000.00
		40	0.1151	8.69
Hot	39.21	38	0.0002	5000.00
		40	0.0312	32.05

Table 1 (Continued)

Species and season	Predicted survival,† %	Test temperature, °C	Mortality rate, number/min	Expected survival time,‡ min
Hypsoblennius ionthas				
Cold	—	37	0.2302	4.34
Hot	0.00	36	0.0000	NLT
		38	0.0767	13.04
Gobionellus hastatus				
Cold	100.00	36	0.0000	NLT
		37	0.2302	4.34
Hot	3.16	35—37	0.0000	NLT
		38	0.0071	140.85
		40	0.1151	8.69
Gobiosoma bosci				
Cold	100.00	32	0.0000	NLT
		34	0.0018	555.56
		35	0.0052	192.31
		36	0.0054	185.19
		38	0.2302	4.34
Cool	100.00	36	0.0000	NLT
		38	0.0767	13.04
		39—40	0.2302	4.34
Hot	3.16	35	0.0000	NLT
		36	0.0000	NLT
		38	0.0210	47.62
		40	0.1151	8.69
Microgobius thalassinus				
Cold	100.00	36	0.0000	NLT
		37—38	0.1151	8.69
Cool	—	37	0.0052	192.31
Hot	0.00	39	0.1151	8.69
Scomberomorus maculatus				
Hot	0.00	34	0.0000	NLT
		35	0.0041	243.90
		36	0.1151	8.69
Peprilus alepidotus				
Hot	0.00	33—35	0.0000	NLT
		36	0.0237	42.19
		38	0.1151	8.69
Paralichthys lethostigma				
Cold	—	35	0.2302	4.34
Hot	0.00	34	0.0000	NLT
		35	0.0012	833.33
		36	0.0020	500.00
		38	0.0575	17.39

(Table 1 continues on following page.)

Table 1 (Continued)

Species and season	Predicted survival,† %	Test temperature, °C	Mortality rate, number/min	Expected survival time,‡ min
Sphoeroides parvus				
Hot	0.00	35	0.0001	10000.00
		36	0.0020	500.00
		38–39	0.2302	4.34

*Exposure was for 30 min above and 150 min below the cooling towers; 54 species were tested at various constant temperatures.

†The dash (—) indicates no observation at discharge-canal temperature; 100% survival is assumed.

‡The abbreviation NLT is normal lifetime.

minnow, *Cyprinodon variegatus* (95%); Gulf killifish, *Fundulus grandis* (96%); mosquitofish, *Gambusia affinis* (62%); sailfin mollie, *Poecilia latipinna* (92%); rough silversides, *Membras martinica* (1%); Atlantic spadefish, *Chaetodipterus faber* (3%); leatherjacket, *Oligoplites saurus* (3%); mojarra, *Eucinostomus* sp. (10%); striped mullet, *Mugil cephalus* (44%); white mullet, *M. curema* (39%); sharptail goby, *Gobionellus hastatus* (3%); and naked goby, *Gobiosoma bosci* (3%). Death rates were zero for all organisms tested during the cool season at 32°C [except the bay anchovy (0.0007)] and during the cold season at 27°C. This indicates that aquatic animals inhabiting the discharge canal and entering from intake water should have little problem with the heated effluent from October to May. Predicted survival probabilities, in terms of average temperatures, were 100% for all animals except bay anchovy (88% in the cool season) from October to May after a 180-min passage through the discharge-canal system (Table 1).

Survival probabilities in the discharge-canal system may be overestimated. The bay anchovy, and probably other least-resistant fishes, died at peak cold-season effluent temperatures. Such nonthermal factors as impingement and mechanical injury from the intake screens and the impact of cascading water in the cooling towers and the thermal factor of cumulative heat dosage were not considered.

Adequacy of the Stochastic Death Model

The model used was previously untried for predicting survival of animal populations passing through the power-plant discharge system. It is based on the model described by Pielou (1969).

This model has two weaknesses. First, death rates were estimated from the laboratory experiments rather than directly from animals passing through the discharge canal. Mortality rates may differ in the discharge because of nonthermal factors. Second, cumulative dosage and acclimation in the discharge canal afferent to the cooling towers (30 min) were not considered. Animals exposed there for 30 min might sustain critical damage and could die in the canal efferent to the towers even though they would have survived if not previously stressed. Laboratory data used to predict mortality past the cooling towers did not consider previous thermal stresses occurring in the canal above the towers. On the other hand, animals might survive longer than expected since they will undergo some upward acclimation in the discharge canal afferent to the towers.

Despite the model's theoretical weaknesses, we accepted it as adequately descriptive of seasonal survival of animal populations in the discharge-canal system because expected survival determined by the stochastic method at various test temperatures (Fig. 2) agreed closely with actual percent of test animals surviving at test temperatures in the laboratory (despite variations for some species tested in small numbers).

Fry, Brett, and Walker (1946) suggested a deterministic model for fish mortality at lethally high temperatures, which, basically, argues that a lethal dose is accumulated as the ratio of exposure time at temperature T to median resistance time at temperature T. When the accumulated dose reaches 1.0, the fish is dead. Their model accurately predicted the results of laboratory tests, but it, too, has not been tested under field conditions.

ACKNOWLEDGMENTS

The research reported here was funded by a grant from Houston Lighting & Power Company to the Department of Wildlife and Fisheries Sciences and the Texas Agricultural Experiment Station (Project 1869-2898).

This manuscript is taken in part from K. S. Chung's thesis prepared in partial fulfillment of the requirement for the degree of Doctor of Philosophy at Texas A. & M. University.

We thank J. M. Matis, Institute of Statistics, Texas A. & M. University, for suggestions in statistical analysis and W. H. Neill, Department of Wildlife and Fisheries Sciences, Texas A. & M. University, and F. G. Schlicht, Houston Lighting & Power Company, for their criticism during the development of the manuscript. Thanks

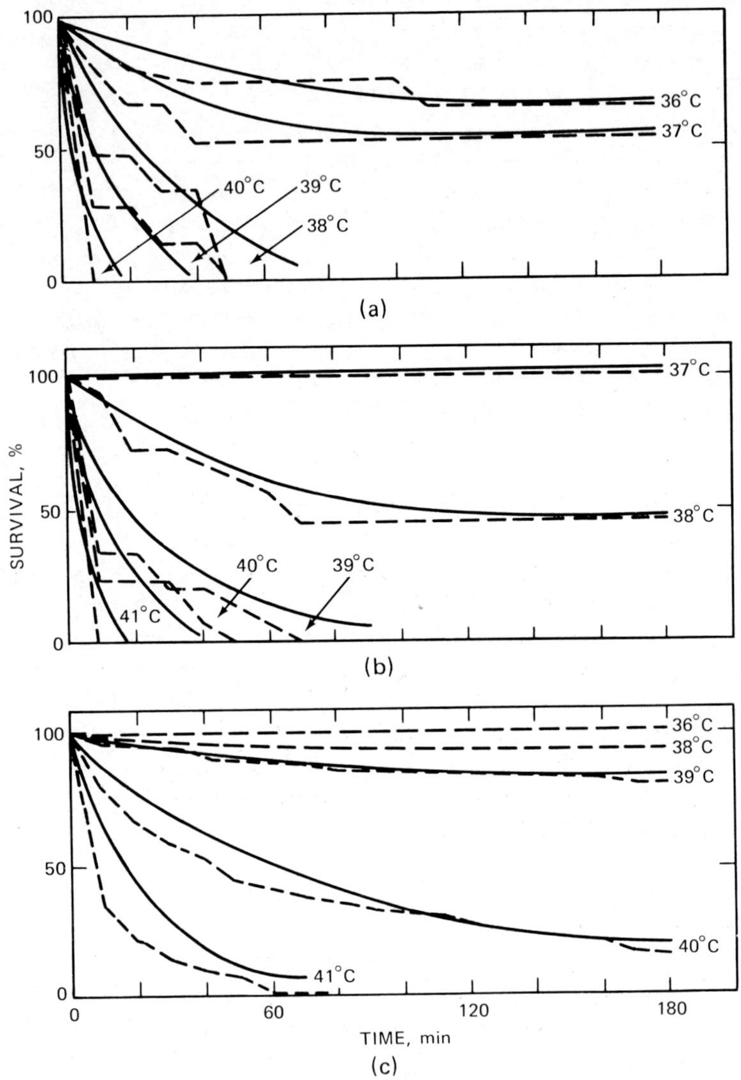

Fig. 2 Seasonal comparisons of mean observed survival (– – –) at 10-min intervals and expected survival (——) calculated by the stochastic method at various constant test temperatures for *Callinectes sapidus* tested for 180 min from June 1974 to September 1975 at the P. H. Robinson Generating Station, Bacliff, Texas. (a) Cold season (November—April). (b) Cool season (October and May). (c) Hot season (June—September).

are also extended to F. J. Margraf, C. L. Chase, D. M. Chase, D. C. McAden, J. A. Palermo, Amzi Crenshaw, and E. N. Cowan for aid in collecting the organisms and performing the experiments.

REFERENCES

Adams, J. R., 1969, Thermal Power, Aquatic Life, and Kilowatts on the Pacific Coast, *Nucl. News*, 12(9): 75-79.

Cairns, J., Jr., 1972, Coping with Heated Waste Water Discharges from Steam-Electric Power Plants, *Bioscience*, 22(7): 411-419.

Chung, K. S., 1977, *Heat Resistance of Crustaceans and Fishes Taken from the Intake Canal of an Estuarine Power Plant and Their Predicted Survival in the Discharge Canal*, Ph.D. Thesis, Texas A. & M. University, College Station.

Churchill, M. A., and T. A. Wojtalik, 1969, Effects of Heated Discharges: The TVA Experience, *Nucl. News*, 12(9): 80-86.

Esch, G. W., and R. W. McFarlane (Eds.), 1976, *Thermal Ecology II*, ERDA Symposium Series, Augusta, Ga., April 2—5, 1975, CONF-750425, NTIS.

French, J. D., 1973, *The Effects of a By-Pass Canal upon the Marine Animals Occurring in the Cooling Canal System of the P. H. Robinson Generating Station*, M.S. Thesis, Texas A. & M. University, College Station.

Fry, F. R. J., J. R. Brett, and K. F. Walker, 1946, *Lethal Temperature Relations for a Sample of Young Speckled Trout* (Salvelinus fontinalis), University of Toronto Studies, Biological Series No. 54, Ontario Fisheries Research Laboratory Publication 66, pp. 5-35.

Gibbons, J. W., and R. R. Sharitz (Eds.), 1974, *Thermal Ecology*, AEC Symposium Series, Augusta, Ga., May 3—5, 1973, CONF-730505, NTIS.

Gross, A. J., and V. A. Clark, 1975, *Survival Distributions: Reliability Applications in the Biomedical Sciences*, John Wiley & Sons, Inc., New York.

Landry, A. M., Jr., and K. Strawn, 1974, Number of Individuals and Injury Rates of Fishes Caught on Revolving Screens at the P. H. Robinson Generating Station, in *Proceedings of the Second Workshop on Entrainment Intake Screening*, L. D. Jensen (Ed.), Cooling Water Discharge Research Report No. 15, Electric Power Research Institute Publication No. 74-049-00-5, Palo Alto, Calif.

McCullough, M. M., 1971, *A Study of the Juvenile Fish Fauna Associated with the Cooling Water of a Steam Electric Generating Station*, M.S. Thesis, Texas A. & M. University, College Station.

Matis, J. H., and O. Hartely, 1971, Stochastic Compartmental Analysis: Model and Least Squares Estimation from Time Series Data, *Biometrics*, 27: 77-102.

Naylor, E., 1965, Effects of Heated Effluents upon Marine and Estuarine Organisms, in *Advances in Marine Biology*, F. S. Russell (Ed.), Vol. 3, pp. 63-103, Academic Press, Inc., New York.

Pielou, E. C., 1969, *An Introduction to Mathematical Ecology*, John Wiley & Sons, Inc., New York.

Reimer, R. D., and K. Strawn, 1973, *Abundance, Distribution and Injury of Four Decapod Crustaceans Collected from Dickinson Bayou and Bay and the*

Revolving Screens, Intake and Discharge Canals of the P. H. Robinson Generating Station, Texas, TAES Project No. 602, Texas A. & M. University, College Station.

Steel, R. G. D., and J. H. Torrie, 1960, *Principles and Procedures of Statistics with Special Reference to the Biological Sciences*, McGraw-Hill Book Company, Inc., New York.

Vernberg, F. J., and W. B. Vernberg, 1969, Thermal Influence on Invertebrate Respiration, *Chesapeake Sci.*, 10(3+4): 234-240.

PATHOGENIC SPECIES OF THERMOPHILIC AND THERMOTOLERANT FUNGI IN REACTOR EFFLUENTS OF THE SAVANNAH RIVER PLANT

M. R. TANSEY* and C. B. FLIERMANS†
*Department of Biology, Indiana University, Bloomington, Indiana; and
†Savannah River Laboratory, Aiken, South Carolina

ABSTRACT

The growth, occurrence, and distribution of thermophilic and thermotolerant fungi were studied in effluents from Savannah River Plant nuclear reactors. Samples of foam, water, microbial mats, plant debris, air, and soils from sites of elevated and ambient temperatures were analyzed by direct microscopic examination and by quantitative plating. Analyses revealed that, except for *Dactylaria gallopava*, populations of thermotolerant and thermophilic fungi were not significantly different at elevated and ambient temperature sites. *Dactylaria gallopava*, a thermotolerant fungus that causes epidemic encephalitis in poultry, is abundant in the microbial mats, in foam, and in soils at the edges of the cooling-water effluents. It is directly associated with effluents that have temperatures of 44°C and higher. The evidence suggests that the thermal effluents are responsible for the occurrence of *D. gallopava*.

Among eukaryotic groups of organisms, evolutionary adaptation to growth at high temperature occurs at the extreme degree in the thermophilic and thermotolerant fungi (Tansey and Brock, 1972). Thermophilic fungi are those having a maximum temperature for growth of 50°C or higher and a minimum temperature for growth of 20°C. Thermotolerant fungi can grow at 50°C or above and grow well below 20°C (Cooney and Emerson, 1964). When high temperatures are relatively constant and coexist with some other environmental extreme, as in acid hot springs, a single fungal species occurs and grows (Tansey and Brock, 1973; Belly, Tansey, and Brock, 1973). In sun-heated soils and in many other habitats where thermal stresses (both high and low temperatures) coexist with moisture stress, additional species of thermophilic and thermotolerant fungi occur and grow (Tansey and Jack, 1976; 1977; Jack and Tansey, 1977; Tansey and Brock, 1978). Thermophilic and thermotolerant fungi thrive at body temperatures, and several species are zoopathogens (Tansey and Brock, 1978). Stresses other than high temperature limit growth of thermophilic and thermotolerant fungi in the body; the immune response of a normal human usually prevents serious infection.

This paper describes the diverse populations of thermophilic and thermotolerant fungi found in foam, water, microbial mats, plant debris, air, and soil at elevated and ambient temperatures at the U. S. Department of Energy's Savannah River Plant (SRP). Savannah River water is heated when it is pumped through heat exchangers to cool nuclear production reactors. These reactors, which are designed to produce nuclear materials for space, civilian, and military applications, differ significantly from nuclear reactors designed for electrical-power production in that effluent temperatures are frequently higher than 70°C. Heated effluents from SRP reactors provide habitats in which pH values are not extreme, and host immune responses do not limit growth of thermophilic and thermotolerant fungi. In these effluents temporal variation in thermal stress occurs over periods of time that are long in comparison with the life cycles of the fungi and thus provide "constant" temperature habitats.

MATERIALS AND METHODS

Samples were collected in late spring and in summer. General sites from which samples were obtained are identified in Figs. 1 and 2. The P Canal, Ponds 2, 4, and 5, and Par Pond receive thermal effluent from P Reactor, and Four Mile Creek receives heated effluent from C Reactor. On the other hand, Steel Creek has not received effluents for the past 9 years, and Upper Three Runs Creek and Intake Canal No. 1 on the Savannah River receive no heated effluents. At the Flowing Streams Laboratory, water from Upper Three Runs Creek is heated to a temperature differential of 25°C before being released through artificial channels (Harvey, 1973).

Water samples from each site were collected by plunging sterile tubes to a depth of 5 to 10 cm with flamed forceps, allowing the tubes to fill completely, and closing them with a screw cap. Water temperature was measured at the collection site, and pH was measured in the laboratory. One milliliter of each sample was spread

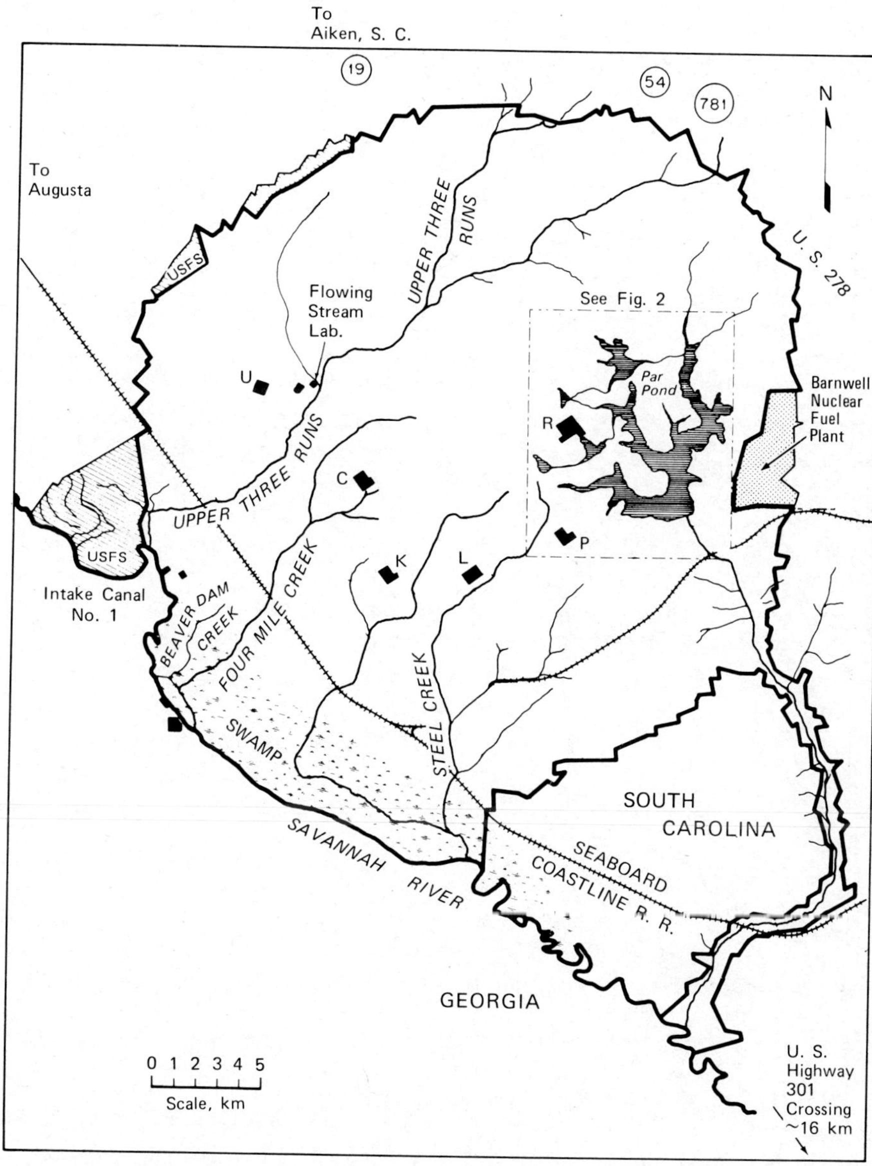

Fig. 1 Map of Savannah River Plant, showing reactor areas, various sampled streams, and the Par Pond system.

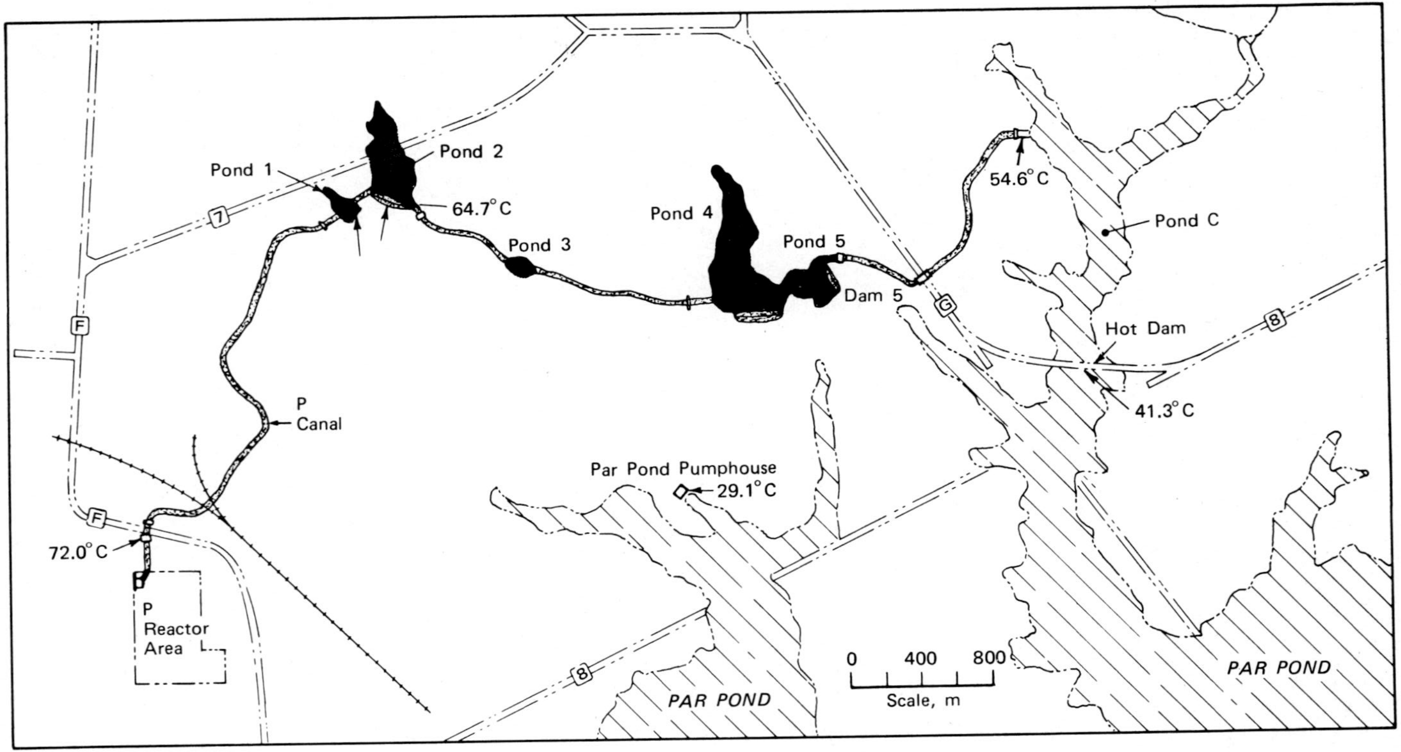

Fig. 2 Thermal reactor canal from P Reactor to Par Pond. Designated pond sampling sites and characteristic summer water temperatures are shown.

on each of four dishes of Emerson Yp Ss Agar (Yp Ss) plus antibiotics and four dishes of Sabouraud Dextrose Agar (SDA) plus antibiotics (Difco Laboratories). A variety of antibacterial antibiotics was used during this study, depending on the resistance and number of bacteria encountered. At the concentrations used, none of the antibiotics affected growth of the fungi isolated. For most samples, adding gentamicin sulfate at 50 μg/ml and penicillin G at 100 units/ml to 50°C Yp Ss or SDA immediately before pouring into dishes provided satisfactory control of bacteria. For some samples it was necessary to use a combination of antibiotics (50 μg/ml of gentamicin sulfate, 100 units/ml of penicillin G, 50 μg/ml of vancomycin hydrochloride, 50 μg/ml of streptomycin sulfate, and 25 μg/ml of chloramphenicol). Dishes of Yp Ss and SDA were incubated at 50 and 45°C, respectively, and observed at least once each 24 hr for 7 days. Colonies were counted, and each was identified by species.

Foam samples were scooped from the water surface with a flamed spoon (Tansey, 1973) and condensed by setting in a sterile plastic bag. Samples were collected in duplicate at most sites. One sample was immediately frozen on dry ice for inorganic and organic carbon determinations; the other was used for microscopic examination, plating, and determining pH. Estimates were made of the amounts of foam present in the field, and the formation, disappearance, and other aspects of foam dynamics were recorded. Condensed foam was spread at dilutions of 10^{-1}, 10^{-2}, 10^{-3}, and 10^{-4} on four dishes at each dilution. Mediums, antibiotics, incubation temperatures, and observation procedures used were the same as those for water samples.

Samples of sediment, microbial mat with subtending soil from the interface of water and shore (water edge), soil, and other materials were quantitatively examined for colony forming units (CFU's) of thermophilic and thermotolerant fungi. After the samples were air dried, the procedures described for water samples were used, except that samples were pour plated. Sites immediately adjacent to water and foam collection sites were sampled in most instances.

Air samples were collected by volumetric impingement, assayed by inoculation of four dishes of SDA + penicillin G + gentamicin sulfate with the impingement liquid from each sample (Tansey and Jack, 1976), and incubated as were the water samples. Approximately 1000 liters of air were impinged onto liquid, which was adjusted to ~20 ml after sampling. Air was also sampled by membrane filtration, with ~2500 liters being drawn through each filter. The filters were halved and placed onto SDA + penicillin

G + gentamicin sulfate and Yp Ss + penicillin G + gentamicin sulfate, as previously described.

RESULTS

Results of isolations from water samples are summarized in Table 1. Water samples from thermal and nonthermal effluents did not contain significant numbers of CFU's of thermophilic and thermotolerant fungi. The few colonies that grew are attributed to spores present in the surface foams accidentally collected as the sample tube was plunged into the water.

Aspergillus fumigatus and *Dactylaria gallopava* were isolated from most of the foam samples taken from thermal effluents (Table 2). No fungi were isolated from foam from the hottest water (66°C) nor from foam on Steel Creek (31.5°C), which had not received hot water for 9 years. Foam on hot water is quite hot; for example, at a depth of 1 cm in a 3-cm-deep pile of foam 30 cm in diameter and on 61°C water, the temperature was 52°C. The amount of foam required to produce 1 ml of liquid varied greatly from sample to sample. In some cases 10 cc of foam yielded 1 ml of liquid; in others ~50 cc were required. Microscopic examination of condensed foam revealed that samples taken from the surface of thermal effluents contained high concentrations of species of algae, fungi, protozoa, and invertebrates that cannot grow at the temperature of the subtending water. Many of these organisms were apparently healthy (e.g., normally pigmented algae, motile protozoa, and intact cytoplasm).

Foam was formed when wind caused waves to lap against the shore. The wind also pushed foam into piles commonly 3 to 5 cm deep and then up onto the shore and the vegetation. Foam frequently formed nearly continuous piles along the edge of parts of Ponds 2, 4, 5, and C, extending for several hundred meters. Foam also accumulated at places where obstructions (riprap, fallen trees, etc.) in canals and ponds caught small patches. On the rapidly flowing canals, foam formed at places where turbulence occurred, e.g., along edges, below spillways, and at obstructions.

The results of plating samples other than water or foam are summarized in Table 3. Except for the hottest sediments, most samples yielded thermophilic and thermotolerant fungi. Qualitatively and quantitatively, samples from thermal effluents did not differ significantly from samples from nonthermal sites or from sun-heated soils (Tansey and Jack, 1976), with one major exception—

(Text continues on page 686.)

TABLE 1

THERMOPHILIC AND THERMOTOLERANT FUNGI ISOLATED FROM WATER SAMPLES COLLECTED AT THE SAVANNAH RIVER PLANT*

Sample site	Temperature, °C	pH	Results of plating on Yp Ss (50°C) and SDA (45°C)
P Canal	72	7.2	No fungi
P Canal	68	7.0	No fungi
P Canal	66	6.8	No fungi
P Canal	65	7.0	No fungi
Pond 2	63	7.3	No fungi
P Canal	60	7.0	No fungi
P Canal	59	7.0	No fungi
Pond 4	56.5	6.9	No fungi
Pond 4	56	7.2	No fungi
Pond 4	54	7.0	No fungi
Pond C	53.5	7.0	No fungi
P Canal	53	7.0	1 colony of *Aspergillus fumigatus*
Pond 2	52	6.6	No fungi
P Canal	51	7.0	1 colony of *Dactylaria gallopava*
Pond C	50	6.7	No fungi
Pond 2	50	6.4	No fungi
Pond C	47.5	6.7	No fungi
Pond 2	47	7.1	No fungi
Pond C	46.5	6.7	No fungi
Pond C	45	6.6	No fungi
Pond C	45	7.0	2 colonies of *A. fumigatus*
Pond 2	44	7.3	No fungi
Pond C	42	7.0	1 colony of *Phanerochaete chrysosporium*
Par Pond	41	6.5	No fungi
Par Pond	38	6.3	No fungi
Par Pond	37.5	7.0	No fungi
Par Pond	37	7.1	2 colonies of *A. fumigatus*
Par Pond	35	6.7	No fungi
Artificial stream	35	6.6	No fungi
Par Pond	34.5	6.6	No fungi
Par Pond	33.5	6.8	No fungi
Par Pond	32.5	7.3	1 colony of *A. fumigatus*
Steel Creek	31.5	6.7	No fungi
Par Pond	31	6.8	No fungi
Par Pond	31	6.8	No fungi
Par Pond	30.5	7.1	No fungi
Artificial stream	30	6.5	No fungi
Artificial stream	27.3	6.5	No fungi
Four Mile Creek	27	6.6	No fungi
Par Pond	26.5	7.2	No fungi
Artificial stream	24.9	6.6	No fungi
Intake Canal No. 1	24	6.7	No fungi
Upper Three Runs Creek	23	6.5	No fungi
Upper Three Runs Creek	22.5	5.6	No fungi
Artificial stream	22.4	6.3	No fungi
Artificial stream	22.3	6.4	1 colony of *A. fumigatus*
Intake Canal No. 1	22	7.2	1 colony of *A. fumigatus*
Upper Three Runs Creek	20	7.0	No fungi

*Samples are listed from hottest to coldest.

TABLE 2

THERMOPHILIC AND THERMOTOLERANT FUNGI ISOLATED FROM FOAM COLLECTED AT THE SAVANNAH RIVER PLANT*

Sample site	Water temperature, °C	pH	Carbon concentration,† mg/ml	Results of plating on Yp Ss (50°C) and SDA (45°C)
P Canal	66	6.4	74, 1290	No fungi
Pond 2	63	6.2	29, 512	*Aspergillus fumigatus, Dactylaria gallopava*
P Canal	61	7.1	304, 1673	*A. fumigatus, D. gallopava*‡
Pond 4	56.5	6.5	98, 3747	*A. fumigatus, D. gallopava, Rhizopus nigricans*
Pond 4	56	6.5	44, 1086	*A. fumigatus, D. gallopava*
Pond C	53.5	6.6	<10, 3410	*A. fumigatus, D. gallopava*
Pond 4	52	6.9	31.7, 519	*A. fumigatus, D. gallopava*‡
Pond C	52	7.5	54.5, 374	*A. fumigatus*‡
Pond 2	48–51	6.3	40, 1112 (foam 10–20 min "old")	*A. fumigatus, D. gallopava*
Pond C	50	6.6	<10, 3416	*A. fumigatus, D. gallopava, Rhizopus rhizopodiformis*
Pond C	47.5	6.4	<10, 2374	*A. fumigatus, D. gallopava, R. nigricans*
Pond C	47	7.4	ND§	*D. gallopava, Emericella (Aspergillus) nidulans*
Pond C	46.5	6.4	<10, 3680	*A. fumigatus, D. gallopava, Rhizopus rhizopodiformis, R. nigricans*
Pond C	45	6.7	60, 2172	*A. fumigatus, D. gallopava*

Table 2 (Continued)

Sample site	Water temperature, °C	pH	Carbon concentration,† mg/ml	Results of plating on Yp Ss (50°C) and SDA (45°C)
Pond C	42.5	6.9	380, 1104	*A. fumigatus, D. gallopava, Humicola lanuginosa, Phanerochaete chrysosporium, Mucor pusillus, Thielavia terricola, Malbranchea sulfurea*
Artificial stream	35	6.5	<10, 2515	*A. fumigatus, D. gallopava, P. chrysosporium, Thielavia terrestris*
Par Pond	32.5	6.9	ND§	*A. fumigatus, D. gallopava, Mucor miehei, Thermoascus thermophilus*
Steel Creek	31.5	6.0	<10, 520	No fungi
Par Pond	31	6.1	<10, 610	*A. fumigatus, H. lanuginosa, P. chrysosporium, Aspergillus niger*
Par Pond	30.5	7.1	<10, 730	*A. fumigatus, Mucor miehei, Mucor pusillus, H. lanuginosa, Thielavia terricola, Thielavia heterothallica*

*Samples are listed from hottest to coldest, in reference to the water on which the foam occurred.

†Carbon is expressed in milligrams of inorganic carbon and organic carbon per milliliter of foam. The first values given are for inorganic and the second for organic carbon.

‡For these samples, it was not possible to control bacterial growth.

§Not determined.

TABLE 3

THERMOPHILIC AND THERMOTOLERANT FUNGI ISOLATED FROM SAMPLES OTHER THAN WATER AND FOAM COLLECTED AT THE SAVANNAH RIVER PLANT*

Sample site	Temperature of nearest water, °C	pH of sample†	Sample description	Results of plating in Yp Ss (50°C) and in SDA (45°C)
P Canal	72	6.1	Bottom sediment	No fungi
P Canal	72	7.6	Green mat in splash zone at water edge	*Aspergillus fumigatus*, *Dactylaria gallopava*, *Thielavia terrestris*, *Talaromyces thermophilus*
P Canal	72	6.8	Greenish-black mat 6 cm above splash zone, bathed by hot vapor	*A. fumigatus*, *D. gallopava*, *Thielavia terrestris*, *Rhizopus rhizopodiformis*
P Canal	72	5.6	Soil beneath herbs and grasses 1 m from water	*A. fumigatus*, *R. rhizopodiformis*, *Humicola lanuginosa*, *Melanocarpus albomyces*
P Canal	72	5.2	Soil beneath pines 10 m from water	*H. lanuginosa*, *Talaromyces thermophilus*, *Thielavia terrestris*
P Canal	68	7.0	Bottom sediment	No fungi

P Canal	68	6.6	Green mat at water edge	*D. gallopava, Thermoascus crustaceus*
P Canal	68	4.9	Green and black mat 3—4 cm above preceding sample	*A. fumigatus, D. gallopava, R. rhizopodiformis*
P Canal	68	6.2	Soil beneath grasses 1 m from water	*Thermoascus crustaceus, R. rhizopodiformis*
P Canal	68	6.2	Soil beneath pines 30 m from water	*A. fumigatus, Thielavia terrestris, R. rhizopodiformis, Thermoascus crustaceus, Scytalidium thermophilum*
Pond 2	63	5.9	Bottom sediment, plant debris, algal mat	*A. fumigatus, Thermoascus crustaceus*
Pond 2	63	6.5	Mat and pine needles at edge of water	*A. fumigatus, D. gallopava, Thielavia terrestris, Thielavia heterothallica*
Pond 2	63	5.7	Soil and plant roots at water edge	*A. fumigatus, R. rhizopodiformis, H. lanuginosa, H. grisea* var. *thermoidea, Talaromyces thermophilus*
Pond 2	63	6.0	Soil beneath grasses 10 cm directly above water (vertical bank)	*A. fumigatus, D. gallopava, H. lanuginosa, Thielavia terrestris, R. rhizopodiformis, R. nigricans*

(Table continues on following page.)

Table 3 (Continued)

Sample site	Temperature of nearest water, °C	pH of sample†	Sample description	Results of plating in Yp Ss (50°C) and in SDA (45°C)
Pond 2	63	4.8	Soil beneath pines 10 m from water	*A. fumigatus, Talaromyces thermophilus, Thermoascus aurantiacus*
P Canal	61	6.7	Drying foam on rocks at water edge	*A. fumigatus, H. lanuginosa, Thielavia terrestris, Thermoascus thermophilus*
P Canal	61	6.7	Drying microbial mat at water edge	*A. fumigatus, Talaromyces thermophilus*
P Canal	61	5.7	Soil beneath grasses 1 m from water	*Thermoascus aurantiacus, H. lanuginosa*
P Canal	61	6.2	Mat at water edge	*D. gallopava*
P Canal	59.5	6.1	Soil, roots, and mat at water edge	*A. fumigatus, D. gallopava*
P Canal	58	6.0	Dry foam and pine needles at water edge	*A. fumigatus, H. lanuginosa, Chaetomium thermophile* var. *dissitum*

Pond 4	54	6.1	Mat and soil at water edge	*A. fumigatus, H. lanuginosa, Thielavia heterothallica, S. thermophilum*
Pond 4	55	6.1	Mat, roots, and soil at water edge	*A. fumigatus, Thermoascus aurantiacus, Chaetomium thermophile* var. *coprophile, R. rhizopodiformis*
Pond 4	55	5.6	Soil beneath grasses 20 m from water	*H. lanuginosa, Thermoascus aurantiacus, C. thermophile* var. *coprophile*
P Canal	54.5	7.0	Plant debris and mat at water edge	*A. fumigatus, H. lanuginosa, Thermoascus aurantiacus, S. thermophilum, Melanocarpus albomyces*
Pond 2	52	6.3	Mat, plant roots, and drying foam at water edge	*A. fumigatus, D. gallopava, R. rhizopodiformis*
Pond 4	52	5.9	Soil, roots, and mat at water edge	*A. fumigatus, H. lanuginosa*
Pond 4	52	5.8	Soil beneath pines 10 m from water edge	*Talaromyces thermophilus*
Pond 4	50.5	5.7	Soil at water edge	*A. fumigatus, H. lanuginosa, S. thermophilum*

(Table continues on following page.)

Table 3 (Continued)

Sample site	Temperature of nearest water, °C	pH of sample†	Sample description	Results of plating in Yp Ss (50°C) and in SDA (45°C)
Pond 2	50	6.0	Mat, plant roots, and drying foam at water edge	*A. fumigatus, D. gallopava, R. rhizopodiformis, Thielavia heterothallica, H. lanuginosa, C. thermophile* var. *dissitum, Thermoascus crustaceus*
Pond C	48	6.2	Soil at water edge	*A. fumigatus, D. gallopava, H. lanuginosa, S. thermophilum*
Pond C	48	5.8	Soil and small plants on shore a few cm from water edge, within reach of waves	*A. fumigatus, D. gallopava, H. lanuginosa*
Pond C	48	4.4	Soil beneath pines 5 m from water	*A. fumigatus, H. lanuginosa, Thielavia terricola*
Pond C	48	4.4	Soil beneath grasses 1–2 m from water	*A. fumigatus, T. terricola, Melanocarpus albomyces*
Pond C	47	7.0	Soil at water edge	*A. fumigatus, H. lanuginosa, H. insolens*

Pond C	47	6.0	Soil on the shore a few cm from water edge, within reach of waves	*A. fumigatus, H. lanuginosa, Melanocarpus albomyces*
Pond C	47	5.7	Soil and small plants on shore a few cm from water edge, within reach of waves	*A. fumigatus, H. lanuginosa, Malbranchea sulfurea*
Pond C	47	4.6	Soil beneath grasses 1—2 m from water	*A. fumigatus, H. lanuginosa, Thielavia terrestris*
Pond C	47	5.4	Soil beneath pines 10 m from water	*Thielavia terrestris, Thielavia terricola, Talaromyces thermophilus*
Pond 2	47	6.6	Mat, plant roots, and drying foam at water edge	*A. fumigatus, R. rhizopodiformis, Thielavia heterothallica*
Pond 2	44	5.7	Bottom sediment and algal mat	*A. fumigatus, H. lanuginosa, R. rhizopodiformis, C. thermophile* var. *coprophile*
Pond 2	44	6.0	Plant roots in water at water edge	*A. fumigatus, C. thermophile* var. *coprophile, Melanocarpus albomyces, H. lanuginosa, Thermoascus aurantiacus, Mucor pusillus*

(Table continues on following page.)

Table 3 (Continued)

Sample site	Temperature of nearest water, °C	pH of sample†	Sample description	Results of plating in Yp Ss (50°C) and in SDA (45°C)
Pond 2	44	6.1	Soil and plant roots at water edge	*A. fumigatus, H. lanuginosa, Melanocarpus albomyces, R. rhizopodiformis*
Pond 2	44	5.8	Soil beneath grasses in area subject to flooding by Pond 2, 1 m from water	*A. fumigatus, D. gallopava, H. lanuginosa, Thielavia terrestris, R. rhizopodiformis, Thielavia heterothallica*
Pond 2	44	5.1	Soil beneath mixed hardwoods and pines, 10 m from water	*A. fumigatus, H. lanuginosa, Thielavia heterothallica, Thielavia terrestris*
Pond C	42.5	5.5	Bottom sediment, algal mat, and plant roots	*A. fumigatus, H. lanuginosa, Melanocarpus albomyces, Talaromyces thermophilus*
Pond C	42.5	6.2	Soil and mat at water edge	*A. fumigatus, H. lanuginosa, Thielavia heterothallica, Thermoascus aurantiacus, Thielavia terrestris*
Pond C	42.5	6.6	Soil a few cm from water edge, within reach of waves	*A. fumigatus, H. lanuginosa, Thermoascus aurantiacus*
Pond C	42.5	5.6	Soil beneath grassy area 1.5 m from water	*A. fumigatus, H. lanuginosa*

Pond C	42.5	6.5	Soil beneath pines and grasses 10 m from water	*A. fumigatus, Thielavia terrestris, H. lanuginosa, Melanocarpus albomyces*
Par Pond	37	6.9	Mat on rocks at water edge	*Thermoascus aurantiacus, Thielavia terrestris*
Par Pond	36	6.9	Bottom sediment and algal mat	*H. lanuginosa*
Par Pond	36	6.2	Soil at water edge	*A. fumigatus, Talaromyces thermophilus*
Par Pond	36	7.1	Soil and plant debris 0.5 m from water edge, within reach of high water	*A. fumigatus, H. lanuginosa*
Par Pond	36	7.2	Soil and small plants 1 m from water	*A. fumigatus*
Par Pond	36	6.4	Soil beneath grasses 2—3 m from water	*H. lanuginosa, Thielavia terrestris, Thermoascus aurantiacus*
Par Pond	36	5.8	Soil beneath pines 10 m from water	*C. thermophile* var. *dissitum, R. rhizopodiformis*
Artificial stream	35	6.5	Algal mat at water edge	*D. gallopava, R. nigricans, Talaromyces thermophilus*
Artificial stream	35	5.8	Bottom sediment and algal mat	*A. fumigatus, Thermoascus aurantiacus, Thielavia heterothallica, Talaromyces emersonii*

(Table continues on following page.)

Table 3 (Continued)

Sample site	Temperature of nearest water, °C	pH of sample†	Sample description	Results of plating in Yp Ss (50°C) and in SDA (45°C)
Par Pond	32.5	6.7	Soil at water edge	*A. fumigatus, H. lanuginosa, Thielavia heterothallica, Thermoascus aurantiacus*
Par Pond	32.5	6.7	Soil within a few cm of water edge, within reach of waves	*A. fumigatus, Talaromyces thermophilus, Mucor miehei*
Par Pond	32.5	6.1	Soil and small plants 0.5 m from water	*A. fumigatus, H. lanuginosa*
Par Pond	32.5	4.6	Soil beneath grasses 1–2 m from water	*A. fumigatus, H. lanuginosa*
Par Pond	32.5	4.6	Soil beneath pines 10 m from water	*A. fumigatus, H. lanuginosa, Thielavia terrestris*
Steel Creek	31.5	5.3	Bottom sediment	*A. fumigatus*
Steel Creek	31.5	5.3	Soil at water edge	*A. fumigatus*
Steel Creek	31.5	6.1	Soil beneath grasses 1 m from water	*A. fumigatus, H. lanuginosa*

Steel Creek	31.5	5.4	Soil beneath pines 20 m from water	*A. fumigatus, H. lanuginosa, Thermoascus aurantiacus, Melanocarpus albomyces*
Artificial stream	30	6.6	Algal mat at water edge	*A. fumigatus, Melanocarpus albomyces*
Artificial stream	30	6.4	Bottom sediment and algal mat	*A. fumigatus, R. nigricans, Melanocarpus albomyces, C. thermophile* var. *dissitum*
Artificial stream	27.3	6.4	Algal mat at water edge	*A. fumigatus, H. lanuginosa, Talaromyces thermophilus, Melanocarpus albomyces, Thermoascus aurantiacus, C. thermophile* var. *dissitum*
Artificial steam	27.3	5.7	Bottom sediment and algal mat	*A. fumigatus, Melanocarpus albomyces, Thermoascus aurantiacus, C. thermophile* var. *dissitum*
Four Mile Creek	27	6.8	Bottom sediment	*A. fumigatus*
Par Pond	27	7.0	Soil at water edge	*A. fumigatus*
Par Pond	27	7.0	Soil 0.5 m from water edge, within reach of high water	*A. fumigatus*

(Table continues on following page.)

Table 3 (Continued)

Sample site	Temperature of nearest water, °C	pH of sample†	Sample description	Results of plating in Yp Ss (50°C) and in SDA (45°C)
Par Pond	27	5.8	Soil and small plants 1–2 m from water	*A. fumigatus, H. lanuginosa, Thermoascus aurantiacus, Melanocarpus albomyces*
Par Pond	27	5.5	Soil beneath grasses 4–5 m from water	*A. fumigatus, H. lanuginosa, C. thermophile* var. *dissitum, Talaromyces thermophilus*
Par Pond	27	4.5	Soil beneath pines 20 m from water	*Thielavia terrestris*
Four Mile Creek	27	6.8	Soil at water edge	*A. fumigatus, D. gallopava, Thermoascus crustaceus*
Four Mile Creek	27	6.1	Soil and algal mat at high water mark	*A. fumigatus, D. gallopava, Thermoascus aurantiacus, Melanocarpus albomyces*

Four Mile Creek	27	6.1	Soil beneath herbs and grasses 2 m from water	*A. fumigatus, Thermoascus aurantiacus, Thielavia terrestris*
Four Mile Creek	27	5.7	Soil beneath trees 10 m from water	*A. fumigatus, Thermoascus aurantiacus, Thielavia heterothallica, Thielavia terrestris, H. lanuginosa*
Artificial stream	24.9	6.5	Algal mat at water edge	*A. fumigatus, R. rhizopodiformis, C. thermophile* var. *dissitum*
Artificial stream	24.9	6.2	Bottom sediment and algal mat	*A. fumigatus, Melanocarpus albomyces, Thielavia heterothallica, C. thermophile* var. *dissitum*
Artificial stream	22.4	6.8	Algal mat at water edge	*A. fumigatus, Melanocarpus albomyces, C. thermophile* var. *dissitum, Thermoascus aurantiacus, Thielavia terrestris, R. rhizopodiformis*
Artificial stream	22.4	6.6	Bottom sediment and algal mat	*A. fumigatus, Thermoascus aurantiacus, Melanocarpus albomyces, Talaromyces thermophilus*

(*Table continues on following page.*)

Table 3 (Continued)

Sample site	Temperature of nearest water, °C	pH of sample†	Sample description	Results of plating in Yp Ss (50°C) and in SDA (45°C)
Artificial stream	22.3	6.6	Algal mat at water edge	*A. fumigatus, Thielavia heterothallica, Thermoascus aurantiacus*
Artificial stream	22.3	6.6	Bottom sediment and algal mat	*A. fumigatus, R. nigricans*
Intake Canal No. 1	22	4.7	Soil at water edge	*A. fumigatus, Melanocarpus albomyces, Thermoascus aurantiacus*
Intake Canal No. 1	22	4.3	Plant-free soil 0.5 m from water	*A. fumigatus, H. lanuginosa, Mucor miehei, Thielavia heterothallica, Talaromyces thermophilus*
Intake Canal No. 1	22	4.3	Soil beneath grasses and herbs 1 m from water	*H. lanuginosa, Thermoascus aurantiacus*
Intake Canal No. 1	22	4.9	Soil beneath grasses 10 m from water	*A. fumigatus, H. lanuginosa, Mucor miehei, S. thermophilum*

Upper Three Runs Creek	21	5.1	Bottom sediment	*Thermoascus aurantiacus, Mucor pusillus*
Upper Three Runs Creek	21	5.4	Soil and plant roots at water edge	*A. fumigatus, Melanocarpus albomyces, Thielavia terrestris*
Upper Three Runs Creek	21	4.2	Soil and plant debris 0.5 m from water	*A. fumigatus, Melanocarpus albomyces, Thermoascus aurantiacus*
Upper Three Runs Creek	21	5.0	Soil and plant debris 10 m from water	*A. fumigatus, Thielavia terrestris*
Thermal Effects Laboratory	11	6.3	Sun-heated mud where condensate from air conditioner drips	*A. fumigatus, H. lanuginosa, Melanocarpus albomyces*
P Canal		4.2	Dry pitch on butts of sun-heated pine stumps	No fungi

*Listed from highest to lowest temperature of nearby water.

†The pH of a 1 : 1 (vol/vol) mixture of sample and distilled water was measured with an electronic pH meter.

D. gallopava was isolated only from thermal sites. The lowest water temperature directly associated with occurrence of *D. gallopava* was 44°C. Samples from cooler sites did not yield this fungus, with one apparent exception; *D. gallopava* was isolated from water edges of Four Mile Creek at a time when water temperature at this site was 27°C. This was a temporary condition since water at this site is higher than 50°C most of the time. In three instances *D. gallopava* was isolated from foam collected on waters that were cooler than 44°C (Table 2), but in each case there was an immediate source having a much higher temperature. The artificial stream received water that had been heated to a higher temperature before mixing with cooler water, and the 42.5°C site on Pond C and the 32.5°C site on Par Pond received current-borne foam formed a few minutes earlier at hotter water—soil interfaces.

Quantitative plating of impingement liquid from 15 air samples yielded five CFU's of *Phanerochaete chrysosporium* and one each of *Rhizopus rhizopodiformis*, *Thielavia terrestris*, and *D. gallopava*. Membrane filters from 17 air samples yielded 22 CFU's of *A. fumigatus*, eight of *P. chrysosporium*, and one each of *Thielavia terricola*, *R. rhizopodiformis*, and *Melanocarpus albomyces*.

Several isolates obtained in this study are believed to represent new species and are not included in the tables; they will be described elsewhere. In no case did these new organisms form many CFU's on assay plates.

DISCUSSION

The following species isolated in this study are zoopathogens: *A. fumigatus*, *A. niger*, *D. gallopava*, *Emericella (Aspergillus) nidulans*, *Thermoascus crustaceus*, *Mucor pusillus*, *M. miehei*, *R. nigricans*, and *R. rhizopodiformis*. There is a single report of disease caused by *Humicola lanuginosa*, and *Phanerochaete chrysosporium* (as *Sporotrichum pruinosum*) caused very localized infections in injected animals (Stretton, 1975). Most disease is initiated by inhalation of spores of these fungi. In view of their relative significance (incidence and severity) as causes of disease, *A. fumigatus* and *D. gallopava* are of greatest interest.

Aspergillus fumigatus is a ubiquitous fungus that is pathogenic, allergenic, and toxigenic. It is a frequent cause of disease of humans and animals (Rippon, 1974; Emmons et al., 1977; Jungerman and Schwartzman, 1972). An important feature of this fungus in terms of our study is that a small amount of fungal biomass can produce immense numbers of readily airborne spores.

Dactylaria gallopava has been reported as the cause of encephalitis in turkey poults in South Carolina and Maryland and in chickens in Australia, Georgia, and Indiana. This fungus has been isolated from samples of sawdust litter from poultry lots, from acid hot springs, from geothermal soils in Wyoming, and from self-heating coal waste piles in Pennsylvania and England (see Tansey and Brock, 1978). *Dactylaria gallopava* is also believed to cause disease in birds other than poultry and has been isolated from a human lung (Weitzman, 1977). Pure cultures of *D. gallopava* isolated from sites having near-neutral pH values grew well between pH ranges of 3.4 and 8.9 (Tansey and Fliermans, unpublished data). The distribution of *D. gallopava* found in our study and the results of the laboratory study indicate that *D. gallopava* does not require especially acidic conditions for establishment but does require elevated temperatures. Failure to isolate this readily recognizable fungus from such well-studied thermal habitats as composts, wood-chip piles, sun-heated soils, and self-heated hay suggests that it requires more moisture than most species of thermophilic and thermotolerant fungi or that it requires a more-constant elevated temperature. These parameters deserve further attention in efforts to determine the factors that limit the occurrence of this pathogen. Many other species of thermophilic and thermotolerant fungi grow in media having an extremely high osmotic concentration (Tansey, unpublished data); this suggests a relative tolerance to low water potential, which would be expected of organisms that characteristically grow in habitats subject to drying. Similarly, many species of thermophilic fungi can grow, sporulate, and exhibit spore germination under conditions in which temperatures fluctuate diurnally between permissive and nonpermissive (Tansey and Jack, 1977; Jack and Tansey, 1977). *Dactylaria gallopava* has not been tested under these conditions.

We must consider sample size when we evaluate the significance of results of plating water samples. For each sample, a total of at least 8 ml was plated (1 ml on each of four plates of two mediums). An additional amount was plated for each of 14 samples for which 10^{-1} and 10^{-2} dilutions were prepared. Five samples were plated one additional time (8 ml each) when initial plates contained too many bacterial colonies to allow reliable counting of fungal colonies. Standards for acceptable levels of zoopathogenic fungi in water intended for various uses are poorly defined (American Public Health Association, 1976). A single spore of *Emmonsiella (Histoplasma) capsulata* will infect a mouse (Ajello and Runyon, 1953); but a normal human may be exposed to millions of spores of *A. fumigatus*

TABLE 4

QUANTITATIVE RESULTS FOR CFU's OF *Dactylaria gallopava* IN SELECTED SAMPLES COLLECTED AT THE SAVANNAH RIVER PLANT

Sample site	Water temperature, °C	Sample description	Results of plating*	
			Yp Ss (50°C)	SDA (45°C)
P Canal	72	Mat and subtending soil 6 cm above splash zone at water edge	1 cc = >30	1 cc = >100
P Canal	68	Mat and subtending soil 3—4 cm above water at edge	1 cc = >100	1 cc = >100 0.1 cc = >100
Pond 4	56.5	Water, foam at water edge	10^{-1} = 9 10^{-3} = 5	10^{-1} = 126 10^{-2} = 28 10^{-3} = 3 10^{-4} = 1
Pond 2	48—51	Water, 10—20 min old foam at water edge	10^{-1} = 8 10^{-2} = 1 10^{-3} = 1	10^{-1} = 24

*Expressed as the sum of numbers of colonies on four plates at the indicated sample dilution (e.g., 10^{-1}) or amount (e.g., 0.1 cc).

without developing disease (Emmons et al., 1977). It is not possible to conclude from our data that the bodies of water studied are mycopathologically safe or that they are hazardous. We can conclude, however, that the concentrations of CFU's of pathogenic thermophilic and thermotolerant fungi are very low in water in comparison with those in the foam occurring on the water and in the microbial mat and the soil at the water's edge.

Our data indicate that the primary place for growth of *D. gallopava* at SRP is the microbial mat and subtending soil at the very edge of thermal effluents and not the water, foam, or other sites (e.g., sun-heated soil distant from the water's edge). Colony forming units of *D. gallopava* are high in number when this mat and associated soil are plated (Table 4). The highest counts are obtained from samples taken at the edge of the hottest water available. These CFU's cannot be coming from water emerging from the reactor (Table 1), the hottest foam (Table 2), the shore distant from the effluent (Table 3), or the air. We could argue that CFU's of *D. gallopava* in water-edge mats and soils associated with cooler foam and water arose from growth of *D. gallopava* in this foam. We do not rule out the possibility that foam at permissive temperatures supports some growth of *D. gallopava.* The foam has a high concentration of organic carbon in comparison with the water, which contains less than 5 ppm (Fliermans, unpublished data), and might support growth (Table 2). Bacteria are abundant in foam (e.g., 1.89×10^8 cells/ml of condensed foam on 50°C water, measured directly in a counting chamber); motile and dividing cells of bacteria are abundant. Germinating conidia of *D. gallopava*, microcolonies, and sporulation were not seen in condensed foam, however. These indicators of growth were readily apparent in condensed foam taken from hot-spring effluents in Wyoming (Tansey and Brock, 1973, and unpublished data). Furthermore, new foam contains CFU's of *D. gallopava* which could not have formed in the few minutes that foam existed (Table 4). Foam is dynamic; individual patches form and disappear in the course of a few minutes or hours. Colony forming units of *D. gallopava* also occur in places where foam cannot accumulate because the mat is well above rapidly flowing water in smooth-walled canals (Table 4).

The photoautotrophic species component of the water-edge microbial mat varies with temperature. At 45°C, it is *Mastigocladus laminosus* and *Oscillatoria* species.

Our data support the conclusion that fungi have not adapted to thermal stress in ways that result in populations of aquatic species comparable to those of prokaryotic microorganisms. No known species of aquatic fungi can grow at 50°C.

ACKNOWLEDGMENTS

The research reported here was developed during the course of work under contract No. AT(07-2)-1 with the U. S. Department of Energy.

We thank I. L. Sauls, J. L. Todd, and A. B. Parrott for technical assistance.

REFERENCES

Ajello, L., and L. C. Runyon, 1953, Infection of Mice with Single Spores of *Histoplasma capsulatum*, *J. Bacteriol.*, 66: 34-40.

American Public Health Association, 1976, *Standard Methods for the Examination of Water and Waste Water*, 14th ed., pp. 982-986, Washington, D. C.

Belly, R. T., M. R. Tansey, and T. D. Brock, 1973, Algal Excretion of ^{14}C-Labeled Compounds and Microbial Interactions in *Cyanidium caldarium* Mats, *J. Phycol.*, 9: 123-127.

Cooney, D. G., and R. Emerson, 1964, *Thermophilic Fungi*, W. H. Freeman & Company, San Francisco.

Emmons, C. W., C. H. Binford, J. P. Utz, and K. J. Kwon-Chung, 1977, *Medical Mycology*, 3rd ed., Lea & Febiger, Philadelphia.

Harvey, R. S., 1973, A Flowing Stream Laboratory, *Assoc. Southeast. Biol. Bull.*, 20: 3-7.

Jack, M. A., and M. R. Tansey, 1977, Growth, Sporulation, and Germination of Spores of Thermophilic Fungi Incubated in Sun-Heated Soil, *Mycologia*, 69: 109-117.

Jungerman, P. F., and R. M. Schwartzman, 1972, *Veterinary Medical Mycology*, Lea & Febiger, Philadelphia.

Rippon, J. W., 1974, *Medical Mycology*, W. B. Saunders Company, Philadelphia.

Stretton, R. J., 1975, Experimentally Induced Mycetoma: Species of *Sporotrichum* and *Sporothrix*, *Mycopathologia*, 55: 83-90.

Tansey, M. R., 1973, Use of Butane Lighter for Sterilization of Soil Sampling Instruments, *Mycologia*, 65: 215-216.

——, and T. D. Brock, 1972, The Upper Temperature Limit for Eukaryotic Organisms, *Proc. Natl. Acad. Sci. U.S.A.*, 69: 2426-2428.

——, and T. D. Brock, 1973, *Dactylaria gallopava*, a Cause of Avian Encephalitis in Hot Spring Effluents, Thermal Soils and Self-Heated Coal Wastes, *Nature*, 242: 202-203.

——, and T. D. Brock, 1978, Microbial Life at High Temperatures: Ecological Aspects, in *Microbial Life in Extreme Environments*, D. J. Kushner (Ed.), pp. 159-216, Academic Press, Inc., London.

——, and M. A. Jack, 1976, Thermophilic Fungi in Sun-Heated Soils, *Mycologia*, 68: 1061-1075.

——, and M. A. Jack, 1977, Growth of Thermophilic Fungi in Soil in situ and in vitro, *Mycologia*, 69: 563-578.

Weitzman, I., 1977, New York City Department of Health, personal communication.

RESPONSES OF THE ALLIGATOR TO INFECTION AND THERMAL STRESS

ARMAND B. GLASSMAN and CAROL E. BENNETT
Department of Laboratory Medicine, Medical University of South Carolina, Charleston, South Carolina

ABSTRACT

The effects of infection by *Aeromonas hydrophila* in *Alligator mississippiensis* were studied. Peripheral blood components and serum protein electrophoretic patterns in normal vs. infected and/or thermally stressed animals are described. Alligators respond to bacterial infection with an increase in white blood cells of the neutrophilic macrophage types. Leech infestations are associated with an increase in eosinophils. Treatment of animals, either with a topical antibiotic ointment or intramuscular injections of an appropriate antibiotic, results in healing of lesions and return of the white blood cell counts to normal levels. Removal of leeches leads to a decrease in eosinophilic counts. Serum protein electrophoretic changes accompanying infection include an increase in total protein and in the gamma and alpha 2 fractions. As alligators age total protein increases, primarily from an increase in the gamma region.

The American alligator plays an important role in shaping and maintaining the animal life in swamps, lakes, and rivers over a widespread area of the southeastern United States (Craighead, 1968). Alligators are aquatic poikilothermic reptiles relying on water as their primary medium for locomotion and their main source of food. Water temperature is intimately involved in their thermoregulation. Alligators remain in the water during hot summer days, almost totally restricting their terrestrial activities until darkness. Increases in ambient water temperature, above 35°C, could pose serious problems by altering their thermoregulatory mechanisms (Colbert, Cowles, and Bogart, 1946).

Coulson and Hernandez (1964) were among the first to study extensively the biochemistry of the alligator. They reported normal chemical values and hemoglobin and hematocrit levels for uninfected animals living at ambient temperatures. Colbert, Cowles, and Bogart (1946) defined the alligator's preferred (32 to 35°C) and lethal (38 to 39°C) temperatures and the precision with which they can regulate body temperature. Immunologic and hematomorphologic data were not reported in either study. Heat energy budgets for alligators were established by Spotila, Soule, and Gates (1972).

Despite all these studies, however, little quantitative information is available for the alligator on a variety of physiological parameters, in particular the host defense mechanisms, the morphology of peripheral blood (Reese, 1917), and the effects of thermal stress on these responses. The purpose of our study was: (1) to examine the effects of temperature alteration on the alligator's capacity to respond to an infection of the bacteria *Aeromonas hydrophila*; (2) to determine the route of infection to which the animal was most susceptible at different temperatures; and (3) to devise methods for quantifying the extent of infection and the capacity of recovery in the alligator under a variety of thermal conditions.

MATERIALS AND METHODS

Phase 1 of the study was designed to generate information regarding possible natural routes of infection and to follow the development and progression of experimental infections. Alligators were exposed to graded doses of *A. hydrophila* (1×10^3 to 5×10^5 *Aeromonas* per milliliter of tank water). Control animals were maintained in a separate tank where levels were from 10 to 25 organisms/ml tank water, and others were exposed by injection of *A. hydrophila* intravenously, intramuscularly, or intraperitoneally. Control animals were given appropriate injections of buffer.

In phase 2 we studied the susceptibility of alligators to *A. hydrophila* as a function of varying levels of thermal stress. Water temperatures of 20, 25, 30, and 35°C were used, and varying *Aeromonas* doses (1×10^3 to 5×10^5/ml) were given at the different temperatures. Hematologic data were collected before exposure to *Aeromonas* and after infection. The effect of temperature on base-line and host-defense effectiveness was appraised.

Aeromonas hydrophila in amounts from 1×10^3 to 5×10^5 organisms/g body weight were pipetted down the animals' gullets in phase 3 of the experiment. Controls were given buffer in an identical manner.

Phase 4 was designed to study the treatment of infected animals that had been thermally stressed. An antibiotic ointment containing (1) polymyxin B sulfate, 5,000 units/gram, (2) zinc bacitracin, 400 units/gram, and (3) neomycin sulfate, 5 mg (equivalent to 3.5 mg neomycin base), was applied topically to lesions. Kanamycin sulfate (10 mg/kg body weight) was given intramuscularly to some of the infected animals daily for 5 days. Injections were stopped for 2 days and then resumed for 5 days with an increased dose, up to 50 mg/kg body weight. Fifty milliliters of oxytetracycline hydrochloride (25 mg/ml) was added to the infected water in the 600-liter tank containing a third set of animals. A fourth tank of animals remained untreated as controls.

Normal values for animals were established beginning with initial capture in their natural habitats off the South Carolina coast. The procedures in studying the alligator blood were:

1. Blood was collected from the dorsal vein of the tail with 1.5-ml ethylenediaminetetraacetic acid tubes to provide an anticoagulated specimen. Slides were prepared, and microhematocrits were determined.

2. Red blood cells were counted in an automated unit that gives total red count, total white count, hemoglobin, hematocrit, and mean corpuscular hemoglobin concentration. Hemoglobin was measured by a cyanomethemoglobin method on a 1 : 500 dilution in an automated hemoglobinometer.

3. Platelet counts and white blood cell counts were performed by adding 20 μl of well-mixed anticoagulated blood to 1.98 ml of 1% ammonium oxalate to make a 1 : 100 dilution. An estimated platelet count was performed using a Miller disk and 50x oil objective (Dacie and Lewis, 1968).

Cytochemical stains; periodic-acid—Schiff (PAS), as described by Lillie (1965); peroxidase, by the Kaplow method (Kaplow, 1965); and leukocyte alkaline phosphatase (LAP) (Kaplow, 1968) were used to determine the identity of certain white blood cells from the chemical content.

The methods for protein studies on serum were as follows:

1. Total protein was determined by the biuret method.

2. Electrophoresis was performed on cellulose acetate gels with a sodium barbital buffer at a pH of 8.6. Migration was performed at 180 V, 11.5 mA for 13 min. The plates were then stained for 3 min in Ponceau S, a general protein stain. The stained gels were scanned and counted by an automated, digital, integrated scanner.

RESULTS AND DISCUSSION

The size and composition of the cellular elements in normal, uninfected animals were determined. Nucleated red blood cells (RBC's) were characterized by size, number, and hemoglobin per deciliter of whole blood (Table 1). No statistically significant changes in red blood cell indexes, hemoglobin, or hematocrit were noted in infected animals vs. wild or captively raised normals. White blood

TABLE 1

RED BLOOD CELL INDEXES FOR HEALTHY, UNINFECTED *Alligator mississippiensis**

Hemoglobin, 7.2 ± 0.4 g/dl
Hematocrit, 19.9% ± 2.0
Red blood cell count, 3.9 ± 0.07 × 10^5/mm^3
Mean corpuscular volume, 522.0 ± 78 μ^3
Mean corpuscular hemoglobin, 190.0 ± 29.8 $\mu\mu$g
Mean corpuscular hemoglobin concentration, 36.4% ± 2.2
Red blood cell size (N = 65)
Long axis, 17.2 ± 1.4 μ
Short axis, 9.1 ± 1.2 μ

*Values are $\bar{x}$ ± standard deviation. Sample size (N) = 8 unless otherwise indicated.

cells were categorized as: (1) granulocytes, (a) eosinophils, (b) basophils; (2) neutrophilic macrophages; (3) lymphocytes; (4) monocytoid cells; and (5) thrombocytes (Tables 2 and 3; Figs. 1 to 3).

Infected animals had increased white blood cell counts, predominantly of the neutrophilic macrophage component. This is consistent with the kind of host immune-defense mechanism seen in higher vertebrates exposed to pyogenic organisms. Neutrophilic macrophages increased from 37 to 70+% in the alligators infected with *A. hydrophila.* Peroxidase staining, which is used in humans as an aid in differentiating cells of the myelocytic systems from those of the monocytic and lymphocyte series, demonstrated that the neutrophilic macrophages were positive. They were also positive with PAS stain, which detects glycogen. There was a high content of LAP in the neutrophilic macrophages.

Modes of exposure resulting in infection were contaminated water in holding pens and injections. No infections were noted in animals given *A. hydrophila* orally. Infected animals died most frequently at 35°C and exhibited a diminished immune response, as measured by ineffective increases of neutrophilic macrophages and

TABLE 2

WHITE BLOOD CELL COUNTS AND PERCENTAGES OF HEALTHY, UNINFECTED *Alligator mississippiensis**

- Normal white blood cell count, 5.5 ± 1.3 × $10^3/mm^3$
- Differential, %
  - Neutrophilic macrophages, 36.6 ± 9.2
  - Eosinophils, 5.1 ± 3.4
  - Basophils, 1.9 ± 2.6
  - Monocytes, 4.0 ± 5.6
  - Lymphocytes, 52.4 ± 10.7
- Thrombocyte count, 2.3 ± 0.8 × $10^4/mm^3$
  - Cell size, 9.5 ± 1.6 μ (N = 16)
  - Thrombocyte nucleus (N = 12)
    - Long axis, 6.6 ± 0.9
    - Short axis, 4.8 ± 1.2

*Values are $\bar{x}$ ± standard deviation. Sample size (N) = 8 unless otherwise noted.

TABLE 3

SIZES OF WHITE BLOOD CELLS (μ) OF HEALTHY, UNINFECTED *Alligator mississippiensis**

- Lymphocytes (N = 15)
  - 12.3 ± 2.3
  - Lymphocyte nucleus
    - Long axis, 9.9 ± 2.1
    - Short axis, 7.3 ± 1.4
- Neutrophilic macrophage (N = 45)
  - 15.3 ± 2.5
  - Neutrophilic macrophage nucleus (N = 37)
    - Long axis, 7.0 ± 2.2
    - Short axis, 5.1 ± 1.6
- Eosinophil (N = 8)
  - 12.4 ± 2.3
  - Eosinophil nucleus (N = 7)
    - Long axis, 7.3 ± 2.8
    - Short axis, 3.9 ± 0.9
- Basophil (N = 7)
  - 11.6 ± 2.2
  - Basophil nucleus (N = 6)
    - 5.4 ± 0.9

*Values are $\bar{x}$ ± standard deviation. N = sample size.

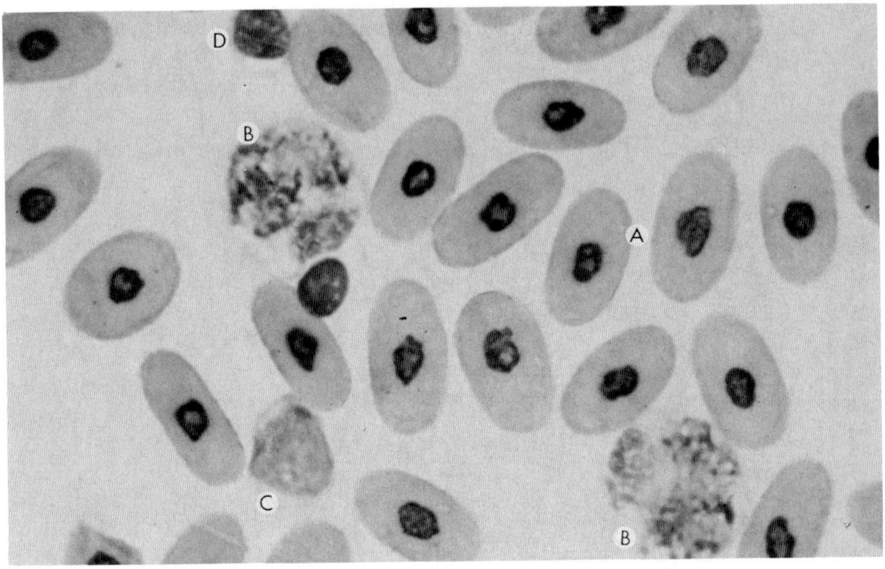

Fig. 1 Wright stained peripheral smear of anticoagulated blood of healthy, uninfected *Alligator mississippiensis.* A, nucleated red blood cells; B, neutrophilic macrophages; C, lymphocytes; and D, thrombocyte.

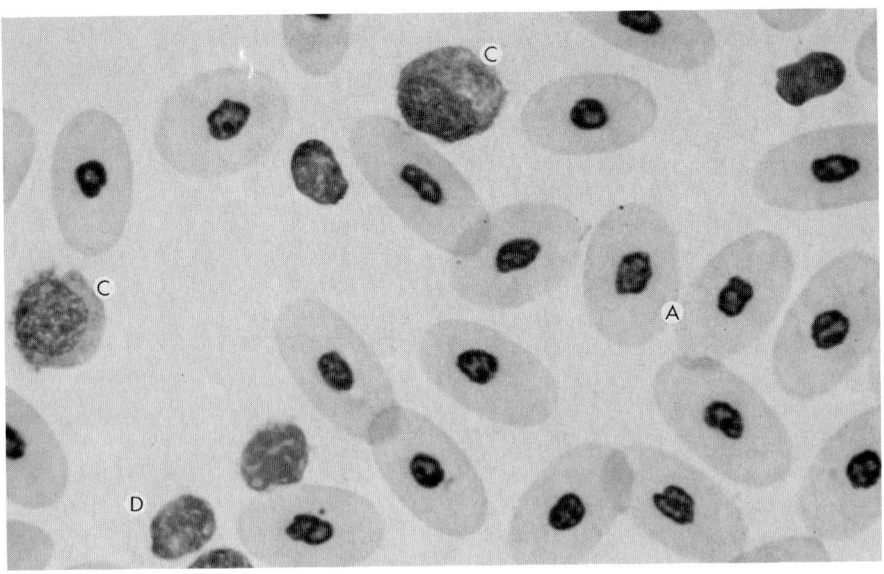

Fig. 2 Wright stained peripheral smear of anticoagulated blood of healthy, uninfected *Alligator mississippiensis.* A, nucleated red blood cells; C, lymphocytes; and D, thrombocytes.

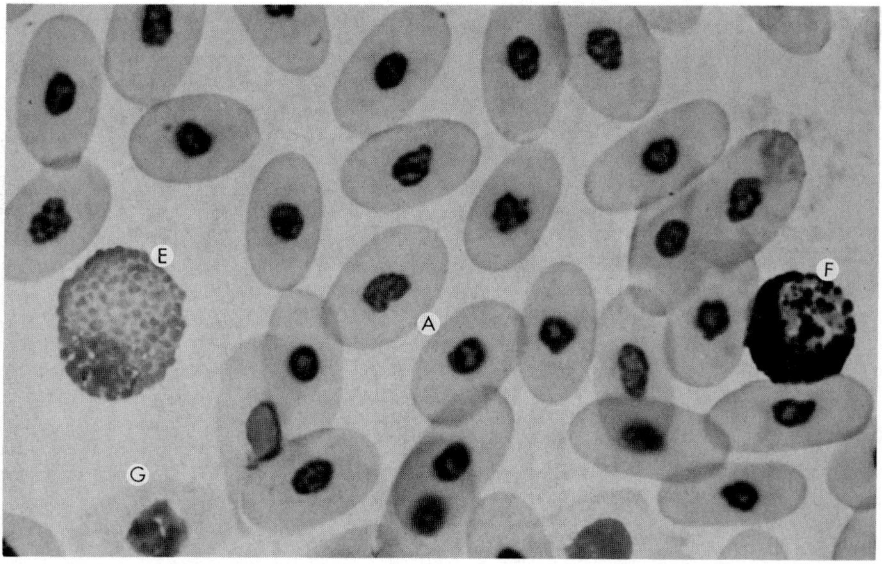

Fig. 3 **Wright stained peripheral smear of anticoagulated blood of healthy, uninfected** *Alligator mississippiensis.* **A, nucleated red blood cells; E, eosinophil; F, basophil; G, immature nucleated red blood cell. Note the loosely packed nuclear material, compared with the compact darker staining nucleus of the mature red blood cell.**

an absence of elevated serum proteins in the alpha 2 region (Fig. 4). Alligators held at 25 to 30°C maintained the best body condition and generated the most striking immune response. Young animals (0.5 to 1.0 years old) died within 3 weeks when placed in tanks at 35°C. This age group also was more susceptible to infection by any route at the higher temperatures. This may represent limited host defense responses of juveniles, increased sensitivity to thermal stress, or some combination of these factors.

Thermal stress seems to have a major effect on the host response of *A. mississippiensis.* Animals kept at various temperatures (20, 25, 30, and 35°C) had similar base-line levels of white blood cell counts (Fig. 6). After infection their white blood cells and neutrophilic macrophages were maximum at 30°C. This temperature corresponded to optimum recovery of infected animals. Exposure in the 35°C tank resulted in slightly lower neutrophilic macrophage numbers and a uniformly fatal outcome (four of four infected animals died). The ineffective host response at 35°C may be the result of complex host agent factors, such as better growth for *Aeromonas* or decreased phagocytosis by neutrophilic macrophages.

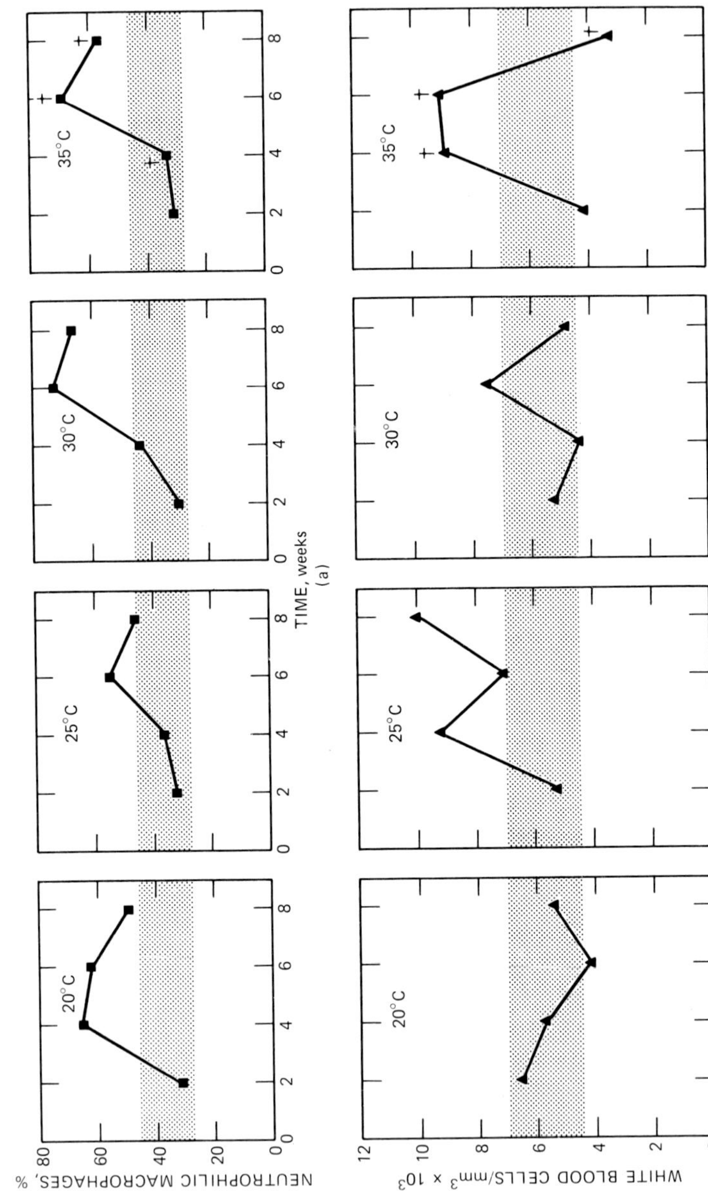

Fig. 4 Response of neutrophilic macrophages (a) and white blood cells (b) in *Alligator mississippiensis* to infection with *Aeromonas hydrophila* at temperatures increasing in 5°C increments. Levels of *Aeromonas* per milliliter of tank water were raised at 2-week intervals, beginning at 1×10^3 for the first 2 weeks and increasing to 5×10^3 for weeks 2 to 4, 5×10^4 for weeks 4 to 6, and 5×10^5 for weeks 6 to 8. Shaded areas indicate normal ranges. Death of animal is indicated by +.

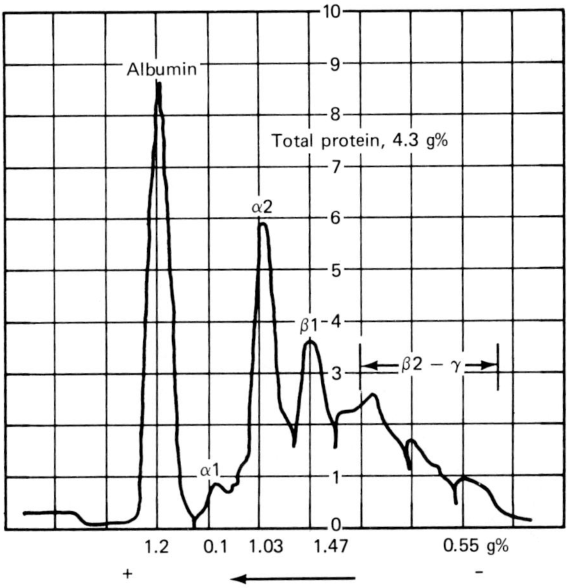

Fig. 5 Normal serum protein electrophoretic pattern for *Alligator mississippiensis.*

Thrombocytes in the alligator are relatively low in number, and their contribution to hemostasis is not immediately known. Thrombocytes are known to participate in hemostasis and host phagocytic defense mechanisms in other animals, however.

Serum protein electrophoresis (SPE) values revealed a total protein of ~4 g in alligators under 1.5 years of age. After 3 years of age, the alligator's total protein values were ~6 g/dl. The normal SPE patterns revealed peaks discernible in the albumin, alpha 1, alpha 2, beta 1, beta 2, and gamma regions (SPE, Fig. 5). In infected animals the SPE showed an elevated alpha 2 globulin peak, with a relative diminution in the albumin regions (Fig. 6). The electrophoretic pattern in the healthy alligators appears similar to that seen in a variety of vertebrates. The alpha 2 peak increases in man in response to acute inflammatory conditions. In older alligators there is an increase in the gamma region above that of laboratory specimens; this indicates increased exposure to a variety of infectious agents and antigens.

The majority of wild animals in this project had the parasite *Haemogregarina* in their red blood cells. The haemogregarines are common intracellular red blood cell parasites with many known invertebrate vectors (Manwell, 1977; Glassman, Holbrook, and

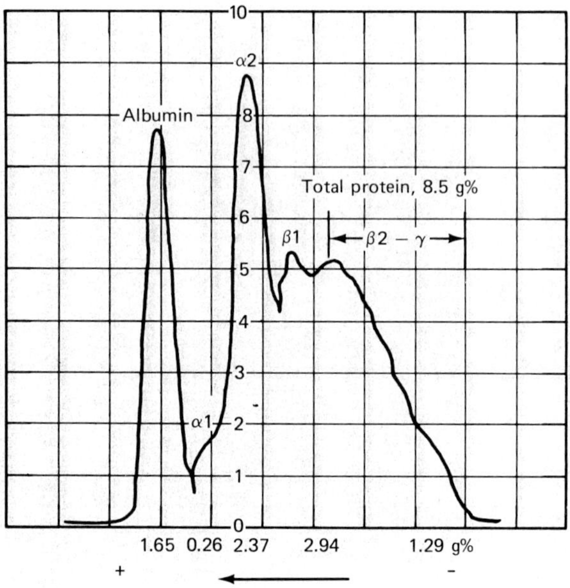

Fig. 6 Serum protein electrophoretic pattern for an alligator with acute infection.

Bennett, 1978). Whether the presence of this parasitic infection altered their red blood cell counts or percentages of white cells is unknown. Animals from the Louisiana Wildlife Preserve, which were raised from eggs in a clean environment, were also studied. They showed no evidence of haemogregarines, and yet their other hematologic values were similar to those of the wild normals.

It was observed in phase 4 of the study that animals captured in the wild (from June 20 to 30) had large numbers of leeches (*Placobdella multilineata*) along their dorsal surface and attached to the upper and lower jaws inside the oral cavity. Upon capture these ten animals all exhibited marked eosinophilia, 60 ± 14%, compared with the normal range of 5 to 10%. Removal of the leeches when the animals were placed in the control tanks resulted in a drop to ~10% over a 1-month span; this suggests that the parasites were responsible for the eosinophilia.

Animals with *A. hydrophila* infections were treated by a variety of methods. An oil-base ointment containing polymyxin B and neomycin, which was applied to lesions topically, was effective in healing severe isolated lesions. Intramuscular injections of kanamycin were given daily at a dose of 10 to 50 mg/kg body weight for 5-day intervals. Kanamycin appeared to halt the progression of the infection, at least temporarily. Oxytetracycline hydrochloride, added to the tank water at a final concentration of 2 mg/ml, had no detectable beneficial effect in these experiments.

The results indicate that *A. mississippiensis* is sensitive to bacteria when heat stressed. The progressive lesions and death that result if the animals are unable to produce adequate defenses may be functions of thermal stress, particularly at temperatures equal to or greater than 35°C.

ACKNOWLEDGMENTS

The research reported here was supported by contract EY-76-S-09-0965 between the Energy Research and Development Administration and Wake Forest University. Robert Gorden, Professor of Microbiology, University of Southern Colorado, served as project coordinator and principal investigator. We are very grateful to Carl Fliermans and the Savannah River Laboratory for providing space, equipment, and assistance. Gerald Esch, Professor and Chairman, and Terry Hazen, Department of Biology, Wake Forest University, made major contributions to overall experimental design. Tom Holbrook, Assistant Professor, Department of Laboratory Medicine, Medical University of South Carolina, aided in identifying *Haemogregarina* sp. and *Placobdella multilineata*. Our special thanks go to Emily Barrineau, our dedicated and competent typist.

REFERENCES

Colbert, E. H., R. B. Cowles, and C. M. Bogart, 1946, Temperature Tolerances in the American Alligator and Their Bearing on the Habits, Evolution and Extinction of the Dinosaurs, *Bull. Am. Mus. Nat. His.*, 86: 329-373.

Coulson, R. A., and T. Hernandez, 1964, *The Biochemistry of the Alligator*, pp. 131-132, Louisiana State University Press, Baton Rouge.

Craighead, F. C., Sr., 1968, The Role of the Alligator in Shaping Plant Communities and Maintaining Wildlife in the Southern Everglades, *Fla. Nat.*, 41: 3-7, 69-74.

Dacie, J. V., and S. M. Lewis, 1968, The Reticulocyte Count, in *Practical Haematology*, 4th ed., p. 30, Grune & Stratton, Inc., New York.

Glassman, A. B., T. Holbrook, and C. E. Bennett, 1978, Eosinophilia Associated with Leech (*Placobdella multilineata*) Infestation of *Alligator mississippiensis*, *Bull. S. C. Acad. Sci.*, 40: 59.

Kaplow, L. S., 1965, Simplified Myeloperoxidase Stain Using Benzidine Dihydrochloride, *Blood*, 26: 215.

——, 1968, Leukocyte Alkaline Phosphatase Cytochemistry: Applications and Methods, *Ann. N. Y. Acad. Sci.*, 155: 911.

Lillie, R. D., 1965, *Histopathologic Technic and Practical Histochemistry*, 3rd ed., p. 495, McGraw-Hill Book Company, New York.

Manwell, R. G., 1977, Gregarines and Haemogregarines, in *Parasitic Protozoa*, J. P. Kreir (Ed.), Vol. 3, pp. 1-32, Academic Press, Inc., New York.

Reese, A. M., 1917, The Blood of *Alligator mississippiensis*, *Anat. Rec.*, 13: 37-44.

Spotila, J. R., O. H. Soule, and D. M. Gates, 1972, The Biophysical Ecology of the Alligator: Heat Energy Budgets and Climate Spaces, *Ecology*, 53: 1094.

ACCLIMATION STATES OF JUVENILE STRIPED BASS HELD IN CONSTANT AND FLUCTUATING TEMPERATURE REGIMES

D. K. COX
Environmental Sciences Division, Oak Ridge National Laboratory,
Oak Ridge, Tennessee

ABSTRACT

Sinusoidal temperature fluctuations of varying periodicities caused juvenile striped bass, *Morone saxatilis,* to have acclimation states higher than the constant average temperatures but lower than the maximum temperatures of the cycles, on the basis of determinations of critical thermal maximums. More rapid cycling caused higher acclimation states. There were no effects of time of day or illumination. Heat tolerance of fish from thermally fluctuating environments will likely be underestimated if average environmental temperatures are assumed to represent the acclimation state. Also, tolerance data developed at constant acclimation temperatures will underestimate survival in fluctuating temperature regimes.

Prediction of power plant impacts and establishment of regulatory standards frequently rely on laboratory data. Thermal responses in the laboratory have traditionally been estimated from constant temperature regimes, but temperatures in a natural aquatic system are characteristically subject to diurnal and seasonal fluctuations. Therefore it is necessary to determine whether reasonable predictions of responses of fish to fluctuating temperature regimes can be made from data developed at constant acclimation temperatures. This investigation compared temperature acclimation patterns developed in fish in fluctuating temperatures with those developed in constant temperatures. Equilibrium loss at high temperatures was the measure of acclimation state, and young-of-the-year striped bass was the test species. Possible differences in thermal tolerance caused by the time of day testing occurred were also investigated. Information on the

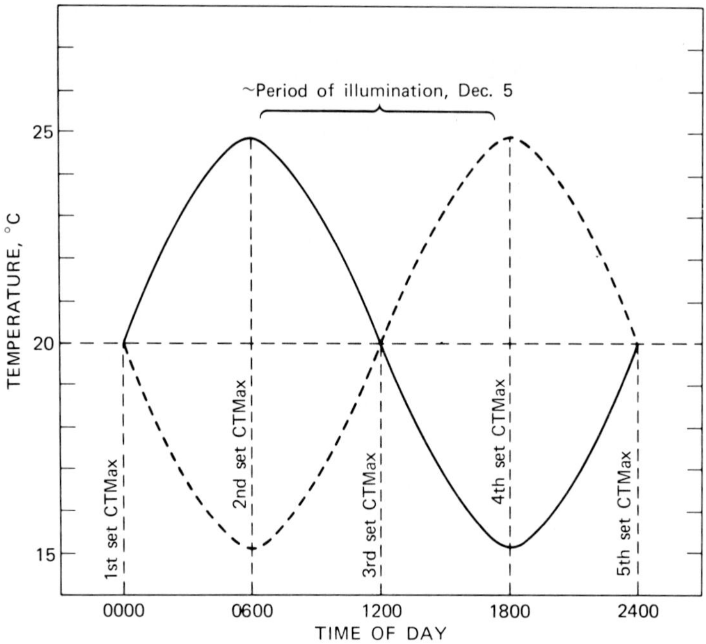

Fig. 1 Fluctuating acclimation temperature regimes for the first-phase experiments indicating the times of day when critical thermal maxima (CTMax) were determined.

effects of cycling acclimation temperature is necessary for predicting more accurately both the impact of entrainment-caused heat shock and the consequences to fish of cold shock induced by power-plant shutdown.

MATERIALS AND METHODS

In the initial phase of the experiment, a total of 270 young-of-the-year striped bass, 6 to 7 g in weight and 8 to 9 cm in total length, were randomly distributed among five circular tanks (1.2 m in diameter and 66 cm deep). Two of the tanks were maintained on a computer-controlled fluctuating temperature regime with a 24-hr period and a sinusoidal curve varying ±5°C around a 20°C base). The temperatures in one tank rose between 0600 and 1800 hr, and, in the other, temperatures fell between 0600 and 1800 hr—a time period extending through the period of illumination (Fig. 1). Illumination was provided by cool white florescent lights controlled by an outside photoswitch to reproduce the seasonal photoperiod. The remaining three tanks of fish were maintained at constant temperatures of 15, 20, and 25°C. The fish were acclimated for 17 days in the fluctuating and constant temperature regimes from Nov. 18 to Dec. 4, 1975. During this period the bass were fed freeze-dried brine shrimp three times a day. Fish were not fed 6 hr before or during the day of testing. On Dec. 5, 1975, critical thermal maximum (CTMax) values were determined for a pair of fish from each tank at 0000, 0600, 1200, 1800, and 2400 hr. This block design provides ten observations for analysis at each temperature and time-of-day treatment. The CTMax values were determined by heating 4 liters of water at a rate of 1.5°C/min until the bass therein permanently lost equilibrium.

In the second phase of the experiment, the fluctuating regimes were modified to provide one, two, and six cycles in each 24-hr period, having the same degree (±5°C) of fluctuation around the same 20°C base temperature. The constant-temperature (15, 20, and 25°C) tanks were retained. The average weight of striped bass used in this phase was 10.8 g, and the average total length was 8.7 cm. After 15 days of acclimation (on Jan. 16, 1976), the test procedure described previously was repeated.

When the small fishes used in the second phase of experimentation were subjected to the most rapid temperature changes used in the study (6 cycles/24 hr), the estimated difference between their internal (gut) and external water temperature was small (Fig. 2). The smaller fish (6 to 7 g) tested in the initial phase are estimated to exhibit even smaller degrees of temperature lag since they were acclimated at 1 cycle/24 hr. These slower rates of change, combined with the smallness of the fishes, could only produce insignificant differences between body and cycling water temperatures.

The test fish were obtained as fry from the Tennessee Wildlife Resource Agency's fish hatchery in Morristown, Tenn., and were raised to experimental size in our laboratory. They were the progeny of spawning stock collected from a landlocked, freshwater population in Cherokee Reservoir, Tennessee.

RESULTS

The mean CTMax values for the initial constant and fluctuating acclimation regimes increased in the order 15, 20, and 15 to 25(—) 1 cycle/24 hr; 15 to 25(+) 1 cycle/24 hr; and 25°C (Table 1). Analysis of variance (ANOVA) indicated significant differences among these acclimation temperature responses ($P < 0.01$), and Duncan's new

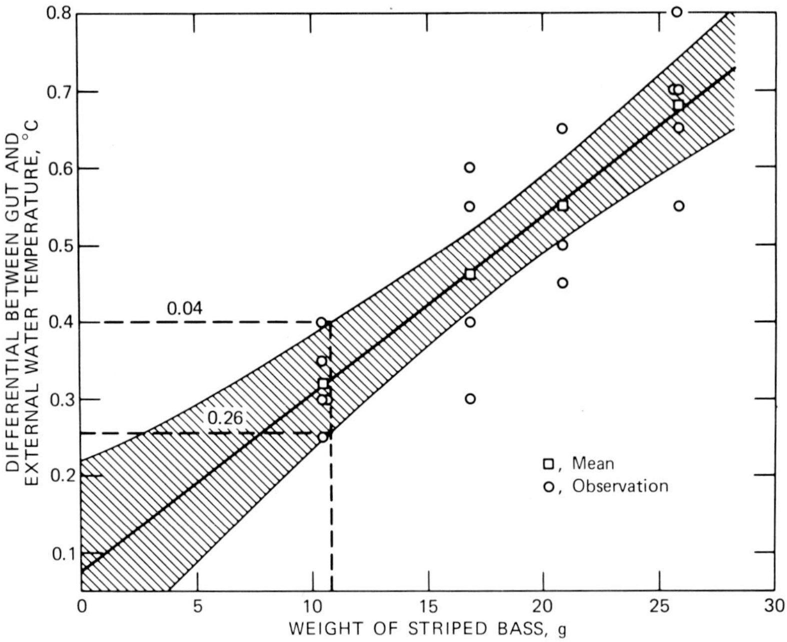

Fig. 2 A linear regression of the difference between gut and water temperature vs. weights of striped bass, *Morone saxatilis*, held in a cycling temperature regime (6 cycles/24 hr) with a sinusoidal curve of ±5°C around a 20°C base. $Y = 0.079 + 0.023X$ and $r = 0.9952$ ($P < 0.001$). The 95% confidence band is for the mean of four new observations at X_R. The mean differences estimated from the confidence band for fish with an average weight of 10.8 g are illustrated.

multiple range test (DMRT) identified differences among the responses. All the mean CTMax values were significantly different ($P < 0.05$) except the two cycling regimes. The mean CTMax values for 0000, 0600, 1200, 1800, and 2400 hr differed only slightly (Table 1), and ANOVA indicated no difference among the time-of-day effects on thermal tolerance ($P < 0.75$).

The mean CTMax values for the second series of constant and fluctuating acclimation regimes increased in the order 15, 20, 15 to 25 for 1, 2, and 6 cycles/24 hr and 25°C (Table 2). Again ANOVA indicated a high degree of significant difference within this group of CTMax values ($P < 0.001$), and DMRT identified the differences. There were no significant differences in mean CTMax ($P < 0.05$) between the 1 and 2 or the 2 and 6 cycles/24-hr temperature regimes, but all other possible combinations of mean CTMax values in this

TABLE 1

ACCLIMATION STATES OF JUVENILE STRIPED BASS, *Morone saxatilis* (DEC. 5, 1975), HELD IN CONSTANT AND FLUCTUATING REGIMES*

Time, hr	Individual	15°C	20°C	15–25°C cycle (+)	15–25°C cycle (–)	25°C	
0000	1	31.5	34.2	35.4	34.6	36.3	$\bar{X}$ = 34.6
	2	32.4	34.6	35.5	34.8	36.5	σ = 1.5
0600	3	32.2	34.5	34.6	35.0	34.8	$\bar{X}$ = 34.5
	4	32.4	34.7	34.9	35.3	36.7	σ = 1.3
1200	5	32.2	34.6	35.0	35.5	36.4	$\bar{X}$ = 34.9
	6	32.5	35.0	35.4	35.7	36.9	σ = 1.5
1800	7	32.3	34.8	35.2	35.1	37.0	$\bar{X}$ = 35.0
	8	32.6	35.2	35.8	35.3	37.2	σ = 1.6
2400	9	31.8	34.4	35.2	34.8	37.2	$\bar{X}$ = 35.0
	10	32.9	34.7	35.8	35.7	37.5	σ = 1.7
†	$\bar{X}$	32.3	34.7	35.3	35.2	36.7	
	σ^2	0.1600	0.0841	0.1444	0.1521	0.5776	
	σ	0.40	0.29	0.38	0.39	0.76	
	$\sigma_{\bar{X}}$	0.13	0.09	0.12	0.12	0.24	

*One fluctuation was with temperature rising (+) and the other was with temperature falling (–) between 0600 and 1800 hr. Fluctuations were for 24-hr periods, with a sinusoidal curve ±5°C around a 20°C base. Acclimation states were compared by determining critical thermal maximum (CTMax) expressed as degrees centigrade. CTMax heating rate was 1.5°C/min, and the end point was permanent loss of equilibrium.

†Duncan's new multiple range test identified differences in these mean CTMax values. Any two means not underscored by the same line are significantly different ($P < 0.05$).

series were different ($P < 0.05$). Test time did not affect CTM (ANOVA $P < 0.75$) (see Table 2).

The acclimation response of young-of-the-year striped bass in fluctuating temperatures fell between the mean (20°C) and maximum (25°C) temperatures of the fluctuation. There were no significant differences, however, between CTMax's obtained for fish tested when temperatures were rising and falling during the period of illumination. The CTMax values determined increased with the frequency of exposure to higher temperatures (Fig. 3).

DISCUSSION

The results of this study differ from those of several previous investigations. Brett (1944) determined that a warm water catfish

TABLE 2

ACCLIMATION STATES OF JUVENILE STRIPED BASS, *Morone saxatilis* (JAN. 16, 1975), HELD IN CONSTANT TEMPERATURES AND 1, 2, AND 6 CYCLES/24 hr TEMPERATURE FLUCTUATIONS*

Time, hr	Individual	15°C	20°C	15—25°C, 1 cycle/24 hr	15—25°C, 2 cycles/24 hr	15—25°C, 6 cycles/24 hr	25°C	
0000	1	31.0	30.9	34.4	34.4	34.9	36.7	$\overline{X}$ = 34.0
	2	31.7	32.0	34.8	34.6	35.1	37.0	σ = 2.08
0600	3	30.8	32.9	34.4		34.0	35.8	$\overline{X}$ = 34.0
	4	31.9	33.8	34.7	34.1	34.9	36.4	σ = 1.62
1200	5	31.4	32.4	33.6	34.2	35.2	35.8	$\overline{X}$ = 34.2
	6	31.6	34.7	34.3	35.0	35.4	37.1	σ = 1.72
1800	7	32.0	31.7	34.7	35.4	35.4	35.0	$\overline{X}$ = 34.4
	8	31.6	33.8	35.2	35.7	35.6	36.4	σ = 1.69
2400	9	31.8	34.3	33.9	33.8	34.8	36.7	$\overline{X}$ = 34.5
	10	32.1	34.4	35.0	35.2	35.0	37.3	σ = 1.59
†	$\overline{X}$	31.6	33.1	34.5	34.7	35.0	36.4	
	σ^2	0.1849	1.6899	0.2401	0.4186	0.1936	0.5041	
	σ	0.43	1.3	0.49	0.65	0.44	0.71	
	$\sigma_{\overline{X}}$	0.14	0.41	0.15	0.22	0.14	0.22	

*Fluctuations were sinusoidal curves ±5°C around a 20°C base. Acclimation states were compared by determining critical thermal maximum (CTMax) expressed as degrees centigrade. CTMax heating rate was 1.5°C/min, and the end point was permanent loss of equilibrium.

†Duncan's new multiple range test identified differences in these mean CTMax values. Any two means not underscored by the same line are significantly different ($P < 0.05$).

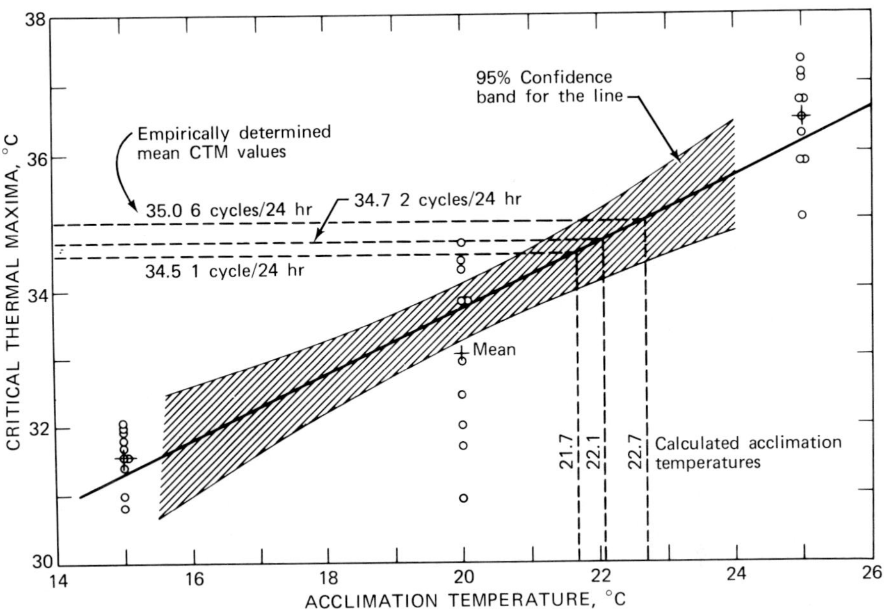

Fig. 3 Acclimation patterns of young-of-the-year striped bass, *Morone saxatilis*. A least squares regression of critical thermal maximum (CTMax) on constant acclimation temperatures ($\hat{Y}$ = 24 + 0.483X, r = 0.901, and $P < 0.001$). Acclimation temperatures were calculated for empirically determined mean CTMax values for bass held in fluctuating temperature regimes. Fluctuations were computer-controlled sinusoidal curves varying ±5°C around a 20°C base at a frequency of 1, 2, and 6 cycles/24 hr.

(Ameirus nebulosus) apparently acclimated to the highest temperature encountered rather than to an average of the environmental cycle. Fry, Hart, and Walker (1946) and Heath (1963) found a similar response for brook trout *(Salvelinus fontinalis)* and sea-run cutthroat *(Salmo clarki clarki)*, respectively. Hubbs (1964) also found similar results for young greenthroat darters *(Etheostoma lepidum)*. In contrast, striped bass in this study did not acclimate to the maximum temperature of the cycle. They did not exhibit greatest heat tolerance for the 24-hr thermoperiod, but tolerance increased as the number of temperature cycles per day increased (Fig. 3).

Other investigators also found exceptions to the tendency for temperature compensation in fishes to approximate the maximum temperature of the fluctuations. Hartwell and Huish (1977) did not find any significant effect of temperature cycles on the thermal shock resistance of either postlarval or juvenile spot, *Leiostomus xanthurus*. Since their intent was to simulate the effect of passage through a power-plant cooling system, latent mortality 96 hr after a 20-min exposure to test temperature was used as the measure of the effect of temperature acclimation to constant and fluctuating temperatures, and the maximum variation of the cycles tested was ±4°C. Hoss et al. (1975) reported that the heat resistance of postlarval pinfish was unaffected by acclimation to 8 to 12°C cycles but was increased by 7 to 13°C cycles. Otto (1974) also found that acclimation to cyclic regimes resulted in CTMax values consistently above the base temperature, but it was always less than the CTMax for fish acclimated to the constant temperature equivalent to the peak of the cycle. The generalization may be affected by the methods used. The critical thermal maximum or minimum or CTM (Cowles and Bogert, 1944; Lowe and Vance, 1955; Lowe and Heath, 1969; Hutchison, 1961) and resistance time to death (Fry, Hart, and Walker, 1946; Brett, 1952) are methods more commonly used to measure the effect of temperature compensation. The results of the differing methods used in these studies may not be directly comparable, however, and the extent of the cyclic temperature range may also be important.

Mechanisms have been proposed to explain increased acclimation in fluctuating regimes. Brett (1944) suggested a series of incremental gains in temperature tolerance during each rise in daily temperature which was not entirely lost when the waters cooled at night. If summation of acclimation responses to the high side of the temperature fluctuation occurs over a sufficient period of time, the eventual gain in heat tolerance should approximate the constant temperature corresponding to the maximum of the cycle. The acclimation to the cycling temperature regimes occurring for more than 2 weeks during both phases of this study may have been insufficient to produce a response equal to the maximum temperature of the cycle. This possibility suggests an unusually slow summation of responses, some other mechanism modifying or acting independently of summation responses, or an inherent "equilibrium" acclimation less than the maximum.

The simple summation mechanism is not adequate to explain all observations. Hutchison and Ferrance (1970) observed that frogs acclimate to the maximum of a daily temperature cycle after exposure to a single 24-hr (15 to 25°C) cycle. Ninety-six hours of acclimation to a constant 25°C was required to achieve a comparable response. The acclimation responses of invertebrates to various constant and fluctuating temperature regimes are considerably less clear than those for fishes (see Costlow and Bookhout, 1971; Thorp and Hoss, 1975).

I suspected a synergistic effect between temperature change and illumination since goldfish *(Carassius auratus)* maintained under constant temperature conditions in the laboratory show relatively greater resistance to heat during the summer and to cold during the winter (Hoar, 1955). Temperature compensation is not responsible for these seasonal alterations of temperature resistance. Alterations of photoperiod can produce some change in thermal resistance similar to the seasonal variations; the neuroendocrine system is assumed to be responsible for this compensation (Hoar, 1959). Therefore it would not have been surprising to find some interaction between temperature change and illumination. I did not observe a synergistic relationship between temperature change and illumination in my study since a rise or fall in temperature during the period of illumination had no apparent effect on either the rate or the extent of temperature acclimation over a 17-day period.

The results of my study make it apparent that predictions for normally fluctuating situations based on static temperatures may be underestimating the acclimation states of striped bass stocks. Such an error would cause an underestimation of either the temperature required to kill a given percentage of fish within a fixed time period or the time required to kill a given percentage at a fixed temperature. Prediction of such power-plant impacts as entrainment of small fish in a power-plant cooling system or cold stress of fishes induced by interrupting the flow of heated condenser cooling water during the cool months of the year rely on thermal tolerance data for which the acclimation state of the fish are known. In most cases the previous thermal histories of the fishes used to develop these data have been a series of constant temperatures. The use of resistance-time-to-death data based on constant acclimation temperatures will, of course, provide a more protective estimate of power-plant effects, but this level of protection may be greater than is justifiable by the biological data.

The effect of artificial temperature fluctuations on the acclimation state of fishes should be applied to the field situation with caution. Temperatures in a natural system are subject to diurnal and seasonal fluctuations, but mobile fish populations exhibit degrees of thermoregulation that may influence the variations of temperature to which they are exposed. Occasionally test fishes can tolerate brief exposures to temperatures approximating or exceeding the upper lethal limit (Lowe and Heath, 1969; Otto, 1974), and fish may actually cycle themselves behaviorally so that near-lethal temperatures are preferred for limited periods of the day (Lowe and Heath,

1969). Ferguson (1958) and Coutant (1977) reported good general agreement between temperature preference determined in the laboratory and in the field, but a wide species-specific temperature distribution range is also observed in the field. Field observations made by Coutant and Carroll (unpublished data) of striped bass tagged with ultrasonic temperature sensors indicated lower and upper avoidance temperatures of only 21 and 24°C despite nearby temperatures of 16 to 32°C. If temperatures in the general range of the preferred temperature are available, fish may be able to avoid the degree of variation tested in this study. The behavioral aspects of the species in the field and the responses to smaller fluctuations at base temperatures different from those tested here must be determined before any realistic predictions of acclimation state caused by general environmental temperature fluctuations can be confidently made.

ACKNOWLEDGMENTS

The research reported here was sponsored by the Division of Biomedical and Environmental Research, U. S. Department of Energy, under contract W-7405-eng-26 with Union Carbide Corporation. This is publication No. 1243, Environmental Sciences Division, Oak Ridge National Laboratory.

REFERENCES

Brett, J. R., 1944, *Some Lethal Temperature Relations of Algonquin Park Fishes*, University of Toronto Studies, Biological Series No. 52, Ontario Fisheries Research Laboratory Publication 63, pp. 1-49.

——, 1952, Temperature Tolerance in Young Pacific Salmon, Genus *Oncorhynchus, J. Fish. Res. Board Can.*, 9: 265-323.

Costlow, J. D., and C. G. Bookhout, 1971, The Effect of Cyclic Temperatures on Larval Development in the Mud Crab, *Rhithropanopeus harrisii*, in *Fourth European Marine Biology Symposium*, D. J. Crisp (Ed.), pp. 211-220, Cambridge University Press, New York.

Coutant, C. C., 1977, Compilation of Temperature Preference Data, *J. Fish. Res. Board Can.*, 34(5): 728-761.

Cowles, R. B., and C. M. Bogert, 1944, A Preliminary Study of the Thermal Requirements of Desert Reptiles, *Bull. Am. Mus. Nat. Hist.*, 83: 265-296.

Ferguson, R. G., 1958, The Preferred Temperature of Fish and Their Midsummer Distribution in Temperate Lakes and Streams, *J. Fish. Res. Board Can.*, 15(4): 607-624.

Fry, F. E. J., J. S. Hart, and K. F. Walker, 1946, *Lethal Temperature Relations for a Sample of Young Speckled Trout* (Salvelinus fontinalis), University of Toronto Studies, Biological Series No. 54, Ontario Fisheries Research Laboratory Publication 66, pp. 5-35.

Hartwell, S. I., and M. T. Huish, 1977, *The Effect of Cycling Acclimation Temperature on the Thermal Shock Resistance of Postlarval and Juvenile*

Spot, Leiostomus xanthurus, report to Carolina Power and Light Company, Raleigh, N. C.

Heath, W. G., 1963, Thermoperiodism in Sea-Run Cutthroat *(Salmo clarki clarki), Science,* 142: 486-488.

Hoar, W. S., 1955, Seasonal Variations in the Resistance of Goldfish to Temperature, *Trans. R. Soc. Can.,* 49: 25-34.

——, 1959, in *Comparative Endocrinology,* A. Gorbman (Ed.), John Wiley & Sons, Inc., New York.

Hoss, D. E., L. C. Coston, J. P. Baptist, and D. W. Engle, 1975, in *Environmental Effects of Cooling Systems at Nuclear Power Plants,* Symposium Proceedings, Oslo, 1974, pp. 519-527, STI/PUB/378, International Atomic Energy Agency, Vienna.

Hubbs, C., 1964, Effects of Thermal Fluctuations on the Relative Survival of Greenthroat Darter Young from Stenothermal and Eurythermal Waters, *Ecology,* 45(2): 376-379.

Hutchison, V. H., 1961, Critical Thermal Maxima in Salamanders, *Physiol. Zool.,* 34: 92-125.

——, and M. R. Ferrance, 1970, Thermal Tolerances of *Rana pipiens* Acclimated to Daily Temperature Cycles, *Herpetologica,* 26(1): 1-8.

Lowe, C. H., Jr., and V. J. Vance, 1955, Acclimation of the Critical Thermal Maximum of the Reptile *Urosaurus ornatus, Science,* 122: 73-74.

——, and W. G. Heath, 1969, Behavior and Physiological Responses to Temperature in the Desert Pupfish *(Cyprinodon macularius), Physiol. Zool.,* 42: 53-59.

Otto, R. G., 1974, The Effects of Acclimation to Cyclic Thermal Regimes on Heat Tolerance of the Western Mosquitofish, *Trans. Am. Fish. Soc.,* 103: 331-335.

Thorp, J. H., and D. E. Hoss, 1975, Effects of Salinity and Cyclic Temperature on Survival of Two Sympatric Species of Grass Shrimp *(Palaemonetes),* and Their Relationship to Natural Distributions, *J. Exp. Mar. Biol. Ecol.,* 18: 19-28.

EFFECTS OF ACUTE AND CHRONIC THERMAL EXPOSURES ON THE EGGS OF THREE HUDSON RIVER ANADROMOUS FISHES

ROBERT L. KELLOGG, JAMES J. SALERNO, and DAVID L. LATIMER
Ecological Analysts, Incorporated, Middletown, New York

ABSTRACT

Atlantic tomcod, alewife, and striped bass eggs were tested for chronic and acute tolerance to elevated temperatures to generate information on potential thermal effects of power-plant operation. Upper lethal temperatures (TL_{50}) for normal hatching success were 26.5°C for alewife eggs, 26.3°C for striped bass eggs, and 6.6°C for Atlantic tomcod eggs. Thermal tolerance of eggs exposed to elevated temperatures for 5 to 60 min increased with advancing stage of embryonic development and decreased with lengthening exposure duration. The applications of the data to the assessment of the impact of power plants are discussed.

One element of the assessment of the environmental impact of steam electric power plants using once-through cooling is the evaluation of the effects of temperature elevations on the embryonic development of fish species. Developing fish eggs are exposed to elevated temperatures in power-plant cooling waters for variable periods of time. Exposures range from a few seconds or minutes, as in the case of eggs entrained into the cooling water at the point of intake or discharge, to several hours or days, as in the case of demersal eggs spawned in the discharge area.

A number of past investigations have considered the effects of temperature elevations on fish embryos. Schubel and Koo (1976), Schubel (1974), and Schubel and Auld (1974) examined the effects of a variety of time—excess temperature histories on the eggs of four estuarine fish species. Typical experimental thermal exposures consisted of 5- to 15-min periods of constant elevated temperatures followed by varying rates of cooling to within 1°C of ambient.

Schubel and Koo (1976) found, for example, that striped bass embryos 24 to 36 hr old could tolerate a 15°C temperature elevation (ΔT) over ambient (19—20°C) for 5 to 15 min followed by varying cooling rates. However, exposure to a 20°C ΔT for 5 min followed by a 90-min cooling period reduced hatching success by 32%, and the same ΔT applied for 10 min followed by a longer cooling period totally eliminated any hatch. Such studies indicate that certain thermal exposures, comparable to those potentially experienced by embryos entrained into once-through cooling systems, can reduce hatching success.

Other investigations have considered the effects of long-term temperature elevations encompassing much or all the embryonic period from fertilization to hatch. Edsall (1970), for example, reported that a temperature of about 26.7°C (80°F) was the upper limit for successful incubation of alewife eggs. Shannon (1969) found no significant difference in the hatching success of striped bass eggs over the temperature range 15.5 to 24.0°C. Such studies as these indicate that certain elevated temperature conditions, comparable to those potentially found in thermal-discharge areas, can reduce hatching success.

In the process of assessing the thermal impact of Hudson River power plants, the need arose for data concerning the effects of elevated temperature on fish eggs. The Hudson River is used extensively for spawning by a number of anadromous fish species during the winter, spring, and early summer months. Some spawning does occur in the general vicinity of the major thermal discharges on the river. Consequently in 1976 a 2-year laboratory study was initiated to determine the effects of both long (chronic) and short (acute) exposures of elevated temperature on the hatching success of the eggs of three major anadromous fishes—striped bass *(Morone saxatilis)*, alewife *(Alosa pseudoharengus)*, and Atlantic tomcod *(Microgadus tomcod)*. The results of this study and the applications of the hatching success data to thermal-impact assessment are presented in this paper.

METHODS AND MATERIALS

The effects of elevated temperature on fish eggs were investigated in a laboratory located on the banks of the Hudson River near Newburgh, N. Y. (river mile 65). The laboratory water supply was river water clarified by flocculation, settling, and filtration. Experimental apparatus included eight flow-through water baths for the incubation of eggs and six static baths for short-duration thermal-tolerance tests. Temperatures in the incubation baths were maintained within about 1°C by thermostatic mixing valves. Temperature in each bath was monitored hourly with thermistor probes and recorded on a computerized data-aquisition system. Thermal-tolerance test-bath temperatures were maintained within 0.1°C by thermostatic immersion heaters.

The source of eggs for study varied by species. Atlantic tomcod and alewife adults were collected from the Hudson River and its tributaries and artificially spawned by the dry fertilization method at ambient temperatures. Striped bass eggs were obtained immediately after fertilization from the Texas Instruments hatchery located at nearby Verplanck, N. Y. These eggs were rushed to the laboratory under controlled temperature conditions and aeration.

The effects of long-term (chronic) temperature elevations on fish eggs were examined by incubating fertilized eggs at ambient and a number of above-ambient temperatures and measuring the hatching success corresponding to each treatment. Eggs fertilized by the dry method were transferred to ambient-temperature baths for a 2- to 12-hr period during which it became possible to distinguish fertilized eggs from those not fertilized. Groups of 100 fertilized eggs from each of one to four replicate adult females were placed in screened-bottom incubation containers in the ambient-temperature baths. Five to seven containers from each replicate female were then assigned to each of the eight experimental incubation baths. The percentage of total hatch and percentage of normal hatch (free of visible deformities) were computed for each replicate female at each temperature. The "optimum hatching range" (as defined by Hokanson, McCormick, and Jones, 1973) was determined as the range of temperatures tested over which the percentage of normal hatch was not significantly different from the highest value recorded [$P = 0.05$, Tukey's multiple range test (Steele and Torrie, 1960)]. Also computed was the upper lethal temperature (TL_{50}) determined as the interpolated temperature at which the percentage of normal hatch is 50% of the highest value recorded.

The effects of short-term (acute) temperature elevations on fish eggs were examined by exposing groups of eggs of several different developmental stages to elevated temperatures for short periods of time and subsequently measuring hatching success. Fertilized eggs were held at ambient temperature before testing. Samples of 50 live eggs were assigned to screened-bottom incubation containers, which were transferred from ambient to several above-ambient temperatures for periods of 5 and 30 min. A control sample accompanied each test and was maintained in an ambient-temperature test bath during the exposure period. Following exposure, all samples were returned to ambient temperature and observed through hatch. The observed percentage of normal hatch for each sample was then computed and corrected for control mortality by the following equation:

$$\text{Corrected hatch, \%} = \frac{\text{observed hatch, \%}}{\text{control hatch, \%}} \times 100\%$$

The corrected hatch percentages were then plotted vs. test temperature, and the TL_{50} was determined by linear regression as the temperature corresponding to 50% corrected hatch.

RESULTS

Chronic Effects

The results of the long-term hatching-success experiments are shown in Table 1. The percentages of total and normal hatch corresponding to each test temperature are shown for each of the three anadromous species examined.

For alewife acclimated to 13 to 14°C, results obtained with eggs from four replicate females were combined. Differences in hatching success between replicate female alewife occurred, but the relative response to temperature between females was similar. The highest percentage of total and normal hatch occurred at 20.8°C (Table 1). Test temperatures from 12.7°C, the lowest test temperature, to 23.9°C were within the "optimum hatching range." The TL_{50} for normal hatch was computed as 26.5°C. The difference between the percentage of total and normal hatch increased at the higher test temperatures, with the greatest deformed hatch for alewife eggs occurring at 26.7°C.

For striped bass acclimated to 18°C, results for eggs from a single female indicated that the highest percentage of total and normal hatch occurred at 22.2°C (Table 1). At temperatures from 18.8 to 25.8°C, normal hatching success remained at levels at least 70% of the highest value recorded. In this case, however, no optimum range was defined since at test temperatures above and below 22.2°C normal hatching success was significantly less than the highest value recorded ($P = 0.05$). The TL_{50} for normal hatch was 26.3°C. At the two highest test temperatures, 27.2 and 29.7°C, the normal hatch percentages were both zero.

TABLE 1

SUMMARY OF RESULTS OF LONG-EXPOSURE HATCHING-SUCCESS EXPERIMENTS

Species	Number of replicate females	Ambient temp., °C	Incubation delay, hr*	Incubation temp., °C ($\bar{x} \pm$ SD)	Mean percentage total hatch	Mean percentage normal† hatch	Time to median hatch,‡ hr
Alewife	4	13–14	8–12	12.7 ± 0.5	68	64	178
				18.1 ± 0.2	68	60	109
				20.8 ± 0.3	77	71	100
				23.9 ± 0.1	72	60	72
				26.7 ± 0.2	51	34	71
				29.7 ± 0.0	0	0	
Striped bass	1	18	8	18.8 ± 0.5	59	59	42
				20.0 ± 0.2	53	53	42
				22.2 ± 0.1	63	63	32
				23.7 ± 0.3	46	46	28
				25.8 ± 0.1	47	47	24
				27.2 ± 0.2	13	0	20
				29.7 ± 0.2	0	0	
Atlantic tomcod	3	2	2–7	2.0 ± 0.5	25	23	39
				3.4 ± 0.4	24	23	33
				4.9 ± 0.5	22	20	28
				7.3 ± 0.4	14	8	24
				8.8 ± 0.4	0	0	

*Duration of period from time of fertilization until the initiation of test incubation.
†Free of visible deformities.
‡Atlantic tomcod times to median hatch reported in days.

For Atlantic tomcod embryos acclimated to 2°C, the optimum hatching range extended from 2.0 to 4.9°C (Table 1). The TL_{50} for normal hatch was 6.6°C. As with alewife, differences in percent hatch between eggs of the three replicate females occurred, but the relative response to temperature between females was similar.

Acute Effects

The results of the short-exposure hatching-success experiments are shown in Table 2. The TL_{50} values for normal hatch at one or more exposure durations are shown for several developmental stages of each of the three anadromous species studied.

For alewife embryos acclimated to 12 to 13°C, short-exposure TL_{50} values for normal hatch ranged from 26.0 to 37.0°C, depending on developmental stage and exposure duration (Table 2). The TL_{50} values for the blastula stage were the lowest, and, as developmental stage advanced, the corresponding TL_{50} values increased. The TL_{50} values for tail-bud and tail-free embryos exceeded those for blastula by 6.0 to 7.9°C. In all cases, TL_{50} values for 30-min exposures were about 3 to 4°C less than those for 5-min exposures. The TL_{50} values varied up to 2.4°C among samples from replicate females; however, no consistent trends were visible among replicates (i.e., there were no consistent differences in thermal tolerance due to female stock).

For striped bass embryos acclimated to 16 to 19°C, short exposure TL_{50} values for normal hatch generally increased with advancing developmental stage and decreased with lengthening exposure duration, as was the case with alewife. Contrary to the general trend, however, striped bass embryos in the late embryo stage were slightly less tolerant of short-term temperature elevations than were embryos in the next younger stage—tail-free embryo. Differences in TL_{50} values as great as 4.5°C with developmental stage were observed. The TL_{50} values for 30-min exposures were up to 5.5°C less than corresponding TL_{50} values for 5-min exposures. The overall range of TL_{50} values observed was 31.5 to 38.5°C.

Short-exposure TL_{50} values for normal hatch for Atlantic tomcod embryos acclimated to 2 to 3°C varied from 19.8 to 31.2°C, depending again on developmental stage and exposure duration. The trends in TL_{50} in relation to these two variables were similar to those for alewife and striped bass. In contrast to the other two species tested, TL_{50} values were available for only relatively advanced embryonic stages of Atlantic tomcod.

TABLE 2

SUMMARY OF RESULTS OF SHORT-EXPOSURE HATCHING-SUCCESS EXPERIMENTS

Species	Developmental stage	Age, hr	Acclimation temp., °C	No. of replicate females	5-min mean TL_{50}, °C	30-min mean TL_{50}, °C	60-min mean TL_{50}, °C
Alewife	Blastula	24	12–13	2	30.3	26.0	
	Tail-bud embryo	76	12–13	3	36.3	32.4	
	Tail-free embryo	105–123	12–13	4*	37.0	33.9	
Striped bass	Blastula	13	16.0	1	33.0		
	Gastrula	11	18.5	1	37.0	31.5	
	Gastrula	16	16.0	1		31.9	
	Tail-bud embryo	28	16.5	1	38.0	34.0	
	Tail-free embryo	32	19.0	1	38.5	35.2	
	Late embryo	49	17.0	1	37.1	34.9	
Atlantic tomcod	Tail-bud embryo	15 days	2–3	1†	26.8‡	23.2	19.8
	Late embryo	30 days	2–3	1†	31.2‡	27.9	24.2

*Three replicates only for 30-min exposure case.
†Eggs from several females were mixed before assignment to incubation containers.
‡Ten-minute TL_{50}.

DISCUSSION

The hatching success data from this study have been used to calculate the maximum temperatures to which eggs can be exposed for given lengths of time without sustaining any measurable reductions in hatching success. Such temperatures, termed "safe levels" or "upper thermal protection limits," were determined, following the convention of the U. S. Environmental Protection Agency (1976) and Mihursky (1969), by subtracting 2.2°C from computed TL_{50} values. The computed safe levels are presented in Table 3. In most cases the 2.2°C safety factor provided estimates of safe levels which are conservative in terms of biological protection.

The chronic-exposure safe levels for alewife and striped bass eggs coincided at 24°C, despite differences in the acclimation temperature ranges for the two species. This safe level exceeds ambient Hudson River temperature during the alewife and striped bass spring spawning seasons by up to 11 and 10°C, respectively (Table 3). In contrast, the 4.4°C safe level estimated for Atlantic tomcod eggs exceeds ambient during the species' midwinter spawning season by only 2.4 to 4.4°C.

Acute-exposure safe levels were determined for 5-min (10-min in the case of Atlantic tomcod) and 30-min exposures for each of the three species tested (Table 3). The estimated safe levels increased with advancing developmental stage and decreased with lengthening exposure duration, as noted above for the TL_{50} data from which the safe levels were derived. The acute safe levels for alewife and striped bass eggs were similar (within 3°C) except in one case, the alewife blastula stage, where the 30-min safe level was 6°C lower than that for striped bass—23.8 vs. 29.7°C (Table 3). Blastula-stage alewife appeared particularly sensitive to short exposures to elevated temperature.

The acute safe levels for Atlantic tomcod eggs were only 3.5 to 8.7°C less than corresponding safe levels for alewife and striped bass (Table 3). This is in contrast to the chronic case, where the Atlantic tomcod safe level was 20°C less than that estimated for the other two species—4.4 vs. 24°C. Atlantic tomcod eggs appear quite tolerant of short exposures to elevated temperatures, whereas they appear very intolerant of long exposures.

Since it was generally observed that safe levels decreased with lengthening exposure duration, the coincidence of the alewife chronic safe level and the alewife blastula-stage 30-min acute safe level, both 24°C, was a cause for question. The source of this coincidence appears to have been an artifact of the experimental design; i.e., the

TABLE 3

COMPUTED SAFE LEVELS (TL_{50} − 2.2°C) OF CHRONIC AND ACUTE EXPOSURES OF FISH EMBRYOS TO ELEVATED TEMPERATURES

Species	Spawning period	Peak spawning temp. range, °C*	Developmental stages	Safe levels (TL_{50}−2.2°C)		
				Chronic	30-min	5-min
Alewife	April—May	13—16	Blastula	24.3	23.8	28.1
			Tail-bud to tail-free embryo		30.2	34.8
Striped bass	May—June	14—18	Blastula to gastrula	24.1	29.7	30.8
			Tail-bud to late embryo		31.8	34.9
Atlantic tomcod	December—February	0—2	Tail-bud embryo	4.4	21.0	24.6
			Eyed-up embryo		25.7	29.0

*Texas Instruments, Inc., 1976.

8- to 12-hr delay following fertilization in the start of the alewife chronic test incubations may have prevented the most thermally sensitive developmental stages from being exposed to elevated temperature. Test incubations should have been initiated as soon after fertilization as possible to keep this condition from occurring. The chronic safe level for alewife eggs reported in Table 3 therefore probably overestimates a safe level appropriate for the entire embryonic period. The same may be true for striped bass eggs, although the delay was only 8 hr in this case. Considering the length of the embryonic period of Atlantic tomcod, up to 39 days at ambient temperatures, the 2-hr delay in the start of the test incubations for this species most likely did not bias the estimated chronic safe level.

Applications

Acute- and chronic-exposure hatching-success data can be used to assess the thermal impact of power plants on fish eggs.

Chronic-exposure safe levels can be used to identify maximum thermal-plume temperatures compatible with successful embryonic development. With knowledge of the spatial characteristics of a thermal plume, such as the surface area enclosed by each plume isotherm, the area of potential habitat in which temperatures exceed the safe level can be determined. Within that area, successful embryonic development is excluded. The assessment of the impact of that exclusion area is completed by considering its biological significance. This may be done by examining a species' spawning habits, spawning locations, egg characteristics, and distribution of eggs in the water body, to determine the relative importance of the excluded area to the reproductive success of the species in the water body.

The chronic-exposure safe levels computed in this study have been used to evaluate the potential impact of the Hudson River thermal discharges. For example, the 4.4°C safe level estimated for Atlantic tomcod (Table 3) indicates that plume temperatures only a few degrees above ambient (0 to 2°C) during the spawning season would exclude successful embryonic development. As a result, the potential for impact within a thermal plume is relatively high. In contrast, the computed safe levels for alewife and striped bass, both about 24°C, considerably exceed river ambient temperatures during the two species' respective spring spawning seasons. The potential for impact is very low in both cases.

Acute safe levels can be used to assess the impact of thermal exposures experienced by fish eggs entrained into power-plant cooling waters at the point of intake or discharge. Since the effects of the acute exposure on fish eggs depend on both the exposure duration and temperature, as noted by this study and by Lauer et al. (1974) and Frank (1974), both these factors must be considered in the acute impact analysis. Once the time—temperature characteristics of a given entrainment exposure are specified, the appropriate acute safe level can be used to determine whether or not a reduction in hatching success would be expected as a result of the thermal exposure.

This approach has been used in the analysis of the thermal impact of Hudson River power plants. The 30-min safe level is applicable to plants with surface outfalls, where the relatively slow rate of temperature decay in the thermal plume may prolong the period of exposure to elevated temperatures. On the other hand, plants with high-velocity diffuser discharges have comparatively short exposure periods, and consequently the 5-min safe level is often applicable. For example, in the former case, blastula-stage alewife eggs, with a 30-min safe level of 23.8°C (Table 3), would be capable of tolerating maximum entrainment temperature elevations of 8 to 11°C, based on ambient temperatures during the spawning season (13 to 16°C). In the latter case, with a 5-min safe level of 28.1°C, the maximum tolerable ΔT values would range from 12 to 15°C. At temperature elevations above these levels, reductions in hatching success would be expected.

The assessment of acute thermal impact requires calculation of the total losses of eggs during a spawning season due to entrainment. Total losses can be determined after computing (1) the total time during the spawning season when discharge temperatures occur in excess of the appropriate safe levels and (2) the average density of eggs passing through the cooling system during those times of excessive temperatures. The biological significance of total losses of eggs must then be determined, one possible approach being the conversion of egg losses to equivalent fish losses.

In conclusion, hatching-success data obtained from acute and chronic thermal-exposure experiments can be used in the assessment of the impact of once-through cooling on fish eggs. When combined with thermal-effects data for other life stages, the impact of a power plant on a given species can be assessed.

ACKNOWLEDGMENTS

Funding for these studies was made available by Central Hudson Gas & Electric Corporation, Consolidated Edison Company of New

York, Niagara Mohawk Power Corporation, and Orange and Rockland Utilities, Inc. Ray van Hoesen, Blair Shepherd, Barry Burgan, and Karen Zseleczky provided valuable assistance in the laboratory. We are particularly indebted to Johnson C. S. Wang for instruction on spawning and incubating techniques and James J. Gift for final review comments.

REFERENCES

Edsall, T. A., 1970, The Effect of Temperature on the Rate of Development and Survival on Alewife Eggs and Larvae, *Trans. Amer. Fish. Soc.*, 99: 376-386.

Frank, M. L., 1974, Relative Sensitivity of Different Developmental Stages of Carp Eggs to Thermal Shock, in *Thermal Ecology*, AEC Symposium Series, Augusta, Ga., May 3—5, 1973, J. W. Gibbons and R. R. Sharitz (Eds.), pp. 171-176, CONF-730505, NTIS.

Hokanson, K. E. F., J. H. McCormick, and B. R. Jones, 1973, Temperature Requirements for Embryos and Larvae of the Northern Pike, *Esox lucius* (Linnaeus), *Trans. Amer. Fish. Soc.*, 102: 89-100.

Lauer, G. J., et al., 1974, Entrainment Studies on Hudson River Organisms, in *Proceedings of the Second Entrainment and Intake Screening Workshop, Baltimore, Maryland, Feb. 5—9, 1973*, L. D. Jensen (Ed.), pp. 37-82, Johns Hopkins University Cooling Water Research Project Report #15.

Mihursky, J. A., 1969, *Patuxent Thermal Studies*, Natural Resources Institute Special Report #1, University of Maryland, College Park.

Schubel, J. R., 1974, Effects of Exposure to Time—Excess Temperature Histories Typically Experienced at Power Plants on the Hatching Success of Fish Eggs, *Estuarine Coastal Mar. Sci.*, 2: 105-116.

——, and A. H. Auld, 1974, Hatching Success of Blueback Herring and Striped Bass Eggs with Various Time vs. Temperature Histories, in *Thermal Ecology*, AEC Symposium Series, Augusta, Ga., May 3—5, 1973, J. W. Gibbons and R. R. Sharitz (Eds.), pp. 164-170, CONF-740505, NTIS.

——, and T. S. Y. Koo, 1976, Effects of Various Time—Excess Temperature Histories on Hatching Success of Blueback Herring, American Shad, and Striped Bass Eggs, in *Thermal Ecology II*, ERDA Symposium Series, Augusta, Ga., April 2—5, 1975, Gerald W. Esch and Robert W. McFarlane (Eds.), pp. 165-170, CONF-750425, NTIS.

Shannon, E. H., 1969, Effect of Temperature Changes upon Developing Striped Bass Eggs and Fry, *Proc. Annu. Conf. Southeast. Assoc. Game Fish Comm.*, 23: 265-274.

Steele, R. G. E., and J. H. Torrie, 1960, *Principles and Procedures of Statistics*, McGraw-Hill Book Co., New York.

Texas Instruments, Inc., 1976, *Fisheries Survey of the Hudson River, March—December 1973*, Vol. IV, Prepared for Consolidated Edison Company of New York, Inc.

U. S. Environmental Protection Agency, 1976, Quality Criteria for Water, Report EPA 440/9-76-0023, Washington, D. C.

MULTIPLE-FACTOR AND SYNERGISTIC STRESSES IN AQUATIC SYSTEMS

F. JOHN VERNBERG
Belle W. Baruch Institute for Marine Biology and Coastal Research,
University of South Carolina, Columbia, South Carolina

ABSTRACT

Most studies on the adaptation of aquatic systems to environmental fluctuations have dealt with the responses of an organism to variation in a single factor, while other factors are kept constant and usually at optimal levels. This is a logical approach and has resulted in a good understanding of response mechanisms of the whole organism, tissues, and cellular components. However, organisms typically are exposed to a complex of factors, with two or more factors varying at the same time. The question is, "How does the complex machinery of either an organism or a multiorganismic system deal with complex environmental fluctuations?" To understand the ability of a biological system to survive in its ecological setting, we must direct our attention to an analysis of multiple factor effects and interactions both under laboratory conditions and in the field. To be ecologically significant, experiments must be designed to simulate conditions of the system's "real world." Unfortunately this is difficult to achieve because, for example, (1) detailed analyses of the microenvironmental conditions experienced by organisms or systems are, with few exceptions, lacking, and (2) simultaneous variation in several distinct factors is both technically difficult and expensive to achieve in the laboratory. Field studies are difficult because of the complexity of biological systems, the lack of undisturbed areas for long-term study, and the need for a multidisciplinary team of researchers. Fortunately some advances have been made with the help of the sophisticated statistical and computer techniques required to analyze the resultant complex data.

Understanding the nature of the functional responses of organisms to environmental changes is one of the challenges facing every new generation of biologists. Aristotle wrote about the adaptive responses of intertidal animals. Pliny the Elder was well schooled in the response physiology of many species. He reported that the simplest way to capture the exotic and wily unicorn was to station a virgin maiden in a field. The unicorn would approach, and, when it was lulled to sleep with its head in her lap, the hunters silently crept up and carted away the helpless beast. Davenport (1897–1899), in his classic two-volume work *Experimental Morphology*, summarized the earlier literature dealing with the influence of environmental factors on the form and function of biological material. He raised many questions concerning physiological adaptation which are still unanswered today, e.g., the influence of multiple environmental factors on organisms. Since the beginning of this century and especially since the end of World War II, the volume of data dealing with the effects of environmental parameters on organisms has been ever increasing. Historically, many studies were restricted to investigating the influence of a single factor on some organismic process. This approach was logical, as it was simpler to vary only one factor at a time rather than many simultaneously. Two problems that discouraged multiple-factor studies were: (1) technological limitations of equipment to control and selectively vary a number of factors and (2) the underdeveloped state of the statistical and computer methodology necessary to handle complex data. Although they have not been completely resolved, these problems are less restrictive at present.

During the past 25 years, laboratory-oriented experimentalists were not the only ones concerned with the influence of the environment on organisms; field-oriented scientists also stressed the probable importance of ecological interactions among organisms. From this interfacing, the field of physiological ecology developed. It is apparent that laboratory and field investigations must be combined if we are to understand organismic–environmental interactions. Many environmental questions are now decided in the courts, with lawyers asking about the relevance of laboratory findings to what is happening in the "real" world. To emphasize that the organism lives in an environment consisting of many factors, each of which must be considered, Billings (1952) reviewed the relationship between the environmental complex and plants. He termed the view that an organism exists in relationship to its entire environment the holocoenotic approach to ecology. Figure 1 is a graphic representation of the interactive relationship between an aquatic animal and its environment. The organism's ambient, or external, environment can be divided into two major subdivisions, abiotic and biotic. Each of these subdivisions includes a number of discrete factors, some of which are indicated in Fig. 1. Obviously the organism is exposed to many factors, each acting independently or in concert with others. Coupling of abiotic and biotic factors can occur as readily as

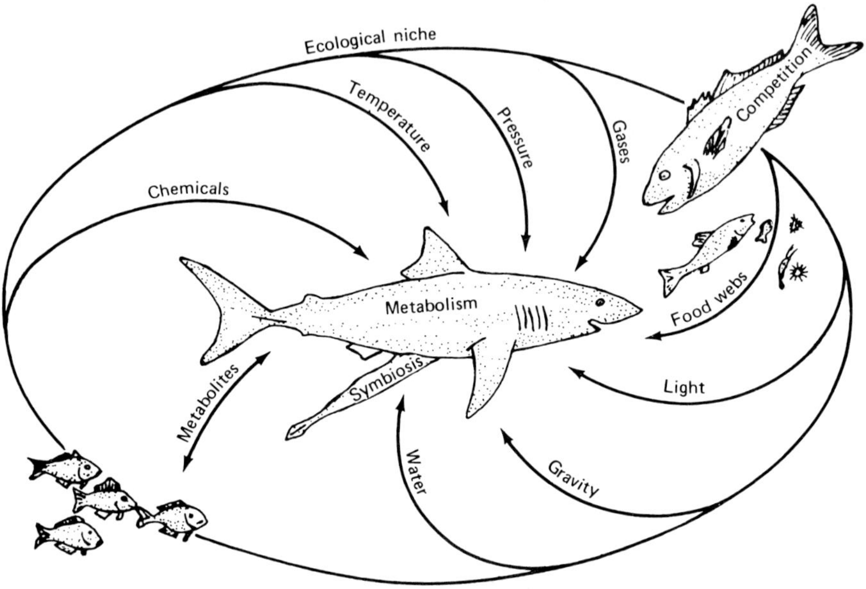

Fig. 1 Interactions of an organism and its environment. (From F. J. Vernberg and W. B. Vernberg, *The Animal and the Environment*, page 2, Holt, Rinehart, and Winston, Inc., New York, 1970.)

coupling of factors within the same subdivision. For example, temperature and predation may be interactive in that a reduced temperature may inhibit feeding in a species, which in turn would reduce the amount of predation on the prey species.

In addition to the external environment, the internal environment (internal milieu) of the organism is also a complex of many factors. In some organisms well-developed homeostatic mechanisms enable the internal environment to be relatively stable and independent of fluctuations in the external environment. For example, some aquatic organisms are excellent osmoregulators and maintain a constant osmotic concentration of their body fluids even when the ambient salinity varies. In contrast, the internal milieu of other organisms changes in direct response to changes in the external milieu. Between these extremes of highly complex regulators and perfect conformers is an array of aquatic species that provide varying degrees of functional regulation. One species, for example, may show different levels of regulation at different stages in its life cycle, at different seasons of the year, or in different parts of its biogeographical range.

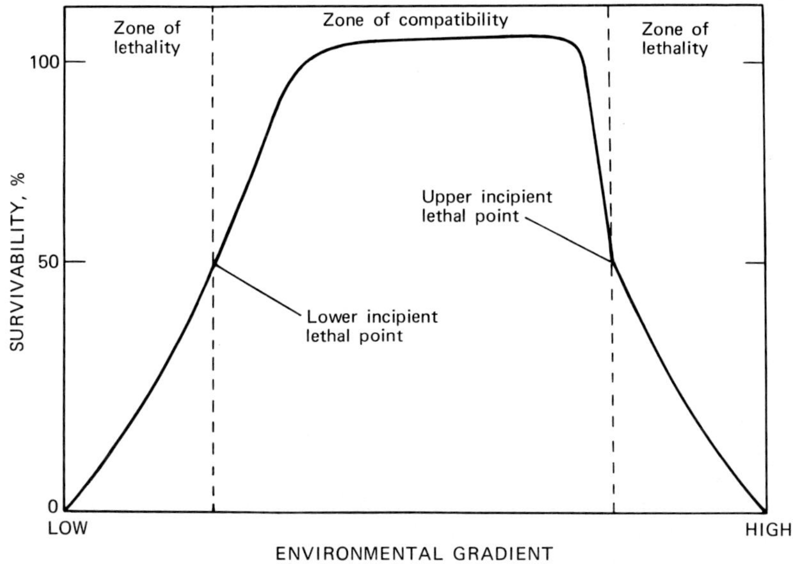

Fig. 2 The zones of lethality and compatibility. (From F. J. Vernberg and W. B. Vernberg, *The Animal and the Environment*, **page 8, Holt, Rinehart, and Winston, Inc., New York, 1970.)**

For any one organism, a balance of external and internal environments must be achieved which permits the functioning of the diverse cellular components required for the intact organism to survive and be successful. When extremes in an environmental factor (or factors) are reached, an organism may cease to function as a viable, integrated entity. Between these extremes, however, it can survive and function at various degrees of effectiveness. With this concept, the zone of lethality (resistance adaptation, zone of resistance) is that portion of the gradient representing the extremes that cause the death of the organism; the zone of compatibility (the biokinetic zone, zone of tolerance) is the central portion of the gradient where life can be sustained (Fig. 2).

Faced with this enormous complexity of the external and internal environments, how can we with our limited knowledge determine which factors are important and initially worthy of study? Mason and Langenbein (1957) suggested the following criteria be applied in assessing the ecological importance of an external factor:

1. It must be operationally significant to an organism.
2. It must be directly effective at some time during the life of the organism.

3. It must be effective as to sequence, as ordered by the ontogeny (development) of the organism.

Just as a decision must be made as to which external factors should be studied, a decision also must be reached as to which internal functions should be investigated. Obviously this decision is based on the nature of the question to be answered. As with external factors, numerous studies have been published dealing with various internal factors, including resistance and capacity adaptations, which have ranged from overt responses of the intact organism to subcellular changes. In addition, the responses of populations and communities to environmental fluctuation have been analyzed.

The purpose of this paper is to review recent research efforts dealing with the influence of multiple factors on organisms in aquatic systems. Since Alderdice (1972), Salisbury (1975), and Vernberg (1975) have reviewed and discussed many papers dealing with this topic, this review will not be comprehensive; rather, it will emphasize recent literature that illustrates some distinct research areas.

Inherent in most scientific disciplines is the problem of definition of terms, and physiological ecology is no exception. Synergism has been variously defined. Only two examples will be cited here. *Gould's Medical Dictionary* defines it as, "the cooperative action of two or more agents or organs." The definition in *Webster's Collegiate Dictionary* is more comprehensive, stating, "synergism is the cooperative action of discrete agencies, such that the total effect is greater than the sum of the two effects taken independently." This latter definition appears to be more widely used. Synergism is not always easy to demonstrate, however, and there are cases where the degree of synergism quantitatively is near zero and the effects of two agencies (environmental factors) appear to be additive and not significantly greater than the sum of the two agencies acting independently. A multiple factor effect is simply the expression of the influence of two or more factors on some response of an organism. The data from this type of study can be statistically analyzed to determine the nature and significance of interactions (see review of Alderdice, 1972). This paper does not deal with the theoretical basis for the mechanism of interaction, but papers dealing with this topic are cited (Lockhart, 1965; Mohr, 1972).

LETHAL LIMITS

One of the classic papers demonstrating the effects of three factors (salinity, temperature, and dissolved oxygen concentration)

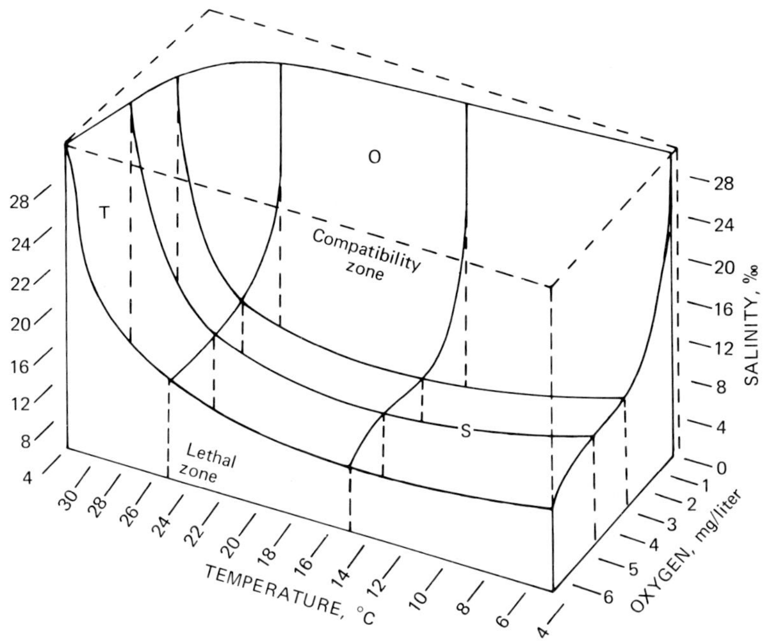

Fig. 3 Boundaries of lethal conditions for the American lobster, *Homarus americanus*. Regions are shown where temperature (T), salinity (S), and dissolved oxygen (O), respectively, act alone as lethal factors. (After McLeese, 1956.)

on the survival of an adult aquatic animal, the American lobster, is that of McLeese (1956) (Fig. 3). A sublethal but stressful exposure to one factor may become lethal when an animal is exposed concurrently to a second sublethal but stressful factor. The net result is a reduction in the size of the zone of compatibility. For example, when salinity is optimal, the lobster can survive at a higher temperature than it can when exposed to low salinity and elevated temperature. Larval stages are also sensitive to multiple factor exposure, as seen in the work of Vernberg, DeCoursey, and O'Hara (1974) (Fig. 4). In this case the multiple effects of temperature, salinity, and cadmium also reduced the size of the compatibility zone for larval fiddler crabs.

Laboratory determination of the response to temperature and salinity does not always correlate with the occurrence of that species in the field. For example, Bradley (1975) reported that the thermal tolerance of the copepod *Eurytemora affinis* determined at various salinities with or without acclimation varied with season. He also

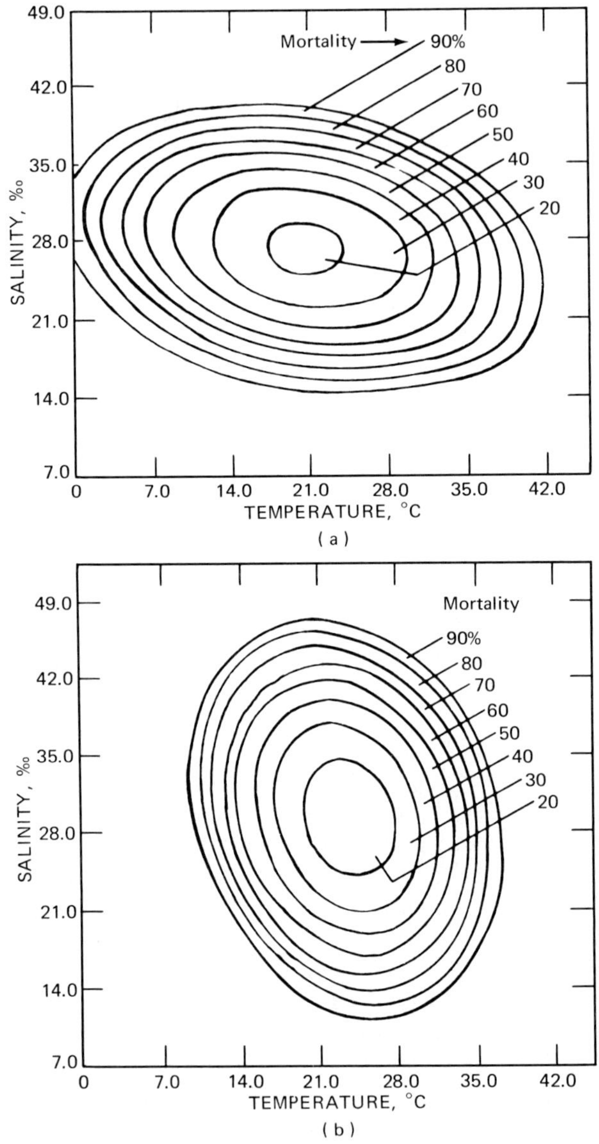

Fig. 4 Estimation of percentage mortality of first-stage *Uca pugilator* zoea based on response surface fitted to observed mortality under 13 combinations of salinity and temperature (a) with and (b) without the addition of 1 ppb cadmium. (From W. B. Vernberg, P. J. DeCoursey, and J. O. O'Hara, Multiple Environment Factor Effects on Physiology and Behavior of the Fiddler Crab, *Uca pugilator*, in *Pollution and Physiology and Marine Organisms*, pp. 381-425, Academic Press, Inc., New York, 1974.

reported that the tolerance level and differences between populations greatly increased with salinity. The seasonal distribution of this copepod, however, is contrary to what would be predicted on the basis of his data. Bradley suggested that other copepod species have a competitive advantage during the summer and fall and thereby restrict populations of *E. affinis*. In contrast, after determining the salinity and temperature tolerances and the salinity preferences of the meiofaunal species *Derocheilocaris typica*, Kraus and Found (1975) found that this species was distributed in nature by active selection of tolerable and avoidance of intolerable salinity—temperature regimes. Distribution patterns were in general agreement with laboratory findings, unlike the work of Bradley (1975).

The levels of various environmental factors may fluctuate with time, and Thorp and Hoss (1975) studied the effects of salinity and cyclic temperatures on the survival of two sympatric species of grass shrimp. Cyclic temperatures appeared to be detrimental in combination with other stresses. Initially, the physiological tolerance to winter conditions of temperature and salinity was determined so as to provide a basis for examining the ecological interaction between these two species. However, neither salinity nor temperature tolerances appear to be of primary importance in habitat partitioning of these sympatric species.

The role of other environmental factors also has been studied in recent years. Roland and Ring (1977) reported on the interrelations of cold, freezing, and desiccation tolerances of the limpet *Acmaea digitalis*. Skoog (1976), working with snails, found that adult and larval *Lymnea peregra* had a greater thermal tolerance than adult and larval *Theodoxus fluviatilis*, a result that could be correlated with the field distribution of the two species. Also, differences in desiccation limits could be correlated with field observations. The freezing tolerance of the intertidal mollusc *Modiolus demissus* increased after acclimation to either low temperature or high salinity (Murphy and Pierce, 1975). It was proposed that low-temperature acclimation does not act to increase freezing tolerance by reducing the amount of tissue water frozen but rather that it increases cellular tolerance to greater levels of dehydration by an unspecified mechanism.

Vargo and Sastry (1977) analyzed the combined effects of two other environmental factors, dissolved oxygen and temperature. They determined the tolerance limits of five zoeal stages and megalops of the crab *Cancer irroratus*. Interstage variation was observed in that the first, second, and fourth zoeal stages showed similar responses, but the responses were different from those of the third and fifth stages, which were similar to each other. Although the larval stages did not show a progressive increase in tolerance to temperature or low dissolved oxygen with development, the megalops is relatively insensitive to changes in oxygen concentration with temperature. In general, these larval stages appear to have the capacity to tolerate a wider range of temperature and oxygen conditions than they encounter in the natural environment.

The interaction of abiotic and biotic factors can influence measures of tolerance. For example, Rippingale and Hodgkin (1977) reported that the brackish-water copepod *Sulcanus conflictus* is rarely found in salinities above 25 ‰. Yet in laboratory studies this species will thrive in higher salinities if excess food is available. It is suggested that the high energy demands for osmoregulation are not met in natural field conditions. Another example of a biotic and abiotic interrelationship was reported by Yocom and Edsall (1974). Whitefish (*Coregonus clupeaformis*) were acclimated to different temperatures and subjected to a thermal shock. The vulnerability of these animals to the biotic factor of predation was assessed. Thermally shocked fry were more vulnerable to capture by perch than were unshocked specimens. Response to multiple factor exposure occurs at the tissue level. This fact has been well demonstrated by a comparison of effect of temperature acclimation and salinity on isolated gill tissue of two estuarine species of bivalves. The gill of the oyster *Crassostrea virginica* is more resistant to low salinity than is the gill from scallops (Vernberg, Schlieper, and Schneider, 1963).

Wallis (1976) has published a simple multifactorial model using response-surface analysis for studies relating to power-station cooling systems. He incorporated salinity, temperature, temperature shock, exposure time, and mortality responses as well as some sublethal effects.

LARVAL DEVELOPMENT AND GROWTH

Various papers have dealt with the effects of environmental factors on the development of embryos and larvae. Alderdice and Forrester (1968) demonstrated that developmental abnormalities occurred in *Parophrys vetulus* eggs when exposed to certain temperature—salinity combinations (Fig. 5). The effects of cadmium and salinity on the larval development of two estuarine crabs, *Rhithropanopeus harrisii* and *Callinectes sapidus*, when subjected to either constant or cycling temperature, were reported by Rosenberg and Costlow (1976). Cycling temperature appeared to lengthen survival times. The megalops were more tolerant of cadmium than were the zoeal stages. Christiansen and Costlow (1975) reported on the effects of salinity and cyclic temperature on *R. harrisii*. Christiansen, Costlow, and Monroe (1977), recently studied the effects of the juvenile hormone mimic ZR-515 (Altosid), salinity,

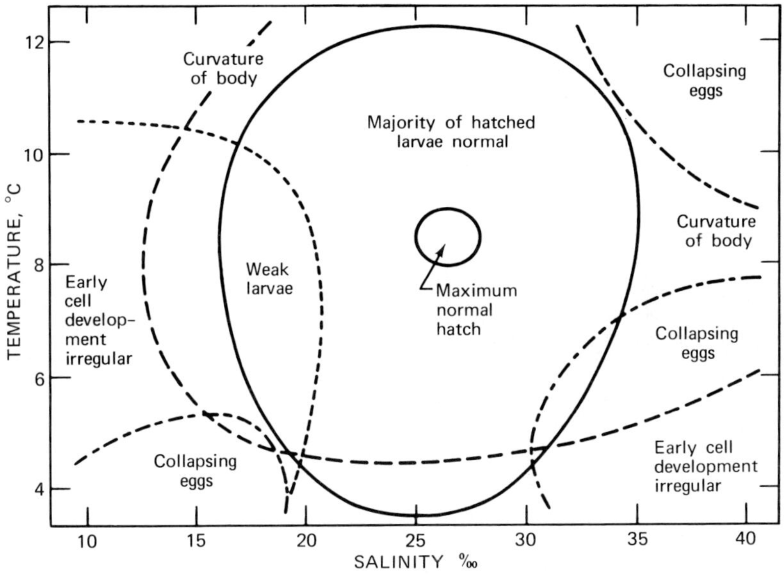

Fig. 5 Effects of incubation salinity and temperature on viability of developing eggs of *Parophrys vetulus*. Various gross developmental anomalies occurred more frequently in certain areas of the salinity–temperature test space, as indicated. (After Alderdice and Forrester, 1968.)

and cyclic temperatures on this same species. No significant synergism between Altosid, salinity, and temperature was reported. The amphipod *Corophium triaenonyx* inhabits parchment tubes in fouling communities on wood and concrete jetties. The eggs are more sensitive to salinity than are adults. The optimum range for development is between 27.5 and 30‰ and a constant temperature of 28°C (Shyamasundari, 1976).

Insects inhabiting freshwater ponds are shown to be affected by experimental photoperiods and temperatures. Larvae of the odonate species *Plathemis lydia* demonstrated stimulated growth rates when subjected to longer photoperiods or higher temperatures. However, the magnitude of these effects varied with the season of the year

(Shepard and Lutz, 1976). Ingram and Jenner (1976) also reported on the influence of photoperiod and temperature on two species of Odonata. Species and seasonal differences were observed, but, generally, development occurred more rapidly under long-day conditions.

In 1975 Lough reevaluated the literature on the combined effects of temperature and salinity on survival and growth of bivalve larvae using response-surface techniques. Generally, tolerance to both temperature and salinity were greater in the late veliger larvae than in the developing embryos. Each species had its own specific characteristic response pattern, which changed as the larvae approached the adult condition. In all species there was a significant temperature—salinity interaction in relation to growth. Cain (1975) subjected larvae of the bivalve *Rangia cuneata* to combinations of temperature and salinity. Embryos were relatively insensitive to thermal shock. In addition, growth was reduced in lower salinities. The combined effects of temperature and salinity resulted in a higher level of mortality than the sum of the mortalities for either of the two factors alone. Recently, Fish and Fish (1977) found that embryonic development of the prosobranch snail *Hydrobia ulvae* was influenced by temperature and salinity. Data were analyzed by fitting surface responses. Hatching was recorded in salinities ranging from 8 to 60‰ and at temperatures up to 35°C, but the minimum incubation time (5 days) was found over a range of 13 to 41‰ salinity and 22.5 to 33°C.

Fish development is also influenced by environmental factors. Hamor and Garside (1976) subjected embryos of the Atlantic salmon, *Salmo salar*, to various combinations of temperature, dissolved oxygen, and water exchange. Survival during embryogenesis and during the hatching period was limited primarily by oxygen supply. Water exchange and temperature were next in importance, but all three were statistically significant in affecting development. May (1975) studied not only the effect of temperature and salinity on fertilization, embryonic development, and hatching of the marine fish *Bairdiella icistia* but also added the factor of parental salinity. Fertilization was blocked at salinities of 10‰ and below. Temperature and salinity interacted, with development at high salinities being most successful at low temperature and development at high temperatures being most successful at low salinities. The overall viability of eggs produced was greatly influenced by the stage of maturity of the spawning female. The salinity tolerance of eggs was not significantly changed when adults were matured sexually in diluted seawater (15‰ salinity).

REPRODUCTION

Asexual reproduction in an intertidal zone sea anemone, *Haliplanella luciae*, was influenced by exposure to both fluctuating temperature and immersion (Johnson and Shick, 1977). Significantly different numbers of fissions were observed under these various test conditions.

Other reproductive phenomena can be influenced by environmental-factor interactions. De Vlaming (1975) reported that at any time during the year a long-photoperiod warm-temperature regime stimulated gonadal development to the prespawning condition or induced spawning in the cyprinid fish *Notemigonus crysoleucas*. Neither photoperiod nor temperature acting alone would stimulate final gonadal maturation. Although spermatocyte formation, proliferation, and the early stages of vitellogenesis occurred independently with respect to imposed conditions, final gonadal maturation and the rate of gametogenesis depend on specific environmental conditions. Rouquette (1976) reported that temperature and light influenced ovary development, formation of periodic sexual characteristics, and egg laying of the crab *Pachygrapsus marmoratus*.

BEHAVIOR

Environmental-factor interactions also influence the behavior of many species of organisms. Both the hard-shell clam *Mercenaria mercenaria* and the Atlantic surf clam *Spisula solidissima* experience extremes of temperature and dissolved oxygen during their burrowing activity. For each species a thermal zone of optimum activity was determined; above and below this zone burrowing declined with changing temperature. Exposing *Mercenaria* to low oxygen conditions (<1 mg/liter) for 3 weeks did not severely impair burrowing (Savage, 1976). Gallepp (1977) observed the behavior patterns (filtering, withdrawal, and case building) of larvae of the caddisfly *Brachycentrus* under different combinations of temperature, food availability, and water velocity. He demonstrated that temperature and food availability were more important than water velocity in influencing behavior. The interaction of these three factors largely controls the bioenergetic state of the larvae and thus influences the behavior of this species.

Another behavioral response, thermal preference, has been studied in relation to thermal acclimation and salinity in the fishes

Fundulus heteroclitus and *F. diaphanus* (Garside and Morrison, 1977) and *Gasterosteus aculeatus* (Garside, Heinze, and Barbour, 1977). Both species of *Fundulus* preferred higher temperatures in a salinity approximating that of the typical habitat. In *Gasterosteus* the preferred temperatures increased through thermal acclimation, ranging from 5 to 25°C. Those for the freshwater tests were about 2°C lower at each acclimation point than those for seawater tests. Differences might be due to differentials in metabolic loading brought on by osmoregulative stresses.

Acute changes in temperature and salinity affect pulsation rates in the ephyrae of the jellyfish *Aurelia aurita* (Dillon, 1977). The rate increases with temperature, although a Q_{10} of 0.97 was found between 20 and 25°C. When salinity is reduced acutely, the pulsation rate is increased for about a 2-day period.

Numerous physiological processes are influenced by interaction of multiple environmental factors (Kinne, 1963; 1964; Vernberg and Vernberg, 1970; 1972; and Alderdice, 1976). Only a few will be cited here.

Water Loss

Humidity and temperature affect the rate of water loss in the shore crab *Carcinus maenas* and the subtidal crab *Portunus marmoreus* (Ahsanullah and Newell, 1977). *Portunus* loses water more readily than *Carcinus*, probably because of differences in relative gill areas.

Metabolic Rate

The results of numerous investigations of the effects of temperature, salinity, and other factors on the metabolic rates of aquatic organisms have been published. The metabolic rate of the goby *Gillichthys mirabilis* increased with increasing temperature, decreased with decreasing ambient oxygen, but remained constant over a wide range of salinities (Courtois, 1976). Nelson et al. (1977) found that the distribution and migration of juvenile *Macrobrachium rosenbergii* in natural habitats could be correlated with their metabolic responses to temperature and salinity. Working with the mysid *Neomysis intermedia*, Simmons and Knight (1975) measured respiration at different combinations of temperature and salinity at three seasons of the year. This species is found in freshwater and in salinities up to 18‰. Unlike the goby, their metabolic rate decreases as salinity is increased. As temperature increases, metabolic rate increases. These workers found seasonal changes that may be related to reproduction and to the type of food available. The metabolic rate of the isopod *Idotea chelipes* from Lake Veere, The Netherlands, is influenced by temperature and salinity in such a manner as to suggest the need for energy for osmoregulatory activities (Vlasblom, Graafsma, and Verhoeven, 1977). Dimock and Groves (1976), working with thermal—salinity effects on the respiration rate of the estuarine crab *Panopeus herbstii*, concluded that these two factors showed significant interaction. The amphipod *Gammarus fossarum* showed partial metabolic regulation in waters of decreasing oxygen content. In the microhabitat of this species, slowly running water is preferred, with a maximal temperature of 22°C (Franke, 1977).

A unique environment in the ocean is the oxygen minimum layer, which may have oxygen levels less than 0.5 ml/liter. The copepod *Gaussia princeps* exhibits a diurnal vertical migration but may spend days below a depth of 400 m in the oxygen minimum layer. Extending some of his earlier work on *G. princeps*, Childress (1977) observed the effects of pressure, temperature, and oxygen consumption of this midwater species. At all temperature—pressure combinations, *G. princeps* displays a very low metabolic rate compared to shallow-living copepods. A higher rate of oxygen consumption was exhibited at the nighttime shallower depths and a much lower rate (partially anaerobic) at deeper daytime depths. Hydrostatic pressure significantly affected oxygen consumption at pressures as low as 28 atm. These responses can be correlated with the ecology of this species. Bivalve molluscs may also be exposed to periods of anoxia. Bayne and Livingstone (1977) reported that *Mytilus edulis* acclimated with respect to rates of oxygen consumption after being maintained at reduced oxygen tensions for more than 5 days. Temperatures of 10 to 22°C did not influence the capacity to acclimate.

Metabolism is also influenced by other environmental factors. Buckingham and Freed (1976) reported that the metabolic rate of *Viviparus contectoides*, a snail, is influenced by both pH, with two optimums at 7.1 and 8.9, and temperature. The oxygen consumption of the freshwater snail *Indoplanorbis exustus* was studied in detail in relation to many factors, such as salinity, temperature, pH, oxygen tension, body size, starvation, exposure to atmospheric air, and diurnal rhythm (Nagabhushanam and Chintawar, 1977). The hemichordate *Ptychodera flava* responded to a lowered salinity by a sudden rise in respiration rate. With respect to tidal changes, the metabolic responses of *P. flava* fall into three classes: (1) a rhythm synchronized with tidal phases, (2) a tidal rhythm with additional peaks during low tides, or (3) no detectable rhythm (Azariah, Ismail, and Najib, 1975).

Metabolic responses of excised gills of the crab *Hemigrapsus nudus* to temperature and salinity adaptation were reported by Hulbert, Schneider, and Moon (1976a). These workers also investigated the effects of temperature and salinity on the intermediary metabolism of excised gill homogenates of this species (Hulbert, Schneider, and Moon, 1976b). At the subcellular level Thomson, Sargent, and Owen (1977) reported on the influence of acclimatization temperature and salinity on Na/K-dependent adenosine triphosphate and fatty acid composition in the gills of the eel *Anguilla anguilla*. They concluded that temperature and not salinity determines the degree of unsaturation of gill lipids. The Arrhenius plot showed a different discontinuity for microsomal Na/K-ATPase from gills of freshwater eels than that for preparations from gills of seawater eels. In another study, temperature and thermal acclimation influenced the osmotic properties and the nonelectrolyte permeability of liver and gill mitochondria from the rainbow trout, *Salmo gairdneri* (Hazel and Schuster, 1976).

Osmoregulation

Temperature and salinity effects have been studied on ion levels in the hemolymph of the blue crab *Callinectes sapidus* (Engel et al., 1974) and on the intracellular osmoregulation and muscle-free amino acids in the fish *Fundulus diaphanus* (Ahokas and Sorg, 1977).

The relative effects of temperature and salinity on the osmoregulatory ability of marine organisms also have been investigated. Two species of shrimp (genus *Crangon*) common to the North Sea vary in their migratory behavior (Spaargaren, 1971). *Crangon crangon* migrates to coastal and inland waters in the spring and returns to the North Sea with the approach of winter; in contrast, *C. allmanni* typically remains in deeper waters throughout the year. The rate of accommodation of the blood concentration to a new salinity is temperature dependent, i.e., the rate increases as the temperature increases, and Spaargaren generalized that marine species show a quicker adaptation rate than estuarine species. The salinity at which maximal regulation in *C. crangon* occurs decreases with increasing temperature as was also shown by Weber and Spaargaren (1970). This shifting of the regulation range to reduced salinities at elevated temperatures appeared to be correlated with the summer migration of this species to brackish water. Spaargaren (1971) reported that, in contrast to the osmoregulatory ability of *C. crangon*, *C. allmanni* showed a very high degree of osmoconformity, and no difference in blood concentration was observed at different temperatures (5 to 15°C). Temperature and salinity did interact to influence the lethal limits of this species. For example, at 5 to 7°C a salinity as low as 10‰ could be tolerated, but at 21°C, salinities below 25‰ were lethal. This response is the reverse of *C. crangon* where lower salinities were tolerated at higher temperatures.

Spaargaren (1972) also investigated the effects of temperature and salinity on the osmoregulatory ability of two species of prawns from the Bay of Naples, a region of high salinity throughout the year. At low temperature and high salinity, *Palaemon serratus* is a strong regulator of osmotic concentration in the body fluids; at low salinities conformity is noted. At high temperatures the regulation range is shifted toward lower salinities.

When comparing the response of these two species with the two species of *Crangon* previously cited, Spaargaren concluded that *L. seticaudata* and *C. allmanni* show similar osmoregulatory responses. However, *C. allmanni* could withstand lower salinities longer at low temperatures, whereas *L. seticaudata* tolerated reduced salinities longer at high temperatures. This response pattern correlated well with conditions normally encountered in their respective habitats. Unlike *C. crangon*, *P. serratus* did not show a causal connection between osmoregulation patterns and migration. Spaargaren suggested that the effect of temperature on osmoregulation is a secondary adaptation to changing seasonal conditions.

In contrast, the differential effects of temperature on the osmoregulatory ability of two closely related species of shrimp influence their seasonal distribution. Both species can hyperosmoregulate in reduced salinity at those temperatures normally encountered during the summer, and they are found in estuaries. Although at low temperature this ability to osmoregulate is impaired in both species, *Penaeus duorarum* is less affected and can overwinter in the cold inshore waters of the Carolina coast (Williams, 1960).

OTHER PHYSIOLOGICAL PROCESSES

Other physiological processes also have been studied when the organism is subjected to multiple environmental factors. Duman and DeVries (1974) found that temperature and photoperiod influenced the production of a macromolecular substance that functions as an antifreeze in the blood serum of a fish. From a biogeographic point of view, it was interesting to note that apparent genetic differences exist between Alaska and California populations of *Anoplarchus purpurescens* in that the California fishes were unable to produce antifreeze when acclimated to cold temperatures.

Endocrine changes can result when levels of environmental factors are changed. For example, McKeown and Peter (1976) found that photoperiod and temperature influenced the release of prolactin from the pituitary gland of the goldfish, *Carassius auratus*. In fiddler crabs, Silverthorn (1975) reported that temperature and thermal acclimation affected the production of a "hormone-like" factor, which influenced oxygen consumption. Regeneration and molting of fiddler crabs are influenced by light, temperature, and salinity (Weis, 1976).

Most of the previous studies stressed the interaction between an organism and its environment. However, an organism is part of a population, which in turn is part of an assemblage of species called a community. In evaluating the potential effects of altered environments, we need to have better techniques to predict effects on communities. Recently Vernberg, McKellar, and Vernberg (1978) proposed a simple population model to predict the influence of the toxicity of a particular pollutant and environmental fluctuation on a natural population.

SUMMARY

In summary, studies of multiple environmental factors and their interactions have involved many diverse organisms inhabiting aquatic systems. These studies have included numerous abiotic factors, both natural and man-induced, and biotic factors. An organism is faced with a complex environment, and it is a natural evolution in our scientific thinking to investigate multiple factor effects and interactions in terms of organismic responses. We need more coordinated field and laboratory studies to answer environmentally oriented questions. A better definition of the characteristics of the microenvironment in which a species lives is required to design laboratory studies that simulate the environment. Experiments need to be designed to help explain and describe the functional mechanisms involved in an organism's responses to multiple factors. As we begin to understand one level of complexity, we encounter an entirely different and higher level of complexity. It is the excitement of trying to understand the next higher level of complexity that forces us to continue research.

REFERENCES

Ahokas, R. A., and G. Sorg, 1977, The Effect of Salinity and Temperature on Intracellular Osmoregulation and Muscle Free Amino Acids in *Fundulus diaphanus*, *Comp. Biochem. Physiol.*, 56A: 101-105.

Ahsanullah, M., and R. C. Newell, 1977, The Effects of Humidity and Temperature on Water Loss in *Carcinus maenas* (L) and *Portunus marmoreus* (Leach), *Comp. Biochem. Physiol.*, 56A: 593-601.

Alderdice, D. F., 1972, Chapter 12 Factor Combination, in *Marine Ecology*, Vol. 1, Part 3, pp. 1659-1722, O. Kinne (Ed.), Wiley-Interscience, Inc., New York.

——, 1976, Some Concepts and Descriptions of Physiological Tolerance: Rate—Temperature Curves of Poikilotherms as Transects of Response Surfaces, *J. Fish. Res. Board Can.*, 33: 299-307.

——, and C. R. Forrester, 1968, Some Effects of Salinity and Temperature on Early Development and Survival of the English Sole (*Parophrys vetulus*), *J. Fish. Res. Board Can.*, 25: 495-521.

Azariah, J., M. M. Ismail, and M. A. Najib, 1975, Investigation on the Ecology and Respiratory Responses of the Hemichordate *Ptychodera flava* to Tidal Cycles and Salinity Changes, *Biol. Bull.*, 149: 455-466.

Bayne, B. L., and D. R. Livingstone, 1977, Responses of *Mytilus edulis* L. to Low Oxygen Tension: Acclimation of the Rate of Oxygen Consumption, *J. Comp. Physiol.*, 114: 129-142.

Billings, W. D., 1952, The Environment Complex in Relation to Plant Growth and Distribution, *Quart. Rev. Biol.*, 27: 251-265.

Bradley, B. P., 1975, The Anomalous Influence of Salinity on Temperature Tolerances of Summer and Winter Populations of the Copepod *Eurytemora affinis*, *Biol. Bull.*, 148: 26-34.

Buckingham, M. J., and D. E. Freed, 1976, Oxygen Consumption in the Prosobranch Snail *Viviparus contectoides* (Mollusca: Gastropoda)—II. Effects of Temperature and pH, *Comp. Biochem. Physiol.*, 53A: 249-252.

Cain, T. D., 1975, Combined Effects of Changes in Temperature and Salinity on Early Stages of *Rangia cuneata*, *Va. J. Sci.*, 25: 30-31.

Childress, J. J., 1977, Effects of Pressure, Temperature and Oxygen on the Oxygen-Consumption Rate of the Midwater Copepod *Gaussia princeps.*, *Mar. Biol.*, 39: 19-24.

Christiansen, M. E., and J. D. Costlow, Jr., 1975, The Effect of Salinity and Cyclic Temperature on Larval Development of the Mud-Crab *Rhithropanopeus harrisii* (Brachyura: Xanthidae) Reared in the Laboratory, *Mar. Biol.* 32: 215-221.

——, J. D. Costlow, Jr., and R. J. Monroe, 1977, Effects of the Juvenile Hormone Mimic ZR-515 (Altosid) on Larval Development of the Mud-Crab *Rhithropanopeus harrisii* in Various Salinities and Cyclic Temperatures, *Mar. Biol.*, 39: 269-279.

Courtois, L. A., 1976, Respiratory Responses of *Gillichthys mirabilis* to Changes in Temperature, Dissolved Oxygen and Salinity, *Comp. Biochem. Physiol.*, 53A: 7-10.

Davenport, C. B., 1897—1899, *Experimental Morphology*, Vols. I and II, Macmillan Publishing Co., London and New York.

De Vlaming, V. L., 1975, Effects of Photoperiod and Temperature on Gonadal Activity in the Cyprinid Teleost, *Notemigonus crysoleucas*, *Biol. Bull.*, 148: 402-415.

Dillon, T. M., 1977, Effects of Acute Changes in Temperature and Salinity on Pulsation Rates in Ephyrae of the Scyphozoan *Aurelia aurita, Mar. Biol.*, 42: 31-35.

Dimock, R. V., Jr., and K. H. Groves, 1976, Interaction of Temperature and Salinity on Oxygen Consumption of the Estuarine Crab *Panopeus herbstii, Mar. Biol.*, 33: 301-308.

Duman, J. G., and A. L. DeVries, 1974, The Effects of Temperature and Photoperiod on Antifreeze Production in Cold Water Fishes, *J. Exp. Zool.*, 190: 89-98.

Engel, D. W., E. M. Davis, D. E. Smith, and J. W. Angelovic, 1974, The Effect of Salinity and Temperature on the Ion Levels in the Hemolymph of the Blue Crab, *Callinectes sapidus,* Rathbun, *Comp. Biochem. Physiol.*, 49A: 259-266.

Fish, J. D., and S. Fish, 1977, The Effects of Temperature and Salinity on the Embryonic Development of *Hydrobia ulvae* (Pennant), *J. Mar. Biol. Ass. U.K.*, 57: 213-218.

Franke, U., 1977, Experimental Investigations of the Respiration of *Gammarus fossarum* Koch 1835 (Crustacea-Amphipoda) in Relation to Temperature, Oxygen Concentration and Water Movement, *Arch. Hydrobiol., Supplement B*, 48: 369-411.

Gallepp, G. W., 1977, Responses of Caddisfly Larvae (*Brachycentrus* spp.) to Temperature, Food Availability and Current Velocity, *Am. Midl. Nat.*, 98: 59-84.

Garside, E. T., D. G. Heinze, and S. E. Barbour, 1977, Thermal Preference in Relation to Salinity in the Threespine Stickleback, *Gasterosteus aculeatus* L., with an Interpretation of Its Significance, *Can. J. Zool.*, 55: 590-594.

——, and G. C. Morrison, 1977, Thermal Preferences of Mummichog, *Fundulus heteroclitus* L., and Banded Killifish, *F. diaphanus* (LeSueus), (Cyprinodontidae) in Relation to Thermal Acclimation and Salinity, *Can. J. Zool.*, 55: 1190-1194.

Hamor, T., and E. T. Garside, 1976, Developmental Rates of Embryos of Atlantic Salmon, *Salmo salar* L., in Response to Various Levels of Temperature, Dissolved Oxygen, and Water Exchange, *Can. J. Zool.*, 54: 1912-1917.

Hazel, J. R., and V. L. Schuster, 1976, The Effects of Temperature and Thermal Acclimation upon the Osmotic Properties and Nonelectrolyte Permeability of Liver and Gill Mitochondria from Rainbow Trout (*Salmo gairdneri*), *J. Exp. Zool.*, 195: 425-438.

Hulbert, W. C., D. E. Schneider, and T. W. Moon, 1976a, Temperature and Salinity Adaptation in the Purple Shore Crab *Hemigrapsus nudus*: An in vitro Physiological Study with Excised Gills, *Mar. Biol.*, 36: 217-222.

——, D. E. Schneider, and T. W. Moon, 1976b, Temperature and Salinity Adaptation in the Purple Shore Crab *Hemigrapsus nudus*: An in vitro Metabolic Flux Study with Excised Gills, *Mar. Biol.*, 36: 223-231.

Ingram, B. R., and C. E. Jenner, 1976, Influence of Photoperiod and Temperature on Developmental Time and Number of Molts in Nymphs of Two Species of Odonata, *Can. J. Zool.*, 54: 2033-2045.

Johnson, L. L. and J. M. Shick, 1977, Effects of Fluctuating Temperature and Immersion on Asexual Reproduction in the Intertidal Sea Anemone *Haliplanella luciae* (Verrill) in Laboratory Culture, *J. Exp. Mar. Biol. Ecol.*, 28: 141-149.

Kinne, O., 1963, The Effects of Temperature and Salinity on Marine and Brackish Water Animals. I. Temperature, *Oceanogr. Mar. Biol., A. Rev.*, 1: 301-340.

——, 1964, The Effects of Temperature and Salinity on Marine and Brackish Water Animals. II. Salinity and Temperature Salinity Combinations, *Oceanogr. Mar. Biol., A. Rev.*, 2: 281-339.

Kraus, M. G., and B. W. Found, 1975, Preliminary Observations on Salinity and Temperature Tolerances and Salinity Preferences of *Derocheilocaris typica*, *Cah. Biol. Mar.*, 16: 751-762.

Lockhart, J. A., 1965, The Analysis of Interactions of Physical and Chemical Factors on Plant Growth, *Annu. Rev. Plant Physiol.*, 16: 37-52.

Lough, R. G., 1975, A Reevaluation of the Combined Effects of Temperature and Salinity on Survival and Growth of Bivalve Larvae Using Response Surface Techniques, *Fish. Bull.*, 73: 86-94.

Mason, H. L., and J. H. Langenbein, 1957, Language Analysis and the Concept Environment, *Ecology*, 38: 325-340.

May, R. C., 1975, Effects of Temperature and Salinity on Fertilization, Embryonic Development, and Hatching in *Bairdiella icistia* (Pisces: Sciaenidae), and the Effect of Parental Salinity Acclimation on Embryonic and Larval Salinity Tolerance, *Fish. Bull.*, 73: 1-22.

McKeown, B. A., and R. E. Peter, 1976, The Effects of Photoperiod and Temperature on the Release of Prolactin from the Pituitary Gland of the Goldfish, *Carassius auratus* L., *Can. J. Zool.*, 54: 1960-1968.

McLeese, D. W., 1956, Effects of Temperature, Salinity and Oxygen on the Survival of the American Lobster, *J. Fish. Res. Board Can.*, 13: 247-272.

Mohr, H., 1972, *Lecture on Photomorphogenesis*, Springer-Verlag New York, Inc.

Murphy, D. J., and S. K. Pierce, Jr., 1975, The Physiological Basis for Changes in the Freezing Tolerance of Intertidal Molluscs, *J. Exp. Zool.*, 193: 313-322.

Nagabhushanam, R., and B. V. Chintawar, 1977, Influence of Environmental Factors on Oxygen Consumption in the Freshwater Snail, *Indoplanorbis exustus, Marathwada Univ. J. Sci., Sect. B*, 15: 209-220.

Nelson, S. G., D. A. Armstrong, A. W. Knight, and H. W. Li, 1977, The Effects of Temperature and Salinity on the Metabolic Rate of Juvenile *Macrobrachium rosenbergii* (Crustacea: Palaemonidae), *Comp. Biochem. Physiol.*, 56A: 533-537.

Rippingale, R. J., and E. P. Hodgkin, 1977, Food Availability and Salinity Tolerance in a Brackish Water Copepod, *Aust. J. Mar. Freshwater Res.*, 28: 1-8.

Roland, W., and R. A. Ring, 1977, Cold, Freezing and Desiccation Tolerance of the Limpet *Acmaea digitalis* (Eschscholtz), *Cryobiology*, 14: 228-235.

Rosenberg, R., and J. D. Costlow, Jr., 1976, Synergistic Effects of Cadmium and Salinity Combined with Constant and Cycling Temperatures on the Larval Development of Two Estuarine Crab Species, *Mar. Biol.*, 38: 291-303.

Rouquette, M. P., 1976, Étude de la fonction de reproduction chez les femelles du crabe *Pachygrapsus marmarotus* (F.) et de différents facteurs qui liu sont liés (I), *Cah. Biol. Mar.*, 17: 387-403.

Salisbury, F. B., 1975, Multiple Factor Effects on Plants, in *Physiological Adaptation to the Environment*, F. J. Vernberg (Ed.), Thomas Y. Crowell Co., New York.

Savage, N. B., 1976, Burrowing Activity in *Mercenaria mercenaria* (L.) and

Spisula solidissima (Dillwyn) as a Function of Temperature and Dissolved Oxygen, *Mar. Behav. Physiol.*, 3: 221-234.

Shepard, L. J., and P. E. Lutz, 1976, Larval Responses of *Plathemis lydia* Drury to Experimental Photoperiods and Temperatures (Odonata: Anisopters), *Am. Midl. Nat.*, 95: 120-130.

Shyamasundari, K., 1976, Effects of Salinity and Temperature on the Development of Eggs in the Tube Building Amphipod *Corophium triaenonyx* Stebbing, *Biol. Bull.*, 150: 286-293.

Silverthorn, S., 1975, Hormonal Involvement in Thermal Acclimation in the Fiddler Crab *Uca pugilator* (Bosc). I. Effect of Eyestalk Extracts on Whole Animal Respiration, *Comp. Biochem. Physiol.*, 50A: 281-283.

Simmons, M. A., and A. W. Knight, 1975, Respiratory Response of *Neomysis intermedia* (Crustacea: Mysidacea) to Changes in Salinity, Temperature and Season, *Comp. Biochem. Physiol.*, 50A: 181-193.

Skoog, G., 1976, Effects of Acclimatization and Physiological State on the Tolerance to High Temperatures and Reactions to Desiccation of *Theodoxus fluviatilis* and *Lymnea peregra, Oikos*, 27: 50-56.

Spaargaren, D. H., 1971, Aspects of the Osmotic Regulation in the Shrimps *Crangon crangon* and *Crangon allmanni, Neth. J. Sea Res.*, 5: 275-333.

——, 1972, Osmoregulation in the Prawns *Palaemon serratus* and *Lysmata* from the Bay of Naples, *Neth. J. Sea Res.*, 5: 416-436.

Thomson, A. J., J. R. Sargent, and J. M. Owen, 1977, Influence of Acclimatization Temperature and Salinity on ($Na^+ + K^+$)-Dependent Adenosine Triphosphatase and Fatty Acid Composition in the Gills of the Eel, (*Anguilla anguilla*), *Comp. Biochem. Physiol.*, 56B: 223-228.

Thorp, J. H., and D. E. Hoss, 1975, Effects of Salinity and Cyclic Temperature on Survival of Two Sympatric Species of Grass Shrimp (*Palaemonetes*), and Their Relationship to Natural Distributions, *J. Exp. Mar. Biol. Ecol.*, 18: 19-28.

Vargo, S. L., and A. N. Sastry, 1977, Acute Temperature and Low Dissolved Oxygen Tolerances of Brachyuran Crab (*Cancer irroratus*) Larvae, *Mar. Biol.*, 40: 165-171.

Vernberg, F. J., and W. B. Vernberg, 1970, *The Animal and the Environment*, Holt, Rinehart, and Winston, New York.

——, C. Schlieper, and D. E. Schneider, 1963, The Influence of Temperature and Salinity on Ciliary Activity of Excised Gill Tissue of Molluscs from North Carolina, *Comp. Biochem. Physiol.*, 8: 271-285.

Vernberg, W. B., 1975, Multiple Factor Effects on Animals, in *Physiological Adaptation to the Environment*, F. J. Vernberg (Ed.), Thomas Y. Crowell Co., New York.

——, and F. J. Vernberg, 1972, *Environmental Physiology of Marine Animals*, Springer-Verlag, New York.

——, P. J. DeCoursey, and J. O'Hara, 1974, Multiple Environment Factor Effects on Physiology and Behavior of the Fiddler Crab, *Uca pugilater*, in *Pollution and Physiology of Marine Organisms*, pp. 381-425, F. J. Vernberg and W. B. Vernberg (Eds.), Academic Press, Inc., New York.

——, H. McKellar, Jr., and F. J. Vernberg, 1978, Toxicity Studies and Environmental Impact Assessment, *Environ. Manage.*, in press.

Vlasblom, A. G., S. J. Graafsma, and J. T. A. Verhoeven, 1977, Survival, Osmoregulatory Ability, and Respiration of *Idotea chelipes* (Crustacea, Isopoda) from Lake Veere in Different Salinities and Temperatures, *Hydrobiologia*, 52: 33-38.

Wallis, R. L., 1976, Some Uses of Multifactorial Response Surface Analysis in Temperature Tolerance Studies, *Aust. J. Mar. Freshwater Res.*, 27: 487-498.

Weber, R. E., and D. H. Spaargaren, 1970, On the Influence of Temperature on the Osmoregulation of *Crangon crangon* and Its Significance Under Estuarine Conditions, *Neth. J. Sea Res.*, 5: 108-120.

Weis, J. S., 1976, Effects of Environmental Factors on Regeneration and Molting in Fiddler Crabs, *Biol. Bull.*, 150: 152-162.

Williams, A. B., 1960, The Influence of Temperature on Osmotic Regulation in Two Species of Estuarine Shrimps (*Penaeus*), *Biol. Bull.*, 119: 560-571.

Yocom, T. G., and T. A. Edsall, 1974, Effect of Acclimation Temperature and Heat Shock on Vulnerability of Fry of Lake Whitefish (*Coregonus clupeaformis*) to Predation, *J. Fish. Res. Board Can.*, 31: 1503-1506.

ORDINATION OF MACROINVERTEBRATE COMMUNITIES IN A MULTISTRESSED RIVER SYSTEM

DAVID C. BECKETT
Department of Biological Sciences, University of Cincinnati, Cincinnati, Ohio

ABSTRACT

The macroinvertebrate communities of the Great Miami River system (southwestern Ohio) were studied over a wide range of environmental conditions and stresses. Stations ranged from relatively undisturbed locations to areas subjected to flood-control modification, power-plant thermal discharges, municipal sewage-treatment-plant discharges and industrial-wastewater outfalls.

Similarity indices and subsequent polar ordination showed that river passage through the city of Dayton did not prohibit high biotic similarity between some sites below Dayton and undisturbed upstream locations. Stations showed varying patterns of affiliation (based on similarity indices), which correlated well with changes in river discharge. Two types of faunal homogeneity were apparent: during relatively high river-discharge periods a high degree of faunal similarity occurred between far upstream and far downstream stations, and during low flow conditions a biotic homogeneity was evident among all downstream stations as a result of the maximization of pollutional influences and the subsequent domination of euryecious species. Ordination of stations during high flow conditions did not produce an ordering along any gradients. During low flow conditions, however, stations ordered well along a gradient of pollutional disturbances. Pollutional effects seemed to overwhelm the effect of normal environmental variables on faunal composition during these low flow conditions.

A comparison of the relatedness of macroinvertebrate communities was used by Burlington (1962) to assess the effect of a point source pollutant on benthic macroinvertebrates. His analysis suffered, however, from a paucity of stations (7) and was taken over just one time period. Crossman, Kaesler, and Cairns (1974) used cluster analysis, based on community similarity to assess the effect of two acute stresses (caused by major industrial spills) on the benthic macroinvertebrates of the Clinch River. Using an ordination procedure (Bray and Curtis, 1957), I have examined the pattern of affinities among a river system's macroinvertebrate communities over a gradient from clean water to multistressed communities. The stressors of significance in this study, municipal and industrial wastewater and power-plant effluents, are fairly common to the large rivers found in the highly populated areas of the United States.

I used 14 stations over three time periods to investigate changes in biotic composition on both a longitudinal and a temporal basis. Modifications in the pattern of relationships between communities owing to changes in river discharge and other physical—chemical parameters were also studied.

METHODS

Study Area

The Great Miami River basin is located in southwestern Ohio. The Miami River flows 257 km in a southwesterly direction from its headwaters in central Ohio, finally entering the Ohio River in the extreme southwestern corner of Ohio. Two major tributaries, the Mad and Stillwater rivers, join the Miami River at Dayton to form the Great Miami River, 129 km above its confluence with the Ohio River. The average Great Miami River discharge is approximately 5100 m^3/min. The population of the Great Miami River basin is approximately 1.5 million; about one-half of this population is concentrated in the Dayton area (Weston, 1967; Little, 1964).

The study area (Fig. 1) consisted of a 53-km section of the Great Miami River and a station on each of the major tributaries, the Mad and Stillwater rivers (for exact station locations see Beckett, 1977). Three upstream stations were established: one each on the Mad River, the Stillwater River, and the Miami River. These upstream stations were in woodland areas. Although there are some small municipal sewage-treatment plants far upstream of these stations, there are no major municipal or industrial discharges near these upstream sample sites.

Eleven stations were established along the Great Miami River within the study area (making a total of 14 stations). A station was established at both the intake and the discharge of the two power plants (the F. M. Tait and the O. H. Hutchings). Each of the power plants has a low-level dam located between the plant's intake and discharge. Station 13 was established below a fairly large dam (Chautauqua Dam). The entire study area on the Great Miami River (stations 4 through 14) underwent some modification for

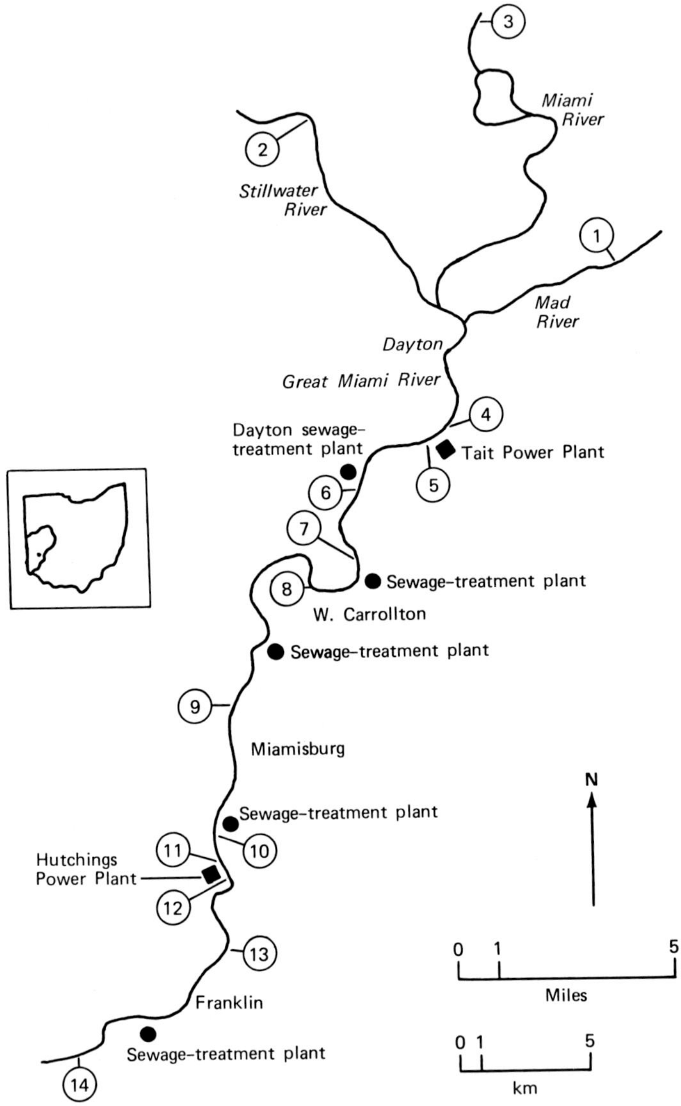

Fig. 1 Map of study area showing numbered station locations. Insert shows the extent of Great Miami River drainage in Ohio; dot within insert indicates Dayton.

flood-control purposes in the early 1920s. The 32-km stretch of the Great Miami River between the intake of the Tait power plant and the final downstream station (14) is subjected to the discharges from two power plants, five municipal sewage-treatment plants, and a number of industrial-wastewater outfalls.

Physical–Chemical Parameters

Several physical–chemical parameters were measured for each of the macroinvertebrate sampling periods. Temperature (°C) was measured with a mercury thermometer. Dissolved oxygen (mg/liter) was measured by means of the azide modification of the Winkler method (American Public Health Association, 1975). Alkalinity was titrated with phenolphthalein and bromcresol green–methyl red indicators (Lind, 1974). Conductivity (μmhos) was measured with a YSI conductivity meter; conductivity readings were corrected to values at a standard temperature of 25.0°C. Current was measured with a General Oceanics digital torpedo-type meter.

The dates for chemical sampling were correlated with the collection dates of macroinvertebrate samplers. Macroinvertebrate samplers and physical–chemical data were collected on the following dates (all dates in 1976): June, macroinvertebrate collection June 21, physical–chemical sampling June 21; August, macroinvertebrate collection August 9, physical–chemical sampling August 16; September, macroinvertebrate collection September 21 and 23; October, physical–chemical sampling October 14. Daily river-discharge data were furnished via monitors of the U. S. Department of Interior Geological Survey.

Macroinvertebrate Sampling and Analysis

Modified Hester–Dendy multiplate samplers similar to those described by Fullner (1971) were used for macroinvertebrate sampling. Each sampler consisted of a series of eight Masonite plates separated by a varying number of washers (spacing varied within the sampler—all samplers were of identical configuration and spacing). Each plate measured 10 by 10 cm and was 0.35 cm thick. Three sampling devices were located at each of the 14 stations. Each device consisted of two samplers connected to a wire suspended from a Styrofoam float. The sampling device was in turn connected to iron and concrete anchors that rested on the river bottom. The samplers were suspended approximately 1 m below the surface at all stations. Sampling devices were placed in quieter waters, i.e., not in riffles, in order to create similar conditions for colonization. The U. S.

Environmental Protection Agency (1973) suggests a 6-week exposure (colonization) period for artificial substrate samplers; my samplers were exposed for 6 weeks for the June and September collections and 7 weeks for the August collection. As colonized samplers were retrieved, new samplers were attached to the sampling devices for later collection. Macroinvertebrates were collected from the washed samples using a U. S. Standard 60 sieve. I used this sieve size rather than the popular 30 sieve so that I could collect smaller or early macroinvertebrate instars for analysis.

Since species diversity and number of taxa are often dependent on sampling effort (Sanders, 1968; Simberloff, 1972), I attempted to collect a uniform number of samplers from each station. Also, since samplers were occasionally lost as a result of vandalism or flooding, I used four samplers per station as my uniform sample size. Pielou (1966) found that diversity reached 95% of the asymptotic level by the fourth sample for stream macroinvertebrates. For the 42 station collections (3 collection dates x 14 stations), 27 collections yielded at least four samplers, 14 yielded two or three samplers, and one station on one date (June 7) yielded no samplers. From the four sampler collections, I empirically determined that two samplers had a mean of 75% of the total number of species for four samplers, and three samplers had a mean of 90% of the total number of species for four samplers. The reported total taxa levels (see Fig. 3) therefore show an adjusted number of species if two or three samplers were recovered instead of the usual four.

In a similar manner two- and three-sampler Shannon–Wiener diversity values (H′) (see below) were compared to four-sampler H′ values. Although four-sampler collections generally had slightly higher H′ values than two- or three-sampler values, this difference was not significant at the 0.05 level using a Wilcoxon signed rank test. Reported H′ values (see Fig. 2) are therefore pooled unadjusted values (Pielou, 1966; Wilhm, 1970b).

I used the Shannon–Wiener diversity index (Shannon, 1948; Wilhm and Dorris, 1968; Wilhm, 1970a) to quantify the species composition of the respective macroinvertebrate communities. To obtain reliable estimates of diversity in this study I have (1) used artificial substrates to reduce the confounding effects of substrate difference (U. S. Environmental Protection Agency, 1973) and (2) pooled samples at each site to approximate an asymptotic level of diversity (Pielou, 1966; Wilhm, 1970b).

Macroinvertebrates were identified in the laboratory to the genus or species level for all taxa except oligochaetes, which were not further identified (Table 1). Oligochaetes were not found on most samplers and when present constituted only a small portion of the

TABLE 1

MACROINVERTEBRATE TAXA COLLECTED IN GREAT MIAMI RIVER SYSTEM (OHIO) STUDY, JUNE–SEPTEMBER 1976

Turbellaria	Zygoptera	*Dicrotendipes* sp.
Dugesia tigrina	*Argia* sp.	*Dicrotendipes* sp. 2
Oligochaeta	*Coenagrion* sp.	*Einfeldia* sp.
Hirudinea	*Enallagma* sp.	*Glyptotendipes senilis*
Helobdella triserialis	*Hetaerina* sp.	*Glyptotendipes* sp.
Mooreobdella microstoma	*Ischnura* sp.	*Larsia* sp.
Placobdella parasitica	Hemiptera	*Microtendipes* sp.
Crustacea	*Sigara* sp.	*Nanocladius* sp.
Amphipoda	Coleoptera	*Nilotanypus* sp.
Crangonyx sp.	*Ancyronyx variegatus*	*Parachironomus abortivus*
Hyalella azteca	*Dineutus* sp.	*P. pectinatellae*
Isopoda	*Laccophilus* sp.	*Phaenopsectra* sp.
Lirceus fontinalis	*Macronychus glabratus*	*Polypedilum fallax*
Decapoda	*Stenelmis* sp.	*Polypedilum* sp.
Orconectes sp.	Trichoptera	*Procladius* sp.
Ephemeroptera	*Ceraclea* sp.	*Pseudochironomus* sp.
Baetis sp.	*Cheumatopsyche* sp.	*Rheotanytarsus* sp.
Caenis sp.	*Cyrnellus marginalis*	*Stenochironomus* sp.
Heptagenia sp.	*Hydropsyche betteni*	*Tanypus* sp.
Isonychia sp.	*H. bronta*	*Tribelos* sp.
Stenacron interpunctatum	*H. frisoni*	Other Diptera
Stenonema integrum	*H. orris*	*Pericoma* sp.
S. pulchellum	*H. simulans*	*Simulium* sp.
Stenonema sp.	*Hydropsyche* sp.	Mollusca
S. terminatum	Diptera	Gastropoda
S. tripunctatum	Chironomidae	*Amnicola* sp.
Tricorythodes sp.	*Ablabesmyia mallochi*	*Ferrissia* sp.
Odonata	*Ablabesmyia* sp.	*Gyraulus* sp.
Anisoptera	*Cardiocladius* sp.	*Physa* sp.
Perithemis tenera	*Chironomus* sp.	*Pleurocera* sp.
	Conchapelopia sp.	
	Cricotopus bicinctus gr.*	
	C. exilis gr.*	
	C. trifasciatus gr.*	
	Cryptochironomus sp.	
	Dicrotendipes nervosus	

*gr., species complex group.

total number of macroinvertebrates found on the samplers. Principal taxonomic keys included Hilsenhoff (1975), insects; Lewis (1974), *Stenonema* mayflies; Holsinger (1972), amphipods; Klemm (1972), leeches; Mason (1973), chironomids; and Beck and Beck (1966), chironomids.

The key portion of this study involved the measuring of relationships among communities at the various stations. I used a similarity index to indicate the distance between communities as a function of their ecological similarity (McIntosh, 1973). This index is a coefficient of community (CC) index variously attributed to Sorenson (1948), Dice (1945), or Kulczynski (1927), and is of the form:

$$CC = \frac{2S_{ab}}{S_a + S_b}$$

where S_a is the number of species in sample A; S_b is the number of species in sample B; and S_{ab} is the number of species found in sample A which are also found in sample B (McIntosh, 1967; Whittaker, 1975). Values of this index can therefore range from 0.0 (perfect dissimilarity) to 1.0 (perfect similarity). This index has been used in determining similarities among plant communities (see review by McIntosh, 1967) and among animal communities (e.g., spiders, Uetz, 1976; benthic macroinvertebrates, Burlington, 1962).

I analyzed my data for each collection date. Using the CC index, I compared each station to each other station, thereby forming a similarity matrix (called a Q matrix or secondary matrix by some authors) for each of the collection periods. I then formulated similarity matrix diagrams (Fig. 4) along an upstream–downstream gradient directly from the similarity matrix. In these diagrams, communities showing similarities equal to or greater than 0.65 are associated by connecting lines. In cases where a number of stations all showed marked similarities ($\geqslant$0.65) to each other, the entire grouping of similar stations was connected to a single similarity line. Those stations not exhibiting marked similarity to any other station were left as single entities. The use of 0.65 as a cutoff value for marked similarity was somewhat arbitrary. However, Hanson (1955) and Hurd (1961) also found 0.65 to be indicative of high biotic similarity in their studies of plant communities.

The similarity matrices were also used as a basis for ordination of data. Ordination provides a graphical summary of the similarity between stations or communities. The ordination technique I used was introduced by Bray and Curtis (1957) and is commonly called Wisconsin comparative or polar ordination. The computational procedure that I used in ordinating the data is presented by Whittaker (1973). In addition, I used ordination as a form of indirect gradient analysis (Whittaker, 1973).

RESULTS AND DISCUSSION

Diversity values (H') showed a number of interesting trends (Fig. 2). Upstream stations (stations 1, 2, and 3) on the Mad, Stillwater, and Great Miami rivers had moderate to high values on all dates. Station 4 (near the F. M. Tait power-plant intake) exhibited a high macroinvertebrate diversity (above 3.0) for all three sampling periods; for two of the three sampling periods the highest value of H'

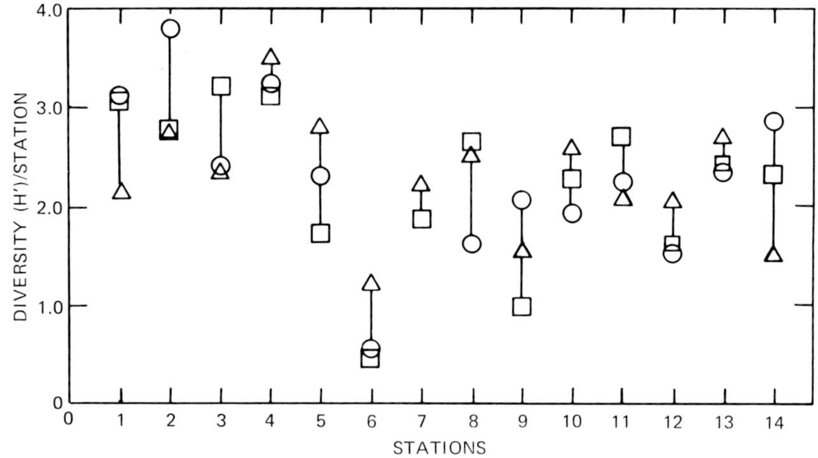

Fig. 2 Macroinvertebrate Shannon—Wiener diversity values (H′/station) for June (○), August (□), and September (△) 1976 collection dates, Great Miami River system, Ohio.

was found at this station. Stations 1 to 4 were the only stations to show H′ values above 3.0, the level supposedly indicative of clean-water areas (Wilhm, 1970a; Wilhm and Dorris, 1968).

Diversity values at station 5 (Tait power-plant discharge) were lower than those of station 4 on all occasions. Macroinvertebrate diversity was quite low below the Dayton sewage-treatment plant (station 6) on all three dates. For two of the three sampling periods H′ was below 1.0 at this point (1 and 22 in September); this is indicative of highly polluted conditions (Wilhm, 1970a; Wilhm and Dorris, 1968). The H′ values at station 7 showed marked increases over those of station 6, whereas moderate diversity values (between 1.0 and 3.0) were observed from stations 8 through 14 for all three collection dates.

Inspection of the number of taxa present per station at the different collection times (Fig. 3) shows some repetition of the diversity data; however, it adds clarity to the general picture of changes in river biota. Upstream stations (1 to 3) exhibited high species richness (S); station 4 (Tait power-plant intake) also had a relatively high number of species present for all three sampling periods. A marked decrease in number of total taxa occurred at station 5 (Tait power-plant discharge) in comparison with station 4; a further decrease occurred below the Dayton sewage-treatment plant. Although S showed some increase at those stations downstream from the Dayton sewage-treatment plant on all three dates, these

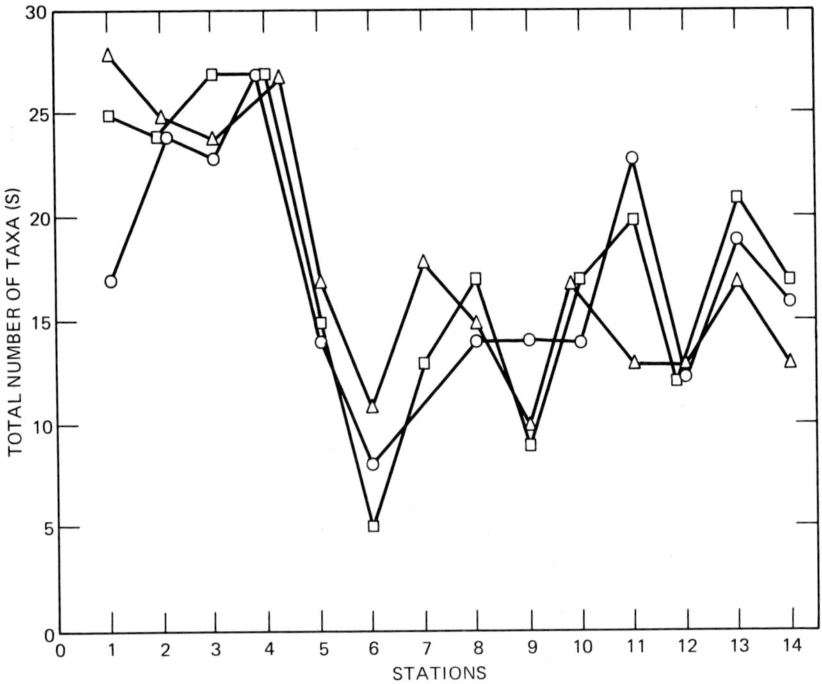

Fig. 3 Total number of macroinvertebrate taxa (S/station) for June (○), August (△), and September (□) 1976 collection dates, Great Miami River system, Ohio.

downstream stations did not show recovery to Tait power-plant intake S levels.

Modifications of the Great Miami's flow (modifications made in the early 1920s for flood-control purposes) and the river's passage through Dayton did not preclude relatively high levels of diversity or species richness below Dayton. The community at the intake of the Tait power plant, located approximately 6 km below the downtown area, had diversity values above 3.0 for all three sampling periods. This station consistently supported the highest or near highest number of macroinvertebrate taxa. Higher diversity and total taxa values at station 4 (in comparison to stations 1 to 3) may be partially due to an increase of stream order (Hynes, 1970; Harrell and Dorris, 1968) with the confluence of the Mad, Stillwater, and Miami rivers. Biotic compositional changes due to longitudinal zonation along my study area were not apparent, however. The CC index values between stations 3 and 4 were quite high in both June (0.72) and September (0.78). No increase in stream order occurred in the remainder of the study area (stations 4 to 14).

The five municipal sewage-treatment plant discharges and the numerous industrial discharges located between stations 6 and 14 probably precluded recovery to Tait intake H′ and S levels. Scott (1969) and Conn (1973) surveyed the macroinvertebrates of the Great Miami River using Beck's (1954) index as a measure of water quality. Although they used a clean-water form index rather than a diversity index, their results were similar to mine. In historical perspective, then, evaluation of river conditions over a number of years seems to give a fairly similar picture.

Physical—chemical parameters measured in this study are presented in Table 2. Alkalinity and pH were within natural limits for rivers such as the Great Miami on all sample dates at all stations. No consistent upstream—downstream pattern was observed for pH and alkalinity. The current was generally greater in June than in August or October owing to greater stream discharge. Because of the channelization of the Great Miami River and the use of low-level dams, current was undetectable at a number of stations.

Observed change in temperature (ΔT) on June 21 across the Tait plant (intake to discharge) was +3.0°C, +4.7°C on August 16, and +6.3°C on October 14. The ΔT for the Hutchings plant was +5.0°C for June 21, +5.2°C on August 16, and +5.5°C for October 14. The greater ΔT values in August and October were due to a decrease in stream discharge in these months, which enhanced plant effects. A summary of stream-discharge data for the Great Miami River over the macroinvertebrate collection periods (Table 3) indicates this steady decrease in stream discharge over the study period.

Observed oxygen concentrations were fairly high at all stations on June 21. However, fairly low O_2 levels were observed at stations 9 through 14 on August 16. Oxygen concentrations at stations 10 and 11 were critically low, less than 1.5 mg/liter on this date. Oxygen levels were higher in October than in August; however, depressed levels at stations 10 and 11 were still apparent.

Conductivity, a measure of the total concentration of dissolved ionic matter in the water, showed a consistent pattern throughout the study. Marked conductivity increases were observed below sewage-treatment plants and below known industrial-wastewater outfalls, resulting in a general pattern of increasing conductivity from upstream to downstream stations. Temperature and conductivity both showed linear negative correlations with total taxa per station (S) ($r_{temp.} = -0.51$, $r_{cond.} = -0.35$). A multiple stepwise regression (computer package BMD02R, Dixon, 1973) showed that temperature, conductivity, and oxygen were all significantly correlated with S [observed $F_{temp.} = 36.95$, $F_{cond.} = 25.39$, $F_{oxy.} = 10.07 >$ critical $F_{0.95\ (1,38)} = 4.10$]. These three variables explained 56.7% of

TABLE 2

PHYSICAL—CHEMICAL PARAMETERS—GREAT MIAMI RIVER SYSTEM, OHIO, JUNE 21 (J), AUG. 16 (A), AND OCT. 14 (O), 1976

Station No.	Date	O_2, mg/liter	Temp., °C	Conductivity, μmhos, at 25°C	pH	Alkalinity, mg $CaCO_3$/liter	Current, m/min
1	J	8.6	21.0	611	7.1	225	42
	A	8.2	20.7	695	7.7	272	28
	O	9.1	13.2	780	7.8	232	38
2	J	4.5	21.5	658	7.3	215	*
	A	5.7	21.4	648	7.5	240	*
	O	8.4	10.3	870	8.4	217	*
3	J	9.6	21.0	767	8.0	247	17
	A	6.0	21.3	749	7.7	248	*
	O	7.4	13.8	812	7.8	232	*
4	J	9.9	23.0	653	7.4	234	*
	A	4.6	21.8	728	7.7	246	*
	O	6.6	13.8	819	7.9	230	*
5	J	10.4	26.0	683	8.0	229	37
	A	6.4	26.5	694	7.8	242	35
	O	8.9	20.1	776	8.1	233	47
6	J	11.2	25.6	768	8.0	209	22
	A	7.0	25.5	840	7.4	217	19
	O	9.1	21.2	939	7.5	192	18
7	J	8.3	22.1	733	8.3	240	28
	A	7.0	25.2	746	7.5	239	*
	O	10.7	20.1	776	8.1	218	*
8	J	8.2	21.3	771	8.3	232	*
	A	5.1	24.8	834	7.4	244	*
	O	8.1	19.8	931	7.9	230	*
9	J	7.4	22.5	704	8.8	232	18
	A	3.6	23.9	894	7.5	250	*
	O	7.2	20.1	946	7.6	229	*
10	J	7.0	22.5	853	8.2	245	6
	A	1.3	24.0	872	7.6	240	*
	O	4.3	18.5	1051	7.6	241	*
11	J	9.3	22.5	725	8.4	229	*
	A	1.2	25.5	859	7.5	233	*
	O	4.8	19.0	929	7.6	229	*
12	J	7.9	27.5	753	8.3	235	*
	A	3.8	30.7	858	7.6	227	*
	O	7.0	24.5	922	7.7	225	*
13	J	9.2	24.6	717	8.3	228	*
	A	3.8	25.2	896	7.5	222	*
	O	7.4	20.5	980	7.7	220	*
14	J	8.8	25.0	705	8.5	222	*
	A	3.7	26.2	883	7.6	233	*
	O	6.7	20.5	935	7.6	220	*

*Undetectable.

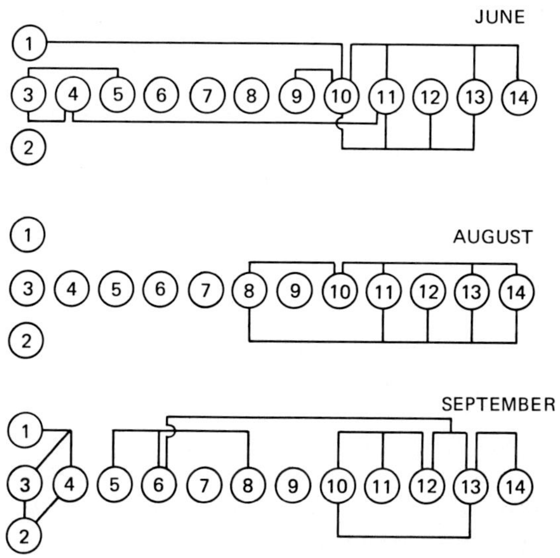

Fig. 4 Diagrammatic representation of similarity matrices (based on coefficient of community) showing strong affinities of macroinvertebrate communities (⩾0.65) for June, August, and September 1976 collection dates, Great Miami River system, Ohio.

the station-to-station variation in S (temperature explained 26.2%, conductivity 19.0%, and oxygen 11.5%).

The strongly negative correlations of conductivity and river temperature with number of species per station was indicative of the stressful effect of discharges (sewage-treatment plants, power plants, and industrial wastewater) on the river's macroinvertebrates. Although conductivity probably did not affect the macroinvertebrates directly, it served as an excellent marker of pollutional disturbances.

Though the use of a diversity index and total taxa levels has proved to be useful in this study, the use of similarity matrices, matrix diagrams, and ordination has provided more interesting and insightful information. The presence—absence data for the 14 stations (see Beckett, 1977) and the diagrams (Fig. 4) generated by these matrices showed that the macroinvertebrates of stations 10 through 14 formed a somewhat continuous association for all three sampling periods. The June similarity matrix diagram also showed marked similarity between various upstream and downstream stations. Stations 1 and 10 and stations 4 and 11 were quite similar in terms of species presence or absence. These distant upstream—downstream affinities were not observed for either the August or September sampling periods.

The cause of these affinities in June (and not in August or September) is most probably changing river discharge. Although the colonization period for the June samplers had a mean discharge (1670 m^3/min) similar to that in August (1713 m^3/min), the minimal flow during these periods is probably very important since pollutional impact would be strongest during periods of least dilution. The minimal flow for the August collection period (569 m^3/min) was 27% lower than that of the June collection period (770 m^3/min). In addition, flow over the last week before collection of samples averaged 758 m^3/min for August compared to 1543 m^3/min for June. Also, depressed dissolved-oxygen concentrations observed in August at downstream stations were not apparent in June (Table 2). The comparatively high flow for the June collection period probably ameliorated pollutional influences, resulting in the existence of markedly similar upstream and downstream communities. An ordering of stations along some type of gradient was not apparent for the June data (Fig. 5).

The August data disclosed no marked similarities between communities upstream of station 8 (Fig. 4). All stations downstream from and including 8 (with the exception of station 9) seemed to be quite similar in terms of macroinvertebrate communities. The August ordination (Fig. 5) showed a polarization tendency with upstream stations at an opposite end of the graph from the downstream stations (8, 10 through 14). Stations 5, 6, 7, and 9 were dissimilar from both these upstream and far downstream communities. This separation of the river into an upstream association and far downstream association is in contrast to the June ordination, which shows a more random distribution of upstream and downstream stations. Again, river flow is thought to be quite significant in forming such a pattern. In August a gradient that correlated with such disturbances as sewage-treatment plant outfalls, industrial effluents, and/or thermal discharges was apparent (Fig. 5).

The August ordination can be viewed as intermediary in a progression caused by pollutional effects when compared to the September ordination (Fig. 5). The separation of the river into an upstream and downstream grouping of stations is accentuated in September. The September similarity matrix diagram (Fig. 4) also displayed this conspicuous clumping of communities. Stations 3 and 4 were quite similar to both stations 1 and 2 for this collection period. The macroinvertebrates of station 5, the station immediately below the Tait power-plant discharge, showed strong association with the macroinvertebrates below both the Dayton sewage-treatment plant (station 6) and the Moraine sewage-treatment plant (station 8). Stations 10 through 14 continued to form a closely related cluster.

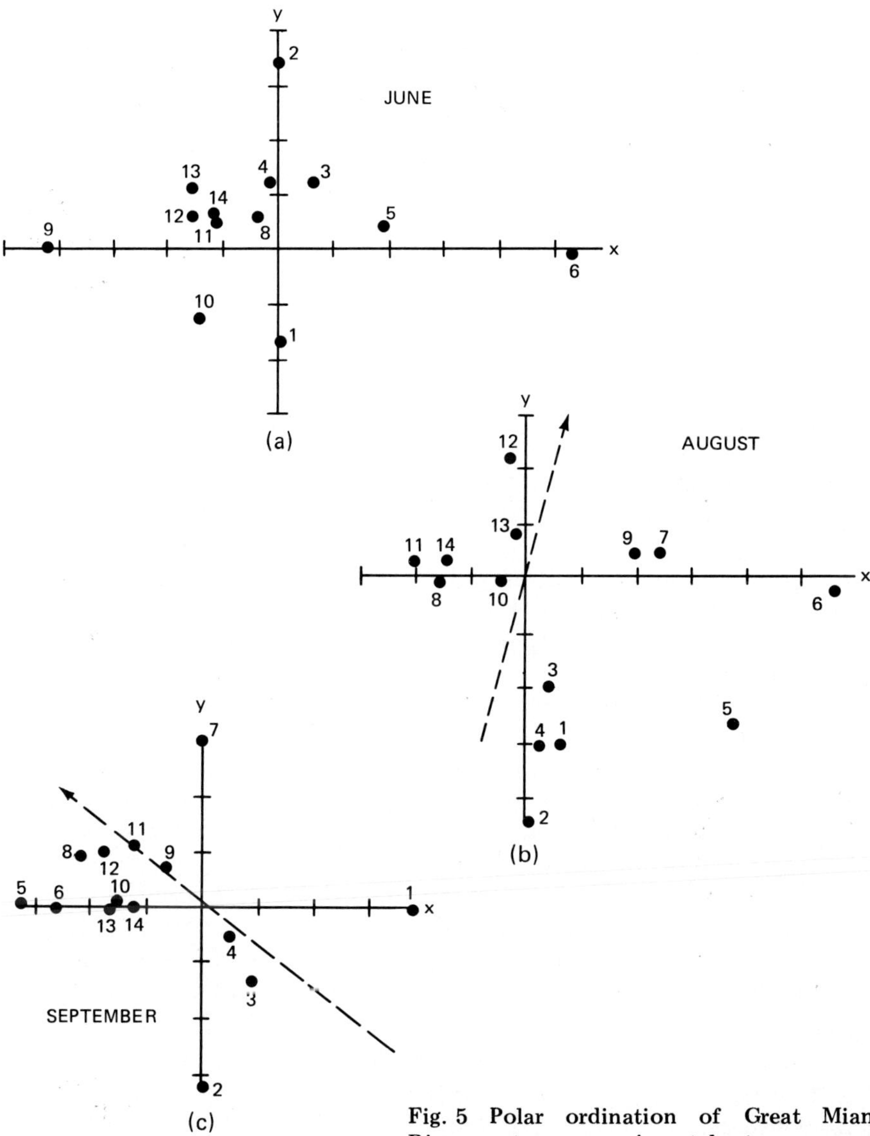

Fig. 5 Polar ordination of Great Miami River system macroinvertebrate communities for June, August, and September 1976. Samples located in relation to two axes by relative similarity (coefficient of community) to end-point samples; units of ecological distance marked on axes are 10% coefficient of community. The first axis is x, and the second axis is y. Dashed lines (August and September) are gradients of pollution disturbance.

TABLE 3

SUMMARY OF GREAT MIAMI RIVER DISCHARGE FOR JUNE, AUGUST, AND SEPTEMBER MACROINVERTEBRATE COLLECTION PERIODS, DAYTON, OHIO

	Sample collection period		
	May 11 to June 21	June 21 to August 9	August 9 to September 22
Mean flow, m^3/min	1670	1713	651
Maximum mean daily flow, m^3/min	5981	6967	970
Minimum mean daily flow, m^3/min	770	569	421
Mean flow last week before sampler collection, m^3/min	1543	758	608

Once again a gradient correlating with wastewater and power-plant discharges seemed apparent (Fig. 5).

The September sampling period featured the lowest mean flow of the three sampling periods, had the lowest minimum daily flow of the stream year (October 1975—September 1976), and had the lowest mean discharge for the week before sample collection (Table 3). It is apparent that pollutional stresses are heightened at this time of the year. With the maximization of pollution influences, the more stenoecious species, e.g., the chironomid *Rheotanytarsus* sp., the mayflies *Baetis* sp., and the *Tricorythodes* sp., have been eliminated, or their numbers significantly reduced, in polluted areas. The downstream areas of the river are then dominated by a few widespread euryecious species, e.g., the chironomids *Einfeldia* sp., *Nanocladius* sp., and *Cricotopus bicinctus*, the planarian *Dugesia tigrina*, and the gastropod *Physa* sp. (see Table 4). This results in a general faunal homogeneity among downstream stations.

Station 5, at the Tait power-plant discharge, showed an interesting change in biotic affinities over the study period. In addition to being influenced by the warm water of the discharge, the macroinvertebrates below the discharge were subjected to some chlorination effects, as chlorine is used to reduce microbial buildup in the plant's condenser tubes. In June, macroinvertebrates at station 5 showed marked similarity to the upstream station (3) (Fig. 4); station 5 was also fairly similar to station 4 (Fig. 5). In August the macroinvertebrate composition of station 5 seemed to be somewhat intermediate between the aggregations of macroinvertebrates at station 4 and station 6 (Dayton sewage-treatment plant) (see Fig. 5). In September, however, station 5 macroinvertebrates showed their strongest affinities with the Dayton sewage-treatment-plant station and the West Carrollton station (8). Both these stations are slightly below the sewage-treatment plants. The probable interpretation of these associations is that, with decreased river flow, a much higher proportion of the water below the plant's discharge comes from the plant discharge itself rather than from directly upstream. Therefore plant effects, such as an increase in river temperature or chlorination, are heightened at this time of year.

Ordinations are usually two-dimensional representations of a multidimensional configuration (Kaesler and Cairns, 1972). By the use of a similarity coefficient, e.g., CC and ordination, a station is reduced to a single multidimensional point in a space defined by axes that reflect changes in species composition (McIntosh, 1973). A factor analysis (Nie et al., 1975) of variations in the macroinvertebrate species and species abundances for the data collected in this study showed 14 common factors to be significant in determining such variations (the factor analysis can estimate the number of significant factors but unfortunately does not identify these factors). It is evident, then, that an interpretation of causal factors may be difficult. This difficulty, due to complexity of factors, can be seen in the June ordination; it was not possible to envision an axis corresponding to an ecocline, known community variables, gradients of disturbance effects, or of community development in the manner of Whittaker (1975). The August and September samples, however, ordered quite well along an oblique axis corresponding to disturbance (shown as dashed lines in Fig. 5). Although other factors may still be somewhat important in variation in species composition during these stress periods, pollution effects become more robust in the ordering of communities and overwhelm the effect of normal variables on faunal composition.

ACKNOWLEDGMENTS

My thanks to Gary Hater, Brad Averill, Joe Redi, Tim Dineen, Tom Federle, Pete Spatt, Dan Strome, and Doug Yeakel for their assistance in the field and in the laboratory. The criticisms and advice of Drs. Michael C. Miller and Thomas C. Kane were very helpful throughout the study.

This study was supported by a grant from Dayton Power and Light Company, Dayton, Ohio.

TABLE 4

DOMINANT (BY NUMBER) MACROINVERTEBRATES BY STATION, GREAT MIAMI RIVER SYSTEM, OHIO, JUNE, AUGUST, AND SEPTEMBER 1976 COLLECTIONS

	Macroinvertebrate taxa*†		
Station	June	August	September
1	*Crangonyx* sp. *Lirceus fontinalis*	*Cheumatopsyche* sp. *Baetis* sp.	*Cheumatopsyche* sp. *Hydropsyche bronta*
	Polypedilum sp. *Phaenopsectra* sp.	*Polypedilum* sp. *Rheotanytarsus* sp.	*Rheotanytarsus* sp. *Polypedilum* sp.
	N = 31	N = 190	N = 443
2	*Cheumatopsyche* sp. *Stenonema pulchellum*	*Stenacron interpunctatum* *Caenis* sp.	*Stenacron interpunctatum* *Dugesia tigrina*
	Conchapelopia sp. *Polypedilum* sp.	*Microtendipes* sp. *Nanocladius* sp.	*Ablabesmyia mallochi* *Glyptotendipes* sp.
	N = 32	N = 67	N = 92
3	*Cheumatopsyche* sp. *Hydropsyche bronta*	*Cheumatopsyche* sp. *Pleurocera* sp.	*Stenacron interpunctatum* *Cheumatopsyche* sp.
	Rheotanytarsus sp. *Nanocladius* sp.	*Nanocladius* sp. *Glyptotendipes* sp.	*Glyptotendipes* sp. *Nanocladius* sp.
	N = 161	N = 324	N = 162
4	*Baetis* sp. *Cheumatopsyche* sp.	*Dugesia tigrina* *Tricorythodes* sp.	*Ischnura* sp. *Dugesia tigrina*
	Nanocladius sp. *Larsia* sp.	*Nanocladius* sp. *Conchapelopia* sp.	*Glyptotendipes* sp. *Procladius* sp.
	N = 86	N = 208	N = 152

Table 4 (Continued)

Station	Macroinvertebrate taxa*†		
	June	August	September
5	*Simulium* sp. *Cheumatopsyche* sp.	*Hydropsyche* sp. *Tricorythodes* sp.	*Physa* sp. *Dugesia tigrina*
	Cricotopus bicinctus *Conchapelopia* sp.	*Cricotopus bicinctus* *Nanocladius* sp.	*Parachironomus pectinatellae* *Cricotopus bicinctus*
	N = 49	N = 197	N = 163
6	*Simulium* sp. *Pericoma* sp.	No nonchironomid macroinvertebrates	*Physa* sp. *Ischnura* sp.
	Cricotopus bicinctus *Nanocladius* sp.	*Cricotopus bicinctus* *Conchapelopia* sp.	*Cricotopus bicinctus* *Conchapelopia* sp.
	N = 87	N = 247	N = 77
7	No samples collected	*Hydropsyche orris* *Hydropsyche* sp.	*Ferrissia* sp. *Argia* sp.
		Cricotopus bicinctus *Conchapelopia* sp.	*Cricotopus bicinctus* *Nanocladius* sp.
		N = 318	N = 328
8	*Physa* sp. *Dugesia tigrina*	*Ferrissia* sp. *Physa* sp.	*Dugesia tigrina* *Physa* sp.
	Nanocladius sp. *Conchapelopia* sp.	*Nanocladius* sp. *Conchapelopia* sp.	*Nanocladius* sp. *Einfeldia* sp.
	N = 256	N = 105	N = 114

(Table continues on following page.)

Table 4 (Continued)

Station	Macroinvertebrate taxa*†		
	June	August	September
9	*Physa* sp. *Dugesia tigrina*	*Dugesia tigrina* *Helobdella triserialis*	*Dugesia tigrina* Oligochaeta
	Nanocladius sp. *Conchapelopia* sp.	*Conchapelopia* sp. *Nanocladius* sp.	*Nanocladius* sp. *Einfeldia* sp.
	N = 169	N = 146	N = 354
10	*Dugesia tigrina* *Crangonyx* sp.	*Ferrissia* sp. *Argia* sp.	*Argia* sp. *Ischnura* sp.
	Conchapelopia sp. *Cricotopus bicinctus*	*Nanocladius* sp. *Cricotopus bicinctus*	*Einfeldia* sp. *Nanocladius* sp.
	N = 211	N = 335	N = 94
11	*Mooreobdella microstoma* *Physa* sp.	*Physa* sp. *Stenacron interpunctatum*	Oligochaeta *Dugesia tigrina*
	Cricotopus bicinctus *Conchapelopia* sp.	*Nanocladius* sp. *Einfeldia* sp.	*Dicrotendipes* sp. *Einfeldia* sp.
	N = 239	N = 165	N = 202
12	*Orconectes* sp. *Physa* sp.	*Argia* sp. *Ischnura* sp.	*Argia* sp. *Ischnura* sp.
	Cricotopus bicinctus *Conchapelopia* sp.	*Nanocladius* sp. *Einfeldia* sp.	*Einfeldia* sp. *Nanocladius* sp.
	N = 132	N = 140	N = 88

Table 4 (Continued)

	Macroinvertebrate taxa*†		
Station	June	August	September
13	*Dugesia tigrina*	*Argia* sp.	*Argia* sp.
	Physa sp.	*Ischnura* sp.	*Perithemis tenera*
	Nanocladius sp.	*Nanocladius* sp.	*Einfeldia* sp.
	Einfeldia sp.	*Einfeldia* sp.	*Dicrotendipes* sp.
	N = 161	N = 317	N = 130
14	*Hydropsyche* sp.	*Physa* sp.	*Dugesia tigrina*
	Argia sp.	*Stenacron interpunctatum*	*Argia* sp.
	Einfeldia sp.	*Glyptotendipes* sp.	*Glyptotendipes* sp.
	Conchapelopia sp.	*Einfeldia* sp.	*Einfeldia*
	N = 150	N = 163	N = 131

*The two most numerous nonchironomid macroinvertebrate taxa are listed first, followed by the two most numerous chironomid taxa.

†N, mean number of macroinvertebrates per sampler.

REFERENCES

American Public Health Association, 1975, *Standard Methods for the Examination of Water and Wastewater*, 14th ed., pp. 440-447, New York.

Beck, W. M., Jr., 1954, Studies in Stream Pollution Biology. I. A Simplified Ecological Classification of Organisms, *J. Fla. Acad. Sci.*, 17: 211-227.

——, and E. C. Beck, 1966, Chironomidae (Diptera) of Florida. 1. Pentaneurini (Tanypodinae), *Bull. Fla. State Mus., Biol. Sci.*, 10: 305-379.

Beckett, D. C., 1977, *Compositional Variation and Ordination of Macroinvertebrate Communities in a Multi-Stressed River System*, M.S. Thesis, University of Cincinnati, Cincinnati, Ohio.

Bray, J. R., and J. T. Curtis, 1957, An Ordination of the Upland Forest Communities of Southern Wisconsin, *Ecol. Monogr.*, 27: 325-349.

Burlington, R. F., 1962, Quantitative Biological Assessment of Pollution, *J. Water Pollut. Control Fed.*, 34: 414-432.

Conn, C. C., 1973, *Biological Survey of the Great Miami River*, Wright State University and Miami Conservancy District, Dayton, Ohio.

Crossman, J. S., R. L. Kaesler, and J. Cairns, Jr., 1974, The Use of Cluster Analysis in the Assessment of Spills of Hazardous Materials, *Am. Midl. Nat.*, 92: 94-114.

Dice, L. R., 1945, Measures of the Amount of Ecological Association Between Species, *Ecology*, 26: 297-302.

Dixon, W. J. (Ed.), 1973, *Biomedical Computer Programs*, pp. 305-330, University of California Press, Berkeley.

Fullner, R. W., 1971, A Comparison of Macroinvertebrates Collected by Basket and Modified Multi-Plate Samplers, *J. Water Pollut. Control Fed.*, 43: 494-499.

Hanson, H. C., 1955, Characteristics of the *Stipa comata-Bouteloua gracilis-Touteloua curtipendula* Association of Northern Colorado, *Ecology*, 36: 269-280.

Harrell, R. C., and T. C. Dorris, 1968, Stream Order, Morphometry, Physico-Chemical Conditions, and Community Structure of Benthic Macroinvertebrates in an Intermittent Stream System, *Am. Midl. Nat.*, 80: 220-251.

Hilsenhoff, W. L., 1975, *Aquatic Insects of Wisconsin, with Generic Keys and Notes on Biology, Ecology, and Distribution*, Wisconsin Technical Bulletin, Vol. 89, pp. 1-82, Department of Natural Resources, Madison, Wisconsin.

Holsinger, J. R., 1972, *The Freshwater Amphipod Crustaceans (Gammaridae) of North America*, pp. 11-15, U. S. Environmental Protection Agency, Cincinnati, Ohio.

Hurd, R. M., 1961, Grassland Vegetation in the Big Horn Mountains, Wyoming, *Ecology*, 42: 459-467.

Hynes, H. B. N., 1970, *The Ecology of Running Waters*, pp. 383-397, University of Toronto Press, Toronto, Ont.

Kaesler, R. L., and J. Cairns, Jr., 1972, Cluster Analysis of Data from Limnological Surveys of the Upper Potomac River, *Am. Midl. Nat.*, 88: 56-67.

Klemm, D. J., 1972, *Freshwater Leeches (Annelida: Hirudinea) of North America*, pp. 1-44, U. S. Environmental Protection Agency, Cincinnati, Ohio.

Kulczynski, S., 1927, Zespoly roslin w Pieninach (Die Pflanzenassociationen der Pieninen.), *Bull. Int. Acad. Polonaise, Sci. Lett., Cl. Sci. Math. Nat., Ser. B*, (Suppl. II): 57-203.

Lewis, P. A., 1974, *Taxonomy and Ecology of Stenonema Mayflies (Heptageniidae: Ephemeroptera)*, pp. 11-55, U. S. Environmental Protection Agency, Cincinnati, Ohio.

Lind, O. T., 1974, *Handbook of Common Methods in Limnology*, pp. 51-54, C. V. Mosby Co., St. Louis.

Little, A. D., 1964, *A Projective Economic Study of the Ohio River Basin*, prepared for the Corps of Engineers by Arthur D. Little Inc., Washington, D. C.

Mason, W. T., Jr., 1973, *An Introduction to the Identification of Chironomid Larvae*, pp. 1-86, U. S. Environmental Protection Agency, Cincinnati, Ohio.

McIntosh, R. P., 1967, The Continuum Concept of Vegetation, *Bot. Rev.*, 33: 130-187.

——, 1973, Ordination and Classification of Communities, in *Handbook of Vegetation Science 5*, R. H. Whittaker (Ed.), pp. 157-191, Dr. W. Junk Publishing Co., The Hague.

Nie, N. H., C. H. Hull, J. G. Jenkins, K. Steinbrenner, and D. H. Bent, 1975, *Statistical Package for the Social Sciences*, pp. 468-514, McGraw-Hill Book Company, New York.

Pielou, E. C., 1966, The Measurement of Diversity in Different Types of Biological Collections, *J. Theor. Biol.*, 13: 131-144.

Sanders, H. L., 1968, Marine Benthic Diversity: A Comparative Study, *Am. Nat.*, 102: 243-282.

Scott, R. D., 1969, *The Macroinvertebrate Biotic Index—a Water Quality Measurement and Natural Continuous Stream Monitor for the Miami River Basin*, Miami Conservancy District, Dayton, Ohio.

Shannon, C. E., 1948, The Mathematical Theory of Communication, in *The Mathematical Theory of Communication*, pp. 3-91, University of Illinois Press, Urbana.

Sorenson, T., 1948, A Method of Establishing Groups of Equal Amplitude in Plant Sociology Based on Similarity of Species Content, and Its Application to Analysis of the Vegetation on Danish Commons, *K. Dan. Vidensk. Selsk. Biol. Skr.*, 5: 1-35.

Simberloff, D., 1972, Properties of the Rarefaction Diversity Measurement, *Am. Nat.*, 106: 414-418.

Uetz, G. W., 1976, Gradient Analysis of Spider Communities in a Streamside Forest, *Oecologia*, 22: 373-385.

United States Environmental Protection Agency, 1973, *Biological Field and Laboratory Methods for Measuring the Quality of Surface Waters and Effluents*, C. Weber (Ed.), Cincinnati, Ohio.

Weston, R. F., 1967, *Water Quality Report—Miami River Basin, State of Ohio*, prepared for Miami Valley Water Quality Committee, Dayton Area Chamber of Commerce, Dayton, Ohio by Roy F. Weston Inc., Environmental Science and Engineering Consultants, West Chester, Pennsylvania.

Whittaker, R. H. (Ed.), 1973, Ordination and Classification of Communities, *Handbook of Vegetation Science 5*, pp. 195-221, Dr. W. Junk Publishing Co., The Hague.

——, 1975, *Communities and Ecosystems*, pp. 112-127, Macmillan Publishing Co., Inc., New York.
Wilhm, J. L., 1970a, Range of Diversity Index in Benthic Macroinvertebrate Populations, *J. Water Pollut. Control Fed.*, 42: R221-R224.
——, 1970b, Effect of Sample Size on Shannon's Formula, *Southwest. Nat.*, 14: 441-445.
——, and T. C. Dorris, 1968, Biological Parameters of Water Quality, *Bioscience*, 18: 447-481.

LONGITUDINAL VARIATION IN THE MACROINVERTEBRATE FAUNA AND WATER CHEMISTRY OF AN EASTERN PENNSYLVANIA TROUT STREAM

PATRICIA T. BRADT, Department of Biology, Lehigh University, Bethlehem, Pennsylvania

ABSTRACT

Chemical and biological parameters in Bushkill Creek, Northampton County, Pennsylvania, were sampled for 17 months to evaluate the benthic macroinvertebrate population and to estimate the future of the stream as habitat for reproducing brown trout. Continuing development in the drainage basin has placed increasing stress on the stream.

Longitudinal zonation was noted both in the macroinvertebrate fauna and in the water chemistry. Total alkalinity, pH, conductivity, nitrate, chloride, and total iron increased downstream. Biomass, total numbers, and number of taxa of benthic macroinvertebrates peaked at a station 10 km above the mouth, then decreased at the downstream station that was located in an urban area. Diversity was highest at the rural upstream station and decreased downstream. Dominant taxa of macroinvertebrates changed from *Stenonema* (Ephemeroptera) at the upstream station, to *Hydropsyche* (Trichoptera) at the two midstream stations, and to Chironomidae (Diptera) at the downstream station. *Hydropsyche* and the Chironomidae are tolerant of a wide range of environmental conditions.

If the stream is to continue to support a reproducing brown trout population in the lower reaches, the stresses on the stream from urbanization must be minimized.

During the past 10 years, Bushkill Creek in Northampton County, Pennsylvania has been subjected to various stresses, including (1) the straightening of 3.2 km of streambed (1968—1969) in conjunction with road building, (2) increased residential and commercial development in the drainage basin, and (3) increasing levels of nutrients.

This study, which covered the period 1972—1974, was undertaken to evaluate the status of the benthic macroinvertebrate community and to estimate the future of the stream as a brown trout habitat.

Bushkill Creek supports a naturally reproducing brown trout (*Salmo trutta*) population. The fish assemblage, however, is dominated by the white sucker, *Catostomus commersoni* (Bradt, 1974). The stream is heavily fished and is also used for swimming and picnicking. Trout habitat is rapidly disappearing from the United States (Needham, 1969), thus Bushkill Creek is a valuable natural resource because it supports reproducing brown trout and is readily accessible to a large numb^r of people.

Road building and attendant rechanneling caused an increase in siltation. Solar energy reaching the stream increased owing to the removal of shade trees and bank cover. The variety of habitats for the benthic macroinvertebrate fauna decreased because the new stream channel was wide and shallow with a uniform substrate (Hynes, 1963). Reconstruction efforts on the stream (1969–1971) involved (1) the installation of gabions (rock current deflectors) to narrow and deepen the stream bed and to increase the flow velocity, (2) the placement of rocks and dams to increase pool and riffle habitat, and (3) the replanting of the banks.

DESCRIPTION OF DRAINAGE BASIN AND SAMPLING STATIONS

The drainage basin of Bushkill Creek covers 206 km^2 (Fig. 1). The main stream runs for 34.2 km and is fed by nine tributaries (Department of the Army, 1972). Originating from springs at an altitude of 420 m on the southern slope of the Kittatiny Mountains, the stream flows southeasterly to its confluence with the Delaware River at Easton. At the base of the mountain (elevation 220 m) more springs enter the stream. The elevation at the mouth of the stream is 54 m above sea level (Department of the Army, 1972).

The stream flows over shale and slate from the base of the Kittatiny Mountains (km 33.6) to the confluence of the Bushkill with the Little Bushkill (km 13). From km 13.0 to the mouth, the stream traverses various limestone formations, according to Young et al. (1972). The upper half of the drainage basin drains forests, meadows, farmland, and a few small towns. The terrain is gently rolling, the stream slope is 14.4 m/km. The lower part of the drainage basin drains the limestone area; this part of the basin is more heavily populated and more developed. The area drained includes agricultural, suburban, and urban areas. The lower quarter of the basin is served by sanitary sewers. Commercial and residential development in the lower part of the drainage basin has been proceeding rapidly with the increased paving of meadowlands.

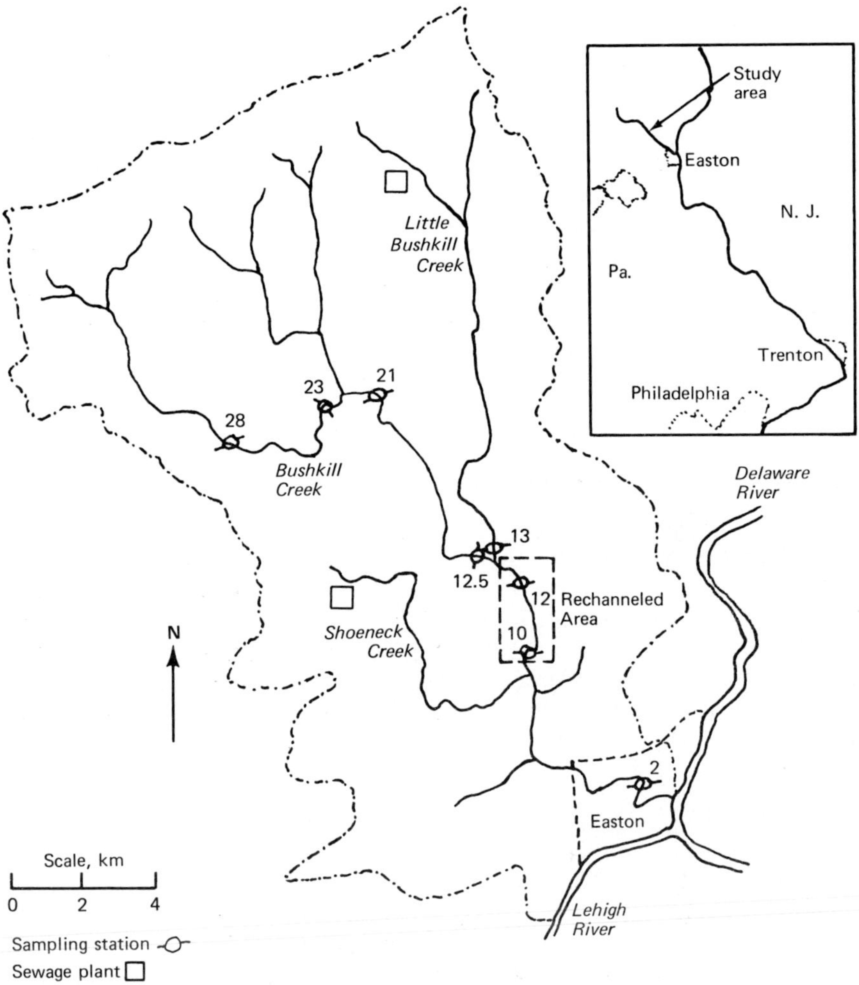

Fig. 1 Bushkill Creek Drainage Basin.

Industries line the banks where the stream enters the urban area of Easton 6.4 km above the mouth.

Stations 28 (km) through 12 (km) are located in rural areas. Station 10 (km) is in an agricultural area, immediately downstream of a 2.0-km section of the rechanneled streambed and the gabion installation. Station 2 (km) is in the city of Easton. Rainfall exerts less impact and stream flow is less variable in the lower part of the drainage basin than in the upper basin. The stream slope averages 9.6 m/km. The lack of overland drainage on the east side of the stream suggests subterranean drainage in the area, and flow studies indicate considerable water entering the stream from limestone springs in the lower basin, according to Young et al. (1972). The slope of the stream increases slightly 4.8 km above the mouth where the stream is rocky and turbulent. The rapid flow and turbulence of this area enables the stream to oxidize industrial pollutants. Two sewage-treatment plants discharge effluents into tributaries. One plant discharges 0.3 million gal/day (secondary treatment) into Little Bushkill Creek 9.6 km above its confluence with the main stem. The other plant discharges 0.5 million gal/day (secondary treatment) into Shoeneck Creek 6.4 km from its entry into the main stem 0.8 km below station 10 (Fig. 1).

Daily average flow, calculated by Young et al. (1972), was 5.41 m^3/sec.

MATERIALS AND METHODS

Water chemistry and the benthic macroinvertebrate fauna were sampled biweekly from March 1972 to September 1973, except during December through March when sampling was monthly. Water chemistry was sampled at eight stations (28, 23, 21, 13, 12.5, 12, 10, and 2). The benthic macroinvertebrates were sampled at four stations (28, 12, 10, and 2). Station numbers represent distance upstream from the confluence of Bushkill Creek and the Delaware River at Easton.

Temperature, pH, total alkalinity, conductivity, orthophosphate, nitrate, chloride, and total iron were measured 31 times at each station over 17 months. Calcium was sampled six times at each station (May to September, 1973). Temperature was measured in the field with a 0°C stem thermometer. A Zeromatic pH meter was used. Total (methyl orange) alkalinity was determined by titration with 0.02*N* sulfuric acid to a pH end point of 4.6, measured with the pH meter (American Public Health Association, 1965). Conductivity was measured with an Industrial Instruments conductivity meter (American Public Health Association, 1965).

Nitrate was measured by the cadmium reduction method (Hach Chemical Company, 1971), and the resulting color was read with a Spectronic 20 spectrophotometer (American Public Health Association, 1965).

Orthophosphate was measured by the modified stannous chloride method (Hach Chemical Company, 1971) and read on the Spectronic 20. Chloride was determined by the Mohr method using titration with silver nitrate (Hach Chemical Company, 1969). Total iron was measured by the 1,10 phenanthroline method (Hach Chemical

Company, 1969), and the resulting color was read on the Spectronic 20. Calcium hardness was determined using titration with disodium dihydrogen-1, 2-cyclohexanediamine-tetra-acetate (CDTA) (Hatch Chemical Company, 1969).

The benthic macroinvertebrate fauna was sampled in riffle areas with a Surber square-foot sampler (Surber, 1936). The mesh size was 9 threads/cm. Two Surber samples were taken at each station, sampling 0.18 m^2 of stream bottom. The mean of the two samples was used in calculating the macroinvertebrate parameters. The sampling stations (28, 12, 10, and 2) were similar in depth (15 to 25 cm) and substrate. The benthic macroinvertebrates were removed from the rocks with a brush, partially sorted from rubble, and preserved in 70% ethyl alcohol. Organisms were identified to genus when possible (Pennak, 1953; Usinger, 1956). Members of the families Elmidae (Coleoptera) and Chironomidae (Diptera) were identified by family. Because Elmidae and Chironomidae were not identified to genus, the diversity index may be biased and slightly low. The same level of resolution was used for all microinvertebrate samples.

The diversity index was derived from the formula of Shannon (Shannon and Weaver, 1963):

$$\bar{d} = \frac{n_i}{n} \log \frac{n_i}{2n}$$

where $\bar{d}$ is diversity, n_i is the number of individuals in the *ith* taxa, and n is the total number of individuals. Wilhm and Dorris (1966; 1968) proposed the use of a diversity index based on information theory to estimate the diversity of aquatic macroinvertebrates. The reliability of a mean diversity index should increase with several collections at each station (Denoncourt and Polk, 1975).

Water quality and macroinvertebrate data were analyzed statistically using the Lehigh Amalgamated Packages for Statistics (LEAPS) Program (Palchak and Koko, 1973). Mean, standard deviation, and correlation matrix were calculated. Significant levels of correlations were calculated from the tables of Edwards (1950).

RESULTS

Chemical and Physical Parameters

The mean temperature at the eight stations on Bushkill Creek did not increase significantly downstream. The mean temperature at station 28 was 12.3°C (SD = 6.4), and the mean temperature at station 2 was 13.2°C (SD = 5.0).

Table 1 lists the means and standard deviations for the chemical parameters. Variation in pH at the eight stations reflects the change in the geological formations which the stream traverses: shale and slate at stations 28, 23, and 21 and limestone at stations 13 through 2. Stations 13 and 12.5 are located at the beginning of the limestone area.

The total alkalinity increases downstream, especially at stations 10 and 2, and was correlated with the downstream station 2 (r = 0.72*).

Conductivity also increased downstream, reflecting an increase in the concentration of the ions of mineral salts. Conductivity was correlated with pH (r = 0.41*), total alkalinity (r = 0.64*), and the downstream station (r = 0.44*), reflecting the relationship between the three parameters and the increase downstream.

The concentration of nitrate in the stream increased downstream and was correlated with alkalinity (r = 0.43*), conductivity (r = 0.36*), and the downstream station 2 (r = 0.39*).

Orthophosphate increased slightly downstream. The mean level at station 28 was 0.05 mg/liter, and the mean level at station 2 was 0.14 mg/liter. The highest mean levels of orthophosphate were recorded at station 13 ($\bar{x}$ = 0.30 mg/liter), a tributary receiving sewage effluent. Orthophosphate levels were correlated with pH (r = 0.23*) and conductivity (r = 0.28*).

An increase in chlorides was also observed as the stream approached the mouth. Chloride levels were correlated with alkalinity (r = 0.43*), conductivity (r = 0.42*), nitrate (r = 0.30*), and the downstream station (r = 0.31*), reflecting the increase in all these parameters toward the mouth.

Total iron levels remained similar at stations 28 through 10 and increased slightly at station 2. The higher levels at station 2 probably reflect the entrance of iron-containing compounds from a mineral-pigment manufacturer 2.4 km upstream of station 2. Total iron was correlated with alkalinity (r = 0.25*), conductivity (r = 0.22*), and the downstream station (r = 0.49*).

Levels of calcium also increased downstream. The mean level at station 28 was 35 mg/liter, at station 12.5, 68.6 mg/liter, and at station 2, 118 mg/liter, reflecting the impact of the limestone areas.

---

*$P < 0.01$, df = 241.

TABLE 1

WATER CHEMISTRY: MEAN, STANDARD DEVIATION, AND RANGE BY STATION (n = 242)

Stations	28	23	21	13	12.5	12	10	2
pH	7.3 ± 0.3 6.6 — 7.7	7.1 ± 0.3 6.3 — 7.7	7.2 ± 0.3 6.6 — 7.7	7.8 ± 0.4 7.2 — 8.5	8.4 ± 0.6 7.3 — 9.3	8.2 ± 0.4 7.2 — 8.9	8.2 ± 0.4 7.3 — 8.8	8.0 ± 0.3 7.5 — 8.4
Alkalinity, $CaCO_3$ mg/liter	25.8 ± 6.0 14.4 — 35.2	27.9 ± 7.4 11.9 — 39.0	26.2 ± 7.3 13.4 — 46.6	49.6 ± 10.1 33.0 — 66.2	44.0 ± 13.5 22.7 — 73.7	46.4 ± 12.9 23.7 — 69.3	87.2 ± 33.4 36.1 — 139.7	125 ± 31.8 57.7 — 167.2
Conductivity, μmho/cm	151 ± 105 75 — 487	194 ± 116 90 — 570	175 ± 101 105 — 456	364 ± 168 220 — 810	284 ± 150 140 — 620	320 ± 161 140 — 736	397 ± 198 180 — 875	529 ± 224 200 — 1050
Nitrate, mg/liter	4.2 ± 2.9 2.0 — 12.1	5 6 ± 3.4 1 0 — 13.0	5.4 ± 2.8 4.0 — 11.8	8.0 ± 3.8 1.2 — 18.6	5.5 ± 2.2 1.1 — 9.8	6.5 ± 3.1 0.8 — 13.0	8.9 ± 2.6 2.6 — 12.6	10.4 ± 2.5 1.3 — 13.0
Chloride, mg/liter	13.3 ± 3.9 5.5 — 21.0	14 2 ± 3.9 7.5 — 23.5	15.0 ± 2.9 9.5 — 20.9	18.4 ± 4.0 11.5 — 27.0	15.7 ± 4.4 7.2 — 27.5	16.4 ± 3.6 9.0 — 24.0	17.7 ± 4.7 11.0 — 30.8	20.0 ± 4.8 11.5 — 34.0
Total iron, mg/liter	0.12 ± 0.11 0.00 — 0.58	0.10 ± 0.08 0.02 — 0.50	0.09 ± 0.08 0.00 — 0.45	0.10 ± 0.11 0.00 — 0.68	0.08 ± 0.06 0.00 — 0.27	0.10 ± 0.05 0.01 — 0.24	0.08 ± 0.07 0.02 — 0.35	0.26 ± 0.13 0.06 — 0.73

Biological Parameters

Total numbers, wet weight, and number of taxa of benthic macroinvertebrates peaked at station 10 (Fig. 2). Diversity was highest at station 28 and lower at stations 12, 10, and 2. According to Wilhm and Dorris (1968), station 28 would be classified as clean water ($\overline{x}$ = 3.42, SD = 0.40), whereas stations 12, 10, and 2 would be classified as mildly polluted (station 12, $\overline{x}$ = 2.89, SD = 0.50; station 10, $\overline{x}$ = 2.97, SD = 0.32; and station 2, $\overline{x}$ = 2.87, SD = 0.55).

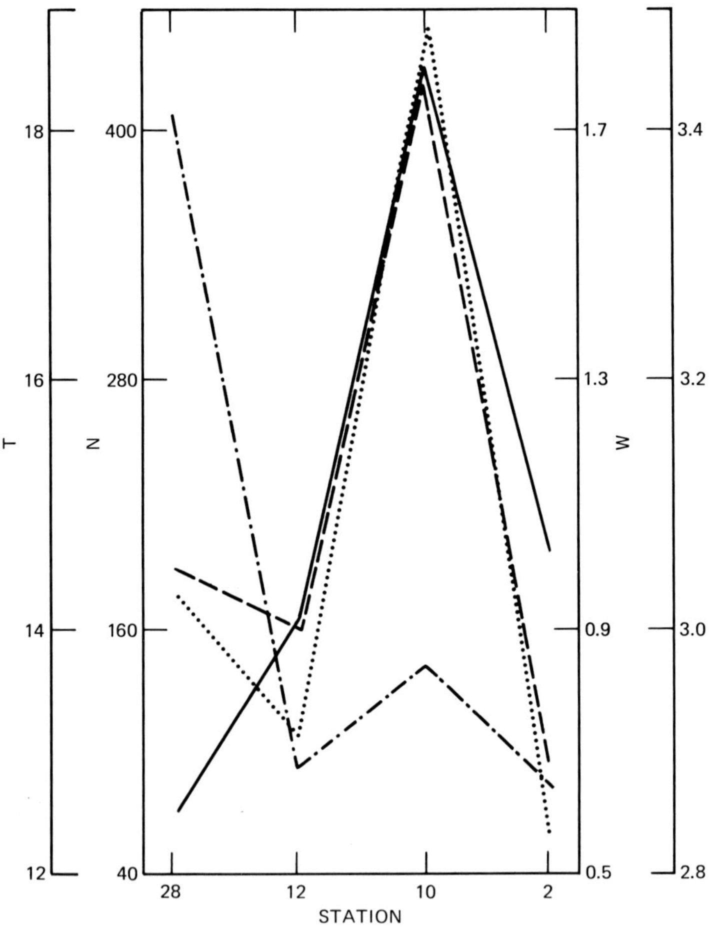

Fig. 2 Mean total numbers, N, ———; wet weight, W, ... ; total taxa, T, — — — —; and diversity of macroinvertebrates, D, —·—·— per 0.092 m^2.

Table 2 shows the different macroinvertebrate families and genera dominating at the four macroinvertebrate stations. In all 50,305 organisms were counted and identified. The mayfly *Stenonema* was dominant at station 28 (18% of the total macroinvertebrates). The caddis fly *Hydropsyche* was dominant at station 12 (30% of the total) and also at station 10 (35% of the total). The family Chironomidae dominated at station 2 (29% of total).

DISCUSSION

Hynes (1972) reports that downstream areas of streams are usually warmer than the headwaters. The reason the temperature in Bushkill Creek does not increase substantially downstream may be that cool water entering the lower drainage basin from limestone springs maintains lower temperatures in the lower reaches. As water from the limestone springs enters the water course, carbon dioxide is lost and the pH rises. The pH also rises in response to carbonates entering the water from the limestone.

The addition of carbonates also contributes to the rise in alkalinity downstream. Industrial effluents in the downstream area

TABLE 2

DOMINANT TAXA OF BENTHIC MACROINVERTEBRATES BY STATION (PERCENT OF TOTAL NUMBERS COLLECTED)

Stations	28	12	10	2
Trichoptera				
Hydropsyche	8	30	35	21
Cheumatopsyche	7	22	7	8
Other Trichoptera	17	2	7	3
Diptera				
Chironomidae	8	10	19	29
Ephemeroptera				
Baetis	1	1	4	13
Ephemerella	8	4	8	2
Stenonema	18	4	1	1
Coleoptera				
Elmidae	11	9	5	3
Total number of organisms collected	4,189	9,601	25,168	11,447

(station 2) may also contribute to the alkalinity. The increase in pH observed as streams flow over limestone areas is accompanied by an increase in alkalinity (Reid, 1961). Hynes (1969) states that the lower reaches of a stream are richer in dissolved nutrients and suspended solids, and this is reflected in the downstream rise in conductivity.

Hynes (1970) suggests that nitrate concentration usually declines downstream as the nutrient is utilized, but nitrate may increase if additions from the surrounding land exceed use by the stream biota. Minckley (1963) also noted a decrease in nitrate downstream. Nitrate, in this study, was highest in downstream suburban and urban areas (stations 10 and 2). Likens and Bormann (1974), however, report that nitrate concentrations are related to the amount of agricultural activity in the drainage basin. Possible sources of nitrate in the downstream area are effluents from the sewage-treatment plants and the limestone springs. Soil percolates may contain more nitrate than surface runoff (Bigga and Corey, 1969).

Minckley (1963) recorded a slight, but consistent, decrease in orthophosphate downstream. The slight increase noted in orthophosphate downstream in this study may be due (1) to orthophosphate concentrations being higher in runoff from urban areas than from rural areas (Owen and Johnson, 1966; Hynes, 1970) and (2) to sewage-plant effluents.

Potential sources of chloride in the stream are minerals in the drainage basin, human or animal sewage, and certain industrial effluents (McKee and Wolf, 1963). The higher mean chloride at station 13 may reflect the effect of the sewage-plant effluent.

The higher iron concentrations at station 2 reflect the influence of the pigment manufacturer and may also result from land runoff and limestone springs (Minckley, 1963).

The plant nutrients, nitrate and orthophosphate, contribute to increased productivity in the stream. Two sewage plants, land runoff, and subterranean drainage probably are the major sources of these nutrients. Algal growths are common in the stream, with diatoms (Chrysophyta) predominating in the winter and *Cladophora* (Chlorophyta) and *Oscillatoria* (Cyanophyta) predominating in the summer (Bradt, 1974). *Cladophora* and *Oscillatoria* are often seen in enriched streams with a high biomass (Hynes, 1969).

The high percentages of the families Hydropsychidae and Chironomidae (Table 2) at Station 12, 10, and 2 parallel the lower diversity at these stations. Members of these two families are ubiquitous and apparently able to cope with the higher nutrient loads in the downstream sections. *Hydropsyche* is an omnivorous net spinner which passively feeds in water rich in organic particulate matter (Hynes, 1970). The Chironomidae are also tolerant of a wide range of environmental conditions (Robach, 1974). When a factor becomes limiting, the most sensitive species disappear, the more tolerant species increase to occupy the vacated niches, and the diversity of the community decreases (Reid, 1961). An increase in the nutrient supply can decrease diversity just as an increase in the intensity of any factor can decrease diversity (Hooper, 1969). Increasing development in the drainage basin has put stress on the lower reaches of the stream. Station 28 is in a rural area with some agricultural lands nearby, but the diversity at station 28 was the highest in the study. Station 12 had diversity ($\bar{x} = 2.89$) similar to station 2 ($\bar{x} = 2.87$). Station 12 is located in an agricultural area. The effluent from the sewage plant on the Little Bushkill may be responsible for the lowered diversity at station 12. The members of the Trichopteran family, Hydropsychidea (*Hydropsyche* and *Cheumatopsyche*), which comprise 55% of the total macroinvertebrate population at station 12 and 42% of the total population at station 10, show preference for open areas with high periphyton density (Hynes, 1972), fast current (Macan, 1962), and high organic loading (Roback, 1974). Station 12 is immediately downstream of a short (183 m) section of rechanneled stream where shading trees were removed and several gabions installed.

Station 10 had the highest total numbers, total taxa, and wet weight. This station is immediately downstream of the 2.0 km of rechanneled streambed and the gabion installation. The stream banks above station 10 were stripped of ground cover, shrubs, and trees during the rechanneling process. The banks were replanted following the rechanneling, but the trees are not yet large enough to shade the stream effectively. Station 10 is the most productive station in terms of biomass ($\bar{x} = 1.87$ g) and has the second highest alkalinity ($\bar{x} = 87.2$ ppm). Armitage (1958) reported that the yearly standing crop of macroinvertebrates and the number of Trichopterans were both correlated with alkalinity. Beyerle and Cooper (1960) found that moderately alkaline streams were more productive than soft-water streams.

Station 2 is influenced by nutrients from the urban area and industrial effluents. The fact that station 2 has a diversity just below 3.0 attests to the effectiveness of rapidly flowing water in maintaining good water quality in the urban area. The mean dissolved-oxygen level at station 2 was 8.9 mg/liter, ranging from 6.0 to 11.8 mg/liter (Bradt, 1974).

Bushkill Creek exhibits longitudinal zonation in chemistry and in the benthic macroinvertebrate community. Longitudinal zonation in streams has been discussed by Minckley (1963), Odum (1971), Hynes (1972), Allen (1975), and Reid and Wood (1976). Total alkalinity, pH, conductivity, nitrate, chloride, total iron, and calcium all increase in the downstream areas. Temperature and orthophosphate levels increase only slightly. The benthic macroinvertebrate community diversity index decreases in the lower reaches of the stream, and the dominant organisms also change. The stream has recovered partially from the effects of rechanneling, but the extent of recovery is difficult to evaluate owing to the lack of studies before the rechanneling.

Urbanization in the form of road building, manufacturing, and land development has stressed the homeostatic mechanisms of terrestrial ecosystems in the drainage basin with the export of increased dissolved and particulate matter into streams. This stress has caused increased rates of eutrophication in aquatic ecosystems (Likens and Bormann, 1974). The longitudinal zonation noted in the stream is evidence of the impact of increased urbanization on the chemistry and biology of the lower stream reaches. If the stream is to continue to support a reproducing brown trout population, the impact of urbanization in the drainage basin must be minimized.

REFERENCES

Allen, J. D., 1975, The Distributional Ecology and Diversity of Benthic Insects in Cement Creek, Colorado, *Ecology*, 56(5): 1040-1053.

American Public Health Association, Inc., 1965, *Standard Methods for the Examination of Water and Wastewater*, 12th ed., American Public Health Association, New York.

Armitage, K. B., 1958, Ecology of Riffle Insects of the Fire Hole River, Wyoming, *Ecology*, 39: 571-580.

Beyerle, G. B., and E. L. Cooper, 1960, Growth of Brown Trout in Selected Pennsylvania Streams, *Trans. Am. Fish. Soc.*, 89: 255-262.

Bigga, J. W., and R. B. Corey, 1969, Agricultural Drainage and Eutrophication, in *Eutrophication: Causes, Consequences, Correctives*, pp. 404-445, National Academy of Sciences, Washington, D. C.

Bradt, P. T., 1974, *The Ecology of the Benthic Macroinvertebrate Fauna of the Bushkill Creek, Northampton County, Pa.*, Ph. D. Thesis, Lehigh University, Bethlehem, Pa.

Cairns, J., Jr., and K. L. Dickson, 1971, A Simple Method for the Biological Assessment of the Effects of Waste Discharges on Aquatic Bottom-Dwelling Organisms, *J. Water Pollut. Control Fed.*, 43(5): 755-772.

Denoncourt, R. F., and J. Polk, 1975, A Five-Year Macroinvertebrate Study with Discussion of Biotic and Diversity Indices as Indicators of Water Quality, Codorus Creek Drainage, York County, Pennsylvania, *Proc. Pa. Acad. Sci.*, 49: 113-120.

Department of the Army, 1972, Philadelphia District, Corps of Engineers, *Flood Plain Information—Bushkill Creek Vicinity of Easton, Pennsylvania*, Philadelphia, Pa.

Edwards, A. L., 1950, *Experimental Design in Psychological Research*, Rinehart and Co., New York.

Hach Chemical Company, 1969, *Water and Wastewater Analysis Procedures.*

——, 1971, *Colorimetric Procedures and Chemical Lists for Water and Wastewater Analysis.*

Hooper, F. F., 1969, Eutrophication Indices and Their Relationship to Other Indices of Ecosystem Change, in *Eutrophication: Causes, Consequences, Correctives*, pp. 225-235, National Academy of Sciences, Washington, D. C.

Hynes, H. B. N., 1963, *The Biology of Polluted Waters*, Liverpool University Press, Liverpool, England.

——, 1969, The Enrichment of Streams, in *Eutrophication: Causes, Consequences, Correctives*, pp. 188-196, National Academy of Sciences, Washington, D. C.

——, 1970, The Ecology of Stream Insects, *Annu. Rev. Entomol.*, 15: 25-42.

——, 1972, *The Ecology of Running Waters*, University of Toronto Press, Toronto.

Likens, G. E., and F. H. Bormann, 1974, Linkages Between Terrestrial and Aquatic Ecosystems, *BioScience*, 24(8): 447-456.

Macan, T. T., 1962, Ecology of Aquatic Insects, *Annu. Rev. Entomol*, 7: 261-288.

McKee, J. E., and H. W. Wolf, 1963, *Water Quality Criteria*, The Resources Agency of California, State Water Resources Control Board.

Minckley, W. L., 1963, The Ecology of a Spring Stream, Doe Run, Meade County, Kentucky, *Wildl. Monogr.*, 11: 1-124.

Needham, J. G., and P. R. Needham, 1962, *A Guide to the Study of Fresh Water Biology*, Holden-Day, Inc., San Francisco.

Needham, P. R., 1969, *Trout Streams*, Revised by C. F. Bond, Charles Scribner's Sons, New York.

Odum, E. P., 1971, *Fundamentals of Ecology*, 3rd ed., p. 320. W. B. Saunders Co., Philadelphia.

Owen, G. E., and M. G. Johnson, 1966, Significance of Some Factors Affecting Yields of Phosphorous from Several Lake Ontario Watersheds, *Mich. Univ. G. Lakes Res. Div. Publ.* 15: 400-410.

Palchak, A. E., and F. W. Koko, Jr., 1973, The LEAPS User's Guide and Manual, Lehigh University Computing Center, prepared by H. J. McFadden, Jr., unpublished.

Pennak, R. W., 1953, *Freshwater Invertebrates of the United States*, The Ronald Press Co., New York.

Reid, G. K., 1961, *Ecology of Inland Waters and Estuaries*, Van Nostrand Reinhold Company, Cincinnati, Ohio.

Reid, G. K., and R. D. Wood, 1976, *Ecology of Inland Waters and Estuaries*, 2nd ed., pp. 393-396, D. Van Nostrand Co., New York.

Roback, S. S., 1974, Insects (Arthropoda: Insecta), in *Pollution Ecology of Freshwater Invertebrates*, C. W. Hart, Jr., and S. L. H. Fuller (Eds.), pp. 313-376, Academic Press, Inc., New York.

Shannon, C. E., and W. Weaver, 1963, *The Mathematical Theory of Communication*, University of Illinois Press, Urbana.

Surber, E. W., 1936, Rainbow Trout and the Bottom Fauna Production in One Mile of Stream, *Trans. Amer. Fish Soc.*, 66: 193-202.

Usinger, R. L., 1956, *Aquatic Insects of California,* University of California Press, Berkeley.

Wilhm, J. L., and T. C. Dorris, 1966, Species Diversity of Benthic Macroinvertebrates in a Stream Receiving Domestic and Oil Refinery Effluents, *Am. Midl. Nat.*, 76(2): 427-449.

——, 1968, Biological Parameters for Water Quality Criteria, *BioScience,* 18: 477-481.

Young, R., et al., 1972, Environmental Baseline Study of the Bushkill Creek, National Science Foundation Sponsored Undergraduate Research Project, Lafayette College, Easton, Pa., unpublished report.

THE EFFECT OF IONIZING RADIATION ON THE THERMAL TOLERANCE OF MOSQUITOFISH

B. G. BLAYLOCK and M. L. FRANK
Oak Ridge National Laboratory, Environmental Sciences Division, Oak Ridge, Tennessee

ABSTRACT

This study determined the effects of acute ionizing radiation on the thermal tolerance of the mosquitofish, *Gambusia affinis. Gambusia* of approximately the same size were given doses of radiation ranging from 500 to 5000 rads from a ^{60}Co source. The irradiated fish were tested 5, 15, and 30 days after irradiation for either lethal temperature with a critical thermal maximum (CTM) type test or mean survival time in a lethal temperature bath (39.5°C). The lethal temperature for *Gambusia* irradiated with 500 rads was not significantly different from that of control fish; neither were lethal temperatures for fish irradiated with 1500 rads and tested at 5 and 30 days after irradiation. At day 15 and 1500 rads, however, lethal temperature was significantly reduced. At a dose of 1500 rads, histopathological effects were observed in the hematopoietic tissues of *Gambusia* 5 days after irradiation; regeneration of the tissue began by day 15 and continued through day 30. The mean survival time of *Gambusia* was not affected by a dose of 500 rads; but, at doses of 1500 rads and above, the effect was significantly different at day 5. A significant difference was also detected at day 15 for the fish receiving a 1500-rad dose. Recovery had taken place by day 30, however, and the mean survival times were not significantly different from controls.

The effects of the interaction of temperature and irradiation on aquatic organisms have been examined by several investigators. Essentially all these studies have been directed toward determining the effects of low sublethal temperatures on the expression of radiation effects. Gros and Bloch (1957) and Gros et al. (1958) hypothesized that low temperature protects against radiation damage. Some subsequent studies (Egami and Etoh, 1966; Etoh and Egami, 1965; Angelovic, White, and Davis, 1969) showed that reducing postirradiation temperatures prolonged the survival of fish, but other investigations (Etoh and Egami, 1967; Aoki, 1964; Aoki, Egami, and Arai, 1966; Egami, 1969) revealed that radiation lesions continued to develop at cold temperatures and were expressed when the fish were returned to a warmer environment. A small change (5°C) in the postirradiation temperature can significantly influence the expression of radiation effects in fish (Blaylock and Mitchell, 1969).

Although all these studies addressed the effects of radiation and temperature on aquatic organisms, only an experiment by Ophel and Judd (1966) investigated the converse; i.e., the effects of radiation on the thermal tolerance of fish. In their study goldfish exposed to ^{90}Sr, ^{131}I, or $^{90}Sr + ^{131}I$ were subjected to a lethal temperature. In 15 of 17 tests, goldfish that had accumulated a maximum estimated dose from internal emitters of 10^4 rads to the bone and tissue and 10^5 rads to the thyroid survived longer than did control fish. In this case the radionuclide burden did not impair the ability of the fish to withstand lethal temperatures. In fact, under the experimental conditions the survival time was increased.

The purpose of our study was to determine the effects of acute external exposures to ionizing radiation on the thermal tolerance of the mosquitofish, *Gambusia affinis*. Our results should help to answer questions arising at public hearings on nuclear power reactors concerning the effects on aquatic biota of the interaction of temperature and radiation. Although the estimated radiation dose accumulated by aquatic biota from routine release of radionuclides at operating nuclear power plants would be insufficient to produce detectable radiation effects (Blaylock and Witherspoon, 1976), the effects of radiation on the thermal tolerance of aquatic organisms should be documented to establish the synergistic effects of radiation and temperature at high and at low temperatures.

MATERIALS AND METHODS

Two experiments were conducted to determine the effect of ionizing radiation on the ability of *Gambusia* to tolerate high temperatures. In the first experiment *Gambusia* were irradiated and subjected to a critical thermal maximum type test (CTM) at intervals up to 30 days postirradiation. Hutchison (1961) described the CTM as the arithmetic means of the collective thermal points at which an animal loses its ability to escape from conditions that will promptly lead to its death when it is heated from a previous acclimation temperature at a constant rate just fast enough to allow body temperature to follow environmental test temperature without a significant time lag. In our study, however, exposure to high temperature continued until opercular beating ceased. In the second experiment irradiated *Gambusia* were exposed to a lethal temperature bath, and their mean survival time was determined.

Fish used in the CTM type experiment were from a *Gambusia* population that had been maintained at 25 ± 1°C under laboratory conditions for over 2 years. The *Gambusia* for the survival-time experiment were collected from a local population and acclimated to laboratory conditions before irradiation. Both experiments made use of female *Gambusia* not obviously laden with young and ranging in length from 2.7 to 3.5 cm.

The fish were irradiated in a Gammacell 200 (Atomic Energy of Canada Limited) containing ^{60}Co at a dose rate of 2.5 rads/sec. Doses of either 500, 1500, 3000, or 5000 rads were given.

After irradiation the fish were maintained at 25 ± 1°C and fed freeze-dried brine shrimp until tested. *Gambusia* for the lethal-temperature-bath experiment were maintained in the same manner, except that after 8 days the acclimation temperature was inadvertently changed to 27 ± 1°C and remained at that temperature until after day 15, when it was returned to 25°C.

Gambusia that had received a dose of either 0, 500, or 1500 rads were subjected to a CTM type test at 5, 15, and 30 days after irradiation. *Gambusia* that had received doses of either 3000 or 5000 rads were tested 3 days after irradiation. Critical thermal maximum tests were conducted on 5 to 10 *Gambusia* at each dose by placing an individual fish in a 1000-ml flask containing 800 ml of water from the tank in which the fish had been maintained after irradiation. The flask was heated by a hemispherical heater controlled by a variable voltage transformer. The rate of temperature elevation was 1.0°C/min. The temperature at which *Gambusia* lost their ability to escape (CTM) and the temperature at which opercular beating ceased were recorded. Cessation of opercular beating was chosen as the biological end point for comparative purposes because less variability was observed in the results.

In the second experiment, *Gambusia* that had received doses of either 0, 1500, 3000, or 5000 rads were exposed to a lethal temperature bath 5, 15, and 30 days after irradiation. During the test the fish were kept in small plastic containers (18 cm in diameter by 14 cm high), which were submerged in 5.5-gal aquariums maintained at 39.5°C. Each container was divided into two compartments by a stainless-steel screen, and groups of either three or four *Gambusia*, the maximum number that could be accurately observed, were confined in each compartment. Portions of the sides and bottom of the containers were replaced with stainless-steel screen to allow for water circulation. Each compartment had a screen cover to keep the fish from swimming to the surface of the water. The temperature of the test compartments was continually monitored. Irradiated and control groups were alternated between test compartments. Cessation of opercular movement was used as the biological end point for survival.

An analysis of variance test was used to compare the results for each dose and time. Duncan's multiple range test was then used to distinguish the significantly different ($P < 0.05$) results.

Three *Gambusia* were taken from the 500- and 1500-rad dose levels at days 5, 15, and 30 for pathological studies. They were fixed in Bouin's fluid, and, after decalcification was completed, routine histological techniques—embedding in paraffin, sectioning on a rotary microtome at 5 μ, and staining with hematoxylin and eosin—were used to prepare the tissue for microscopy.

RESULTS

The results of the CTM type tests for irradiated *Gambusia* are shown in Table 1. Only at 1500 rads and day 15 was the value significantly different from the controls ($P < 0.05$). The greatest range of individual CTM values (38.1 to 40.6°C) was also found at this dose and time. The CTM values were very consistent, with a majority of the fish having CTMs of 40°C or above, except for those irradiated at 1500 rads at day 15. For this treatment 7 of the 10 fish had CTM values of less than 40°C.

TABLE 1

LETHAL TEMPERATURES (°C) FOR *Gambusia affinis* AFTER ACUTE DOSES OF IONIZING RADIATION

Dose, rads	Days after irradiation*				
	3	5	10	15	30
0	40.3 (0.11)	40.1 (0.15)	40.2 (0.12)		40.3 (0.24)
500		40.2 (0.05)	40.2 (0.16)	40.3 (0.17)	40.4 (0.10)
1500		40.1 (0.09)		39.5† (0.24)	40.2 (0.08)
3000	40.2 (0.08)				
5000	40.4 (0.07)				

*Values in parentheses are standard errors.
†This CTM differed significantly ($P < 0.05$) from the others.

In the CTM type test, either 5 or 10 *Gambusia* were scheduled to be tested for each dose and time, but injury to fish during testing or loss during time after irradiation resulted in less than 10 fish at each dose for the 30-day test.

Results of the mean survival times after exposures to lethal temperatures for irradiated *Gambusia* are given in Table 2. Ten fish were tested at each dose and time period, except for the 30-day time period, when 6 fish were tested. Thirty fish were irradiated at doses of 3000 and 5000 rads, but none survived for the 15- and 30-day tests. Many significant differences were observed which can be explained by dose response and recovery patterns.

TABLE 2

MEAN SURVIVAL TIME (MIN) OF IRRADIATED *Gambusia affinis* AFTER EXPOSURE TO LETHAL TEMPERATURE (39.5°C)

Dose, rads	Days after irradiation*		
	5	15	30
0	47.8^{c} (3.8)	80.4^{a} (4.0)	$38.2^{c,d}$ (6.2)
1500	24.6^{d} (2.1)	65.2^{b} (5.3)	23.3^{d} (7.0)
3000	25.5^{d} (3.2)		
5000	7.6^{e} (3.2)		

*Values in parentheses are standard errors. Means with different superscripts (a to e) differ significantly ($P < 0.05$).

Doses up to 5000 rads tested at either 3 or 5 days after irradiation did not significantly change the lethal-temperature values of *Gambusia* (Table 1). However, a dose effect on the mean survival time was observed at day 5 for doses of 1500 rads and above (Table 2). The survival times for the 1500- and 3000-rad fish at day 5 were significantly different from the controls, but they were not significantly different from each other. When the dose was increased to 5000 rads, however, the survival time was significantly reduced. The mean survival time for a dose of 1500 rads was significantly different from the controls at days 5 and 15 but not at day 30. Apparently after 30 days the fish had recovered sufficiently from radiation injury so that the mean survival times of the irradiated fish and controls were not significantly different.

The mean survival time of the control fish at day 15 was significantly different from that of the control fish at days 5 and 30. This difference can be attributed to an acclimation temperature of

27°C, instead of 25°C, inadvertently maintained for a period of 1 week before the 15-day test. The effect of a different acclimation temperature was also observed in the irradiated fish at the 1500-rad dose level, in which the 15-day mean survival time was also significantly greater than the 5- and 30-day mean survival time. The acclimation temperature was returned to 25°C after the 15-day tests, and the mean survival time of the controls at 30 days returned to a level that was not significantly different from the controls at 5 days.

The tissues of three *Gambusia* irradiated with 500 or 1500 rads were examined for pathological changes at 5, 15, and 30 days after irradiation. No pathological effects were observed at any time period at the 500-rad dose level. At 1500 rads after 5 days, atrophy of the hematopoietic tissues and gonads was observed, but, as expected at this dose level, no effects were observed on the intestinal epithelium. At day 15 postirradiation, atrophy of the gonads was complete, and some regeneration of the hematopoietic tissues was observed. At day 30 regeneration of the hematopoietic tissues was continuing.

DISCUSSION

Our experiment demonstrated by use of CTM and mean-survival-time tests that acute doses of ionizing radiation significantly reduce the thermal tolerance of *Gambusia*. These results are in opposition to those of Ophel and Judd (1966), who reported that irradiation from internal emitters increased the thermal tolerance of goldfish. Ophel and Judd suggested that impairment of the heat-transfer properties of the gills or reduction in blood circulation caused by irradiation injury would result in a slower increase of body temperature, thereby prolonging the mean survival time.

In our experiment *Gambusia* suffered heat shock immediately after being placed in a temperature bath at 39.5°C, but they recovered after a short time before succumbing to the lethal temperature. After the initial recovery from heat shock, the fish showed a great deal of variation in their activity, and this appeared to influence their survival time. Very active fish died in a shorter time than did fish that remained inactive until death. In the experiment of Ophel and Judd (1966), a high body burden of ^{131}I or ^{90}Sr could have possibly influenced the response of goldfish to lethal temperature; however, it was difficult to relate increase in mean survival time of the fish to high doses of radiation. Conversely, the results of our experiments, in which an acute dose of 1500 rads significantly decreased the mean survival time, can be related to radiation effects.

Except for those receiving a 1500-rad dose at day 15, the CTMs for *Gambusia* in our experiment were very consistent. Acclimation temperatures can influence the CTM of *Gambusia* (Otto, 1973); however, in our experiment the irradiated and control fish were maintained at the same acclimation temperature. Thus the difference in CTM values was attributed to radiation injury.

Irradiation death of fish receiving doses less than 2000 rads is usually attributed to damage to the hematopoietic tissues (Aoki, 1964; Aoki, Egami, and Arai, 1966; Egami and Etoh, 1966; Etoh, 1969; Cosgrove et al., 1975). An early and rapid depletion of hematopoietic cells takes place in proportion to radiation dose, with recovery from sublethal doses beginning in about 2 weeks. Depression of blood cells in the circulatory system reportedly lags behind the loss of cells in the hematopoietic tissues (Cosgrove et al., 1975). The period, 2 weeks after irradiation, appears to be a crucial time for the survival of irradiated fish kept at about 25°C (Egami, 1969). Fish that survive this period usually show regeneration of tissues and have a good chance of recovering from radiation damage.

An acute dose of 500 rads did not influence the CTM values of *Gambusia* at 5, 15, or 30 days. Pathological studies of the hematopoietic tissues support these results; no damage to the tissues was observed at a dose of 500 rads. Atrophy of the hematopoietic tissues occurred 5 days after a dose of 1500 rads, but regeneration of the tissue had started by day 15 and continued at day 30. Only at day 15 was the CTM influenced by radiation effects. Apparently radiation lesions had not developed sufficiently after 5 days to produce an effect, and by the end of 30 days regeneration of tissue and recovery from radiation effects had taken place.

Even doses of 3000 and 5000 rads produced no effect on the CTM values 3 days after irradiation. At these dose levels the majority of deaths are attributed to intestinal damage (Hyodo, 1964; Egami and Etoh, 1966). The development of radiation lesions in fish is temperature and time dependent, however, and apparently development of the lesions had not progressed sufficiently to influence CTM at 3 days. Mean survival time and CTM type tests were not conducted at 15 and 30 days after irradiation for fish that had received doses of 3000 and 5000 rads because insufficient numbers of *Gambusia* survived.

The effect of radiation on the temperature tolerance of *Gambusia* was more evident when mean survival time was used as a test. Although the lethal temperature obtained with the CTM type test was affected only at day 15 postirradiation, the effect on the mean survival time was expressed at both the 5- and 15-day intervals.

Although the difference in survival time between doses of 1500 and 3000 rads was not significant, survival time was considerably reduced at a dose of 5000 rads (Table 2).

It is evident from this study that an acute dose of ionizing radiation influences the thermal tolerance of *Gambusia*. A dose of 1500 rads, however, which was required to affect the thermal tolerance of *Gambusia*, is much higher than the doses expected for aquatic biota in the vicinity of a nuclear power station. On the other hand, 500 rads, which is also much higher than doses expected near a nuclear power station, did not cause a significant reduction of thermal tolerance. Blaylock and Witherspoon (1976) estimated doses of less than 1 rad/day to aquatic biota from routine releases of radionuclides from nuclear power stations. Such doses could not be expected to produce any effect on the thermal tolerance of fish.

ACKNOWLEDGMENTS

The research reported here was sponsored by the U. S. Department of Energy under contract with Union Carbide Corporation. Environmental Sciences Division Publication No. 1208.

We wish to express our appreciation to G. E. Cosgrove, Biology Division, Oak Ridge National Laboratory (ORNL), for histological studies; to J. J. Beauchamp, Computer Sciences Division, ORNL, for his advice on the statistical analysis of the data; and to Carol P. Allen, Environmental Sciences Division, ORNL, for technical assistance.

REFERENCES

Angelovic, J. W., J. C. White, Jr., and E. M. Davis, 1969, Interactions of Ionizing Radiation, Salinity, and Temperature on the Estuarine Fish, *Fundulus heteroclitus*, in *Symposium on Radioecology*, D. J. Nelson and F. C. Evans (Eds.), USAEC Report CONF-670503, pp. 131-141, NTIS.

Aoki, Kazuko, 1964, The Effect of Temperature on the Development of Histological Damage of Hematopoietic Tissue Following Whole-Body X-Irradiation in the Goldfish, *Carassius auratus*, *Annual Report, 1963*, Report NIRS-3, pp. 35-36, National Institute of Radiological Sciences, Chiba, Japan.

——, Nobuo Egami, and Ryoichi Arai, 1966, Histological Changes in Interrenal Tissue of the Goldfish, *Carassius auratus*, Following X-Irradiation, *J. Radiat. Res.*, 7(2): 83-90.

Blaylock, B. G., and T. J. Mitchell, 1969, The Effect of Temperature on the Dose Response of *Gambusia affinis affinis* from Two Natural Populations, *Radiat. Res.*, 40: 503-511.

——, and J. P. Witherspoon, 1976, Radiation Doses and Effects Estimated for Aquatic Biota Exposed to Radioactive Releases from LWR Fuel-Cycle Facilities, *Nucl. Saf.*, 17: 351-361.

Cosgrove, G. E., B. G. Blaylock, G. U. Ulrikson, and P. H. Cohan, 1975, Radiation-Induced Hematopoietic Lesions in Fish, in *The Pathology of Fishes*, W. E. Ribelin and G. Migaki (Eds.), pp. 463-476, The University of Wisconsin Press, Madison.

Egami, Nobuo, 1969, Kinetics of Recovery from Injury after Whole-Body X-Irradiation of the Fish *Oryzias latipes* at Different Temperatures, *Radiat. Res.*, 37: 192-201.

——, and Hisami Etoh, 1966, Effect of Temperatures on the Rate of Recovery from Radiation-Induced Damage in the Fish *Oryzias latipes*, *Radiat. Res.*, 27: 630-637.

Etoh, Hisami, 1969, Changes in ^{3}H-Thymidine Incorporation into Hematopoietic Cells of Goldfish During Recovery Period from Radiation Injury, *Annot. Zool. Jpn.*, 42(3): 159-165.

——, and Nobuo Egami, 1965, Effect of Temperature on Survival Period of the Fish, *Oryzias latipes*, Following Irradiation with Different X-Ray Doses, *Annot. Zool. Jpn.*, 38(3): 114-121.

——, and Nobuo Egami, 1967, Damage Accumulation and Recovery in the Fish *Oryzias latipes* Exposed to Fractionated or Protracted Radiation at Different Temperatures, *Radiat. Res.*, 32: 884-891.

Gros, C. M., and J. Bloch, 1957, Influence de la température sur la survie d'un poisson *Carassius carassium* (L) après irradiation totale, *C. R. Soc. Biol.*, 151: 602-604.

——, R. Keiling, J. Bloch, and J. P. Villain, 1958, Action combince du froid et des rayons X sur poisson, *Radiobiol. Lat.*, 1: 361-375.

Hutchison, V. H., 1961, Critical Thermal Maxima in Salamanders, *Physiol. Zool.*, 34: 92-125.

Hyodo, Y., 1964, Effect of X-Irradiation on the Intestinal Epithelium of the Goldfish *Carassius auratus*. I. Histological Changes in the Intestine of Irradiated Fish, *Annot. Zool. Jpn.*, 37: 104-111.

Ophel, I. L., and J. M. Judd, 1966, Effects of Internally Deposited Radionuclides on the Thermal Tolerance of Fish, *Disposal of Radioactive Wastes into Seas, Oceans and Surface Waters*, Symposium Proceedings, Vienna, 1966, pp. 825-831, STI/PUB/126, International Atomic Energy Agency, Vienna.

Otto, R. G., 1973, Temperature Tolerance of the Mosquitofish, *J. Fish. Biol.*, 5: 575-585.

RESPONSES OF ICHTHYOPLANKTON TO STRESSES SIMULATING PASSAGE THROUGH A POWER-PLANT CONDENSER TUBE

GERALD V. POJE, THOMAS C. GINN,* and JOSEPH M. O'CONNOR
New York University Medical Center, Institute of Environmental Medicine, Tuxedo, New York

ABSTRACT

A power-plant condenser-tube simulator that can subject entrainable organisms to temperature changes, pressure and flow variation, biocides, and fluid-induced stresses similar to those at existing power-plant condenser tubes was designed and constructed. Larvae of carp (*Cyprinus carpio*) and striped bass (*Morone saxatilis*) were exposed for 5 min total duration to a range of entrainment stresses in three experimental designs: (1) Temperature was varied while a flow velocity of 1.22 m/sec and a minimum pressure of 0.51 atm was maintained; (2) flow was varied, and a constant, sublethal temperature increase ($\Delta T = 0$ and 8.2°C for bass or 16.7°C for carp) was maintained; and (3) biocide concentrations (sodium hypochlorite) were varied at constant velocity (1.22 m/sec) and ΔT (0 and 8.3°C or 11.1°C). Carp larvae had significantly higher latent mortalities at 18.1°C ΔT (36.1°C final temperature), and striped bass were most sensitive at 15.5°C ΔT (33.6°C). A complicated survival pattern developed because of the combined stresses.

The past quarter century has seen a rapid proliferation of steam-electric power plants using once-through cooling. Among the stresses encountered by organisms entrained in the cooling-water flow are thermal shock, turbulent flow and velocity-induced shear forces, subatmospheric pressures, and variations in dissolved oxygen concentrations (Coutant, 1970). In addition, many installations treat cooling-water systems with biocides to defoul the surfaces of heat exchangers (condenser tubes; Beauchamp, 1969).

*Present Address: Tetra Tech, Inc., Lafayette, California.

It was assumed that the mechanical effect of entrainment could be assessed by comparing samples taken at plant intakes with those taken at discharges. Similarly, the stresses of temperature change and biocide concentration could be isolated by sampling during plant outages and during chlorination. Marcy (1973) used these techniques to estimate entrainment mortality as 100% for clupeid larvae at the Connecticut Yankee nuclear power plant; but Lauer and co-workers (1974) showed significant mean differences in survival of *Morone* sp. larvae of only 5 to 39% between intake and discharge samples at the Indian Point nuclear power plant.

Widely divergent estimates of entrainment mortality are caused by several factors. First, standard methods for mortality evaluation have not been established; second, different power-plant designs result in varying stress profiles; and third, the ability of an organism to tolerate stress is seasonal, age specific, and species specific. In addition, the inherent temporal and spatial "patchiness" of natural populations frequently precludes statistically adequate samples of each species for in-plant comparison (Wrenn, 1976). Most importantly, collection-induced mortality (death caused by impingement on collection nets) may be high enough to totally eliminate the calculation of plant-induced mortality (O'Connor and Schaffer, 1977).

These problems prompted laboratory investigations that subjected plankton to simulations of the individual stresses encountered during power-plant passage. Schubel, Koo, and Smith (1976) exposed striped bass eggs and larvae to several thermal stress regimes. Thermally induced mortalities were apparent after short-term exposures to temperatures 15°C above an ambient of 20°C. Morgan and co-workers (1976) developed a system for exposing striped bass eggs and larvae to shear stress. Although their LD_{50} values exceeded those occurring during power-plant passage, the possibility of shear-induced damage to entrained bass was considered real (Ulanowicz, 1975).

Rapid pressure changes, especially exposure to subatmospheric pressure, may be lethal to striped bass eggs and larvae (Beck, Poje, and Waller, 1975). Gas bubble disease, caused by thermal and/or pressure changes, is recognized as problematic at several existing power-generation facilities (Harvey, 1975).

Biocides, particularly the various dissolved forms of chlorine, have measurable lethal and sublethal effects on phytoplankton, zooplankton, and fishes (Brungs, 1973; Eppley, Renger, and Williams, 1976; Zillich, 1972).

As in field studies, problems are encountered in assessing entrainment impact by these laboratory data. The possible synergistic relationship of the stresses must be quantified.

We have investigated the responses of striped bass and carp larvae to simulated power-plant condenser-tube passage by using a unique experimental apparatus that closely approximates the thermal, hydraulic, mechanical, and biocide stresses of condenser tubes.

MATERIALS AND METHODS

Simulator Operation

The power-plant condenser-tube simulator can subject entrainable-sized organisms to variable flow velocities, temperature changes (ΔT), biocide concentrations, and pressure changes. All stresses occur simultaneously during passage through a condenser tube of a length, diameter, and construction comparable to that in a power plant. The apparatus consists of a source-tank system, circulation pump, injection station, hypochlorite pumping system, dual condenser tubes to which heat is applied, and collection tanks for each condenser tube (Fig. 1).

A variable-speed feed pump (pump 1 in the figure) supplies water from any available water source (the simulator is portable) to a 950-liter reservoir (A), which is held at a constant static head (±50 mm of water) by an overflow drain. The centrifugal circulation pump (pump 2) propels water through the system at precisely controlled flow velocities over the range from 0.5 to 3.0 m/sec.

Organisms are injected into the flow of the simulator in an isolation loop (B), and loop pressure is equalized to that of the circulating water. Organisms are entrained and accelerated instantaneously to the selected flow by operating ball valves that shunt the entire flow through the isolation loop.

The water then enters a vertical section of pipe which divides at a y junction (C). Pressure, temperature, and flow are monitored by a gauge before the split. Dual sections of clear plastic piping allow the investigator to view organisms after the split. The biocide system feeds into the main simulator flow above the split. The biocide injection system (D) consists of a 200-liter reservoir tank that feeds a variable speed diaphragm pump (pump 3). The flow of biocide can be manipulated to supply either one or both condenser loops.

The vertical piping constricts into 25.65-mm ID condenser tubes above the inlet from the biocide system. The tubes are composed of

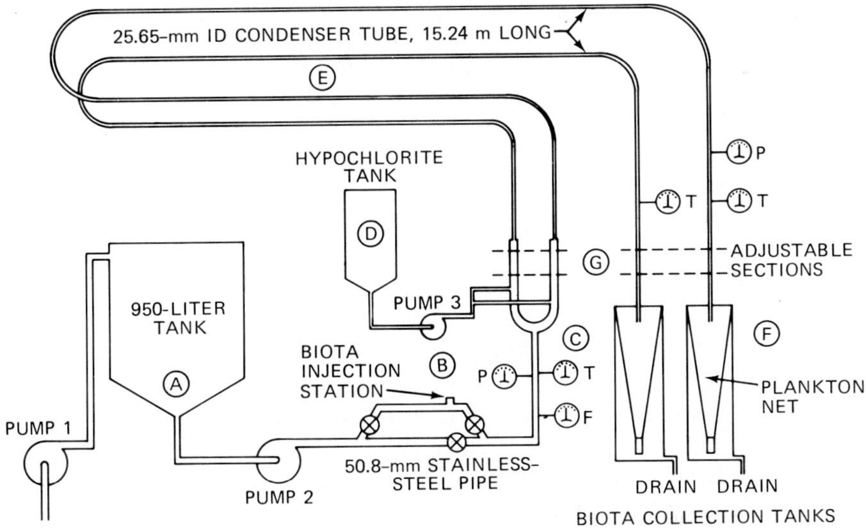

Fig. 1 Schematic diagram of the power-plant condenser-tube simulator. A, source tank; B, injection station for organisms; C, flow-splitting unit; D, biocide injection system; E, condenser tubes; F, fiber-glass collection-tank system with plankton nets; G, area of the simulator where height adjustments are made. Various gauge positions are noted.

15.24-m lengths of type-304 stainless steel arranged in two horizontally displayed, U-shaped loops (E); the inner loop has a 1.38-m radius and the outer, a 1.69-m radius. Pressure and temperature gauges are mounted on the discharge ends of the condenser tubes.

Water exiting the condenser tubes flows downward through vertical pipes positioned centrally over a collection system (F) consisting of dual 200-liter fiber-glass cylinders containing 0.5-m conical plankton nets (571-μ mesh). The vertical piping has manually operated valves that can cause the flow to by-pass the collection tanks and move directly to the drain to facilitate removal of the nets without additional buffeting from the flow.

Heating System

As in power-plant condenser systems, water is heated during a single pass through the simulator. Each tube is embedded in machined, octagonal, aluminum blocks with grooves cut into their surfaces to accommodate electrical resistance heaters. The power to the heaters is supplied independently to each condenser tube. At maximum flows through the simulator, a ΔT of 0 to 25°C can be achieved rapidly on one or both loops and maintained at tolerances of ±0.25°C. Temperature is monitored before the y junction and at the end of each condenser tube.

Pressure

As the speed of the circulating pump increases, the pressure of water flowing through the simulator increases. In addition, the weight of the column of water draining the condenser tube exerts a subatmospheric pressure on the system. Varying the length of these sections (G) alters the magnitude of the suction pressure. The elevation of the condenser tubes can be adjusted to three heights above the base skid, 3.05, 4.39 and 5.49 m. By varying flow and height of the condenser loops, we can regulate pressures at the condenser inlet over the range from 0.5 to 1.5 atm (Fig. 2). Pressure is monitored by a gauge before the y junction and at the entrance and exit of the condenser tubes.

Experimental Design

Although the invertebrates *Gammarus tigrinus*, *Chaoborus* sp., and *Daphnia magna* have also been tested, only experiments with striped bass and carp larvae are described here. They were chosen because of their availability and their susceptibility to entrainment and because they represent species with contrasting physiologies. Carp are eurythermal physostomes (with open swim bladders), and striped bass are more stenothermal and are physoclists (with closed swim bladders). Carp were wild spawned, and striped bass were hatchery spawned. Both species were hatched and reared in quarry water from Verplanck, N. Y., the same water supply used during simulator operation and in larval holding facilities.

For each simulator run at least 60 organisms were placed in and passed through the simulator and held in the collection net for 5-min total exposure. The split flow divided the organisms to each condenser loop. The frequency distribution of the ratio of organisms collected from each tank was a normal approximation of the binomial distribution ($P = 0.5$ and $\alpha = 0.05$).

For all experiments duplicate sets of 30 larvae were counted and placed in flow-through holding containers as absolute controls. They indicated overall mortality in the population and mortality from handling and observational stress. Static "barrel" controls, consisting of 30 organisms placed in flow-through containers suspended in each collection tank, were run in all experiments. These groups were

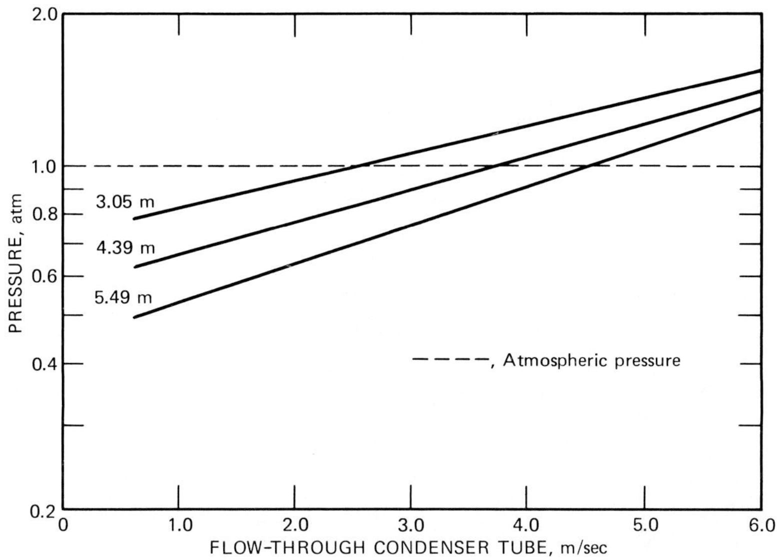

Fig. 2 Pressure at the inlet to the condenser tube vs. flow velocity at three elevations above the simulator skid.

exposed to the biocide concentrations and thermal stresses applied to the condenser loop. Neither the fluid-induced stresses of condenser passage nor the stress of net collection was experienced by barrel controls. Flow controls were used in experiments where flow and temperature were the variables examined. These consisted of 30 organisms poured into each collection net and held for a 5-min exposure. The stresses encountered included the thermal changes in the condenser loop and the mechanical damage of net collection but not the fluid-induced damage of condenser passage.

After the 5-min exposure period, all organisms were rinsed in quarry water and returned immediately to ambient conditions. All test groups were held at ambient temperature in flow-through containers equipped with individual quarry water supplies. Saturation of dissolved oxygen did not fall below 90% in the source-tank (intake) and collection-tank (discharge) water. Mortalities were determined immediately after exposure, and latent effects were observed at 24 and 48 hr.

Both species were subjected to a range of exposures for each variable examined (Table 1). For temperature experiments the lethal limits of exposure regimes and ΔT for each species were the objectives. For flow and biocide experiments, we assumed that the

TABLE 1

EXPERIMENTAL REGIMES TESTED

Organism	$\bar{x}$ size (range)	Variable examined	Exposures	Flow,* m/sec	Temp., °C
Striped bass	8.52 mm (5.2—8.6)	ΔT, °C	0, 8.3, 11.1, 13.9, 15.5	1.22	18.1 ambient
Striped bass	7.35 mm (4.3—8.8)	Flow, m/sec	0.61, 1.22, 1.83		17.2 ambient and 28.3 (ΔT = 11.1°C, sublethal)
Striped bass	8.69 mm (5.3—12.5)	Chlorine, ppm	0, 0.31, 0.59, 1.09, 2.47	1.22	21.1 ambient and 29.4 (ΔT = 8.3°C, sublethal)
Carp	7.60 mm (6.3—8.7)	ΔT, °C	0, 11.1, 13.9, 16.7, 18.1, 18.6, 19.4	1.22	18.1 ambient
Carp	8.02 mm (6.6—9.3)	Flow, m/sec	0.61, 1.22, 1.83		18.3 ambient and 35.0 (ΔT = 16.7°C, sublethal)
Carp	9.54 mm (6.9—11.5)	Chlorine, ppm	0, 0.42, 0.87, 1.90, 3.75, 7.81	1.22	19.3 ambient and 30.4 (ΔT = 11.1°C, sublethal)

*Flow in the condenser tube.

synergies of fluid-induced stresses and ΔT or of biocide and ΔT would be most apparent at sublethal temperatures near the lethal limit for the species. Therefore, striped bass flow experiments were run at ambient temperature and at 8.3°C ΔT, and carp received 16.6°C ΔT. In the biocide experiments, bass were exposed at ambient temperature and at 8.3°C ΔT. Carp were run at ambient temperature and at 11.1°C ΔT. Chlorine (as total residual chlorine) was measured by a Wallace and Tiernan amperometric titrator. Free available chlorine was the most prevalent species.

RESULTS

Temperature Experiments

At 24 hr after exposure to short-term thermal stress, striped bass larvae showed a mortality pattern that depended on exposure regime and on the magnitude of thermal stress [Fig. 3(a)]. Larvae exposed to turbulent flow in the collection net (simulator-run and flow-control groups) had similar mortalities, which were higher than those exposed to static-run thermal stresses (barrel controls) ($P < 0.012$). By 48 hr the mean mortality of all static controls had increased from 27.8 to 47.9%, and the analysis of variance disclosed no differences among exposure regimes ($P < 0.33$).

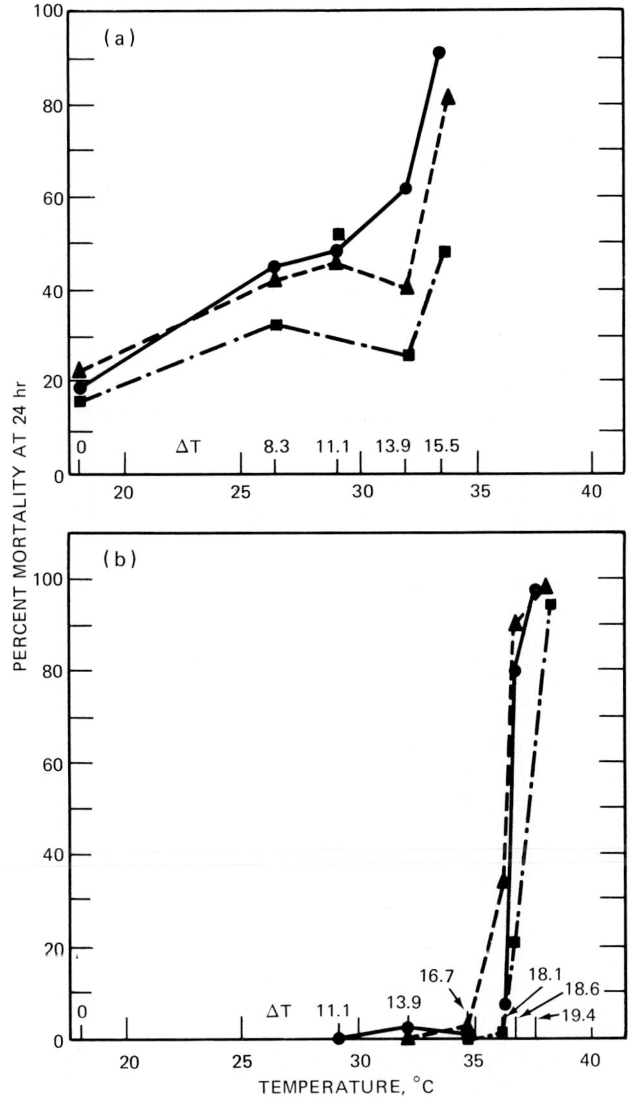

Fig. 3 **Percent mortality of (a) striped bass,** ***Morone saxatilis,*** **and (b) carp,** ***Cyprinus carpio,*** **larvae, 24 hr after 5-min exposures to several temperature increases and three exposure regimes in the condenser tube simulator. ●——●, simulator. ▲——▲, flow control. ■—·—■, barrel control.**

Analysis of the various temperature exposures showed significant differences in their mortalities at 24 hr ($P \ll 0.001$). The data were arranged in five groups according to magnitude of ΔT and subjected to a multiple range test by the Student—Newman—Keuls procedure to identify homogeneous subsets. Bass exposed to 15.5°C ΔT (final temperature, 33.6°C) were isolated from other groups by their higher mortalities. At ambient temperature and at 8.3°C ΔT (final temperature, 26.4°C), survival of bass was greater than at any other ΔT exposure. The results showed no roe-specific variance. In addition, there were no interactions among variables (roe, batch, test regime, and ΔT) which generated variances indicative of synergistic behavior.

The mortality of the barrel-control striped bass, 42.7% at 15.5°C ΔT (33.6°C), compared favorably with that reported in the literature. Schubel, Koo, and Smith (1976) showed that short-term static exposure ~4 to 6 min of bass to a ΔT of 15.0°C (35.0°C) resulted in 48% mortality after 24 hr.

Carp larvae demonstrated a much greater tolerance to ΔT [Fig.3(b)]. Absolute controls showed 100% survival to 48 hr. Increased mortality occurred only at temperatures above 16.7°C ΔT (34.8°C). At 19.4°C ΔT (37.8°C), 24-hr survival was 1.7% for simulator runs, 1.6% for flow controls, and 3.3% for barrel controls. Over a 1.6°C rise (36.1 to 37.8°C), barrel-control mortality increased by 90%.

At the 24-hr observation, the test regime to which the larvae were exposed contributed no significant variance to survival ($P < 0.194$). At 48 hr, however, the carp exposed to turbulent flow in the collection net (simulator runs and flow controls) had a higher mortality rate than those in the barrel controls ($P < 0.052$).

Flow Experiments

Kedl and Coutant (1976) exposed striped bass larvae to passage through a simulated power-plant condenser at 2.1 m/sec and low ΔT. Their study and ours did not identify any mortality attributable to condenser-tube passage at typical power-plant velocities and low ΔT. We pursued this line of investigation by additional exposures of bass larvae at ambient temperature, a typical power-plant ΔT of 11.1°C (28.3°C), and flows of 0.61, 1.22, and 1.83 m/sec. Schubel (1975) calculated that the general condenser-tube velocity was in the range of 1.5 to 2.5 m/sec.

Responses of striped bass larvae to changing flow velocities varied with temperature exposure. At 24 hr after exposure to ambient temperature and low flows (0.61 and 1.22 m/sec), simulator-run groups had mean mortalities of 39.7 and 37.2% above the ambient-exposed static group (32.7% mortality), and flow controls had mean increases in mortality of 24.3 and 28.7% above the static group [Fig. 4(a)]. There were no significant differences between the two groups. At 11.1°C ΔT, mean mortalities of simulator runs at 0.61 and 1.22 m/sec were 43.2 and 44.2%, respectively; mean flow-control mortalities were 3.8 and 14.1% above the thermal-exposed static controls (44.3% mortality) [Fig. 4(b)]. Survivals for the two test regimes were compared at the same velocity by Student's t-test. At 0.61 m/sec the difference was significant ($P < 0.025$) and at 1.22, $P < 0.067$.

When flow velocities increased to 1.83 m/sec, the mortality caused only by the collection net increased beyond simulator-induced mortality with or without ΔT [Figs. 4(a) and 4(b)].

Carp had 100% survival at 48 hr under all test regimes with a flow velocity of 0.61 m/sec and at either ambient (18.3°C) or 16.7°C-ΔT (35.0°C) exposure (Table 2). Increasing flow velocity to 1.22 m/sec resulted in a maximum mortality of 5.2%. At a velocity of 1.83 m/sec and 16.7°C ΔT, simulator-run and flow-control groups had mortality of 22.0% at 48 hr.

TABLE 2

SURVIVAL OF CARP LARVAE AFTER EXPOSURE TO THREE CONDENSER FLOW VELOCITIES

		Survival, %					
		Ambient temp. (18.3°C)			35.0°C (16.7°C ΔT)		
Sample	Flow velocity, m/sec	Immediate	24 hr	48 hr	Immediate	24 hr	48 hr
Simulator run	0.61	100	100	100	100	100	100
Flow control	0.61	100	100	100	100	100	100
Barrel control	No flow	100	100	100	100	100	100
Simulator run	1.22	100	100	98.1	100	98.5	98.5
Flow control	1.22	100	98.3	98.3	100	98.3	94.8
Simulator run	1.83	100	100	96.7	100	89.8	78.0
Flow control	1.83	100	98.3	98.3	100	78.0	78.0

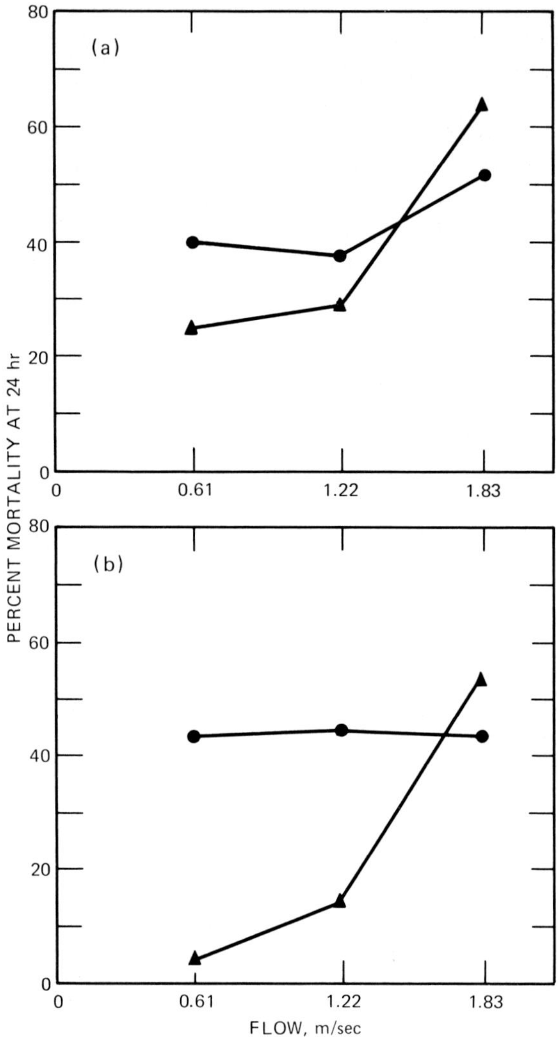

Fig. 4 Percent mortality of striped bass larvae 24 hr after exposure to three flow velocities in the condenser tubes. (a) Ambient temperature, 17.2°C. (b) 11.1°C ΔT. ▲, flow control. ●, simulator exposed. Mortality is that in excess of respective barrel controls, 32.7% for (a) and 44.3% for (b).

Chlorine Experiments

Striped bass larvae exposed to different chlorine concentrations with and without simultaneous temperature elevations of 8.3°C ΔT (29.4°C) displayed a complex pattern of mortality at 24 hr (Fig. 5). There were no differences in mortality among barrel controls exposed at ambient temperature at all chlorine doses. Barrel controls exposed to elevated temperatures had significantly higher mortalities ($P < 0.05$) at free chlorine concentrations of 1.09 and 2.47 ppm. At free chlorine concentrations of 0.31, 0.59, 1.09, and 2.47 ppm, all simulator-run groups had higher mortalities than barrel controls at ambient temperature.

Carp exposed to chlorine doses at ambient temperature displayed increased mortality (>10%) at 24 hr and at doses above 1.90 ppm. Simultaneous exposure to 11.1°C ΔT (30.4°C) and 1.90 ppm chlorine produced 90.2% mortalities in simulator-exposed organisms and 52.0% mortalities in barrel controls. Variable response among replicates prevented statistical comparisons among groups.

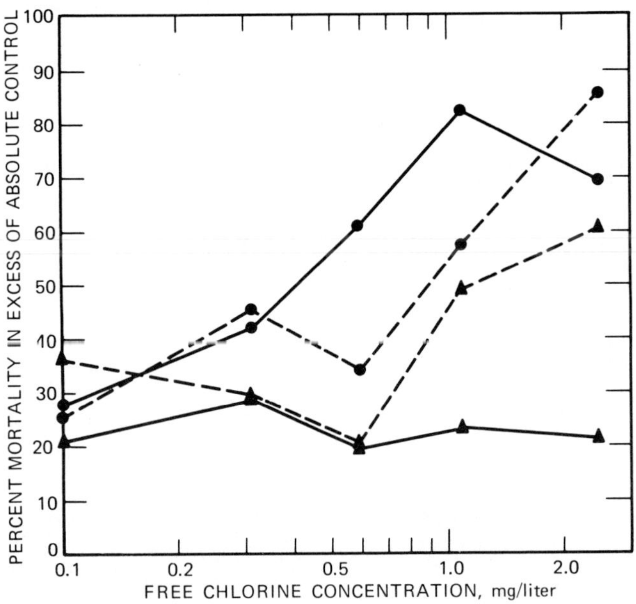

Fig. 5 Percent mortality of striped bass larvae at 24 hr after a 5-min exposure to several doses of free chlorine. ●, simulator exposed. ▲, barrel control. — — —, exposure at 8.3°C ΔT (29.4°C actual temperature). ———, exposure at ambient temperature (21.1°C).

DISCUSSION

The condenser-tube simulator represents an important development in the field of entrainment research. This apparatus provides a controlled experimental system that can test individual and combined stresses encountered by biota entrained in power-plant condensers. The split flow design permits simultaneous exposure of test organisms to different concentrations of biocides and changes in temperature at a variety of flow velocities. The research presented here is preliminary in nature and seeks primarily to demonstrate the fact that testing of condenser-related stresses is feasible in a simulator. In addition, relevant data on synergistic stress phenomena can be obtained with a controlled-variable testing device. The results from these experiments are consistent with data from laboratory studies with the same species and shed new light on power-plant impact studies conducted in the field with conventional techniques.

In experiments under conditions similar to those expected in operating steam-electric stations (Coutant, 1970), the power-plant condenser simulator confirms that short-term exposure to thermal stresses in the range of $\sim 10^{\circ}C$ ΔT in the condensers is most likely not the major cause of ichthyoplankton mortality at power stations. Thermal-induced mortality of striped bass in the condenser simulator was similar to that derived by Schubel, Koo, and Smith (1976) in a model system and by Lauer et al. (1974) in static tests.

The simulator tests verified the hypotheses of Marcy (1975) and of Kedl and Coutant (1976) that mechanical damage to organisms in the circulating water system of power plants was of significance in entrainment impact. Fluid-induced stresses (see, e.g., Morgan et al., 1976) were identified as contributing to the mortality of striped bass but not of carp.

Synergism between temperature and fluid-induced stress was identified in the simulator as contributing to collection mortality. O'Connor and Schaffer (1977) suggested that the differential effects of collection mortality may preclude valid impact assessment at operating power stations. The condenser tube simulator has demonstrated that collection mortality is affected by both flow velocity and thermal exposure. O'Connor and Schaffer tested striped bass collection mortality at different velocities but did not test the effect of thermal or biocide shock. Our results demonstrate that true estimates of cross-plant mortality may be most accurate under near-lethal conditions for the species in question. Under moderate conditions of ΔT, fluid-induced stress, and chlorination, such as those seen at most operating stations, collection mortality may be as great as or greater than mortality caused by plant passage. Since the questions of collection mortality, mechanical damage, and thermal shock have yet to be resolved, further simulator tests will certainly yield valuable results for detailed analysis.

Collection mortality in the simulator system is of great significance. Future testing will be conducted, not only with standard collection devices but also with the newly developed LMS larval table (McGroddy and Wyman, 1977).

Several controversies presently reign with regard to power-plant operation in biologically active areas. Views on the resolution of these controversies differ widely and tend to be supported with great tenacity and emotion, despite an overall paucity of relevant data. At least some of these factors, e.g., fluid-induced stress and various physical—chemical synergies, lend themselves to analysis with the simulator and other devices like it.

ACKNOWLEDGMENTS

The research reported here was funded in part by New York State Energy Research and Development Authority and in part by Grant No. ES from the National Institute of Environmental Health Sciences.

We would like to thank the staff of Consolidated Edison Company of New York, Inc., for logistic support and Texas Instruments, Inc., for supplying fish larvae from the Consolidated Edison hatchery at Verplanck, N. Y. Gary Jinks, Stacey Rigdon, and Cindy Griffin of New York University Medical Center provided valuable assistance in carrying out the research.

REFERENCES

Beauchamp, R. S. A., 1969, The Use of Chlorine in the Cooling Water System of Coastal Power Station, *Chesapeake Sci.*, 10: 280.

Beck, A. P., G. Poje, and W. Waller, 1975, A Laboratory Study on the Effects of Exposure of Some Entrainable Hudson River Biota to Hydrostatic Pressure Regimes Calculated for the Proposed Cornwall Pumped-Storage Plant, in *Fisheries and Energy Production: A Symposium*, S. B. Saila (Ed.), pp. 167-204, Lexington Books, D. C. Heath & Company, Lexington, Mass.

Brungs, W. A., 1973, Effects of Residual Chlorine on Aquatic Life, *J. Water Pollut. Control Fed.*, 45: 2180-2193.

Coutant, C. C., 1970, Biological Aspects of Thermal Pollution. I. Entrainment and Discharge Canal Effects, *CRC Crit. Rev. Environ. Control*, 1(3): 341-381.

Eppley, R. W., E. H. Renger, and P. M. Williams, 1976, Chlorine Reactions with Seawater Constituents and the Inhibition of Photosynthesis of Natural Marine Phytoplankton, *Estuarine Coastal Mar. Sci.*, 4: 147-161.

Harvey, H. H., 1975, Gas Disease in Fishes—A Review, in *Chemistry and Physics of Aqueous Gas Solutions,* W. A. Adams (Ed.), pp. 450-485, Electrochemical Society, Inc., Princeton, N. J.

Kedl, R. J., and C. C. Coutant, 1976, Survival of Juvenile Fishes Receiving Thermal and Mechanical Stresses in a Simulated Power-Plant Condenser, in *Thermal Ecology II,* ERDA Symposium Series, Augusta, Ga., Apr. 2—5, 1975, G. W. Esch and R. W. McFarlane (Eds.), pp. 394-400, CONF-750425, NTIS.

Lauer, G. J., et al., 1974, Entrainment Studies on Hudson River Organisms, in *Proceedings of the Second Workshop on Entrainment and Intake Screening,* L. D. Jensen (Ed.), pp. 37-82, Report No. 15, Electric Power Research Institute Publication No. 74-049-00-5, Palo Alto, Calif.

McGroddy, P. M., and R. L. Wyman, 1977, Efficiency of Nets and a New Device for Sampling Living Fish Larvae, *J. Fish. Res. Board Can.*, 34: 571-574.

Marcy, B. C., Jr., 1973, Vulnerability and Survival of Young Connecticut River Fish Entrained at a Nuclear Power Plant, *J. Fish. Res. Board Can.*, 30(8): 1195-1203.

——, 1975, Entrainment of Organisms at Power Plants with Emphasis on Fishes—An Overview, in *Fisheries and Energy Production: A Symposium,* S. B. Saila (Ed.), pp. 89-106, Lexington Books, D. C. Heath & Company, Lexington, Mass.

Morgan, R. P., II, R. E. Ulanowicz, V. J. Rasin, Jr., Linda A. Noe, and G. B. Gray, 1976, Effects of Shear on Eggs and Larvae of the Striped Bass, *Morone saxatilis,* and the White Perch, *M. americana, Trans. Am. Fish. Soc.*, 105(1): 149-154.

O'Connor, J. M., and S. Schaffer, 1977, Survival of Striped Bass Ichthyoplankton in Nets, *Chesapeake Sci.*, 18(3): 312-315.

Schubel, J. R., 1975, Some Comments on the Thermal Effects of Power Plants on Fish Eggs and Larvae, in *Fisheries and Energy Production: A Symposium,* S. B. Saila (Ed.), pp. 31-54, Lexington Books, D. C. Heath & Company, Lexington, Mass.

——, T. S. Y. Koo, C. F. Smith, 1976, Thermal Effects of Power Plant Entrainment on Survival of Fish Eggs and Larvae: A Laboratory Assessment, in Special Report 52, PPRP-13, Chesapeake Bay Institute, The Johns Hopkins University, Baltimore, Md.

Ulanowicz, R. E., 1975, The Mechanical Effects of Water Flow on Fish Eggs and Larvae, in *Fisheries and Energy Production: A Symposium,* S. B. Saila (Ed.), pp. 77-88, Lexington Books, D. C. Heath & Company, Lexington, Mass.

Wrenn, W. B., 1976, Preliminary Assessment of Larval Fish Entrainment, Colbert Steam Plant, Tennessee River, in *Thermal Ecology II,* ERDA Symposium Series, Augusta, Ga., Apr. 2—5, 1975, G. W. Esch and R. W. McFarlane (Eds.), pp. 381-385, CONF-750425, NTIS.

Zillich, J. A., 1972, Toxicity of Combined Chlorine Residuals to Fresh Water Fish, *J. Water Pollut. Control Fed.*, 44: 212-220.

EFFECTS OF SIMULATED ENTRAINMENT ON THE BIOLOGY OF A FRESHWATER CLADOCERAN

ARTHUR L. BUIKEMA, JR., SARA R. SHERBERGER,
GREGORY W. KNAUER, LINDA A. NEWBERN,
JEFFREY T. READING, and JOHN CAIRNS, JR.
Biology Department and Center for Environmental Studies, Virginia
Polytechnic Institute and State University, Blacksburg, Virginia

ABSTRACT

Under simulated entrainment conditions *Daphnia pulex* were exposed separately and in combination to temperature (10 and 14°C above 20°C ambient), chlorine (0.35 to 0.50 ppm), and mechanical stress (4 liters/min through a 12.5-mm tube). In a 12-min period, maximum exposure occurred for less than 2 min, followed by dilution and/or cooling. After exposure the animals were monitored for mortality, adenosine triphosphate (ATP), metabolism, feeding, growth, molting, reproduction, and behavior. Chlorine and combined stress significantly reduced the levels of ATP. A 14°C stress significantly depressed metabolism. Mechanical stress significantly increased metabolism, and the combined stress had no significant effect on metabolism. Stresses of 10 and 14°C caused a decrease in feeding. Mechanical stress accounted for 30 to 80% acute mortality. Survivors of stress were observed for latent effects. Chlorine reduced growth rates. No significant effects on reproduction were observed when the animals were fed high-quality food. Poor food quality caused a significant reduction in the number and vitality of young produced by females exposed to 10°C stress or to combined stress. There were no significant effects of entrainment on molting, behavior, or long-term survival.

Electric power generating facilities may use large quantities of water for cooling purposes. During passage through a cooling system, zooplankton populations may be exposed to various combinations of thermal, chlorine, and mechanical stress. The few studies available on entrained zooplankton suggest that the factors affecting zooplankton populations are site specific (e.g., Coutant and Talmadge, 1977).

Heinle (1976) and Lanza et al. (1975) noted that copepod mortality was caused primarily by chlorine and that elevated temperature and mechanical stress had little effect on mortality. Davies, Hanson, and Jensen (1976) also found that elevated temperature and entrainment had little effect on copepod mortality. However, Carpenter, Peck, and Anderson (1974) noted that hydraulic stress was the major factor contributing to copepod mortality, accounting for 70% mortality.

Astrauskas and Rachyunas (1975) noted ~50% reduction in a zooplankton population after entrainment, but Buss (1975) and Benda and Gulvas (1976) stated that entrainment accounted for less than 15% mortality in Lake Michigan zooplankton. Most of this mortality was attributed to mechanical stress. Storr (1974) also noted that mechanical stress accounted for a major part of total mortality. Davies and Jensen (1975), summarizing a study of three mid-Atlantic power plants, concluded that zooplankton mortality was not affected by mechanical stress per se but by a combination of ambient temperature, temperature rise, mechanical stress, and chlorination levels.

Freshwater cladocera appear to be the most sensitive of the zooplankters to entrainment (Bunting, 1974); marine cladocera seemingly are not affected (Markowski, 1960; 1962). Mechanical stress appears to be the most important factor for large cladoceran mortality (Harmsworth, 1974) although temperature may be a limiting factor for some species (Whitehouse, 1971). The effect of chlorination is well known (see, e.g., Brungs, 1973), and its principal effects were on mortality and reproduction of *Daphnia magna.*

The primary purpose of our research was to look at the interactions of thermal, chlorine, and mechanical stresses in the biology of a representative cladoceran, *Daphnia pulex.* The results may be of value in identifying research tools to quantify the impact of entrainment on zooplankton.

MATERIALS AND METHODS

Daphnia pulex cultures were maintained at 20°C and 150 lux intensity under cool-white fluorescent bulbs. Culture and test water was soft (40 ppm hardness), carbon-dechlorinated, Blacksburg, Virginia, tap water. *Daphnia* cultures were fed ad libitum with log phase *Chlamydomonas reinhardii* (wild, − strain), and the algae were cultured in Bold's basal medium.

Simulated entrainment exposure was modeled after that of Coutant (1970; 1974) (Fig. 1). In a 12-min period, the animals were

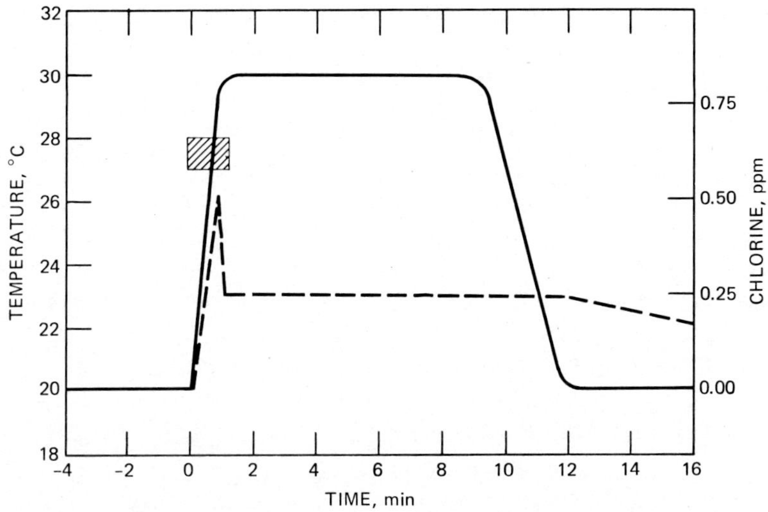

Fig. 1 Time course for thermal, chlorine, and mechanical stresses applied to *Daphnia pulex*. Temperature, ———. Chlorine, — — —. Pumping, ▨.

exposed separately and in combination to temperature, chlorine, and mechanical stresses. The Coutant model for time of exposure to various stresses was as realistic a choice as any model designed for a specific power plant would have been. Coutant (1970; 1974) said that the details do vary among power plants. The effects of pressure were not studied because Beck, Poje, and Waller (1975) demonstrated that survival of larger crustaceans was not significantly affected by pressure changes. After exposure, survival, oxygen consumption, growth, molting, filtering rate, reproduction, adenosine triphosphate (ATP), and behavior were monitored. Unless otherwise noted, animals were held and studied at 20°C after exposure. Although most studies were conducted with adult *Daphnia* (8 day), some experiments were made with first instar organisms.

With different temperature water baths, thermal stress (10 and 14°C above ambient, 20°C) was applied within 1 min and maintained for 8 additional min (Fig. 1). After 9 min, the temperature was cooled to ambient over a 3-min period. Total thermal exposure was 12 min. These thermal stresses were selected as realistic estimates of the temperature changes that would occur for entrained summer populations of cladocera. Maximum summer populations of most cladocera occur at temperatures near 20°C.

Chlorine, introduced as calcium hypochlorite, was monitored with an Amperometer Titrator (Wallace & Tiernan Division, Pennwalt

Corp.). Preliminary bioassays were conducted to determine the relationship between chlorine concentration and exposure time. Mortality was monitored for up to 48 hr after exposure. In the final studies the animals were initially exposed in a continuous-flow system to 0.5 ppm total chlorine for 30 sec. Then the total chlorine concentration was diluted to 0.25 ppm and maintained for 11.5 min. After 12 min of exposure, the chlorine was diluted over a period of ~20 min until there was no chlorine. The final chlorine concentration selected, 0.5 ppm, is lower than the maximum pulses of chlorine reported for power plants. Chlorine levels above 0.5 ppm were lethal after short exposures, however, and 0.5 ppm was the maximum concentration that did not affect survival in chronic studies.

Mechanical stress was applied by pumping the animals through a submersible pump and 46 m of 12.5-mm-ID Tygon tubing (Norton Company) at a rate of 4 liters/min. Total time of passage through the system was approximately 1 min. Total time of mechanical stress in the pump–tube system and turbulence in the receiving tank was 30 min. The mechanical stress emphasis of this research was on passage of the organism through a 12.5-mm tube. The flow velocity, 46 m/min, was low for data presented for some power plants, and the pressure, 4.84 psi, was within the minimum range reported for the Indian Point power station (Lauer et al., 1974). The calculated shear stress was 0.0485 psf.

In the combined stress study, the *Daphnia* were simultaneously exposed to a 10°C thermal, 0.5 ppm chlorine, and mechanical stresses in a continuous-flow system. The times of exposure for each component of the combined stress were as outlined and as presented in Fig. 1. Mechanical stress and turbulence in the receiving tank lasted for 30 min, the time required to dilute the total chlorine to zero.

In studies where the *Daphnia* were monitored for their entire life, the animals were usually fed an ad libitum concentration of algae after exposure. The effects of food quality were monitored in three experiments. In two studies, thermally stressed animals were fed trout chow and scenescent algae, and in the third, animals exposed to the combined stress were fed algae at a rate of 20,000 cells/ml. In this study, 10 animals were each fed healthy log-phase algae, stressed log-phase algae, or a 50 : 50 mixture of stressed and unstressed log-phase algae. The algae were preexposed to 10°C thermal and 0.5 ppm total chlorine stresses for the time periods discussed (see Fig. 1).

Acutely exposed *Daphnia* were observed for mortality. The criterion for death was no body movement after probing or swirling of the test container. For organisms exposed to mechanical stress, the mortality values include lack of recovery of test animals after exposure.

In the long-term studies, 10 *Daphnia* were raised in 10-ml vials until the fifth instar and then placed in 50-ml containers. The animals were fed and transferred into clean culture water daily. Notes were made daily on mortality, physiological condition, molting, presence or absence of eggs, number of live and dead young, and incremental growth. Body size was measured with an ocular micrometer from the head to the base of the spine in 0.1-mm increments.

Oxygen consumption, filtering rate, ATP, and behavior measurements were begun immediately after exposure to the stresses. The number of replicates for each test are identified in the appropriate tables. Oxygen consumption was measured by placing five unfed organisms in each 60-ml respirometer containing aerated, dechlorinated water at 20°C. After 24 hr, the oxygen content was determined using the modified Winkler technique (American Public Health Association, 1976) with a 0.005*N* thiosulfate solution. The difference in oxygen content between experimental and control organisms after 24 hr was considered the amount of oxygen consumed.

Filtering rate was determined by placing three *Daphnia* in each replicate containing 3.0 x 10^5 algal cells/ml at 20°C. At 0 and 24 hr, algal concentration was determined with an Electrozone Celloscope particle analyzer (Particle Data, Inc.). Filtering rate (F) was determined by

$$F = v \frac{\log C_0 - \log C_t}{\log e} \quad (1)$$

where v is volume of water per animal, C_0 is algal concentration at time zero, and C_t is algal concentration after 24 hr. Data were expressed as filtering rate (F) in milliliters per *Daphnia* per day.

Levels of ATP were determined by boiling five exposed animals per replicate in tris buffer and determining the ATP with the luciferin—luciferase assay using a Lab-Line Instruments, Inc., 1940 photometer.

Behavior was observed by placing 20 stressed animals into 50-cm vertical columns and noting changes in the vertical distribution with changes in light intensity from an incandescent bulb.

T-tests were conducted on all experiments except the *Daphnia* fed stressed and unstressed algae. For this study, a one-way factorial analysis of variance was conducted, and differences were determined using Duncan's new multiple range test. In all cases, the level of significance was $P \leqslant 0.05$.

RESULTS

After 24 hr, survival of first instar and adult organisms was not affected by a 10°C stress. When adults were exposed to 34°C, they also survived, but 35°C (a 15°C stress) was lethal. The effect of chlorine was dependent on concentration and length of exposure (Table 1). A 1-min exposure to 0.1 ppm chlorine killed 50% of the population within 48 hr. If food was provided immediately after a 22-sec chlorine exposure, the tolerance level increased to 0.5 ppm. Mechanical stress accounted for 30 to 60% mortality in adults and 60 to 80% in first instars within 1 hr after exposure. When all three stresses were combined, the percent mortality was similar to that obtained for mechanical stress.

The daphnids that survived the mechanical and combined stress were held for the duration of their lives, and adult survival was not significantly different from the controls (Table 2). First instar organisms, however, exhibited 50% mortality within 24 hr and rarely survived for more than 3 days. Chlorine and 10°C temperature stress did not affect long-term survival.

Surviving first instar and adult organisms were monitored after exposure for the duration of their lives. As long as healthy food was provided, there was no significant effect of individual or combined stresses on survival, molting rate, number of live young per brood (Table 2), or number of deteriorated eggs. Growth was not affected by stress in adult animals but was significantly reduced in chlorine-stressed first instar organisms.

If food was limiting in any way, however, latent effects of stress were observed. Significant depressions in reproduction occurred if the daphnids were fed trout chow or scenescent algae. Figure 2 shows the effect of scenescent algae fed between the seventh and eleventh instars. The thermally shocked daphnids had significantly lower numbers of live young immediately after being fed scenescent algae and recovered more slowly after being fed log-phase algae. During this period both groups of animals had increased numbers of deteriorated eggs per brood, and in instar eight the shocked animals had significantly higher numbers of deteriorated eggs (18 vs. 4 for the controls). In a subsequent study, *Daphnia* exposed to the combined stresses were fed unstressed log-phase algae, algae stressed by temperature and chlorine, and a 50 : 50 mixture of stressed and unstressed algae (Table 3). In most instances stressed food signifi-

TABLE 1

ACUTE LETHALITY EFFECTS OF CHLORINE CONCENTRATION AND EXPOSURE TIME ON ADULT *Daphnia pulex*

Original chlorine level, ppm	Exposure time, sec	Chlorine level after dilution, ppm	Exposure time, min	% survival	Time, hr
0.5	60	0.25	9	60	0.5
	60			0	1.0
	22			83	0.5
	22			67	1.0
	22			33	2.0
0.25	60	0.13	9	100	0.5
				100	1.0
				100	2.0
				60	24.0
	22	0.13	9	100	0.5
				100	1.0
				100	2.0
				83	24.0
				50	48.0
0.1	60	0.05	9	100	0.5
				100	1.0
				100	2.0
				83	24.0
				50	48.0
	22	0.05	9	100	0.5
				100	1.0
				100	2.0
				100	24.0
				83	48.0
	30	0.05	9.5	100	0.5
				100	1.0
				100	2.0
				83	24.0
				83	48.0
0.0	less than 10% mortality in all experiments				

TABLE 2

LONG-TERM BIOLOGICAL EFFECTS OF ENTRAINMENT STRESSES ON *Daphnia pulex*

Stress	Number of replicates	Mean survival, days	Molts per day	Live young per brood	Growth per instar, mm
Thermal, 10°C					
Adult	10	*	0.39*	7.85*	0.17*
First instar	10	2.6	0.34	11.4†	0.15†
Chlorine, 0.5 ppm					
Adult	10	32.1	0.31	24.25†	0.18†
First instar	10	41.1	0.42	21.21†	0.18†
Mechanical					
Adult	10	33.0	0.37	26.48†	0.13†
Combined					
Adult	10	41.0	0.38	24.36†	0.13†
Control					
Adult	40‡	41.9	0.35	26.13†	0.12†
First instar	20‡	33.7	0.38	17.06†	0.16†

*No significant differences noted; test terminated after 8 days.
†Average over first 15 instars after stress.
‡Control data for all experiments were combined.

cantly depressed individual growth, number of eggs produced, and number of live young.

A 10°C thermal stress caused all young cladocerans to become primaparous one instar earlier, whereas only 40% of the controls were primaparous one instar early. Consequently, thermally shocked animals produced an average of one more brood than did controls during the observation period.

Adult animals exposed to temperature and chlorine stress had depressed rates of oxygen consumption, but animals exposed to mechanical and combined stress had increased rates of oxygen consumption. Only a 14°C thermal shock significantly reduced oxygen consumption in adults, but mechanical stress significantly increased oxygen consumption (Table 4). A 10°C thermal shock caused a significant reduction in oxygen consumption of first instar animals, and the effect was significant 48 hr after exposure.

Within 24 hr after thermal stress, there were significant decreases in filtering rate (Table 5). The negative filtering rates were obtained because algal growth rate during the test exceeded removal rate. Chlorine and mechanical stresses had no significant effects, and no data were available for the effect of combined stress.

Chlorine and combined stresses significantly reduced ATP levels in adult organisms, and other stresses had no significant effect

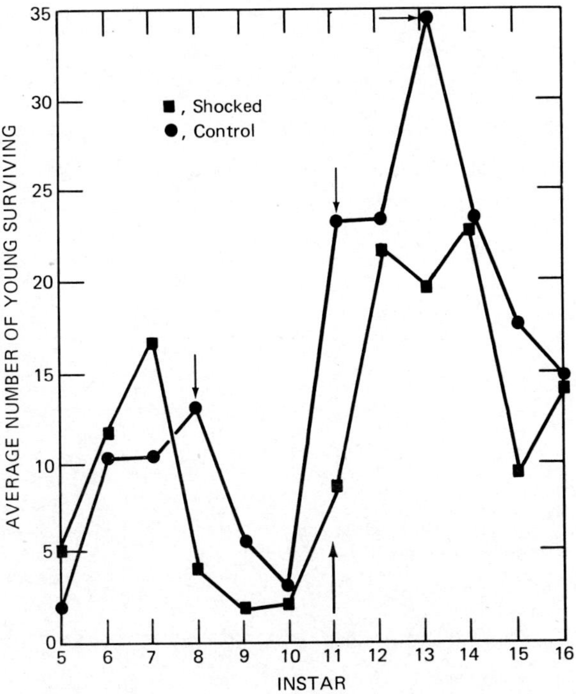

Fig. 2 Effect of food stress on reproduction of thermally stressed (■) and control (●) *Daphnia pulex*. Small arrows refer to significant differences among control and test organisms. Large arrow refers to instar when healthy food was reintroduced.

(Table 6). The reduction noted after combined stress was probably caused by chlorine.

No significant effects of individual stress on behavior were noted.

DISCUSSION

Mechanical stress had the greatest effect on cladoceran survival. This response generally was not in agreement with studies of copepods (Lanza et al., 1975; Davies, Hanson, and Jensen, 1976; Heinle, 1976) but agreed with Bunting's (1974) observation that entrainment "crops-off" cladoceran species susceptible to mechanical damage. Fish eggs and larvae also were susceptible to mechanical effects (Ulanowicz, 1975). Ulanowicz discussed the major mechanical effects on survival, pressure changes, acceleration, shear stress, and abrasion. All these factors, except pressure changes, would be

TABLE 3

EFFECTS OF FOOD STRESS ON GROWTH AND REPRODUCTION OF *Daphnia pulex**

Instar after shock	Mean variable	Food types†		
2	Length, mm	Control	50 : 50	Stressed
		2.23	2.06	2.04
	Number of eggs	Control	50 : 50	Stressed
		18.6	14.1	11.6
	Number of live young	Control	Stressed	50 : 50
		18.3	9.4	7.5
3	Length, mm	Control	50 : 50	Stressed
		2.39	2.26	2.16
	Number of eggs	Control	50 : 50	Stressed
		21.7	17.9	14.7
	Number of live young	Control	50 : 50	Stressed
		6.7	14.0	14.4
4	Length, mm	Control	50 : 50	Stressed
		2.43	2.31	2.25
	Number of eggs	Control	Stressed	50 : 50
		14.1	8.9	7.5
	Number of live young	Control	50 : 50	Stressed
		1.6	3.7	0.1
5	Length, mm	Control	50 : 50	Stressed
		2.52	2.43	2.31
	Number of eggs	Control	50 : 50	Stressed
		17.4	13.3	8.6
	Number of live young	Control	50 : 50	Stressed
		17.4	6.7	6.7
6	Length, mm	Control	50 : 50	Stressed
		2.54	2.46	2.16
	Number of eggs	Control	50 : 50	Stressed
		19.7	19.5	10.7
	Number of live young	Control	50 : 50	Stressed
		15.2	17.0	5.7

*See text for detailed discussion of study.

†Significance determined by Duncan's new multiple range test. Food types not underscored by continuous line are significantly different at the 0.05 level.

TABLE 4

EFFECTS OF ENTRAINMENT STRESSES ON OXYGEN CONSUMPTION OF ADULT *Daphnia pulex*

Stress and sample	Number of replicates	O_2 consumption, $\mu l\ mg^{-1}\ hr^{-1}$ (± 95% CI)
Thermal, 10°C		
Test	6	6.60 ± 0.84
Control	5	7.55 ± 0.68
Thermal, 14°C		
Test	4	4.85 ± 0.61*
Control	3	6.04 ± 0.42
Chlorine, 0.5 ppm		
Test	4	11.41 ± 1.16
Control	3	11.71 ± 1.06
Mechanical		
Test	6	10.68 ± 1.96*
Control	5	6.96 ± 0.34
Combined		
Test	5	15.56 ± 8.79
Control	4	13.50 ± 3.88

*Significant at the 0.05 level.

expected to cause mortality in cladocera and other crustaceans (Beck, Poje and Waller, 1975).

Adult animals that survived the mechanical stress had higher rates of oxygen consumption, and preadults did not survive after 3 days. The exact effect of mechanical stress on oxygen consumption is not known, but it may be related to increased energy expenditure in reorientation behavior and tissue repair. This increase in oxygen consumption may have in part accounted for the poor survival of preadult *Daphnia.*

Temperature was the second most important parameter affecting cladoceran biology. A 14°C stress significantly reduced oxygen consumption, and 10 to 14°C stresses reduced filtering rates. In no case did thermal stress significantly affect growth, molting, or reproduction when there was an ad libitum supply of algae. The depressant effect of thermal shock on oxygen consumption may be the result of prolonged undershoot in response. Kinne (1964), citing studies of others, noted that, when *Daphnia magna* encountered a sharp thermal gradient, it showed a characteristic overshoot or

TABLE 5

EFFECTS OF ENTRAINMENT STRESSES ON THE FILTERING RATE OF ADULT *Daphnia pulex*

Stress and sample	Number of replicates	O_2 consumption, ml animal^{-1} day^{-1} (± 95% CI)
Thermal, 10°C		
Test	3	−0.69 ± 10.69*
Control	3	0.65 ± 4.62
Thermal, 14°C		
Test	3	−2.37 ± 7.09*
Control	3	3.71 ± 1.32
Chlorine, 0.5 ppm		
Test	4	5.23 ± 0.92
Control	4	3.50 ± 2.29
Mechanical		
Test	5	3.49 ± 2.32
Control	4	3.50 ± 2.29
Combined	no data	

*Algal growth exceeded rate of uptake.

undershoot in response intensity. If the animal survived the initial shock, it quickly adjusted its metabolism to a new level and then acclimated to the new environmental condition. This rapid type of response may have occurred in 10°C stressed animals, but the 14°C stress may have caused physiological damage that resulted in a reduced rate of oxygen consumption. Theoretically, oxygen consumption should have increased as temperature increased (see, e.g., Schindler, 1968).

The depressant effect of 10°C stress on filtering rate was also unexpected since filtering rate usually increases with temperature. Kibby (1971) and Hayward and Gallup (1976) demonstrated that *Daphnia* filtered best at an optimum temperature dependent on prior acclimation. At temperatures above the optimum, filtering rates decreased. The lower filtering rate at 30°C for acclimated *D. schødleri* was similar to the response observed for *D. pulex* stressed at 30°C for 12 min. The possible mode of action for this response is not known.

If nutrition was a limiting factor, then subtle latent thermal effects significantly decreased the number of live young per brood. Temperature may have affected reproduction because of the reduced

TABLE 6

EFFECTS OF ENTRAINMENT STRESSES ON THE ATP LEVELS OF ADULT *Daphnia pulex*

Stress and sample	Number of replicates	ATP level, μg ATP/organism (± 95% CI)
Thermal, 10°C		
Test	5	0.23 ± 0.09
Control	4	0.28 ± 0.10
Thermal, 14°C		
Test	5	0.02 ± 0.01
Control	4	0.03 ± 0.02
Chlorine, 0.5 ppm		
Test	5	0.01 ± 0.01*
Control	5	0.04 ± 0.02
Mechanical		
Test	5	0.10 ± 0.03
Control	5	0.11 ± 0.07
Combined		
Test	3	0.01 ± 0.01*
Control	4	0.04 ± 0.03

*Significant difference at the 0.05 level.

filtering rates of the cladocerans. Less food means less young will be produced (Richman, 1958). Thermal stress may also have depressed digestion, assimilation rates, and gonadal functions. Enzyme inactivation may be a causal factor.

Because of its daily intermittent presence in field situations, chlorine was concluded to be less important than mechanical and thermal stress. The effect of chlorine as a major factor in entrainment mortality of zooplankton was varied (see, e.g., Heinle, 1976; Lanza et al., 1975; Davies and Jensen, 1975). In this laboratory simulation higher concentrations of chlorine would have been lethal, but concentrations ⩽0.5 ppm were not. The effect of chlorine dose exposure was dependent not only on time but also on the availability of food after exposure. Chlorine significantly reduced the level of ATP of surviving animals, and the mode of action may have been inhibition of carbohydrate oxidation (Green and Stumpf, 1946; Eaton et al., 1973). Chlorine retarded the growth of exposed first instar *Daphnia* but not of adult animals. This effect may have a nominal impact on egg production because the number of eggs produced is size dependent. Because there were no significant effects on reproduction, the effect of chlorine on body size was considered insignificant.

Even though we did not study the interactions of two specific stresses, we felt that the effects of combined stress on cladoceran functions probably reflected only the action of specific stress. For example, the significant decrease in ATP levels may be caused by chlorine, and the poor survival of first instars may be caused by mechanical stress. That there was no significant effect of combined stress on oxygen consumption suggests that the stimulatory effect of mechanical stress was negated by the inhibitory effect of thermal stress.

Generally, entrainment caused a decrease in field populations of zooplankton (see, e.g., Astrauskas and Rachyunas, 1975), probably because of the "predatory" impact of mechanical stress (Bunting, 1974). On the basis of our data, adult *Daphnia pulex* have a greater chance of surviving entrainment stress than do young *D. pulex* unless the thermal shock approaches 35°C, the incipient lethal for *D. pulex* (Goss and Bunting, 1976), or unless exposure to chlorine exceeds 0.5 ppm for more than 30 sec without subsequent dilution.

Evaluating the response of surviving individual organisms to entrainment, we find no measurable effect on their biology if food is healthy and in sufficient quantities. If food was limiting in any way, latent effects of thermal stress to the *Daphnia* became apparent, and these effects contribute to the depressed population numbers observed below the discharge areas. The effects are further increased because damaged and scenescent algae inhibit feeding responses of cladocera (Ryther, 1954).

The significant increase in zooplankton abundance in heated discharges observed by Goryajnova (1975) and others may be caused not by the thermal effects on entrained animals but by a combination of recruitment of organisms not previously entrained and the effects of variable temperatures on reproductive potential. Bunting (1974) and Swain, Wilson, and Neri (1975) showed that higher temperatures decreased development time and longevity while increasing reproduction rates. Feeding rates also increased with increasing temperature (Duval and Geen, 1976; Hayward and Gallup, 1976), and this would affect reproduction. Buikema and Loeffelman (1977), Halbach (1973), and Ruttner-Kolisko (1964) demonstrated that fluctuating temperatures increased the reproductive potential of rotifers, and similar responses would be expected for other zooplankton.

Except for acute survival, means for assessing the impact of entrainment on cladoceran zooplankton in the field appear to be limited. Significant effects on ATP levels appear to be restricted to the effects of chlorination. Because chlorination is intermittent, ATP levels would not be affected when there is no chlorine present. Oxygen consumption and long-term biological studies also are insignificant, especially if the animals are cultured in the laboratory under ideal conditions. Filtering rate may be an indicator, but, in all probability, it will not provide meaningful data. Behavior was the most unreliable method investigated.

ACKNOWLEDGMENTS

The research reported here was funded by a contract from the Energy Research and Development Administration.

REFERENCES

American Public Health Association, 1976, *Standard Methods for the Examination of Water and Waste Water*, 14th ed., pp. 443-449, Washington, D. C.

Astrauskas, A. S., and L. A. Rachyunas, 1975, Hydrological Conditions in the Reservoir Cooler of the Lithuanian State Regional Electric Power Station, *Gidrobiol. Zh.*, 11: 19.

Beck, A. P., G. V. Poje, and W. T. Waller, 1975, A Laboratory Study on the Effects of the Exposure of Some Entrainable Hudson River Biota to Hydrostatic Pressure Regimes Calculated for the Proposed Cornwall Pumped-Storage Plant, in *Fisheries and Energy Production: A Symposium*, S. B. Saila (Ed.), pp. 167-204, D. C. Heath & Company, Lexington, Mass.

Benda, R. S., and J. Gulvas, 1976, Effects of the Palisades Nuclear Power Plant on Lake Michigan, in *Thermal Ecology II*, ERDA Symposium Series, Augusta, Ga., Apr. 2–5, 1975, G. W. Esch and R. W. McFarlane (Eds.), pp. 243-250, CONF-750425, NTIS.

Brungs, W. A., 1973, Effects of Residual Chlorine on Aquatic Life, *J. Water Pollut. Control Fed.*, 45: 2180-2193.

Buikema, A. L., Jr., and P. J. Loeffelman, 1977, Effects of Pumped-Storage Operations on Rotifer Populations, Societas Internationalis Limnologiae, XX Congress, Copenhagen.

Bunting, D., 1974, Zooplankton: Thermal Regulation and Stress, in *Energy Production and Thermal Effects*, B. J. Gallagher (Ed.), Ann Arbor Science Pubs., Inc., Ann Arbor, Mich.

Buss, D. F., 1975, An Environmental Study of the Ecological Effects of the Thermal Discharge from the Point Beach Nuclear Plant, *Trans. Am. Micros. Soc.*, 22: 640.

Carpenter, E. J., B. B. Peck, and S. J. Anderson, 1974, Survival of Copepods Passing Through a Nuclear Power Station on Northeastern Long Island Sound, U.S.A., *Mar. Biol.*, 24: 49-55.

Coutant, C. C., 1970, Biological Aspects of Thermal Pollution. I. Entrainment and Discharge Canal Effects, *CRC Crit. Rev. Environ. Control*, 1(3): 341-381.

——, 1974, Evaluation of Entrainment Effect, in *Proceedings of the Second Workshop on Entrainment and Intake Screening,* L. D. Jensen (Ed.), pp. 1-12, Cooling Water Discharge Research Project Report No. 15, Electric Power Research Institute Publication No. 74-049-00-5, Palo Alto, Calif.

——, and S. S. Talmadge, 1977, Thermal Effects, *J. Water Pollut. Control Fed.*, 49: 1369-1425.

Davies, R. M., C. H. Hanson, and L. D. Jensen, 1976, Entrainment of Estuarine Zooplankton into a Mid-Atlantic Power Plant: Delayed Effects, in *Thermal Ecology II,* ERDA Symposium Series, Augusta, Ga., Apr. 2—5, 1975, G. W. Esch and R. W. McFarlane (Eds.), pp. 349-357, CONF-750425, NTIS.

——, and L. D. Jensen, 1975, Zooplankton Entrainment at Three Mid-Atlantic Power Plants, *J. Water Pollut. Control Fed.*, 47: 2130-2142.

Duval, W. S., and G. H. Geen, 1976, Diel Feeding and Respiration Rhythms in Zooplankton, *Limnol. Oceanogr.*, 21: 823-829.

Eaton, J. W., C. F. Kolpin, H. S. Swofford, C. M. Kjellstrand, and H. S. Jacob, 1973, Chlorinated Urban Water: A Cause of Dialysis-Induced Hemolytic Anemia, *Science,* 181: 463-464.

Goryajnova, L. F., 1975, The Effect of the Norossijsk Power Station Heated Discharge on Zooplankton, *Gidrobiol. Zh.*, 11: 28.

Goss, L. B., and D. L. Bunting, 1976, Thermal Tolerance of Zooplankton, *Water Res.*, 10: 387-398.

Green, D. E., and P. K. Stumpf, 1946, The Mode of Chlorine, *J. Am. Water Works Assoc.*, 38: 1301-1305.

Halbach, U., 1973, Life Table Data and Population Dynamics of the Rotifer *Brachionus calyciflorus* Pallas as Influenced by Periodically Oscillating Temperature, in *Effects of Temperature on Ectothermic Organisms: Ecological Implications and Mechanisms of Compensation,* W. Wieser (Ed.), Springer-Verlag, New York.

Harmsworth, R. V., 1974, Artificial Cooling Lakes as Unique Aquatic Ecosystems, in *Energy Production and Thermal Effects,* B. J. Gallagher (Ed.), Ann Arbor Science Pubs., Inc., Ann Arbor, Mich.

Hayward, R. S., and D. N. Gallup, 1976, Feeding, Filtering and Assimilation in *Daphnia schødleri* Sars as Affected by Environmental Conditions, *Arch. Hydrobiol.*, 77: 139-163.

Heinle, D. R., 1976, Effects of Passage Through Power Plant Cooling Systems on Estuarine Copepod Species, *Environ. Pollut. (London),* 11: 39-58.

Kibby, H. V., 1971, Effect of Temperature on the Feeding Behavior of *Daphnia rosea, Limnol. Oceanogr.*, 16: 580-581.

Kinne, O., 1964, Non-Genetic Adaptation to Temperature and Salinity, *Helgol. Wiss. Meeresunters.*, 9: 433-458.

Lanza, G. R., G. J. Lauer, T. C. Ginn, P. C. Storm, and I. Zubarik, 1975, Biological Effects of Simulated Discharge Plume Entrainment at Indian Point Nuclear Power Station, Hudson River Estuary, U.S.A., in *Combined Effects of Radioactive, Chemical and Thermal Releases to the Environment,* Symposium Proceedings, Stockholm, 1975, pp. 95-126, STI/PUB/404, International Atomic Energy Agency, Vienna.

Lauer, G. J., et al., 1974, Entrainment Studies on Hudson River Organisms, in *Proceedings of the Second Workshop on Entrainment and Intake Screening,* L. D. Jensen (Ed.), pp. 37-82, Cooling Water Discharge Research Project Report No. 15, Electric Power Research Institute, Publication No. 74-049-00-5, Palo Alto, Calif.

Markowski, S., 1960, Observations on the Response of Some Benthic Organisms to Power Station Cooling Water, *J. Anim. Ecol.*, 29: 349-357.

——, 1962, Faunistic and Ecological Investigations in Cavendish Dock, Barrow-in-Furness, *J. Anim. Ecol.*, 31: 43-52.

Richman, S., 1958, The Transformation of Energy by *Daphnia pulex*, *Ecol. Monogr.*, 28: 273-291.

Ruttner-Kolisko, A., 1964, Uber die labile periode in fortpflanzungszyclus der radertiere, *Int. Rev. Gesam. Hydrobiol. Hydrogr.*, 49: 473-482.

Ryther, J. H., 1954, Inhibitory Effects of Phytoplankton upon the Feeding of *Daphnia magna* with Reference to Growth, Reproduction and Survival, *Ecology*, 35: 522-533.

Schindler, D. W., 1968, Feeding, Assimilation and Respiration Rates of *Daphnia magna* Under Various Environmental Conditions and Their Relation to Production Estimates, *J. Anim. Ecol.*, 37: 369-385.

Storr, J. F., 1974, Plankton Entrainment by the Condenser Systems of Nuclear Power Stations on Lake Ontario, in *Thermal Ecology*, AEC Symposium Series, May 3—5, 1973, Augusta, Ga., J. W. Gibbons and R. R. Sharitz (Eds.), pp. 281-295, CONF-730505, NTIS.

Swain, W. R., R. M. Wilson, and R. P. Neri, 1975, *Studies on the Effects of Thermal Additions on Selected Zooplankton Populations*, Water Resources Research Center Bulletin 84, Minneapolis, Minn.

Ulanowicz, R. E., 1975, The Mechanical Effects of Water Flow on Fish Eggs and Larvae, in *Fisheries and Energy Production: A Symposium*, S. B. Saila (Ed.), pp. 77-88, D. C. Heath & Company, Lexington, Mass.

Whitehouse, J. W., 1971, Some Aspects of the Biology of Lake Trawsfynydd: A Power Station Cooling Pond, *Hydrobiologia*, 38: 253-288.

THE AVOIDANCE RESPONSE OF THE COMMON SHINER TO TOTAL AND COMBINED RESIDUAL CHLORINE IN THERMALLY INFLUENCED DISCHARGES

DONALD S. CHERRY, STEPHAN R. LARRICK, JAMES D. GIATTINA,
KENNETH L. DICKSON, and JOHN CAIRNS, Jr.
Biology Department and Center for Environmental Studies, Virginia Polytechnic Institute and State University, Blacksburg, Virginia

ABSTRACT

Chlorine avoidances of the common shiner [*Notropis cornutus* (Mitchell)] were determined in a trough containing a steep horizontal gradient at a field laboratory located at the Glen Lyn Power Plant in southwestern Virginia. Responses were evaluated by treating New River water with doubling increments of total residual chlorine (TRC) from 0.025 to 0.40 mg/liter. Free residual chlorine (FRC) was measured; combined residual chlorine (CRC) and the hypochlorous acid content (HOCl) were calculated; and each of the fractions comprising the TRC was correlated with fish behavior. Responses were also evaluated according to ammonia nitrogen concentrations, chloramine solutions with no FRC, and chlorination at preferred temperatures. Tests were carried out at acclimation temperatures of 12, 18, 24, and 30°C for the unheated TRC treatments and at 12 and 24°C for all other tests. The first significant or threshold avoidance to TRC (represented by a significant decrease in residence time in treated vs. untreated water) occurred at levels varying from 0.10 to 0.20 mg/liter, depending on the acclimation temperature tested. The CRC and FRC comprising the TRC at the threshold avoidance level varied from 0.077 to 0.159 and 0.035 to 0.122 mg/liter, respectively. Variation in threshold avoidance concentrations for fractions within the TRC was lowest for HOCl, 0.010 to 0.028 mg/liter. In chloramine trials conducted at 12 and 24°C, avoidances first occurred at 0.207 and 0.205 mg/liter, respectively. When fish were acclimated at 12 and 24°C and tested in preferred temperatures for these acclimation temperatures (17.6 and 26.8°C, respectively), TRC threshold avoidances occurred at 0.211 and 0.209 mg/liter. Chloramine avoidance thresholds were 0.399 mg/liter at 12°C and 0.195 mg/liter at 24°C. Avoidance of ammonia was not a factor in the chloramine tests since the fish did not avoid ammonia nitrogen concentrations of 0.015 to 0.275 mg/liter in separate tests. The avoidance responses to chlorine in ambient and preferred temperature conditions suggest that the common shiner was less influenced by the presence of chloramine and was more sensitive to a low threshold concentration of HOCl in TRC treatments.

Fish populations inhabiting thermally influenced waters near power-plant installations often encounter chlorine perturbations resulting from antifouling practices. Residual chlorine concentrations for certain regions of a plant's discharge vary from 0.1 to 0.5 mg/liter (Becker and Thatcher, 1973), with a recommended maximum weekly average of free residual of 0.2 mg/liter (Environmental Protection Agency, 1973). Although these chlorine levels are potentially hazardous to most aquatic biota, including fish, species that actively detect and avoid sublethal or lethal residuals may not suffer deleterious effects. Field studies have demonstrated the ability of fish to avoid polluted water (Katz and Gaufin, 1952; Sprague, Elson, and Saunders, 1965; Tsai, 1968; 1970). Power-production discharges are unique, however, in that many fish species are attracted to the thermal plumes, which periodically contain toxic chlorine residuals.

Many earlier investigations dealing with chlorine toxicity (see reviews by Brungs, 1973; 1976) did not adequately consider the constituents comprising total residual chlorine (TRC). When relatively high levels of nitrogenous compounds are present, most of the chlorine residual may be in a combined (CRC) or chloramine form (Clark, Viessmann, and Hammer, 1971). In relatively unpolluted water the residual may be in an uncombined or free form (FRC), which, depending on pH and temperature, is comprised of hypochlorous acid (HOCl) and hypochlorite ion (OCl^-). Hypochlorous acid is considered to be the most toxic of these components (Becker and Thatcher, 1973), but each component differentially influences fish physiology and avoidance behavior. Heath (1977) supported the hypothesis that the toxic actions of CRC and FRC are based on separate physiological mechanisms. In behavioral studies, Fava and Tsai (1976) reported that the blacknose dace (*Rhinichthys atratulus*) avoided FRC to a lesser degree than it did CRC. *Micropterus punctulatus* and *Notropis rubellus* were shown to avoid very low concentrations of HOCl (Cherry et al., 1977). In addition, avoidance responses to chlorine have been species specific. Rainbow trout (*Salmo gairdneri*) demonstrated a very sensitive response to chlorine concentrations of 0.001 mg/liter (Sprague and Drury, 1969), but rockbass (*Ambloplites rupestris*) failed to significantly avoid doses $\leqslant$0.80 mg/liter TRC (Cherry et al., 1978). Threshold or first avoidance responses of four other species were within a narrower range, generally from 0.005 to 0.500 mg/liter CRC (Fava and Tsai, 1976; Bogardus, 1977). The objectives of our study were to evaluate the avoidance responses of the common shiner (*Notropis cornutus*) to separate TRC and chloramine treatments at several acclimation temperatures and to compare these responses with responses to the same chlorinated treatments at the fish's preferred temperature conditions.

MATERIALS AND METHODS

Common shiners from the Ohio River drainage were obtained by air freight, in cooperation with the Center for Great Lakes Studies, University of Wisconsin, Milwaukee. Holding facilities, feeding procedures, and temperature acclimation rates were reported earlier (Cherry, Dickson, and Cairns, 1975a). All fish tested were generally less than 2 years old and ranged from 45 to 65 mm fork length.

Preference temperatures were determined in a horizontal trough heated underneath by a battery of infrared lamps. Two groups of four fish were monitored separately in a 20°C temperature gradient. Fish were allowed 2 to 3 hr to adjust to the environmental conditions, and then temperature selection was determined for fish acclimated to 12 and 24°C.

Chlorine dosing procedures and equipment used for chlorine avoidance trials were the same as those reported by Cherry et al. (1977). Two water baths fed opposite ends of each avoidance trough. A calcium hypochlorite solution was added to one bath with a peristaltic pump; this created a steep gradient ranging from chlorinated to untreated water in each trough. After a 60- to 90-min orientation period, and before chlorine was added, fish behavior was continuously monitored for 10 min. The residence time in the prechlorinated side of the trough was the reference for the subsequent chlorine exposures. Target TRC concentrations were established in increments from 0.025 mg/liter in doubling sequence to 0.400 mg/liter. Residual concentrations were measured by amperometric titration (American Public Health Association et al., 1976) before and after each dose. Eight replications were obtained, with each fish being timed for 10-min intervals at each target chlorine concentration at acclimation temperatures of 12, 18, 24, and 30°C. The CRC and HOCl concentrations were calculated; the pH was measured with a portable Markson pH meter; and ammonia nitrogen was measured by an Orion specific ion electrode. These treatments, the TRC trials, contained CRC and FRC fractions in

TABLE 1

CHLORINE AVOIDANCE TRIALS AT VARIOUS ACCLIMATION TEMPERATURES

Type of avoidance trial	Major components	Acclimation temperature, °C
TRC	30–70% CRC–FRC, depending on river water quality	12, 18, 24, 30
CRC	97% chloramine, mainly monochloramine	12, 24
TRC in preferred temperature	30–70% CRC–FRC, depending on river water quality	12, 24
CRC in preferred temperature	97% chloramine, mainly monochloramine	12, 24
NH_3	Ammonium ion	12, 24

various proportions depending on the New River water quality each day (Table 1).

The second type of chlorine avoidance trial, chloramine avoidance, consisted of total residual treatments comprised mainly of CRC. These trials were carried out at 12 and 24°C in the same manner as the TRC trials, except that the original stock solution of calcium hypochlorite used to chlorinate one water bath was also dosed with ammonium chloride. Dissociation of the ammonium chloride resulted in the ammonia ions' combining with FRC to form a solution comprised primarily of CRC. More than 90% of the solution was monochloramine (determined amperometrically), and pH levels ranged from 7.5 to 8.4. Ammonia levels were measured at the beginning and the end of each 10-min time interval. To ascertain the possible interference of excess ammonia with fish responses, we carried out a series of trials using an ammonium chloride solution separately from the chloramine tests. Ammonia solutions from 0.015 to 0.275 mg/liter were established in one-half of each trough, and fish responses at 12 and 24°C were evaluated as in chlorinated trials.

To evaluate fish responses relative to power-plant thermal discharges, we carried out additional trials in which the chlorinated water was heated. After the control period was completed, the temperature in one water bath was raised above the acclimation temperature of 12 or 24°C to the preference temperature. This created two distinct temperature regions in the avoidance trough. A second 10-min timing was then conducted to ensure preference for the warmer water, and then chlorine was added to the heated side in the same dosing increments as in the previous chlorine trials. In all avoidance trials the residence time in the control or reference period was compared with that during each chlorinated treatment with Duncan's new multiple range test (Steel and Torrie, 1960). The first significant ($P \leqslant 0.05$) reduction in residence time after each chlorinated increment, as compared with the reference period, represented the avoidance threshold.

RESULTS

Chlorinated Trials

The first significant avoidance to TRC varied from 0.10 mg/liter when the fish were acclimated and tested at 18°C to 0.20 mg/liter at 12, 24, and 30°C (Fig. 1). Residence times declined steadily when chlorine exposures surpassed 0.10 mg/liter and were generally lowest at the highest residual encountered (0.40 mg/liter). At 12 and 24°C, the threshold avoidance to chloramine residuals (0.20 mg/liter) was the same as that for TRC (Fig. 1), but, unlike those in TRC trials, residence times declined steadily after the first exposure to chloramine treatments (0.025 mg/liter).

In the TRC avoidance trials, the range of variation at the first significant avoidance between acclimation temperatures was highest for TRC concentrations (0.112 to 0.209 mg/liter), intermediate for CRC (0.077 to 0.159 mg/liter) and FRC (0.035 to 0.122 mg/liter) levels, and least variable for HOCl (0.010 to 0.028 mg/liter) determinations (Table 2). Avoidance thresholds for the chloramine trials (0.205 to 0.207 mg/liter at 12 and 24°C, respectively) were considerably higher than for CRC fractions in the TRC trials (0.087 to 0.106 mg/liter).

Chlorinated–Preferred-Temperature Trials

The TRC avoidance thresholds for common shiners at preferred temperatures of 17.6 or 26.8°C (when acclimated at 12 and 24°C, respectively) were 0.20 mg/liter (Fig. 2). Avoidance thresholds for chloramine trials, however, were 0.40 mg/liter at 17.6°C and 0.20 mg/liter at 26.8°C. Thus the threshold concentration for chloramine trials at 12°C was approximately three times as great as the CRC threshold level in the TRC trials (0.399 to 0.138 mg/liter). At 24°C, avoidance thresholds in chloramine trials were also higher

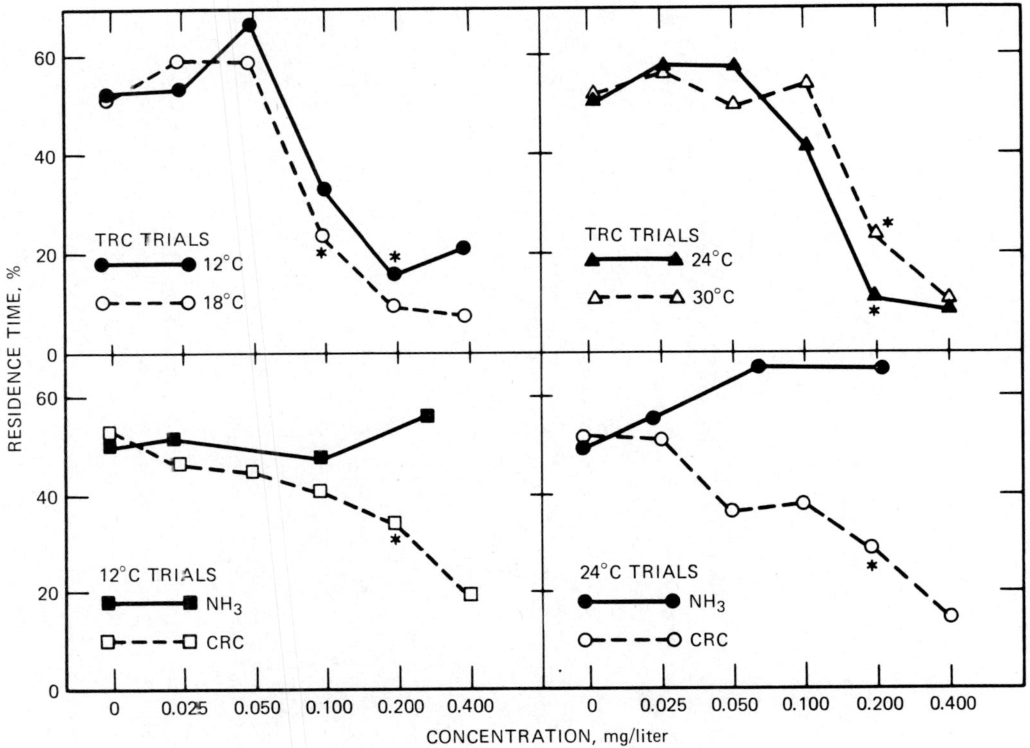

Fig. 1 Residence time and first significant avoidance ($P \leqslant 0.05$), indicated by an asterisk (*), of the common shiner (*Notropis cornutus*) to several concentrations of total residual chlorine (TRC), ammonia nitrogen (NH_3), and chloramine (CRC) at several acclimation temperatures.

TABLE 2

FIRST STATISTICAL (OR THRESHOLD) AVOIDANCES BY THE COMMON SHINER (*NOTROPIS CORNUTUS*) SUBJECTED TO VARIOUS TREATMENTS AND TEMPERATURE CONDITIONS

Acclimation temperature, °C	Concentration at first significant avoidance,* mg/liter				
	TRC	CRC	FRC	HOCl	NH_3
	TRC Avoidance at Ambient Temperature				
12	0.209	0.087	0.122	0.025	
18	0.112	0.077	0.035	0.011	
24	0.202	0.106	0.096	0.028	
30	0.203	0.159	0.044	0.010	
	Chloramine Avoidance at Ambient Temperature				
12	0.213	0.207	0.006	0	0.006
24	0.207	0.205	0.002	0	0.012
	TRC Avoidance at Preferred Temperature				
12	0.211	0.138	0.073	0.029	
24	0.209	0.163	0.046	0.011	
	Chloramine Avoidance at Preferred Temperature				
12	0.401	0.399	0.002	0	0.013
24	0.195	0.195	0	0	0.002

*Abbreviations are TRC, total residual chlorine; CRC, combined residual chlorine; and FRC, free residual chlorine.

than the CRC fraction of TRC treatments (0.195 to 0.163 mg/liter, respectively).

The common shiner's response to ammonia nitrogen resulted in no significant avoidances in the range tested, ~0.015 to 0.275 mg/liter (Fig. 1). In most treatments the species tended to spend greater time in the ammonia-treated water. At the highest concentration tested (0.275 mg/liter), which greatly exceeded ammonia nitrogen levels in chloramine trials (Table 2), residence time was nearly the same as during the control period.

DISCUSSION

The common shiner actively responded to chlorinated water by avoiding TRC and CRC exposures after a minimal concentration level

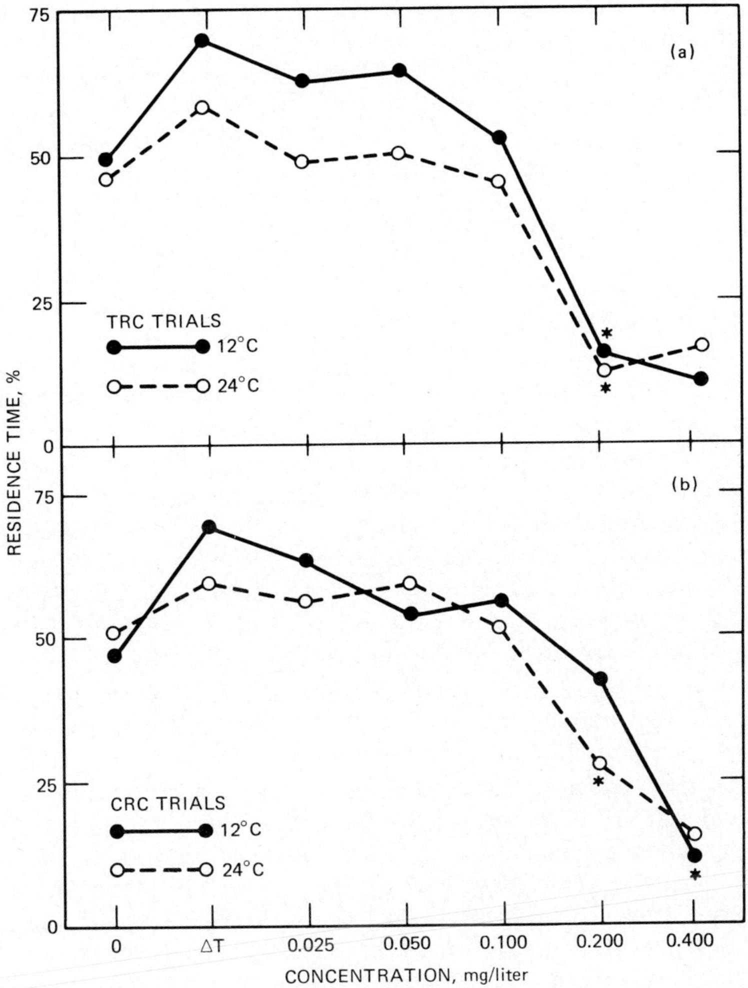

Fig. 2 Residence time and first significant avoidance ($P \leqslant 0.05$), indicated by an asterisk (*), of the common shiner to several concentrations of (a) total residual chlorine (TRC) and (b) chloramine (CRC) at preferred (ΔT) temperature conditions when acclimated at 12 and 24°C.

was attained (0.10 mg/liter TRC or 0.20 mg/liter CRC), depending on the acclimation temperature (Fig. 1). As these levels increased in magnitude, the shiner showed greater avoidance behavior, with as low as 6.4% residence time at 18°C for 0.40 mg/liter TRC. Cherry et al. (1977) reported similar avoidance trends to TRC (0.05 to

0.20 mg/liter) for the rosyface shiner (*Notropis rubellus*) and spotted bass *(Micropterus punctulatus)*, and Cherry et al. (1978) reported the same trends for smallmouth bass *(Micropterus dolomieui)* and the Coho salmon *(Oncorhynchus kisutch)*.

A greater concentration of chloramine (0.205 mg/liter), in comparison with TRC levels comprised of CRC and FRC, was required to initiate the threshold avoidance behavior of the common shiner at 12 and 24°C (Table 2). In TRC concentrations at 12°C, avoidance first occurred when the CRC level was 0.087 mg/liter, with a FRC fraction of 0.122 mg/liter. At 24°C, CRC and FRC levels comprising the TRC were nearly equal. The presence of FRC at certain minimal concentrations may initiate avoidance behavior sooner since the CRC fraction in the TRC trials was approximately half the amount of the chloramine avoidance threshold.

Excess ammonia concentrations were minimal (0.006 to 0.013 mg/liter) at the chloramine avoidance thresholds in both unheated and heated trials and did not significantly influence avoidance behavior (Table 2). At higher ammonia nitrogen concentrations (0.069 to 0.207 mg/liter in heated trials), fish resided in the ammonia-treated water more often than in the untreated river water (Fig. 1). Similar trends were reported by Jones (1948).

Comparing avoidance thresholds for chloramine vs. TRC trials indicated that higher chloramine than FRC concentrations were required to produce an avoidance response (Table 2). For example, at 12°C, the common shiner showed a stronger preference for the warmer water (17.6°C) than at the 24°C trial (preference of 26.8°C). A chloramine concentration at least twice as high as the amount needed in the other chloramine or TRC trials (0.399 mg/liter) was necessary to force the shiner from the preferred water. Although the FRC avoidance thresholds in preferred temperature trials were less than those in unheated conditions, the HOCl components of both were generally the same. Therefore, a specific minimal HOCl fraction may influence threshold avoidance behavior more than the CRC components regardless of temperature. Fava and Tsai (1976) reported, however, that chloramine concentrations were more instrumental in avoidance behavior than free residual chlorine for blacknose dace, but HOCl was not considered. Differences in experimental designs and the fact that domestic wastewater was used could account for these ambiguities.

In our studies much of the variability in threshold avoidances to TRC and its components was reduced when responses were evaluated as a function of the HOCl fraction. Cherry et al. (1977) reported that the most consistent patterns of chlorine avoidance by the spotted bass and rosyface shiner occurred when avoidance thresholds were correlated with the HOCl fraction in the TRC. Larrick et al. (1978) reported that the first significant avoidance by the golden shiner (*Notemigonus crysoleucas*), tested at 12, 18, 24, and 30°C, correlated highly with the HOCl fraction in the TRC. Although the golden shiner first avoided TRC levels from 0.20 to 0.40 mg/liter, depending on the acclimation temperature, the HOCl concentration at these threshold avoidances ranged from 0.015 to 0.017 mg/liter.

The common shiner demonstrated an ability to actively avoid lethal and sublethal chloramine concentrations. The first significant avoidance (0.207 mg/liter at 12°C) by the shiner of chloramine in unheated treatments was less than half the 48-hr LC_{50} obtained by Brooks and Seegert (1978) at 10 and 20°C (0.54 and 0.50 mg/liter, respectively). The avoidance threshold for preferred temperature trials at 12°C (0.399 mg/liter) was also less than the LC_{50} values obtained at 10 and 20°C, and the avoidance threshold at 24°C (0.195 mg/liter) was approximately half the lethal chloramine concentration found at 30°C (0.38 mg/liter).

General standards for intermittent chlorine discharges have been established. Recommended weekly FRC levels are 0.20 mg/liter, and the maximum is 0.50 mg/liter FRC (Environmental Protection Agency, 1973); these levels are applicable to the Glen Lyn Power Plant. Although the lethal threshold of FRC for the common shiner has not been reported, nitrogenous compounds inherent in New River water may form chloramine concentrations in potentially hazardous levels. Whether or not the chlorinated residuals are considered acceptable or toxic, the common shiner appears to have demonstrated the ability to actively perceive and avoid sublethal chloramine residuals.

ACKNOWLEDGMENT

The research reported here was supported by the American Electric Power Service Corporation, Canton, Ohio.

REFERENCES

American Public Health Association, American Water Works Association, Water Pollution Control Federation, 1976, *Standard Methods for the Examination of Water and Wastewater*, 14th ed., Washington, D. C.

Becker, C. D., and T. O. Thatcher (Comps.), 1973, *Toxicity of Power Plant Chemicals to Aquatic Life*, USAEC Report WASH-1249, Atomic Energy Commission, Washington, D. C.

Bogardus, R. B., 1977, Wapora, Inc., Washington, D. C., personal communication.

Brooks, A. S., and G. L. Seegert, 1978, A Preliminary Look at the Effects of Intermittent Chlorination on Selected Warm Water Fish, in *Conference on Water Chlorination: Environmental Impact and Health Effects*, Gatlinburg, Tenn., Oct. 31—Nov. 4, 1977, Vol. 2, pp. 95-110, Ann Arbor Science Pubs., Inc., Ann Arbor, Mich.

Brungs, W. A., 1973, Effects of Residual Chlorine on Aquatic Life, *J. Water Pollut. Control Fed.*, 45: 2180-2193.

——, 1976, *Effects of Wastewater and Cooling Water Chlorination on Aquatic Life*, Report No. EPA-600/3-76-098, Environmental Protection Agency, Duluth, Minn.

Cherry, D. S., K. L. Dickson, and J. Cairns, Jr., 1975a, Temperatures Selected and Avoided by Fish Populations at Various Acclimation Temperatures, *J. Fish. Res. Board Can.*, 32: 485-491.

——, K. L. Dickson, and J. Cairns, Jr., 1975b, The Use of a Mobile Laboratory to Study Temperature Response of Fish, in *Twenty-Ninth Industrial Waste Conference*, Symposium Proceedings, Purdue University, Lafayette, Ind., May 7—9, 1974, Engineering Bulletin, Engineering Extension Service No. 145, pp. 129-145, Purdue University.

——, S. R. Larrick, K. L. Dickson, R. C. Hoehn, and J. Cairns, Jr., 1977, Significance of Hypochlorous Acid in Free Residual Chlorine to the Avoidance Response of Spotted Bass (*Micropterus punctulatus*) and Rosyface Shiner (*Notropis rubellus*), *J. Fish. Res. Board Can.*, 34: 1365-1372.

——, S. R. Larrick, K. L. Dickson, and J. Cairns, Jr., 1978, Response of Eurythermal and Stenothermal Fish Species to Chlorinated Discharges, in *Eleventh Annual Conference on Trace Substances in Environmental Health*, Columbia, Mo., June 6—9, 1977, University of Missouri Press, Columbia.

Clark, J. W., W. Viessman, Jr., and M. J. Hammer, 1971, *Water Supply and Pollution Control*, 2nd ed., Intext Education Publishers, Scranton, Pa.

Environmental Protection Agency, 1973, *Water Quality Criteria 1972*, Report EPA-R3-73-033, Washington, D. C.

Fava, J. A., Jr., and C. Tsai, 1976, Immediate Behavioral Reactions of Blacknose Dace, *Rhinichthys atratulus*, to Domestic Sewage and Its Toxic Constituents, *Trans. Am. Fish. Soc.*, 105: 430-441.

Heath, A. G., 1977, Toxicity in Intermittent Chlorination to Freshwater Fish: Influence of Temperature and Chlorine Form, *Hydrobiologia*, 56: 39-47.

Jones, J. R. E., 1948, A Further Study of the Reactions of Fish to Toxic Substances, *J. Exp. Biol.*, 25: 22-34.

Katz, M., and A. R. Gaufin, 1952, The Effects of Sewage Pollution on the Fish Populations of a Midwestern Stream, *Trans. Am. Fish. Soc.*, 82: 156-165.

Larrick, S. R., D. S. Cherry, K. L. Dickson, and J. Cairns, Jr., 1978, The Use of Various Avoidance Indices to Evaluate the Behavioral Response of the Golden Shiner to Components of Total Residual Chlorine, in *Conference on Water Chlorination: Environmental Impact and Health Effects*, Gatlinburg, Tenn., Oct. 31—Nov. 4, 1977, Ann Arbor Science Pubs., Inc., Ann Arbor, Mich.

Sprague, J. B., and D. E. Drury, 1969, Avoidance Reactions of Salmonid Fish to Representative Pollutants, in *Advances in Water Pollution Research*, Proceedings of the 4th International Conference on Water Pollution Research, Pergamon Press, Inc., Elmsford, N. Y.

——, P. F. Elson, and R. L. Saunders, 1965, Sublethal Copper and Zinc Pollution in a Salmon River—A Field and Laboratory Study, *Int. J. Air Water Pollut.*, 9: 531-543.

Steel, R. G. D., and J. H. Torrie, 1960, *Principles and Procedures of Statistics*, McGraw-Hill Book Company, Inc., New York.

Tsai, C., 1968, Effects of Chlorinated Sewage Effluents on Fishes in Upper Patuxent River, Maryland, *Chesapeake Sci.*, 9: 83-93.

——, 1970, Changes in Fish Populations and Migration in Relation to Increased Sewage Pollution in Little Patuxent River, Maryland, *Chesapeake Sci.*, 11: 34-41.

AUTHOR INDEX

SUBJECT INDEX*

*Page numbers indicate only the first appearance of that subject in the article. Readers are encouraged to refer to the remainder of the study for additional information on the indexed subject.